Methods in Enzymology

Volume 406
REGULATORS AND EFFECTORS OF SMALL GTPASES: RHO FAMILY

METHODS IN ENZYMOLOGY

EDITORS-IN-CHIEF

John N. Abelson   Melvin I. Simon

DIVISION OF BIOLOGY
CALIFORNIA INSTITUTE OF TECHNOLOGY
PASADENA, CALIFORNIA

FOUNDING EDITORS

Sidney P. Colowick and Nathan O. Kaplan

Methods in Enzymology

Volume 406

Regulators and Effectors of Small GTPases: Rho Family

EDITED BY

William E. Balch

DEPARTMENT OF CELL BIOLOGY
THE SCRIPPS RESEARCH INSTITUTE
LA JOLLA, CALIFORNIA

Channing J. Der

DEPARTMENT OF PHARMACOLOGY
THE UNIVERSITY OF NORTH CAROLINA
CHAPEL HILL, NORTH CAROLINA

Alan Hall

CRC ONCOGENE AND SIGNAL TRANSDUCTION GROUP
MRC LABORATORY FOR MOLECULAR CELL BIOLOGY
UNIVERSITY COLLEGE LONDON
LONDON, ENGLAND

AMSTERDAM • BOSTON • HEIDELBERG • LONDON
NEW YORK • OXFORD • PARIS • SAN DIEGO
SAN FRANCISCO • SINGAPORE • SYDNEY • TOKYO
Academic Press is an imprint of Elsevier

ELSEVIER

Elsevier Academic Press
525 B Street, Suite 1900, San Diego, California 92101-4495, USA
84 Theobald's Road, London WC1X 8RR, UK

This book is printed on acid-free paper.

Permissions may be sought directly from Elsevier's Science & Technology Rights
Department in Oxford, UK: phone: (+44) 1865 843830, fax: (+44) 1865 853333,
E-mail: permissions@elsevier.com. You may also complete your request on-line
via the Elsevier homepage (http://elsevier.com), by selecting
"Support & Contact" then "Copyright and Permission" and then "Obtaining Permissions."

For all information on all Elsevier Academic Press publications
visit our Web site at www.books.elsevier.com

ISBN-13: 978-0-12-182811-0
ISBN-10: 0-12-182811-5

PRINTED IN THE UNITED STATES OF AMERICA
06 07 08 09 10 9 8 7 6 5 4 3 2 1

Table of Contents

Contributors to Volume 406

Article numbers are in parentheses following the names of contributors.
Affiliations listed are current.

REZA AHMADIAN (1), *Max-Planck Institute for Molecular Physiology, Department of Structural Biology, Dortmund, Germany*

HUZOOR AKBAR (43), *Division of Experimental Hematology, Childrens Hospital Research Foundation, Cincinnati, Ohio*

KLAUS AKTORIES (10), *Albert Ludwigs Universitat Freiburg, Institut fur Experimentelle & Klinische Pharmakologie & Toxikologie, Freiburg, Germany*

BRUNO ANTONNY (6), *CNRS-Institut de Pharmacologie, Moleculaire et Cellulaire, Valbonne, France*

WILLIAM T. ARTHUR (31), *Cytoskeleton, Denver, Colorado*

DANIEL BAIRD (5), *Department of Chemistry, and Chemical Biology, Cornell University, Ithaca, New York*

HOLGER BARTH (10), *Abteilung Pharmakologie und Toxikologie, Universität Ulm, Ulm, Germany*

ANASTACIA C. BERZAT (2), *Department of Radiation Oncology, University of North Carolina, Chapel Hill, North Carolina*

LARS BLUMENSTEIN (1), *Max-Planck Institute for Molecular Physiology, Department of Structural Biology, Dortmund, Germany*

GARY M. BOKOCH (7), *Department of Immunology and Cell Biology, The Scripps Research Institute, La Jolla, California*

GARY G. BORISY (57), *Cell and Molecular Biology, Northwestern University Medical School, Chicago, Illinois*

ROHIT BOSE (32), *Mount Sinai Hospital, University of Toronto, Toronto, Ontario, Canada*

DAVID M. BOURDON (37), *Department of Pharmacology, University of North Carolina School of Medicine, Chapel Hill, North Carolina*

LAURENT BOYER (33), *INSERM, Faculte de Medecine, Nice, France*

VANIA M. M. BRAGA (29), *Molecular and Cellular Medicine Division, Division of Biomedical Sciences, Imperial College, London, United Kingdom*

KEITH BURRIDGE (31), *Department of Cell and Developmental Biology, Lineberger Cancer Center, University of North Carolina, Chapel Hill, North Carolina*

JOSE CANCELAS (43), *Division of Experimental Hematology, Children's Hospital Research Foundation, Cincinnati, Ohio*

RICHARD A. CERIONE (5, 48), *Departments of Molecular Medicine and Chemistry and Chemical Biology, Cornell University, Ithaca, New York*

LOUISE CHANG (55), *Life Sciences Institute, University of Michigan, Ann Arbor, Michigan*

XINYU CHEN (26), *Center for Cell Signalling, University of Virginia School of Medicine, University of Virginia, Charlottesville, Virginia*

ZHUI CHEN (35), *Department of Pharmacology, University of Texas Southwestern Medical Center, Dallas, Texas*

EMILY J. CHENETTE (2), *Lineberger Comprehensive Cancer Center, University of North Carolina, Chapel Hill, North Carolina*

SHIAN-HUEY CHIANG (55), *Life Sciences Institute, University of Michigan, Ann Arbor, Michigan*

MELANIE H. COBB (35), *Department of Pharmacology, University of Texas Southwestern Medical Center, Dallas, Texas*

LUDOVIC COLLIN (27), *MRC Laboratory for Molecular Cell Biology, University College London, London, United Kingdom*

JEAN-FRANCOIS CÔTÉ (4), *Institut de Recherches, Cliniques de Montréal, Montréal, Canada*

ADRIENNE D. COX (2), *Department of Pharmacology, University of North Carolina, Chapel Hill, North Carolina*

DANIEL R. CROFT (42), *The Beatson Institute for Cancer Research, Glasgow, United Kingdom*

CHANNING J. DER (2), *Department of Pharmacology, University of North Carolina, Chapel Hill, North Carolina*

CELINE M. DERMARDIROSSIAN (7), *Department of Immunology and Cell Biology, The Scripps Research Institute, La Jolla, California*

XUEMEI DONG (47), *Department of Genetics & Developmental Biology, University of Connecticut Health Center, Farmington, Connecticut*

ANNE DOYE (33), *INSERM, Faculte de Medecine Nice, France*

RADOVAN DVORSKY (1), *Max-Planck Institute for Molecular Physiology, Department of Structural Biology, Dortmund, Germany*

EARL B. EDWARDS (37), *Department of Pharmacology, University of North Carolina School of Medicine, Chapel Hill, North Carolina*

SHAWN M. ELLERBROEK (31), *Department of Chemistry, Wartburg College, Waverly, Iowa*

YUKINORI ENDO (25), *Craniofacial Development Biology & Regeneration Branch, National Institute of Dental and Craniofacial Research, National Institutes of Health, Bethesda, Maryland*

JENNIFER C. ERASMUS (29), *Molecular and Cellular Medicine Section, Division of Biomedical Sciences, Imperial College, London, United Kingdom*

SANDRINE ETIENNE-MANNEVILLE (44), *Morphogeneses et Signalisation Cellulaire, Institut Curie, Paris, France*

QIYU FENG (5), *Department of Molecular Medicine, Cornell University, Ithaca, New York*

DENNIS FIEGEN (1), *Max-Planck Institute for Molecular Physiology, Department of Structural Biology, Dortmund, Germany*

RAFAEL GARCIA-MATA (31), *Department of Cell and Developmental Biology, Lineberger Comprehensive Cancer Center, University of North Carolina, Chapel Hill, North Carolina*

ANNETTE GÄRTNER (27), *MRC Laboratory for Molecular Cell Biology, University College London, London, United Kingdom*

SVETLANA GERSHBURG (20), *Department of Pharmacology, University of North Carolina, Chapel Hill, North Carolina*

EDGAR R. GOMES (45), *Department of Anatomy & Cell Biology, Columbia University, New York, New York*

BRUCE GOODE (16), *Rosenstiel Center, Brandeis University, Waltham, Massachusetts*

LLOYD GREENE (36), *Columbia University, Center for Neurobiology & Behavior, New York, New York*

GREGG G. GUNDERSEN (45), *Department of Anatomy & Cell Biology, Columbia University, New York, New York*

RAYMOND (38), *Department of Biochemistry, University of Medicine and Dentistry of New Jersey, Piscataway, New Jersey*

LARS C. HAEUSLER (1), *Max-Planck Institute for Molecular Physiology, Department of Structural Biology, Dortmund, Germany*

KLAUS HAHN (12), *Department of Pharmacology, University of North Carolina, Chapel Hill, North Carolina*

ALAN HALL (52), *CRC Oncogene and Signal Transduction Group, MRC Laboratory for Molecular Cell Biology, University College London, London, United Kingdom*

T. KENDALL HARDEN (19, 20, 37), *Department of Pharmacology, University of North Carolina School of Medicine, Chapel Hill, North Carolina*

ELIZABETH S. HARRIS (15), *Dartmouth Medical School, Department of Biochemistry, Hanover, New Hampshire*

GERD HAUG (10), *Albert Ludwigs Universitat Freiburg, Institut fur Experimentelle & Klinische Pharmakologie & Toxikologie, Freiburg, Germany*

PHILLIP T. HAWKINS (8), *Inositide Laboratory, The Babraham Institute, Cambridge, United Kingdom*

XI HE (38), *Neurobiology Program, Children's Hospital, Harvard Medical School, Boston, Massachusetts*

LARS HEMSATH (1), *Max-Planck Institute for Molecular Physiology, Department of Structural Biology, Dortmund, Germany*

ULRIKE HERBRAND (1), *Max-Planck Institute for Molecular Physiology, Department of Structural Biology, Dortmund, Germany*

HENRY N. HIGGS (15), *Dartmouth Medical School, Department of Biochemistry Hanover, New Hampshire*

KIRSTI HILL (3), *Inositide Laboratory, The Babraham Institute, Cambridge, United Kingdom*

LOUIS HODGSON (12), *Department of Pharmacology, University of North Carolina, Chapel Hill, North Carolina*

HSIN-YI HENRY HO (13, 14), *Division of Neuroscience, Children's Hospital, Boston, Massachusetts*

STEVEN HOOPER (49), *Tumour Cell Biology Laboratory, Cancer Research United Kingdom London Research Institute, London, United Kingdom*

JUSTYNA JANAS (46), *Cold Spring Harbor Laboratory, Cold Spring Harbor, New York*

MARK R. JEZYK (20), *Department of Pharmacology, University of North Carolina, Chapel Hill, North Carolina*

TZUU-SHUH JOU (53), *Department of Internal Medicine, National Taiwan University College of Medicine, Taipei, Taiwan*

MARC W. KIRSCHNER (13, 14), *Harvard Medical School, Boston, Massachusetts*

ETSUKO KIYOKAWA (23), *Department of Signal Transduction, Research Institute for Microbial Diseases, Osaka University, Osaka, Japan*

SONJA KRUGMANN (8), *Inositide Laboratory, The Babraham Institute, Cambridge, United Kingdom*

KAZUO KUROKAWA (23), *Department of Signal Transduction, Research Institute for Microbial Diseases, Osaka University, Osaka, Japan*

GIOVANNA LALLI (27), *MRC Laboratory for Molecular Cell, Biology, University College London, London, United Kingdom*

DANA LASKO (39), *BioAxone Therapeutic Inc., Montreal, Quebec, Canada*

ANDRES M. LEBENSOHN (13, 14), *Harvard Medical School, Boston, Massachusetts*

EMMANUEL LEMICHEZ (33), *INSERM, Faculte de Medecine de Nice, Nice, France*

DAISY W. LEUNG (21), *Department of Biochemistry, University of Texas Southwestern Medical Center at Dallas, Dallas, Texas*

YINGHUA LI (56), *Institut fur Allgemeine Zoologie and Genetik, Westfalische Wilhelms Universitat, Munster, Germany*

ERZSÉBET LIGETI (9), *Semmelweis University, Department of Physiology, Budapest, Hungary*

QIONG LIN (48), *Weis Center for Research, Geisinger Clinic, Danville, Pennsylvania*

MINGJIAN LU (28), *Carter Immunology Center, University of Virginia, Charlottesville, Virginia*

LE MA (13), *Department of Biological Sciences, Stanford University, Stanford, California*

IAN G. MACARA (26), *Center for Cell Signalling, University of Virginia School of Medicine, University of Virginia, Charlottesville, Virginia*

SANKAR MAITI (16), *Rosenstiel Center, Brandeis University, Waltham, Massachusetts*

JEAN-BAPTISTE MANNEVILLE (40), *Institut Curie—CNRS, Paris, France*

JOHN F. MARSHALL (49), *Tumour Cell Biology Laboratory, Cancer Research United Kingdom London Research Institute, London, United Kingdom*

MICHIYUKI MATSUDA (23), *Department of Signal Transduction, Research Institute for Microbial Diseases, Osaka University, Osaka, Japan*

KAZUE MATSUMOTO (25), *Craniofacial Development Biology & Regeneration Branch, National Institute of Dental and Craniofacial Research, National Institutes of Health, Bethesda, Maryland*

LISA MCKERRACHER (39), *Department Pathologie et Biologie Cellulaire, Université de Montreal/BioAxone Therapeutic Inc, Montreal, Quebec, Canada*

BRUNO MESMIN (6), *CNRS-Institut de Pharmacologie, Moleculaire et Cellulaire, Valbonne, France*

AMEL METTOUCHI (33), *INSERM, Faculte de Medecine, Nice, France*

DAVID MICHAELSON (22), *Cell Biology and Pharmacology, NYU School of Medicine, New York, New York*

JAIME MILLÁN (50), *Ludwig Institute for Cancer Research, University College London, London, United Kingdom*

W. TODD MILLER (18), *Department of Physiology and Biophysics, SUNY at Stony Brook, Stony Brook, New York*

DAVID M. MORGAN (21), *Department of Biochemistry, University of Texas, Southwestern Medical Center at Dallas, Dallas, Texas*

JAMES B. MOSELEY (16), *Rosenstiel Center, Brandeis University, Waltham, Massachusetts*

KEITH MOSTOV (53), *Departments of Anatomy, Biochemistry, and Biophysics, University of California School of Medicine, San Francisco, California*

HIDEYUKI MUKAI (17), *Biosignal Research Center, Kobe University, Kobe, Japan*

SENTHIL K. MUTHUSWAMY (54), *Cold Spring Harbor Laboratory, Cold Spring Harbor, New York*

TAKESHI NAKAMURA (23), *Department of Signal Transduction, Research Institute for Microbial Diseases, Osaka University, Osaka, Japan*

PERIHAN NALBANT (12), *Department of Immunology, The Scripps Research Institute, La Jolla, California*

SHUH NARUMIYA (24, 51), *Department of Pharmacology, Kyoto University, Faculty of Medicine, Kyoto, Japan*

NICOLE K. NOREN (31), *The Burnham Institute, La Jolla, California*

LUCY ERIN O'BRIEN (53), *Department of Molecular and Cell Biology, University of California, Berkeley, California*

FABIAN OCEGUERA-YANEZ (24), *Department of Pharmacology, Kyoto University, Faculty of Medicine, Kyoto, Japan*

HISAKAZU OGITA (30), *Department of Molecular Biology and Biochemistry, Osaka University Graduate School of Medicine, Osaka, Japan*

ABIODUN A. OGUNJIMI (32), *Mount Sinai Hospital, University of Toronto, Toronto, Ontario, Canada*

MICHAEL OLSON (11, 42), *Molecular Cell Biology Laboratory, The Beatson Institute for Cancer Research, Glasgow, United Kingdom*

YOSHITAKA ONO (17), *Biosignal Research Center, Kobe University, Kobe, Japan*

BARISH OZDAMAR (32), *Mount Sinai Hospital/University of Toronto, Toronto, Ontario, Canada*

ROUMEN PANKOV (25), *Department of Cytology, Histology, and Embryology, Faculty of Biology, Sofia University, Sofia, Bulgaria*

JOHN PELOQUIN (57), *Cell and Molecular Biology, Northwestern University Medical School, Chicago, Illinois*

MARK R. PHILIPS (22), *Cell Biology and Pharmacology, NYU School of Medicine, New York, New York*

ANDREAS W. PÜSCHEL (56), *Institut fur Allgemeine Zoologie and Genetik, Westfalische Wilhelms Universitat, Munster, Germany*

SHARONA EVEN-RAM (25), *Craniofacial Development Biology & Regeneration Branch, National Institute of Dental and Craniofacial Research, National Institutes of Health, Bethesda, Maryland*

KODI S. RAVICHANDRAN (28), *Carter Immunology Center, University of Virginia, Charlottesville, Virginia*

ANNE J. RIDLEY (41, 50), *Ludwig Institute for Cancer Research, University College London, London, United Kingdom*

KIRSTI RIENTO (41), *Ludwig Institute for Cancer Research, University College London, London, United Kingdom*

RAJAT ROHATGI (14), *Department of Oncology, Stanford University Medical Center, Stanford, California*

MICHAEL K. ROSEN (21), *Howard Hughes Medical Institute, Department of Biochemistry, University of Texas Southwestern Medical Center at Dallas, Dallas, Texas*

ERIK SAHAI (11, 49), *Tumour Cell Biology Laboratory, Cancer Research United Kingdom London Research Institute, London, United Kingdom*

ALAN R. SALTIEL (55), *Life Sciences Institute, University of Michigan, Ann Arbor, Michigan*

JENS C. SCHWAMBORN (56), *Institut fur Allgemeine Zoologie and Genetik, Westfalische Wilhelms Universitat, Munster, Germany*

JASON P. SEIFERT (19), *Department of Pharmacology, University of North Carolina School of Medicine, Chapel Hill, North Carolina*

JEFFREY SETTLEMAN (9), *Harvard Medical School, MGH Cancer Center, Charlestown, Massachusetts*

FEIMO SHEN (12), *Department of Pharmacology, University of North Carolina, Chapel Hill, North Carolina*

ADAM SHUTES (2), *Lineberger Comprehensive Cancer Center, University of North Carolina, Chapel Hill, North Carolina*

JACEK SKOWRONSKI (46), *Cold Spring Harbor Laboratory, Cold Spring Harbor, New York*

JASON T. SNYDER (19, 20), *Department of Pharmacology, University of North Carolina School of Medicine, Chapel Hill, North Carolina*

JON SONDEK (19, 20, 37), *Department of Pharmacology, University of North Carolina School of Medicine, Chapel Hill, North Carolina*

PATRICIA STEGE (1), *Max-Planck Institute for Molecular Physiology, Department of Structural Biology, Dortmund, Germany*

LEN STEVENS (8), *Inositide Laboratory, The Babraham Institute, Cambridge, United Kingdom*

HIDEKI SUMIMOTO (34), *Medical Institute of Bioregulation, Kyushu University, Fukuoka, Japan*

YOSHIMI TAKAI (30), *Department of Molecular Biology and Biochemistry, Osaka University Graduate School of Medicine, Osaka, Japan*

RYU TAKEYA (34), *Medical Institute of Bioregulation, Kyushu University, Fuku-oka, Japan*

KITTY TANG (53), *Departments of Anatomy and Biochemistry and Biophysics, University of California School of Medicine, San Francisco, California*

LAURA TURNER (52), *Eisai London Research Laboratories, University College London, London, United Kingdom*

NORIKO UENO (34), *Medical Institute of Bioregulation, Kyushu University, Fukuoka, Japan*

LINDA VAN AELST (46), *Cold Spring Harbor Laboratory, Cold Spring Harbor, New York*

DANIJELA VIGNJEVIC (57), *Institut Curie—Section Recherche, Paris, France*

KRISTIINA VUORI (4), *Cancer Center, The Burnham Institute, La Jolla, California*

HONG-RUI WANG (32), *Mount Sinai Hospital, University of Toronto, Toronto, Ontario, Canada*

HEIDI C. E. WELCH (3), *Inositide Laboratory, The Babraham Institute, Cambridge, United Kingdom*

KRISTER WENNERBERG (31), *Cytoskeleton, Denver, Colorado*

DAVID A. WILLIAMS (43), *Division of Experimental Hematology, Children's Hospital Research Foundation, Cincinnati, Ohio*

LYNN WILLIAMS (50), *The Kennedy Institute of Rheumatology, Imperial College London, London, United Kingdom*

MICHELE R. WING (37), *Department of Pharmacology, University of North Carolina School of Medicine, Chapel Hill, North Carolina*

JEFF WRANA (32), *Mount Sinai Hospital, University of Toronto, Toronto, Ontario, Canada*

DIANQING WU (47), *Department of Genetics & Developmental Biology, University of Connecticut Health Center, Farmington, Connecticut*

BIN XIANG (54), *Cold Spring Harbor Laboratory, Cold Spring Harbor, New York*

XHIHENG XU (36), *Institute of Genetics and Developmental Biology, Chinese Academy of Sciences, Beijing, China*

KENNETH M. YAMADA (25), *Craniofacial Development Biology & Regeneration Branch, National Institute of Dental and Craniofacial Research, National Institutes of Health, Bethesda, Maryland*

WANNIAN YANG (48), *Weis Center for Research, Geisinger Clinic, Danville, Pennsylvania*

SHINGO YASUDA (51), *Department of Pharmacology, Kyoto University, Faculty of Medicine, Kyoto, Japan*

NORIKO YOKOYAMA (18), *Department of Physiology and Biophysics, SUNY at Stony Brook, Stony Brook, New York*

WEI YU (53), *Departments of Anatomy and Biochemistry and Biophysics, University of California School of Medicine, San Francisco, California*

MIRJAM M. P. ZEGERS (53), *Department of Epithelial Pathobiology, College of Medicine, University of Cincinnati, Cincinnati, Ohio*

YUE ZHANG (32), *Mount Sinai Hospital, Toronto/University of Toronto, Ontario, Canada*

JIE ZHENG (43), *Department of Structural Biology, St. Jude's Children's Research Hospital, Memphis, Tennessee*

YI ZHENG (43), *Children's Hospital Research Foundation, Cincinnati, Ohio*

Preface

The Ras superfamily (>150 human members) encompasses Ras GTPases involved in cell proliferation, Rho GTPases involved in regulating the cytoskeleton, Rab GTPases involved in membrane targeting/fusion, and a group of GTPases including Sar1, Arf, Arl and dynamin involved in vesicle budding/fission. These GTPases act as molecular switches, and their activities are controlled by a large number of regulatory molecules that affect either GTP loading (guanine nucleotide exchange factors or GEFs) or GTP hydrolysis (GTPase activating proteins or GAPs). In their active state, they interact with a continually increasing and functionally complex array of downstream effectors.

In this new series of *Methods in Enzymology*, we have striven to bring together the latest thinking, approaches, and techniques in this area. Two volumes (403 and 404) focus on membrane regulating GTPases, the first dedicated to those involved in budding and fission (Sar1, Arf, Arl and dynamin), and the second focused on those that control targeting and fusion (Rabs). Volumes 406 and 407 focus on the Rho and Ras families, respectively. It is important to emphasize that while each of these volumes deals with a different GTPase family, both contain a wealth of common methodologies. As such, the techniques and approaches pioneered with respect to one class of GTPase are likely to be equally applicable to other classes. Furthermore, the functional distinctions that have been classically associated with the distinct branches of the superfamily are beginning to blur. There is now considerable evidence for biological and biochemical interplay and crosstalk among seemingly divergent family members. The compilation of a database of regulators and effectors of the whole superfamily by Bernards (Vol. 407) reflects some of these complex interrelationships. In addition to fostering cross-talk among investigators who study different GTPases, these volumes will also aid the entry of new investigators into the field.

Since the last *Methods in Enzymology* volume on this topic in 2000, Rho GTPases have continued to receive a huge amount of attention. The human genome sequence has revealed the full extent of the Rho GEF and Rho GAP families (over 80 members for each), and the challenge of identifying the molecular interactions and cellular pathways influenced by each of these regulators is a daunting prospect. This new volume describes some of the methods currently being used to examine Rho family GTPase regulation at the biochemical and cellular level. The number of downstream targets of Rho GTPases also continues to grow, and since there is no single motif that defines a target, we do not yet know what the final number might be. As might be expected, this large number of targets is reflected in the diverse cellular effects elicited by Rho GTPases. Some, such as actin filament assembly, are becoming well characterized at the biochemical level, and the influence of Rho GTPases on Arp2/3 and the Diaphanous-related formins is covered in this volume. Other cellular responses to Rho GTPases are not as well characterized molecularly, but these too are included here, since they contribute to fundamentally important aspects of cell behavior. The last five years have witnessed a transformation in the way that cellular pathways can be analyzed in multi-cellular organisms through the use of interfering RNA technology. Some of these approaches that have been used to analyze Rho GTPase pathways are included. Finally, new imaging techniques are revolutionizing our ability to visualize GTPase activities, both spatially and temporally, within a single cell.

We are extremely grateful to the many investigators who have generously contributed their time and knowledge to bring a wealth of technical expertise into this and other volumes comprising the Ras superfamily series.

ALAN HALL
WILLIAM E. BALCH
CHANNING J. DER

METHODS IN ENZYMOLOGY

VOLUME 176. Nuclear Magnetic Resonance (Part A: Spectral Techniques and Dynamics)
Edited by NORMAN J. OPPENHEIMER AND THOMAS L. JAMES

VOLUME 177. Nuclear Magnetic Resonance (Part B: Structure and Mechanism)
Edited by NORMAN J. OPPENHEIMER AND THOMAS L. JAMES

VOLUME 178. Antibodies, Antigens, and Molecular Mimicry
Edited by JOHN J. LANGONE

VOLUME 179. Complex Carbohydrates (Part F)
Edited by VICTOR GINSBURG

VOLUME 180. RNA Processing (Part A: General Methods)
Edited by JAMES E. DAHLBERG AND JOHN N. ABELSON

VOLUME 181. RNA Processing (Part B: Specific Methods)
Edited by JAMES E. DAHLBERG AND JOHN N. ABELSON

VOLUME 182. Guide to Protein Purification
Edited by MURRAY P. DEUTSCHER

VOLUME 183. Molecular Evolution: Computer Analysis of Protein and Nucleic Acid Sequences
Edited by RUSSELL F. DOOLITTLE

VOLUME 184. Avidin-Biotin Technology
Edited by MEIR WILCHEK AND EDWARD A. BAYER

VOLUME 185. Gene Expression Technology
Edited by DAVID V. GOEDDEL

VOLUME 186. Oxygen Radicals in Biological Systems (Part B: Oxygen Radicals and Antioxidants)
Edited by LESTER PACKER AND ALEXANDER N. GLAZER

VOLUME 187. Arachidonate Related Lipid Mediators
Edited by ROBERT C. MURPHY AND FRANK A. FITZPATRICK

VOLUME 188. Hydrocarbons and Methylotrophy
Edited by MARY E. LIDSTROM

VOLUME 189. Retinoids (Part A: Molecular and Metabolic Aspects)
Edited by LESTER PACKER

VOLUME 190. Retinoids (Part B: Cell Differentiation and Clinical Applications)
Edited by LESTER PACKER

VOLUME 191. Biomembranes (Part V: Cellular and Subcellular Transport: Epithelial Cells)
Edited by SIDNEY FLEISCHER AND BECCA FLEISCHER

VOLUME 192. Biomembranes (Part W: Cellular and Subcellular Transport: Epithelial Cells)
Edited by SIDNEY FLEISCHER AND BECCA FLEISCHER

VOLUME 244. Proteolytic Enzymes: Serine and Cysteine Peptidases
Edited by ALAN J. BARRETT

VOLUME 245. Extracellular Matrix Components
Edited by E. RUOSLAHTI AND E. ENGVALL

VOLUME 246. Biochemical Spectroscopy
Edited by KENNETH SAUER

VOLUME 247. Neoglycoconjugates (Part B: Biomedical Applications)
Edited by Y. C. LEE AND REIKO T. LEE

VOLUME 248. Proteolytic Enzymes: Aspartic and Metallo Peptidases
Edited by ALAN J. BARRETT

VOLUME 249. Enzyme Kinetics and Mechanism (Part D: Developments in Enzyme Dynamics)
Edited by DANIEL L. PURICH

VOLUME 250. Lipid Modifications of Proteins
Edited by PATRICK J. CASEY AND JANICE E. BUSS

VOLUME 251. Biothiols (Part A: Monothiols and Dithiols, Protein Thiols, and Thiyl Radicals)
Edited by LESTER PACKER

VOLUME 252. Biothiols (Part B: Glutathione and Thioredoxin; Thiols in Signal Transduction and Gene Regulation)
Edited by LESTER PACKER

VOLUME 253. Adhesion of Microbial Pathogens
Edited by RON J. DOYLE AND ITZHAK OFEK

VOLUME 254. Oncogene Techniques
Edited by PETER K. VOGT AND INDER M. VERMA

VOLUME 255. Small GTPases and Their Regulators (Part A: Ras Family)
Edited by W. E. BALCH, CHANNING J. DER, AND ALAN HALL

VOLUME 256. Small GTPases and Their Regulators (Part B: Rho Family)
Edited by W. E. BALCH, CHANNING J. DER, AND ALAN HALL

VOLUME 257. Small GTPases and Their Regulators (Part C: Proteins Involved in Transport)
Edited by W. E. BALCH, CHANNING J. DER, AND ALAN HALL

VOLUME 258. Redox-Active Amino Acids in Biology
Edited by JUDITH P. KLINMAN

VOLUME 259. Energetics of Biological Macromolecules
Edited by MICHAEL L. JOHNSON AND GARY K. ACKERS

VOLUME 260. Mitochondrial Biogenesis and Genetics (Part A)
Edited by GIUSEPPE M. ATTARDI AND ANNE CHOMYN

VOLUME 261. Nuclear Magnetic Resonance and Nucleic Acids
Edited by THOMAS L. JAMES

VOLUME 352. Redox Cell Biology and Genetics (Part A)
Edited by CHANDAN K. SEN AND LESTER PACKER

VOLUME 353. Redox Cell Biology and Genetics (Part B)
Edited by CHANDAN K. SEN AND LESTER PACKER

VOLUME 354. Enzyme Kinetics and Mechanisms (Part F: Detection and Characterization of Enzyme Reaction Intermediates)
Edited by DANIEL L. PURICH

VOLUME 355. Cumulative Subject Index Volumes 321–354

VOLUME 356. Laser Capture Microscopy and Microdissection
Edited by P. MICHAEL CONN

VOLUME 357. Cytochrome P450, Part C
Edited by ERIC F. JOHNSON AND MICHAEL R. WATERMAN

VOLUME 358. Bacterial Pathogenesis (Part C: Identification, Regulation, and Function of Virulence Factors)
Edited by VIRGINIA L. CLARK AND PATRIK M. BAVOIL

VOLUME 359. Nitric Oxide (Part D)
Edited by ENRIQUE CADENAS AND LESTER PACKER

VOLUME 360. Biophotonics (Part A)
Edited by GERARD MARRIOTT AND IAN PARKER

VOLUME 361. Biophotonics (Part B)
Edited by GERARD MARRIOTT AND IAN PARKER

VOLUME 362. Recognition of Carbohydrates in Biological Systems (Part A)
Edited by YUAN C. LEE AND REIKO T. LEE

VOLUME 363. Recognition of Carbohydrates in Biological Systems (Part B)
Edited by YUAN C. LEE AND REIKO T. LEE

VOLUME 364. Nuclear Receptors
Edited by DAVID W. RUSSELL AND DAVID J. MANGELSDORF

VOLUME 365. Differentiation of Embryonic Stem Cells
Edited by PAUL M. WASSAUMAN AND GORDON M. KELLER

VOLUME 366. Protein Phosphatases
Edited by SUSANNE KLUMPP AND JOSEF KRIEGLSTEIN

VOLUME 367. Liposomes (Part A)
Edited by NEJAT DÜZGÜNEŞ

VOLUME 368. Macromolecular Crystallography (Part C)
Edited by CHARLES W. CARTER, JR., AND ROBERT M. SWEET

VOLUME 369. Combinational Chemistry (Part B)
Edited by GUILLERMO A. MORALES AND BARRY A. BUNIN

VOLUME 370. RNA Polymerases and Associated Factors (Part C)
Edited by SANKAR L. ADHYA AND SUSAN GARGES

VOLUME 391. Liposomes (Part E)
Edited by NEJAT DÜZGÜNEŞ

VOLUME 392. RNA Interference
Edited by ENGELKE ROSSI

VOLUME 393. Circadian Rhythms
Edited by MICHAEL W. YOUNG

VOLUME 394. Nuclear Magnetic Resonance of Biological Macromolecules
(Part C)
Edited by THOMAS L. JAMES

VOLUME 395. Producing the Biochemical Data (Part B)
Edited by ELIZABETH A. ZIMMER AND ERIC H. ROALSON

VOLUME 396. Nitric Oxide (Part E)
Edited by LESTER PACKER AND ENRIQUE CADENAS

VOLUME 397. Environmental Microbiology
Edited by JARED R. LEADBETTER

VOLUME 398. Ubiquitin and Protein Degradation (Part A)
Edited by RAYMOND J. DESHAIES

VOLUME 399. Ubiquitin and Protein Degradation (Part B)
Edited by RAYMOND J. DESHAIES

VOLUME 400. Phase II Conjugation Enzymes and Transport Systems
Edited by HELMUT SIES AND LESTER PACKER

VOLUME 401. Glutathione Transferases and Gamma Glutamyl Transpeptidases
Edited by HELMUT SIES AND LESTER PACKER

VOLUME 402. Biological Mass Spectrometry
Edited by A. L. BURLINGAME

VOLUME 403. GTPases Regulating Membrane Targeting and Fusion
Edited by WILLIAM E. BALCH, CHANNING J. DER, AND ALAN HALL

VOLUME 404. GTPases Regulating Membrane Dynamics
Edited by WILLIAM E. BALCH, CHANNING J. DER, AND ALAN HALL

VOLUME 405. Mass Spectrometry: Modified Proteins and Glycoconjugates
Edited by A. L. BURLINGAME

VOLUME 406. Regulators and Effectors of Small GTPases: Rho Family
Edited by WILLIAM E. BALCH, CHANNING J. DER, AND ALAN HALL

VOLUME 407. Regulators and Effectors of Small GTPases: Ras Family
(in preparation)
Edited by WILLIAM E. BALCH, CHANNING J. DER, AND ALAN HALL

VOLUME 408. DNA Repair (Part A) (in preparation)
Edited by JUDITH L. CAMPBELL AND PAUL MODRICH

VOLUME 409. DNA Repair (Part B) (in preparation)
Edited by JUDITH L. CAMPBELL AND PAUL MODRICH

[1] Purification and Biochemical Properties of Rac1, 2, 3 and the Splice Variant Rac1b

By Lars Christian Haeusler, Lars Hemsath, Dennis Fiegen, Lars Blumenstein, Ulrike Herbrand, Patricia Stege, Radovan Dvorsky, and Mohammad Reza Ahmadian

Abstract

Rac proteins (Rac1, 1b, 2, 3) belong to the GTP-binding proteins (or GTPases) of the Ras superfamily and thus act as molecular switches cycling between an active GTP-bound and an inactive GDP-bound form through nucleotide exchange and hydrolysis. Like most other GTPases, these proteins adopt different conformations depending on the bound nucleotide, the main differences lying in the conformation of two short and flexible loop structures designated as the switch I and switch II region. The three distinct mammalian Rac isoforms, Rac1, 2 and 3, share a very high sequence identity (up to 90%), with Rac1b being an alternative splice variant of Rac1 with a 19 amino acid insertion in vicinity to the switch II region. We have demonstrated that Rac1 and Rac3 are very closely related with respect to their biochemical properties, such as effector interaction, nucleotide binding, and hydrolysis. In contrast, Rac2 displays a slower nucleotide association and is more efficiently activated by the Rac-GEF Tiam1. Modeling and normal mode analysis corroborate the hypothesis that the altered molecular dynamics of Rac2, in particular at the switch I region, may be responsible for different biochemical properties. On the other hand, our structural and biochemical analysis of Rac1b has shown that, compared with Rac1, Rac1b has an accelerated GEF-independent GDP/GTP-exchange and an impaired GTP-hydrolysis, accounting for a self-activating GTPase. This chapter discusses the use of fluorescence spectroscopic methods, allowing real-time monitoring of the interaction of nucleotides, regulators, and effectors with the Rac proteins at submicromolar concentrations and quantification of the kinetic and equilibrium constants.

Introduction

The GTP-binding Rac-like proteins act as tightly regulated molecular switches that cycle between an inactive GDP-bound and an active GTP-bound state in response to a variety of extracellular stimuli. The interconversion between both states is controlled by two intrinsically

METHODS IN ENZYMOLOGY, VOL. 406
0076-6879/06 $35.00
DOI: 10.1016/S0076-6879(06)06001-0

slow biochemical reactions, namely the GDP/GTP exchange and the GTP-hydrolysis. Both reactions can be accelerated by several orders of magnitude by guanine nucleotide exchange factors (GEFs) and GTPase activating proteins (GAPs), respectively (Vetter and Wittinghofer, 2001). The crystal structures of several GTPases in either state revealed that the switching mechanism depends on the conformational change of two regions, termed switch I and switch II (Dvorsky and Ahmadian, 2004; Ihara et al., 1998). In the GTP-bound state, the switch regions provide a platform for the selective interaction with effector proteins (Bishop and Hall, 2000) and thereby initiate downstream signaling. The three distinct mammalian Rac isoforms (Rac1, 2 and 3) being encoded by different genes share between 89–93% amino acid sequence identity (Didsbury et al., 1989; Haataja et al., 1997; Polakis et al., 1989; Wherlock and Mellor, 2002). Rac1, the best-investigated isoform, regulates gene expression, cell cycle progression, and rearrangement of the actin cytoskeleton (Michiels and Collard, 1999). Rac2 is proposed to be responsible for the regulation of the oxidative burst in hematopoietic cells (Dinauer et al., 2003). Rac3 has been shown to be serum-inducible and hyperactive in breast cancer cells, where it controls proliferation by a PAK-dependent pathway (Haataja et al., 1997; Mira et al., 2000). Rac1 and Rac3 are ubiquitously expressed and therefore regulate a wide variety of cellular processes, whereas Rac2 is predominantly expressed in cells of the hematopoietic lineage. Rac1b was discovered in human tumors as an alternative splice variant of Rac1 containing a 19 amino acid insertion (between codons 75 and 76) next to the switch II region (Jordan et al., 1999; Schnelzer et al., 2000). It has been suggested that this insertion may create a novel effector-binding site in Rac1b and thus participates in signaling pathways related to the neoplastic growth of the intestinal mucosa (Jordan et al., 1999). Recently, it has been shown that Rac1b neither interacts with Rho-GDI, PAK1 nor induces lamellipodia formation but is able to promote growth transformation of NIH3T3 cells in a similar way as a constitutive active Rac1 mutant (Matos et al., 2003; Singh et al., 2004). In this chapter, we outline the methods that allowed us to quantitatively measure the physical interaction between the Rac proteins and nucleotides, GEFs, effectors and GAPs (Fiegen et al., 2004; Haeusler et al., 2003), providing a more detailed insight into the biochemical properties of the four mammalian Rac proteins (Table I).

Materials and Methods

Buffers

All buffers were filtered (0.2 μm) and degassed.

TABLE I
BIOCHEMICAL PROPERTIES OF RAC1, RAC1B, RAC2, AND RAC3

	Rac1	Rac1b	Rac2	Rac3
MantGDP-binding				
k_{on} $(M^{-1} s^{-1})^a$	2.5×10^6	1.1×10^6	0.17×10^6	2.2×10^6
k_{off} $(s^{-1}$; no Tiam1$)^a$	0.7×10^{-4}	18×10^{-4}	1.2×10^{-4}	0.9×10^{-4}
K_{off} $(s^{-1}$; 5 μM Tiam1$)^a$	36×10^{-4}	19×10^{-4}	165×10^{-4}	32×10^{-4}
K_d (pM)	28	1640	705	41
PAK-GBD binding				
K_d $(\mu M$; 0–7.5 μM GBD$)^b$	0.49	3.55^d	0.13	0.61
GTP-hydrolysis				
rate (min^{-1}; no GAP$)^c$	0.11	0.0035	0.16	0.18
rate (min^{-1}; 8 μM GAP$)^c$	2.36	0.194	n.d.	n.d.

The dissociation constant (K_d) for the nucleotide binding values has been calculated from the association and the dissociation rate constants of the respective nucleotides ($K_d = k_{off}/k_{on}$). 0.1 μM^a, 0.2 μM^b, and 80 μM^c Rac proteins were used for these measurements. ^{d}Rac1b measurements were carried out at 10° due to the much faster nucleotide dissociation rate. n.d.: not determined.

Buffer A: 30 mM Tris/HCl, pH 7.5, 1 mM MgCl$_2$, 3 mM dithioerythritol (DTE), 50 mM NaCl
Buffer B: 30 mM Tris/HCl, pH 7.5, 1 mM MgCl$_2$, 10 mM Na$_2$HPO$_4$/NaH$_2$PO$_4$, 3 mM DTE
Buffer C: 100 mM K$_2$HPO$_4$/KH$_2$PO$_4$, pH 6.5, 10 mM tetrabutylammonium bromide, 7.5–25% acetonitrile.

Proteins

All proteins (i.e., Rac1, Rac1b, Rac2, Rac3), the catalytic domains of Tiam1 (DHPH; aa 1033–1404) and p50RhoGAP (GAP; aa 198–439), and the GTPase-binding domain (GBD) of αPAK (aa 57–141) were synthesized as glutathione-S-transferase (GST) fusion proteins in *Escherichia coli*. Bacterial lysates were applied to a glutathione-sepharose column (Amersham Pharmacia, Freiburg), and the GST-fusion proteins were eluted in buffer A containing 20 mM glutathione. Cleavage of GST was performed in batch by incubation with 1–2 units of thrombin (Serva) per milligram proteins for 4–16 h at 4°. Subsequently, the cleaved proteins were subjected to two additional chromatographic steps, a gel filtration (Superdex 75, Pharmacia) and a second glutathione-sepharose affinity chromatography, to obtain proteins of interest at high purity and to remove residual GST.

Gpp(NH)p-Bound and mantGpp(NH)p-Bound Rac Proteins

Gpp(NH)p-bound and mantGpp(NH)p-bound Rac proteins were prepared using the enzymatic activity of alkaline phosphatase (Roche Diagnostics), thus degrading bound GDP. Herein, 0.1–1 of the enzyme were incubated overnight at 4° with 1 mg GDP-bound GTPase in buffer A containing a 1.2 molar excess of Gpp(NH)p or mantGpp(NH)p, 200 mM ammonium sulfate, and 1 mM ZnCl$_2$. The GDP degradation can be readily monitored by high-performance liquid chromatography (HPLC) analysis. After complete GDP degradation, the protein solution was applied to a prepacked NAP-5 gel filtration column (Pharmacia, Uppsala, Sweden) in buffer B. The concentration of the (mant-)Gpp(NH)p-bound GTPase was determined by HPLC as described later. The proteins were stored at –80°.

Preparation of the Nucleotide-Free GTPases

Preparation of the nucleotide-free GTPases was carried out in two steps as originally described for H-Ras (John *et al.*, 1990). In the first step, the bound GDP was degraded by alkaline phosphatase and replaced by Gpp(CH2)p (another non-hydrolyzable GTP-analog, which is resistant to alkaline phosphatase) under the same conditions as described previously for the preparation of Gpp(NH)p-bound GTPases. After GDP was completely degraded, 1–2 snake venom phosphodiesterase (Sigma-Aldrich) per milligram GTPase was added to the solution to cleave Gpp(CH2)p between the α- and β-phosphate. The reaction was monitored by HPLC measurements, and after complete degradation of Gpp(CH2)p, both enzymes (alkaline phosphatase and phosphodiesterase) were inactivated by two cycles of snap freezing in liquid nitrogen and quick defrosting. In the presence of GMP or guanosine (the products of the enzymatic reactions), most GTPases are quite stable and can be stored at –80° for several months.

MantGDP-Bound GTPases

Fluorescent GDP-bound GTPases were prepared by loading nucleotide-free proteins with a 1.2-fold molar excess of mantGDP. Unbound nucleotides were separated from nucleotide-bound GTPases in buffer B on prepacked NAP-5 columns (Pharmacia). The concentrations of the respective nucleotide-bound GTPases were determined by HPLC as described later. The mantGDP-bound proteins were stored at –80°.

Analysis of Free or Protein-Bound Fluorescent Nucleotides by Reversed-Phase HPLC

Free- or protein-bound nucleotides (non-labeled or mant-labeled nucleotide) were analyzed by reversed-phase HPLC using buffer C, a C-18 column (ODS-Hypersil 5 μm, Bischoff, Leonberg, Germany), and a prefilter (Nucleosil 100 C18, Bischoff), which separates protein–nucleotide complexes by adsorbing the denatured protein. A reasonable elution time ranging between 2 and 7 minutes on a C-18 column is achieved in buffer C containing 7.5% acetonitrile for non-labeled nucleotides and 25% acetonitrile for fluorescently labeled nucleotides. The HPLC system is calibrated with a solution having a clearly defined nucleotide concentration. The molar extinction coefficients at 252 nm of 13,700 M^{-1} cm^{-1} for nonlabeled and 22,600 M^{-1} cm^{-1} for mant-labeled guanine nucleotides were used. The linear relationship between peak area and nucleotide concentration allows calculating the concentration of nucleotide samples by integration of the respective peak areas.

Fluorescence Spectroscopic Methods

The stopped-flow instrument is routinely used for the analysis of rapid kinetics, such as guanine nucleotide association and GEF-catalyzed nucleotide dissociation (single turnover conditions) and the interaction between GTPases and their interacting partners (Ahmadian et al., 2002). The instrument (Applied Photophysics SX16MV) used had a fully automated operation modus and allowed collection of up to 1000 data points within a time window of 100 msec up to 1000 sec. In the fully automated measuring step, equal volumes of two previously prepared solutions are rapidly injected into a mixing chamber where the fluorescence can be detected directly after the rapid mixing has occurred. The reaction time of the apparatus is approximately 2 msec. To obtain high accuracy, several identical measurements are recorded and averaged. For mant-nucleotides, the used excitation wavelength is 366 nm, with the fluorescence being detected by a photomultiplier behind a cutoff-filter at wavelengths above 408 nm. Reactions that are not completed within the 1000-sec timeframe, such as nucleotide dissociation and indirect detection of GTPase-effector interactions using the guanine nucleotide dissociation inhibition (GDI) assay, can be monitored with a fluorescence spectrometer (Perkin-Elmer LS50B and SPEX Instruments FluoroMax II). With a temperature-controlled four-position turret, four experiments can be performed simultaneously, providing a better comparability of the single setups. The fluorescent component is provided in four quartz cuvettes (Hellma) that are thermally equilibrated until a constant fluorescence signal is reached. After addition of the reaction partner into the cuvette, a time-resolved fluorescence change can be monitored. For mant-nucleotides, maximal fluorescence is obtained at excitation and emission wavelengths of 366 and 450 nm, respectively.

Assays

Nucleotide Binding

The equilibrium dissociation constants (K_d) of high-affinity interactions can be obtained by the kinetic approach from the ratio of the dissociation rate constant (k_{off}) and the association rate constant (k_{on}). Taking advantage of the large change in fluorescence intensity, the time course of the association of the nucleotide-free Rac proteins and the mant-labeled nucleotide can be monitored after rapidly mixing the two components in a stopped-flow instrument. The association of 0.1 μM mantGDP with increasing concentrations of nucleotide-free Rac (0.5–10 μM) is measured in buffer B at 25°, which leads to an incremental increase in fluorescence intensity. The association kinetics exhibit a single exponential behavior under pseudo first-order conditions. The association rate constant (k_{on}) is determined from the slope of the linear regression of the observed rate constants (k_{obs}) plotted versus the concentration of nucleotide-free Rac. Because these experiments are performed under pseudo first-order conditions, the Rac proteins need to be in high molar excess. In principle, these experiments can be carried out for all other GTPases. However, the most important limiting factor is the accessibility and stability of the nucleotide free form of the respective GTPase. Measurements of the intrinsic nucleotide dissociation rate of the Rac proteins are carried out in a competitive displacement experiment by preincubating 0.1 μM mantGDP-bound Rac in buffer B at 25° and a final cuvette volume of 600 μl. The reaction is initiated by adding a large molar excess of unlabeled GDP (20 μM), leading to a very slow single exponential decrease in fluorescence. The presence of a large excess of unlabeled nucleotide ensures both a constant concentration for the exchange and a negligible signal of the reverse reaction, namely the reassociation of mantGDP. To make sure that the dissociation reaction has been completed, 20 mM EDTA was added to the sample, depleting Rac-bound magnesium ions and leading to a complete spontaneous nucleotide release. The exponential curve fitted to the obtained data yields the dissociation rate constant (k_{off}).

The dissociation constant (K_d), calculated from the kinetic parameters of nucleotide dissociation and association reactions ($K_d = k_{off}/k_{on}$) of the Rac proteins demonstrates the highest mantGDP affinity for Rac1 and the lowest for Rac1b with the following order Rac1>Rac3>Rac2>Rac1b (Table I). The overall affinity of Rac proteins for mantGDP is, particularly in the case of Rac1 and 3, in the same range as reported before for the members of other small GTPase families such as Ras (John et al., 1990), Ran (Klebe et al., 1995), and Rab (Simon et al., 1996). It is important to note that we obtained similar data for mantGTP binding to the nucleotide-free Rac proteins, suggesting that the presence of the γ-phosphate has no influence on nucleotide affinity.

GEF-Catalyzed Nucleotide Dissociation (GEF) Assay

A number of techniques are available for investigating the guanine nucleotide exchange of GTPases. A more detailed analysis of the interaction of the exchange factor with the GTPase–nucleotide complex is provided by fluorescence measurements in a stopped-flow instrument as described for Ran/RCC1 (Klebe et al., 1995), Ras/Cdc25 (Lenzen et al., 1998), Rap/C3G (van den Berghe et al., 1999), Rab/Vps9 (Esters et al., 2001), Rho/GDS (Hutchinson and Eccleston, 2000), and Rho/p190 (van Horck et al., 2001). The Tiam1-catalyzed mantGDP dissociation reaction of the Rac proteins is measured under the same conditions as described for the intrinsic mantGDP dissociation. These fast reactions are monitored in a stopped-flow instrument by rapid mixing of 0.1 μM Rac·mantGDP and 5 μM DHPH domain of Tiam1 with 20 μM GDP in buffer B at 25°. Using increasing GEF concentrations allows us to quantitatively determine the maximal GEF-catalyzed nucleotide dissociation rate (k_{max}) and the apparent K_d value for the interaction between GEF and nucleotide-bound Rac. A maximal Tiam1-activity of 6000-fold acceleration of intrinsic nucleotide dissociation was obtained for Rac2, which was 30-fold higher than that for Rac1 (Haeusler et al., 2003). We also have obtained an overall rate enhancement of 40,000-fold for the DHPH domain of the Rho-specific p190GEF (Haeusler et al., in preparation), suggesting that the DH-containing proteins can be in principle very efficient GEFs.

Effector-Induced Nucleotide Dissociation Inhibition (GDI) Assay

GTPase–effector interactions can be quantitatively analyzed on the basis of the observation that effector GBDs bind close to the nucleotide binding region of the GTPase and, thereby, in contrast to the GEFs, inhibit the nucleotide dissociation (GDI effect). This method is used to determine the binding affinity of effector domains for their GTPases in an indirect manner, as described for Ras/Raf kinase (Herrmann et al., 1995, 1996), Ras/RalGDS (Linnemann, 2002; Rudolph et al., 1999), Ras/Byr2 (Scheffzek et al.,

2001), and Rho/Rhotekin, Rho/PKN, and Rho/Rho kinase interactions (Blumenstein and Ahmadian, 2004). Increasing concentrations of PAK-GBD result in an incremental inhibition of the mantGppNHp dissociation from Rac, using 0.2 μM Rac-mantGppNHp and 40 μM GppNHp in buffer B at 25°. Individual observed rate constants (k_{obs}) are calculated from the respective single exponential decay and plotted against the PAK concentrations to obtain the K_d values according to an equation described by Hermann *et al.* (1995, 1996).

Intrinsic and GAP-Stimulated GTP-Hydrolysis Reaction

The intrinsic and GAP-stimulated GTP-hydrolysis reaction of the Rac proteins can be measured using various methods as previously described (Ahmadian *et al.*, 2002). A generally useful and accurate method is HPLC. Hereby, 80 μM nucleotide-free GTPase and 70 μM GTP are incubated at 25° in a final volume 200 μl buffer B containing 10 mM MgCl$_2$ in the absence and presence of 8 μM GAP. Samples of 25 μl are taken at different time intervals and snap frozen in liquid nitrogen to halt the GTP–hydrolysis reaction. The samples (25 μl) are then applied to a RP-18 HPLC column equilibrated with buffer C containing 7.5% acetonitrile and isocratically eluted at a flow rate of 1.8 ml/min. Concentrations of GDP and GTP are determined from the area of the elution peaks of GTP and GDP to calculate the relative GTP content, represented by the ratio [GTP]/ ([GTP]+[GDP]), to describe the reaction progress. By plotting the relative GTP content against the time, the hydrolysis rate k_{cat} can be determined with a single exponential fit of the data points.

Most recently, we synthesized a new fluorescent GTP, which is sensitive toward the GTP hydrolysis–induced conformational changes of the Rho GTPases and thus allows direct monitoring of the kinetics of the intrinsic and GAP-catalyzed GTP–hydrolysis reactions (Eberth *et al.*, submitted).

Biochemical Evaluation of the Rac Proteins

The biochemical assays described previously together with structural studies have shown that the Rac isoforms exhibit different properties concerning ligand– and protein–protein interactions (Table I; Fiegen *et al.*, 2004; Haeusler *et al.*, 2003). Whereas Rac1 and Rac3 behave almost identically, Rac2 revealed (1) a 25-fold lower nucleotide affinity because of a decreased nucleotide association rate, (2) a slightly higher PAK binding affinity, and (3) a significant increase in Tiam1-catalyzed nucleotide disso-ciation. These aberrant properties are the consequence of different confor-mational flexibilities in the switch I region (Haeusler *et al.*, 2003). In contrast, Rac1b showed (1) a dramatic increase in its intrinsic nucleotide dissociation, which cannot be further enhanced by Tiam1; (2) a drastic decrease in intrinsic GTP-hydrolysis rate, which can be restored by GAP; and (3) reduced affinity for PAK-GBD. The altered characteristics of Rac1b are based on an open switch I conformation and a highly mobile switch II, which are induced by the 19 amino acid insertion (Fiegen *et al.*, 2004).

Acknowledgments

Work in the authors' laboratory is supported, in part, by a European Community Marie Curie Fellowship, the Volkswagen-Stiftung, the Deutsche Forschungsgemeinschaft, the Max Planck Society, the Verband der Chemischen Industrie and the Bundesministerium für Bildung und Forschung.

References

Ahmadian, M. R., Wittinghofer, A., and Herrmann, C. (2002). Fluorescence methods in the study of small GTP-binding proteins. *Methods Mol. Biol.* **189,** 45–63.

Bishop, A. L., and Hall, A. (2000). Rho GTPases and their effector proteins. *Biochem. J.* **348,** 241–255.

Blumenstein, L., and Ahmadian, M. R. (2004). Models of the cooperative mechanism for Rho-effector recognition: Implications for RhoA-mediated effector activation. *J. Biol. Chem.* **279,** 53419–53426.

Didsbury, J., Weber, R. F., Bokoch, G. M., Evans, T., and Snyderman, R. (1989). Rac, a novel Ras-related family of proteins that are botulinum toxin substrates. *J. Biol. Chem.* **264,** 16378–16382.

Dinauer, M. C. (2003). Regulation of neutrophil function by Rac GTPases. *Curr. Opin. Hematol.* **10,** 8–15.

Dvorsky, R., and Ahmadian, M. R. (2004). Always look on the bright site of Rho—structural implications for a conserved intermolecular interface. *EMBO Rep.* **5,** 1130–1136.

Esters, H., Alexandrov, K., Iakovenko, A., Ivanova, T., Thoma, N., Rybin, V., Zerial, M., Scheidig, A. J., and Goody, R. S. (2001). Vps9, Rabex-5 and DSS4: Proteins with weak but distinct nucleotide-exchange activities for Rab proteins. *J. Mol. Biol.* **310,** 141–156.

Fiegen, D., Haeusler, L. C., Blumenstein, L., Herbrand, U., Dvorsky, R., Vetter, I. R., and Ahmadian, M. R. (2004). Alternative splicing of Rac1 creates a self-activating GTPase. *J. Biol. Chem.* **279,** 4743–4749.

Haataja, L., Groffen, J., and Heisterkamp, N. (1997). Characterization of RAC3, a novel member of the Rho family. *J. Biol. Chem.* **272,** 20384–20388.

Haeusler, L. C., Blumenstein, L., Stege, P., Dvorsky, R., and Ahmadian, M. R. (2003). Comparative functional analysis of the Rac GTPases. *FEBS Lett.* **555,** 556–560.

Herrmann, C., Martin, G. A., and Wittinghofer, A. (1995). Quantitative analysis of the complex between p21ras and the Ras-binding domain of the human Raf-1 protein kinase. *J. Biol. Chem.* **270,** 2901–2905.

Herrmann, C., Horn, G., Spaargaren, M., and Wittinghofer, A. (1996). Differential interaction of the ras family GTP-binding proteins H-Ras, Rap1A, and R-Ras with the putative effector molecules Raf kinase and Ral-guanine nucleotide exchange factor. *J. Biol. Chem.* **271,** 6794–6800.

Hutchinson, J. P., and Eccleston, J. F. (2000). Mechanism of nucleotide release from Rho by the GDP dissociation stimulator protein. *Biochemistry* **39,** 11348–11359.

Ihara, K., Muraguchi, S., Kato, M., Shimizu, T., Shirakawa, M., Kuroda, S., Kaibuchi, K., and Hakoshima, T. (1998). Crystal structure of human RhoA in a dominantly active form complexed with a GTP analogue. *J. Biol. Chem.* **273,** 9656–9666.

John, J., Sohmen, R., Feuerstein, J., Linke, R., Wittinghofer, A., and Goody, R. S. (1990). Kinetics of interaction of nucleotides with nucleotide-free H-ras p21. *Biochemistry* **29,** 6058–6065.

Jordan, P., Brazao, R., Boavida, M. G., Gespach, C., and Chastre, E. (1999). Cloning of a novel human Rac1b splice variant with increased expression in colorectal tumors. *Oncogene* **18,** 6835–6839.

Klebe, C., Prinz, H., Wittinghofer, A., and Goody, R. S. (1995). The kinetic mechanism of Ran–nucleotide exchange catalyzed by RCC1. *Biochemistry* **34,** 12543–12552.

Lenzen, C., Cool, R. H., Prinz, H., Kuhlmann, J., and Wittinghofer, A. (1998). Kinetic analysis by fluorescence of the interaction between Ras and the catalytic domain of the guanine nucleotide exchange factor Cdc25Mm. *Biochemistry* **37,** 7420–7430.

Linnemann, T., Kiel, C., Herter, P., and Herrmann, C. (2002). The activation of RalGDS can be achieved independently of its Ras binding domain. Implications of an activation mechanism in Ras effector specificity and signal distribution. *J. Biol. Chem.* **277,** 7831–7837.

Matos, P., Collard, J. G., and Jordan, P. (2003). Tumor-related alternatively spliced Rac1b is not regulated by Rho-GDP dissociation inhibitors and exhibits selective downstream signaling. *J. Biol. Chem.* **278,** 50442–50448.

Michiels, F., and Collard, J. G. (1999). Rho-like GTPases: Their role in cell adhesion and invasion. *Biochem. Soc. Symp.* **65,** 125–146.

Mira, J. P., Benard, V., Groffen, J., Sanders, L. C., and Knaus, U. G. (2000). Endogenous, hyperactive Rac3 controls proliferation of breast cancer cells by a p21-activated kinase-dependent pathway. *Proc. Natl. Acad. Sci. USA* **97,** 185–189.

Polakis, P. G., Weber, R. F., Nevins, B., Didsbury, J. R., Evans, T., and Snyderman, R. (1989). Identification of the Ral and Rac1 gene products, low molecular mass GTP-binding proteins from human platelets. *J. Biol. Chem.* **264,** 16383–16389.

Rudolph, M. G., Wittinghofer, A., and Vetter, I. R. (1999). Nucleotide binding to the G12V-mutant of Cdc42 investigated by X-ray diffraction and fluorescence spectroscopy: Two different nucleotide states in one crystal. *Protein Sci.* **8,** 778–787.

Scheffzek, K., Grunewald, P., Wohlgemuth, S., Kabsch, W., Tu, H., Wigler, M., Wittinghofer, A., and Herrmann, C. (2001). The Ras-Byr2RBD complex: Structural basis for Ras effector recognition in yeast. *Structure* **9,** 1043–1050.

Schnelzer, A., Prechtel, D., Knaus, U., Dehne, K., Gerhard, M., Graeff, H., Harbeck, N., Schmitt, M., and Lengyel, E. (2000). Rac1 in human breast cancer: Overexpression, mutation analysis, and characterization of a new isoform, Rac1b. *Oncogene* **19,** 3013–3020.

Simon, I., Zerial, M., and Goody, R. S. (1996). Kinetics of interaction of Rab5 and Rab7 with nucleotides and magnesium ions. *J. Biol. Chem.* **271,** 20470–20478.

Singh, A., Karnoub, A. E., Palmby, T. R., Lengyel, E., Sondek, J., and Der, C. J. (2004). Rac1b, a tumor associated, constitutively active Rac1 splice variant, promotes cellular transformation. *Oncogene* **23,** 9369–9380.

van den Berghe, N., Cool, R. H., and Wittinghofer, A. (1999). Discriminatory residues in Ras and Rap for guanine nucleotide exchange factor recognition. *J. Biol. Chem.* **274,** 11078–11085.

van Horck, F. P., Ahmadian, M.R, Haeusler, L. C., Moolenaar, W. H., and Kranenburg, O. (2001). Characterisation of p190RhoGEF: A RhoA-specific guanine nucleotide exchange factor that interacts with microtubules. *J. Biol. Chem.* **276,** 4948–4956.

Vetter, I. R., and Wittinghofer, A. (2001). The guanine nucleotide-binding switch in three dimensions. *Science* **294,** 1299–1304.

Wherlock, M., and Mellor, H. (2002). The Rho GTPase family: A Racs to Wrchs story. *J. Cell. Sci.* **115,** 239–240.

[2] Biochemical Analyses of the Wrch Atypical Rho Family GTPases

By ADAM SHUTES, ANASTACIA C. BERZAT, EMILY J. CHENETTE, ADRIENNE D. COX, and CHANNING J. DER

Abstract

The Rho family of GTPases comprises a major branch of the Ras superfamily of small GTPases. To date, at least 22 human members have been identified. However, most of our knowledge of Rho GTPase function comes from the study of the three classical Rho GTPases, RhoA, Rac1, and Cdc42. These Rho GTPases function as GDP/GTP-related binary switches that are activated by diverse extracellular signal–mediated stimuli. The activated GTPases then interact with downstream effectors to regulate cytoplasmic signaling networks that in turn regulate actin organization, cell cycle progression, and gene expression. Recently, studies have begun to explore the regulation and function of some of the lesser-known members of the Rho GTPase family. Wrch-1 (*Wnt*-regulated *Cdc42 homolog-1*) and the closely related Chp (*Cdc42 homologous protein*)/Wrch-2 protein comprise a distinct branch of the mammalian Rho GTPase family. Although both share significant sequence and functional similarities with Cdc42, Wrch proteins possess additional N- and C-terminal sequences that distinguish them from the classical Rho GTPases (Cdc42, RhoA, and Rac1). We have determined that Wrch-1 and Wrch2 exhibit unusual GDP/GTP binding properties and undergo posttranslational lipid modifications distinct from those of the classical Rho GTPases. In this chapter, we summarize our experimental approaches used to characterize the biochemical properties of these atypical Rho GTPases.

METHODS IN ENZYMOLOGY, VOL. 406
0076-6879/06 $35.00
DOI: 10.1016/S0076-6879(06)06002-2

Introduction

The Wnt-1 regulated Cdc42 homolog-1 (Wrch-1) was first identified as a gene up-regulated in Wnt-1–transformed C57MG mouse mammary cells (Tao et al., 2001). Together with the Cdc42 homologous protein Chp, also called Wrch-2, these related GTPases comprise a distinct branch of the human Rho family of small GTPases (Wennerberg and Der, 2004). Wrch-1 is a 258-amino acid protein (23.9 kDa) that shares 52% amino sequence identity with Cdc42 and contains conserved residues in the G-domain, which are required for GTP hydrolysis (G58 and Q107 in Wrch-1, G12 and Q61 in Cdc42). However, Wrch-1 and Chp contain unique sequence extensions at their N- and C-termini whose roles seem to be important in the negative regulation of Wrch-1 and Chp GTPase activity (Shutes et al., 2004) and in the subcellular localization and membrane association essential for their biological activity (Berzat et al., 2005; Chenette et al., 2005).

In this chapter, we describe the preparation and purification of recombinant glutathione S-transferase (GST)-Wrch-1 and hexa-histidine (His_6)-tagged Wrch-1 protein from *Escherichia coli* bacteria and demonstrate the measurement of its basic GTPase characteristics: GTP hydrolysis and nucleotide exchange rates. In addition, we describe a cell-based labeling assay to monitor the covalent modification of C-terminal cysteine residues found in Wrch proteins by the fatty acid palmitate, as well as a method to evaluate the effects of pharmacologically inhibiting this modification on Wrch subcellular localization.

Experimental Methods to Evaluate Wrch-1 GDP/GTP Regulation

Molecular Constructs of Wrch-1

To generate bacterial expression vectors encoding human Wrch-1, we used the polymerase chain reaction (PCR) DNA amplification was done using oligonucleotide primers to the 5' (GGA TCC ATG CCC CCG CAG) and 3' (CTC GAG TCA TTC TTT GCA TTT GTC C) ends of Wrch-1. These primers contain *Bam*HI and *Xho*I restriction sites, respectively. The fragments were then ligated into the vectors pGEX 4T1 (27–4580-01, Amersham Pharmacia Biotech) or pPROEX HTb (Invitrogen) that had been previously digested with *Bam*HI and *Xho*I. The sequence-verified plasmids were transformed into the BL21 strain of *E. coli* (Promega).

For expression in mammalian cells, PCR-mediated DNA amplification was used to introduce 5' and 3' *Bam*HI sites flanking Wrch-1(WT) for subcloning into the 5' *Bgl*II and 3' *Bam*HI sites of the pEGFP-C1 (MCS) to form an N-terminal green fluorescent protein-tagged Wrch-1 chimeric protein.

Preparation of Bacterially Expressed GST-Wrch-1 and His$_6$-Wrch-1

Both GST- and His$_6$-tagged forms of Wrch-1 protein can be expressed in the BL21 strain of *E. coli* bacteria (Promega). The preparation of Wrch-1 is a straightforward process, although the affinity purification steps for either GST-Wrch-1 or His$_6$-Wrch-1 differ, and as such, each is described later. In contrast, we found that Wrch-2/Chp protein was not easily amenable to expression in bacteria; consequently, biochemical analyses have focused on Wrch-1.

Place a stab of the relevant glycerol stock (stored at $-80°$) into 50 ml of LB medium and incubate shaking overnight at $37°$. On the following day, place 20 ml of the overnight culture into 1 L of warmed LB broth in a 2-L baffled shaker flask and incubate at $37°$ in a gyratory incubator. Take hourly OD_{600} readings, and when the culture reaches an $OD_{600} \approx 0.8$ (normally approximately 2 h), induce each 1-L culture with a final concentration of 1 mM isopropyl-beta-D-thiogalactopyranoside (IPTG) (e.g., 1 ml of 1 M IPTG stock). Remember to take a preinduction sample of the bacteria: 200 μl of cells, centrifuge at 14,000 rpm for 1 min in a benchtop centrifuge, and resuspend the pellet in 40 μl of protein sample buffer. Store at $-20°$. This will be used as a control for induction efficiency of the Wrch-1 protein.

Induce the bacteria for 3 h at $37°$, at which time take a postinduction sample (200 μl of cells, centrifuge, and resuspend the pellet in 40 μl of protein sample buffer). Harvest the bacteria by pouring the cultures into 250-ml wide-mouth centrifuge bottles (e.g., Nalgene) and centrifuge these at 4000 rpm for 10 min at $4°$ in a Sorvall RSS-15 refrigerated centrifuge or similar. Remove the final supernatant, leaving a cell pellet at the bottom of the centrifuge tube. Resuspend the pellets, pooling them into a 50-ml conical centrifuge tube (e.g., BD Falcon) in a total of 20 ml of Tris/GDP buffer (20 mM Tris-HCl, pH 7.5, 50 mM NaCl, 1 mM MgCl$_2$, 1 μM GDP), for GST-Wrch-1, or 20 ml of Tris/GDP/imidazole buffer (20 mM Tris-HCl, pH 7.5, 50 mM NaCl, 1 mM MgCl$_2$, 1 μM GDP, 1 μM imidazole) for His$_6$-Wrch-1. This is best performed by adding the entire 20-ml buffer to the first centrifuge bottle, resuspending that pellet, and then moving the entire suspension to the next centrifuge bottle. Snap-freeze the final resuspended bacteria (normally approximately 25 ml total final volume) in liquid nitrogen and store at $-80°$ until ready to perform the purification step.

Purification of GST-Wrch-1

GST-Wrch-1 can be purified by the batch-loading technique, as well as by affinity chromatography on a fast protein liquid chromatography (FPLC) machine. Batch loading is more convenient for GTPase studies, whereas FPLC purification is more effective and best suited to producing pure, soluble protein (for exchange assays, for example). Thaw the bacterial suspension on ice, and then sonicate on ice for 1 min at 14 W (RMS). The suspension should appear less viscous than it was before, and small white membranous bodies should appear in the suspension during a productive sonication. Pour the sonicate into a chilled 25-ml centrifuge tube (Falcon) and centrifuge in a Sorvall RC-5B centrifuge with an SA-300 rotor (or comparable combination) for 30 min at 10,000 rpm and 4°. The resulting supernatant should be straw colored, and there should also be a significantly sized pellet. At this stage, take 20 μl of supernatant and place it into 40 μl of protein sample buffer; store at –80° with the other sequential monitoring samples.

For batch-loading, take the supernatant into a cold 50-ml conical tube (BD Falcon) and add 150 μl of a 50% slurry of either glutathione-agarose or glutathione-Sepharose, preequilibrated in 20 mM Tris-HCl, pH 7.5, 50 mM NaCl, 1 mM MgCl$_2$. Allow this to agitate at 4° for 30 min so the protein can bind to the beads. After the 30 min binding step, centrifuge the 50-ml Falcon tube at 4000 rpm in a swinging bucket-style centrifuge at 4° for 5 min, so as to pellet the bead-bound Wrch-1 protein. Remove the supernatant and resuspend the bead-bound GST-Wrch-1 pellet in 1 ml of cold Tris/GDP buffer and move to a 1.5-ml microcentrifuge tube. Centrifuge in a benchtop centrifuge at 14,000 rpm for 1 min at 4° to pellet the glutathione-agarose beads. Remove the supernatant and resuspend the pellet in another 1 ml of Tris/GDP buffer. Repeat these washing steps a total of three times. When washing is complete, remove the supernatant and add an equal amount of 100% glycerol to the beads (i.e., 150 μl). Take a sample of the beads into 40 μl protein sample buffer. The bead-bound protein can either be used immediately or aliquoted into 0.5-ml microcentrifuge tubes, snap frozen in liquid nitrogen, and stored in a –80° freezer. A –20° freezer is insufficient, and the Wrch-1 will lose activity over the period of a week when stored at this temperature. It is convenient to keep Wrch-1 on the beads for as long as possible, because this facilitates nucleotide loading for GTPase and nucleotide exchange assays, and the protein appears quite stable in this state. Shown in Fig. 1A is a 12% SDS-polyacrylamide gel (PAGE) separation of a batch GST-Wrch-1 protein preparation.

For purification on an FPLC, load the supernatant from the 10,000 rpm centrifugation step onto a 1-ml GSTrap FF column (Amersham 17–5130-01). Use the built in "GST purification" protocol, with a wash buffer containing 20 mM Tris-HCl (pH 7.5), 50 mM NaCl, 1 mM MgCl$_2$, 1 μM GDP, and an elution buffer containing 20 mM Tris-HCl (pH 7.5), 50 mM NaCl, 1 mM MgCl$_2$, 1 μM GDP, 100 mM glutathione. Collect the fractions and run the samples (20-μl sample combined with 40 μl of protein sample

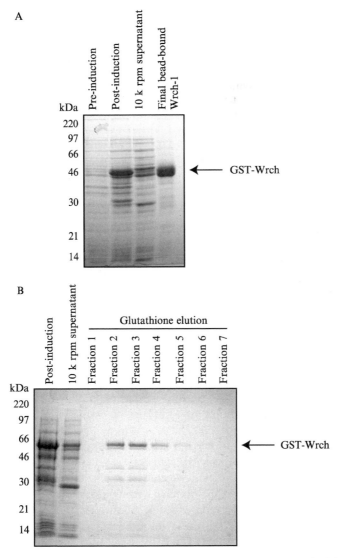

FIG. 1. Purification of GST-Wrch-1. (A) GST-Wrch-1 purification. A 12% SDS-PAGE gel of GST-Wrch-1 purification by batch loading onto glutathione-agarose beads. (B) A 12% SDS-PAGE gel of GST-Wrch-1 purification through affinity chromatography using a GSTrap column attached to an FPLC machine.

buffer) on an SDS-PAGE gel. Wrch-1 will elute primarily in the second and third fractions (Fig. 1B). To remove the glutathione, dialyze overnight in 20 mM Tris-HCl (pH 7.5), 50 mM NaCl, 1 mM MgCl$_2$, 1 μM GDP. The fractions containing Wrch-1 should be pooled and placed into a hydrated Slide-A-Lyzer dialysis cassette (Pierce 66380 or similar) overnight (or for a minimum of 4 h) at 4° with stirring. We also find that this is an excellent point, if desired, to perform an exchange reaction, by simply replacing the GDP in the dialysis buffer with GMPPNP. Because of the rapid exchange rate of Wrch-1, the exchange will be complete within the minimum 4 h of dialysis.

If desired, the protein solution can be concentrated further. We use centrifugal concentrators (Vivascience VS2022) over nitrogen-pressure–driven concentrators, which often have problems with clogging of proteins onto the membrane. A 4-L preparation and purification performed using the FPLC will often produce between 0.5–1 mg of purified Wrch-1 protein.

We have found that purification by both batch-loading and FPLC produces full-length protein, which does not suffer from the C-terminal truncation so often a problem with preparation of small GTPases. This can be observed by the use of a rabbit polyclonal antiserum that we have made against the C-terminus of Wrch-1 (Shutes and Der, unpublished) in a Western blot of a protein preparation gel.

Purification of His$_6$-Wrch-1

Recombinant His$_6$-Wrch protein is purified using a 5-ml HiTrap Chelating HP affinity column (Amersham 17–0409-01). Before use, the column needs to be prepared. Using a 20-ml syringe, attached to the correct syringe-adaptor, apply each of the wash steps: add 10 ml of N1/GDP buffer (10 mM NaH$_2$PO$_4$, 10 mM Na$_2$HPO$_4$, 100 mM NaCl, 10 mM imidazole, 1 μM GDP, pH 7.5) and 1 mM EDTA to the column and let the column sit for 1 min. Wash the column with 20 ml of H$_2$O, and then with 10 ml NaOH to remove any nonspecific bound proteins or membranes. Wash again with 20 ml H$_2$O. Add 10 ml of 0.1 M NiSO$_4$—the column will become a pale green. Then wash again with 20 ml H$_2$O to remove any excess NiSO$_4$. Finally, wash the column with 10 ml N1/GDP buffer. The column should now be attached to an FPLC (the Pharmacia Äkta Prime is a capable system at a reasonable price).

Thaw the bacterial suspension on ice and then sonicate on ice for 1 min at 14,000 rpm. Pour the sonicate into a 25-ml centrifuge tube and centrifuge in a Sorvall RC-5B in an SA-300 rotor for 30 min at 10,000 rpm and 4°. The resulting supernatant should be straw colored, and there should also be a reasonably sized pellet. At this stage, remove 20 μl of supernatant and add together and mix with 40 μl of protein sample buffer; store at –80° until ready to monitor purification steps. The remainder of the supernatant should be loaded onto the FPLC's loop system. The "wash" tubing should be connected to a reservoir of filtered buffer N1/GDP, and the "elution" tubing should be connected to a reservoir of filtered N2/GDP elution buffer (10 mM NaH$_2$PO$_4$, 10 mM Na$_2$HPO$_4$, 100 mM NaCl, 1 M imidazole, 1 μM GDP, pH 7.5). On the Äkta Prime system, select "His-tag Purification" from its preset protocol menu, enter the volume of supernatant to be loaded onto the column, and press "Start."

When purification is complete (which takes approximately 90 min), 20-μl samples can be taken from every third fraction (into 40 μl protein sample buffer) and resolved by 12% SDS-PAGE to match to the trace of the recorder (if one is connected to the FPLC) with protein concentrations. Pool the desired His$_6$-Wrch-1 fractions into a 15-ml conical centrifuge tube (BD Falcon) and take a 20-μl sample into 40 μl of protein sample buffer. We find that His$_6$-Wrch-1 elutes approximately 8–11 fractions. To remove the high concentrations of imidazole, the protein solution must be dialyzed. Place the protein solution in 3-ml Slide-a-Lyzer dialysis cassettes (Pierce 66380 or similar) and dialyze overnight (or for a minimum of 4 h) at 4° against 20 mM Tris-HCl, 50 mM NaCl, 1 mM MgCl$_2$, 1 μM GDP. The protein is stable at 4° in the presence of Mg^{2+} and GDP.

The final stage involves a concentration of the purified His$_6$-Wrch-1 protein. Using centrifugal concentrators, we have achieved final concentrations of up to 1.5 mM. The protein should be dispensed in 25–100-μl aliquots, snap frozen in liquid nitrogen, and stored at –80° until required. A 12% SDS-PAGE gel from a preparation of His$_6$-ΔC Wrch-1 is shown in Fig. 2.

Nucleotide Loading of Wrch-1

The procedure described previously for the preparation and purification of Wrch-1 produces GDP-bound Wrch-1. The method of loading nucleotide onto Wrch-1 depends on the form of Wrch-1 prepared—either bead-bound GST-Wrch-1 or free His$_6$-Wrch-1. Bead-bound GST-Wrch-1 facilitates the washing stages in the nucleotide loading procedure, whereas the His$_6$-Wrch-1 requires separation of excess nucleotide from the protein–nucleotide complex through a PD-10 column (Amersham Biosciences 17–0851-01). Unlike other small GTPases, we have found that it is important to have no EDTA present for the loading reaction. Not only is it unnecessary to accelerate the exchange reaction, but it also causes loss of nucleotide from Wrch-1 and loss of protein activity (as shown by anisotropy measurements in Shutes et al., 2004).

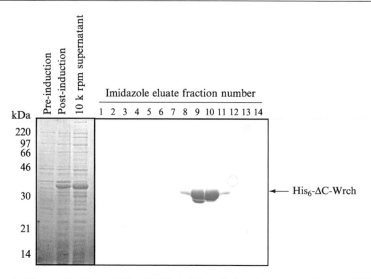

FIG. 2. The purification of His$_6$-ΔC-Wrch-1 by affinity chromatography on a HisTrap column attached to an FPLC.

GST-Wrch-1

This procedure can be used for loading any nucleotide (e.g., mant-nucleotides, available from Molecular Probes) onto the bead-bound GST-Wrch-1, although this particular example illustrates the loading of GTP.

Take freshly prepared, or freshly thawed, bead-bound GST-Wrch-1 in a 0.5-ml microcentrifuge tube and quickly wash 3 × with 200 μl of cold Tris buffer (20 mM Tris-HCl, pH 7.5, 50 mM NaCl, 1 mM MgCl$_2$) using a glass Hamilton syringe to remove the supernatant and a benchtop picofuge to precipitate the beads. Be sure to wash out any glycerol used to help stabilize the protein when frozen, and on the last wash, remove all the remaining supernatant. Resuspend the Wrch-1 beads in 100 μl of Wrch-1·GTP loading buffer (20 mM Tris-HCl, pH 7.5, 50 mM NaCl, 1 mM MgCl$_2$, 100 μM GTP) and incubate at room temperature for 1 min with gentle agitation. Place on ice and quickly wash the beads twice in 200 μl cold Tris buffer (20 mM Tris-HCl, pH 7.5, 50 mM NaCl, 1 mM MgCl$_2$), and then add 30 μl of elution buffer (20 mM Tris-HCl, pH 7.5, 50 mM NaCl, 1 mM MgCl$_2$, 1 mM glutathione). Incubate on ice for 10 min, again with gentle agitation. Remove the supernatant, which now contains free GST-Wrch-1·GTP, and measure the protein concentration. The GST-Wrch-1·GTP should be used immediately, because GTP hydrolysis occurs, albeit at a much slower rate, on ice.

His_6-Wrch-1

Loading of His_6-Wrch-1 with mant-nucleotide is more intricate than that of GST-Wrch-1 and requires separation of protein–nucleotide complex from nucleotide. Incubate His_6-Wrch-1 in buffer (20 mM Tris-HCl, pH 7.5, 50 mM NaCl, 1 mM MgCl$_2$, and 100 μM mant-nucleotide) in a total of 500 μl for 1 min. Immediately apply the incubation mix down a PD-10 column (Amersham Bioscience 17–0851-01), which has been equilibrated in 20 mM Tris-HCl, pH 7.5, 50 mM NaCl, and 1 mM MgCl$_2$. Wash through by applying further 500-μl aliquots of cold buffer and collect 500-μl fractions as they come off the column. This should be performed at 4° (a cold room is best). The fractions with the highest protein concentration should be isolated, measured, and snap frozen in 50 μl aliquots at –80°.

Measuring GTPase Activity of Wrch-1

To monitor the hydrolysis of small GTPases *in vitro*, we use the phosphate-binding protein probe (PBP) coupled to the MDCC fluorophore (MDCC-PBP) (Shutes and Der, 2005). For this method, we find that 2 μM of small GTPase provides a sufficient signal to measure the GTP hydrolysis in the presence of GAPs or effectors. To move between micrograms/ microliters of protein and molar concentrations, use the formula:

$$M = (\mu g/\mu l)/gmol^{-1} \qquad (1)$$

which represents:

$$\text{Molar concentration} = (\text{g/L concentration})/\text{Molecular weight} \qquad (2)$$

Prepare a black 96-well microtiter plate (Falcon) by adding 50 μl 2× GTP buffer (40 mM Tris-HCl, 100 mM NaCl, 2 mM MgCl$_2$) and 15 μM MDCC-PBP to a total of 100 μl (with fresh Milli-Q water). The MDCC-PBP needs to equilibrate to room temperature, which usually takes approximately 5 min, and the changes in fluorescence this equilibration causes can be followed on the spectrofluorimeter (λ_{ex} = 425 nm, λ_{em} = 465 nm). When a steady fluorescence is observed, add 2 μM Wrch-1·GTP and begin a new fluorescence measurement. We use the SpectroMax Gemini spectrofluorimeter (Gemini Instruments), which is a reasonable cost and capable system.

The process of GTP hydrolysis can be represented as:

$$\text{Wrch} \cdot \text{GTP} \cdot \text{Mg}^{2+} \overset{k_{+1}}{\rightarrow} \text{Wrch} \cdot \text{GDP} \cdot \text{P}_i \cdot \text{Mg}^{2+} \overset{k_{+2}}{\rightarrow} \text{Wrch} \cdot \text{GDP} \cdot \text{Mg}^{2+} + \text{P}_i$$

$$(3)$$

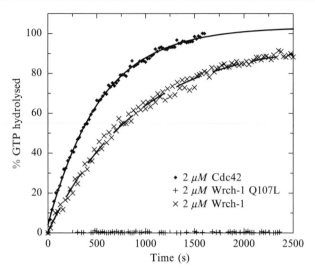

FIG. 3. GTP hydrolysis of Wrch-1; 2 μM GST-Wrch-1, GST-Cdc42 or GST-Wrch-1 (Q107L) was incubated in 20 mM Tris, pH 7.5, 1 mM MgCl$_2$, 50 mM NaCl, and 15 μM MDCC-PBP at 25°. The changes in fluorescent intensity were followed at λ_{ex} = 425 nm, λ_{em} = 465 nm, from which observed hydrolysis rates could be calculated.

It is the release of Wrch-bound P$_i$ (increasing the free P$_i$), k_{+2}, that is measured in this assay. The process can be fitted to a single exponential process. Data can be best normalized by a spreadsheet application such as Microsoft Excel and then analyzed (single exponential curve fitting, etc.) by applications such as ProFit (http://www.quansoft.com/) or Kaleidagraph (http://www.synergy.com/) for Mac OS X or Windows. Figure 3 shows the GTP hydrolysis rates of Wrch-1 and Cdc42. 2 μM Wrch-1 shows a similar rate of GTP hydrolysis (k_{obs} = 0.0011 s^{-1}) to that of 2 μM Cdc42 (k_{obs} = 0.0018 s^{-1}), but constitutively active Wrch-1 Q107L shows no GTPase activity.

Measuring the Nucleotide Exchange Rate of Wrch-1

The rate of Wrch-1 nucleotide exchange can be measured using the established fluorescent mant nucleotide (Leonard *et al.*, 1994). Nucleotide exchange can be simplified to a one-step process:

$$GTP + Wrch \cdot GDP \cdot Mg^{2+} \underset{k_{-1}}{\overset{k_{+1}}{\rightleftarrows}} GDP + Wrch \cdot GTP \cdot Mg^{2+} \quad (4)$$

It is often assumed that GTP and GDP will show similar on/off rates; however, this is not necessarily the case (Shutes *et al.*, 2002), and indeed

Wrch-1 loads GTP-analogs more rapidly than GDP analogs (data not shown). Strict kinetic analysis is not the purpose of this chapter, and a comparison of the rates of mantGDP association between small GTPases is usually sufficient to give an idea of the exchange rate for a particular small GTPase.

In a 96-well black microtiter plate and in the same spectrofluorometer as used in the previously described GAP assay, incubate 2 μM Wrch-1·GDP with excess (100 μM) mantGDP in 20 mM Tris-HCl, pH 7.5, 50 mM NaCl, 1 mM MgCl$_2$ (all in a total volume of 100 μl). Before addition of the Wrch-1, allow the mantGDP to equilibrate to the temperature of the solution (in this case room temperature). This normally takes 2–5 min. Wrch-1 possesses a remarkably rapid intrinsic nucleotide exchange rate and requires no EDTA to stimulate exchange (unlike Cdc42), so exchange will begin immediately. Increases in fluorescent intensity (λ_{ex} = 360 nm, λ_{em} = 440 nm) represent an increase in loading of mant nucleotide to Wrch-1 from solution. Figure 4 shows MantGDP exchange onto His$_6$-ΔC Wrch-1, with a rate of 0.012 s^{-1}.

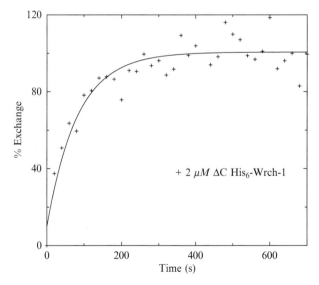

FIG. 4. Nucleotide exchange of Wrch-1; 2 μM of His$_6$-ΔC-Wrch-1 was incubated in 20 mM Tris, pH 7.5, 1 mM MgCl$_2$, 50 mM NaCl, and 100 μM mant-GDP at 25°. The changes in fluorescent intensity were followed at λ_{ex} = 365 nm, λ_{em} = 440 nm, from which exchange rates could be calculated.

Determination of Palmitoylation of Wrch Proteins

Labeling of Palmitoylated Cysteine Residues in Wrch-1/2 Proteins

We have applied a biotin-BMCC labeling protocol (Drisdel and Green, 2004) for direct analyses of protein palmitoylation to evaluate the modification of wild-type and mutant Wrch-1 and Chp proteins expressed in mammalian cells (Berzat *et al.*, 2005; Chenette *et al.*, 2005). It is important to note that this assay will not detect palmitate groups added to serine and threonine residues by means of oxyester linkages, because these bonds are not hydrolyzed by hydroxylamine. This assay involves blocking all available free cysteines in a protein with *N*-ethylmaleimide (NEM), which alkylates thiol groups. Cysteines with bound palmitate are not sensitive to NEM-induced alkylation; hence, after the thioester bond holding palmitate to a cysteine is cleaved with hydroxylamine, only the free thiol groups at the sites of palmitate linkage are recognized and bound by the 1-biotinamido-4-[4'-(maleimidomethyl)cyclohexanecarboxamido] butane (BMCC) compound. This method, therefore, provides a sensitive and efficient way of detecting palmitoylation of cysteines, although a definitive demonstration of Wrch protein modification by palmitoylation requires analysis by mass spectroscopy.

For this assay, seed 10^6 293T cells in 100-mm dishes and then transfect cells with 5–10 μg of the desired construct using a calcium phosphate precipitation transfection technique. Twenty hours after transfection, wash the complexes off the cells and feed with fresh growth medium. Forty-eight hours after transfection, wash cells once in cold 1 × phosphate-buffered saline (PBS) and lyse in 700 μl lysis buffer (150 mM NaCl, 5 mM EDTA, 50 mM Tris, pH 7.4, 0.02% NaN$_3$, 2% Triton-X 100, protease inhibitors). Incubate lysate with 5 μg of the appropriate primary antibody at 4° for 1 h and then add 20 μg protein G or protein A to the lysates and incubate at 4° for 30 min. Wash bound protein twice with lysis buffer and incubate with lysis buffer containing 50 mM NEM (Sigma) for 48 h at 4°. Change NEM solution once during incubation. NEM is a labile compound, so it is especially important to make fresh NEM solution for each experiment and to keep all solutions containing NEM cold and protected from light. We also find that ordering fresh NEM compound every 3 to 4 months helps keep background low.

Wash bound protein with lysis buffer and treat with 1 M hydroxylamine, pH 7.4, to cleave thioester bonds for 1 h at 25°. Wash bound protein again, and treat with 1 μM 1-biotinamido-4-[4'-(maleimidomethyl)cyclohexane-carboxamido] butane (biotin-BMCC, Pierce), which recognizes free sulfhydryl groups, for 2 h at 25°. Wash bound protein two to three times using lysis

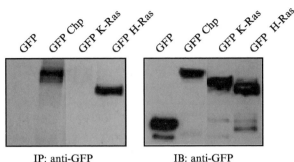

IP: anti-GFP   IB: anti-GFP

FIG. 5. Use of biotin-BMCC–labeling to detect palmitate modification of human Chp/Wrch-2. 293T cells were transiently transfected with the indicated pEGFP expression plasmids encoding GFP alone or GFP fused to the N-terminus of the indicated proteins. Forty-eight hours after transfection, the cells were lysed and used for the biotin-BMCC–labeling assay.

buffer, resuspend in 20–100 μl 2 × protein sample loading buffer, boil for 5 min, and resolve protein by 12% SDS-PAGE. Transfer protein to polyvinylidene difluoride membrane. After transfer is complete, block membrane in 5% bovine serum albumin in TBS with 0.1% Tween-20 (TBST) at 25° for 1 h. To detect biotin-labeled protein, incubate the membrane with streptavidin-horseradish peroxidase (Pierce) at 25° for 2 h, wash membrane three times in TBST, expose to ECL reagents (Pierce), and expose to X-ray film.

Among suitable controls for this assay are the nonpalmitoylated K-Ras4B and the palmitoylated H-Ras protein (Cox and Der, 2002). With this assay, labeling of H-Ras, but not of K-Ras4B, was seen (Fig. 5). Similarly, labeling of Chp (Chenette *et al.*, 2005) and of Wrch-1 (Berzat *et al.*, 2005) was also detected, and mutation of the cysteine residue that is the third residue from the C-termini of both proteins abolished this labeling (data not shown).

Inhibition of Palmitoylation-Mediated Localization of Wrch-1

We have also used a pharmacological approach to demonstrate Wrch protein modification by palmitoylation and to determine whether palmitoylation of Wrch-1 protein is critical for its proper localization to cellular membranes. We used a recently described method of monitoring lipid-modified protein subcellular distribution by using a pharmacological inhibitor of palmitoylation "2-bromopalmitate (2-BP)" to alter the localization patterns of enhanced green fluorescent protein (EGFP)–tagged proteins (Keller *et al.*, 2005). When expressed in cells, EGFP alone localizes diffusely throughout the cytosol and nucleus, but not to cellular membranes, in part because of its lack of membrane targeting signals (i.e., lipid modifications) and to a putative nuclear localization signal (NLS) (Keller *et al.*, 2005) (BD Biosciences Clontech). However, when EGFP is used as a tag to track localization patterns of lipid-modified proteins such as Wrch-1, the palmitate lipid moiety attached to Wrch-1 is dominant over the putative NLS of EGFP. Notably this causes nuclear exclusion of EGFP-tagged Wrch-1 and directs the protein to cellular membranes instead of only the cytosol and nucleus (Berzat *et al.*, 2005). Disruption of Wrch-1 palmitoylation by 2-BP results in a redistribution of EGFP-tagged Wrch-1 with a subcellular localization pattern similar to that of unmodified EGFP.

2-BP, which is a noncompetitive inhibitor of palmitoyl-acyltransferase activity, has been used extensively to show that the subcellular membrane association of members of the Ras and Rho subfamilies are dependent on palmitate modification (Michaelson *et al.*, 2001; Varner *et al.*, 2003; Webb *et al.*, 2000). It is thought to function by blocking the transfer of palmitate lipid moieties from palmitoyl CoA to palmitoylatable protein substrates (Varner *et al.*, 2003). Therefore, when combined with the previously described biotin-BMCC labeling assay, this method provides a simple and visual way to provide independent verification of Wrch modification by palmitoylation and to assess whether palmitoylation is necessary for Wrch subcellular localization.

In this assay, 10^5 NIH 3T3 fibroblasts are seeded in 60-mm dishes containing uncoated glass square coverslips (Corning No. 11/2, 0.16-mm thick, 18-mm). For some cell types, it may be necessary to precoat the glass coverslips with extracellular matrix proteins or poly-L-lysine to facilitate adherence of cells to coverslips. The next day, transiently transfect cells with 1 μg of pEGFP expression vector encoding Wrch-1 using a standard transfection method such as LipofectAMINE Plus (Invitrogen). Before transfection, remove the growth medium from the NIH 3T3 culture and replace with fresh growth medium supplemented with 150 μM 2-BP final concentration (Sigma). If using the LipfectAMINE Plus transfection method, culture medium containing 150 μM 2-BP can be added after the 3 to 5 h incubation period. The concentration of 2-BP used for this experiment depends on the cell type and should be determined empirically. Because 2-BP is soluble in dimethylsulfoxide (DMSO), cells treated with 2-BP should be compared with parallel cultures treated with the same final concentration of DMSO to control for vehicle effects.

Twenty-four hours after transfection, cells express the EGFP fluorophore and are ready for fluorescent microscopy analysis. To visualize GFP-tagged Wrch-1, coverslips should be rinsed in PBS. Then, add 1 drop of PBS to a clean glass slide and gently invert glass coverslip, cell side down, onto the PBS. Remove any excess PBS with a paper tissue and then view the EGFP-tagged Wrch-1 localization patterns in the presence or absence

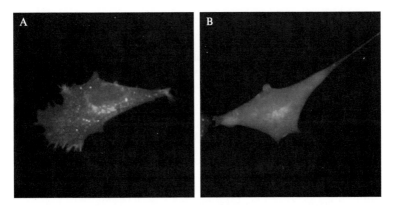

FIG. 6. Inhibition of palmitoylation with 2-BP to disrupt Wrch-1 localization patterns. NIH 3T3 fibroblasts were transiently transfected with pEGFP-Wrch-1 constructs and treated with either DMSO vehicle (A) or 2-BP (B) for 24 h. Changes in EGFP-tagged Wrch-1 localization after 2-BP treatment were captured using FITC filter on fluorescent microscope.

of 2-BP on a fluorescent microscope containing a fluorescein isothiocyanate (FITC) filter. When EGFP-tagged Wrch-1 is treated with 2-BP, a dramatic redistribution of Wrch-1 protein from cellular membranes to the cytosol and nucleus is observed (Fig. 6B) compared with vehicle-treated EGFP-tagged Wrch-1. Vehicle-treated EGFP-tagged Wrch-1 localizes properly to plasma and internal membranes and is distinctly excluded from the nucleus (Fig. 6A). Mislocalization of EGFP-tagged Chp on treatment with 2-BP has also been observed (Chenette *et al.*, 2005). Image software such as MetaMorph (Universal Imaging Corp.) is required to capture images for publication.

Concluding Remarks

In this review we have described the preparation method of both GST- and His_6-Wrch-1, followed by examples of the assays used in the biochemical characterization of both its GTP hydrolysis rate and its nucleotide exchange rate. In addition, we describe an *in vitro* labeling assay to evaluate the *in vivo* modification of the C-terminal sequences of Wrch-1 and Chp by palmitoylation.

References

Berzat, A. C., Buss, J. E., Chenette, E. J., Weinbaum, C. A., Shutes, A., Der, C. J., Minden, A., and Cox, A. D. (2005). Transforming Activity of the Rho Family GTPase, Wrch-1, a Wnt-regulated Cd_c42 Homolog, is dependent on a novel carboxyl-terminal palmitoylation motif. *J. Biol. Chem.* **280,** 33055–33069.

Chenette, E., Abo, A., and Der, C. (2005). Critical and distinct roles of amino- and carboxyl-terminal sequences in regulation of the biological activity of the Chp atypical Rho GTPase. *J. Biol. Chem.* **280,** 13784–13792.

Cox, A., and Der, C. (2002). Ras family signaling: Therapeutic targeting. *Cancer Biol. Ther.* **1,** 599–606.

Drisdel, R. C., and Green, W. N. (2004). Labeling and quantifying sites of protein palmitoylation. *Biotechniques* **36,** 276–285.

Keller, P. J., Fiordalisi, J. J., Berzat, A. C., and Cox, A. D. (2005). *Methods* In Press.

Leonard, D. A., Evans, T., Hart, M., Cerione, R. A., and Manor, D. (1994). Investigation of the GTP-binding/GTPase cycle of Cdc42Hs using fluorescence spectroscopy. *Biochemistry* **33,** 12323–12328.

Michaelson, D., Silletti, J., Murphy, G., D'Eustachio, P., Rush, M., and Philips, M. R. (2001). Differential localization of Rho GTPases in live cells: Regulation by hypervariable regions and RhoGDI binding. *J. Cell Biol.* **152,** 111–126.

Shutes, A., Phillips, R. A., Corrie, J. E., and Webb, M. R. (2002). Role of magnesium in nucleotide exchange on the small G protein Rac investigated using novel fluorescent guanine nucleotide analogues. *Biochemistry* **41,** 3828–3835.

Shutes, A., Berzat, A. C., Cox, A. D., and Der, C. J. (2004). Atypical mechanism of regulation of the Wrch-1 Rho family small GTPase. *Curr. Biol.* **14,** 2052–2056.

Shutes, A., and Der, C. J. (2005). Real time *in vitro* measurement of GTP hydrolysis. *Methods* **37,** 183–189.

Tao, W., Pennica, D., Xu, L., Kalejta, R., and Levine, A. J. (2001). Wrch-1, a novel member of the Rho gene family that is regulated by Wnt-1. *Genes Dev.* **15,** 1796–1807.

Varner, A., Ducker, C., Xia, Z., Zhuang, Y., De Vos, M. L., and Smith, C. D. (2003). Characterization of human palmitoyl-acyl transferase activity using peptides that mimic distinct palmitoylation motifs. *Biochem. J.* **373,** 91–99.

Webb, Y., Hermida-Matsumoto, L., and Resh, M. (2000). Inhibition of protein palmitoylation, raft localization, and T cell signaling by 2-bromopalmitate and polyunsaturated fatty acids. *J. Biol. Chem.* **275,** 261–270.

Wennerberg, K., and Der, C. (2004). Rho-family GTPases: It's not only Rac and Rho (and I like it). *J. Cell Sci.* **117,** 1301–1312.

[3] Purification of P-Rex1 from Neutrophils and Nucleotide Exchange Assay

By KIRSTI HILL and HEIDI C. E. WELCH

Abstract

The P-Rex family of guanine-nucleotide exchange factors (GEFs) are activators of the small GTPase Rac (Donald *et al.,* 2004; Rosenfeldt *et al.,* 2004; Welch *et al.,* 2002). They are directly regulated *in vitro* and *in vivo* by the lipid second messenger phosphatidylinositol (3,4,5)-triphosphate (PtdIns(3,4,5)P$_3$) and by the $\beta\gamma$ subunits of heterotrimeric G proteins

0076-6879/06 $35.00
DOI: 10.1016/S0076-6879(06)06003-4

(Donald *et al.*, 2004; Rosenfeldt *et al.*, 2004; Welch *et al.*, 2002). Activation by PtdIns(3,4,5)P$_3$ occurs by means of the PH domain of P-Rex1 and activation by G$\beta\gamma$ subunits by means of the catalytic DH domain (Hill *et al.*, 2005). P-Rex1 and P-Rex2 also contain two DEP and two PDZ protein interaction domains, as well as homology over their COOH-terminal half to inositol polyphosphate 4-phosphatase (Donald *et al.*, 2004; Welch *et al.*, 2002). These domains, although not necessary for P-Rex1 activity *in vitro*, influence its basal and/or stimulated Rac-GEF activity, suggesting that their interaction with the DH/PH domain tandem is important for P-Rex1 function (Hill *et al.*, 2005). P-Rex2B, a splice variant of P-Rex2, lacks the C-terminal half (Rosenfeldt *et al.*, 2004). P-Rex1 was originally identified during a search for PtdIns(3,4,5)P$_3$–dependent activators of Rac in neutrophils and purified to homogeneity from pig leukocyte cytosol, in which it is the major such activity (Welch *et al.*, 2002). P-Rex1 is mainly expressed in neutrophils and regulates reactive oxygen species formation in these cells (Welch *et al.*, 2002), whereas P-Rex2 is expressed in a wide variety of tissues but not in neutrophils (Donald *et al.*, 2004), and P-Rex2B is expressed in the heart (Rosenfeldt *et al.*, 2004).

This Chapter describes our methods for (1) the purification of endogenous P-Rex1 from pig leukocyte cytosol, (2) the production and purification of recombinant P-Rex proteins and their substrate GTPase Rac from Sf9 cells, and (3) the *in vitro* assay for measuring the GEF activities of native or recombinant P-Rex proteins.

Purification of P-Rex1 Protein from Pig Leukocyte Cytosol

Native P-Rex1 is purified from pig leukocyte cytosol through a series of chromatography columns (Welch *et al.*, 2002). All columns (from Pharmacia) are packed beforehand, and purification solutions are prepared, degassed, and stored cold. Purification is done in the cold room and monitored by measuring PtdIns(3,4,5)P$_3$-stimulated P-Rex1 Rac-GEF activity, using the *in vitro* assay described in "*In Vitro* Rac GEF Activity Assay." Typically, approximately 0.5 mg of pure P-Rex1 protein can be obtained from 120 l of pigs' blood. For us, one run through the entire process, including preparation, took 4 weeks once conditions were established.

Preparation of Pig Leukocyte Cytosol

Prepare pig leukocyte cytosol in batches of 30 l of blood obtained from a local slaughterhouse. Collect only first 1–2 l of arterial blood per pig immediately after slaughter, mixing each 2 l with 280 ml anticoagulant (80 m*M* sodium citrate, 18 m*M* NaH$_2$PO$_4$, 17 m*M* citric acid, 161 m*M*

D-glucose). To sediment red blood cells, pool blood into two containers of 15 l, mix each with 3 l of PVP solution (3% polyvinyl-pyrrolidone, avg. MW = 360,000, Sigma PVP-360, 0.9% NaCl), and wait until red blood cells have settled to approximately 50% of the volume. Syphon supernatant into centrifuge bottles and spin at 500g for 8 min at 15°. To wash leukocyte pellets, resuspend in a total of 1.3 l calcium-free Hanks at RT (made from 10× Hanks (see later) with added 10 mM glucose, 0.5 mM MgCl$_2$, 0.4 mM MgSO$_4$), and repeat spin. Lyse remaining red blood cells by resuspending pellet in 700 ml of ice-cold H$_2$O for exactly 30 sec, stop by addition of 77 ml 10× Hanks (33 mM KCl, 4.4 mM KH$_2$PO$_4$, 1.37 M NaCl, 42 mM NaHCO$_3$, 35 mM Na$_2$HPO$_4$), and repeat spin. Wash cells once in ice-cold Hanks and once in ice-cold leukocyte lysis buffer (30 mM Tris, pH 7.8, at 4°, 0.1 M NaCl, 4 mM EGTA). Resuspend pellet in 4–5 volumes of ice-cold leukocyte lysis buffer containing 1× protease inhibitors (10 µg/ml each of antipain, aprotinin, pepstatin, leupeptin), 0.1 mM PMSF, and 1 mM DTT. Sonicate on ice for 4 × 15 sec using a 1.2-cm probe sonicator (Heat Systems, on 40% output), then sediment nuclei and debris at 900g for 10 min at 4°. Clear the supernatant containing the cytosol by spinning at 100,000g for 1 h at 4°, freeze in 50-ml aliquots in liquid N$_2$, and store at –80°. Typically, each preparation yields 200 ml cytosol at 9 mg/ml protein.

Q-Sepharose Fast-Flow Column

Thaw enough pig leukocyte cytosol to have 7.5 g of protein, clear by spinning at 100,000g for 1 h at 4°, filter through a 0.2-µm filter, then dilute 1:6 in 10 mM Tris, pH 7.8, at 4°, 1 mM EGTA, 10% ethylene glycol, 1% betaine, 0.01% Na azide, 0.2× protease inhibitors, 50 µM PMSF, 1 mM DTT. Load diluted cytosol onto a 400-ml Q-Sepharose fast-flow column (5-cm diameter, 20-cm high) equilibrated in 30 mM Tris, pH 7.8, at 4°, 0.5 mM EGTA, 0.1 mM EDTA, 10% ethylene glycol, 1% betaine, 0.01% Na azide, 0.2× protease inhibitors, 50 µM PMSF, 1 mM DTT. Load sample at 20 ml/min, using an empty column as a prefilter. Wash in the same buffer, then elute protein with a 3 l linear NaCl gradient of 0.1–0.6 M at 16 ml/min, collecting 48-ml fractions. Record protein and salt concentration traces. Measure PtdIns(3,4,5)P$_3$-stimulated P-Rex1 Rac-GEF activity in load, flow through, and fractions (diluted 1:2). Expect two discrete peaks of PtdIns (3,4,5)P$_3$-dependent Rac-GEF activities (see Fig. 1). Peak A, the major of the two peaks that elutes between 0.43 and 0.52 M NaCl and makes up approximately 65% of the total activity, is P-Rex1. Pool P-Rex1 fractions, which will be roughly 400 mg of protein distributed over seven fractions, for further use.

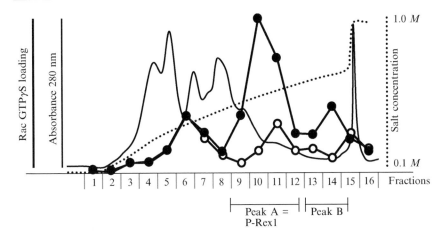

FIG. 1. P-Rex1 is the major PtdIns(3,4,5)P$_3$-stimulated Rac-GEF activity in pig leukocyte cytosol fractionated on a Q Sepharose Fast Flow column. Pig leukocyte cytosol was separated by salt gradient on a Q Sepharose fast flow column as described in "Purification of P-Rex Protein from Pig Leukocyte Cytosol, Q-Sepharose Fast-flow Column," and the fractions were assayed for Rac-GEF activity either in the presence (solid circles) or the absence (empty circles) of PtdIns(3,4,5)P$_3$, as described in "*In Vitro* Rac GEF Activity Assay." This revealed two discrete peaks of PtdIns(3,4,5)P$_3$-stimulated Rac-GEF activity. Peak A is P-Rex1.

Desalting Column 1 and SP-Sepharose HP Column

Before the next purification step, the sample needs to be desalted. Use a 1.4-l G25 fine column (5 cm diameter, 70 cm high) equilibrated with desalting buffer 1 (20 mM Hepes, pH 6.8, at 4°, 10% ethylene glycol, 1% betaine, 10 mM NaCl, 0.5 mM EGTA, 0.01% azide, 0.2× protease inhibitors, 50 μM PMSF, 1 mM DTT). Load sample at 10 ml/min, wash with desalting buffer, record protein and salt traces, and collect eluting protein. Load desalted sample straight onto a 50-ml SP Sepharose HP column (2.6-cm diameter, 10-cm high), equilibrated in desalting buffer 1, at 3 ml/min. Wash, then elute protein with a 600-ml linear 0.25–0.75 M KCl gradient at 0.8 ml/min, collecting 12-ml fractions. Expect P-Rex1 protein to elute between 0.31 and 0.375 M KCl. Test fractions (diluted 1:3) for PtdIns(3,4,5)P$_3$-dependent P-Rex1 Rac-GEF activity and pool active ones (about 20 mg of protein over seven fractions) for next step.

Desalting Column 2 and Heparin Sepharose Column

Desalt the sample on a 300-ml G25 fine column (5-cm diameter, 15-cm high) equilibrated with desalting buffer 2 (20 mM Hepes, pH 7.2, at 4°, 10% ethylene glycol, 1% betaine, 10 mM KCl, 0.5 mM EGTA, 0.01% azide,

0.2× protease inhibitors, 50 μM PMSF, 1 mM DTT). Load at 5 ml/min, wash, collect desalted protein, and load straight onto a 12-ml heparin Sepharose column (1.6 cm diameter, 6 cm high) equilibrated in desalting buffer 2. Load at 1 ml/min, wash, then elute protein with a 150-ml linear 0.1–0.7 M KCl gradient at 0.2 ml/min, collecting 3 ml fractions. P-Rex1 protein will elute between 0.5 and 0.69 M KCl. Test fractions (diluted 1:10) for PtdIns(3,4,5) P$_3$-dependent P-Rex1 Rac-GEF activity, and pool those giving best purification (corresponding to 0.6–0.65 M KCl, about 1.5 mg of protein) for further use.

Gel Filtration

Before loading the sample onto the next column, adjust its pH to below 7.0 and its Hepes concentration to 50 mM, by addition of an appropriate volume of 250 mM Hepes, pH 6.8, at 4°. Then, concentrate the sample in a Centriprep (Amicon) device to a volume of less than 1 ml and preclear by spinning in a Microfuge at maximal speed for 30 min. Load onto a 200-ml FPLC gel filtration column equilibrated in 20 mM Hepes, pH 6.9, at 4°, 10% ethylene glycol, 1% betaine, 120 mM NaCl, 0.5 mM EGTA, 0.01% azide, 0.2× protease inhibitors, 50 μM PMSF, 1 mM DTT at 0.3 ml/min, wash, and collect fractions of 2 ml. Test fractions (diluted 1:8) for PtdIns (3,4,5)P$_3$-dependent P-Rex1 Rac-GEF activity. P-Rex1 protein elutes at 100–108 ml, in about 0.6 mg of protein (peak is at 104 ml, corresponding to an apparent MW of 203 kDa).

Mono S Column

Dilute the sample 1:3 in 20 mM Hepes, pH 7.0, at 4°, 10% ethylene glycol, 1% betaine, 0.5 mM EGTA, 0.01% azide, 0.2× protease inhibitors, 50 μM PMSF, 1 mM DTT. Load onto a 1-ml Mono S FPLC column at 0.4 ml/min, wash, and elute with a 54-ml linear gradient of 0.1–0.7 M KCl at 0.2 ml/min, collecting 0.4-ml fractions. Test fractions (diluted 1:80) for P-Rex1 activity. Pure P-Rex1 protein elutes between 0.375 and 0.425 M KCl, about 530 μg in 1.2 ml.

Assessment of Purity

Purity can be assessed by running a few microliters of the active fractions from the last two columns on SDS-PAGE followed by silver staining. P-Rex1 runs with an apparent MW of 196 kDa. A minor second band of apparent MW 142 kDa is almost certainly a proteolytic fragment of P-Rex1 (we did tryptic digests of both proteins followed by MALDI-TOF analysis and N-terminal sequencing of the resulting peptides).

Storage

The purified protein can be stored at 4° for a week at least. For long-term storage, add fatty-acid–free BSA to 2 mg/ml (essential to recover activity after thawing), snap freeze in aliquots in liquid N_2, and store at –80°. Samples can be repeatedly thawed.

Production of Human Recombinant P-Rex and Rac Proteins in Sf9 Cells

To test the Rac-GEF activity of purified native P-Rex1, we use as substrate recombinant GDP-loaded Rac produced in and purified from Sf9 cells. We have also made recombinant human wild-type P-Rex1 and P-Rex2 and several mutants of human P-Rex1 in Sf9 cells (Fig. 2). The production and purification of recombinant Rac and P-Rex family proteins is described here.

We have chosen the Sf9 cell/baculovirus system because it produces posttranslationally modified recombinant proteins in appreciable quantity. Typically, we obtain about 250 μl of pure recombinant Rac or P-Rex proteins at concentrations of 1 mg/ml from one 400-ml culture of Sf9 cells (2 l for Rac). The entire procedure (excluding plasmid preparation) takes approximately 4–6 weeks, but high titer viral supernatants can be stored for at least 1 year for further use.

Constructs in Sf9 Cell Expression Vector

Our constructs for the small GTPase Rac1 and for human wild type P-Rex1, P-Rex2, as well as the deletion, truncation, and point mutants of P-Rex1 are all in the pAcOG1 insect cell expression vector, which gives the resulting recombinant proteins an NH_2-terminal EE-epitope tag, which does not interfere with the GEF activity assay (Donald *et al.,* 2004; Hill *et al.,* 2005; Welch *et al.,* 2002).

Sf9 Cell Culture

Maintain Sf9 cells (*Spodoptera frugiperda*; European Tissue Culture Collection) at 27° without gassing in suspension, at a density of 0.5–2.0 × 10^6 cells/ml, in 0.5 or 1 l spinner flasks (Techne). For routine cell culture, grow Sf9 cells in TMN-FH insect medium (Sigma) supplemented with 11% FBS, 1 unit/ml penicillin, and 0.1 mg/ml streptomycin. To prepare TNM-FH insect medium (Sigma), take 5 1-l powder aliquots, add 1.75 g $NaHCO_3$, 87.5 ml 1 *M* NaOH, and H_2O to 5 l, pH to 6.3, filter through a 0.2-μm filter, and store at RT in the dark. Use Sf9 cells for protein production between 1 and 10 weeks in culture. When required, grow Sf9 cells as adherent monolayers in tissue culture flasks. During the lipofection procedure and for cloning of colonies, grow cells in Grace's insect medium (Invitrogen).

Lipofection of Sf9 Cells

Seed 2.5×10^6 cells into a 25-cm^2 tissue culture flask and leave to adhere. Take 100 μl of serum-free Grace's insect medium, add 5 μg of plasmid DNA, 0.5 μl of linear baculogold viral DNA (PharMingen), and 20 μl of Cellfectin reagent (Invitrogen), vortex for 10 sec, and incubate for 40 min at RT for liposomes to form. Wash Sf9 cells twice with serum-free Grace's insect medium and aspirate. Take the liposome mix, add 1 ml of serum-free Grace's insect medium, pipette this to the aspirated cells, and incubate for 5 h with gentle rocking at RT. The plasmid and baculogold DNA will be taken up by the Sf9 cells and will recombine to form the complete viral genome. Add 1 ml of complete TMN-FH medium, then incubate for 3 days at 27° to allow for the formation and expansion of viral particles in the cells. Remove the supernatant, clear by centrifugation at 300g for 5 min at RT, and store at 4° in the dark.

Cloning of Viruses in Soft Agar

For each Sf9 cell lipofection, seed five 10-cm tissue culture dishes with 7×10^6 cells and leave to adhere. Infect them with 10-fold serial dilutions of lipofection supernatant in Grace's insect medium for 3 h at 27°. For overlaying with soft agar, autoclave 1 g of Seaplaque™ agarose (BioWhittaker Molecular Applications) in 15 ml of Grace's insect medium. When cooled to 40°, add a further 45 ml of warm Grace's insect medium and keep this mixture at 40°. Immediately before use, add 50 ml of complete TMN-FH at RT, and overlay aspirated, infected Sf9 cells with 10 ml of the agar mix. Transfer the cells to a humidified 27° incubator. After 5–10 days, white round plaques of dead cells can be observed. Use glass Pasteur pipettes to pick 5–10 plaques for each protein, and incubate overnight at 4° in the dark in 0.5 ml Grace's insect medium to allow for diffusion of viral particles out of the agar plugs.

Amplification of Viral Particles

Amplify suspensions of viral particles three times before protein production to produce low, medium, and high titer supernatants. For the first round of amplification, use an entire agar plug in 0.5 ml of Grace's insect medium to infect 2.5×10^6 cells in a 25-cm^2 tissue culture flask for 2 days.

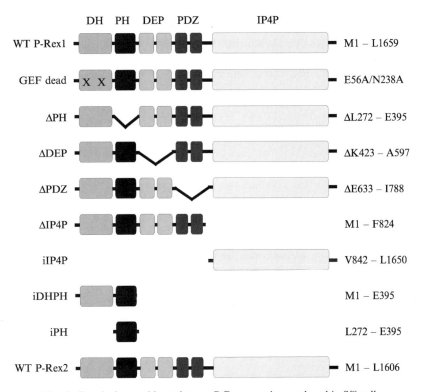

Fig. 2. Panel of recombinant human P-Rex proteins produced in Sf9 cells.

Harvest the supernatant by centrifugation at 300g for 15 min at RT (low titer supernatant). For medium titer supernatant, use 0.75 ml of low-titer supernatant to infect 1.5×10^7 cells in 80-cm^2 tissue culture flasks for 2.5 days. Remove the supernatant from the cells, centrifuge at 300g for 15 min at RT, and store at 4° in the dark (medium titer supernatant). At this point, analyze cells for protein expression. Wash the cells with 1 ml of Sf9 wash solution (0.7% KCl, 2.66% sucrose, 7 mM NaH$_2$PO$_4$, pH 6.2, at RT, 20 mM MgCl$_2$), collect cells using cell scrapers, and centrifuge at 5000g for 10 min at 4°. Aspirate the supernatant, snap freeze the pellets, and analyze for protein expression as described later. Amplify viral suspensions that produce a high level of expression of the protein of interest a third time to produce high-titer supernatant. For high-titer supernatant, use 1.2 ml of medium-titer supernatant to infect two 175-cm^2 tissue culture flasks containing 3.5×10^7 cells for 2 days. Remove the high-titer supernatant, centrifuge at 300g for 15 min at RT, and store at 4° in the dark.

Small-Scale Immunoprecipitations from Sf9 Cells

Small-scale immunoprecipitations can be performed to test viral clones for protein expression or identify optimal infection times and viral titers for protein expression. Resuspend infected Sf9 cell pellets (or drops) by vortexing into ice-cold lysis buffer (1% Triton X-100, 0.12 M NaCl, 20 mM Hepes, pH 7.4, at 4°, 1 mM EGTA), incubate for 5 min on ice. Clear lysates by centrifugation at 9700g for 12 min at 4°, add to prewashed anti-EE beads (anti-EE epitope tag antibody coupled to protein G sepharose beads; Onyx Pharmaceuticals), and incubate at 4° with end-over-end rotation for 60–90 min. Wash beads four times in ice-cold lysis buffer, aspirate, and boil in SDS sample buffer. Proteins can be separated by SDS-PAGE and visualized by Coomassie staining.

Optimal Conditions for P-Rex and Rac Protein Production in Sf9 Cells

To establish conditions for optimal protein expression, put 150 ml of Sf9 cells at a density of 1×10^6 cells/ml into 500-ml spinner flasks and infect with 0.3–3% (v/v) of high-titer supernatant. Harvest 15 ml of cells 1.5, 2, 2.25, 2.5, and 3 days after infection, centrifuge at 300g for 5 min at RT, and snap freeze the pellets. Protein production can be analyzed by immunoprecipitation as described in "Small-scale Immunoprecipitations from Sf9 Cells." Figure 3 shows ΔDEP-P-Rex1 as a typical example of such an analysis.

Using similar experiments as that shown in Fig. 3, we have identified optimal infection times and viral titers to obtain the highest levels of

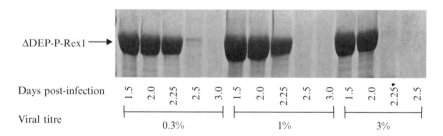

FIG. 3. Optimal infection time and viral titer for production of recombinant human ΔDEP-P-Rex1 protein in Sf9 cells. Sf9 cells were infected with 0.3–3% (v/v) of high titer viral supernatant for ΔDEP-P-Rex1 as described in "Production of Human Recombinant P-Rex and Rac Proteins in Sf9 Cells, Optimal Conditions for P-Rex and Rac Protein Production in Sf9 Cells" Aliquots of the cells were harvested at 1.5, 2, 2.25, 2.5. or 3 days after infection, as indicated, and protein production was analyzed by immunoprecipitation, SDS-PAGE, and Coomassie staining, as described in "Production of Human Recombinant P-Rex and Rac Proteins in Sf9 Cells, Small-Scale Immunoprecipitations from Sf9 Cells."

TABLE I
Optimal Infection Times and Viral Titers for Production of Recombinant
Human P-Rex and Rac Proteins in Sf9 Cells

Protein	Optimal infection time (days after infection)	Optimal viral titer% (v/v)
WT P-Rex1	2.2	1.5
GEF dead	2.5–3.0	0.8
ΔPH	2.0	0.8
ΔDEP	1.5	1.0
ΔPDZ	1.5	2.5
ΔIP4P	1.5	0.8
iIP4P	0.3	2.0
iDHPH	2.25–2.5	0.8
iPH	1.5	0.8
WT P-Rex2	2.0	1.0
Rac1	2.5	1.5

Sf9 were infected with 0.3–3% (v/v) of high-titer viral supernatants for the indicated P-Rex or Rac proteins as described in "Production of Human Recombinant P-Rex and Rac Proteins in Sf9 Cells, Optimal Conditions for P-Rex and Rac Protein Production in Sf9 Cells." Aliquots of the cells were harvested 1.5, 2, 2.25, 2.5, or 3 days after infection as indicated, and protein production was analyzed as described in "Production of Human Recombinant P-Rex and Rac Proteins in Sf9 Cells, Small-scale Immunoprecipitations from Sf9 Cells."

production for all of our P-Rex and Rac proteins. These are summarized in Table I.

Production of Recombinant Protein

Once optimal infection times have been established for the production of each protein (see Table I), Sf9 cell cultures can be scaled up for protein purification. Infect 2 l of cells at 1×10^6 cells/ml in a 5-l spinner flask with the optimal concentration of high-titer supernatant for the optimal length of time. Harvest cells by centrifugation at $500g$ for 25 min at 4°, wash once in Grace's insect medium, centrifuge at $300g$ for 7 min at 4°, and resuspended in a small volume of Grace's insect medium. Pipette cells drop-wise into liquid N_2 and store the resulting Sf9 cell drops at –80° until they are ready for protein purification.

Purification of Recombinant P-Rex1 Proteins from Sf9 Cells

Thaw drops of infected Sf9 cells (the equivalent of approximately 400 ml of culture) on ice and add 25 ml of ice-cold lysis buffer (PBS, 1% Triton X-100, 5 mM EGTA, 1 mM EDTA, 25 mM NaF, 20 mM β-glycerophosphate, 1 mM DTT, 0.1 mM PMSF, and 10 μg/ml each of antipain, pepstatin A, leupeptin, aprotinin). Incubate on ice for 5 min and centrifuge at 118,000g for 1 h at 4° to obtain the cytosolic fraction. Add the supernatant to 800 μl of prewashed anti-EE beads and incubate with end-over-end rotation for 90 min at 4°. After immunoprecipitation, wash the beads three times in ice-cold wash buffer (2× PBS, 0.1% Triton X-100, 1 mM EGTA) and then wash five times in ice-cold wash/elution buffer (PBS, 10% glycerol, 1 mM DTT, 1 mM EGTA, 0.01% Na azide), and aspirate the beads close to the pellet. To elute the EE-tagged proteins from the anti-EE beads, add an equal volume of wash/elution buffer containing 150 μg/ml EY peptide (EEYMPME; NH2-terminus acetylated). Incubate on ice for 20 min with occasional agitation, then collect the supernatant by centrifugation at 500g for 2 min at 4°. Repeat the elution twice more and pool the supernatants. Concentrate the protein using a Centricon™ device (Millipore) with the appropriate molecular weight cutoff. For some of the P-Rex proteins, the purification protocol needs to be modified slightly: Elute the ΔIP4P protein from the anti-EE beads with EY peptide in one single step, because further concentration steps lead to its loss. For iDHPH and iPH proteins, increase the amount of NaCl in the elution step to 0.3 M to improve their elution. For storage of all purified recombinant P-Rex proteins, add fatty-acid–free BSA to 2 mg/ml and glycerol to 50% (also store some aliquots without BSA). Snap freeze aliquots and store at –80°. BSA-containing stocks of P-Rex proteins can be thawed repeatedly without loosing GEF activity. Typical yields of all our P-Rex proteins are summarized in Table II.

TABLE II
HILL AND WELCH: TYPICAL YIELDS OF HUMAN RECOMBINANT P-REX AND RAC PROTEINS
PURIFIED FROM S19 CELLS. RAC AND THE VARIOUS P-REX PROTEINS WERE PRODUCED IN SF9
CELLS USING OPTIMAL CONDITIONS FOR EACH, AND WERE PURIFIED AS DESCRIBED IN PART II,
SECTIONS J AND K, ABOUT 250 μl OF EACH PURIFIED RAC OR P-REX PROTEIN, AT THE INDICATED
CONCENTRATIONS, WERE OBTAINED FROM 400 μl SF9 CULTURE (2 LITRES FOR RAC)

Protein	Concentration (mg/ml)	Concentration (μM)
WT-P-Rex1	1	5.4
GEF dead	0.3	1.6
ΔPH	1	5.8
ΔDEP	1	6
ΔPDZ	0.3	1.8
ΔIP4P	0.2	2.2
iIP4P	0.5	5.4
iDHPH	1	22.4
iPH	0.1	67.4
WT-P-Rex2	1	5.4
Rac1	1	47.6

Purification of Recombinant Rac Protein from Sf9 Cells

For use in GEF activity assays, purify EE-Rac1 from Sf9 cells in the GDP-bound state. Thaw droplets of infected Sf9 cells (from of 2 l of culture) on ice. Add five volumes of lysis buffer (PBS, 40 mM Hepes, pH 7.5, at RT, 10 mM EGTA, 5 mM MgCl$_2$, 300 μM GDP, 1 mM DTT, 0.2 mM PMSF, and 20 μg/ml each of antipain, pepstatin A, leupeptin, aprotinin) and sonicate using a probe sonicator for 3 × 15 sec. Add 11% Triton X-114 (in 0.15 M NaCl, 10 mM Tris, pH 7.5, 0.01% Na azide) to give a final concentration of 1% (v/v), mix well, and leave on ice for 5 min. Obtain the cytosolic fraction by centrifugation at 118,000g for 40 min at 4°. Warm the supernatant to 37° for 30–45 sec until it goes cloudy, and centrifuge at 1000g for 90 sec at RT to obtain a phase-split. Discard the upper phase containing non-lipid modified Rac1. Add fresh ice-cold lysis buffer to the lower phase to the original volume and repeat the warming and centrifugation steps twice. Wash the lower phase containing lipid-modified Rac1 with ice-cold lysis buffer and centrifuge at 100,000g for 40 min at 4°. Incubate the supernatant with 1 ml of prewashed anti-EE-beads for 90 min at 4° with end-over-end rotation. Wash the anti-EE-beads three times in wash buffer (PBS, 5 mM MgCl$_2$, 1 mM DTT, 10 μM GDP, 1 mM EGTA, 0.2% Triton X-100), followed by five washes in wash/elution buffer (1 mM EGTA, 40 mM Hepes, pH 7.5, at RT, 0.15 M NaCl, 5 mM MgCl$_2$, 1 mM DTT, 10 μM GDP). To elute EE-Rac1 from the EE-beads, add wash/elution buffer containing 300 μg/ml EY peptide and 1% cholate, and incubate on ice for 20 min with occasional agitation. Repeat the elution step twice and pool the supernatants. Concentrate the protein in a 10-kDa molecular weight cutoff Centricon™ device, snap freeze aliquots, and store at –80°.

In Vitro Rac GEF Activity Assay

This assay measures GTPγS-loading of Rac *in vitro*. In principle, liposomes that either contain or not PtdIns(3,4,5)P$_3$ and/or G$\beta\gamma$ subunits are incubated with GDP-bound EE-Rac1 as substrate, with [^{35}S]-GTPγS, and with P-Rex to catalyze the reaction. After the incubation, Rac is immunoprecipitated using anti-EE-beads (see earlier), and its GTPγS-loading is assessed by β-counting.

As substrate, we routinely use Sf9-cell–derived purified GDP-loaded recombinant EE-Rac1 (see "Production of Human Recombinant P-Rex and Rac Proteins in Sf9 Cells"). The source of P-Rex can be either native P-Rex1 at varying stages of purity (see "Purification of P-Rex Protein from Pig Leukocyte Cytosol") or any recombinant Sf9-cell–derived purified P-Rex protein with a DH domain (see "Production of Human Recombinant P-Rex and Rac Proteins in Sf9 Cells"). Final assay concentrations of the key ingredients are 100 nM Rac1, 50 nM P-Rex (when using recombinant protein), 200 μM each of phosphatidylcholine, phosphatidylserine, and phosphatidylinositol, 5 μM GTPγS including 1 μCi of [^{35}S]-GTPγS per sample, 0.0024% cholate (from Rac storage buffer), and typically 10 μM PtdIns(3,4,5)P$_3$ or 0.3 μM G$\beta\gamma$ subunits as stimuli. When assaying for synergistic activation of P-Rex proteins by PtdIns(3,4,5)P$_3$ and G$\beta\gamma$ subunits, submaximal concentrations of 0.2 μM PtdIns(3,4,5)P$_3$ and 0.03 μM G$\beta\gamma$ subunits are used.

Materials

As sources of PtdIns(3,4,5)P$_3$, we routinely use either D-D-(stearoyl/arachidonoyl)-PtdIns(3,4,5)P$_3$ synthesized by Piers Gaffney (Gaffney and Reese, 1997), or D-D-dipalmitoyl PtdIns(3,4,5)P$_3$ synthesized by Gavin Painter (Painter *et al.*, 1999). A commercially available source would be dipalmitoyl-PtdIns(3,4,5)P$_3$ from Echelon. Store PtdIns(3,4,5)P$_3$ lyophilized at –80°, but keep a working stock (625 μM in H$_2$O) at –20° for several months. Briefly bath-sonicate the working stock before each use. Phosphatidylcholine and phosphatidylserine from bovine brain and phosphatidylinositol from bovine liver are all from Sigma. Store these at 10 mg/ml (approximately 12.5 mM) in CHCl$_3$ at –80°. As a source of G$\beta\gamma$ subunits, we routinely use recombinant EE-Gβ1,γ2 subunits purified from Sf9 cells. They are stored in 1% cholate, 1 mM DTT, 20 mM Hepes/NaOH, pH 8.0, at 4°, and 1 mM EDTA (Welch *et al.*, 2002).

Preparation of Liposomes

For a typical assay, prepare fresh on the day 3 different tubes of 5× liposomes, the first with PtdIns(3,4,5)P$_3$, the second without, and the third with EDTA. EDTA is the positive control. It chelates the Mg^{2+} in the assay, causing GDP to dissociate from Rac, and, in conditions of excess free GTPγS, allowing GTPγS to bind, thus showing how much of the Rac in the assay can be activated. To make the liposomes, make a 1:1:1 master-mix of phosphatidylcholine, phosphatidylserine, and phosphatidylinositol. Pipette 12 μl of the master-mix into each of three Microfuge tubes and dry down the lipids by leaving at RT. When dry, add 4 μl of 625 μM D-D-(S/A)-PtdIns(3,4,5)P$_3$ to the first tube, together with 21 μl H$_2$O, and 25 μl 2× lipid buffer (40 mM Hepes, pH 7.5, at 4°, 200 mM NaCl, 2 mM EGTA). To the second tube, add 50 μl of lipid buffer (20 mM Hepes, pH 7.5, at 4°, 100 mM NaCl, 1 mM EGTA), and to the third tube add 50 μl of lipid buffer containing 10 mM EDTA. Vortex (liquid goes cloudy), and then bath-sonicate the tubes until liposome preparation goes clear. The liposomes can be left at RT while the other assay components are prepared.

Preparation of 10× Assay Buffer

To a stock of 10× assay buffer (20 mM Hepes, pH 7.5, at 4°, 10 mM MgCl$_2$, 100 mM NaCl, 1 mM EGTA), add fatty-acid–free BSA to 100 mg/ml and DTT to 10 mM. Keep on ice until the assay.

Preparation of GTPγS

Prepare a 5× working solution of GTPγS by diluting a 0.5 mM GTPγS stock (tetralithium salt, from ICN, kept at –20°) to 25 μM in lipid buffer, adding 1 μCi of [^{35}S]-GTPγS (1250 Ci/mmol, from NEN) for each 2 μl of 5× stock. Keep on ice until use.

Preparation of P-Rex Proteins

When monitoring the purification of native P-Rex1, dilute column fractions shortly before the assay, as described in "Purification of P-Rex1 Protein from Pig Leukocyte Cytosol," in ice-cold P-Rex dilution buffer (20 mM Hepes, pH 7.0, at 4°, 1% betaine, 0.01% Na azide, 0.5 mM EGTA, 200 mM KCl, 10% ethylene glycerol), and keep on ice until use. When using recombinant P-Rex proteins, thaw these shortly before the assay and dilute in P-Rex dilution buffer to 125 nM, to give a 2.5× working stock, except in assays with G$\beta\gamma$ subunits, where P-Rex working stocks must be 5×. Immediately snap-freeze recombinant P-Rex proteins again for storage at –80°.

Preparation of Rac

Prepare a 10× working stock of Rac immediately before the assay by diluting the recombinant GDP-loaded lipid-modified EE-Rac1 purified from Sf9 cells (see "Production of Human Recombinant P-Rex and Rac Proteins in Sf9 Cells") to 1 μM in lipid buffer. Immediately snap-freeze Rac source again for storage at –80°.

Preparation of Gβγ Subunits

For assays with G$\beta\gamma$ subunits, dilute Sf9-cell–derived purified recombinant lipid-modified EE-Gβ1,γ2 subunits in ice-cold P-Rex dilution buffer to 1.5 μM, to give a 5× working stock. Do this only after having aliquoted the Rac/assay buffer mix into the assay tubes (see later), because G$\beta\gamma$ subunits do not retain their activity for long when diluted from their storage buffer, even at 4°. Immediately snap-freeze G$\beta\gamma$ subunits again for storage at –80°.

Assay

This protocol is for a typical assay with 20 samples performed in duplicates. Mix equal volumes of 10× Rac stock and 10× assay buffer, and aliquot 2 μl of the mix into precooled Microfuge tubes in an ice-slush bath. Stagger all subsequent additions by 30 sec between samples to ensure equal incubation times. Add 2 μl of 5× liposomes (containing or not PtdIns (3,4,5)P$_3$ or EDTA) per sample, vortex, incubate 10 min on ice, then add 4 μl of the 2.5× P-Rex1 stock, immediately followed by 2 μl 5× GTPγS, to give a final assay volume of 10 μl. Vortex and incubate for 10 min in a 30° water bath. Stop the reaction by adding 400 μl of ice-cold lysis buffer (1× PBS, 1 m*M* EGTA, 10 m*M* MgCl$_2$, 1% Triton X-100, 0.1 m*M* GTP). Immunoprecipitate Rac using prewashed and pre-aliquoted anti-EE-beads (see earlier), with end-over-end rotation for 60 min at 4°. Wash the anti-EE-beads four times with 400 μl of ice-cold lysis buffer. After the last wash, aspirate tight to the beads, add 400 μl of Ultima Gold scintillation fluid (Packard), and vortex well. Transfer open tubes into scintillation vials (Packard), add 5 ml of scintillation fluid, vortex again, and quantitate [^{35}S] GTPγS-loading of Rac by scintillation β-counting.

For assays with Gβγ subunits, add 2 μl of 5× Gβγ subunits per sample to the liposome/Rac mix before the 10 min incubation on ice. After the 10 min, add 2 μl of 5× stock P-Rex proteins instead of the 2.5× stock. The Gβγ subunit storage buffer contains 1% cholate that strongly inhibits the Rac-GEF activity of P-Rex enzymes. Hence, the effects of Gβγ subunits in the GEF assay are the combined effects of the stimulation by Gβγ subunits and the inhibition by the cholate. To account for this, perform controls for each Gβγ subunit concentration, using an equivalent dilution of Gβγ subunit storage buffer.

References

Donald, S., Hill, K., Lecureuil, C., Barnouin, R., Krugmann, S., Coadwell, W. J., Andrews, S. R., Walker, S. A., Hawkins, P. T., Stephens, L. R., and Welch, H. C. (2004). P-Rex2, a new guanine-nucleotide exchange factor for Rac. *FEBS Lett.* **572**, 172–176.

Gaffney, P. R. J., and Reese, C. B. (1997). Synthesis of 1-O-stearoyl-2-O-arachidonoyl-sn-glycer-3-yl-D-myo-inositol 3,4,5-trisphosphate and its stereoisomers. *Bioorg. Med. Chem. Lett.* **7**, 3171–3176.

Hill, K., Krugmann, S., Andrews, S. R., Coadwell, W. J., Finan, P., Welch, H. C., Hawkins, P. T., and Stephens, L. R. (2005). Regulation of P-Rex1 by phosphatidylinositol (3,4,5)-trisphosphate and Gβγ subunits. *J. Biol. Chem.* **280**, 4166–4173. Epub 2004.

Painter, G. F., Grove, S. J. A., Gilbert, I. H., Holmes, A. B., Paithby, P. R., Hill, M. L., Hawkins, P. T., and Stephens, L. R. (1999). General synthesis of 3-phosphorylated myo-inositol phospholipids and derivatives. *J. Chem. Soc. Perkin Trans.* **1**, 923–935.

Rosenfeldt, H., Vazquez-Prado, J., and Gutkind, J. S. (2004). P-Rex2, a novel PI-3-kinase sensitive Rac exchange factor. *FEBS Lett.* **572,** 167–171.

Welch, H. C., Coadwell, W. J., Ellson, C. D., Ferguson, G. J., Andrews, S. R., Erdjument-Bromage, H., Tempst, P., Hawkins, P. T., and Stephens, L. R. (2002). P-Rex1, a PtdIns (3,4,5)P$_3$- and G$\beta\gamma$-regulated guanine-nucleotide exchange factor for Rac. *Cell* **108,** 809–821.

[4] *In Vitro* Guanine Nucleotide Exchange Activity of DHR-2/DOCKER/CZH2 Domains

By Jean-François Côté and Kristiina Vuori

Abstract

Rho family GTPases regulate a large variety of biological processes, including the reorganization of the actin cytoskeleton. Like other members of the Ras superfamily of small GTP-binding proteins, Rho GTPases cycle between a GDP-bound (inactive) and a GTP-bound (active) state, and, when active, the GTPases relay extracellular signals to a large number of downstream effectors. Guanine nucleotide exchange factors (GEFs) promote the exchange of GDP for GTP on Rho GTPases, thereby activating them. Most Rho-GEFs mediate their effects through their signature domain known as the Dbl Homology-Pleckstrin Homology (DH-PH) module. Recently, we and others identified a family of evolutionarily conserved, DOCK180-related proteins that also display GEF activity toward Rho GTPases. The DOCK180-family of proteins lacks the canonical DH-PH module. Instead, they rely on a novel domain, termed DHR-2, DOCKER, or CZH2, to exchange GDP for GTP on Rho targets. In this chapter, the experimental approach that we used to uncover the exchange activity of the DHR-2 domain of DOCK180-related proteins will be described.

Introduction

Dynamic regulation of the actin cytoskeleton is critical for many cellular processes, including phagocytosis and cell migration. The Rho GTPases, such as Rho, Rac, and Cdc42, are central players in virtually all aspects of actin regulation (Raftopoulou and Hall, 2004). Like all GTPases, Rho proteins cycle between an inactive GDP-bound and an active GTP-bound state. When active, Rho proteins elicit their effects by activating specific effectors that transduce the signals to the cytoskeleton. Several

METHODS IN ENZYMOLOGY, VOL. 406
0076-6879/06 $35.00
DOI: 10.1016/S0076-6879(06)06004-6

classes of regulatory proteins that control the nucleotide state of Rho GTPases have been identified, among which guanine nucleotide exchange factors (GEFs) catalyze the exchange of GDP for GTP in response to extracellular signaling. A large family of Rho-GEFs consisting of more than 50 members has been described (Rossman *et al.*, 2005). It is now known that two signaling motifs, always arranged in tandem to form the DH-PH module, are responsible for the catalytic activity of these GEFs. The DH-PH module functions by creating a transitional interaction with the Rho-GDP target that results in the dissociation of the GDP. This transitional intermediate state, composed of the nucleotide-free Rho bound to the DH-PH, is then disrupted by the loading of the Rho protein with GTP.

DOCK180 was originally identified as a binding protein for the SH3 domain of the proto-oncogene product CrkII (Hasegawa *et al.*, 1996). Subsequent genetic screens in *C. elegans* and *Drosophila* suggested that the DOCK180 orthologs in these organisms were acting upstream of the Rho-family member Rac in a spectrum of biological events, including phagocytosis of apoptotic cells, migration of gonad cells, and myoblast fusion (Rushton *et al.*, 1995; Wu and Horvitz, 1998). Independent studies in mammalian cells demonstrated that the CrkII/DOCK180 complex regulates several Rac-dependent signaling pathways, including activation of the JNK kinase cascade and cell migration (Cheresh *et al.*, 1999; Dolfi *et al.*, 1998). Further studies uncovered that overexpression of DOCK180 leads to GTP loading of Rac1 in mammalian cells and that mice lacking DOCK2, a DOCK180 homolog exclusively expressed in hematopoietic cells, are deficient in lymphocyte migration and Rac activation in response to chemokines (Fukui *et al.*, 2001; Kiyokawa *et al.*, 1998a). Because DOCK180 lacks the canonical DH-PH module, it was first assumed that DOCK180 activates Rac by means of an unidentified exchange factor. Unexpectedly, we and others found that DOCK180 is a direct activator of Rac and that the exchange activity resides in a novel domain, termed DOCK Homology Region (DHR)-2, DOCKER, or CZH2, which displays no apparent sequence homology to the canonical DH-PH module (Brugnera *et al.*, 2002; Cote and Vuori, 2002; Meller *et al.*, 2002). Similar to the DH-PH module, the DHR-2 domain forms a stable complex with nucleotide-free Rac and promotes the exchange of GDP for GTP *in vitro*. Furthermore, we identified several proteins with homology to DOCK180 that harbors the DHR-2 domain. We will later describe the methods we used to uncover and measure the guanine exchange activity in the DHR-2 domain of the evolutionarily conserved DOCK180 superfamily of novel Rho GEFs.

Identification of the DOCK180 Superfamily and the DHR-2 Domain

We initially used several bioinformatics approaches to gain further insight into the DOCK180 function. As such, standard BLAST searches using the human DOCK180 protein as a query indicated that DOCK180 belongs to a protein superfamily, which consists of at least 11 human members. On the basis of BLAST scores and phylogenetic data, we classified proteins in four subfamilies: DOCK-A, -B, -C, and -D (Fig. 1A). Common in the C-terminal of all these proteins is a region of high sequence homology, which we termed the DHR-2 domain (aas 1111–1636 of human DOCK180). Furthermore, sequence alignments of full-length proteins revealed the presence of high homology in these proteins that we termed DHR-1 (Fig. 1B). We demonstrated recently that the DHR-1 domain functions as a novel lipid binding domain, and targets the DOCK 180 signaling complex to the plasma membrane by specifically recognizing the major lipid product of PI 3-kinases, PtdIns(3,4,5)P$_3$ (Cote *et al.*, 2005). For full details on the identification of the DOCK180 superfamily, please see Cote and Vuori (2002).

In Vitro GTPase Binding Assays Demonstrate That the DHR-2 Domains Form a Stable Complex with Nucleotide-free Rho GTPases

Matsuda and colleagues previously reported a role for the C-terminal region of DOCK180 in Rac GTP-loading on DOCK180 overexpression in

A

	Homo sapiens	D. melanogaster	C. elegans	D. discoideum	A. thaliana	S. cerevisiae
DOCK-A	DOCK180 DOCK2 DOCK5	Myoblast city	CED-5	DocA		
DOCK-B	DOCK3/MOCA DOCK4	CG11754			SPIKE1	YLR422W
DOCK-C	DOCK6 DOCK7 DOCK8	CG11376	F46H5.4	Unnamed (#AAL92252)		
DOCK-D	DOCK9/zizimin1 DOCK10/zizimin3 DOCK11/zizimin2	CG6630	F22G12.5	Unnamed (#AAM08471)		

B

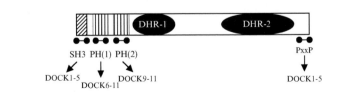

FIG. 1. (*continued*)

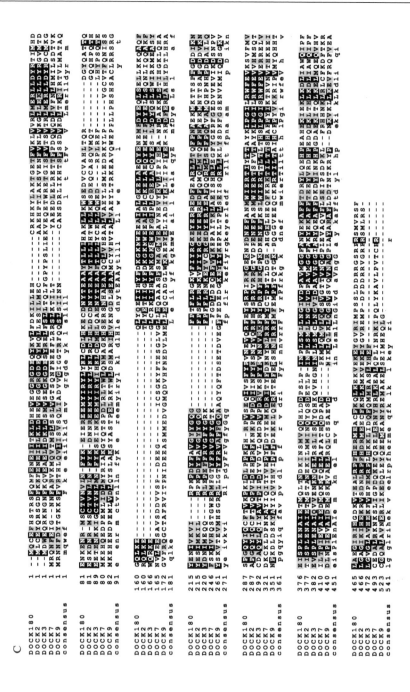

mammalian cells (Kobayashi *et al.*, 2001); this region overlaps with the DHR-2 domain identified previously. Furthermore, earlier reports by the Matsuda and Settleman groups suggested that DOCK180 could form a complex with nucleotide-free Rac in cell lysates (Kiyokawa *et al.*, 1998a; Nolan *et al.*, 1998). Importantly, GEFs can be distinguished from other GTPase-interacting proteins by their ability to bind to the nucleotide-free state of GTPase (Cherfils and Chardin, 1999). On the basis of these findings, we decided to examine the potential role of the DHR-2 domain in direct binding to and activation of Rac GTPase by DOCK180 and subsequently expanded our studies to the other DHR-2 domains. Three *in vitro* approaches were undertaken to examine the capability of the DHR-2 domains, and the full-length DOCK180-family proteins, to directly interact with a nucleotide-free Rho GTPase. First, GST-fusion proteins encoding the various DHR-2 domains of DOCK180-proteins were produced in bacteria, and their ability to precipitate nucleotide-free forms of either endogenously or exogenously expressed Rho GTPases in mammalian cell lysates was examined. In a reverse experiment, immobilized GST-fusion proteins of Rho GTPases were tested for their capability to pull down *in vitro* transcribed and translated ^{35}S-methionine–labeled DHR-2 domains. As a third line of investigation, immobilized GST-fusion proteins of Rho GTPases were tested for their capability to precipitate exogenously expressed, full-length forms of DOCK180 proteins in mammalian cells.

Bacterial Expression and Purification of GST-Tagged DHR-2 Domains and Rho GTPases

By using conventional DNA methods, the appropriate cDNA portions of the DHR-2 domains of DOCK180, DOCK2, DOCK3/MOCA, DOCK7, and DOCK9/zizimin1 were cloned in pGEX-4T-1 (see Fig. 1C for an

---

FIG. 1. (A) Identification and classification of the DOCK180-related proteins. The proteins were identified by Blast searches using DOCK180 protein sequence as a query. Classification was mainly based on phylogenetic analyses and complemented by Blast identity scores. (B) Schematic diagram of the DOCK180-related proteins. All the DOCK proteins are characterized by the presence of the DHR-1 and DHR-2 domains. An amino-terminal SH3 domain is found in DOCK1–5 (DOCK1=DOCK180). A PH domain (PH1) is found in DOCK6–11 (our unpublished data), whereas an additional PH domain (PH2) is characteristic to DOCK9–11 proteins. Furthermore, DOCK1–5 proteins harbor carboxyl terminal proline-rich regions that serve as binding sites for, among others, the SH3 domain of Crk proteins. (C) Multiple sequence alignment of DHR-2 domains. The multiple sequence alignment of the DHR-2 domains was performed using the ClustalW algorithm, and the final output was generated by Boxshade. The boundaries of the DHR-2 domains in the alignment correspond exactly to the boundaries of the DHR-2 proteins generated in the biochemical studies reported in this chapter. Modified after Cote and Vuori (2002).

alignment of the exact boundaries of the DHR-2 domains used in this study). To produce the corresponding GST fusion proteins, single colonies of *Escherichia coli* BL21(DE3) harboring the plasmids of interest were inoculated in 5 ml of LB media supplemented with 100 μg/ml of ampicillin and cultured overnight at 37°. These minicultures were subsequently used to inoculate 250 ml of LB media supplemented with 100 μg/ml of ampicillin and cultured overnight at 37° in 2-l flasks. The next morning, flasks were cooled on ice for 15 min, and 250 ml of LB media (supplemented with 100 μg/ml of ampicillin and with 2 mM IPTG) was added to each culture before moving them to 30° for 2 h for protein induction. We found that inducing saturated bacterial cultures at 30° for short periods resulted in a better yield of the various DHR-2 fragments. Bacteria were harvested by centrifugation at 5000 rpm in a GSA rotor. Bacterial pellets were resuspended in 10 ml of PBS-1% Triton X-100 supplemented with 1× complete protease inhibitor cocktail (Roche) and lysed by sonication on ice (three cycles of 30 sec with continuous pulse, Branson Sonifier 250). The bacterial lysates were cleared by centrifugation (12,000 rpm in an SS-34 rotor for 10 min). GST fusion proteins in the cleared lysates were affinity absorbed on glutathione sepharose (100 μl of packed beads) for 30 min under gentle agitation in the cold room. Beads were collected by low-speed centrifugation (1000 rpm) and washed extensively in PBS-0.1% Triton X-100 and stored at 4° in the same buffer. Protein concentration was approximately 0.5 μg/ml of packed beads.

GST Rho GTPases were expressed and purified similarly to the DHR-2 described previously with the following modifications. For protein production, 50 ml of overnight cultures was diluted to 500 ml (in LB media with 100 μg/ml of ampicillin) and grown for 2 h at 37° before induction with 1 mM IPTG (final concentration) for 3 h at 37°. Proteins were extracted from the bacteria and purified as previously. The yield of the Rho GTPases was typically high (approximately 5 μg/500 μl of culture), with the exception of RhoG and CHP (500 μg/500 ml). The purified proteins were stored at 4° for several days as described previously for DHR-2 domains. The constructs for GST-Rac, GST-Cdc42, and GST-RhoA were obtained from A. Hall. The cDNAs for Rac2 and Rac3 were amplified by PCR and cloned in the *Bam*HI and *Eco*RI sites of pGEX-4T-1. The constructs for GST-RhoG and GST-TCL were a gift from Dr. P. Fort. The TC10 cDNA was obtained from Dr. A. Saltiel and subcloned in the *Bam*HI and *Eco*RI sites of pGEX-4T-1. The CHP cDNA was obtained from H. Mellor and subcloned in the *Eco*RI and *Xho*I sites of pGEX-4T-1. The cDNAs of Rnd1 and Rnd2 were obtained from H. Katoh and were digested with *Kpn*1 and *Eco*RV, blunt-ended with T4 polymerase and ligated in the *Sma*I site of pGEX-4T-1.

*Precipitation of Rho GTPases in Mammalian Cells with the Bacterially
Produced GST-DHR-2 Domains in Nucleotide-Free Conditions*

To examine the binding of the DHR-2 domains to an exogenously expressed Rho GTPase, COS-1 cells cultured on 10-cm dishes were transfected (4 μg of pRK5 plasmids, Lipofectamine Plus) to express either Myc-Rac1 or Myc-Cdc42 (obtained from A. Hall). Forty-eight hours after transfection, the cells were lysed in 1 ml per dish of an EDTA-containing buffer (buffer A: 25 mM Tris-HCl, pH 7.5, 150 mM NaCl, 5 mM EDTA, 0.5% Triton X-100, 10 mM NaF, 0.1 mM Na3VO4, 1× complete inhibitor cocktail) to stabilize the nucleotide-free form of Rac and Cdc42. When binding to endogenous Rho GTPases was studied, untransfected, confluent COS-1 cell cultures were directly lysed as previously. The clarified lysates (100 μg of proteins in 500 μl of buffer A) were then incubated with the GST-DHR-2 fusion proteins immobilized on glutathione sepharose (10 μl of packed beads, see previously) for 90 min under gentle agitation at 4°. The beads were washed three times in buffer A, and the bound material was resolved by SDS-PAGE. The exogenously expressed GTPases that bound to the DHR-2 fragments were detected by immunoblotting with the anti-Myc antibody (9E10, Santa Cruz), whereas the bound endogenous Rac or Cdc42 proteins were detected by immunoblotting using anti-Rac (Upstate) or anti-Cdc42 (Pharmingen) monoclonal antibodies, respectively.

As shown in Fig. 2A, the DHR-2 domain of DOCK180, but not its deletion mutants, was capable of precipitating Myc-Rac from cell lysates. This binding was specific, because no precipitation of Myc-Cdc42 with the DOCK180 GST DHR-2 protein was detected. As shown in Fig 2B, the DHR-2 domain of DOCK2, but not of DOCK3/MOCA, interacted robustly with the endogenous Rac protein. No binding to Rac was detected by the GST DHR-2 domains of either DOCK7 or DOCK9/zizimin1 (not shown). In these conditions, the DHR-2 domains of the DOCK2, DOCK3/MOCA, DOCK7, and DOCK9/zizimin1 proteins failed to demonstrate any interaction with the endogenous Cdc42 protein (not shown).

In Vitro *Transcription and Translation of* [35]*S-Labeled DHR-2
Domains and their Binding to Immobilized GST-Fusion
Proteins of Rho GTPases*

To produce the radiolabeled DHR-2 domains of DOCK180, DOCK3/MOCA, DOCK7, or DOCK9/zizimin1, we cloned the corresponding cDNAs in pcDNA3.1-Myc (see Fig. 1C for boundaries) and took advantage of the T7 RNA polymerase priming site upstream of the Myc epitope to

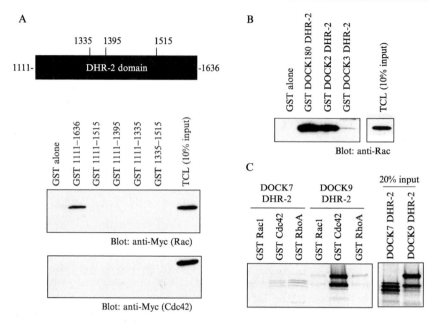

FIG. 2. (A) The DHR-2 domain of DOCK180 binds to Rac in nucleotide-free conditions. A schematic representation of the DHR-2 domain and of the boundaries of the various truncation mutants of DOCK180 is shown. COS cells were transfected to express either Myc-Rac or Myc-Cdc42 and lysates (containing EDTA to promote the nucleotide-free intermediate of GTPases) were incubated with various immobilized GST fusion proteins of the DHR-2 domain of DOCK180. The bound material was detected by immunoblotting with the 9E10 anti-Myc antibody. (B) The DHR-2 domain of DOCK2 binds to nucleotide-free Rac. The DHR-2 domains of DOCK180, DOCK2, and DOCK3 were tested for their ability to precipitate endogenous Rac from COS cell lysate as described in (A). The bound Rac was detected by immunoblotting with an anti-Rac antibody. (C) The DHR-2 domain of DOCK9 binds to nucleotide-free Cdc42. Immobilized GST fusion proteins of Rac, Cdc42, and Rho were tested for their ability to precipitate the DHR-2 domains of DOCK7 or DOCK9 in nucleotide-free conditions. The DHR-2 domains were generated by coupled *in vitro* transcription and translation in the presence of ^{35}S-methionine. The bound material was detected by autoradiography. Modified after Cote and Vuori (2002).

generate the DHR-2 domains by *in vitro* transcription and translation in the presence of ^{35}S-methionine. We used the Promega TnT kit for coupled transcription and translation of the DHR-2 domains (1 μg of plasmid, 2 μl of translation-grade ^{35}S-methionine in a 50-μl total reaction volume for 90 min at 30°). The various labeled DHR-2 domains (10 μl) were then diluted in 500 μl of buffer A and incubated with 5 μg of immobilized GST-Rac, GST-Cdc42, or GST-RhoA for 90 min at 4°. After extensive washing, the bound material was resolved by SDS-PAGE and detected by autoradiography. By use of this approach, we detected binding of the DHR-2 domains of DOCK180 to GST-Rac (not shown) and of DOCK9/zizimin1 to GST-Cdc42 (Fig. 2C). No binding of the DHR-2 domains of DOCK3/MOCA or DOCK7 to Rac, Cdc42, or RhoA GST-fusion proteins was detected (Fig. 2C and not shown).

Precipitation of Full-Length DOCK180 Proteins in Mammalian Cells with the Bacterially Produced GST-Rho GTPases

The studies described previously successfully identified DOCK180 and DOCK2 (members of the DOCK-A subfamily) and DOCK9/zizimin1 (member of the DOCK-D subfamily) as binding proteins for Rac and Cdc42, respectively. However, no Rho GTPase binding was detected for the DHR-2 domains of DOCK3/MOCA (member of the DOCK-B subfamily) and DOCK7 (member of the DOCK-C subfamily). This could be due to the requirement of sequences adjacent to the DHR-2 domain for the binding activity, and/or members of the DOCK-B and DOCK-C subfamilies may recognize Rho GTPases other than the ones we had studied (there are 22 known members in the Rho GTPase family). To address these issues, we tested the ability of full-length DOCK180, DOCK3/MOCA, DOCK6 (member of the DOCK-C subfamily), and DOCK9/zizimin1 proteins to interact with a larger panel of GST-fusions of Rho GTPases, namely Rac1, Rac2, Rac3, RhoG, Cdc42, TC10, TCL, CHP, RhoA, Rnd1, and Rnd2 (see earlier). Flag-DOCK180 (a gift from Dr. Matsuda) and HA-DOCK3/MOCA (a gift from Dr. Schubert) have been described elsewhere (Kashiwa *et al.*, 2000; Kiyokawa *et al.*, 1998b). The full-length human DOCK6 cDNA (obtained from the Kazuza DNA Research Institute, clone KIAA1395) was ligated in pcDNA3 Myc-His. A portion of the cDNA of DOCK9/zizimin1 (obtained from the Kazuza DNA Research Institute, clone KIAA1058), coding for amino acids 43–2069, was ligated in pEGFP-C2. Five micrograms of each of the plasmids was transfected in HEK293-T cells using Lipofectamine 2000. The cells were lysed in buffer A 48 h after transfection, and the clarified lysates were incubated with 5 μg of each of the Rho GTPases immobilized on glutathione sepharose for 90 min at 4°. After several washes of the beads, the bound material was resolved on SDS-PAGE and detected by immunoblotting with the appropriate antibodies. As shown in Fig. 3, we detected binding of Flag-DOCK180 to Rac1, 2, and 3, and of EGFP-DOCK9/zizimin1 to Cdc42 but failed to detect binding of DOCK3/MOCA or DOCK6 to any of the GTPases tested. Thus, additional work is clearly required to clarify the biochemical properties of the DHR-2 domains of the DOCK-B and DOCK-C subfamilies.

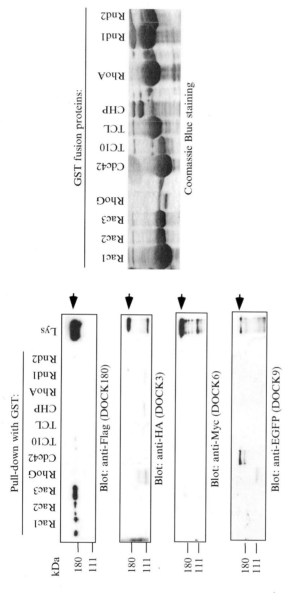

FIG. 3. Binding of RhoGTPases to full-length DOCK180-related proteins in nucleotide-free conditions. HEK 293-T cells were transfected to express Flag DOCK180, HA-DOCK3, DOCK6-Myc-His, or EGFP-DOCK9. A panel of immobilized GST-Rho GTPases was tested for the ability to precipitate the DOCK180-related proteins in nucleotide-free conditions. Bound material was detected using anti-Flag, anti-HA, anti-Myc, or anti-EGFP monoclonal antibodies (left panels). The Rho GTPases used in this experiment were fractionated by SDS-PAGE, and the gel was stained with Coomassie Blue to ascertain the integrity of the proteins.

The DHR-2 Domains Contain Rho-GEF Activity *In Vitro*

That the DHR-2 domains are capable of interacting with Rho GTPases *in vitro* under nucleotide-free conditions suggests that these domains could have direct catalytic exchange activity toward Rho GTPases. Both *in vitro* and *in vivo* GEF activity studies, described in this and in the following sections, were undertaken to study this possibility.

Expression and Purification of Recombinant DHR-2 Domains for In Vitro *GEF Activity Assays*

The DHR-2 domains of DOCK180, DOCK2, DOCK3/MOCA, DOCK7, and DOCK9/zizimin1 were produced as GST-fusion proteins in bacteria (2 l of final IPTG-induced volume for each construct) as described previously. GST alone was also prepared to be used as a negative control. The proteins were eluted from the glutathione beads by three successive incubations with 500 μl of 50 mM Tris-HCl (pH 8.0)/150 mM NaCl/10 mM glutathione. The proteins were then buffer-exchanged against PBS-10% glycerol and concentrated to approximately 200 μl using centrifugation concentrating units (10-kDa cutoff, Millipore). Yields were typically low for the eluted DHR-2 domains (50–100 μg per 2 l of culture). Proteins were stored in 10-μl aliquots (2.5–5 μg) at −80°, and they remained stable (and active) for several weeks under these storage conditions. We initially noted that the GST-DHR-2 domain of DOCK180, although capable of binding to Rac, did not exchange GDP for GTP in a GEF assay (described later). We found that removal of the GST moiety by thrombin cleavage was necessary to uncover the GEF activity of this DHR-2 domain, and, thus, we subjected all DHR-2 domains used in our studies to a similar cleavage when performing exchange assays. To remove the GST-tag, 1 μl of thrombin (1 NIH Unit, from bovine plasma, Sigma) was added per the 10-μl aliquot containing the DHR-2 domain, and the mixture was incubated for 20 min at room temperature. The digestion reaction was stopped by adding 1 μl of 50× complete protease inhibitor cocktail. GST alone was treated identically, as a control. Importantly, we noted a complete cleavage of the GST tag for all the DHR-2 domains tested and the concomitant appearance of a cleaved product of the anticipated size when analyzed by SDS-PAGE (not shown). The digested DHR-2 domains were used immediately in GEF assays (see later).

Preparation of ^{3}H-GDP–Loaded Rac and Cdc42

GST-Rac and GST-Cdc42 were produced in bacteria and purified as described in section "*In Vitro* GTPase Binding Assay Demonstrate That the DHR-2 Domains Form a Stable Complex with Nucleotide-free Rho

GTPases (Bacterial Expression and Purification of GST-tagged DHR-2 Domains and Rho GTPases)"and eluted from the affinity matrix as depicted in "The DHR-2 Domains Contain Rho-GEF Activity *In Vitro* (Expression and Purification of Recombinant DHR-2 Domains for *In Vitro* GEF Activity Assays)." These proteins were then concentrated to 1 mg/ml and buffer-exchanged to PBS before storage at $-80°$. The ^{3}H-GDP loading of the purified GTPases was performed as follows: 10 μg (10 μl) of each GTPase was diluted in 50 μl of $2\times$ loading buffer (20 mM Hepes, pH 7.4, 200 mM NaCl, 4 mM EDTA, 0.4 mM DTT, 10 mM GDP, and 10 μM ^{3}H-GDP [Amersham]), and the mixture was brought to 100 μl with water. The samples were incubated for 15 min at room temperature, and the GDP-loaded GTPases were stabilized in this state by adding magnesium chloride to a final concentration of 5 mM (1 μl of a 0.5 M stock). The loaded GTPases were kept on ice for a maximum of 4 h until using them in exchange assays.

In Vitro *GEF Assays with the DHR-2 Domains*

^{3}H-GDP Rac or Cdc42 molecules were incubated with the purified recombinant DHR-2 domains in the presence of cold GTP, and the dissociation of the ^{3}H-GDP was monitored. To perform this assay, we diluted 20 μl of the ^{3}H-GDP loaded Rho GTPases (2 μg) with 30 μl of water and 50 μl of $2\times$ exchange buffer (20 mM Hepes, pH 7.4, 200 mM NaCl, 20 mM MgCl$_2$, 1 mg/ml lipid-free BSA, 0.4 mM DTT, and 2 mM GTP) at room temperature. To initiate the exchange reactions, 10 μl of the DHR-2 domains of DOCK180, DOCK2, DOCK3/MOCA, DOCK7, and DOCK9/zizimin1 (prepared as described in "The DHR-2 Domains Contain Rho-GEF Activity *In Vitro* [Expression and Purification of Recombinant DHR-2 Domains for *In Vitro* GEF Activity Assays])" or as a control, GST alone, were added to the reactions. Aliquots (30 μl) were taken at different time points (0, 15, 30 min) and diluted in 1 ml of ice-cold STOP buffer (10 mM Hepes, pH 7.4, 100 mM NaCl, and 5 mM MgCl$_2$). The Rho GTPases were trapped on nitrocellulose membranes (Millipore) by filtration under gentle vacuum and washed with 5 ml of ice-cold STOP buffer. Dried membranes were subjected to liquid scintillation counting for quantification of the bound ^{3}H-GDP.

The amount of bound ^{3}H-GDP was expressed as a percentage of the 0 min time point. The GEF domain of Vav2 was used as a positive control in our studies (not shown but details can be found in Cote and Vuori [2002]). In these experiments, the DHR-2 domains of DOCK180 and DOCK2 exchanged GDP for GTP on Rac, whereas the DHR-2 domain of DOCK9/zizimin1 was active toward Cdc42 (Fig. 4). We failed to detect exchange activity toward either Rac or Cdc42 by the DHR-2 domains of DOCK3/MOCA or DOCK7, and as expected, by GST alone (Fig. 4).

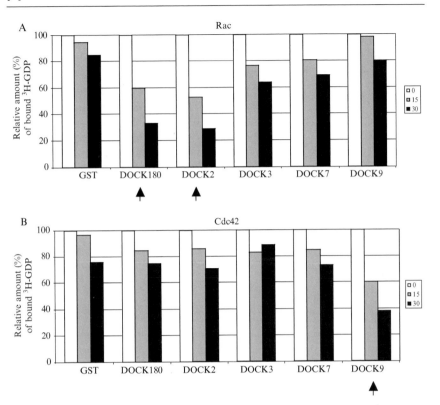

FIG. 4. Purified DHR-2 domains of various DOCK180-related proteins have GEF activity toward the small GTPases Rac (A) and Cdc42 (B) *in vitro*. The indicated DHR-2 domains, or GST alone as a control, were tested in an *in vitro* GEF assay toward either ^{3}H-GDP-loaded Rac (A) or Cdc42 (B). Reactions were stopped at 0, 15, 30 min time points. Radiolabeled GDP remaining bound to the GTPases was measured by use of a filter-binding assay. Arrows indicate the samples with statistically significant GEF activity.

Importantly, the exchange activity of the DHR-2 domains correlated with their ability to form a molecular complex with their cognate GTPases under nucleotide-free conditions (Figs. 2 and 4).

The DHR-2 Domains of DOCK180 and DOCK2 Contain Rho-GEF Activity *In Vivo*

We next investigated whether the DHR-2 domain of DOCK180 was sufficient and necessary to activate Rac in cells. For these experiments, we generated a mammalian expression vector coding for DOCK180 lacking the DHR-2 domain (DOCK180▲DHR-2) and a construct that expresses the sequences of the DHR-2 domain of DOCK180 (aas 1111–1636,

Myc-DOCK180 DHR-2). As positive and negative controls, respectively, we used a flag-tagged, full-length form of DOCK180 and a truncated version of the DHR-2 domain (aas 1111–1335), which lacks Rac-binding and exchange activity *in vitro* (see Fig. 2A). COS-1, HEK293-T, and LR73 cells were used to overexpress these various constructs, and similar results were obtained in all lines. The classical method in which the PBD domain of p21PAK, immobilized on glutathione sepharose, is used to affinity-purify the GTP-loaded Rac from cell lysates was used here. The PAK-PBD construct, as well as its expression and purification, has been described in detail in (Benard and Bokoch, 2002). The cells transfected with the various plasmids were lysed 48 h after transfection in 500 μl of

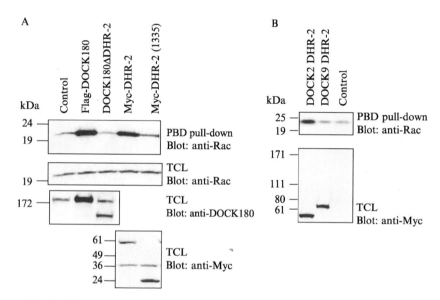

FIG. 5. (A) The DHR-2 domain of DOCK180 is both necessary and sufficient for Rac activation in cells. HEK 293-T cells were transfected with a control vector or vectors coding for Flag-DOCK180, DOCK180 ▲ DHR-2, Myc-DHR-2 (aas 1111–1636 of DOCK180), or Myc-DHR-2 (1395), which is a truncated version of the DHR-2 domain (aas 1111–1335 of DOCK180). Rac-GTP was pulled down from cell lysates using the GST-PAK-PBD immobilized to beads. The amount of bound Rac was detected by immunoblotting with an anti-Rac antibody. Expression levels of the exogenous proteins were analyzed by immunoblotting with anti-DOCK180 or anti-Myc antibodies. Lysates were also immunoblotted with an anti-Rac antibody to confirm equal loading between samples. (B) The DHR-2 domain of DOCK2 activates Rac *in vivo*. HEK 293-T cells were transfected with a control plasmid or vectors coding for either the Myc-DHR-2 domains of DOCK2 or DOCK9. Activated Rac was pulled down from lysates as described in (A). Bound material was detected by immunoblotting with an anti-Rac antibody. Expression levels of exogenous proteins were verified by immunoblotting the lysates with an anti-Myc antibody. Modified after Cote and Vuori (2002).

MLB buffer (25 mM Hepes (pH 7.5), 150 mM NaCl, 1% NP-40, 10 mM MgCl$_2$, 1 mM EDTA, 10% glycerol and 1× complete protease inhibitor). The clarified cell lysates were then incubated with 10 μg of immobilized PAK-PBD for 30 min under gentle agitation. After three washes with MLB buffer, the bound material was resolved by SDS-PAGE, and the amount of activated Rac was detected by immunoblotting with an anti-Rac antibody. Equal volumes of each lysates were also separated by SDS-PAGE, and equal loading (anti-Rac) and exogenous protein expression (anti-DOCK180 or anti-Myc) was verified by immunoblotting. As shown in Fig. 5A, DOCK180 readily stimulated the Rac GTP-loading by 4.4-fold in cells, whereas the DHR-2 deletion mutant protein exhibited no activity. Similarly, expression of the Myc-tagged DHR-2 domain of DOCK180 led to a 3.8-fold Rac activation, but a truncated form of this domain was devoid of any exchange activity (Fig. 5A). Similar to what we found for DOCK180, the DHR-2 domain of DOCK2, but not the one of DOCK9, was potent in activating Rac in cells (Fig. 5B). Together, these results demonstrate that the DHR-2 domain of DOCK180 and DOCK2 is both sufficient and necessary for Rac activation in cells.

Discussion

Our studies by us and those of others have revealed the existence of a superfamily of evolutionarily conserved DOCK180-related proteins that function as GEFs for Rho GTPases (Brugnera *et al.*, 2002; Cote and Vuori, 2002; Meller *et al.*, 2002). As described here, we have demonstrated that members of the DOCK-A subfamily (DOCK180, DOCK2) and DOCK-D subfamily (DOCK9/zizimin1) function as GEFs toward Rac and Cdc42, respectively. In Table I, an up-to-date summary of what is known with respect to the GEF activity of mammalian DOCK180 proteins toward small GTPases is provided.

The GEF activity in the DOCK proteins resides in the DHR-2 domain, which seems to function in a manner similar to the DH-PH module found in classical Rho-GEFs. Thus, the DHR-2 forms a complex with nucleotide-free GTPases and catalyzes the exchange of GDP for GTP on Rho GTPases both *in vitro* and *in vivo*. At present, the details as to how the DHR-2 domain functions to catalyze the exchange reaction remain unknown. Thus, atomic level structural characterization of the DHR-2 domain is required to clarify the molecular basis of the DHR-2–mediated GEF activity. At the same time, biochemical analyses, such as those presented here, will be essential to validate the specificity and exchange activity of the various DHR-2 domains toward GTPases. This information will bring

TABLE I
SUMMARY OF THE SPECIFICITY AND ACTIVITY OF DOCK180-RELATED
PROTEINS TOWARD GTPASES

DOCK	Members	*In vitro* binding to nucleotide-free GTPases[a]	*In vitro* GEF activity on GTPases	*In vivo* GEF activity on GTPases
A	DOCK180	Rac1[b,c]	Rac1[b]	Rac1[b]
	DOCK2	Rac1[b,c]	Rac1[b]	Rac1[b]
	DOCK5	Rac2[d], Rac3[d]	?	?
B	DOCK3	Rac1[e,f]	?	Rac1[e,f]
	DOCK4	?	?	Rap1[g], Rac[h]
C	DOCK6	?	?	?
	DOCK7	?	?	?
	DOCK8	?[i]	?	?
D	DOCK9	Cdc42[b,j]	Cdc42[b]	Cdc42[j]
	DOCK10	Cdc42[k], TCL[k]	?	?
	DOCK11	Cdc42[k]	?	Cdc42[k]

[a] Denotes binding between the DHR-2 of the DOCK180-family member and the nucleotide-free GTPase, unless otherwise indicated.
[b] Côté and Vuori, *J. Cell Sci.* (115) 2002:p4901–4913.
[c] Brugnera *et al.*, *Nat. Cell. Biol.* (4) 2002:p574–582.
[d] Our unpublished results.
[e] Binding and activation detected with full-length DOCK3, but not with the isolated DHR-2 domain.
[f] Grimsley *et al.*, *J. Biol. Chem.* (279) 2004:p6087–6097.
[g] Yajnik *et al.*, *Cell* (112) 2003:p673–684.
[h] Lu *et al.*, *Curr. Biol.* (15) 2005:p371–377.
[i] Interaction between FL DOCK8 and Rac1, Cdc42, TC10 and TCL has been detected in a yeast two-hybrid assay.
[j] Meller *et al.*, *Nat. Cell. Biol.* (4) 2002:p639–647.
[k] Nishikimi *et al.*, *FEBS Lett.* (579) 2005:p1039–1046.

guidance to future studies to understand the physiological functions of this novel superfamily of signaling molecules.

Acknowledgments

We are indebted to the many colleagues mentioned in the text who have shared reagents in these studies. We also apologize for not citing important articles because of space limitation. Our studies have been supported by the Terry Fox Foundation/National Cancer Institute of Canada (Fellowship to J.-F.C.) and by the National Institutes of Health (to K.V.).

References

Benard, V., and Bokoch, G. M. (2002). Assay of Cdc42, Rac, and Rho GTPase activation by affinity methods. *Methods Enzymol.* **345,** 349–359.

Brugnera, E., Haney, L., Grimsley, C., Lu, M., Walk, S. F., Tosello-Trampont, A. C., Macara, I. G., Madhani, H., Fink, G. R., and Ravichandran, K. S. (2002). Unconventional Rac-GEF activity is mediated through the Dock180-ELMO complex. *Nat. Cell Biol.* **4**, 574–582.

Cheresh, D. A., Leng, J., and Klemke, R. L. (1999). Regulation of cell contraction and membrane ruffling by distinct signals in migratory cells. *J. Cell Biol.* **146**, 1107–1116.

Cherfils, J., and Chardin, P. (1999). GEFs: Structural basis for their activation of small GTP-binding proteins. *Trends Biochem. Sci.* **24**, 306–311.

Cote, J. F., and Vuori, K. (2002). Identification of an evolutionarily conserved superfamily of DOCK180-related proteins with guanine nucleotide exchange activity. *J. Cell Sci.* **115**, 4901–4913.

Cote, J. F., Motoyama, A. B., Bush, J. A., and Vuori, K. (2005). A novel and evolutionarily conserved PtdIns(3,4,5)P$_3$-binding domain is necessary for DOCK180 signaling. *Nat. Cell Biol.* **7**, 797–807.

Dolfi, F., Garcia-Guzman, M., Ojaniemi, M., Nakamura, H., Matsuda, M., and Vuori, K. (1998). The adaptor protein Crk connects multiple cellular stimuli to the JNK signaling pathway. *Proc. Natl. Acad. Sci. USA* **95**, 15394–15399.

Fukui, Y., Hashimoto, O., Sanui, T., Oono, T., Koga, H., Abe, M., Inayoshi, A., Noda, M., Oike, M., Shirai, T., and Sasazuki, T. (2001). Haematopoietic cell-specific CDM family protein DOCK2 is essential for lymphocyte migration. *Nature* **412**, 826–831.

Hasegawa, H., Kiyokawa, E., Tanaka, S., Nagashima, K., Gotoh, N., Shibuya, M., Kurata, T., and Matsuda, M. (1996). DOCK180, a major CRK-binding protein, alters cell morphology upon translocation to the cell membrane. *Mol. Cell Biol.* **16**, 1770–1776.

Kashiwa, A., Yoshida, H., Lee, S., Paladino, T., Liu, Y., Chen, Q., Dargusch, R., Schubert, D., and Kimura, H. (2000). Isolation and characterization of novel presenilin binding protein. *J. Neurochem* **75**, 109–116.

Kiyokawa, E., Hashimoto, Y., Kobayashi, S., Sugimura, H., Kurata, T., and Matsuda, M. (1998a). Activation of Rac1 by a Crk SH3-binding protein, DOCK180. *Genes Dev.* **12**, 3331–3336.

Kiyokawa, E., Hashimoto, Y., Kurata, T., Sugimura, H., and Matsuda, M. (1998b). Evidence that DOCK180 up-regulates signals from the CrkII-p130(Cas) complex. *J. Biol. Chem.* **273**, 24479–24484.

Kobayashi, S., Shirai, T., Kiyokawa, E., Mochizuki, N., Matsuda, M., and Fukui, Y. (2001). Membrane recruitment of DOCK180 by binding to PtdIns(3,4,5)P$_3$. *Biochem. J.* **354**, 73–78.

Meller, N., Irani-Tehrani, M., Kiosses, W. B., del Pozo, M. A., and Schwartz, M. A. (2002). Zizimin1, a novel Cdc42 activator, reveals a new GEF domain for Rho proteins. *Nat. Cell Biol.* **4**, 639–647.

Nolan, K. M., Barrett, K., Lu, Y., Hu, K. Q., Vincent, S., and Settleman, J. (1998). Myoblast city, the *Drosophila* homolog of DOCK180/CED-5, is required in a Rac signaling pathway utilized for multiple developmental processes. *Genes Dev.* **12**, 3337–3342.

Raftopoulou, M., and Hall, A. (2004). Cell migration: Rho GTPases lead the way. *Dev. Biol.* **265**, 23–32.

Rossman, K. L., Der, C. J., and Sondek, J. (2005). GEF means go: Turning on RHO GTPases with guanine nucleotide-exchange factors. *Nat. Rev. Mol. Cell Biol.* **6**, 167–180.

Rushton, E., Drysdale, R., Abmayr, S. M., Michelson, A. M., and Bate, M. (1995). Mutations in a novel gene, myoblast city, provide evidence in support of the founder cell hypothesis for *Drosophila* muscle development. *Development* **121**, 1979–1988.

Wu, Y. C., and Horvitz, H. R. (1998). *C. elegans* phagocytosis and cell-migration protein CED-5 is similar to human DOCK180. *Nature* **392**, 501–504.

[5]   Biochemical Characterization of the
Cool (Cloned-Out-of-Library)/Pix (Pak-Interactive
Exchange Factor) Proteins

By DANIEL BAIRD,* QIYU FENG,* and RICHARD A. CERIONE

Abstract

The Cool (Cloned out of Library)/Pix (Pak interactive exchange factor) proteins have been implicated in a diversity of biological activities, ranging from pathways initiated by growth factors and chemoattractants to X-linked mental retardation. Initially discovered through yeast two-hybrid and biochemical analyses as binding partners for the Cdc42/Rac-target/effector, Pak (p21 activated kinase), the sequences for the Cool/Pix proteins revealed a DH (Dbl homology) domain. Because the DH domain is the limit functional unit for stimulating guanine nucleotide exchange on Rho family GTP-binding proteins, it was assumed that the Cool/Pix proteins would act as guanine nucleotide exchange factors (GEFs) for the Rho proteins. Of the three known isoforms, (p50Cool-1, p85Cool-1/β-Pix, and 90Cool-2/α-Pix), only Cool-2/α-Pix has exhibited significant GEF activity. A number of experimental techniques have been used to characterize Cool-2, and in vitro analysis has revealed that its GEF activity is under tight control through intramolecular interactions involving several binding partners. Here we describe the biochemical methods used to study the Cool/Pix proteins and, in particular, the regulation of the GEF activity of Cool-2/α-Pix.

Introduction

Members of the Rho-family of GTP-binding proteins function as molecular switches in biological response pathways that result in changes in the actin cytoskeletal architecture, the stimulation of cell cycle progression and gene transcription, and the regulation of intracellular trafficking activities (Bar-Sagi and Hall, 2000; Erickson and Cerione, 2001; Hall, 1998; Van Aelst and D'Souza-Schorey, 1997). Rho-family guanine nucleotide exchange factors (Rho GEFs) serve as critical upstream activators of these molecular switches. The GEFs activate their Rho-family substrates by accelerating the rate at which they release GDP and thereby enabling them

---

* Both authors have contributed equally to this chapter.

METHODS IN ENZYMOLOGY, VOL. 406
0076-6879/06 $35.00
DOI: 10.1016/S0076-6879(06)06005-8

to rapidly bind GTP in cells (Hoffman and Cerione, 2002; Whitehead *et al.*, 1997). Given the key regulatory roles that Rho-GEFs play in the signaling cascades mediated by Rho-family GTP-binding proteins, detecting and quantifying the activities of potential GEFs have occupied a significant amount of investigative effort.

The Cool/Pix (for *C*loned-*o*ut *o*f *l*ibrary/*P*ak-*i*nteractive *ex*change factor) proteins were originally identified through their abilities to bind to members of the Pak (*p*21 *a*ctivated *k*inase) family of proteins and represent one subgroup of the Dbl (*D*iffuse *B*-cell *l*ymphoma) family of Rho-GEFs (Bagrodia *et al.*, 1998; Manser *et al.*, 1998; Oh *et al.*, 1997). Three different isoforms of the Cool/Pix proteins have been extensively studied; these include Cool-2 or α-Pix and two alternative-splice variants designated p50Cool-1 and p85Cool-1/β-Pix. The Cool/Pix proteins have been shown to exhibit a variety of functional activities and to elicit a diversity of cellular responses (Albertinazzi *et al.*, 2003; Kim *et al.*, 2001; Ku *et al.*, 2001; Kutsche *et al.*, 2000; Koh *et al.*, 2001; Lee *et al.*, 2001; Ramakers, 2002; Turner *et al.*, 1999, 2001). The GEF activity of the Cool/Pix proteins is tightly controlled. Thus far, only Cool-2/α-Pix has been shown to exhibit GEF activity, at least as measured in GDP-dissociation assays, and to stimulate Pak activity in cells. It is becoming clear that the GEF activity of the Cool/Pix proteins is likely to be influenced by a number of factors, including protein dimerization, various key interacting domains (e.g., the SH3 domain), and an important regulatory region designated as T1, as well as through a number of binding partners such as Pak, Cbl, and GTP-bound Cdc42 (Baird *et al.*, 2005; Feng *et al.*, 2002, 2004).

In this chapter, we describe various methods used to biochemically characterize this interesting family of GEFs. In particular, we describe methods for the expression and purification of Cool/Pix proteins from *Escherichia coli*, insect cells, and mammalian cells, as well as outline assays for measuring their GEF activities and methods for analyzing their regulation.

Preparation of Proteins

Expression and Purification of Cdc42/Rac from E. coli
Expression Systems

Cdc42 and Rac can be expressed as glutathione-s-transferase (GST) fusion proteins in *E. coli*. Cdc42 or Rac, which has been cloned into the *Bam*HI/*Eco*RI site of the pGEX-2T (Amersham) vector, is expressed in BL21 (DE3) cells (Novagen). The cells are grown to an OD_{600} of 0.8–1.0 and induced with 1 mM IPTG at 37°, shaking at 250 rpm. After 3 h of induction, the cells are centrifuged for 10 min at 5500 rcf, washed with HMA buffer (20 mM HEPES, pH 8.0, 5 mM MgCl$_2$, and 1 mM sodium azide) and recentrifuged. The pellets are then frozen at $-80°$ or lysed immediately.

Cell lysis is performed at 4° using a French Pressure Cell Press (SLM-Aminco) at 1500 psi in HMA, in the presence of 1 mM sodium orthovanadate, 10 μg/ml leupeptin, 10 μg/ml aprotinin, 1 mM phenylmethylsulfonyl fluoride (PMSF), and DNAse. The lysed cells are centrifuged at 40,000 rpm for 40 min in a Ti45 rotor (Beckman). The particulate fractions are discarded, and the supernatant fractions are incubated with glutathione-conjugated Sepharose beads (Amersham). The supernatant/bead mixture is continuously rotated at 4° for 1 h. The beads are then loaded onto a gravity column and washed with a significant amount of HMA containing 250 mM NaCl. Protein is eluted from the beads with HMA and 10 mM glutathione. The eluant is then concentrated to a volume of ~10 ml. To liberate the protein from its GST tag, an appropriate amount of thrombin (Amersham) is added, followed by overnight dialysis against HMA and 2 mM CaCl$_2$ at 4°.

As a final purification step, Cdc42 or Rac can be separated from its GST tag by loading the dialyzed sample on to a HiTrap Q Column (Amersham). Both Rac and Cdc42 flow directly through the column using HMA buffer that lacks NaCl, whereas the GST tag and any remaining protein contaminants adhere to the column. At this point, the purified GTP-binding protein is concentrated to ~1 mg/ml, snap-frozen, and stored in usable aliquots at $-80°$.

Expression and Purification of Cool/Pix and Pak Proteins in Insect Cells

We have found that the Cool-2/α-Pix dimer is a Rac-specific GEF (Feng et al., 2004). Moreover, the ability of activated forms of Cdc42 to stimulate the Rac-specific GEF activity of Cool-2/α-Pix (Baird et al., 2005) requires that Cool-2/α-Pix is in the dimeric state, because Cdc42 seems to bind to the DH domain of one of the monomers making up the dimer, whereas Rac binds to the DH domain of the adjoining monomer. Because the dimerization of Cool-2/α-Pix occurs by way of a leucine zipper located at the carboxyl-terminal end of the protein, it becomes necessary to express the full-length Cool-2/α-Pix protein, rather than simply the DH and PH domain fragments, to study many aspects of its Rac-specific GEF activity. We have had very little success expressing full-length Cool-2/α-Pix in a bacterial system and thus turned to insect cells (Sf21). We have found that this eukaryotic cell system allows large-scale production of highly purified, full-length Cool-2/α-Pix protein. Likewise, because we have found that the binding of Pak3 to the SH3 domain of monomeric Cool-2/α-Pix activates its GEF activity toward Cdc42 (Feng *et al.*, 2004), we have been interested in obtaining purified recombinant Pak3 for structure-function studies. Here again, it is advantageous to use insect cells to generate suitable preparations of Pak3.

The insect (*Sf21*) cells are maintained in Grace's Insect Media (Invitrogen), supplemented with 10% fetal bovine serum. Typically, cells are cultured in T150 tissue culture flasks (Fisher) at 27° and then grown to confluency before being split 1:4 (every third day). A virus for full-length, 6-histidine(His)–tagged Pak3 or a virus expressing His-tagged Cool-2 can be produced using the Bac-to-Bac Baculovirus Expression Kit (Invitrogen). *Sf21* cells (1.8×10^7) are infected with their respective virus at a MOI of 5 for 72 h. After 3 days of infection, the cells are washed once with PBS and centrifuged at 500 rcf. The pellet is then either snap-frozen and stored at $-80°$ or lysed immediately in 20 mM Tris, pH 8.0, 20 mM imidazole, 0.2% CHAPS, plus 1 mM sodium orthovanadate, 10 μg/ml leupeptin, 10 μg/ml aprotinin, and 1 mM PMSF. The lysate is centrifuged at 12,000g for 10 min, and the supernatant is then incubated at 4° for 30 min with NTA-agarose conjugated to nickel. The agarose beads are washed three times with lysis buffer and then eluted in 20 mM Tris, pH 8.0, 200 mM imidazole, and 0.2% CHAPS. The protein is concentrated to 1 μg/μl. The typical yield per infection of 1.8×10^7 insect cells is 50 μg of purified protein.

Expression and Immunoprecipitation of Cool/Pix and its Interacting Proteins in Mammalian Cells

The cDNAs encoding Cool/Pix, as well as their interacting proteins Pak3 and Cbl, have been subcloned into Myc- or HA-tagged mammalian expression vectors (Bagrodia *et al.*, 1995, 1998; Feng *et al.*, 2002, 2004). These cDNAs can be expressed in NIH3T3 or COS-7 cells that are maintained in Dulbecco's modified Eagle's medium plus 10% fetal bovine serum (Invitrogen). The plasmids are transfected into these cells using LipofectAMINE (Invitrogen). The cells are lysed after 24–48 h in 20 mM Hepes, pH 7.4, 1% Nonidet P-40, 150 mM NaCl, 1 mM EDTA, 1 mM sodium orthovanadate, 10 mg/ml leupeptin, and 10 mg/ml aprotinin, at 4° for 20 min. The lysates are prepared by centrifugation at 12,000g for 10 min at 4°.

To immunoprecipitate the Myc-tagged Cool/Pix proteins, cell lysates are incubated with anti-Myc primary antibody (Covance) for 1.5 h on ice, followed by mixing with protein-G Sepharose beads (Invitrogen) for 1 h.

The beads are washed three times with lysis buffer. The Cool/Pix proteins, now adherent to the beads, are ready for GDP-dissociation or PAK assays.

Preparing Activated Forms of Recombinant Cdc42

We have found that the Cool/Pix proteins are able to bind Cdc42 in its activated state (Wu *et al.*, 2003) and that the binding of activated Cdc42 enhances the Rac-specific GEF activity of dimeric Cool-2/α-Pix (Baird *et al.*, 2005). To examine the regulatory effects of Cdc42 on Cool-2/α-Pix, activated forms of recombinant Cdc42 can be prepared as follows.

Either GTPγS, GMP-PNP, or GMP-PCP can be loaded on to Cdc42 to generate an activated form of the protein, although GMP-PCP has been found to be the most stable nonhydrolyzable GTP analog and therefore is most often used. To ensure the complete activation of Cdc42, GDP that is initially bound to the wild-type protein is dissociated and then degraded with alkaline phosphatase.

Ammonium sulfate is added (final concentration of $\sim$200 mM) to the stock of purified, recombinant GDP-bound Cdc42. The ammonium sulfate acts in a similar manner to EDTA, disrupting essential Mg^{2+} coordination between the nucleotide and the GTP-binding protein, thereby facilitating GDP dissociation. A fourfold molar excess of GMP-PCP (Sigma), relative to Cdc42, is then added, followed by alkaline phosphatase conjugated to acrylic beads (Sigma) (4 U/mg Cdc42). This mixture is rotated continuously for 4 h at 4°. After the phosphatase has had sufficient time to react, the mixture is centrifuged to remove the beads. As a final purification step, Cdc42 is loaded on to a Superdex-75 column (Amersham), resolved from excess nucleotide, and equilibrated in HMA buffer.

GDP Dissociation Assay

Because the rate-limiting step for the guanine nucleotide exchange reaction is the dissociation of GDP from GTP-binding proteins like Cdc42, it is this step that is catalyzed by GEFs. Thus, GEF activity is typically assessed by assaying the ability of a candidate GEF (e.g., Cool-2/α-Pix) to stimulate the release of [^{3}H]GDP from a specific GTP-binding protein. This assay is composed of three steps: loading the GTP-binding protein (recombinant Cdc42 or Rac) with [^{3}H]GDP, assaying the release of [^{3}H]GDP as stimulated by the GEF, and measuring the [^{3}H]GDP that has remained associated with the GTP-binding protein during the time period of the assay.

Loading Cdc42/Rac with [³H]GDP

Cdc42·[³H]GDP or Rac·[³H]GDP is prepared by incubating ~1 μM recombinant (*E. coli*) Cdc42 or Rac with ~0.5 μM [³H]GDP (14.2 Ci/mmol, from Amersham) in loading buffer (10 m*M* HEPES, 100 m*M* NaCl, 7.5 m*M* EDTA, pH 7.5) at room temperature for 30 min. Under these conditions, the [³H]GDP is incorporated rapidly into Cdc42/Rac. The exchange of [³H]GDP for GDP is terminated by increasing the concentration of Mg²⁺ to 20 m*M*.

Stimulation of [³H]GDP Release by the GEF

The recombinant Cool/Pix proteins (~4 μM) purified from insect cells or Cool/Pix proteins immunoprecipitated from mammalian cells are incubated in the presence or absence of their binding partners (Pak3 or Cbl; expressed and purified from insect cells) in exchange buffer (14 m*M* HEPES, 140 m*M* NaCl, 7 m*M* MgCl₂, 1.4 m*M* DTT, and 0.7 m*M* GTPγS, pH 7.5). The exchange reaction is initiated by mixing the Cool/Pix proteins (in exchange buffer) with Cdc42/Rac preloaded with [³H]GDP (in loading buffer), typically in a 1:1 molar ratio, because complete nucleotide exchange is not attainable when lower amounts of Cool/Pix are used relative to the GTP-binding protein.

Measurement of [³H]GDP Remaining Associated with the GTP-Binding Protein During the Exchange Assay

At different time points (typically ranging from 0.5–25 min), equal amounts of reaction mixture are removed and diluted into 1 ml of ice-cold termination buffer (20 m*M* Tris-HCl, 10 m*M* MgCl₂, and 100 m*M* NaCl, pH 7.4) to stop the reaction. The samples are then filtered over nitrocellulose membranes (Schleicher & Schuell, BA85). The membranes are washed three times with 2 ml of ice-cold termination buffer to remove free [³H] GDP. Protein-bound [³H]GDP on the filter is quantitated by scintillation counting using a LS 6500 Scintillation System (Beckman).

As indicated previously, both recombinant Cool/Pix proteins purified from insect cells and Cool/Pix proteins immunoprecipitated from mammalian cells can be used in GEF assays. Figure 1 shows the ability of wild-type Cool-2/α-Pix to exhibit GEF activity toward Cdc42 and Rac. In this experiment, Cool-2/α-Pix was transiently expressed in COS-7 cells and immunoprecipitated using anti-Myc antibody. The amount of [³H]GDP that dissociated from Cdc42 or Rac, measured as cpms by scintillation counting, was converted into the percentage of [³H]GDP remaining bound to Cdc42/Rac. As shown in Fig. 1, no detectable Cdc42-GEF activity was measured

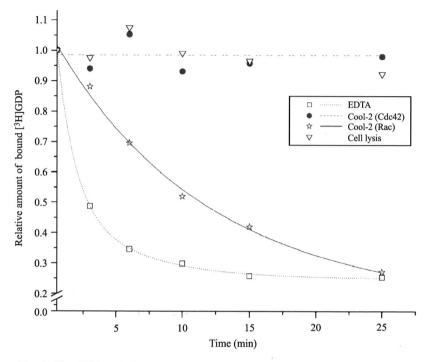

FIG. 1. The abilities of wild-type Cool-2 to exhibit GEF activity toward Cdc42 and Rac. Cool-2 stimulates the dissociation of [³H]GDP from Rac but not Cdc42. COS-7 cells were transiently transfected with a plasmid expressing Myc-tagged Cool-2, and the Cool-2 proteins were immunoprecipitated using an anti-Myc antibody. Recombinant (*E. coli*) Cdc42 or Rac was preloaded with [³H]GDP. The exchange of bound [³H]GDP for GTPγS, as monitored by the dissociation of the radiolabeled nucleotide from Cdc42 or Rac, was assayed in the presence of immunoprecipitated Myc-Cool-2 or in the presence of lysates from cells that were not transfected with the cDNA encoding Myc-Cool-2 (i.e., control cells). The data shown are representative of three experiments and were taken from Feng *et al.* (2004). (See color insert.)

for Cool-2/α-Pix as indicated by the fact that little dissociation of [³H]GDP from Cdc42 occurred over 20–25 min. On the other hand, Cool-2 stimulated the dissociation of [³H]GDP from Rac, indicating that full-length Cool-2 acts as a specific Rac-GEF. Using the same types of assays of [³H]GDP dissociation, we can determine how the GEF activity of the Cool/Pix proteins is influenced by dimerization or by interacting proteins such as Pak, Cbl, and GTP-bound activated Cdc42 (Baird *et al.*, 2005; Feng *et al.*, 2004). For example, Fig. 2 shows how GMP-PCP-bound Cdc42 is able to stimulate the Rac-GEF activity of Cool-2/α-Pix.

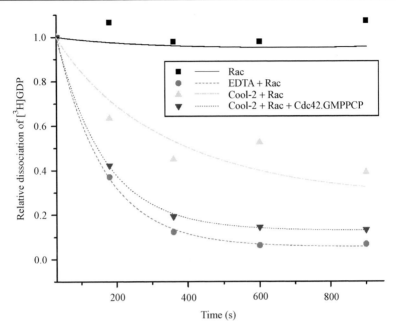

FIG. 2. GMP-PCP–bound Cdc42 stimulates the Rac-GEF activity of Cool-2/α-Pix. The GEF activity of Cool-2 toward Rac is specifically enhanced by GMP-PCP–bound Cdc42. Relative dissociation of preloaded [^{3}H]GDP from Rac1 or Cdc42 in the presence of excess GTPγS and insect cell expressed His-Cool-2 was measured as outlined in this chapter. The data were taken from Baird et al. (2005). (See color insert.)

Pak Assay

Because Paks (p21-activated kinases) are the downstream targets of Cdc42 and Rac, we can indirectly measure the GEF activity of Cool/Pix proteins by assaying Pak activity (Feng et al., 2002, 2004). COS–7 cells transiently expressing Cool/Pix proteins and hemagglutinin (HA)-tagged Pak3 are maintained in Dulbecco's modified Eagle's medium plus 10% fetal bovine serum for 24–36 h and then starved for 18–24 h. The cells are lysed, and Pak3 is immunoprecipitated as described under "Preparation of Proteins." After the Pak3 proteins are bound to protein-G beads, the beads are washed three times with lysis buffer, twice with 2× phosphorylation buffer (10 mM MgCl$_2$ and 40 mM Tris-HCl, pH 7.4), and mixed with 1.0 mg/sample of the substrate myelin basic protein (Sigma). Pak assays are initiated by adding 25 μM ATP and 10 mCi of [γ-^{32}P] ATP(3000 Ci/mmol, NEN) to the immunoprecipitation mixture. The assays are typically performed for 10–30 min at 25° and then stopped by adding 2× SDS-PAGE

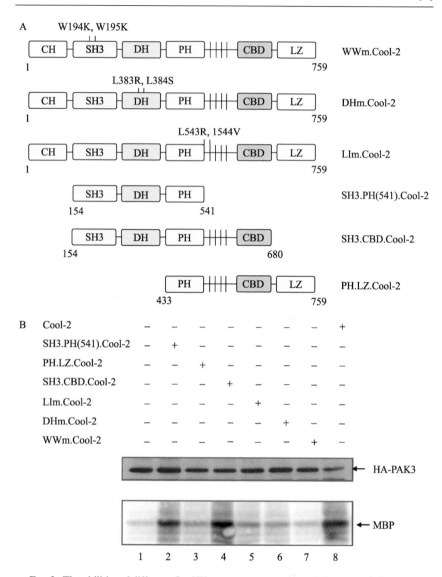

FIG. 3. The abilities of different Cool/Pix constructs to activate Pak *in vivo*. (A) Depiction of different Cool/Pix constructs used in the experiment. (B) SH3.PH.Cool-2 (lane 2), SH3. CBD.Cool-2 (lane 4), and wild-type Cool-2 (lane 8) can activate Pak *in vivo*, whereas PH.LZ. Cool-2 (lane 3), LIm.Cool-2 (lane 5), DHm.Cool-2 (lane 6), and WWm.Cool-2 (lane 7) are not able to activate Pak. The Pak assays (shown in the lower panel) were performed using myelin basic protein (MBP) as a phosphosubstrate. Upper panel shows the relative amounts of Pak3 assayed under the different conditions. Data were taken from Feng *et al.* (2004). (See color insert.)

sample buffer containing 20 mM EDTA. The phosphoproteins are resolved by SDS-PAGE (10 or 12% gel) and visualized by PhosphorImager analysis (Amersham Biosciences) before immunoblotting.

A typical Pak assay is shown in Fig. 3B. Different Cool/Pix constructs (depicted in Fig. 3A) have varying abilities to stimulate the Pak3-catalyzed phosphorylation of myelin basic protein. This type of assay initially indicated that the different Cool/Pix constructs were capable of distinct functional activities (Bagrodia et al., 1998; Feng et al., 2002). We have found that it is also possible to assay Pak activity by directly mixing recombinant Cool/Pix proteins (purified from insect cells) with Pak3 that has been immunoprecipitated from mammalian cells. However, recombinant Pak3 often displays a high background, suggesting that it is apparently purified in its active form from insect cells.

Discussion

The Cool/Pix proteins contain tandem DH and PH domains like other members of the Dbl family and were originally assumed to function as traditional Rho-GEFs. However, it turns out that the members of this family exhibit an interesting array of regulatory features and functional activities. First, both p85Cool-1/β-Pix and Cool-2/α-Pix form stable complexes with the GTP-bound activated form of Cdc42, which has important functional consequences. Specifically, p85Cool-1/β-Pix has been shown to interface GTP-bound Cdc42 with the ubiquitin E3 ligase c-Cbl, resulting in the sequestration of Cbl away from EGF receptors and the negative regulation of EGF receptor ubiquitination and down-regulation (Wu et al., 2003). In the case of Cool-2/α-Pix, the binding of activated Cdc42 strongly stimulates its Rac-GEF activity (Baird et al., 2005).

Perhaps even more interesting are the roles of protein oligomerization and protein–protein interactions in the regulation of Cool-2/α-Pix's GEF activity and its specificity for GTP-binding protein substrates. The monomer-dimer equilibrium of Cool-2/α-Pix directly impacts GEF specificity (Feng et al., 2004). The dimeric form of Cool-2/α-Pix is a specific GEF for Rac, in which case, the DH domain from one monomer and the carboxyl-terminal end of the PH domain from the adjoining monomer are essential for binding and activating Rac. GTP-bound forms of Cdc42 can only regulate the GEF activity of dimeric Cool-2/α-Pix, because the activated Cdc42 molecule binds to the available DH domain that is not being used by Rac. When the Cool-2/α-Pix dimer dissociates into a monomer, it then is able to act as a GEF for either Cdc42 or Rac, but only when the SH3 domain of monomeric Cool-2/α-Pix is occupied by one of its binding partners (i.e., Pak or Cbl). Thus, extracellular signals that influence the monomer–dimer equilibrium of Cool-2/α-Pix will dictate what down-stream GTP-binding proteins are activated. Indeed, we and others have obtained evidence showing that G protein–coupled receptors, through their ability to activate large G proteins and generate $G\beta\gamma$ complexes, can trigger the dissociation of the Cool-2/α-Pix dimer and promote its Cdc42-GEF activity (Feng *et al.*, 2004; Li *et al.*, 2003). Reconstitution assays using purified recombinant proteins, as outlined in the previous sections, demonstrated that this was the outcome of the binding of $G\beta\gamma$ to Pak3, with the Pak3-$G\beta\gamma$ complex then interacting with the SH3 domain of Cool-2/α-Pix and causing it to dissociate into monomeric GEFs.

Given the complexity of their regulation, and the many functional consequences of their actions, there are a number of important questions regarding where and how the Cool/Pix proteins participate in different cellular activities that await future studies. For example, what cellular signaling pathways use p85Cool-1/β-Pix as a GEF and what types of regulatory events are necessary to enable p85Cool-1/β-Pix to exhibit GEF activity as read-out by stimulated [^{3}H]GDP dissociation from Rac or Cdc42? What other types of regulatory signals (e.g., phosphorylation) influence the monomer–dimer equilibrium of Cool-2/α-Pix and its GEF activities? How does the binding of Pak or Cbl to monomeric Cool-2/α-Pix induce the necessary changes in its DH domain to stimulate its GEF activity? The answers to these questions will require structure-function studies in well-defined systems, as well as the preparation of sufficient quantities of purified, recombinant Cool/Pix proteins for high resolution structural studies. This in turn will likely involve the procedures for generating Cool/Pix proteins and assaying their GEF activities that we have outlined previously.

References

Albertinazzi, C., Za, L., Paris, S., and de Curtis, I. (2003). ADP-ribosylation factor 6 and a functional PIX/p95-APP1 complex are required for Rac1B-mediated neurite outgrowth. *Mol. Biol. Cell* **14**, 1295–1307.

Bagrodia, S., Taylor, S. J., Creasy, C. L., Chernoff, J., and Cerione, R. A. (1995). Identification of a mouse p21Cdc42/Rac activated kinase. *J. Biol. Chem.* **270**, 22731–22737.

Bagrodia, S., Taylor, S. J., Jordon, K. A., Van Aelst, L., and Cerione, R. A. (1998). A novel regulator of p21-activated kinases. *J. Biol. Chem.* **273**, 23633–23636.

Baird, D., Feng, Q., and Cerione, R. A. (2005). The Cool-2/alpha-Pix protein mediates a Cdc42-Rac signaling cascade. *Curr. Biol.* **15**, 1–10.

Bar-Sagi, D., and Hall, A. (2000). Ras and Rho GTPases: A family reunion. *Cell* **103**, 227–238.

Erickson, J. W., and Cerione, R. A. (2001). Multiple roles for Cdc42 in cell regulation. *Curr. Opin. Cell Biol.* **13**, 153–157.

Feng, Q., Albeck, J. G., Cerione, R. A., and Yang, W. (2002). Regulation of the Cool/Pix proteins: Key binding partners of the Cdc42/Rac targets, the p21-activated kinases. *J. Biol. Chem.* **277**, 5644–5650.

Feng, Q., Baird, D., and Cerione, R. A. (2004). Novel regulatory mechanisms for the Dbl family guanine nucleotide exchange factor Cool-2/alpha-Pix. *EMBO J.* **23**, 3492–3504.

Hall, A. (1998). Rho GTPases and the actin cytoskeleton. *Science* **279**, 509–514.

Hoffman, G. R., and Cerione, R. A. (2002). Signaling to the Rho GTPases: Networking with the DH domain. *FEBS Lett.* **513**, 85–91.

Kim, S., Lee, S. H., and Park, D. (2001). Leucine zipper-mediated homodimerization of the p21-activated kinase-interacting factor, beta Pix. Implication for a role in cytoskeletal reorganization. *J. Biol. Chem.* **276**, 10581–10584.

Koh, C. G., Manser, E., Zhao, Z. S., Ng, C. P., and Lim, L. (2001). Beta1PIX, the PAK-interacting exchange factor, requires localization via a coiled-coil region to promote microvillus-like structures and membrane ruffles. *J. Cell Sci.* **114**, 4239–4251.

Ku, G. M., Yablonski, D., Manser, E., Lim, L., and Weiss, A. (2001). A PAK1-PIX-PKL complex is activated by the T-cell receptor independent of Nck, Slp-76 and LAT. *EMBO J.* **20**, 457–465.

Kutsche, K., Yntema, H., Brandt, A., Jantke, I., Nothwang, H. G., Orth, U., Boavida, M. G., David, D., Chelly, J., Fryns, J. P., Moraine, C., Ropers, H. H., Hamel, B. C., van Bokhoven, H., and Gal, A. (2000). Mutations in ARHGEF6, encoding a guanine nucleotide exchange factor for Rho GTPases, in patients with X-linked mental retardation. *Nat. Genet.* **26**, 247–250.

Lee, S. H., Eom, M., Lee, S. J., Kim, S., Park, H. J., and Park, D. (2001). BetaPix-enhanced p38 activation by Cdc42/Rac/PAK/MKK3/6-mediated pathway. Implication in the regulation of membrane ruffling. *J. Biol. Chem.* **276**, 25066–25072.

Li, Z., Hannigan, M., Mo, Z., Liu, B., Lu, W., Wu, Y., Smrcka, A. V., Wu, G., Li, L., Liu, M., Huang, C. K., and Wu, D. (2003). Directional sensing requires G beta gamma-mediated PAK1 and PIX alpha-dependent activation of Cdc42. *Cell* **114**, 215–227.

Manser, E., Loo, T. H., Koh, C. G., Zhao, Z. S., Chen, X. Q., Tan, L., Tan, I., Leung, T., and Lim, L. (1998). PAK kinases are directly coupled to the PIX family of nucleotide exchange factors. *Mol. Cell* **1**, 183–192.

Oh, W. K., Yoo, J. C., Jo, D., Song, Y. H., Kim, M. G., and Park, D. (1997). Cloning of a SH3 domain-containing proline-rich protein, p85SPR, and its localization in focal adhesion. *Biochem. Biophys. Res. Commun.* **235**, 794–798.

Ramakers, G. J. (2002). Rho proteins, mental retardation and the cellular basis of cognition. *Trends Neurosci.* **25**, 191–199.

Turner, C. E., Brown, M. C., Perrotta, J. A., Riedy, M. C., Nikolopoulos, S. N., McDonald, A. R., Bagrodia, S., Thomas, S., and Leventhal, P. S. (1999). Paxillin LD4 motif binds PAK and PIX through a novel 95-kD ankyrin repeat, ARF-GAP protein: A role in cytoskeletal remodeling. *J. Cell Biol.* **145**, 851–863.

Turner, C. E., West, K. A., and Brown, M. C. (2001). Paxillin-ARF GAP signaling and the cytoskeleton. *Curr. Opin. Cell Biol.* **13**, 593–599.

Van Aelst, L., and D'Souza-Schorey, C. (1997). Rho GTPases and signaling networks. *Genes Dev.* **11**, 2295–2322.

Whitehead, I. P., Campbell, S., Rossman, K. L., and Der, C. J. (1997). Dbl family proteins. *Biochim. Biophys. Acta* **1332**, F1–F23.

Wu, W. J., Tu, S., and Cerione, R. A. (2003). Activated Cdc42 sequesters c-Cbl and prevents EGF receptor degradation. *Cell* **114**, 715–725.

[6] GEF and Glucosylation Assays on Liposome-Bound Rac

By BRUNO MESMIN and BRUNO ANTONNY

Abstract

Rac binds tightly to lipid membranes through a lipid modification. The influence of the lipid membrane environment on the multiple interactions of Rac has not been well documented. In this chapter, we detail a method to prepare geranyl-geranylated Rac bound to liposomes of defined composition. With this method, one can dissect some lipid–protein interactions that facilitate the interaction of Rac with other proteins such as guanine nucleotide exchange factors and bacterial toxins.

Introduction

The small G protein Rac is modified at its C-terminus by a geranyl–geranyl group, a long 20-carbon isoprene, which binds avidly to lipid membranes. Experiments with model lipopeptides have shown that the membrane/water-partitioning coefficient of the geranyl–geranyl group is in the range of 10^{-6} M (Silvius and l'Heureux, 1994). This implies that at the millimolar concentration of lipids "seen" by a protein in a cell, a geranyl–geranylated protein, should be membrane associated.

If geranyl-geranylated Rac and other Rho proteins are intrinsically nonsoluble, they can form a cytosolic soluble complex with GDI, a protein that masks the lipid modification (Hoffman *et al.*, 2000). Importantly, GDI also masks the switch regions, which are essential for the interaction of small G proteins with other proteins such as guanine nucleotide exchange factors (GEFs), effectors, and GTPase activating proteins (GAPs). Rac in complex with GDI is, therefore, functionally "inert" in contrast to membrane-associated Rac. For practical reasons, however, most *in vitro* studies are carried out in solution using nonprenylated Rac expressed in *Escherichia coli*. Although such studies provide detailed mechanistic information, particularly with the resolution of protein structures, it is likely that the lipid membrane environment influences the cross-talk between Rac and its partners.

Here we describe a protocol to prepare liposome-bound geranyl-geranylated Rac in the desired conformation (GDP or GTP) and on liposomes of defined composition. Two examples will be presented that

METHODS IN ENZYMOLOGY, VOL. 406
Copyright 2006, Elsevier Inc. All rights reserved.

0076-6879/06 $35.00
DOI: 10.1016/S0076-6879(06)06006-X

illustrate the dramatic effect of the lipid membrane environment on the interaction of prenylated Rac with other proteins.

Liposome Preparation

It should be noted that a recent volume of *Methods in Enzymology* (vol. 372) is dedicated to the preparation and characterization of liposomes.

Azolectin Liposomes

Azolectin is a soybean lipid extract (type II S, Sigma). Its exact composition is not known, but it is much cheaper than pure lipids. Many membrane-associated proteins readily adsorb on azolectin liposomes. To determine whether a membrane environment provides a kinetic advantage in the interaction between two proteins, it is a good idea to perform pilot experiments with azolectin liposomes before dissecting the contribution of various lipids using liposomes of defined composition. Azolectin liposomes are prepared by the phase-reversion method (Duzgunes, 2003; Szoka and Papahadjopoulos, 1978). Azolectin is dissolved in diethylether (6 ml/20 mg of azolectin lipids) in a 100-ml round-bottom glass flask. Addition of 1 ml sucrose buffer (20 mM HEPES–KOH, pH 7.5, 180 mM sucrose), nonmiscible in ether, leads to the formation of two phases. The mixture is sonicated at 4° during 2 min in a bath sonicator to obtain an emulsion in which small droplets of water are separated from the ether phase by a lipid monolayer (inverted micelles). Ether is then gently evaporated in a rotary evaporator (100 rpm) under vacuum at 25° to promote the coalescence of the inverted micelles into unilamellar liposomes. After 5 min, a viscous gel forms and falls down when most solvent is evaporated. The suspension is collected in an Eppendorf tube and incubated in a vacuum chamber for 15 min to remove residual traces of solvent. The solution can be completed up to 1 ml with sucrose buffer. Liposomes are finally sized by extrusion.

Liposomes of Defined Lipid Composition

Lipids are purchased as chloroform solutions from Avanti polar lipids (www.avantilipids.com) or from Sigma and stored at −20° under argon in small (2-ml) glass vials with a Teflon cap. To prepare liposomes of defined lipid composition, lipids are mixed at the desired molar ratio in a peershaped glass flask. The fluorescent lipid nitrobenzoxadiazol-PE (www.probes.com) is added at 0.2–1 mol% as a tracer. The total amount of lipid ranges from 1–5 mg. The solvent is evaporated in a rotatory evaporator (500 rpm) under vacuum at 25°. A lipid film rapidly forms at the glass surface. After 15 min, the evaporator is purged with argon, and the dried lipid film is placed in a vacuum chamber for 30 min. The film is gently resuspended at 1–4 mg/ml in sucrose buffer to obtain multilamellar vesicles, and the suspension is collected in an Eppendorf tube. After five cycles of freezing/thawing in liquid nitrogen and hot water (40°), the liposome suspension is stored at −20°.

Liposome Extrusion

The procedure for the extrusion of liposomes has been described elsewhere (MacDonald *et al.*, 1991; Mayer *et al.*, 1986; Mui *et al.*, 2003). There are two commercially available hand extruders with similar design, one from Avanti polar lipids (www.avantilipid.com) and one from Avestin (www.avestin.com). A polycarbonate filter (diameter, 19 mm; pore size, 0.4 μm) is tightly sandwiched between two Teflon parts in a metal holder. Using Hamilton syringes, one manually forces the liposome suspension to pass 19 times through the filter. After extrusion, external sucrose can be removed by diluting the suspension fivefold in an iso-osmotic buffer (20 mM HEPES, pH 7.5, 100 mM KCl) and centrifuged at 100,000 rpm (400,000g) for 15 min in a TLA100.3 rotor. The liposome pellet is resuspended at 1–5 mg/ml in the same buffer. Liposomes are stored at room temperature under argon and used within 3 days.

Buffers and Proteins

Guanine nucleotide exchange reactions are performed in 20 mM HEPES–KOH (pH 7.5), 100 mM NaCl, 1 mM MgCl$_2$, and 1 mM dithiothreitol (DTT) (buffer A). Glucosylation assays are performed in 20 mM HEPES–KOH (pH 7.5), 100 mM KCl, 1 mM MgCl$_2$, 1 mM DTT (buffer B), supplemented with 1 mM MnCl$_2$. For unknown reason, the substitution of KCl for NaCl increases the binding of lethal toxin to lipid membranes. The assay buffer should be iso-osmotic with the sucrose buffer to protect liposomes from leakage (100 mM NaCl is iso-osmotic with 180 mM sucrose). If the ionic strength of the assay buffer has to be changed, we recommend adjusting the concentration of sucrose in the buffer used for the preparation of liposomes accordingly. A table for the osmolarity of saline and sucrose solutions can be found in the *Handbook of Chemistry and Physics* (Weast, 1989). It should be noted that MgCl$_2$ and MnCl$_2$, which are present in the assay buffer, are omitted in the liposome buffer, because divalent cations are known to promote long-term fusion of liposomes. Mn^{++} is a cofactor of glucosyltransferase enzymes, whereas Mg^{++} is critical for efficient binding of guanine nucleotides to small G proteins. This feature is used to promote or block guanine nucleotide exchange in small G proteins (see later).

Geranyl-geranylated Rac1 (with an N-terminal hexahistidine tag) is coexpressed with RhoGDI-1 (containing an N-terminal FLAG) in *Saccharomyces cerevisiae* and purified as described for RhoA/GDI (Read and Nakamoto, 2000; Read *et al.*, 2000). Unprenylated Rac1-GDP fused to glutathione S-transferase (GST), His-tagged Tiam DH-PH (amino-acids 1033–1406), and His tagged lethal toxin LT_{cyt} (a.a 1–546) are expressed in *E. coli* and purified on glutathione–agarose beads (Amersham Biosciences) or on Ni^{2+}–agarose beads (Qiagen) according to the manufacturer protocols.

GDI-Removal Protocol

General Description

The GDI-removal protocol enables preparation of liposome-bound geranyl-geranylated Rac in the absence of detergent. This protocol is based on the reduced affinity of GDI for Rac-GTP compared with the GDP-bound form (Robbe *et al.*, 2003; Sasaki *et al.*, 1993).

The Rac-GDP/GDI complex (1 μM) is mixed with liposomes (6 mM lipids) and with 20 μM GTP in buffer A or B, depending on the assay chosen. EDTA (2 mM) is added from a 100× stock solution at pH 7 to give 1 μM free Mg^{2+}, and the suspension is incubated for 40 min at 30° in a Thermomixer (800 rpm; Eppendorf) (Fig. 1, stage 1). At this low Mg^{2+} concentration, GDP dissociates within minutes and is replaced by the

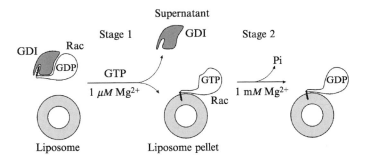

FIG. 1. GDI-removal protocol. During the first stage, the Rac-GDP-GDI complex is incubated with liposomes and with GTP and EDTA to promote the formation of membrane-bound Rac-GTP. Liposomes are then collected by centrifugation and washed to separate membrane-bound Rac-GTP from soluble GDI and excess GTP. In the second stage, the lipid pellet is resuspended in a buffer containing 1 mM $MgCl_2$ and incubated at 30° to allow Rac to hydrolyze the bound GTP. Adapted from Robbe *et al.* (2003).

added nucleotide (here GTP). The binding of GTP shifts the partitioning equilibrium of Rac toward the liposome-bound form at the expense of the GDI-bound form. After incubation, the sample is centrifuged at 100,000 rpm (400,000g) in a Beckman TLA 100.1 rotor for 20 min at 4°. The supernatant is removed, and the liposome pellet is gently washed (but not resuspended) with cold buffer A or B, supplemented with 2 mM EDTA ([Mg^{2+}]$_{free}$ = 1 μM). The sample is centrifuged again for 5 min at 4°, and the pellet is finally resuspended in buffer A or B ([Mg^{2+}]$_{free}$ = 1 mM) and incubated for 30 min at 30° to promote GTP hydrolysis (Fig. 1, stage 2). The half time of GTP hydrolysis in Rac at 1 mM Mg^{2+} and at 30° is about 8 min. The final liposome suspension, which contains membrane-bound Rac-GDP, is stored on ice before use. To assess the efficiency of the GDI-removal protocol, aliquots of the initial lipid/protein suspension, of the supernatant, and of the final liposome suspension are analyzed by SDS-PAGE. This analysis can be performed with Coomassie staining, but we recommend the use of the fluorescent dye Sypro Orange (Molecular probes; www.probes.com). This probe can be visualized and quantified in a fluorescence imaging system such as the LAS 3000 (Fuji), which consists of a black box with blue diodes for excitation (460 nm), a filter wheel (with a 515 nm DI filter for sypro orange and NBD fluorescence), and a CCD camera. The digitalized images are analyzed using the Aida software (Fuji). This method is more sensitive and more linear than Coomassie staining. Typically, almost 90% of GDI are found in the first supernatant after the GTP-binding stage, whereas 50–80 % of Rac are found associated with liposomes after the GTP hydrolysis stage (Robbe and Antonny, 2003). The lipid probe (NBD-PE excitation 465 nm; emission 535 nm) can be also visualized and quantified in the fluorescence imaging system. For this, we deposit small drops (10 μl) of the various fractions on a black plastic plate and take a picture of the plate.

The GDI-removal protocol can be modified in various ways. First, a nonhydrolyzable analog of GTP (such as GTPγS) can be used to obtain liposome-bound Rac in a permanently active form. Second, the concentration of liposome can be varied. However the efficiency of Rac translocation strongly decreases at lipid concentration <0.5 mM. Third, and as detailed later, the GDI-removal protocol can be performed with liposomes of various compositions.

Optimal Lipid Composition

GDI contains not only a hydrophobic pocket that shields the geranyl–geranyl group from the solvent but also a cluster of acidic residues that interact electrostatically with the C-terminal polybasic region of Rac

(Hoffman *et al.*, 2000). Therefore, the presence of anionic lipids is required to displace Rac from GDI in the GDI-removal protocol (Mesmin *et al.*, 2004). Figure 2A shows that under conditions that favor GDP–GTP exchange (low Mg^{2+} concentration), Rac accumulates strongly and with similar efficiency to all liposomes containing 30 mol% anionic lipids but not to neutral liposomes. Dose-response curves for the anionic lipids phosphatidylserine (PS) and phosphatidylglycerol (PG) are shown in Fig. 2B.

In summary, the partitioning of Rac between GDI and liposomes in the GDI-removal protocol is governed by three parameters: the conformation of Rac, the concentration of lipids, and the percentage of anionic lipids. For efficient translocation of Rac, it is thus critical (1) to use high lipid concentration (in the millimolar range); (2) to add GTP and to transiently lower the concentration of Mg^{++} in the micromolar range using EDTA; and (3) to include anionic lipids in the liposome formulation. However, the nature of the anionic lipid is not important, because the interaction with the polybasic tail of Rac is electrostatic and nonspecific (Fig. 2A). This leaves the choice of the lipid composition quite open for studying the influence of the lipid membrane environment (Mesmin *et al.*, 2004). If one lipid species is suspected to play a role in the interaction of Rac with a given target, one has to take into account its contribution to the percentage of anionic lipids,

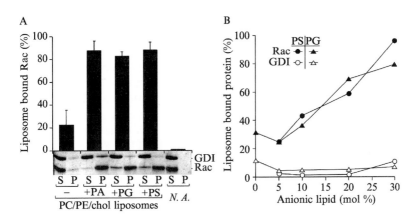

FIG. 2. Effect of anionic lipids on the partitioning of Rac between GDI and liposomes. (A) The Rac-GDP/GDI complex (1 μM) is incubated for 40 min at 30° with GTP (20 μM) and with or without PC/PE/cholesterol liposomes (30–60/20/20 mol%) and supplemented with 30 mol% PA, PG, or PS as indicated. The concentration of free Mg^{2+} is 1 μM. After incubation, the sample is centrifuged, and the supernatant (*S*) and the pellet (*P*) are analyzed by SDS-PAGE and Sypro Orange staining. *N. A.*, No addition of liposomes. (B) Dose-response experiments using liposomes containing increasing amounts of anionic lipids (PS or PG) at expense of PC. (A) from Mesmin *et al.* (2004).

so as to keep this percentage constant. A neutral lipid should be added at the expense of PC, whereas an anionic lipid should be added at the expense of PS (or PG).

Rac accumulates on the inner leaflet of the plasma membrane. Ideally, the background composition of the liposome should imitate this leaflet, notably the abundance of PS and phosphatidylethanolamine (PE). Our standard liposome composition consists of egg PC (30 mol%), liver PE (20 mol%), brain PS (30 mol%), and cholesterol (20 mol%). This composition can be considered as a reasonable starting mixture, which can be modified according to the molecular mechanism studied. Of note, lipid chain unsaturation has strong impact on the adsorption of some proteins, because it modulates the penetration of hydrophobic residues. Experiments in which dioleolyl lipids (C18:1–C18:1) are substituted for lipids from natural sources can be informative.

Guanine Nucleotide Exchange on Liposome-Bound Rac

GEFs for Rho proteins contain a catalytic module, the DH-PH tandem, which promotes the release of GDP (Schmidt and Hall, 2002). This tandem is quite inefficient in catalyzing nucleotide exchange in solution, and it is believed that the membrane environment could favor better positioning of the DH–PH tandem toward membrane-associated Rho proteins (Baumeister et al., 2003). This hypothesis is supported by the observation that the DH–PH tandem of the exchange factor Tiam1 catalyses GDP to GTP exchange on gerenyl-geranylated Rac bound to azolectin liposomes much more efficiently than on nonprenylated Rac in solution (Robbe et al., 2003).

Liposome-bound Rac-GDP, which is obtained by the GDI removal protocol, is diluted twofold in buffer A supplemented with 20 μM [^{35}S] GTPγS (2000 cpm/pmol) to give a final lipid concentration of 3 mM and a final protein concentration of $\sim$0.25 μM (as estimated by SDS-PAGE). The sample is incubated at 30° and at time zero, Tiam DH-PH (0–4 μM) is added. At the indicated times, 20-μl aliquots are removed and immediately diluted into 2 ml of ice-cold buffer (buffer C; 20 mM HEPES, pH 7.5, 100 mM NaCl, and 10 mM MgCl$_2$), and filtered on nitrocellulose filter disks (Schleicher & Schüll) mounted on a holder (Millipore, ref XX1002530) connected to a vacuum pump. The filter-disk is washed twice with 2 ml of buffer C, dried, and counted. The same protocol is used to measure the binding of [^{35}S]GTPγS on unprenylated-Rac in solution. Figure 3A compares the time courses of [^{35}S]GTPγS binding on prenylated Rac bound to azolectin liposomes and on soluble nonprenylated Rac catalyzed by 0.08 μM Tiam1 DH-PH. A control experiment performed in the absence of DH–PH shows that the spontaneous rate of nucleotide exchange on Rac

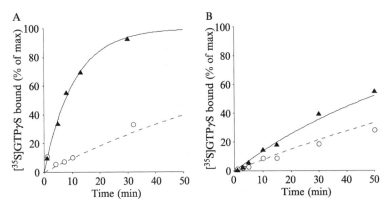

FIG. 3. Nucleotide exchange assays. Comparison between geranyl-geranylated Rac-GDP associated with azolectin liposomes (A) and unprenylated Rac-GDP in solution (B) for Tiam 1 DH-PH-catalyzed GDP-to-GTPγS exchange. All experiments are performed at 1 mM free Mg^{2+} with 20 μM [^{35}S]GTPγS with or without 0.08 μM Tiam1 DH-PH. Adapted from Robbe *et al.* (2003).

is not influenced by the membrane environment. In marked contrast the DH-PH–catalyzed reaction is accelerated 10-fold at the liposome surface.

Glucosylation of Liposome-Bound Rac

Large clostridial toxins such as lethal toxin (LT) are glucosyltransferases, which catalyze the transfer of a glucosyl group from UDP-glucose to a conserved threonine residue within the switch I region of small G proteins (Busch and Aktories, 2000). With the help of the GDI-removal protocol, we have recently shown that LTcyt, the fragment that is liberated in the cytosol of infected cells, takes advantage of the lipid membrane environment to glucosylate efficiently membrane-bound prenylated Rac (Mesmin *et al.*, 2004). Indeed, LTcyt binds specifically to PS. Two protocols are detailed here: one to assess the binding of LTcyt to liposomes and a second to follow the glucosylation of liposome-bound Rac.

The binding of LT to liposome is determined by sedimentation experiments using the protocol developed by McLaughlin (Buser and McLaughlin, 1998). LT, Lt$_{cytr}$, or other purified toxin fragments are mixed in buffer B with sucrose-loaded liposomes of defined composition obtained by extrusion through 0.4-μm polycarbonate filters. The protein concentration should be in the micromolar range, and the lipid concentration in the millimolar range. The mixture (50 μl) is prepared directly in small polycarbonate centrifuge tubes and centrifuged at 100,000 rpm (400,000g)

in a TL100.1 centrifuge (Beckman) for 20 min at 4°. One control experiment should be performed in the absence of liposome. The supernatant and the pellet, which is resuspended in the same volume of buffer, are analyzed by SDS-PAGE. The protein gels are stained with the fluorescent dye sypro-orange, digitalized, and quantified using the fluorescent imaging system (FUJI LAS 3000).

If the sedimentation assay with sucrose-loaded liposomes is straightforward, it should be noted that it is not adapted to the study of large protein complexes (which tend to pellet in the absence of liposomes) or when very small liposomes are used to study the effect of membrane curvature. In these cases, it is preferable to collect the liposomes and bound proteins by flotation on dense sucrose cushions (see Bigay and Antonny, chapter 10, *Methods in Enzymology*, vol. 404).

The radioactive assay used to follow Rac glucosylation by LT is very similar to the guanine nucleotide exchange assay described previously. Liposome-bound Rac-GDP is diluted twofold in buffer B supplemented with 1 mM MnCl$_2$ and 10 μM [^{14}C]UDP-glucose (732.6 dpm/pmol) to give a final lipid concentration of 3 mM and a final protein concentration of $\sim$0.25 μM (as estimated by SDS-PAGE). The glucosylation reaction is performed

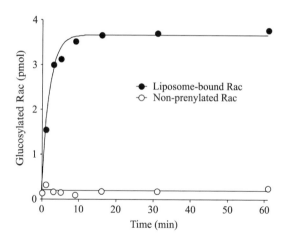

FIG. 4. Glucosylation experiments. Comparison between geranyl-geranylated Rac-GDP bound to PC/PE/cholesterol liposomes supplemented with 30 mol% PS (*filled circles*), or unprenylated Rac-GDP in solution (*open circles*) for LT$_{cyt}$-catalyzed glucosylation. Glucosylation is initiated by the addition of 0.25 nM of toxin on membrane bound prenylated Rac-GDP ($\sim$0.25 μM) or on unprenylated Rac-GDP (0.5 μM) in the presence of [^{14}C]-UDP-glucose (10 μM). Protein-bound radioactivity was determined on 20-μl aliquots ($\sim$5 pmol of Rac per aliquot). Continuous traces represent best single exponential fits. Adapted from Mesmin *et al.* (2004).

at 30° and is initiated by the addition of the LT or other LT fragments. At the indicated times, 20-μl aliquots are withdrawn, diluted in 2 ml of ice-cold buffer D (20 mM HEPES, pH 7.5, 100 mM KCl, and 10 mM MgCl$_2$), and filtered on 25-mm BA85 nitrocellulose filters (Schleicher & Schüll) mounted on a holder connected to a vacuum pump. After two additional washes with 2 ml of buffer D, the filters are dried and counted. The same protocol is used to measure the unprenylated Rac glucosylation in solution.

Figure 4 compares the time courses of glucosylation of liposome-bound geranyl–geranylated Rac-GDP and of unprenylated Rac with 0.25 nM LTcyt. At the concentration of LTcyt used here, the glucosylation of soluble unprenylated Rac is not detectable, whereas liposome-bound pre-nylated Rac is glucosylated within minutes. The liposomes used in these experiments contain 30% PC, 20% PE, 30% PS, and 20% cholesterol. Fast glucosylation occurs because LTcyt binds to PS.

In conclusion, the GDI-removal protocol is a straightforward method to obtain pure prenylated Rac in any conformation (GDP or GTP) and bound to liposomes that can imitate the cell environment. In contrast, experiments in which lipids and notably phosphoinositides are simply added in solution as emulsions obtained by sonication lead to conflicting results, because phosphoinositides are so highly charged that they can promote nonspecific effects (Chabre *et al.*, 1998). The protocol described here should be useful for studying various molecular events controlled by Rac in a minimal and controlled lipid environment.

References

Baumeister, M. A., Martinu, L., Rossman, K. L., Sondek, J., Lemmon, M. A., and Chou, M. M. (2003). Loss of phosphatidylinositol 3-phosphate binding by the C-terminal Tiam-1 pleckstrin homology domain prevents *in vivo* Rac1 activation without affecting membrane targeting. *J. Biol. Chem.* **278**, 11457–11464.

Buser, C. A., and McLaughlin, S. (1998). Ultracentrifugation technique for measuring the binding of peptides and proteins to sucrose-loaded phospholipid vesicles. *Methods Mol. Biol.* **84**, 267–281.

Busch, C., and Aktories, K. (2000). Microbial toxins and the glycosylation of rho family GTPases. *Curr. Opin. Struct. Biol.* **10**, 528–535.

Chabre, M., Antonny, B., and Paris, S. (1998). PIP2: Activator ... or terminator of small G proteins? *Trends Biochem. Sci.* **23**, 98–100.

Duzgunes, N. (2003). Preparation and quantitation of small unilamellar liposomes and large unilamellar reverse-phase evaporation liposomes. *Methods Enzymol.* **367**, 23–27.

Hoffman, G. R., Nassar, N., and Cerione, R. A. (2000). Structure of the Rho family GTP-binding protein Cdc42 in complex with the multifunctional regulator RhoGDI.PG-345–56. *Cell* **100**, 345–356.

MacDonald, R. C., MacDonald, R. I., Menco, B. P., Takeshita, K., Subbarao, N. K., and Hu, L. R. (1991). Small-volume extrusion apparatus for preparation of large, unilamellar vesicles. *Biochim. Biophys. Acta* **1061**, 297–303.

Mayer, L. D., Hope, M. J., and Cullis, P. R. (1986). Vesicles of variable sizes produced by a rapid extrusion procedure. *Biochim. Biophys. Acta* **858,** 161–168.

Mesmin, B., Robbe, K., Geny, B., Luton, F., Brandolin, G., Popoff, M. R., and Antonny, B. (2004). A phosphatidylserine-binding site in the cytosolic fragment of Clostridium sordellii lethal toxin facilitates glucosylation of membrane-bound Rac and is required for cytotoxicity. *J. Biol. Chem.* **279,** 49876–49882.

Mui, B., Chow, L., and Hope, M. J. (2003). Extrusion technique to generate liposomes of defined size. *Methods Enzymol.* **367,** 3–14.

Read, P. W., Liu, X., Longenecker, K., Dipierro, C. G., Walker, L. A., Somlyo, A. V., Somlyo, A. P., and Nakamoto, R. K. (2000). Human RhoA/RhoGDI complex expressed in yeast: GTP exchange is sufficient for translocation of RhoA to liposomes. *Protein Sci.* **9,** 376–386.

Read, P. W., and Nakamoto, R. K. (2000). Expression and purification of Rho/RhoGDI complexes. *Methods Enzymol.* **325,** 15–25.

Robbe, K., and Antonny, B. (2003). Liposomes in the study of GDP/GTP cycle of Arf and related small G proteins. *Methods Enzymol.* **372,** 151–166.

Robbe, K., Otto-Bruc, A., Chardin, P., and Antonny, B. (2003). Dissociation of GDP dissociation inhibitor and membrane translocation are required for efficient activation of Rac by the Dbl homology-pleckstrin homology region of Tiam. *J. Biol. Chem.* **278,** 4756–4762. Epub 2002 Dec 5.

Sasaki, T., Kato, M., and Takai, Y. (1993). Consequences of weak interaction of rho GDI with the GTP-bound forms of rho p21 and rac p21. *J. Biol. Chem.* **268,** 23959–23963.

Schmidt, A., and Hall, A. (2002). Guanine nucleotide exchange factors for Rho GTPases: Turning on the switch. *Genes Dev.* **16,** 1587–1609.

Silvius, J. R., and l'Heureux, F. (1994). Fluorimetric evaluation of the affinities of isoprenylated peptides for lipid bilayers. PG-3014–22. *Biochemistry* **33,** 3014–3022.

Szoka, F., Jr., and Papahadjopoulos, D. (1978). Procedure for preparation of liposomes with large internal aqueous space and high capture by reverse-phase evaporation. *Proc. Natl. Acad. Sci. USA* **75,** 4194–4198.

Weast, R. C. (1989). *Handbook of Chemistry and Physics.* CRC Press, Boca Raton, FL.

[7] Phosphorylation of RhoGDI by p21-Activated Kinase 1

By CELINE M. DERMARDIROSSIAN and GARY M. BOKOCH

Abstract

Rho GTPase activation is partially regulated at the level of guanine nucleotide dissociation inhibitors, or GDIs. The binding of Rho GTPases to GDIs has been shown to dramatically reduce the action of guanine nucleotide exchange factors (GEFs) to initiate Rho GTPase activation. The GDI–GTPase complex thus serves as a major point of regulation of Rho GTPase activity and function. It is likely that specific mechanisms exist to dissociate individual members of the Rho GTPase family from

METHODS IN ENZYMOLOGY, VOL. 406
Copyright 2006, Elsevier Inc. All rights reserved.

0076-6879/06 $35.00
DOI: 10.1016/S0076-6879(06)06007-1

cytosolic Rho GDI complexes to facilitate the activation process. Such dissociation would likely be tightly coupled to GEF-mediated guanine nucleotide exchange and membrane association of the activated GTPase, resulting in effector binding and functional responses.

Accumulating evidence suggests that the phosphorylation of either the Rho GTPases themselves and/or phosphorylation of GDIs might serve as a mechanism for regulating the formation and/or dissociation of Rho GTPase–GDI complexes. Indeed, the selective release of Rac1 from RhoGDI complexes induced by the p21-activated kinase-regulated phosphorylation of RhoGDI has been reported. We describe here methods for the analysis of RhoGDI phosphorylation and regulation by p21-activated kinase 1 (Pak1).

Introduction

The low-molecular weight Rho GTPases are involved in the regulation of a variety of biological pathways. These GTPases function as molecular switches, regulating cellular signaling by alternating between an inactive, primarily cytosolic, GDP-bound state and an active GTP-bound state usually associated with membranes where effector targets reside. Conversion of inactive GTPases to an active form requires the action of guanine nucleotide exchange factors (GEFs) that catalyze the exchange of bound GDP for ambient GTP. GTPase inactivation involves the catalysis of GTP hydrolysis, which is intrinsically slow, through the action of GAPs (GTPase-activating proteins) that convert the GTPase to the inactive GDP-bound state. An additional level of regulation exists for GTPases of the Rho family because of their association with a third class of protein, the GDP dissociation inhibitors (GDIs).

Three human Rho GDIs have been identified: the ubiquitously expressed RhoGDI (or GDIα/GDI1) (Fukumoto et al., 1990; Ueda et al., 1990), the hematopoietic cell–selective Ly/D4GDI (or GDIβGDI2) (Lelias et al., 1993; Scherle et al., 1993), and RhoGDIγ (or GDI3), specifically expressed in lung, brain, and testis (Adra et al., 1997; Zalcman et al., 1996). Both RhoGDI and D4GDI are cytosolic and form 1:1 complexes with Rho family GTPases. Three distinct biochemical activities have been described for Rho GDIs (DerMardirossian and Bokoch, 2005; Olofsson, 1999). First, they inhibit the dissociation of GDP from Rho proteins, maintaining the GTPase in an inactive form and preventing GTPase activation by GEFs. Second, they are able to interact with the GTP-bound form of the Rho GTPase to inhibit GTP hydrolysis, blocking both intrinsic and GAP-catalyzed GTPase activity and preventing interactions with effector targets. A third biochemical activity of GDIs is to modulate the cycling of Rho GTPases between cytosol and membranes. GDIs maintain Rho GTPases as soluble cytosolic proteins by forming high-affinity complexes in which the geranyl-geranyl membrane-targeting moiety present at the C terminus of the Rho GTPases is shielded from the solvent by its insertion into the hydrophobic pocket formed by the immunoglobulin-like β sandwich of GDI (Hoffman *et al.*, 2000; Longenecker *et al.*, 1999). When Rho proteins are released from GDIs, they are able to insert into the lipid bilayer of the plasma membrane through their isoprenylated C terminus. In this uncomplexed form, the Rho GTPases can then interact with, and are activated by, membrane-associated GEFs, thereby initiating the association with effector targets at the membrane. An extraction from the membrane because of the reassociation with GDI, possibly initiated by GTP hydrolysis, is postulated to induce recycling of the GTPase back into the cytosol (DerMardirossian and Bokoch, 2005; Olofsson, 1999).

The binding of Rho GTPases to GDI has been shown to dramatically reduce the action of GEFs, such as Dbl (Yaku *et al.*, 1994) and Tiam1 (Robbe *et al.*, 2003), to catalyze nucleotide exchange. The GDI–GTPase complex is thus a major point of regulation of Rho GTPase activity and function. It is likely that specific mechanisms exist to dissociate individual members of the Rho GTPase family from cytosolic RhoGDI complexes to facilitate the activation process. This dissociation would likely be tightly coupled to GEF-mediated guanine nucleotide exchange and membrane association of the activated GTPase, resulting in effector binding and functional responses.

DerMardirossian *et al.* (2004) recently described the binding and phosphorylation of RhoGDI, both *in vitro* and *in vivo*, by p21-activated kinase 1 (Pak1). Pak1 is a serine/threonine kinase whose activity is regulated by the binding of active Rac or Cdc42, as well as a number of other mediators (Bokoch, 2003). Phosphorylation was shown to occur on two sites (Ser101 and Ser174) in RhoGDI on the external surface of the hydrophobic cleft in which the GTPase prenyl group binds. Both of these sites lie adjacent to hydrophobic residues that directly line the RhoGDI geranylgeranyl-binding pocket. Phosphorylation of these two sites resulted in the selective release of Rac1, but not RhoA, from the GDI complex, leading to its subsequent activation by exchange factors. Rac1 dissociation from RhoGDI and subsequent Rac1 activation induced by the growth factors EGF and PDGF required phosphorylation of S101 and S174 by Pak1. The phosphorylation of RhoGDI by Pak1 might serve as a positive feed-forward mechanism to account for sustained Rac activation during processes such as cell motility (Fig. 1).

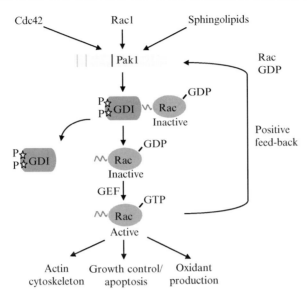

FIG. 1. Model for Pak1-mediated positive feedback activation of Rac GTPase. The interaction of active Cdc42, Rac1, or sphingolipids with Pak1 stimulates the phosphorylation of Rac-RhoGDI complexes at Ser101 and Ser174. This induces the dissociation of Rac1GDP, which is then activated to Rac1GTP through interaction with GEF(s). Active Rac1 feeds back to further activate Pak1, leading to more free Rac1 in a positive feedback cycle that enhances subsequent biological activities. (See color insert.)

Materials and Methods

In Vitro *Phosphorylation of RhoGDI by Pak1*

To examine whether pure, recombinant Pak1 could directly phosphorylate RhoGDI, we performed an *in vitro* kinase assay using GST-Pak1 as a constitutively active form of Pak1 and recombinant RhoGDI as a substrate. The preparation of purified GST-Pak1 and RhoGDI have been described elsewhere in detail (Chuang *et al.*, 1993; King *et al.*, 2000a). We note that to have highly active GST-Pak1, it is better to retransform pGEX-GST-Pak1 DNA in *Escherichia coli* strain BL21–competent cells, plated on LB broth, Miller (Fisher) plus ampicillin, and incubate overnight at 30°. Do not incubate the plates at 37°, because spontaneous mutations in Pak1 may occur. Kinase reactions (King *et al.*, 2000b) contained recombinant GST-Pak1 (1 μg per reaction) in kinase buffer (50 mM HEPES/NaOH, pH 7.5, 10 mM MgCl$_2$, 2 mM MnCl$_2$, 0.2 mM DTT) in a 70-μl reaction mixture, with 2 μg of pure recombinant RhoGDI, 20 μM ATP and 0.5 μCi/reaction radiolabeled [γ-32P]ATP (specific activity 4500 mCi/mmol, from ICN, Costa Mesa, CA). As a positive control, perform the kinase reaction in presence of 1 μg of myelin basic protein (Sigma M-1891). The reactions were incubated for 30 min at 30° and stopped by addition of 15 μl of 4× concentrated Laemmli's SDS-PAGE sample buffer. Phosphorylation reactions were resolved by 12% SDS-polyacrylamide gel electrophoresis (SDS-PAGE), then gels were stained with Coomassie Brilliant Blue R-250, destained, dried, and subjected to autoradiography for 1–24 h (Fig. 2).

In Vivo *Phosphorylation of RhoGDI by Pak1*

HEK293T cells were maintained in Dulbecco's modified Eagle's medium (DMEM, Life Technologies, Inc.) containing 8% heat-inactivated fetal bovine serum, 10 mM HEPES, pH 7.0, 2 mM glutamine, 100 units/ml penicillin G, and 100 μg/ml streptomycin at 37° in 5% CO_2. For transfection experiments, cells were seeded on 10-cm cell culture dishes at 50–70% confluency and transfected using appropriate concentrations of LipofectAmine (Life Technologies, Inc.) as a transfection agent. Cells were transfected with 1 μg His-tagged RhoGDI constructs (wt, S101A, S174A, S101/174A) and either 3 μg Myc-tagged Pak1 constructs (wt, constitutively active Pak1^{T423E,L107F} or kinase-dead Pak1^{K299A}) or 3 μg empty pCMV6-vector DNA control and 20 μl of LipofectAmine (Invitrogen, Carlsbad, CA) per dish. The transfection protocol was essentially followed according to the manufacturer's guidelines. At 16 h after transfection, cells were gently washed once with phosphate-buffered saline (PBS 1×) and incubated at 37°, 5% CO_2 for 1 h with phosphate-free DMEM; 0.4 mCi [^{32}P] orthophosphate (ICN) was then added to the plates, and cells were incubated overnight at 37°, 5% CO_2. Phosphate-buffered saline was used to wash once the plates (be very gentle, as cells may detach) and cells were

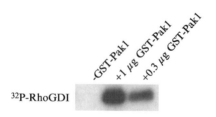

FIG. 2. Phosphorylation of RhoGDI by Pak1 in *in vitro* kinase assay. Purified recombinant RhoGDI (0.3 and 1 μg) were phosphorylated in an *in vitro* kinase assay with purified recombinant GST-Pak1, as described in the text. A concentration-dependent phosphorylation of RhoGDI by Pak1 was observed.

scraped into 0.5 ml of lysis buffer (25 mM Tris/HCl, pH 7.5, 150 mM NaCl, 1 mM EDTA, 1 mM DTT, 10% glycerol, and 1% NP-40 supplemented with 1 mM phenylmethylsulfonyl fluoride, 1 mM sodium orthovanadate, 10 μg/ml of aprotinin, 10 μg/ml leupeptin, 10 μg/ml pepstatin, and 10 nM microcystein).

To verify protein expression, whole-cell lysates from nonradioactive transfected cells were scraped into 0.5 ml of lysis buffer. After 30 min on ice, the homogenate was spun down 10 min at 14,000 rpm at 4° in a benchtop Eppendorf Microfuge to remove insoluble cells. A protein assay was carried out on the clarified homogenate using a BCA reagent protein assay (Pierce). Equivalent amounts of protein (25 μg) of the clarified homogenate were run on 12% polyacrylamide gels to resolve proteins. After electrophoresis, the proteins were transferred to nitrocellulose (Amersham) and probed with 9E10 anti-myc antibody (dilution 1/5,000, generated in-house) and with anti-His antibody (dilution 1/2,000, Babco) to detect the expression of Myc-tagged Pak1 and His-tagged RhoGDI proteins, respectively.

For precipitation experiments, His-tagged RhoGDI-expressing lysates were preincubated with equilibrated protein G-Sepharose (Amersham-Pharmacia) for 1 h at 4°, then the beads were pelleted to reduce the background. Precleared lysates were then incubated with monoclonal anti-His antibody (4 μl/0.5 ml lysates, Babco) for 2 h at 4°, followed by 2 h incubation at 4° with equilibrated Protein G-sepharose beads (1:1 slurry/IP). The bead fraction was washed four times with lysis buffer and 4× concentrated Laemmli's SDS-PAGE sample buffer was added to the beads, which were boiled 5 min at 110°. Samples were resolved by 12% SDS-polyacrylamide gel electrophoresis and RhoGDI phosphorylation was detected by autoradiography, as described previously.

Binding of Pak1 to RhoGDI

To evaluate interactions between endogenous RhoGDI and Pak1, we carried out coimmunoprecipitation experiments. As described previously, HEK293T cells grown to approximately 70% confluence on 100-mm tissue culture dishes were transiently transfected with 20 μl LipofectAmine reagent and 3 μg of either empty vector control (pCMV6M) or Myc-tagged Pak1 (wt, Pak1^{K299A}, Pak1^{T423E,L107F} in pCMV6M vector) (King et al., 2000b). The cells were allowed to express the protein for 20 h after transfection and were then washed in PBS and scraped into 0.5 ml of lysis buffer. Clarified lysates were subjected to protein concentration determined by BCA assay, and protein expression analyzed by immunoblotting using enhanced chemiluminescence (Pierce). To immunoprecipitate endogenous

RhoGDI, we used 10 μl of RhoGDI antibody (Santa Cruz SC-360 A-20) per 0.5 ml of lysates. After overnight incubation at 4°, 20 μl of pre-equilibrated protein A-Sepharose beads (Repligen, Waltham, MA) were added to the lysates and incubated for 2 h at 4°. Precipitated proteins were washed four times with lysis buffer, suspended in SDS-PAGE sample buffer, boiled for 5 min, resolved by 12% SDS-PAGE, and transferred to nitrocellulose membranes. The immunoprecipitated proteins were analyzed by immunoblotting the membrane with either Myc (dilution 1/5,000) or RhoGDI (dilution 1/2,000) antibodies using enhanced chemiluminescence (Pierce). Because RhoGDI has the same molecular weight as the antibody light chain, to avoid any background on the immunoblot we use an anti-rabbit secondary antibody that does not recognize the light chain (Pierce #31463).

We observed substantial coimmunoprecipitation of the active form of Pak1 with RhoGDI. In contrast, there was only slight interaction of Pak1 wt with RhoGDI, whereas kinase-dead Pak1 failed to interact with RhoGDI. These data indicate the formation of a relatively stable complex between Pak1 and RhoGDI is dependent on Pak1 being in an active state capable of substrate binding and phosphorylation. To investigate the regions of Pak1 that were important for the interaction with RhoGDI, we transfected in HEK293T cells either the N-terminal regulatory domain (amino acids 1-205) or the C-terminal catalytic domain of Pak1 (amino acids 206-545) following the protocol described previously. Only the catalytically active C terminus, but not the N-terminal regulatory domain, of Pak1 binds RhoGDI.

Analysis of Phosphorylation Sites on RhoGDI

To analyze the phosphorylation sites on RhoGDI, we performed a phosphoamino acid analysis according to the procedure of Boyle et al. (1991) with minor modifications. HEK293T cells were transfected with His-tagged RhoGDI (wt S101A, S174A, S101/174A) and active Pak1 constructs, and cells were labeled with [^{32}P] orthophosphate as described previously. Proteins were immunoprecipitated with anti-His mouse antibody and protein G-Sepharose, resolved by 12% SDS-PAGE, blotted onto nitrocellulose membrane, and phosphorylated RhoGDI proteins revealed by autoradiography. The radiolabeled band corresponding to RhoGDI was excised and blocked with 0.5% polyvinylpyrrolidone (PVP-10 Sigma) in 100 mM acetic acid for 30 min at 37°. After extensive washes in H$_2$O, the excised samples were incubated for 15 h at 37° with 1.5 μg trypsin in 50 mM NH$_4$HCO$_3$, pH 8.3. Samples were then lyophilized, resuspended in 30 μl of 5.7 N HCl, and hydrolyzed at 110° for 1 h. The hydrolysate was concentrated using a Speed-Vac concentrator, and the resulting pellet was washed twice in 200 μl of distilled water and finally resuspended in 3 μl of pH 1.9 electrophoresis buffer along with phosphoamino acid standards (phosphoserine, phosphotyrosine, phosphothreonine, at 1 mg/ml), and spotted onto cellulose-coated plates (EM Science 25 TLC plate 20 × 20 cm Cellulose OB124186). Separation of the proteolytic fragments was accomplished by electrophoresis in the first dimension for 20 min at 1500 V and by another electrophoresis in the second dimension (plates are turned 90 degrees counterclockwise) for 16 min at 1300 V in a Multiphor II horizontal electrophoresis unit (Amersham Pharmacia Biotech). The radioactive tryptic fragments were located by autoradiography. Phosphoamino acid standards were developed using 0.2% ninhydrin solution spray.

To determine the phosphorylated residues on RhoGDI, we performed two-dimensional phosphopeptide mapping of RhoGDI that had been phosphorylated *in vivo* by Pak1 as described previously. Proteolyzed samples were lyophilized and resuspended in 2 μl of pH 1.9 electrophoresis buffer (2.2% formic acid, 8% acetic acid), spotted onto 20-cm cellulose-coated plates (EM Science 25 TLC plate 20 × 20 cm Cellulose OB124186), and electrophoresed for 40 min at 1300 V in pH 1.9 electrophoresis buffer. Plates were air-dried and chromatographed in buffer containing 62.5% isobutyric acid, 1.9% n-butanol, 4.8% pyridine, and 2.9% glacial acetic acid for 14 h. The radioactive phosphopeptides were located by autoradiography (Boyle *et al.*, 1991) (Fig. 3).

Analysis for Rac-RhoGDI Complexes In Vitro *and* In Vivo

In Vitro. To evaluate the effects of phosphorylation on GTPase–GDI complexation *in vitro*, we used an assay based on the binding of [^{35}S] GTPγS to Rho GTPase. The basis for this assay lies in the fact that nucleotide binding only occurs to free GTPase, because nucleotide exchange is inhibited when the GTPase is bound with RhoGDI (Chuang *et al.*, 1993; Ueda *et al.*, 1990). Lipid-modified RhoA and Rac1 were purified from baculovirus membranes and preloaded with GDP as in Chuang *et al.* (1993) and Cerione *et al.* (1995). GST-GDP-Rac or RhoA immobilized on glutathione beads and precomplexed with RhoGDIwt or Rho GDI S101/174A mutant were obtained by incubating the GDP-RhoGTPase with 4–10-fold excess RhoGDI proteins overnight at 4°. The resulting complexes were washed three times with 25 mM Tris-HCl, pH 7.5, 1 mM DTT, 5 mM MgCl$_2$, 1 mM EDTA, and 0.1% Chaps to remove free RhoGDI. We verified the formation of the complex by the lack of significant [^{35}S]GTPγS binding in the absence of added detergent and a recovery of 100% of GTPase binding activity on the addition of 2% sodium cholate, which totally disrupts the complex and serves as a positive control. The

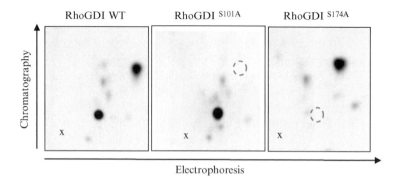

FIG. 3. Analysis of Pak1 phosphorylation sites on RhoGDI. 293T cells expressing active Pak1 and RhoGDI constructs were metabolically labeled with $[\gamma\text{-}^{32}P]$ orthophosphoric acid and RhoGDI immunoprecipitated from the whole cell lysates and subjected to phosphopeptide mapping, as in text. RhoGDI constructs that were used are indicated at the top of each map, and the application point of each sample is indicated in the lower left-hand corner by a cross. An arrow shows the direction of electrophoresis and chromatography migration. On the electrophoresis axis, the cathode is on the right. Two distinct radiolabeled phosphopeptide spots were detected with RhoGDI wild type (left panel). Missing phosphopeptides in RhoGDI mutants are circled by dots (middle and right panel). (See color insert.)

binding of $[35_S]$GTPγs to the RhoGTPases was monitored by filtration on BA85 nitrocellulose filters, using the method described in Chuang *et al.* (1993) and Knaus (1992); 50 n*M* of the complex was subjected to *in vitro* kinase assay with 1 m*M* unlabeled ATP for 20 min at 30° as described previously in the presence of 1 µg/ml of GST-Pak1 or 1 µg/ml of GST protein as a control. The sample was then pelleted and washed twice with 25 m*M* Tris-HCl, pH 7.5, 1 m*M* DTT, 5 m*M* $MgCl_2$, 1 m*M* EDTA, and 0.1% Chaps. The binding of $[35_S]$GTPγS was determined by the filter binding method as described by Chuang *et al.* (1993) and Knaus (1992).

In Vivo. Growth factors such as PDGF and EGF activate Rac to modulate cell growth and cytoskeletal remodeling. We examined the role of Pak1 in hormone-initiated dissociation of Rac from RhoGDI complexes during cell activation. HeLa Tet-off cells (Gossen and Bujard, 1992) were stably transfected with pTRE2hyg-EGFP-tagged Pak PID (inhibits specifically Pak kinase activity) or pTRE2hyg-EGFP-tagged Pak PIDL107F (the corresponding inactive mutant) by transfection with lipofectAMINE (Invitrogen). Resistant clones were isolated in the presence of hygromycin (500 µg/ml), screened for inducible expression and then cultured in the presence of doxycycline (1 µg/ml) according to the manufacturer's instructions. HeLa cells, which contained a Tet-off vector (Clontech), were maintained in DMEM supplemented with 10% fetal bovine serum, 10 m*M*

HEPES pH 7.0, 2 mM glutamine, 100 μg/ml G-418, 100 μg/ml hygromycin B, and 1 μg/ml doxycycline at 37° in 5% CO_2. To induce EGFP-Pak-PID or EGFP-Pak-PIDL107F protein expression, doxycycline has to be removed from the medium for at least 30 h. Cells should be washed twice with PBS, trypsinized, reseeded on plates, and washed again twice with PBS after 3–12 h. Separate experiments verified that expression of PIDwt resulted in >90% inhibition of serum-induced Pak1 activity.

For stimulation experiments, the HeLa cells stably transfected with inducible EGFP-tagged Pak PID wt or EGFP-tagged Pak PID L107F were cultured in 6-well plates in normal serum without doxycycline. At confluency, the cells were serum-starved 18 h and then treated with or without human recombinant PDGF (10 ng/ml) or EGF (100 ng/ml) (Fisher Scientific, Tustin, CA) for 0, 5, and 10 min. For transient transfection, HeLa cells were seeded on 6-well culture dishes at 0.5×10^6 cells and the next day were transfected using LipofectAMINE as described previously. At 18 h after transfection, cells were serum starved and treated with growth hormone as described. HeLa cells were then harvested in lysis buffer, and the whole-cell lysates were subjected to protein quantitation by BCA assay. For immunoprecipitation experiments, endogenous RhoGDI was immunoprecipitated from whole-cell lysates with RhoGDI antibody (Santa Cruz SC-360 A-20) overnight at 4°, followed by 2 h incubation with protein-A Sepharose beads. The proteins were transferred to PVDF membranes, and the level of Rac or RhoA proteins, as well as the immunoprecipitated RhoGDI, was assessed by Western blotting analysis using anti-Rac1 (#23A8, #05–389, Upstate Biotechnology Inc., Lake Placid, NY), anti-RhoA (#119 Sc-179 or #26C4 Sc-418, Santa Cruz Biotechnology Inc., Santa Cruz, CA), or anti-RhoGDI antibodies (Sc-360 A-20, Santa Cruz Biotechnology Inc.), respectively. Densitometry analysis was used to compare the level of Rho GTPase binding to RhoGDI in the various samples, and results were averaged over a minimum of three experiments.

References

Adra, C. N., Manor, D., Ko, J. L., Zhu, S., Horiuchi, T., Van Aelst, L., Cerione, R. A., and Lim, B. (1997). RhoGDIgamma: A GDP-dissociation inhibitor for Rho proteins with preferential expression in brain and pancreas. *Proc. Natl. Acad. Sci. USA* **94,** 4279–4284.
Bokoch, G. M. (2003). Biology of the p21-Activated Kinases. *Annu. Rev. Biochem.* **72,** 743–781.
Boyle, W. J., van der Geer, P., and Hunter, T. (1991). Phosphopeptide mapping and phosphoamino acid analysis by two-dimensional separation on thin-layer cellulose plates. *Methods Enzymol.* **201,** 110–149.
Cerione, R. A., Leonard, D., and Zheng, Y. (1995). Purification of baculovirus-expressed Cdc42Hs. *Methods Enzymol.* **256,** 11–15.

Chuang, T. H., Xu, X., Knaus, U. G., Hart, M. J., and Bokoch, G. M. (1993). GDP dissociation inhibitor prevents intrinsic and GTPase activating protein-stimulated GTP hydrolysis by the Rac GTP-binding protein. *J. Biol. Chem.* **268,** 775–778.

DerMardirossian, C., and Bokoch, G. M. (2005). GDI's: A central regulatory point in Rho GTPase activation. *Trends Cell Biol.* **15,** 356–363.

DerMardirossian, C., Schnelzer, A., and Bokoch, G. M. (2004). Phosphorylation of RhoGDI by Pak1 mediates dissociation of Rac GTPase. *Mol. Cell* **15,** 117–127.

Fukumoto, Y., Kaibuchi, K., Hori, Y., Fujioka, H., Araki, S., Ueda, T., Kikuchi, A., and Takai, Y. (1990). Molecular cloning and characterization of a novel type of regulatory protein (GDI) for the rho proteins, ras p21-like small GTP-binding proteins. *Oncogene* **5,** 1321–1328.

Gossen, M., and Bujard, H. (1992). Tight control of gene expression in mammalian cells by tetracycline-responsive promoters. *Proc. Natl. Acad. Sci. USA* **89,** 5547–5551.

Hoffman, G. R., Nassar, N., and Cerione, R. A. (2000). Structure of the Rho family GTP-binding protein Cdc42 in complex with the multifunctional regulator RhoGDI. *Cell* **100,** 345–356.

King, C. C., Reilly, A. M., and Knaus, U. G. (2000a). Purification and *in vitro* activities of p21-activated kinases. *Methods Enzymol.* **325,** 155–166.

King, C. C., Sanders, L. C., and Bokoch, G. M. (2000b). *In vivo* activity of wild-type and mutant PAKs. *Methods Enzymol.* **325,** 315–327.

Knaus, U. G., Heyworth, P. G., Kinsella, B. Therese, Curnutte, J. T., and Bokoch, G. M. (1992). Purification and characterization of Rac2. *J. Biol. Chem.* **267,** 23575–23580.

Lelias, J. M., Adra, C. N., Wulf, G. M., Guillemot, J. C., Khagad, M., Caput, D., and Lim, B. (1993). cDNA cloning of a human mRNA preferentially expressed in hematopoietic cells and with homology to a GDP-dissociation inhibitor for the rho GTP-binding proteins. *Proc. Natl. Acad. Sci. USA* **90,** 1479–1483.

Longenecker, K., Read, P., Derewenda, U., Dauter, Z., Liu, X., Garrard, S., Walker, L., Somlyo, A. V., Nakamoto, R. K., Somlyo, A. P., and Derewenda, Z. S. (1999). How RhoGDI binds Rho. *Acta Crystallogr. Dev. Biol. Crystallogr.* **55**(Pt. 9), 1503–1515.

Olofsson, B. (1999). Rho guanine dissociation inhibitors: Pivotal molecules in cellular signalling. *Cell Signal* **11,** 545–554.

Robbe, K., Otto-Bruc, A., Chardin, P., and Antonny, B. (2003). Dissociation of GDP dissociation inhibitor and membrane translocation are required for efficient activation of Rac by the Dbl homology-pleckstrin homology region of Tiam. *J. Biol. Chem.* **278,** 4756–4762.

Scherle, P., Behrens, T., and Staudt, L. M. (1993). Ly-GDI, a GDP-dissociation inhibitor of the RhoA GTP-binding protein, is expressed preferentially in lymphocytes. *Proc. Natl. Acad. Sci. USA* **90,** 7568–7572.

Ueda, T., Kikuchi, A., Ohga, N., Yamamoto, J., and Takai, Y. (1990). Purification and characterization from bovine brain cytosol of a novel regulatory protein inhibiting the dissociation of GDP from and the subsequent binding of GTP to rhoB p20, a ras p21-like GTP-binding protein. *J. Biol. Chem.* **265,** 9373–9380.

Yaku, H., Sasaki, T., and Takai, Y. (1994). The Dbl oncogene product as a GDP/GTP exchange protein for the Rho family: Its properties in comparison with those of Smg GDS. *Biochem. Biophys. Res. Commun.* **198,** 811–817.

Zalcman, G., Closson, V., Camonis, J., Honore, N., Rousseau-Merck, M. F., Tavitian, A., and Olofsson, B. (1996). RhoGDI-3 is a new GDP dissociation inhibitor (GDI). Identification of a non-cytosolic GDI protein interacting with the small GTP-binding proteins RhoB and RhoG. *J. Biol. Chem.* **271,** 30366–30374.

[8] Purification of ARAP3 and Characterization of GAP Activities

By Sonja Krugmann, Len Stephens, and Phillip T. Hawkins

Abstract

ARAP3 is a dual Arf and Rho GTPase activating protein (GAP) that was identified from pig leukocyte cytosol using a phosphatidylinositol-(3,4,5)-trisphosphate (PtdIns[3,4,5]P_3) affinity matrix in a targeted proteomics study. ARAP3's domain structure includes five PH domains, an Arf GAP domain, three ankyrin repeats, a Rho GAP domain, and a Ras association domain. ARAP3 is a PtdIns(3,4,5)P_3-dependent GAP for Arf6 both *in vitro* and *in vivo*. It acts as a Rap-GTP–activated RhoA GAP *in vitro*, and this activation depends on a direct interaction between ARAP3 and Rap-GTP; *in vivo* PtdIns(3,4,5)P_3 seems to be required to allow ARAP3's activation as a RhoA GAP by Rap-GTP. Overexpression of ARAP3 in pig aortic endothelial (PAE) cells causes the PI3K-dependent loss of adhesion to the substratum and interferes with lamellipodium formation. This overexpression phenotype depends on ARAP3's intact abilities to bind PtdIns(3,4,5)P_3, to interact with Rap-GTP, and to be a catalytically active RhoA and Arf6 GAP.

Introduction

Rho family small GTPases are involved in the regulation of cellular actin dynamics, whereas Arf family members are involved in vesicular trafficking events. All small GTPases act as molecular switches, cycling between an active, GTP-bound, and an inactive, GDP-bound state. The endogenous GTPase activity of the small GTPases is often slow but can be substantially activated by GAPs; similarly, exchange of GDP for GTP is catalyzed by guanine nucleotide exchange factors (GEFs) (Takai *et al.*, 2001). Although the functioning and mode of activation of GEFs has been studied to a considerable extent, GAPs are less understood. The ARAP1-3 protein family (Krugmann *et al.*, 2002; Miura *et al.*, 2002) was recently identified as dual GAPs for Arf and Rho family small GTPases. These large multidomain proteins contain five PH domains, Arf- and Rho-GAP domains, ankyrin repeats, and a Ras association domain. ARAP3 is the most studied family member. ARAP3 is activated by PtdIns(3,4,5)P_3, the lipid second messenger produced by class I phosphoinositide 3-OH kinases (PI3K) both as an Arf6 GAP and as a RhoA GAP; furthermore,

METHODS IN ENZYMOLOGY, VOL. 406   0076-6879/06 $35.00

the RhoA GAP activity can be activated by the Ras family small GTPase Rap (Krugmann *et al.*, 2002, 2004). ARAP3 is therefore ideally placed to mediate cross-talk between PI3K and Rho, Arf, and Ras family members. This article describes the assays used in the identification and characterization of ARAP3.

Materials

Dipalmitoyl forms of phosphoinositides (and their nonbiological stereoisomers) were synthesized, and PtdIns(3,4,5)P$_3$ was covalently coupled to affigel beads as described (Painter *et al.*, 1999, 2001).

Competition-Based Phosphoinositide Binding Assay

This assay (see also Fig. 1) was used to identify PtdIns(3,4,5)P$_3$-binding proteins from pig leukocyte cytosol (Krugmann *et al.*, 2002) and to assay lipid binding specificities of recombinant or overexpressed proteins in cell lysates or buffer.

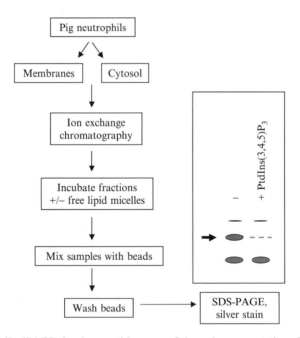

FIG. 1. PtdIns(3,4,5)P$_3$ bead competition assay. Schematic representation of the identification of PtdIns(3,4,5)P$_3$-binding proteins using covalently coupled PtdIns(3,4,5)P$_3$ affigel beads as an affinity matrix.

Analysis of Lipid Binding Specificities of Recombinant Proteins or Cell Lysates Containing Overexpressed Proteins

Recombinant, tagged proteins were diluted into ice-cold $1\times$ PBS, $1mM$ $MgCl_2$, 0.1% Triton X-100, 0.5 mg/ml fatty, acid free BSA to a concentration of 0.2–2 μM. Alternately, COS-7 cells overexpressing a tagged protein of interest were scraped into 10-ml/15-cm T/C dish of ice-cold lysis buffer (1% NP_40, 20 mM HEPES, pH 7.4, at $4°$, 5 mM EDTA, 5 mM EGTA, 10 mM NaF, 1 mM sodium orthovanadate, 120 mM NaCl, 5 mM β-glycerophosphate, 10 μg/ml each of antipain, aprotinin, leupeptin, and pepstatin A, and 1 mM PMSF. Preliminary experiments determined it is important to include phosphatase inhibitors (EDTA, EGTA, vanadate, NaF) at this stage to preserve the identity of the $PtdIns(3,4,5)P_3$ on the beads and in the solution during the incubation period. Lysates were centrifuged (100,000g at 4C for 1 h) to pellet insoluble material.

Two to 10 μl per sample of packed $PtdIns(3,4,5)P_3$ beads were washed three times in lysis buffer and pellets aspirated carefully before assays. Recombinant proteins or lysates were preincubated on ice for 10 min in the presence or absence of 5–50 μM $PtdIns(3,4,5)P_3$ or another phosphoinositide of choice ($PtdIns(3,4)P_2$, $PtdIns(3,5)P_2$, $PtdIns(3)P$, $PtdIns(4,5)P_2$), to allow subsequent competition with the lipid on the beads. Samples (0.5–1.5 ml) were then added to the beads and left to incubate for 30 min under gentle end-over-end movement at $4°$. Beads were then washed three times with wash buffer (20 mM HEPES, pH 7.2, at $4°$, 0.2 M NaCl, 5 mM EDTA, 5 mM EGTA, 5 mM β-glycerophosphate, 10 mM NaF, 0.1% sodium azide, 0.1% Nonidet P40) once in 5 mM HEPES, pH 7.2, at $4°$ and aspirated tightly. Samples were eluted for 5 min at $95°$ in SDS-PAGE sample buffer; the eluted protein was separated by SDS-PAGE and transferred onto PVDF membrane for Western blotting. Proteins of interest were identified according to their tags using appropriate antibodies. Beads were reused for up to five experiments after washing. For this, they were washed four times in $1\times$ PBS, 0.1% SDS, and twice more in $1\times$ PBS, 0.01% sodium azide in which they were stored at $4°$ also.

Identification of ARAP3

The preceding assay was carried out with pig leukocyte cytosol (in 0.1 M NaCl, 0.25% Nonident P-40, 2.5 mM EDTA, 2.5 mM EGTA, 2.5 mM sodium orthovanadate, 25 mM β-glycerophosphate, 25 mM NaF, 125 mM HEPES, pH 7.2, at $4°$) that had been prefractionated by anion or by cation exchange chromatography. For analytical experiments, 1 ml of lysate was used with 10 μl beads; for preparative experiments, 30- to 100 ml fractions were used with 0.1–0.3 ml beads. Beads were washed, and proteins were eluted off the beads as described previously. For analytical experiments, proteins were visualized by silver staining. This identified 25 proteins that bound to PtdIns(3,4,5)P$_3$ beads and were effectively competed off with free PtdIns(3,4,5)P$_3$. For a preparative experiment, proteins were separated by SDS-PAGE, transferred to nitrocellulose, stained with ponceau S, and bands of interest excised. These proteins were digested with trypsin and processed for mass spectrometric fingerprinting (Erdjument-Bromage *et al.*, 1998). Mass values from the MALDI-TOF experiments were used to search nonredundant protein databases. This identified known proteins known to bind to PtdIns(3,4,5)P3, known proteins unknown to bind to PtdIns(3,4,5)P3, and novel proteins; most PtdIns(3,4,5)P3-binding proteins identified contained at least one PH domain. At the time of its sequencing, no complete mRNA had been described for ARAP3, and it was cloned from cDNA libraries.

Purification of Recombinant ARAP3

ARAP3 was expressed in the Sf9 baculovirus system with a C-terminal Glu-Glu epitope tag (EEEEYMPME) using the pAcSG2 baculovirus transfer vector (BD Biosciences) and an epitope tag (GAGGAGGAGGAGTT-CATGCCCATGGAA) incorporated into the C-terminal PCR primer. Initial expression experiments revealed ARAP3 is very susceptible to proteolysis from the C-terminus. Preparations of N-terminally tagged protein contained high amounts of degradation products, whereas C-terminally tagged protein was full length. ARAP3-EE was purified from Sf9 cells in large batches. Sf9 cells from 2-3 l of culture infected for 1.85 days with baculovirus driving expression of ARAP3-EE were pelleted, washed once in 0.7% KCl, 2.66% sucrose, 7 mM sodium phosphate buffer, pH 6.2, 20 mM MgCl$_2$, and dropwise snap frozen in liquid nitrogen. For purification, the droplets were warmed quickly until they started to thaw, vortexed into 100 ml of ice-cold 1% Triton X-100, 0.15 M NaCl, 40 mM HEPES (pH 7.4 at 4°), 1 mM DTT, 0.01% sodium azide, 1 mM PMSF, and 10 μg/ml each of pepstatin A, antipain, leupeptin, and aprotinin and left to lyse on ice for 15 min. The Triton X-100 insoluble phase was pelleted by centrifugation (100,000g for 45 min at 4°). In the meantime, 1.5 ml of packed, covalently coupled anti-EE sepharose was equilibrated into lysis buffer (without protease inhibitors). ARAP3-EE was then immunoprecipitated out of the clarified lysates in two 50 ml tubes for 2.5 h with slow end-over-end rotation at 4°. Beads were then recombined into one tube and washed five times with 50 ml of ice-cold 0.3 M NaCl, 1% Triton X-100, 25 nM HEPES (pH 7.4 at 4°), 1 mM DTT, 0.01% sodium azide followed by further five washes in ice-cold 0.14 M NaCl, 25 mM HEPES (pH 7.4 at 4°), 1 mM DTT, 10% glycerol, and 0.01% sodium azide. Between washes, beads were pelleted in a swinging bucket centrifuge with slow deceleration so as not to disturb the fragile bead pellet; supernatants were aspirated carefully. ARAP-EE was eluted off the EE-sepharose with four times 1.5 ml of the same buffer supplemented with 150 μg/ml EY peptide (EYMPTD, NH2-terminus acetylated; twice that concentration was added in the first step to equilibrate the buffer volume contained between the beads) on ice over the course of 1.5 h. During the elution steps, tubes were flicked gently to mix every few minutes. All eluates were pooled and protein concentration determined by Bradford protein assay. Attempts to concentrate ARAP3 using Centriplus concentrators (Amicon) failed beyond 0.8 mg/ml. After stepwise addition under constant agitation of glycerol to 50% of the final volume, ARAP3-EE was snap frozen in aliquots and stored at –80°. The purity of the ARAP preparation was determined on a Coomassie-stained gel. See Fig. 2B for an example.

Purification of Recombinant Rho or Ras GTPases from E. coli

GST-fusions of Rho or Ras family small GTPases were expressed in *E. coli* (BL-21 strain). Typically, overnight cultures were diluted 1/100 into LB and grown at 37° until $OD_{600} = 0.5$; at this stage, cells were induced with 0.1 mM IPTG for 3 h. Rap proteins, which expressed very poorly, were grown in 2× TY until $OD_{600} = 0.3$ and induced with 0.01 mM IPTG

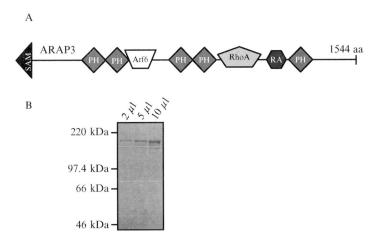

FIG. 2. ARAP3 domain structure and purification of recombinant protein. (A) ARAP3 domain structure. Domains are abbreviated as follows: SAM, sterile alpha motive; PH, pleckstrin homology domain; Arf6, Arf6-GTPase activating domain; RhoA, RhoA-GTPase activating domain; RA, Ras association domain. Ankyrin repeats are C-terminal to the Arf GAP domain but not indicated. (B) 2 μl, 5 μl, and 10 μl of a ARAP3-EE large-scale preparation were assessed for their purity on a Coomassie-stained 7% SDS-PAGE gel.

for 24 h at 32° (Herrmann *et al.*, 1996). Cells were pelleted by centrifugation, washed once in PBS, centrifuged again, and pellets snap frozen and stored until further use at –80°. Thawed cell pellets were vortexed into 25 ml ice-cold 20 mM Tris-HCl, pH 8.0, 150 mM NaCl, 10 μM GDP, 1 mM MgCl$_2$, 1 mM PMSF. After four bursts of sonication (10 sec each with 30-sec intervals on ice) using a probe sonicator, Triton X-100 was added to 1% w/v), and cells were left to lyse on ice for 30 min before Triton-insoluble material was pelleted by centrifugation for 1 h (100,000g at 4°). The supernatant was then poured onto 0.5 ml of packed glutathione sepharose 4B (Amersham Biosciences) that had been pre-equilibrated in the lysis buffer supplemented with Triton X-100. Protein was allowed to bind to beads for 2 h under slow end-over-end rotation at 4°; beads were then washed five times in 15 ml each of lysis buffer + 1% Triton-X100, twice in lysis buffer lacking Triton X-100, and a final time in the same buffer containing only 0.5 mM MgCl$_2$. Beads were transferred to a gravity-fed Polyprep chromatography column (BioRad) during the final wash, and GST-fusion proteins were eluted in 0.5-ml fractions in lysis buffer supplemented with freshly prepared 20 mM reduced glutathione, pH 8.0. For Ras proteins, "second" washes were in 20 mM Tris-HCl, pH 8.0, at 4°, 10% glycerol, 1 mM DTT, 150 mM NaCl, 5 mM β-glycerophosphate, 10 μM GDP; in the final wash, the MgCl$_2$ concentration was reduced to 0.5 mM; elution was in 20 mM Tris-HCl. pH 8.0. at 4°, 20% glycerol, 1 mM DTT, 0.5 mM MgCl$_2$, 1 μM GDP, and 20 mM reduced glutathione, pH 8.0. Protein peaks were determined by small-scale protein assay (Bradford), and the corresponding fractions were pooled. Rho family small GTPases were concentrated using Centricon concentrators (10-kDa cutoff; Amicon) until protein concentration was >2 mg/ml. Proteins were snap-frozen in small aliquots and stored at –80°C. Ras family proteins were concentrated using Slidealyzer dialysis devices in Slidealyzer solution (both Pierce) until c >1 mg/ml, or snap frozen immediately. Purity and concentration of proteins were confirmed on a Coomassie-stained protein gel. The active concentration of the GTPases (i.e., the amount that is able to bind nucleotide) was determined by loading with [^{3}H]-GTP as previously described (Self and Hall, 1995).

In Vitro *Rho GAP Assays*

Pilot experiments revealed that ARAP3 is inhibited by chelating agents. The reason is likely to lie in ARAP3's binding to Zn^{++} via its N-terminal Zinc finger motif and the fact that the affinity for Zn^{++} of EDTA/EGTA is higher than that for Mg^{++}. To avoid chelating agents in the assay, GTPases were loaded with [γ-^{32}P]GTP in the absence of EDTA and Mg^{++}. A 62.5-nM (active) stock of GTPases was prepared with 5 μCi [γ-^{32}P]GTP (3000 Ci/mmol)/10 μl loading reaction in 20 mM HEPES (pH 7.8 at 30°), 50 mM NaCl, 1 mg/ml BSA, 0.1 mM DTT for 10 min at 30°, followed by addition of $MgCl_2$ to 5 mM mixing and placing on ice. Stocks of GTPγS (or GDPβS)-loaded Ras proteins were prepared in the same way by loading 60 mM (active) Ras protein with 0.1 mM GTPγS (or GDPβS). As control, a GTPγS (or GDPβS) blank solution lacking the Ras protein was prepared. For assays containing lipid vesicles, lipids were dried to a film at the bottom of a minifuge tube and bath sonicated into GAP buffer (150 mM NaCl, 20 mM HEPES, pH 7.4, at 30°, 0.1 mM DTT, 10 μM GTP, 0.2 mM $MgCl_2$, 1 mg/ml BSA) to make a stock into which all other ingredients were diluted. The final lipid concentration in the GAP reaction was 200 μM each of PtdSer, PtdIns, PtdCho, and 10 μM PtdIns(3,4,5)P_3. To start the assay, the [γ-^{32}P]GTP-preloaded Rho GTPases were diluted into GAP reactions on ice, which already contained ARAP3 or its buffer, Ras-GTPγS or its blank and lipid vesicles or not, tubes vortexed briefly, and placed in a 30° water bath. For ease of handling, individual assay tubes were staggered by 20 sec. A control reaction was carried through on ice. The total volume of each GAP reaction was 20 μl; it contained 0.85–0.93 mM final $MgCl_2$ and 2 μl of the [γ-^{32}P]GTP-Rho stock; 5 μl aliquots were taken out of the GAP reactions after regular intervals and placed into 125 μl ice-cold 1 M perchloric acid and 1 mM inorganic phosphate. To extract the released, inorganic $^{32}P_i$ out of the reaction mixture, 125 μl 2% ammonium molybdate and 500 μl isobutanol/toluene (1:1) were added, the mix was vortexed and spun, and the upper phase quantified by scintillation counting in the presence of scintillant. To keep degradation of GDP/GTP to a minimum during these manipulations, all reagents and tubes were kept on ice throughout, and centrifugation was in a cooled benchtop centrifuge. The counts obtained from the control reaction account for inorganic phosphate released during the loading reaction or because of degradation during extraction procedures; they were subtracted from the other counts. When setting up these assays, we experimented also with filter paper assays, in which loaded GTPases are trapped on nitrocellulose at different stages throughout the reaction; the filters are washed in a vacuum manifold, and loss of counts, rather than increase of counts, is measured. In our hands, duplicates were tighter, reactions easier to handle, and more assays possible per day when extracting free phosphate as opposed to trapping protein, but the general idea of the assay is identical in each case. Figure 3B shows an example of an *in vitro* Rho GAP assay with Rho-GTP as substrate, ARAP3 as GAP, and RalA-GTPγS as Ras family protein.

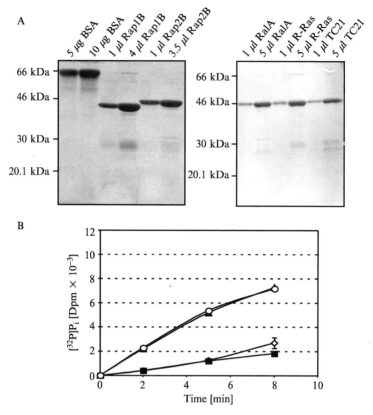

FIG. 3. Purification of Ras family proteins and *in vitro* Rho GAP assay. (A) GST-Ras family protein fusions were expressed in and purified from *E. coli* as described in the "Methods." Concentrated preparations were assessed for purity on a Coomassie-stained 12% SDS-PAGE gel. (B) In this *in vitro* Rho GAP assay, $[\gamma$-^{32}P]GTP-preloaded GST-RhoA served as substrate for ARAP3, and GTPγS preloaded GST-RalA was assessed for its ability to activate ARAP3 as a RhoA GAP. Aliquots of $[\gamma$-^{32}P]GTP-RhoA were incubated with ARAP3 or its buffer and with GTPγS-RalA or "GTPγS blank" as described in "Methods" for the indicated lengths of time. Then samples were placed on ice, and ^{32}P$_i$ was extracted for scintillation counting. Legend: diamonds, RhoA + GTPγS blank; solid squares, RhoA + GTPγS-RalA; circles, Rho + ARAP3 + GTPγS blank; solid triangles, RhoA + ARAP3 + RalA-GTPγS. This experiment was carried out twice in duplicate. Where error bars cannot be seen, they fall within the size of the symbols.

In Vivo *Rho GAP Assay*

We used a slight modification of previously published "pull-down" assays (Nimnual *et al.*, 2003; Sander *et al.*, 1998; Ren *et al.*, 1999). These rely on identification of the fraction of a total GTPase that is GTP-loaded inside cells at a given time. Most Rho GTPases are GDP-loaded in resting cells. Therefore, for analysis of a GEF, endogenous GTPases can be assessed because of the large signal-to-noise created by a small increase in the proportion of GTP-loaded GTPases. However, for the analysis of a GAP, it has generally been found necessary to overexpress its substrate. This has previously been done in transiently transfected mammalian cells (Aresta *et al.*, 2002). The disadvantage with this approach is that cotransfections with several plasmids (in our case RhoA, Rap, ARAP3, and a catalytic PI3K subunit) leads to uneven protein expression, making results difficult to interpret. To allow fine control of protein expression, combined with the certainty that small G protein signaling, but not endogenous ARAP3 exists, we used baculovirus-driven expression of proteins in *Spodoptera frugiperda* (Sf9) insect cells; 5×10^6 Sf9 cells were seeded and left to adhere in 6-cm tissue culture dishes and then infected with appropriate viral titers (which were determined by trial and error) of baculoviruses encoding a Rho GTPase, alone or together with ARAP3 and/or the p110γ PI3K catalytic subunit and a Ras GTPase. Cells were incubated at 27° for 1.8 days to express proteins. Then, dishes were placed on ice, medium was aspirated, cells washed quickly with 6 ml ice-cold PBS, and aspirated tightly; cells were then scraped into 1.3 ml of lysis buffer (for RhoA we used 50 mM Tris-HCl, pH 7.4, at 4°, 1% Triton X-100, 500 mM NaCl, 10 mM MgCl$_2$, 1 mM PMSF, 10 μg/ml each of aprotinin, leupeptin, antipain, and pepstatin A; for Rac as in Sander *et al.* (1998). Lysates were transferred into cooled 2-ml minifuge tubes, and insoluble proteins were pelleted for 5 min in a refrigerated minifuge; 2% of lysates were then used for a Western blot to determine the total amount of the GTPase in question (and expression of all other overexpressed proteins in the samples). The remaining 98% of lysate was subjected to pull-down assay using 10 μl packed glutathione sepharose 4B beads carrying GST-PAK-CRIB or GST-rhotekin Rho binding domain, which had been prepared on the preceding day as previously described (Nimnual *et al.*, 2003; Sander *et al.*, 1998). Proteins were allowed to bind at 4° for 40 min under slow end-over-end rotation and then washed five times very quickly using ice-cold lysis buffer without PMSF and antiproteases. Beads were then aspirated tightly, boiled in sample buffer, spun down, and sample buffer loaded on a gel for Western blotting with an anti-RhoA or anti-Rac1 antibody (Santa Cruz and BD Transduction Laboratories, respectively). To visualize expression of all proteins in lysates, Western blots were probed with anti-tag antibodies for Rap and the p110γ PI3K catalytic subunit and with a sheep anti-ARAP3 antiserum that has been previously described (Krugmann *et al.*, 2002).

In Vitro *Arf GAP Assay*

HA-tagged Arf1, Arf5, and Arf6 were transiently expressed by electro-poration in COS7 cells. COS7 cells were cultured in DMEM medium supplemented with 10% fetal bovine serum and Pen/Strep (all Gibco). Cells were trypsinized, washed twice in PBS, resuspended at 1×10^7 cells per 450 μl ice-cold electroporation buffer (30.8 mM NaCl, 120 mM KCl, 8.1 mM Na_2HPO_4, 1.46 mM KH_2PO_4, 5 mM $MgCl_2$), aliquoted into cold electroporation cuvettes (0.4 mm gap), and mixed with 50 μl electropora-tion buffer without $MgCl_2$ and 20 μg plasmid DNA. Cuvettes were placed in a gene pulser (BioRad) and cells electroporated at 0.25 V, 960 μFd. Cells were resuspended in 40 ml of $37°$ prewarmed DMEM + 10% FBS and plated into a large tissue culture flask (175 cm^2). After 8 h, medium was replaced by growth medium supplemented with Pen/Strep. After 40 h, cells were harvested by trypsinization, washed once in PBS, pelleted by centri-fugation, cell pellets snap frozen in liquid nitrogen, and stored at $-80°$. For Arf GAP assays, frozen cell pellets were vortexed into ice-cold lysis buffer (150 mM NaCl, 40 mM HEPES, pH 7.4, at $4°$, 1% Triton X-100, 1 mM $MgCl_2$, 1 mM EGTA, 20 μM GTP, 1 mM PMSF, 10 μg/ml each of antipain, aprotinin, pepstatin A, and leupeptin), thawed for 10 min on ice, and subjected to centrifugation (15 min, 20,000g, $4°$) to pellet the insoluble fraction. The supernatant was transferred onto 20 μl packed, covalently coupled anti-HA protein G sepharose beads that had previously been equilibrated in lysis buffer. Arf proteins were left to bind to the beads for 2.5 h at $4°$ under slow end-over-end rotation, beads were then washed five times with 1.5 ml lysis buffer (without antiproteases) and three times in loading buffer (0.1% triton X-100, 1 mM EDTA, 2.5 mM $MgCl_2$, 25 mM KCl, 100 mM NaCl, 40 mM HEPES, pH7.4, at $30°$, 1 mM DTT) to equilibrate Arf proteins for loading with $[\alpha\text{-}^{32}P]$-GTP. To prepare for loading with GTP and subsequent GAP assays, appropriate quantities of lipids were dried down and vesicles generated by bath sonicating into the appropriate buffers. Loading was performed for 1 h in a $30°$ waterbath with occasional gentle mixing in a total volume of 65 μl in the presence of 20–40 μCi of $[\alpha\text{-}^{32}P]$-GTP, 3 mM phosphoenol pyruvate, 200 μM PtdEtn, and 1.25 U/ml pyruvate kinase (Roche) as an ATP-GTP regenerating system to ensure GTP was not being degraded. In the meantime, lipid stocks (PtdEtn and $PtdIns(3,4,5)P_3$ or other phosphoinositides to be tested) were dried down at the bottom of a minifuge tube and bath sonicated into GAP buffer to make a concentrated stock to be diluted in the GAP reactions. Beads were then washed once in GAP buffer (40 mM HEPES, pH 7.4, at $30°$, 2.5 mM $MgCl_2$, 100 mM NaCl, 0.5 mM GTP, 1 mM DTT, 0.1% Triton X-100), mixed with 20 μl pre-equilibrated "blank" beads (for ease of handling), and the mixture aliquoted onto eight tubes for GAP reactions. GAP reaction tubes contained 5 μl packed beads in 35 μl GAP buffer containing lipid vesicles (typically 200 μM PtdEtn with or without 1–20 μM PtdIns(3,4,5)P$_3$) and 0.05 μM ARAP3-EE or its vehicle. Assays were left for 15 min at 30°. Reactions were stopped by addition of 1 ml ice-cold 50 mM HEPES, pH 7.4, 5 mM MgCl$_2$, 500 mM NaCl, 0.1% Triton X-100, and placing on ice. Beads were subsequently washed four more times with the same buffer to discard nonincorporated label and twice with 10 mM HEPES, pH 7.4, at 4°, 1 mM MgCl$_2$. Nucleotides were eluted off the beads by 10 min incubation and vortexing with 2 M formic acid, 50 μM GTP and 50 μM GDP. Samples were spun briefly, supernatants trans-ferred to new tubes, and freeze dried. Dry samples were taken up in 2 μl 50 mM NaPO$_4$, 10 mM NaP$_2$O$_7$, and spotted onto PEI-CEL thin-layer chromatography plates (Schleicher and Schuell); once spots were dried, tubes were "cleaned" out with a further 1 μl 50 mM NaPO$_4$, 10 mM NaP$_2$O$_7$ that was applied to the same spot. As a standard, [^{32}P]-GTP was applied. Nucleotides were resolved by thin-layer chromatography in 1 M NaH$_2$PO$_4$. Plates were dried down and spots visualized by autoradio-graphy. GDP/GTP ratios were calculated after scanning on a Phos-phorImager (Molecular Dynamics). Figure 4 shows an example of *in vitro* Arf1 loading with [α-^{32}P]GTP and an *in vitro* Arf1 GAP assay using the well-characterized ARF-GAP (Cukierman *et al.*, 1995) as GAP.

In Vivo *Arf6 GAP Assay*

This assay takes advantage of the fact that Arf6-GTP localizes to the plasma membrane and, when overexpressed, accumulates there in places, causing very characteristic actin and membrane-rich "bumps" to appear. In contrast, Arf6-GDP does not cause these structures and localizes in a granular fashion inside the cells, which has been described as an endotheli-al department (Peters *et al.*, 1995). Overexpression of a GTP-locked (L67) Arf6 and a GDP-locked (N27) Arf6 construct in pig aortic endothelial (PAE) cells confirmed these previous observations; overexpressed wild-type Arf6 shows pixillated staining throughout the cytoplasm and causes plasma membrane protrusions. Simultaneous overexpression of ARAP3, but not of its Arf6GAP point mutation, caused disappearance of the characteristic protrusions. PAE cells were propagated in F12 medium supplemented with 10% fetal bovine serum; 1×10^7 PAE cells were trans-fected by electroporation with 20 μg HA-tagged wt Arf6 construct together with 20 μg pEGFP as a control or with a pEGFP-ARAP3 wild-type or point mutant construct. Cells were plated onto coverslips in 6-well dishes, left to recover for 20 h, fixed for 10 min in 4% paraformaldehyde in PBS,

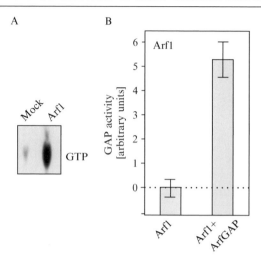

FIG. 4. *In vitro* Arf GAP assay. (A) Arf1-HA was immunoprecipitated from Arf1-HA or mock transiently transfected COS-7 cell lysates using covalently coupled HA-sepharose. The immunoprecipitated material was loaded with $[\alpha\text{-}^{32}\text{P}]$GTP in a 30° waterbath in the presence of an ATP-GTP regenerating system. Nucleotides were extracted off the sepharose beads and resolved by thin-layer chromatography. (B) $[\alpha\text{-}^{32}\text{P}]$GTP–loaded HA-tagged Arf1 was incubated for 15 min in a 30° waterbath alone or in the presence of the Arf1 GAP ARF-GAP (Cukierman *et al.*, 1995). GAP reactions were stopped, nucleotides extracted, separated by thin-layer chromatography, and scanned on a PhosphorImager for calculations of GDP/GTP ratios.

permeabilized for 5 min with 0.2% Triton X-100 in PBS, and stained with an anti-HA antibody (clone 12CA5) followed by a TRITC anti-mouse coupled secondary antibody (Jackson). Coverslips were mounted onto slides using Aquamount (Polysciences Inc). Cells were viewed on a Zeiss Axiovert fluorescent microscope, and photographs were taken using a digital camera and spot software.

Conclusion

We have described the isolation of ARAP3 on the basis of its specific binding to PtdIns(3,4,5)P$_3$. We have developed assays to characterize its Arf- and Rho-GAP activities *in vitro* and *in vivo*.

Acknowledgments

This work was supported by a Deutsche Forschungsgemeinschaft research fellowship to SK and Cancer Research UK project grant. SK is currently a BBSRC David Phillip's Fellow.

References

Aresta, S., de Tand-Heim, M. F., Beranger, F., and de Gunzgurg, J. (2002). A novel Rho GTPase-activating-protein interacts with GEM, a member of the Ras superfamily of GTPases. *Biochem. J.* **367,** 57–65.

Cukierman, E., Huber, I., Rotman, M., and Cassel, D. (1995). The ARF1 GTPase-activating protein: Zinc finger motif and Golgi complex localization. *Science* **270,** 1999–2002.

Erdjument-Bromage, H., Lui, M., Lacomis, L., Grewal, A., Annan, A. S., McNulty, D. E., Carr, S. A., and Tempst, P. (1998). Examination of a micro-tip reversed-phase liquid chromatographic extraction of peptide pools for mass spectrometric analysis. *J. Chromatogr. A.* **826,** 167–181.

Herrmann, C., Horn, G., Spaargarden, M., and Wittinghofer, A. (1996). Differential interaction of the Ras family GTP-binding proteins H-Ras, Rap1A, and R-Ras with the putative effector molecules Raf kinase and Ral-guanine nucleotide exchange factor. *J. Biol. Chem.* **271,** 6794–6800.

Krugmann, S., Anderson, K. E., Ridley, S. H., Risso, N., McGregor, A., Coadwell, J., Davidson, K., Eguinoa, A., Ellson, C. D., Lipp, P., Manifava, M., Ktistakis, N., Painter, G., Thuring, J. W., Cooper, M. A., Lim, Z. Y., Holmes, A. B., Dove, S. K., Michell, R. H., Grewal, A., Nazarian, A., Erdjument-Bromage, H., Tempst, P., Stephens, L. R., and Hawkins, P. T. (2002). Identification of ARAP3, a novel PI3K effector regulating both Arf and Rho GTPases, by selective capture on phosphoinositide affinity matrices. *Mol. Cell* **9,** 95–108.

Krugmann, S., Williams, R., Stephens, L., and Hawkins, P. T. (2004). ARAP3 is a PI3K- and Rap- regulated GAP for RhoA. *Curr. Biol.* **14,** 1380–1384.

Miura, K., Jacques, K. M., Stauffer, S., Kubosaki, A., Zhu, K., Hirsch, D. S., Resau, J., Zheng, Y., and Randazzo, P. A. (2002). ARAP1: A point of convergence for Arf and Rho signaling. *Mol. Cell* **9,** 109–119.

Nimnual, A. S., Taylor, L. J., and Bar-Sagi, D. (2003). Redox-dependent downregulation of Rho by Rac. *Nat. Cell Biol.* **5,** 236–241.

Painter, G. F., Grove, S. J. A., Gilbert, I. H., Holmes, A. B., Raithbly, P. R., Hill, M. L., Hawkins, P. T., and Stephens, L. R. (1999). General synthesis of 3-phosphorylated myo-inositol phospholipids and derivatives. *J. Chem. Soc. Perkin Trans.* **1,** 923–935.

Painter, G. F., Thuring, J. W., Lim, Z.Y, Holmes, A. B., Hawkins, P. T., and Stephens, L. R. (2001). Synthesis and biological evaluation of a PtdIns(3,4,5)P3 affinity matrix. *Chem. Commun.* **7,** 645–646.

Peters, P. J., Hsu, V. W., Ooi, C. E., Finazzi, D., Teal, S. B., Ooschot, V., Donaldson, J. G., and Klausner, R. D. (1995). Overexpression of wt and mutant Arf1 and Arf6: Distinct perturbations of nonoverlapping membrane compartments. *J. Cell Biol.* **128,** 1003–1017.

Ren, X. D., Kiosses, W. B., and Schwartz, M. A. (1999). Regulation of the small GTP-binding protein Rho by cell adhesion and the cytoskeleton. *EMBO J.* **18,** 578–585.

Sander, E. E., van Delft, S., ten Kloster, J. P., van der Kammen, R. A., Michiels, F., and Collard, J. G. (1998). Matrix-dependent Tiam1/Rac signalling in epithelial cells promotes either cell-cell adhesion or cell migration and is regulated by phosphatidylinositol 3-kinase. *J. Cell. Biol.* **143,** 1385–1398.

Self, A. J., and Hall, A. (1995). Measurement of intrinsic nucleotide exchange and GTP hydrolysis rates. *Methods Enzymol.* **156,** 67–76.

Takai, Y., Sasaki, T., and Matozaki, T. (2001). Small GTP-binding proteins. *Physiol. Rev.* **81,** 153–208.

[9] Regulation of RhoGAP Specificity by Phospholipids and Prenylation

By ERZSÉBET LIGETI and JEFFREY SETTLEMAN

Abstract

Among the key protein regulators of the various and numerous small GTPases are the GTPase activating proteins (GAPs). Experimental studies of some of the ~170 GAPs predicted by the human genome indicate that their catalytic GAP activity is regulated by a variety of mechanisms, including phosphorylation, protein–protein interactions, proteolysis, and interactions with lipids. Most reported biochemical studies to address the specificity of GAPs for particular GTPases have been conducted *in vitro* with bacterially produced GTPases. Thus, the potential influence of these various regulatory mechanisms in the context of GAP–GTPase specificity may be overlooked in such assays. Here, we present experimental studies that highlight the role of lipids in modulating the GTPase specificity for some of the Rho GAPs. We find that particular phospholipids can substantially alter the substrate "preference" for the p190 GAPs. We find that C-terminal prenylation of GTPases can influence the specificity of GAP interactions as well. These observations emphasize the limitations of standard *in vitro* GAP assays in definitively establishing the physiologically relevant GTPase targets for particular GAPs.

Introduction

GTPase activating proteins (GAPs) accelerate the rate of GTP hydrolysis on small GTPases and thereby promote the down-regulation or termination of GTPase-mediated cellular signaling pathways. On the basis of sequence similarities, it has been predicted that the human genome includes ~170 genes whose products may function as GAPs (Bernards, 2003). The largest family of these potential GAP genes, about 70, contain the consensus sequence typical of proteins that specifically regulate various members of the Rho subfamily of small GTPases as substrates (Bernards, 2003). The large number of predicted Rho-family GAPs suggests that each individual GAP is likely to be involved in very specific molecular interactions with substrate GTPases and at localized subcellular domains. From the large variety of potential Rho-family GAPs individual cells express a

METHODS IN ENZYMOLOGY, VOL. 406
0076-6879/06 $35.00
DOI: 10.1016/S0076-6879(06)06009-5

unique "RhoGAP repertoire" that is likely to contribute to the specificity of GTPase-mediated biological functions carried out by a particular cell type.

The molecular weight of Rho-family GAPs varies across a broad range (approximately 35–210 kDa), whereas the functionally essential domain ("GAP domain") consists of only approximately 200 amino acids (20 kDa). The GAP catalytic domain contains a conserved arginine that protrudes into the nucleotide-binding pocket of the small GTPase and by stabilizing its transition state promotes the hydrolysis of GTP (Scheffzek *et al.*, 1997). The peptide sequences outside of the GAP domain include a wide variety of domains of predicted functions that provide the possibility for numerous molecular interactions and mechanisms for regulation of the GAP catalytic activity (Bernards and Settleman, 2004).

Most of the proteins with documented Rho-GAP activity have the capacity to regulate all three major subdivisions of the Rho subfamily (i.e., Rho, Rac, and Cdc42), whereas some Rho GAPs have been reported to regulate only selected members of the family (Table I). In most cases, substrate specificity has been determined in *in vitro* GAP assays, using the predicted GAP domain of the relevant GAP and small GTPases expressed as fusion proteins in bacteria. There are two potential problems with this approach. First, bacteria are not able to carry out the posttranslational

TABLE I
Substrate Preference of Some Rho Family GAPs

GAP	Smg recognized as substrate	Smg not recognized as substrate	References
RalBP1	Cdc42 Rac	Rho	Jullien-Flores *et al.*, 1995; Park and Weinberg, 1995
BCR	Cdc42 Rac	Rho	Chuang *et al.*, 1995
GRAF GRAF2	Cdc42 RhoA	Rac1	Hildebrand *et al.*, 1996; Ren *et al.*, 2001; Shibata *et al.*, 2001
RICH-1 RICH-2	Cdc42 Rac1	RhoA	Richnau and Aspenstrom, 2001
p190A		"Preferential Rho"	Ridley *et al.*, 1993
mgcRacGAP	Rac Cdc42	Rho	Agnel *et al.*, 1992; Toure *et al.*, 1998
srGAP1	Rho Cdc42	Rac1	Wong *et al.*, 2001
RhoGAP1	RhoA	Rac1 Cdc42	Prakash *et al.*, 2000
p50RhoGAP	Preferential Cdc42		Bourne *et al.*, 1990, 1991

modifications of small GTPases that typically occur in mammalian cells: farnesylation, proteolytic cleavage, and methylation. Second, because most GAPs are complex multidomain proteins, the modifying effect of protein domains outside of the GAP domain may be neglected.

In this Chapter, we provide examples that demonstrate that both the lipid environment and the prenylation state of the small GTPases are important factors that can modify the substrate specificity of several GAPs in *in vitro* assays. Our data suggest that in the context of intracellular signaling, the substrate specificity of any given GAP may differ considerably from the specificity determined under typical *in vitro* conditions using purified recombinant proteins.

Expression and Purification of Proteins

Expression of p190RhoGAP in Insect Cells

Full-length p190A and p190B RhoGAPs and a C-terminal fragment of p190A (amino acids 1135–1513) were expressed with a hexahistidine tag in Sf9 insect cells using the baculovirus system. Recombinant viruses were produced in the Bac-to-Bac system following the manufacturer's protocol. Sf9 insect cells were transfected with the amplified viral stocks and harvested after 3 days. The protein fragment was affinity purified on nickel beads (Invitrogen), and the purified proteins were found to be essentially homogeneous as determined by Coomassie-Blue staining of samples resolved by SDS-PAGE. For unknown reasons, the yield of p190A RhoGAP is consistently substantially greater than that of p190B, which seems to be a less stable protein both *in vitro* and *in vivo*.

Expression of p50RhoGAP as GST-Fusion Protein in Escherichia coli

An *E. coli* clone expressing the full-length p50RhoGAP protein in pGEX-2T vector was obtained from Dr. Alan Hall. For purification of the protein, bacteria are grown overnight at 37°. Expression of the protein is induced by addition of 0.5 mM IPTG to the exponentially growing culture. Bacteria are harvested after 4 h, and purification of the protein is carried out using glutathione-Sepharose beads, as described by Self and Hall (Self and Hall, 1995).

Expression of Rac and Rho

E. coli clones transformed with plasmids expressing Rac1 and RhoA GTPases as GST fusion protein were a generous gift of Dr. Alan Hall. The nonprenylated small GTPases were prepared as described by Self and Hall

(Self and Hall, 1995). Prenylated Rac and Rho were expressed and purified from Sf9 insect cells as described by Di-Poi *et al.* (2001). Basically, the cDNA corresponding to the coding sequence of human Rac1 and RhoA was transferred from the pGEX-2T vector into the baculoviral pBacPA-KHis1 vector using the *Bam*HI and *Eco*RI restriction sites. To obtain recombinant baculoviruses, Sf9 insect cells were transduced using the protocol provided with the Baculogold kit (Pharmingen). After 5 days, the supernatants were harvested and used for transfection of fresh Sf9 cells. Repeated cycles of amplification were carried out, and the virus-containing supernatants were collected and stored. For production of proteins, Sf9 cells (2×10^6/ml) were infected at a concentration of 100 μl of high titer virus per 10^6 cells. After 3 days, cells were pelleted, disrupted by ultrasonication, and the membrane fraction was extracted by 1% CHAPS. His-tagged small GTPases were affinity purified on nickel beads (Invitrogen) as described in detail by Di-Poi *et al.* (2001).

Preparation of Lipid Vesicles

The investigated lipids were dissolved in chloroform or a chloroform/methanol (95:5) mixture, aliquoted in 200-μg fractions, and dried in nitrogen atmosphere. To prepare liposomes, lipids were rehydrated for at least 1 h at room temperature in 100 μl buffer1 (25 mM HEPES, 50 mM NaCl, 1 mM MgCl$_2$, pH = 7.5) followed by extensive vortexing until the mixture became homogenously opalescent. Alternately, the vortexed lipid suspension was sonicated in a bath-type sonicator for 20 min at 4°. Phospholipids were used in a concentration of 0.1 mg/ml (approximately 200 μM).

GAP Assays

Required Solutions

Buffer 1 (wash buffer): 25 mM HEPES, 50 mM NaCl, 1 mM MgCl$_2$, pH = 7.5

Buffer 2 (assay buffer): 25 mM HEPES, 50 mM NaCl, 1 mM MgCl$_2$, 0.1 mM ATP, 0.1 mM GTP, 0.1 mM DTT, and 1 mg/ml BSA, pH = 7.5

Nucleotide depletion buffer: 50 mM HEPES, 50 mM NaCl, 0.1 mM DTT, 0.1 mM ATP, 0.1 mM EGTA, 5 mM EDTA, and 1 mg/ml BSA, pH = 7.4

Acidic Na-phosphate solution: 0.1 M Na-phosphate, pH = 2.0, containing 5% charcoal.

Loading of Rac and Rho with [^{32}P]GTP

An identical procedure was used for loading of prenylated and non-prenylated proteins with nucleotide. The GTPase (1–2 μg) was incubated with 10 μCi [γ-^{32}P]GTP (NEN, 6000 Ci/mmol) or [α-^{32}P]GTP (NEN, 800 Ci/mmol) in 100 μl nucleotide depletion buffer for 10 min at 30° or 20° in the case of Rho or Rac, respectively. After incubation, MgCl$_2$ was added to a final concentration of 10 mM followed by dilution of the loaded proteins with 2–4 ml of buffer 2 on ice. At 4°, hydrolysis of GTP by Rho was not significant, thus Rho–GAP assays could be carried out on a large number of samples within 1–2 h after nucleotide loading. In contrast, significant GTP hydrolysis (~15%) by Rac proteins was detected even on ice within 20 min. Consequently, in the Rac–GAP assays, additional controls were included for each analysis (12 samples handled at the same time).

GAP Assay Using Separation on Nitrocellulose

The GAP assay was initiated by adding 100 μl of loaded small GTPase to 10 μl of buffer 1 containing the investigated GAP (in a volume of 1 or 2 μl) and the lipid vesicles (in a volume of 1–5 μl). In Rho-GAP assays incubation was carried out for 15 min at 30°, whereas Rac-GAP assays were carried out for 5 min at 20°.

At the end of the incubation period, GTP hydrolysis was stopped by the rapid addition of 150 μl ice-cold buffer 1, and the sample was immediately deposited onto nitrocellulose filters of 0.45-μm pore size (Schleicher-Schull) on a 12-place vacuum manifold (Millipore). The filters were washed with 5 ml ice-cold buffer1 and placed into scintillation vials. Filter-retained radioactivity was determined in a Beckman LS 5000TD liquid scintillation spectrometer.

Potential Experimental Error Associated with Separation on Nitrocellulose

The pore size of the nitrocellulose filters is relatively large compared with the size of the investigated proteins. Retention is based on electrostatic interactions that may be altered by changes in the charge density of particular protein. The presence of hydrophobic substances, such as lipid vesicles, in the filtered solution may in principle alter the retention efficiency of the filters. A decrease in protein retention could, therefore, potentially mimic an increased rate of GTP hydrolysis.

One possible means of ruling out any potential alteration in the separation efficiency is to carry out the GAP assay with small GTPases loaded with [α-^{32}P]GTP. In this case, GTP hydrolysis does not influence the amount of protein-bound radioactivity. In our experiments, neither the presence of GAPs nor lipid vesicles was determined to exert any influence on the amount of $[\alpha\text{-}^{32}P]$GTP radioactivity retained on filters in a Rho- or a RacGAP assay.

GAP Assay Using Charcoal Precipitation

Another means of avoiding potential errors associated with separation on nitrocellulose is the use of charcoal precipitation. This technique is based on the fact that active charcoal is able to bind organic phosphates, but it does not adsorb inorganic phosphate. Thus, hydrolyzed inorganic ^{32}P-phosphate can be measured in the separated supernatant.

At the end of the incubation period of the GAP assay, active charcoal suspended in 0.6 ml ice-cold acidic Na-phosphate solution was added to samples, which were then vortexed rapidly and maintained at 4° until the end of the experiment. Thereafter, samples were centrifuged at 12,000g for 5 min, and radioactivity within an aliquot of the clear supernatant was counted in a Beckman LS 5000TD liquid scintillation spectrometer on the basis of the Cerenkov effect.

Modification of the Substrate Specificity by the Lipid Environment

P190RhoGAP is able to accelerate the rate of GTP hydrolysis both on RhoA and Rac1 proteins. However, when p190 GAP is examined for activity against these small GTPases in the presence of phosphatidylserine vesicles, a characteristic change is consistently observed in the substrate preference; specifically, the RhoGAP activity is inhibited and the RacGAP activity is enhanced (Fig. 1) (Ligeti *et al.*, 2004). Under conditions in which the concentration of the small GTPase is in 10–100 fold excess relative to the GAP, an almost complete switch of the substrate specificity could be observed: a potent RhoGAP became a weak RhoGAP but an effective RacGAP. The experiments summarized in Fig. 1 demonstrate that the same results were obtained independently of the applied method (i.e., whether a decrease in the substrate $[[\gamma\text{-}^{32}P]$GTP in A and B] or an increase in the product [hydrolyzed inorganic ^{32}P-phosphate in C and D] was followed).

Phosphatidylserine was not the only phospholipid that affects the substrate specificity of p190RhoGAP. Two more acidic phospholipids, phosphatidylinositol and phosphatidylinositol bisphosphate (PIP2), exerted a similar effect on p190 GAP activity (Fig. 2). Interestingly, phosphatidic acid only inhibited RhoGAP activity but did not influence the RacGAP activity of p190RhoGAP. This difference may indicate that p190RhoGAP

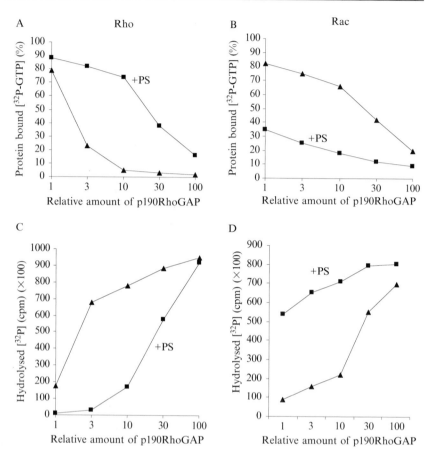

FIG. 1. Effect of phosphatidylserine (PS) on the Rho- and Rac-GAP activity of p190. (A) and (B) Protein-bound $[\gamma\text{-}^{32}P]GTP$ was retained on nitrocellulose filters, whereas in (C) and (D) hydrolyzed ^{32}P-phosphate was separated by charcoal precipitation. Reproduced from Ligeti *et al.* (2004).

possesses more than one lipid-binding site with different affinities for various phospholipids. A similar behavior was previously reported for ArfGAPs (Brown *et al.*, 1998). It is important to note that all the modifying effects of lipids are seen only when GAP activity is assayed using prenylated small GTPases derived from insect cells. Thus, the effects of these lipids on GAP specificity are dependent on the presence of a C-terminal modification of the GTPase. Moreover, the isolated C-terminal region of p190RhoGAP, which contains the catalytic domain, was determined to be

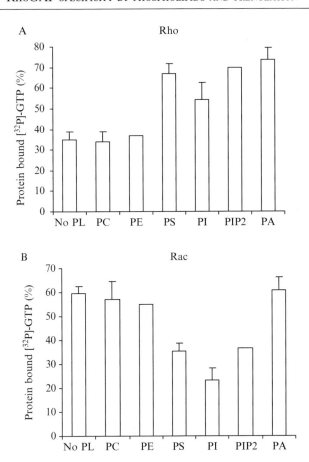

Fɪɢ. 2. Relative effects of various phospholipids on the Rho- and Rac-GAP activity of p190 using prenylated GTPases and the isolated C-terminal catalytic domain of p190. No PL, no phospholipids; PC, phosphatidylcholine; PE, phosphatidylethanolamine; PS, phosphatidylserine; PI, phosphatidylinositol, PIP2, phosphatidylinositol-4,5-bisphosphate; PA, phosphatidic acid. Reproduced from Ligeti *et al.* (2004).

sufficient for the lipid-mediated switch in GTPase preference, suggesting that this region of the protein contains all of the lipid-sensitive regulatory elements.

Significantly, alteration of the GAP substrate preference as a function of the composition of the lipid environment seems not to be a completely generalizable property that applies to all RhoGAPs. Thus, p50RhoGAP, which, like p190, also recognizes both Rho and Rac GTPases as substrate, is not influenced by any of the tested phospholipids (Ligeti *et al.*, 2004).

Effect of the Posttranslational Modification of the Small GTPase on the
  Substrate Specificity of GAPs

As summarized in Table I, several GAPs have been reported to exhibit
an apparent preference for one or more of the small GTPases belonging to
the Rho subfamily. We show here a few examples indicating that the
substrate preference of GAPs may differ depending on whether prenylated
(expressed in Sf9 insect cells) or nonprenylated (expressed in bacteria)
small GTPases are used for testing.

BCR is the gene involved in translocation on the Philadelphia chromosome that occurs in cases of chronic myelogenous leukemia. The BCR gene
product, a 160-kDa protein that consists of a N-terminal serine/threonine
kinase domain, contains a central Dbl-homology domain that catalyzes
nucleotide exchange on Rho family small GTPases and a C-terminal
GAP domain for the Rho GTPases. Using Bcr-null mutant mice, a functional role of Bcr has been verified in neutrophilic granulocytes (Voncken
et al., 1995), but the exact molecular interactions of this protein are still not
known. The GAP domain of Bcr was shown to accelerate the GTP hydrolysis of nonprenylated Rac1, Rac2, and Cdc42, but it was inactive with
RhoA (Fig. 3A). This same biochemical finding has been obtained by two
different groups (Chuang *et al.*, 1995; Diekmann *et al.*, 1991) and was

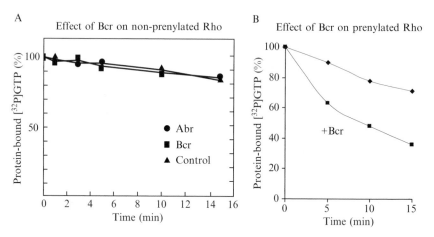

FIG. 3. Comparison of Bcr GAP activity using non-prenylated (A) or prenylated (B) Rho
as substrate. (A) Reproduced from Chuang *et al.* (1995) with permission. (B) Ligeti and
Dagher, unpublished observation.

confirmed in our own laboratory. However, when the effect of Bcr was tested with prenylated Rho, we consistently observed a significant increase in GTP hydrolysis by RhoA (Fig. 3B). Both the previously published data on Bcr's substrate specificity and its role in neutrophils implicated Bcr as a RacGAP. However, our findings indicate that a potential regulatory effect of Bcr on Rho-dependent cellular processes should also be considered.

P190RhoGAP has been reported to react "preferentially with Rho." As shown in Fig. 4A, this is indeed the case when nonprenylated small GTPases are investigated: at a relatively high concentration, p190GAP reacts largely with Rho but not with Rac. However, the efficiency is significantly increased and the substrate preference is altered if prenylated GTPases are studied (Fig. 4B). The RhoGAP activity is evident when p190 GAP is present at relatively low concentration, and significantly, the RacGAP activity becomes dominant at the lowest p190 GAP concentration tested. Thus, when prenylated small GTPases are considered, the conclusion that p190 GAP reacts preferentially with Rac would seem to be more appropriate.

Finally, the 50-kDa GAP, alternately referred to as p50Cdc42GAP or p50RhoGAP, was reported to predominantly affect GTP hydrolysis by Cdc42 and to react only marginally with RhoA and Rac1 (Barfod *et al.*,

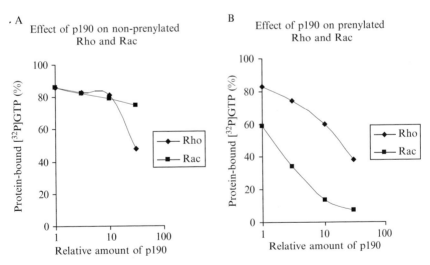

FIG. 4. Comparison of p190RhoGAP using nonprenylated (A) or prenylated (B) RhoA and Rac1 GTPases as substrates. (Ligeti and Settleman, unpublished observation).

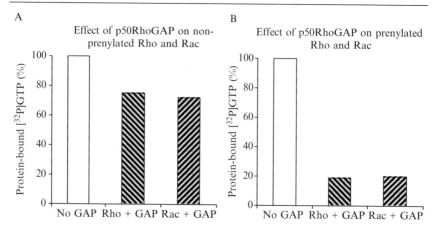

FIG. 5. Comparison of p50RhoGAP activity using nonprenylated (A) or prenylated (B) RhoA and Rac1 GTPases as substrates. (Ligeti and Moskwa, unpublished observation).

1993; Lancaster *et al.*, 1994). These early data were obtained with GST-fusion proteins expressed in *E. coli*. Indeed, when nonprenylated small GTPases were investigated, this protein proved to have a relatively weak effect on GTP hydrolysis by both RhoA and Rac1. As shown in Fig. 5A, only 20% of the protein-bound GTP is hydrolyzed in 15 (RhoA) or 5 (Rac1) min, respectively. In contrast, p50GAP exhibits strong GTPase activating activity both on prenylated Rho and Rac: 80% of the protein-bound GTP is hydrolyzed in 15 (RhoA) or 5 (Rac1) min, respectively (Fig. 5B). In the case of p50GAP, the difference in its reactivity with prenylated or nonprenylated small GTPases is due to the autoinhibitory conformation of the protein that limits the access of the small GTPase to the critical amino acids of the GAP domain and that is relieved by the prenyl moiety (Moskwa *et al.*, 2005). Further experimentation is needed to determine whether this mechanism is uniquely relevant to the p50GAP protein or is a more general property of GAPs for the Rho family small GTPases. In any case, in the cellular environment, where most of the small GTPase molecules are prenylated, p50GAP should be considered as a GAP with potential broad substrate specificity.

Conclusions

The GTPase-activating action of the various Rho-family GAPs resides in their highly homologous "GAP-domain" and involves the catalytically essential arginine finger (2). However, interaction of specific GAPs with specific small GTPases is also influenced by the nonhomologous amino acids within, and sequences outside of the GAP-domain, resulting in a wide variation of substrate preference. The results summarized previously emphasize the fact that the substrate preference of at least some GAPs is not constant but depends on several factors, such as the lipid environment and the posttranslational modification of the GTPase. Recently, phosphorylation and autoinhibition have also been identified as modifying factors in the cases of mgcRacGAP (Minoshima *et al.*, 2003) and p50RhoGAP (Moskwa *et al.*, 2005), respectively. Modification of the phosphorylation state of proteins, release of autoinhibition by protein–protein or protein–lipid interactions, changes in the composition of the lipid environment, or translocation of a protein to a different lipid environment are all well-known factors that influence intracellular signaling. It is less well known that in mammalian cells the activity of prenyl transferases is hormonally regulated, and significant changes do occur in the prenylation state of small GTPases (Finlayson *et al.*, 2003; Goalstone *et al.*, 2001). In conclusion, we suggest that considerable care should be taken when speculating on the substrate specificity of particular GAPs simply on the basis of *in vitro* assays, especially when using nonprenylated GTPases.

Acknowledgments

The authors acknowledge Drs. Marie-Claire Dagher, Patryk Moskwa, and Samuel Hernandez for reagents and for technical and intellectual contributions. Experimental work in the authors' laboratories was supported by grants from the Hungarian Research Fund (OTKA T37755 and Ts040865) to E. L., and by NIH grants R03 TW006421 and RO1 CA62142 to J. S.

References

Agnel, M., Roder, L., Vola, C., and Griffin-Shea, R. (1992). A *Drosophila* rotund transcript expressed during spermatogenesis and imaginal disc morphogenesis encodes a protein which is similar to human Rac GTPase-activating (racGAP) proteins. *Mol. Cell Biol.* **12**, 5111–5122.

Barfod, E. T., Zheng, Y., Kuang, W. J., Hart, M. J., Evans, T., Cerione, R. A., and Ashkenazi, A. (1993). Cloning and expression of a human CDC42 GTPase-activating protein reveals a functional SH3-binding domain. *J. Biol. Chem.* **268**, 26059–26062.

Bernards, A. (2003). GAPs galore! A survey of putative Ras superfamily GTPase activating proteins in man and *Drosophila*. *Biochim. Biophys. Acta* **1603**, 47–82.

Bernards, A., and Settleman, J. (2004). GAP control: Regulating the regulators of small GTPases. *Trends Cell Biol.* **14**, 377–385.

Bourne, H. R., Sanders, D. A., and McCormick, F. (1990). The GTPase superfamily: A conserved switch for diverse cell functions. *Nature* **348**, 125–132.

Bourne, H. R., Sanders, D. A., and McCormick, F. (1991). The GTPase superfamily: Conserved structure and molecular mechanism. *Nature* **349**, 117–127.

Brown, M. T., Andrade, J., Radhakrishna, H., Donaldson, J. G., Cooper, J. A., and Randazzo, P. A. (1998). ASAP1, a phospholipid-dependent arf GTPase-activating protein that associates with and is phosphorylated by Src. *Mol. Cell Biol.* **18,** 7038–7051.

Chuang, T. H., Xu, X., Kaartinen, V., Heisterkamp, N., Groffen, J., and Bokoch, G. M. (1995). Abr and Bcr are multifunctional regulators of the Rho GTP-binding protein family. *Proc. Natl. Acad. Sci. USA* **92,** 10282–10286.

Di-Poi, N., Faure, J., Grizot, S., Molnar, G., Pick, E., and Dagher, M. C. (2001). Mechanism of NADPH oxidase activation by the Rac/Rho-GDI complex. *Biochemistry* **40,** 10014–10022.

Diekmann, D., Brill, S., Garrett, M. D., Totty, N., Hsuan, J., Monfries, C., Hall, C., Lim, L., and Hall, A. (1991). Bcr encodes a GTPase-activating protein for p21rac. *Nature* **351,** 400–402.

Finlayson, C. A., Chappell, J., Leitner, J. W., Goalstone, M. L., Garrity, M., Nawaz, S., Ciaraldi, T. P., and Draznin, B. (2003). Enhanced insulin signaling via Shc in human breast cancer. *Metabolism* **52,** 1606–1611.

Goalstone, M. L., Leitner, J. W., Berhanu, P., Sharma, P. M., Olefsky, J. M., and Draznin, B. (2001). Insulin signals to prenyltransferases via the Shc branch of intracellular signaling. *J. Biol. Chem.* **276,** 12805–12812.

Hildebrand, J. D., Taylor, J. M., and Parsons, J. T. (1996). An SH3 domain-containing GTPase-activating protein for Rho and Cdc42 associates with focal adhesion kinase. *Mol. Cell Biol.* **16,** 3169–3178.

Jullien-Flores, V., Dorseuil, O., Romero, F., Letourneur, F., Saragosti, S., Berger, R., Tavitian, A., Gacon, G., and Camonis, J. H. (1995). Bridging Ral GTPase to Rho pathways. RLIP76, a Ral effector with CDC42/Rac GTPase-activating protein activity. *J. Biol. Chem.* **270,** 22473–22477.

Lancaster, C. A., Taylor-Harris, P. M., Self, A. J., Brill, S., van Erp, H. E., and Hall, A. (1994). Characterization of rhoGAP. A GTPase-activating protein for rho-related small GTPases. *J. Biol. Chem.* **269,** 1137–1142.

Ligeti, E., Dagher, M. C., Hernandez, S. E., Koleske, A. J., and Settleman, J. (2004). Phospholipids can switch the GTPase substrate preference of a GTPase-activating protein. *J. Biol. Chem.* **279,** 5055–5058.

Minoshima, Y., Kawashima, T., Hirose, K., Tonozuka, Y., Kawajiri, A., Bao, Y. C., Deng, X., Tatsuka, M., Narumiya, S., May, W. S., Jr., Nosaka, T., Semba, K., Inoue, T., Satoh, T., Inagaki, M., and Kitamura, T. (2003). Phosphorylation by aurora B converts MgcRacGAP to a RhoGAP during cytokinesis. *Dev. Cell* **4,** 549–560.

Moskwa, P., Paclet, M. H., Dagher, M. C., and Ligeti, E. (2005). Autoinhibition of p50 Rho GTPase-activating protein (GAP) is released by prenylated small GTPases. *J. Biol. Chem.* **280,** 6716–6720.

Park, S. H., and Weinberg, R. A. (1995). A putative effector of Ral has homology to Rho/Rac GTPase activating proteins. *Oncogene* **11,** 2349–2355.

Prakash, S. K., Paylor, R., Jenna, S., Lamarche-Vane, N., Armstrong, D. L., Xu, B., Mancini, M. A., and Zoghbi, H. Y. (2000). Functional analysis of ARHGAP6, a novel GTPase-activating protein for RhoA. *Hum. Mol. Genet.* **9,** 477–488.

Ren, X. R., Du, Q. S., Huang, Y. Z., Ao, S. Z., Mei, L., and Xiong, W. C. (2001). Regulation of CDC42 GTPase by proline-rich tyrosine kinase 2 interacting with PSGAP, a novel pleckstrin homology and Src homology 3 domain containing rhoGAP protein. *J. Cell Biol.* **152,** 971–984.

Richnau, N., and Aspenstrom, P. (2001). Rich, a rho GTPase-activating protein domain-containing protein involved in signaling by Cdc42 and Rac1. *J. Biol. Chem.* **276,** 35060–35070.

Ridley, A. J., Self, A. J., Kasmi, F., Paterson, H. F., Hall, A., Marshall, C. J., and Ellis, C. (1993). Rho family GTPase activating proteins p190, Bcr and RhoGAP show distinct specificities *in vitro* and *in vivo*. *EMBO J.* **12,** 5151–5160.

Scheffzek, K., Ahmadian, M. R., Kabsch, W., Wiesmülller, L., Lautwein, A., Schmitz, F., and Wittinghofer, A. (1997). The Ras-RasGAP complex: Structural basis for GTPase activation and its loss in oncogenic Ras mutants. *Science* **277,** 333–338.

Self, A. J., and Hall, A. (1995). Purification of recombinant Rho/Rac/G25K from *Escherichia coli*. *Methods Enzymol.* **256,** 3–10.

Shibata, H., Oishi, K., Yamagiwa, A., Matsumoto, M., Mukai, H., and Ono, Y. (2001). PKNbeta interacts with the SH3 domains of Graf and a novel Graf related protein, Graf2, which are GTPase activating proteins for Rho family. *J. Biochem. (Tokyo)* **130,** 23–31.

Toure, A., Dorseuil, O., Morin, L., Timmons, P., Jegou, B., Reibel, L., and Gacon, G. (1998). MgcRacGAP, a new human GTPase-activating protein for Rac and Cdc42 similar to Drosophila rotundRacGAP gene product, is expressed in male germ cells. *J. Biol. Chem.* **273,** 6019–6023.

Voncken, J. W., van Schaick, H., Kaartinen, V., Deemer, K., Coates, T., Landing, B., Pattengale, P., Dorseuil, O., Bokoch, G. M., Groffen, J., and Heisterkamp, N. (1995). Increased neutrophil respiratory burst in Bcr-null mutants. *Cell* **80,** 719–728.

Wong, K., Ren, X. R., Huang, Y. Z., Xie, Y., Liu, G., Saito, H., Tang, H., Wen, L., Brady-Kalnay, S. M., Mei, L., Wu, J. Y., Xiong, W. C., and Rao, Y. (2001). Signal transduction in neuronal migration: Roles of GTPase activating proteins and the small GTPase Cdc42 in the Slit-Robo pathway. *Cell* **107,** 209–221.

[10] Purification and Activity of the Rho ADP-Ribosylating Binary C2/C3 Toxin

By Gerd Haug, Holger Barth, and Klaus Aktories

Abstract

C3 exoenzyme from *Clostridium limosum,* specifically ADP-ribosylates and inactivates Rho GTPases, but not or much less than Rac and Cdc42. To bypass the poor cell accessibility of the exoenzyme, a chimeric fusion toxin was constructed consisting of C3 exoenzyme and the N-terminal adaptor domain of the enzyme component C2I of the actin–ADP-ribosylating *Clostridium botulinum* C2 toxin. This fusion toxin C2IN-C3 is transported into cells by interaction with the binding and translocation component (C2II) of C2 toxin. Purification and activity of the chimeric toxin is reported.

Introduction

Rho proteins encompass a family of approximately 20 GTPases, which act as molecular switches to control a large array of cellular processes, including organization of the actin cytoskeleton, cell polarity, transcriptional activation,

METHODS IN ENZYMOLOGY, VOL. 406
0076-6879/06 $35.00
DOI: 10.1016/S0076-6879(06)06010-1

cell cycle progression, and many other cellular functions (Bishop and Hall, 2000; Burridge and Wennerberg, 2004; Etienne-Manneville and Hall, 2002; Wennerberg and Der, 2004). Rho proteins are targets of various bacterial protein toxins, which specifically activate or inactivate the GTPases (Lerm et al., 2000). These bacterial toxins turned out to be extremely helpful to study the cellular functions of Rho GTPases. Whereas clostridial glucosylating toxins like Clostridium difficile toxins A and B are used to inactivate many members of the Rho family, including Rho, Rac, and Cdc42 subtypes; the ADP-ribosylating exoenzyme of the C3 family especially exhibits a high selectivity for Rho subtypes GTPases (Aktories et al., 2004).

C3 exoenzymes are ~25 kDa proteins with 35–75% sequence identity, which are produced by C. botulinum, C. limosum, Staphylococcus aureus, and Bacillus cereus (Aktories et al., 1987, 2004; Rubin et al., 1988). They interrupt Rho signaling by mono-ADP-ribosylation of Rho proteins at position asparagine-41 (Sekine et al., 1989). ADP-ribosylated RhoA is biologically inactive, because it is trapped in the Rho–GDI complex, and GEF-induced activation is blocked (Genth et al., 2003; Sehr et al., 1998).

Exotoxins that modify cytosolic target proteins usually consist of an enzyme domain (A domain) and a binding/translocation domain (B domain), facilitating cell entry of the A domain. So far, no B domain was identified for C3 transferases. Accordingly, entry of the exoenzymes into eukaryotic cells is poor. Until recently, inhibition of Rho in living cells by C3 transferases was achieved by either using microinjection protocols or by long-term incubation (at least several hours) of cells with high amounts of C3 protein. To increase the cell permeability of C3 transferases, the ability of other protein toxins was exploited to enter cells readily. To this end, chimeric fusion toxins have been constructed, consisting of the binding/translocation domain of a highly cell permeable toxin (e.g., diphtheria toxin) and the Rho modifying exoenzyme C3 (Aullo et al., 1993). Particularly useful was the design of a cell-permeable C3 fusion toxin on the basis of the delivery system of the binary C. botulinum C2 toxin. This article focuses on the purification and application of the C2IN-C3 fusion toxin in cell culture experiments.

The C2 Toxin as a Cell Delivery System

The binary actin ADP-ribosylating C2 toxin from C. botulinum consists of the enzyme component C2I (49.4 kDa) that ADP-ribosylates G-actin and the binding/translocation component C2II (80.5 kDa), which enables

C2I to enter the cytosol of target cells. C2II is activated by proteolytic cleavage (Ohishi, 1987). An ~20-kDa fragment is released from the N terminus. The resulting C2IIa protein (59.8 kDa) forms heptamers (Barth *et al.*, 2000) that bind to its cellular receptor, which represents complex and hybrid carbohydrates (Eckhardt *et al.*, 2000). C2I assembles to C2IIa heptamers, and the complex is taken up by means of receptor-mediated endocytosis. In acidic endosomes, C2IIa inserts into the membrane of endosomes, forms pores (Blocker *et al.*, 2003), and facilitates translocation of C2I into the cytosol (Fig. 1A). The enzyme activity of C2I is harbored within the C-terminal domain of C2I (Barth *et al.*, 1998b). The N-terminal 225 amino acid residues of the C2I protein are required for interaction with the binding/transport component C2IIa. To exploit cell entry of C2 toxin for uptake of C3, a fusion protein was constructed (Fig. 1B) consisting of the amino acid residues 1–225 of C2I (C2IN) and the C3 transferase of *C. limosum* (Barth *et al.*, 1998a).

Expression and Purification of Recombinant C2IN-C3 and C2II Proteins

Cloning of the C2IN-C3 fusion toxin was described in detail (Barth *et al.*, 1998a). Recombinant GST fusion proteins are produced in *E. coli* (TG1 for C2IN-C3 and BL21 for C2II) transformed with the respective DNA fragment inserted in the pGEX2T vector. *Escherichia coli* harboring plasmid pGEX2T-C2IN-C3 or pGEX2T-C2II is grown at 37° in Luria Bertani (LB) medium, containing ampicillin (final concentration 100 μg/l) to an optical density of 0.8 at 600 nm. Isopropyl-β-D-thiogalactopyranoside (IPTG) is added to a final concentration of 0.2 mM and cultures are incubated at 29° for an additional 20 h. Cells are sedimented (10 min, 4°, 6000g), pellet is resuspended in lysis buffer (10 ml/l culture) and disrupted by sonication (strokes of 3 × 1 min, cycl. 70, 80% power, interrupted by incubation on ice for 1 min; Sonopuls HD60, Bandelin Berlin). Cellular debris is sedimented (20 min, 4°, 20,000g), and the resulting supernatant is incubated (end-over-end rotation) for 60 min at room temperature with glutathione-Sepharose 4B (2-ml bead volume per 100 ml of suspension) to allow binding of GST fusion proteins to glutathione. The suspension is centrifuged (5 min, 1000g), and the beads are subsequently washed two times with washing buffer (10 bed volumes) and three times with PBS (5 bed volumes). To cleave GST fusion proteins from GST and to release recombinant proteins from the matrix, beads are incubated (end-over-end rotation) in PBS with thrombin (3.25 NIH units/ml bead suspensions) for 1 h at room temperature. Then, the suspension is centrifuged, and the supernatant is incubated (end-over-end rotation) with benzamidine beads

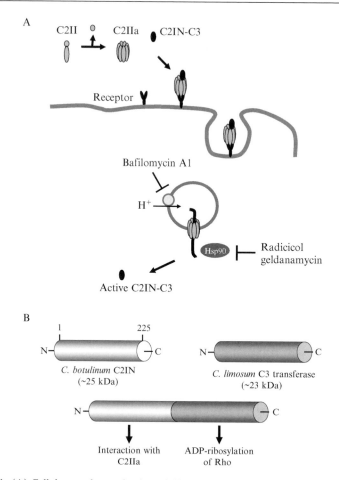

FIG. 1. (A) Cellular uptake mechanism of *Clostridium botulinum* C2 toxin and C2IN-C3. After proteolytic activation, C2IIa forms heptamers and binds to the receptor on the cell surface. C2I assembles and the toxin-receptor complex is taken up by means of receptor-mediated endocytosis. In acidic endosomes, C2IIa inserts into the membrane, forms pores, and C2I is translocated into the cytosol. Bafilomycin A1 is an inhibitor of the v-ATPase in the endosomal membrane and prevents acidification of the endosomes. Translocation is also dependent on the cellular chaperone Hsp90, which can be inhibited or delayed with the specific inhibitors geldanamycin or radicicol. (B) The fusion toxin C2IN-C3. The N-terminal domain of C2I (C2IN; amino acid residues 1-225) was used as an adaptor to deliver the C3 exoenzyme from *Clostridium limosum* into cells. C2IN interacts with C2IIa and is sufficient for translocation of fusion proteins into the cytosol.

(100 μl bead volume, 10 min, 4°) to inactivate and remove thrombin. Thrombin treated GST beads are incubated for another two times with PBS (without further thrombin supplementation) to elute remaining bead-associated protein, and the resulting fractions are also incubated together with the benzamidine beads that were used before. After centrifugation, the resulting supernatant is analyzed by SDS-PAGE. The resulting recombinant proteins do not require any further purification steps. The proteins should be stored in small aliquots at −20 or −80° to prevent inactivation. The enzyme activity of the C2IN-C3 protein decreases when stored at 4° for more than 3 days.

Activation of C2II

The binding and translocation component C2II is not able to mediate cellular uptake of C2I without proteolytic activation. For activation *in vitro*, C2II is incubated with trypsin (50 μg/ml C2II protein solution) for 30 min at 37° and then kept on ice. Complete activation of C2II is examined in a subsequent SDS-PAGE, because C2II and C2IIa proteins show different molecular masses. When all C2II has been cleaved to its activated form C2IIa ($\sim$60 kDa), soybean trypsin inhibitor (75 μg/ml C2II/ trypsin protein solution) is added to the protein solution to prevent further protein degradation by trypsin. C2IIa should immediately be stored in aliquots at −20 or −80°. When stored at 4° for up to 3 days, precipitation of C2IIa may occur, but precipitation can be dissolved without any detectable loss of activity.

[^{32}P]Labeling of Rho by the C3 Fusion Toxin *In Vitro*

Similar to C3 exoenzyme, Rho GTPases can be radioactively labeled by ADP-ribosylation with the C2N/C3 fusion toxin. The assay protocol for C3 was described in detail earlier (Aktories and Just, 1995). Recombinant Rho protein (0.5–1 μg) or cell lysate (20–150 μg) is incubated in a buffer (total assay volume 50 μl), containing 50 mM triethanolamine hydrochloride (pH 7.5), 2 mM MgCl$_2$, 1 mM EDTA, 1 mM dithiothreitol (DTT), 0.2 mM PMSF, 0.2–1 μM [^{32}P]NAD (approximately 0.2–0.5 μCi), and 0.15–1 μg/ml C3 or C2IN-C3 at 37° for 5–30 min. Labeled protein is analyzed by SDS-PAGE and autoradiography. Alternately, the use of a buffer containing

50 mM HEPES, 2 mM MgCl$_2$, and 100 μg/ml bovine serum albumin for ADP-ribosylation was described (Wilde *et al.*, 2001).

The specificity of the fusion toxin seems to be the same as that from the respective C3 wild-type exoenzyme. C3 from *C. limosum*, which is most often used as the enzyme part for the fusion toxin, exhibits strong activity toward RhoA, RhoB, and RhoC. Cdc42 is poorly modified by C3lim *in vitro* (Just *et al.*, 1992). Rac, which is a poor substrate for C3 exoenzyme from *C. botulinum*, is not modified by C3lim (Just *et al.*, 1992). RhoE and Rnd3, which are modified by C3staul-3 from *S. aureus*, are also no substrates for C3lim (Wilde *et al.*, 2001). Recently, it was reported that C3bot and C3lim interact with high affinity with Ral (K$_D$ ~10 nM) *in vitro* without ADP-ribosylating the GTPase (Wilde *et al.*, 2002). So far, it is not clear whether this interaction has any biological consequence for studies with intact cells.

Inactivation of Rho in Cultured Monolayer Cells

For intoxication and inactivation of Rho, subconfluent monolayer cells are incubated with C2IIa (200 ng/ml) plus C2IN-C3 (100 ng/ml) in cell-specific medium. At these concentrations, nearly all cells are intoxicated after an incubation period of 2–3 h. When the intoxication efficiency of the combination of C2IN-C3 and C2IIa is tested in comparison with C3 exoenzyme in Chinese-Hamster-Ovary cells, no morphological changes are observed within 3 h with up to 30 μg/ml of *C. limosum* C3 exoenzyme, whereas CHO cells treated with the combination of C2IN-C3 and C2IIa start to round up after 60 min. Approximately 120 min after toxin addition, more than 50% of cells exhibit the typical "C3-morphology" (i.e., contraction of cell bodies and prolonged extensions) (Barth *et al.*, 1998a). At early time points, HeLa cells seem to be less sensitive to the toxin, but after 3 h of incubation, 80% of the cells are affected (Fig. 2A). The potency of C2IN-C3 is increased more than 300-fold compared with C3 exoenzyme of *C. limosum* (Barth *et al.*, 1998a). For most cell types, 100 ng/ml C2IN-C3 in combination with 200 ng/ml C2IIa are sufficient to obtain an efficient intoxication of subconfluently growing monolayer cells within 2–3 h (Barth *et al.*, 1999, 2002b). Usually, the toxin is more efficient when cell culture medium is used without fetal calf serum.

In most cells (e.g., Vero cells, Hela cells, NIH-3T3 fibroblasts), the inactivation of Rho by C3 can be followed by monitoring the morphological changes of cultured cells. Because of inhibition of Rho signaling, the actin cytoskeleton is disrupted, and drastic morphological changes ("C3-morphology") occur (Fig. 2B). Intoxication of cells by C2IN-C3/

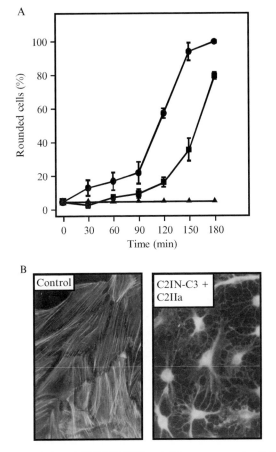

FIG. 2. Cytotoxic activity of C2IN-C3/C2IIa. A Time course of cell intoxication with C2IN-C3/C2IIa (Barth *et al.*, 1998a). CHO or HeLa cells were treated with C2IN-C3 (200 ng/ml) together with C2IIa (200 ng/ml) or with *Clostridium limosum* C3 exoenzyme (30 μg/ml) and incubated at 37°. Every 30 min, the number of rounded cells was determined (●, CHO + C2IN-C3/C2IIa; ■, HeLa + C2IN-C3/C2IIa; ▲, CHO/HeLa + C3). (B) Cytotoxic activity of C2IN–C3 fusion toxin on rat astrocytes (Barth *et al.*, 1999). Rat astrocytes were incubated without (control) or with C2IN-C3 (100 ng/ml) together with C2IIa (200 ng/ml) for 90 min at 37°. Cells were fixed and actin filaments were stained with rhodamine-phalloidin.

C2IIa is reversible when the fusion toxin is removed from the medium, and cells regenerate because of both neosynthesis of Rho protein in the cells and proteasomal degradation of the fusion toxin (Barth *et al.*, 1999). This has to be considered when data are compared with those obtained with wild-type C3 transferases. Depending on the concentration, the fusion toxin acts as a transient inactivator of Rho and does not lead to cell death but to a complete recovery of the actin cytoskeleton and cell morphology.

The level of ADP-ribosylated and, thereby, inactivated cellular Rho can be determined in an *in vitro* assay by "sequential ADP-ribosylation." After incubation of cells with or without C2IN-C3/C2IIa (100 ng/ml and 200 ng/ml) for 2 h, cells are washed and lysed in ADP-ribosylation buffer. Cell lysate (50 μg of protein) is ADP-ribosylated in the presence of C2IN-C3 (300 μg/ml) and [^{32}P]NAD (0.5 μM; 1 nCi/μl). Lysate of control cells contains nonmodified RhoA, which serves as substrate for C2IN-C3 and is radioactively labeled with [^{32}P]ADP-ribose *in vitro*. By contrast, in lysate from cells previously treated with C2IN-C3 in combination with C2II, Rho proteins are already ADP-ribosylated during incubation of intact cells and, therefore, are not radioactively modified (Barth *et al.*, 1998a). Moreover, ADP-ribosylated Rho can be monitored by an

TABLE I
STUDIES USING C2IN-C3 FUSION TOXIN AS A TOOL FOR STUDYING THE ROLE OF RHO GTPASE IN CELLULAR PROCESSES (BARTH *ET AL.*, 2002A)

Role of Rho in activation of volume-regulated anion channels	Nilius *et al.*, 1999
Role of Rho in Ephrin-A5–induced neuronal growth cone collapse	Wahl *et al.*, 2000
Role of Rho in secretion of van Willebrand factor	Vischer *et al.*, 2000
Role of Rho in aquaporin 2 translocation	Klussmann *et al.*, 2001
Role of Rho GTPases in IL-1 receptor-induced signaling	Dreikhausen *et al.*, 2001
Role of Rho GTPases in cerebellar granule neuronal survival	Linseman *et al.*, 2001
Role of Rho GTPases in uptake and degradation of lipoproteins by macrophages	Sakr *et al.*, 2001
Role of Rho GTPases in cyclo-oxygenase-2-gene expression	Hahn *et al.*, 2002
Role of Rho in monocyte transendothelial migration	Strey *et al.*, 2002
Role of Rho GTPases in formation of branched dendrites	Leemhuis *et al.*, 2004

upwards shift in 12.5% SDS-PAGE detected with anti-RhoA antibody (Barth *et al.*, 1999).

Concluding Remarks

C3 exoenzymes have been widely used as cell biological tools to study the functions of Rho GTPases. However, their applications are hampered by low cell accessibility. Use of C3 exoenzymes as C2IN-C3 fusion toxins together with the binding and translocation component C2II is able to circumvent this disadvantage. Because C2II binds to all mammalian cells tested so far, this delivery system is applicable to a wide range of cells. Studies using the C2IN-C3 fusion toxin are listed in Table I. Compared with the treatment of cells with C3 exoenzymes, one has to consider that the C2IN-C3 fusion delivery system may result in reversible effects.

Appendix

Reagents

1. Lysis buffer: 10 mM NaCl, 20 mM Tris-HCl, pH 7.4, 0.1% (v/v) Triton X-100 and 1 mM phenylmethylsulfonyl fluoride (PMSF).
2. Washing buffer: 20 mM Tris, pH 7.4 and 150 mM NaCl
3. Phosphate buffered saline (PBS): 8 g/l NaCl, 0.2 g/l KCl, 1.15 g/l Na_2HPO_4 and 0.2 g/l KH_2PO_4.
4. ADP-ribosylation buffer: 50 mM HEPES pH 7.5, 1 mM DTT, 10 mM thymidine and "Complete" (protease inhibitor cocktail).

References

Aktories, K., and Just, I. (1995). *In vitro* ADP-ribosylation of Rho by bacterial ADP-ribosyltransferases. *Methods Enzymol.* **256,** 184–195.

Aktories, K., Weller, U., and Chatwal, G. S. (1987). *Clostridium botulinum* type C produces a novel ADP-ribosyltransferase distinct from botulinum C2 toxin. *FEBS Lett.* **212,** 109–113.

Aktories, K., Wilde, C., and Vogelsgesang, M. (2004). Rho-modifying C3-like ADP-ribosyltransferases. *Rev. Physiol Biochem. Pharmacol.* **152,** 1–22.

Aullo, P., Giry, M., Olsnes, S., Popoff, M. R., Kocks, C., and Boquet, P. (1993). A chimeric toxin to study the role of the 21 kDa GTP binding protein rho in the control of actin microfilament assembly. *EMBO J.* **12,** 921–931.

Barth, H., Blöcker, D., and Aktories, K. (2002a). The uptake machinery of clostridial actin ADP-ribosylating toxins - a cell delivery system for fusion proteins and polypeptide drugs. *Naunyn-Schmiedeberg's Arch. Pharmacol.* **366,** 501–512.

Barth, H., Blöcker, D., Behlke, J., Bergsma-Schutter, W., Brisson, A., Benz, R., and Aktories, K. (2000). Cellular uptake of *Clostridium botulinum* C2 toxin requires oligomerization and acidification. *J. Biol. Chem.* **275**, 18704–18711.

Barth, H., Hofmann, F., Olenik, C., Just, I., and Aktories, K. (1998a). The N-terminal part of the enzyme component (C2I) of the binary *Clostridium botulinum* C2 toxin interacts with the binding component C2II and functions as a carrier system for a Rho ADP-ribosylating C3-like fusion toxin. *Infect. Immun.* **66**, 1364–1369.

Barth, H., Olenik, C., Sehr, P., Schmidt, G., Aktories, K., and Meyer, D. K. (1999). Neosynthesis and activation of Rho by *Escherichia coli* cytotoxic necrotizing factor (CNF1) reverse cytopathic effects of ADP-ribosylated Rho. *J. Biol. Chem.* **274**, 27407–27414.

Barth, H., Preiss, J. C., Hofmann, F., and Aktories, K. (1998b). Characterization of the catalytic site of the ADP-ribosyltransferase *Clostridium botulinum* C2 toxin by site-directed mutagenesis. *J. Biol. Chem.* **273**, 29506–29511.

Barth, H., Roebling, R., Fritz, M., and Aktories, K. (2002b). The binary *Clostridium botulinum* C2 toxin as a protein delivery system: Identification of the minimal protein region necessary for interaction of toxin components. *J. Biol. Chem.* **277**, 5074–5081.

Bishop, A. L., and Hall, A. (2000). Rho GTPases and their effector proteins. *Biochem. J.* **348**, 241–255.

Blocker, D., Pohlmann, K., Haug, G., Bachmeyer, C., Benz, R., Aktories, K., and Barth, H. (2003). *Clostridium botulinum* C2 toxin: Low pH-induced pore formation is required for translocation of the enzyme component C2I into the cytosol of host cells. *J. Biol. Chem.* **278**, 37360–37367.

Burridge, K., and Wennerberg, K. (2004). Rho and Rac take center stage. *Cell* **116**, 167–179.

Dreikhausen, U., Varga, G., Hofmann, F., Barth, H., Aktories, K., Resch, K., and Szamel, M. (2001). Regulation by rho family GTPases of IL-1 receptor induced signaling: C3-like chimeric toxin and *Clostridium difficile* toxin B inhibit signaling pathways involved in IL-2 gene expression. *Eur. J. Immmunol.* **31**, 1610–1619.

Eckhardt, M., Barth, H., Blöcker, D., and Aktories, K. (2000). Binding of *Clostridium botulinum* C2 toxin to asparagine-linked complex and hybrid carbohydrates. *J. Biol. Chem.* **275**, 2328–2334.

Etienne-Manneville, S., and Hall, A. (2002). Rho GTPases in cell biology. *Nature* **420**, 629–635.

Genth, H., Gerhard, R., Maeda, A., Amano, M., Kaibuchi, K., Aktories, K., and Just, I. (2003). Entrapment of Rho ADP-ribosylated by *Clostridium botulinum* C3 exoenzyme in the Rho-guanine nucleotide dissociation inhibitor-1 complex. *J. Biol. Chem.* **278**, 28523–28527.

Hahn, A., Barth, H., Kress, M., Mertens, P. R., and Goppelt-Strübe, M. (2002). Role of Rac and Cdc42 in lysophosphatidic acid-mediated cyclooxygenase-2 gene expression. *Biochem. J.* **362**, 33–40.

Just, I., Mohr, C., Schallehn, G., Menard, L., Didsbury, J. R., Vandekerckhove, J., van Damme, J., and Aktories, K. (1992). Purification and characterization of an ADP-ribosyltransferase produced by *Clostridium limosum*. *J. Biol. Chem.* **267**, 10274–10280.

Klussmann, E., Tamma, G., Lorenz, D., Wiesner, B., Maric, K., Hofmann, F., Aktories, K., Valenti, G., and Rosenthal, W. (2001). An inhibitory role of Rho in the vasopressin-mediated translocation of aquaporin-2 into cell membranes of renal principal cells. *J. Biol. Chem.* **276**, 20451–20457.

Leemhuis, J., Boutillier, S., Barth, H., Feuerstein, T. J., Brock, C., Nürnberg, B., Aktories, K., and Meyer, D. K. (2004). Rho GTPases and Phosphoinositide 3-kinase organize formation of branched dendrites. *J. Biol. Chem.* **279**, 585–596.

Lerm, M., Schmidt, G., and Aktories, K. (2000). Bacterial protein toxins targeting Rho GTPases. *FEMS Microbiol. Lett.* **188**, 1–6.

Linseman, D. A., Laessig, T., Meintzer, M. K., McClure, M., Barth, H., Aktories, K., and Heidenreich, K. A. (2001). An essential role for Rac/Cdc42 GTPases in cerebellar granule neuron survival. *J. Biol. Chem.* **276**, 39123–39131.

Nilius, B., Voets, T., Prenen, J., Barth, H., Aktories, K., Kaibuchi, K., Droogmans, G., and Eggermont, J. (1999). Role of Rho and Rho kinase in the activation of volume-regulated anion channels in bovine endothelial cells. *J. Physiol.* **516**, 67–74.

Ohishi, I. (1987). Activation of botulinum C2 toxin by trypsin. *Infect. Immun.* **55**, 1461–1465.

Rubin, E. J., Gill, D. M., Boquet, P., and Popoff, M. R. (1988). Functional modification of a 21-Kilodalton G protein when ADP- ribosylated by exoenzyme C3 of *Clostridium botulinum. Mol. Cell. Biol.* **8**, 418–426.

Sakr, S., Eddy, R., Barth, H., Xie, B., Greenberg, S., Maxfield, F. R., and Tabas, I. (2001). An *in-vitro* model for the uptake and degradation of lipoproteins by arterial-wall macrophages during atherogenesis. *J. Biol. Chem.* **276**, 37649–37658.

Sehr, P., Joseph, G., Genth, H., Just, I., Pick, E., and Aktories, K. (1998). Glucosylation and ADP-ribosylation of Rho proteins—Effects on nucleotide binding, GTPase activity, and effector-coupling. *Biochemistry* **37**, 5296–5304.

Sekine, A., Fujiwara, M., and Narumiya, S. (1989). Asparagine residue in the rho gene product is the modification site for botulinum ADP-ribosyltransferase. *J. Biol. Chem.* **264**, 8602–8605.

Strey, A., Janning, A., Barth, H., and Gerke, V. (2002). Endothelial Rho signaling is required for monocyte transendothelial migration. *FEBS Lett.* **517**, 261–266.

Vischer, U. M., Barth, H., and Wollheim, C. B. (2000). Regulated von Willebrand factor secretion is associated with agonist-specific patterns of cytoskeletal remodeling in cultured endothelial cells. *Arterioscler. Thromb. Vasc. Biol.* **20**, 883–891.

Wahl, S., Barth, H., Ciossek, T., Aktories, K., and Mueller, B. K. (2000). Ephrin-A5 induces collapse of growth cones by activating Rho and Rho kinase. *J. Cell Biol.* **149**, 263–270.

Wennerberg, K., and Der, C. J. (2004). Rho-family GTPases: It's not only Rac and Rho (and I like it). *J. Cell Sci.* **117**, 1301–1312.

Wilde, C., Barth, H., Sehr, P., Han, L., Schmidt, M., Just, I., and Aktories, K. (2002). Interaction of the Rho-ADP-ribosylating C3 exoenzyme with RalA. *J. Biol. Chem.* **277**, 14771–14776.

Wilde, C., Chatwal, G. S., Schmalzing, G., Aktories, K., and Just, I. (2001). A novel C3-like ADP-ribosyltransferase from *Staphylococcus aureus* modifying RhoE and Rnd3. *J. Biol. Chem.* **276**, 9537–9542.

[11]   Purification of TAT-C3 Exoenzyme

By ERIK SAHAI and MICHAEL F. OLSON

Abstract

The *Clostridium botulinum* C3 exoenzyme has been an invaluable tool for the study of the biological functions of Rho GTPases. The C3 enzyme selectively catalyzes the ADP-ribosylation, and consequent inactivation, of RhoA, RhoB, and RhoC of the Rho GTPase protein family. Through the experimental use of C3, it has been possible to determine the contributions made by these signaling proteins to processes including the regulation of cell morphology, cell cycle progression, and gene transcription. Unlike bacterial toxins that have some means to attach to and/or enter cells, C3 does not have an element that facilitates efficient entry. As a result, numerous methods have been used to effectively deliver C3 into cells. One approach has been to engineer a recombinant C3 with an HIV TAT leader sequence that permits transduction of the protein across the plasma membrane. In this chapter, the purification and characterization of the recombinant TAT-C3 protein is described.

Introduction

A number of proteins produced by bacteria are capable of disrupting the actin cytoskeleton in target cells (Barbieri *et al.*, 2002). One way that actin filament disruption can be achieved is through the modification and inhibition of Rho GTPases (Aktories *et al.*, 2004). The *Clostridium botulinum* C3 exoenzyme (unlike a bacterial toxin, C3 lacks a cell-binding domain that would facilitate efficient entry) targets RhoA, RhoB, and RhoC and catalyzes the transfer of ADP-ribose from NAD onto asparagine at position 41. Although other members of the Rho family, including Rac and Cdc42, contain asparagine at the analogous position, C3 catalyzed ADP-ribosylation is extremely inefficient unless the protein has been denatured, suggesting that target protein conformation influences specificity. Once a Rho GTPase has been ADP-ribosylated, not only is the activity of the modified protein compromised, but it seems to exert a "dominant-negative" effect on Rho signaling pathways (Paterson *et al.*, 1990).

The C3 exoenzyme (also called C3 transferase) has been extensively used to examine the role of Rho GTPases in biological process such as proliferation (Olson *et al.*, 1995), gene transcription (Hill *et al.*, 1995; Sahai

METHODS IN ENZYMOLOGY, VOL. 406
0076-6879/06 $35.00
DOI: 10.1016/S0076-6879(06)06011-3

et al., 1998), and apoptosis (Coleman *et al.*, 2001). However, given that C3 does not possess the intrinsic ability to enter cells, it has been necessary to use active methods to deliver recombinant protein including microinjection (Paterson *et al.*, 1990), electroporation (Beltman *et al.*, 1999), or by means of liposomes (Borbiev *et al.*, 2000). To overcome this shortcoming, we engineered a bacterial expression plasmid that encodes for C3 with a HIV TAT protein transduction domain (PTD) (Coleman *et al.*, 2001; Sahai *et al.*, 2001). The TAT PTD promotes protein entry by means of macropinocytosis (Wadia *et al.*, 2004) and has been used to deliver a large variety of proteins and peptides into cells (Wadia and Dowdy, 2005). The TAT PTD allows for virtually uniform transduction of target cells, which permits biochemical studies of Rho GTPase function that previously would not have been possible. TAT-C3 protein has been used to assess the contribution of Rho to a variety of biological functions, including apoptotic membrane blebbing (Coleman *et al.*, 2001; Sebbagh *et al.*, 2001), oncogenic transformation (Sahai *et al.*, 2001), adherens junction formation (Sahai and Marshall, 2002), smooth muscle contraction and proliferation (Sauzeau *et al.*, 2001), neurite growth (Monnier *et al.*, 2003), cardiac myocyte morphology (Grounds *et al.*, 2005), vascular neointimal formation (Guerin *et al.*, 2005), ezrin phosphorylation (Croft *et al.*, 2004), and tumor cell invasion (Sahai and Marshall, 2003; Vial *et al.*, 2003).

In this Chapter, we outline the purification and characterization of TAT-C3. The increased intracellular entry of the modified exoenzyme should increase the range of experiments possible, making the TAT-C3 protein an even more useful tool in the analysis of Rho GTPase function.

Methods

Generation of TAT-MYC-C3 Expression Vector

To add the TAT protein transduction domain to recombinant C3 exoenzyme, a pGEX-KG MYC-C3 plasmid was modified to include the nucleotide sequence 5′-GGAGGATACGGCCGAAAGAAGCGACGACAG CGACGCCGTGGAGGA-3′ at a position 5′ to the C3 coding sequence (Fig. 1). C3 sequence corresponds to nucleotides 1–780 of EMBL accession X51464. Complete plasmid sequence is available on request.

Recombinant Protein Preparation

To determine the optimum concentration of isopropyl β-D-thiogalactopyranoside (IPTG; Sigma) for induction of GST TAT-MYC-C3 expression, a single bacterial colony of pGEX-KG TAT-MYC-C3 in BL21 (DE3)

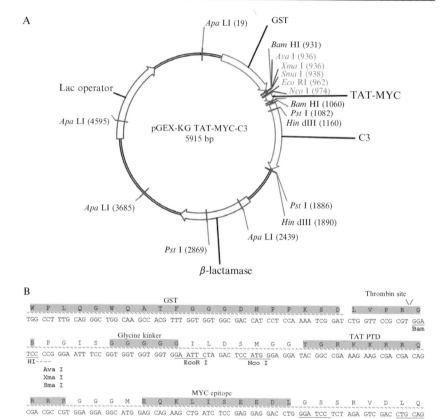

FIG. 1. Map of pGEX-KG TAT-MYC-C3 with detail of region between GST and C3 moieties. (A) Based on the pGEX-KG plasmid (Guan *et al.*, 1991), an expression plasmid encoding a glutathione-*S*-transferase (GST) fusion with the C3 exoenzyme was made. (B) The region between the GST and C3 domains is shown in greater detail. At the amino terminus of GST, a thrombin cleavage site allows for proteolytic cleavage and release of the C3 domain. A five amino acid "glycine kinker" facilitates access of thrombin to the cleavage site. The TAT protein transduction domain (PTD) was added to improve transduction of the C3 protein into cells. The MYC epitope sequence allows for detection of the C3 protein using anti-MYC epitope antibodies such as the mouse monoclonal 9E10. C3 sequence corresponds to nucleotides 1–780 of EMBL accession X51464.

pLys^s cells (Novagen) is cultured in 5 ml of Terrific broth (1.2% [w/v] bacto tryptone, 2.4% [w/v] yeast extract, 0.4% [v/v] glycerol, 17 mM KH2PO4, 72 mM K$_2$HPO$_4$) with 20 μg/ml chloramphenicol (Sigma) and 100 μg/ml ampicillin (Sigma). The overnight culture is diluted 1:10 into 50 ml of Terrific broth including 100 μg/ml ampicillin and grown to OD$_{600}$ 0.6 at 37° before removing 1 ml aliquots of culture to tubes containing IPTG concentrations ranging from 1–1000 μM. After continued incubation at 37° for 2 h, bacterial cells are pelleted by centrifugation, lysed with an equal volume of 2× Laemmli sample buffer (Sigma), and sonicated with two 10-sec rounds at 20% intensity using a Branson Digital Sonifier. After boiling samples at 100° for 5 min, equal volumes from each condition are run on 10% SDS-polyacrylamide gels, transferred to nitrocellulose, blocked with 5% (w/v) dried milk in TBST (137 mM NaCl, 5 mM KCl, 25 mM Tris-HCl [pH 7.4], 0.5% [v/v] Tween 20) before immunoblotting with 9E10 anti-MYC epitope mouse monoclonal antibody (Institute of Cancer Research, London) and then goat anti-mouse secondary antibody conjugated with Alexa Fluor 680 (Molecular Probes). Molecular weights are determined by running 10 μl of prestained SDS-PAGE broad-range standards (Bio-Rad). Detection and quantitation of GST TAT-MYC-C3 expression is performed with a Li-Cor Odyssey Infrared Imaging System.

For large-scale protein production, single bacterial colonies of pGEX-KG TAT-MYC-C3 in BL21 (DE3) pLys^s cells are cultured in six 50-ml batches of Terrific broth with 20 μg/ml chloramphenicol and 100 μg/ml ampicillin. Overnight cultures are each diluted 1:10 into one of six flasks containing 500 ml Terrific broth including 100 μg/ml ampicillin and grown to OD$_{600}$ 0.6–1.0 at 37° before inducing with 100 μM IPTG for 3 h at 37°. Cells are pelleted by centrifugation, resuspended in 10 ml of Tris-buffered saline (TBS) lysis buffer (137 mM NaCl, 5 mM KCl, 25 mM Tris-HCl [pH 7.4] with 1 mM dithiothreitol [DTT] and 1 mM phenylmethylsulfonyl fluoride [PMSF]) and snap frozen in a dry ice/ethanol solution. At this point, samples may be stored at −80° indefinitely. After thawing in a 37° water bath, samples are kept on ice while being disrupted by three 1-min rounds of sonication at 20% intensity using a Branson Digital Sonifier, followed by removal of debris by centrifugation at 10,000g for 20 min at 4°. Clarified supernatants are incubated with 0.5 ml glutathione-Sepharose (Sigma) bead slurry for 2 h at 4° to bind GST fusion protein. Beads are washed six times with 50 ml of TBS lysis buffer, followed by two washes with 50 ml of thrombin cleavage buffer (1 mM MgCl$_2$, 1 mM CaCl$_2$, 1 mM DTT in TBS). The beads are resuspended in 0.5 ml of thrombin cleavage buffer and TAT-MYC-C3 released from the GST moiety by incubation with 25 units of bovine thrombin (Sigma) overnight at 4°. The supernatant is removed, beads are washed twice with 0.5 ml thrombin cleavage buffer, and the collected supernatants are incubated with 30 μl of TBS-washed p-aminobenzamide beads (Sigma) for 1 h to remove thrombin, before snap freezing aliquots and storage at $-80°$.

To monitor expression and purification of TAT-MYC-C3, samples from each stage and 10 μl of prestained SDS-PAGE broad-range standards (Bio-Rad) are run on 10% SDS-polyacrylamide gels, then stained with Brilliant Blue G (Sigma), and destained with one wash in 10% (v/v) acetic acid, 25% (v/v) methanol for 1 min, and several washes with 25% (v/v) methanol until background staining has been reduced. Gel images are acquired at 700 nm with a Li-Cor Odyssey Infrared Imaging System.

Cell Culture and Western Blotting

NIH 3T3, HCT 116, and LS174T cell lines are routinely maintained in Dulbecco's modified Eagles medium (DMEM) (Invitrogen) supplemented with 10% (v/v) donor calf serum (DCS) (Invitrogen) and 100 U/ml penicillin-streptomycin (Invitrogen). For experiments, LS174T (1×10^6 per 10-cm dish), HCT116 (1×10^6 per 10-cm dish), or NIH 3T3 cells (0.6×10^6 per 10-cm dish) are plated in 10% DCS/DMEM. The following day, cells are transferred to serum-free medium and treated with TAT-MYC-C3 at indicated concentrations for 16 h. Dishes are placed on ice, rinsed with ice-cold phosphate-buffered saline (PBS), and then cells are lysed with 300 μl of RIPA buffer (10 mM Tris-HCl [pH 7.5], 5 mM EDTA, 1% [v/v] Nonidet P-40, 0.5% [w/v] sodium deoxycholate, 40 mM Na PP$_i$, 1 mM Na$_3$VO$_4$, 50 mM NaF, 1 mM PMSF, 0.025% [w/v] SDS, 150 mM NaCl, and 1$\times$ Complete protease inhibitors [Roche]). The cell lysates are scraped and collected, then centrifuged at 13,000g at 4° for 15 min. Equal aliquots of 100 μg of each lysate are run on 10% SDS-polyacrylamide gels, transferred to nitrocellulose, and blocked with 5% (w/v) dried milk in TBST before immunoblotting with 9E10 anti-MYC epitope mouse monoclonal antibody, anti-RhoA mouse monoclonal antibody (Santa Cruz Biotechnology), or rabbit anti-ERK2 polyclonal antibody (C. J. Marshall, Institute of Cancer Research London). Goat anti-mouse and goat anti-rabbit horseradish peroxidase secondary antibodies (Pierce) are used to visualize immunoreactivity with ECL Plus Western Blotting Detection Reagents (Amersham Pharmacia) or SuperSignal West Femto Maximum Sensitivity Substrate (Pierce) using Biomax MR autoradiography film (Kodak).

Immunofluorescence

To examine how TAT-MYC-C3 affects the organization of the actin cytoskeleton, 1×10^4 NIH 3T3 cells are plated onto glass coverslips (12 mm diameter) in 10% DCS/DMEM for 24 h. Cells are placed in serum-free medium, either without or with 1 μM TAT-MYC-C3 for 16 h. Cells are washed once with PBS and then fixed with 4% (w/v) p-formaldehyde in PBS for 10 min at room temperature. Cells are washed three times with PBS and then permeabilized in 0.5% (v/v) Triton X-100 in PBS for 10 min. Actin structures are visualized by staining with 2 μg/ml Texas Red–conjugated phalloidin (Molecular Probes). Coverslips are then washed twice with PBS and once with water before mounting onto glass slides with 10 μl Mowiol (Calbiochem) containing 0.1% (w/v) p-phenylenediamine (Sigma) as an anti-quenching agent. Confocal laser-scanning microscopy is performed using a Leica SP2 and Leica confocal software.

Rho-GTP Pull-Down

HCT 116 cells are seeded at 1×10^6 per 150-mm dish and grown overnight. The following day, cells are either untreated or treated with 0.5 μM TAT-MYC-C3 for 16 h, before rinsing in ice-cold PBS and lysis in 1 ml of pull-down buffer (50 mM Tris-HCl [pH 7.2], 500 mM NaCl, 1% [v/v] Triton X-100, 5 mM MgCl$_2$, 1 mM DTT, 1 mM PMSF, and 1× Complete protease inhibitors [Roche]). Lysates are centrifuged at 10,000g for 15 min at 4° before removing 30 μl of supernatant for Western blotting; the remainder is used for the pull-down assay. Supernatants are mixed with 10 μg GST-Rhotekin RBD (amino acids 7–89) bound to glutathione-Sepharose beads and rotated at 4° for 30 min. Beads are then washed twice in 1 ml wash buffer (50 mM Tris-HCl [pH 7.2], 150 mM NaCl, 1% [v/v] Triton X-100, 5 mM MgCl$_2$ and 1 mM DTT) before eluting protein complexes with 1× Laemmli buffer. Samples are run on 10% SDS-polyacrylamide gels, transferred to nitrocellulose, and blocked with 5% (w/v) dried milk in TBST before immunoblotting with RhoA mouse monoclonal antibody (Santa Cruz Biotechnology) and horseradish peroxidase conjugated goat anti-mouse antibody (Pierce). Immunoreactivity is visualized with SuperSignal West Femto Maximum Sensitivity Substrate (Pierce) using Biomax MR autoradiography film (Kodak).

Results and Discussion

Recombinant C3 exoenzyme produced in *E. coli* has been used in many studies to examine the biological functions of Rho GTPases. Given that we wished to more efficiently deliver C3 into cells, an expression plasmid that was already in use for the production of MYC-epitope–tagged C3 was modified by inserting a linker sequence corresponding to the HIV TAT protein transduction domain (PTD) (Vocero-Akbani *et al.*, 2000) to generate pGEX-KG TAT-MYC-C3 (Fig. 1). The resulting construct has several features, β-lactamase expression allows for selection with ampicillin while the lac operator drives isopropyl β-D-thiogalactopyranoside (IPTG)-responsive protein expression. Proteins are expressed as glutathione-S-transferase (GST) fusions, which allows for rapid one-step batch purification using glutathione-Sepharose beads. Proteins can either be eluted with reduced glutathione, or recombinant protein can be released from the GST moiety by thrombin-mediated cleavage at an engineered site (Fig. 1B). Efficient cleavage at this site may sometimes be hindered by fused proteins; to overcome this, five glycine residues were added in the form of the "glycine kinker" to improve the accessibility of the thrombin cleavage site (Guan et al., 1991). The TAT protein transduction domain (PTD) consists of nine amino acids (YGRKKRRQRRR) that facilitate the passage of peptides and small proteins into cells (Wadia et al., 2005). The MYC epitope allows for delivery of C3 into cells to be monitored by Western blotting (e.g., Fig. 4A) or by immunohistochemistry.

We determined the optimal concentration for GST TAT-MYC-C3 induction to be 100 μM IPTG, with protein expression induced approximately 90–100 fold above uninduced levels (Fig. 2A). When induced, GST TAT-MYC-C3 expression represents approximately 5% of total soluble protein (Fig. 2B) as estimated by Brilliant Blue G staining and quantitation at 700 nm by infrared imaging. A one-step affinity purification with glutathione (GSH) Sepharose beads gave a single dominant band with only traces of contaminating bands (Fig. 2B), the two most prominent likely being incomplete translation products also seen in the anti-MYC Western blot (Fig. 2A). Although the GST-fusion protein was not completely cleaved by thrombin treatment, efficiency of cleavage was approximately 80% (Fig. 2B). Most of the GST was retained on the GSH-Sepharose beads, giving a final TAT-MYC-C3 preparation that was approximately 60% pure (Fig. 2B). Using gel electrophoresis and mass spectrometry, the two largest contaminants were identified as a lower molecular weight band consisting of a mixture of incompletely translated C3, GST, and E. coli ribosomal protein S4, and a higher molecular weight band, also observed with glutathione-eluted GST, consisting of heat shock protein 70 (HSP70) and ribosomal protein S1. These contaminants do not seem to have any biological effects. For example the addition of GST, with the contaminating HSP70/S1 and E. coli ribosomal proteins, to tissue culture cells does not affect the actin cytoskeleton (data not shown). We factor in TAT-MYC-C3 purity when calculating molar concentrations used in experiments.

For the TAT modification to be experimentally useful, it should improve the intracellular delivery of C3. We compared the abilities of MYC-C3 and TAT-MYC-C3 to enter NIH 3T3 cells by Western blotting cell extracts with anti-MYC epitope antibody 16 h after treatment with

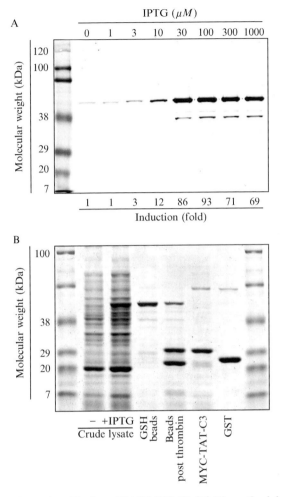

FIG. 2. Expression and purification of TAT-MYC-C3. (A) The optimal dose of isopropyl
β-D-thiogalactopyranoside (IPTG) for TAT-MYC-C3 induction was determined by Western
blotting crude *E. coli* lysates that had been treated with IPTG concentrations from 1–1000 μM
for 2 h. The fold induction was determined as the average of two experiments using Alexa
Fluor 680 secondary antibody and infrared imaging (Li-Cor Odyssey). (B) Purification of
TAT-MYC-C3 was monitored by running samples from each stage on SDS-polyacrylamide.
Crude lysates of untreated or IPTG-treated *E. coli*, protein associated with glutathione (GSH)
Sepharose beads, protein associated with GSH Sepharose beads, and eluted TAT-MYC-C3
after thrombin treatment are each shown. For comparison purposes, glutathione-*S*-transferase
(GST) expressed in *E. coli*, purified and eluted from GSH Sepharose beads with an excess of
GSH, is also shown. Molecular weights are as reported for the batch of prestained markers.

recombinant protein (Fig. 3A). The conventional MYC-C3 was undetectable at 40 nM and slightly detectable at 320 nM in cell extracts. However, treatment of cells with 80 nM TAT-MYC-C3 resulted in dramatically higher levels of intracellular protein, indicating that the addition of the 9 amino acid protein transduction domain significantly increased the delivery of C3 protein into cells (Fig. 3A). Although we typically use overnight incubations with TAT-MYC-C3 for convenience, biological effects are usually evident after 4–8 h (data not shown).

The C3 exoenzyme was a critical reagent in the determination of the role of Rho GTPases in the regulation of the actin cytoskeleton (Paterson et al., 1990). However, given that uptake of recombinant C3 protein by many cell lines, including murine fibroblasts, from tissue culture medium is inefficient, it was necessary that the recombinant protein be actively introduced by methods such as microinjection. Therefore, we wished to determine whether TAT-MYC-C3 added to tissue culture medium would affect actin cytoskeletal structures. NIH 3T3 fibroblasts were placed in serum-free medium, either without or with TAT-MYC-C3 or an equal amount of GST. After 16 h, cells were stimulated for 20 min with medium containing 10% serum to induce typical Rho-dependent actin stress fibers (Fig. 3B, left panel). Cells treated with TAT-MYC-C3 were marked by their lack of stress fibers, instead many displayed peripheral membrane ruffles and long axon-like projections (Fig. 3B, right panel). GST treatment was without effect (data not shown). These data indicate that the TAT-MYC-C3 protein modified actin cytoskeletal structures in fibroblast cells, consistent with the previously demonstrated actions of microinjected C3 protein.

ADP-ribosylation of RhoA is sufficient to alter the electrophoretic mobility of the protein on SDS-polyacrylamide gels (Morii et al., 1995). Therefore, to monitor the effectiveness of transduced TAT-MYC-C3, cell extracts can be Western blotted with RhoA antibody and protein mobility compared against untreated cell extracts. Treatment of LS174T cells with TAT-MYC-C3 overnight effectively resulted in complete modification of RhoA to a slower mobility form (Fig. 4A). In addition, a reduction in RhoA protein level has been consistently observed after TAT-MYC-C3 treatment, which may also reflect a component of the mechanism of action in vivo. Although treatment of some cell lines with TAT-MYC-C3 results in complete RhoA modification, less than complete modification has been observed in other cell lines despite significant inhibition of RhoA biological activity (e.g., reduction in actin stress fibers). One explanation for this phenomenon is that ADP-ribosylated RhoA may act in a "dominant-negative" manner to block downstream signaling (Paterson et al., 1990); however, the mechanism for this effect has not been determined. We have observed that the incompletely modified RhoA may act in a

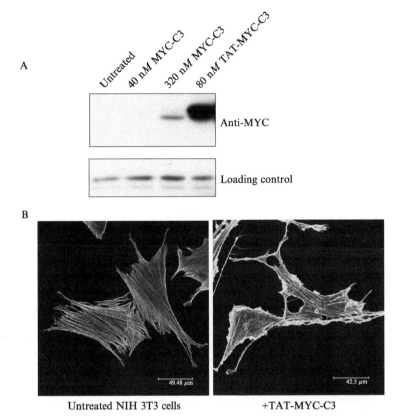

FIG. 3. The TAT protein transduction domain improves delivery of C3 into NIH 3T3 fibroblasts, resulting in disruption of actin stress fibers. (A) NIH 3T3 cells were left untreated or treated with MYC-C3 or TAT-MYC-C3 as indicated for 16 h. Cell extracts were prepared and Western blotted with anti-MYC epitope antibody 9E10 to assess protein entry into cells. Although 320 nM MYC-C3 did permit some protein to enter cells, treatment with only 80 nM TAT-MYC-C3 resulted in dramatically greater protein delivery, indicating that the TAT protein transduction domain significantly enhanced protein entry. (B) NIH 3T3 cells were put in serum-free medium without (left panel) or with TAT-MYC-C3 (right panel) as indicated. After 16 h, cells were stimulated with medium containing 10% serum for 20 min then fixed and stained for filamentous actin structures with Texas Red conjugated phalloidin. Untreated cells displayed typical actin stress fibers; however, TAT-MYC-C3–treated cells were devoid of stress fibers, and, instead, they showed increased peripheral membrane ruffling and long axon-like extensions consistent with the inhibition of RhoA by the transduced C3 exoenzyme. Scale bars represent 49.5 μm (left panel) and 42.5 μm (right panel).

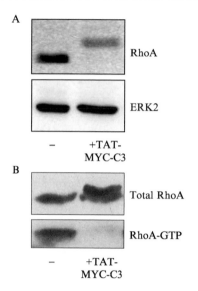

FIG. 4. Shift in electrophoretic mobility of ADP-ribosylated RhoA induced by TAT-MYC-C3 also correlates with inhibition of RhoA activation. (A) LS174T cells were untreated or treated with TAT-MYC-C3 for 16 h as indicated. Cell extracts were prepared and Western blotted for RhoA (upper panel) and ERK2 as a loading control (lower panel). Treatment with TAT-MYC-C3 resulted in decreased RhoA levels and a shift in electrophoretic mobility, which can be used as a biochemical readout for TAT-MYC-C3 entry and function. (B) Treatment of HCT 116 cells with TAT-MYC-C3 resulted in incomplete ADP-ribosylation of RhoA (upper panel) but blocks the GTP-loading and consequent activation of the unmodified RhoA as determined by a pull-down assay using the Rho binding domain of Rhotekin (lower panel).

dominant-negative fashion by inhibiting the GTP-loading of the unmodified RhoA population (Fig. 4B). These results suggest that ADP-ribosylated RhoA may bind and sequester Rho guanine nucleotide exchange factors (GEFs) in nonfunctional "dead-end" complexes. Therefore, functionally critical RhoGEFs potentially could be identified by purification of ADP-ribosylated RhoA after treatment of cells with TAT-MYC-C3.

Conclusions

The C3 exoenzyme has been an invaluable research tool in signal transduction research. By enhancing uptake of C3 from tissue culture medium through the addition of a protein transduction domain, we have improved its effectiveness and, therefore, usefulness. We will provide plasmid DNA and full sequence to our academic colleagues on request.

Acknowledgments

This work was supported by Cancer Research UK and an NIH/NCI (R01 CA030721) grant to M. F. O. We wish to acknowledge Richard Treisman (CRUK London Research Institute) for the gift of pGEX-KG MYC-C3, and the technical assistance of Steve Hooper (CRUK London Research Institute), Chris Ward, and Dan Croft (Beatson Institute, Glasgow).

References

Aktories, K., Wilde, C., and Vogelsgesang, M. (2004). Rho-modifying C3-like ADP-ribosyltransferases. *Rev. Physiol. Biochem. Pharmacol.* **152**, 1–22.

Barbieri, J. T., Riese, M. J., and Aktories, K. (2002). Bacterial toxins that modify the actin cytoskeleton. *Annu. Rev. Cell Dev. Biol.* **18**, 315–344.

Beltman, J., Erickson, J. R., Martin, G. A., Lyons, J. F., and Cook, S. J. (1999). C3 toxin activates the stress signaling pathways, JNK and p38, but antagonizes the activation of AP-1 in rat-1 cells. *J. Biol. Chem.* **274**, 3772–3780.

Borbiev, T., Nurmukhambetova, S., Liu, F., Verin, A. D., and Garcia, J. G. (2000). Introduction of C3 exoenzyme into cultured endothelium by lipofectamine. *Anal. Biochem.* **285**, 260–264.

Coleman, M. L., Sahai, E. A., Yeo, M., Bosch, M., Dewar, A., and Olson, M. F. (2001). Membrane blebbing during apoptosis results from caspase-mediated activation of ROCK I. *Nat. Cell Biol.* **3**, 339–345.

Croft, D. R., Sahai, E., Mavria, G., Li, S., Tsai, J., Lee, W. M., Marshall, C. J., and Olson, M. F. (2004). Conditional ROCK activation *in vivo* induces tumor cell dissemination and angiogenesis. *Cancer Res.* **64**, 8994–9001.

Grounds, H. R., Ng, D. C., and Bogoyevitch, M. A. (2005). Small G-protein Rho is involved in the maintenance of cardiac myocyte morphology. *J. Cell Biochem.* **95**, 529–542.

Guan, K. L., and Dixon, J. E. (1991). Eukaryotic proteins expressed in *Escherichia coli*: An improved thrombin cleavage and purification procedure of fusion proteins with glutathione S-transferase. *Anal. Biochem.* **192**, 262–267.

Guerin, P., Sauzeau, V., Rolli-Derkinderen, M., Al Habbash, O., Scalbert, E., Crochet, D., Pacaud, P., and Loirand, G. (2005). Stent implantation activates RhoA in human arteries: Inhibitory effect of rapamycin. *J. Vasc. Res.* **42**, 21–28.

Hill, C. S., Wynne, J., and Treisman, R. (1995). The Rho family GTPases RhoA, Rac1, and CDC42Hs regulate transcriptional activation by SRF. *Cell* **81**, 1159–1170.

Monnier, P. P., Sierra, A., Schwab, J. M., Henke-Fahle, S., and Mueller, B. K. (2003). The Rho/ROCK pathway mediates neurite growth-inhibitory activity associated with the chondroitin sulfate proteoglycans of the CNS glial scar. *Mol. Cell Neurosci.* **22**, 319–330.

Morii, N., and Narumiya, S. (1995). Preparation of native and recombinant *Clostridium botulinum* C3 ADP-ribosyltransferase and identification of Rho proteins by ADP-ribosylation. *Methods Enzymol.* **256**, 196–206.

Olson, M. F., Ashworth, A., and Hall, A. (1995). An essential role for Rho, Rac, and Cdc42 GTPases in cell cycle progression through G1. *Science* **269**, 1270–1272.

Paterson, H. F., Self, A. J., Garrett, M. D., Just, I., Aktories, K., and Hall, A. (1990). Microinjection of recombinant p21rho induces rapid changes in cell morphology. *J. Cell Biol.* **111**, 1001–1007.

Sahai, E., Alberts, A. S., and Treisman, R. (1998). RhoA effector mutants reveal distinct effector pathways for cytoskeletal reorganization, SRF activation and transformation. *EMBO J.* **17**, 1350–1361.

Sahai, E., and Marshall, C. J. (2002). ROCK and Dia have opposing effects on adherens junctions downstream of Rho. *Nat. Cell Biol.* **4**, 408–415.

Sahai, E., and Marshall, C. J. (2003). Differing modes of tumour cell invasion have distinct requirements for Rho/ROCK signalling and extracellular proteolysis. *Nat. Cell Biol.* **5**, 711–719.

Sahai, E., Olson, M. F., and Marshall, C. J. (2001). Cross-talk between Ras and Rho signalling pathways in transformation favours proliferation and increased motility. *EMBO J.* **20**, 755–766.

Sauzeau, V., Le Mellionnec, E., Bertoglio, J., Scalbert, E., Pacaud, P., and Loirand, G. (2001). Human urotensin II-induced contraction and arterial smooth muscle cell proliferation are mediated by RhoA and Rho-kinase. *Circ. Res.* **88**, 1102–1104.

Sebbagh, M., Renvoize, C., Hamelin, J., Riche, N., Bertoglio, J., and Breard, J. (2001). Caspase-3-mediated cleavage of ROCK I induces MLC phosphorylation and apoptotic membrane blebbing. *Nat. Cell Biol.* **3**, 346–352.

Vial, E., Sahai, E., and Marshall, C. J. (2003). ERK-MAPK signaling coordinately regulates activity of Rac1 and RhoA for tumor cell motility. *Cancer Cell* **4**, 67–79.

Vocero-Akbani, A., Lissy, N. A., and Dowdy, S. F. (2000). Transduction of full-length Tat fusion proteins directly into mammalian cells: Analysis of T cell receptor activation-induced cell death. *Methods Enzymol.* **322**, 508–521.

Wadia, J. S., and Dowdy, S. F. (2005). Transmembrane delivery of protein and peptide drugs by TAT-mediated transduction in the treatment of cancer. *Adv. Drug Deliv. Rev.* **57**, 579–596.

Wadia, J. S., Stan, R. V., and Dowdy, S. F. (2004). Transducible TAT-HA fusogenic peptide enhances escape of TAT-fusion proteins after lipid raft macropinocytosis. *Nat. Med.* **10**, 310–315.

[12] Imaging and Photobleach Correction of Mero-CBD, Sensor of Endogenous Cdc42 Activation

By LOUIS HODGSON, PERIHAN NALBANT,
FEIMO SHEN, and KLAUS HAHN

Abstract

This chapter details quantitative imaging of the Mero-CBD biosensor, which reports activation of endogenous Cdc42 in living cells. The procedures described are appropriate for imaging any biosensor that uses two different fluorophores on a single molecule, including FRET biosensors. Of particular interest is an algorithm to correct for fluorophore photobleaching, useful when quantitating activity changes over time. Specific topics

METHODS IN ENZYMOLOGY, VOL. 406   0076-6879/06 $35.00
DOI: 10.1016/S0076-6879(06)06012-5

include procedures and caveats in production of the Mero-CBD sensor, image acquisition, motion artifacts, shading correction, background subtraction, registration, and ratio imaging.

Introduction

We recently described a novel biosensor to visualize and quantify Cdc42 nucleotide state in living cells (Nalbant *et al.*, 2004). This biosensor (Mero-CBD) is based on a fragment of Wiskott–Aldrich syndrome protein (CBD, the Cdc42 binding domain) that binds only to activated GTP-bound Cdc42. The fragment is covalently labeled with a merocyanine dye whose fluorescence responds to the polarity of its solvent environment (Toutchkine *et al.*, 2003). When the biosensor encounters and binds to activated Cdc42, dye fluorescence greatly increases. We use a GFP fusion of the WASP domain to enable ratio imaging; the ratio of dye/GFP fluorescence provides a quantitative readout of Cdc42 activation level. This approach is advantageous in that it monitors the activation of endogenous, untagged Cdc42. It provides a very bright signal resulting from direct excitation of a dye, enabling many time points to be monitored using low biosensor concentrations. This review first briefly expands on the published protocol for Mero-CBD preparation (Nalbant *et al.*, 2004) but focuses primarily on a general procedure for quantitative imaging of Mero-CBD, including corrections for photobleaching and caveats in image-processing steps. These procedures will also be useful for other biosensors, including those based on FRET, which incorporate two different fluorophores. Of special interest is an algorithm to correct for photobleaching. We illustrate software operations with Metamorph, but the procedures are available in most commercial image processing packages.

Mero-CBD Preparation

To avoid pitfalls that have been encountered by laboratories attempting to follow the published procedure for Mero-CBD production, the following notes are included:

1. It is important to use bacteria strain BL21(DE3) (Stratagene). Most problems encountered with producing the sensor have resulted from attempts to use similar strains (i.e., do not use BL21(DE3)pLysS). The protein should be induced and expressed at room temperature (26°) to increase the proportion of correctly folded, soluble CBD-EGFP.

2. Use Talon resin (Co^{2+} affinity, Clontech Inc.) rather than Ni-NTA resin. Use 2 ml Talon resin (dry volume) for 3 g of cell pellet.

3. Use the suggested buffers. Apparently, unimportant modifications here have also caused problems in several cases.

4. Be sure to use enough buffer during lysis (i.e., for 3-g cell pellet use 35 ml total lysis buffer). CBD-EGFP needs to be in 50 mM NaH$_2$PO$_4$ buffer, pH 7.5, at a protein concentration of 100 μM. The reactive dye cannot be kept as a DMSO solution for more than 12 h. The dye/protein ratio in the reaction mixture should be 5:1 or 6:1.

5. When labeling, the reaction time is critical. Because EGFP contains internal cysteines, longer reaction times result in overlabeling and loss of EGFP signals. Optimize reaction times to produce a final dye/protein ratio of 0.7–0.9 in the purified product (starting point for optimization: 1 h at room temperature). Dye/protein is measured as described elsewhere (Nalbant et al., 2004). The final Mero-CBD concentration after gel filtration is usually 50–60 μM.

Cell Injection and Image Acquisition

Cells are microinjected on coverslips and allowed to recover for 30 min in an incubator before imaging. Mero-CBD can form puncta in some cells in about 3 h, possibly because of autophagocytosis, so timing these steps is important. During image acquisition, a dye image and a GFP image must be obtained at each time point for later dye/GFP division to obtain the final ratio image. It is important to show that localized activation is not an artifact caused by motion of the cell between acquisitions of the two images. For example, retraction of the edge can lead portions of an image to be divided by regions containing no cell. This will produce what seems to be very high activity at the cell edge. To control for motion artifacts, the order in which the images are taken can be reversed. Alternately, an EGFP picture can be taken both at the beginning and end of each sequence. Division of the earlier EGFP by the later EGFP image will reveal regions where movement has occurred (Fig. 1).

In our experiments, images are acquired at 30 sec–1 min time intervals using 2 by 2 binning on a full-field 1.3 k by 1.0 k CCD camera. The excitation light from a 100 W Hg arc lamp is too bright, necessitating use of neutral density filters. These are valuable also to decrease bleaching. Bleaching is much less for samples exposed over longer time to weak illumination, rather than briefly to bright light, even when the total light exposure is the same. This is also beneficial to cell health. We routinely use 25–36% transmittance (ND 1.4–1.0) neutral density filters. These numbers are provided only as a rough example, because exposure times will depend on the quantum efficiency and noise characteristics of the

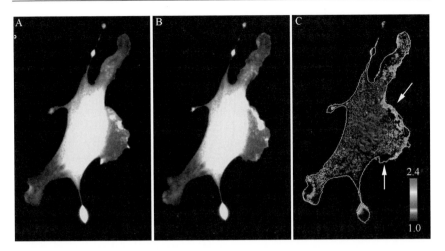

FIG. 1. Effects of motion artifacts on ratiometric measurements. (A and B) EGFP images taken before and after I-SO acquisition. Dividing the second EGFP image by the first shows where motion has occurred during the acquisition sequence. In this example, (B)—second EGFP acquisition—was divided by (A)—first EGFP acquisition—to produce ratiometric (C). Areas that underwent protrusion generated artificial high ratios (upper white arrow) and low ratios (lower white arrow). The outline of the cell is shown in white in (C). (See color insert.)

camera. Twelve-bit cameras have an intensity range of 0–4096; to obtain sufficient dynamic range, we target roughly 1500–3000 as the maximum pixel value in the field of view. We carefully monitor the pixel intensities near the cell periphery where the signals are usually low, to be at least 100–150 over background values.

In Mero-CBD, we used two fluorophores with very different wavelengths to minimize bleed through of one fluorescence channel into the other. Even with lenses well corrected for chromatic aberration, switching between two such very different wavelengths necessitates moving the objective to maintain the same focal plane. In our system, the objective must be moved approximately 200 nm. We determine the appropriate offset in control experiments using commercially available beads that fluoresce over a wide range of wavelengths (Molecular Probest, Eugene Oregon).

Bandpass Filters and Dichroic Selection

Motion-induced artifacts in activity levels can be eliminated by taking the two images simultaneously, either using two cameras or a device that obtains two images on different portions of the same CCD chip (hardware for both is commercially available). Either approach requires

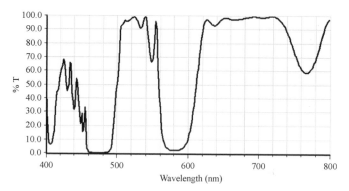

Fɪɢ. 2. Transmittance spectra of the custom multiband pass dichroic mirror designed by Chroma Technology for simultaneous imaging of I-SO and EGFP.

use of appropriate dichroic mirrors to send different wavelengths to the two images; such dichroics and matching excitation and emission filters have been designed for Mero-CBD by the Chroma Company (Chroma Technology, Rockingham, VT) (Fig. 2; bandpass filters = HQ470/40, HQ525/50 for EGFP, and HQ580/30, HQ630/40 for I-SO). Because the I-SO dye and EGFP have different brightness, the relative amounts of excitation light at each wavelength must be controlled by the filters if images are to be obtained simultaneously. Dyes are being developed that can be excited by two different laser lines. This will simplify adjustment of excitation intensities.

One of the key concerns with biosensors using two fluorescent components is the bleed through of emission from one fluorophore into the image of the other. When both fluorophores are on the same molecule, this is not so critical, because the two fluorophores are equally distributed throughout the cell. In such cases, bleed through affects the linearity of response to physiological stimuli, but this has been ignored in most experiments, where it is considered an acceptable error. When the two fluorophores are on separate molecules, it is essential to correct for bleed through; otherwise, differences in subcellular distribution of the two fluorophores can profoundly affect results. This has been covered in detail elsewhere (Chamberlain and Hahn, 2000; Chamberlain et al., 2000).

Shading Correction

The first step of image analysis is to correct for uneven illumination across the field of view. This is present in almost all images, including those taken with Plan-corrected objectives, but can be greatly reduced using fiberoptic light scramblers. Users can check their images by taking a diagonal line-scan of an empty field. To correct for shading, obtain images without any sample present, using the integration times and illumination conditions applied to actual samples. This should be part of each experiment, using either cell-free areas within the coverslip or mounting a fresh coverslip with identical media. In the latter case, it is easier to focus if a mark is made on the side of the coverslip where the cells would have been. Usually, 20–30 shading images are acquired and averaged to produce a single shade correction image for each fluorescence channel. This averaging reduces stochastic camera noise associated with image acquisition (Fig. 3). Once the averaged shading images are acquired, shading correction is simply a matter of dividing the sample image by the shading image. Here, care must be taken to prevent floating point errors. Because many common software programs do not use a "floating" decimal place in specifying pixel intensities, there are effectively rounding-off errors in the calculations. Metamorph, which does have this problem, offers a convenient plug-in module for shade correction; images are multiplied by a scaling factor to increase pixel intensity values before division, thus effectively eliminating these errors. One either scales so that the maximum pixel value in the image remains on scale or simply specifies a number high enough to eliminate the error (i.e., 1000 for 12-bit images acquired as

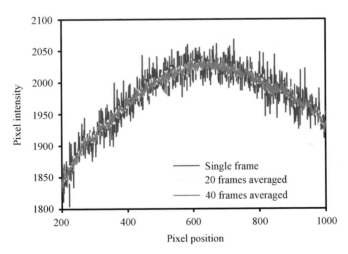

FIG. 3. The effect of averaging on shading correction images. A sample-free field of view was imaged under identical illumination conditions once (blue), 20 times (yellow), and 40 times (red) and averaged. Averaging of many frames reduces the stochastic noise in the image. (See color insert.)

described previously). The same scaling factor should be specified for both fluorescence images before division, so that final ratios are not affected; this is important when a set of final images are to be compared with one another, as in a time course. Matlab (Mathworks, Natick, MA) codes can be written to process images using floating-point procedures.

Background Subtraction

Because of shade correction, cell-free areas within the image show uniform intensity. Background subtraction can, therefore, be accomplished by choosing a small area within the cell-free background and subtracting the average intensity from the entire image. It is important to choose the same background region in all image pairs to minimize artifacts. Metamorph offers a convenient plug-in module ("Use region as background") to expedite this; one selects a background region in the first fluorescence image of the ratio pair, which can then be pasted onto the second fluorescence image. When using such a utility for an entire series of images, it is important that the average background from the first image pair not be used for the entire image stack. The same background region can be applied to each image, but average background must be determined and subtracted anew for each time point in the image stack.

Image Registration

For ratiometric sensors, one image is divided by another to obtain the final map of protein activity distribution (i.e., for Mero-CBD, I-SO image divided by EGFP to obtain Cdc42 activation). One of the most challenging aspects of this procedure is properly registering the two images before division. It is very difficult to change the microscope so that pixel misalignment does not occur, because it stems from so many causes (varying thickness and alignment of dichroics and filters; movement of many microscope components including filter wheels, turrets, and stages; temperature fluctuations; ambient vibrations; chromatic aberration within the objective lens). Figure 4 illustrates the effects of misalignment on ratiometric analysis. These include characteristic artifacts that warn of misalignment: "edge artifacts" where ratio values are high on one side of the cell and low on the opposite side, and bright shading on one side of objects within the cell such as the nucleus. Automated methods for image registration are available (Shen et al., 2004), but manual methods can be used successfully and remain more accurate in some cases. A Metamorph "Color align" plug-in module can be used to determine the relative pixel shifts of up to three fluorescence channels. One fluorescence image channel is held stationary, whereas the other is shifted manually while following the overlay of red and

Correct   Incorrect

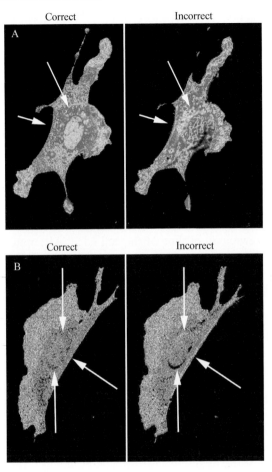

Correct   Incorrect

FIG. 4. Image registration artifacts. (A) Example of misregistration producing artifactual edge effects and spurious intracellular features (white arrows). The images were misaligned by 5 whole pixels in both the x and y directions. (B) More subtle effects of misalignment resulting in edge effects around the nucleus (white arrows) plus an acute, high ratio region at the cell edge. In this case, the shifts were less than 1 pixel (0.1 pixel in the x direction, −0.8 pixel in the y direction). A telltale sign of misalignment is high values on one side of the misaligned object and low values on the opposite side. (See color insert.)

green pseudocolor images. The resulting x and y pixel shift values can be applied to other images from the same experiment using the "Subpixel shift" module, which can make corrections at subpixel accuracy. If using two cameras, rotational alignment must be performed before image acquisition (a grid-type micrometer can be used). This is critical, because rotational correction is much more difficult to perform computationally. Off-line methods to correct for such issues are available (Danuser, 1999).

Image Masking

Dividing two images will introduce noise and hot spots wherever the intensity in the either image is too low to use. Division of small numbers can greatly distort variations from pixel to pixel. Although intensities within sufficiently bright portions of the image may be well over 100, values can be below 10 in the background or dim portions of the cell. This produces large fluctuations even though the actual difference in brightness between neighboring pixels is small (compare 5/1 versus 1/5). To limit ratio calculations to image regions with sufficient intensity, various masking operations are used. The simplest is to set areas outside of the cell to zero by manually drawing an outline of the cell edge. Alternately, masking can be based on exclusion of pixels below a user-specified minimum intensity level. Masking can be based on volumetric markers such as fluorescent dextrans or membrane markers to unambiguously specify the true cell boundary. In the case of Mero-CBD, distribution of the biosensor is sufficiently close to the cell edge to use biosensor signaling in defining the cell boundaries. To produce the mask of the cell edge, the fluorescence image of the brighter signal (EGFP) is thresholded by histogram intermodal localization. This is a straightforward process ("Threshold Image" command in Metamorph). After proper shading correction and background subtraction, the intensity versus pixel number histogram usually will show a large distinct peak for low intensity pixels and a spread of high intensity pixels comprising the image (Fig. 5). The low-end threshold value is determined by trial and error. The upper bound needs to be set at the maximum possible value for the image (65,535 for a 16-bit image). When processing a series of images in a stack, such as from a time-lapse experiment, it is important to compare the low-end threshold value selected for the first and last images. The photobleaching of biosensor will shift the low end of the intensity distribution histogram downward over time. Therefore, using a threshold based on the first image would eliminate important areas in cells from later time points. Threshold values are usually determined for later time points. A mask can be produced based on the selected thresholds.

In Metamorph, the "Clip" tab within the "Threshold Image" module is used to produce a binary mask in 16 bit. This will produce a binary mask, where regions outside the selected area are uniformly zero, whereas inside is 65,535. The "Arithmetic" function is then used to divide the binary mask by a constant (65,535) to produce a true binary mask containing the pixel value of one inside the masked region and zero outside. This binary mask is

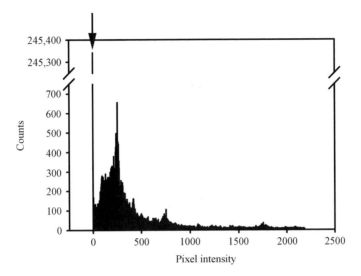

FIG. 5. A representative histogram from a shading corrected, background subtracted image. The prominent histogram spike at zero intensity (arrow) is followed by a continuous distribution of non-zero intensity peaks.

multiplied by the fluorescence images before ratio calculation. It is important that the color scheme used for image display uses a specific color, usually black, for pixels in which the numerator or denominator is set at zero.

Ratio Calculation

As with shading correction, floating point errors must be considered during the final division that produces the ratio image (see preceding). In the Metamorph "Arithmetic" module, a scaling factor of 1000 is specified as a multiplication factor during the ratio calculation. Smaller factors (10–100) tend to produce a stepped histogram distribution (Fig. 6). The same scaling factor needs to be applied for each ratio calculation if quantitative comparison of ratio images is desired. The pseudocolor display of the ratio image is scaled to emphasize or study certain intensity ranges, bringing out different features in the image. An image without the accompanying pseudocolor scale is almost meaningless.

Photobleach Correction

Time-lapse imaging of fluorescence is complicated by photobleaching of the biosensor. Real activation changes are superimposed on a steady decrease in fluorescence intensity, at different rates for each fluorophore in the sensor. There is too much variability to determine a standard bleaching

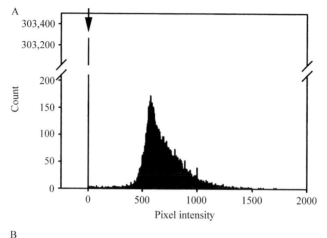

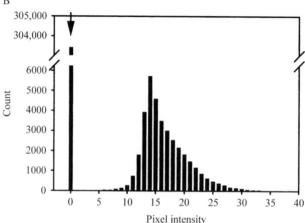

FIG. 6. Effect of multiplication factor during ratiometric calculation. In software that does not use floating point operations, ratiometric division requires a scaling factor. (A) A smoothly varying histogram distribution of a ratio image produced using a scaling factor of 1000. (B) The same ratio image with a scaling factor of 25. Although the general trend of the data is similar, the histogram in (A) retains higher resolution and produces a smoother ratio image. In both panels, background value of zero is marked with an arrow.

curve in separate control experiments. In real experiments, fitting curves to the data is complicated by the fact that bleaching is superimposed on the fluorescence changes the biosensor was designed to undergo.

Photobleach correction for Rho family sensors is simplified by the fact that only ∼5% of GTPases are active (Del Pozo *et al.*, 2002; Ren *et al.*, 1999), and that even this ∼5% is localized to specific regions (i.e., cell periphery, membrane ruffles). We can assume that averaging all the low-activity regions in the cell provides a reasonable approximation of the intensity of biosensor in the inactive state. Even if we include all the activated material, the average for Mero-CBD would be too high by only 10% (Mero-CBD shows a 3× change in I-SO intensity between the active and inactive state. Therefore, the whole cell average would be at most 5(3) + 95(1)/100 = 1.1. Using only the portions of the cell where there is minimal, activated Cdc42 should reduce this error greatly.

To determine the bleaching kinetics, changes in I-SO and EGFP intensities were measured. Intensities can be measured using the masked fluorescence images, because the background is at or near zero and can be readily removed from the calculations. A Metamorph journal can be constructed to automate these procedures after the initial determination of the intensity threshold values from the EGFP and I-SO image stacks. Background subtraction removes any signal from dye and fluorophores outside the cell, which may bleach differently in the extracellular environment.

Because relatively little change in ratio occurs near the cell nucleus, and the thick nuclear regions are brighter than the cell periphery, it is straightforward to set a threshold value that includes only the regions near the nucleus. This is better than using the Region tool to select a small area within the nucleus for measurement, because such measurements would be skewed by cell movement.

The average intensity values determined for EGFP and I-SO at each time point are normalized; each is divided by the initial value (t = 0 time point) to produce a normalized ratio change as a function of time. This curve takes the form of an exponential decay function (Patterson and Piston, 2000) and is fitted to a double exponential function:

$$y = A \cdot e^{-B\times} + C \cdot e^{-D\times} \qquad (1)$$

Curve fitting provides a robust and objective means to produce a bleach correction function. For example, if a physiological stimulus produces a large, transient change in ratio values, it is possible to analyze only portions of the curve before stimulation and after effects of stimulation have subsided. When provided with a segment at the beginning of the curve and a few ratio values near the end of the assay, the double exponential function can approximate the general trend of the photobleach curve. Once the photobleaching function is determined, it can be used to obtain a function for correcting the data: a simple inverse of the photobleach function. For convenient application to Metamorph ratio image stacks, a simple Matlab program has been provided (Appendix A). A version with graphical user interface can be downloaded from http://www.med.unc.edu/pharm. The program reads consecutively numbered image files in tiff format, reads an

MS-Excel data sheet containing the raw intensity values of the I-SO and EGFP channels, calculates the photobleach correcting function, and applies this to the original images to output the corrected and consecutively numbered tiff images.

Conclusion

We hope that this description of image-processing procedures, including common pitfalls, will be valuable for those wishing to use the many biosensors that require ratio imaging of two separate fluorophores. In the future, ratiometric sensors of protein activity that are based on shifts in the fluorescence maxima of a single dye should further simplify ratio imaging, eliminating altogether the need for bleaching corrections and enabling precise quantification without bleed through correction.

Appendix A

% %

% Program to do ratio photobleach correction.
%
% Need 16-bit individual tiff images in a folder with consecutive numbers starting with "xxx1" for data up to 9 planes, "xxx01" for up to 99 planes, and "xxx001" for up to 999 planes.
% Plus MS-Excel "XXXX.xls" file with Plane/Numerator/Denominator in 3 consecutive columns, with first cell in each column containing descriptions (i.e., "Plane", "I-SO", "GFP" etc.
% Plane numbering should start from 1 to however many images.
% Intensity data for I-SO plane number 1 should never be Zero.
% I-SO intensity can be set to zero in cells after the plane number 1 to prevent them from being used in the curve fitting, if justified.
%
% author: Louis Hodgson 2004

% %
clear all; clc;
% getting the ratio image file names and intensities file names
I = input ('Enter number of ratio images: '); % This sets the number of corrections to loop
Fname = input ('Enter file name without running numbers and file extension: ','s'); % This is the name of the ratio file to be corrected

Dname = input('Enter Excel file name for the intensities, Plane#/ Num/Denom; without Excel file extensions: ','s'); % This sets file name for intensities

DNAMEIN = sprintf ('%s.xls', dname);

data_all = xlsread(DNAMEIN);

Plane_initial = data_all(1:I);   % Plane array goes from 1 to I

num_Int = data_all(I+1:2*I);   % absolute value intensities for the numerator

denom_Int = data_all(2*I+1:3*I);   % absolute value intensities for the denominator

num_Int_norm = num_Int ./ num_Int(1);   % calculates normalized intensity values for numerator, normalized to t = 0

denom_Int_norm = denom_Int/denom_Int(1);   % calculates normalized intensity values for denominator, normalized to t = 0

Plane = (Plane_initial - 1).';   % makes plane array start from zero instead of 1 for data fitting purposes

Ratio = (num_Int_norm/denom_Int_norm).';   % calculates the ratio of the normalized intensities

outliers = ~excludedata (Plane, Ratio, 'range', [0 0.1]); % sets fit exclusion range in Y values [fresult,gof] = fit(Plane, Ratio, 'exp2', 'exclude', outliers); % double exponential fit, Plane numbers as X and normalized Ratio as Y

display (fresult);   % displays fitted parameters and coefficients

display (gof);   % displays goodness of fit parameters

Fit_ratio = fresult(Plane); % producing fitted curve using the plane numbers

CF = 1/Fit_ratio;   % correction factor is 1 over the ratio decay function

a = num2str(fresult.a); b = num2str(fresult.b); c = num2str(fresult.c); d = num2str(fresult.d); rsq = num2str(gof.rsquare);

coef = sprintf('a = %s ; b = %s', a, b);

coef2 = sprintf('c = %s ; d = %s', c, d);

coef3 = sprintf('r^2 = %s', rsq);

plot(Plane, Ratio, 'rd', Plane, fresult(0:I-1), 'b-'); % Plots results

title('Double exponential fit to Photobleach data');

xlabel('Plane Number');

ylabel('Normalized Ratio Decay');

legend('Data', 'Fit');

axis([0, I, 0, 1.2]);

text (2, 0.4, 'y = a exp(b*x) + c exp(d*x)');

text (2, 0.33, coef);

text (2, 0.26, coef2);

```
text (2, 0.19, coef3);
% Writes data files for output and plotting off-line
fid = fopen('Fit_Decay_curve.txt', 'a');
fprintf(fid, '%12.6f \r', Fit_ratio);
fclose(fid);
fid = fopen('Actual_Ratio_data.txt', 'a');
fprintf(fid, '%12.6f \r', Ratio);
fclose(fid);
fid = fopen('goodness_of_fit.txt', 'a');
fprintf(fid, '%12.6f \r', gof);
fclose(fid);
% Image processing starting at this point, using the CF as the correction
if I < 100;
if I < 10;
for mult_rep = 1:I;
counter = mult_rep;
FNAMEIN = sprintf('%s%i.tif', fname, counter);
ratio = imread(FNAMEIN, 'tif');
R = double(ratio);
CR = R .* CF(counter); %Array element by element multiplication
CR_out = uint16(CR);
FNAMEOUT = sprintf('out_%i.tif',counter);
imwrite(CR_out, FNAMEOUT, 'tif', 'compression', 'none'); % write
    output file
end;
else;
for mult_rep = 1:I;
counter = mult_rep;
if counter < 10;
FNAMEIN = sprintf('%s0%i.tif', fname, counter);
else;
FNAMEIN = sprintf('%s%i.tif', fname, counter);
end;
ratio = imread(FNAMEIN, 'tif');
R=double(ratio);
CR = R .* CF(counter); %Array element by element multiplication
CR_out = uint16(CR);
if counter < 10; FNAMEOUT = sprintf('out_0%i.tif',counter);
else;
FNAMEOUT = sprintf('out_%i.tif',counter);
end;
```

```
imwrite(CR_out, FNAMEOUT, 'tif', 'compression', 'none'); % write
   output file
end;
end;
elseif I > = 100;
for mult_rep = 1:I;
counter = mult_rep;
if counter < 10;
FNAMEIN = sprintf('%s00%i.tif', fname, counter);
ratio = imread(FNAMEIN, 'tif');
R = double(ratio);
CR = R .* CF(counter); %Array element by element multiplication
CR_out = uint16(CR);
FNAMEOUT = sprintf('out_00%i.tif',counter);
imwrite(CR_out, FNAMEOUT, 'tif', 'compression', 'none'); % write
   output file
elseif counter < 100;
FNAMEIN = sprintf('%s0%i.tif', fname, counter);
ratio = imread(FNAMEIN, 'tif');
R = double(ratio);
CR = R .* CF(counter); %Array element by element multiplication
CR_out = uint16(CR);
FNAMEOUT = sprintf('out_0%i.tif',counter);
imwrite(CR_out, FNAMEOUT, 'tif', 'compression', 'none'); % write
   output file
elseif counter > = 100;
FNAMEIN = sprintf('%s%i.tif', fname, counter);
ratio = imread(FNAMEIN, 'tif');
R = double(ratio);
CR = R .* CF(counter); %Array element by element multiplication
CR_out = uint16(CR);
FNAMEOUT = sprintf('out_%i.tif',counter);
imwrite(CR_out, FNAMEOUT, 'tif', 'compression', 'none'); % write
   output file
end;
end;
end;
```

Acknowledgment

This work was supported by The National Institutes of Health (GM057464 and GM064346).

References

Chamberlain, C., and Hahn, K. M. (2000). Watching proteins in the wild: Fluorescence methods to study protein dynamics in living cells. *Traffic* **1**, 755–762.

Chamberlain, C. E., Kraynov, V., and Hahn, K. M. (2000). Imaging spatiotemporal dynamics of Rac activation *in vivo* with FLAIR. *Methods Enzymol.* **325**, 389–400.

Danuser, G. (1999). Photogrammetric calibration of a stereo light microscope. *J. Microscopy* **193**, 62–83.

Del Pozo, M. A., Kiosses, W. B., Alderson, N. B., Meller, N., Hahn, K. M., and Schwartz, M. A. (2002). Integrins regulate GTP-Rac localized effector interactions through dissociation of Rho-GDI. *Nat. Cell Biol.* **4**, 232–239.

Nalbant, P., Hodgson, L., Kraynov, V., Toutchkine, A., and Hahn, K. M. (2004). Activation of endogenous Cdc42 visualized in living cells. *Science* **305**, 1615–1619.

Patterson, G. H., and Piston, D. W. (2000). Photobleaching in two-photon excitation microscopy. *Biophys. J.* **78**, 2159–2162.

Ren, X. D., Kiosses, W. B., and Schwartz, M. A. (1999). Regulation of the small GTP-binding protein Rho by cell adhesion and the cytoskeleton. *EMBO J.* **18**, 578–585.

Shen, F., Hodgson, L., Rabinovich, A., Pertz, O., Hahn, K. M., and Price, J. H. (2004). Functional proteometrics for cell migration. *Cytometry*. In press.

Toutchkine, A., Kraynov, V., and Hahn, K. (2003). Solvent-sensitive dyes to report protein conformational changes in living cells. *J. Am. Chem. Soc.* **125**, 4132–4145.

[13] Cdc42 and PI(4,5)P₂-Induced Actin Assembly in *Xenopus* Egg Extracts

By ANDRES M. LEBENSOHN, LE MA,
HSIN-YI HENRY HO, and MARC W. KIRSCHNER

Abstract

Xenopus egg cytoplasmic extracts have been used to study a variety of complex cellular processes. Given their amenability to biochemical manipulation and physiological balance of regulatory proteins, these extracts are an ideal system to dissect signal transduction pathways leading to actin assembly. We have developed methods to study Cdc42 and PI(4,5) P₂-induced actin assembly in *Xenopus* egg extracts. In this chapter, we describe detailed procedures to prepare *Xenopus* egg extracts, Cdc42, and PI(4,5)P₂ for use in actin assembly experiments. We also describe a fluorometric pyrene actin assay for quantitative kinetic analysis of actin polymerization and a microscopic rhodamine actin assay for quick measurement of actin rearrangements in extracts. Finally we provide a protocol for immunodepletion of proteins and discuss the use of immunodepletion

METHODS IN ENZYMOLOGY, VOL. 406
0076-6879/06 $35.00
DOI: 10.1016/S0076-6879(06)06013-7

and rescue experiments for functional analysis of components in the extracts.

Introduction

Cytoplasmic extracts prepared from the eggs of the African clawed frog *Xenopus laevis* have proven to be a powerful cell-free system for dissecting a variety of complex biological processes, including nuclear assembly and disassembly (Newport and Spann, 1987), chromosome condensation (Ohsumi *et al.*, 1993), spindle assembly (Sawin *et al.*, 1992), DNA replication (Hutchison and Kill, 1989), and the cell cycle (Murray and Kirschner, 1989). A unique property of *Xenopus* egg extracts is that they recapitulate these complex processes in a manner very similar to that observed in intact cells. This is because they are prepared at high protein concentrations and maintain a physiological balance of regulatory proteins. Yet, unlike intact cells or whole organisms, extracts are amenable to extensive biochemical manipulation. Proteins can be removed by immunodepletion and added back at known concentrations or in mutant forms, and the extracts can be fractionated to purify components required for an activity of interest.

The utility of *Xenopus* egg cytoplasmic extracts for studying actin assembly was first established by experiments demonstrating that the intracellular pathogen *Listeria monocytogenes* moved in extracts propelled by actin comet tails very similar to those formed in infected somatic cells (Marchand *et al.*, 1995; Theriot *et al.*, 1994). These results suggested that the extracts contained the machinery required for actin assembly. Our own experiments then demonstrated that extracts also contained the components required to couple this machinery to cellular signaling intermediates such as phosphoinositides and Cdc42 (Ma *et al.*, 1998a), making them an ideal system to dissect signal transduction pathways leading to actin assembly.

Our initial studies showed that Cdc42 was required for phosphatidylinositol-4,5-biphosphate (PI[4,5]P₂)-induced actin assembly (Ma *et al.*, 1998a). Then, using Cdc42 as the agonist and microscopic and kinetic assays as activity readouts, we fractionated *Xenopus* egg extracts to purify downstream components required to modulate actin assembly. This strategy, combined with immunodepletion experiments, led us to discover that the Arp2/3 complex and N-WASP mediate actin polymerization downstream of Cdc42 (Ma *et al.*, 1998b; Rohatgi *et al.*, 1999), demonstrating for the first time that the actin nucleating activity of the Arp2/3 complex can be modulated by a cellular signaling pathway. The finding that N-WASP directly links Cdc42 to the Arp2/3 complex also helped establish the role of WASP family proteins as important nodes connecting upstream signals to actin nucleation. Furthermore, purification from *Xenopus* egg extracts revealed that the native N-WASP protein is almost entirely in a complex with WIP (WASP interacting protein) and led to the identification of Toca-1 (transducer of Cdc42-dependent actin assembly) as an essential modulator of this complex (Ho *et al.*, 2004). Unlike recombinant N-WASP, which can be activated by Cdc42 alone, activation of the N-WASP-WIP complex requires both Cdc42 and Toca-1.

The results of these studies, along with the work from a number of other laboratories, delineated a pathway from PI(4,5)P$_2$ through Cdc42, Toca-1, and the N-WASP-WIP complex regulating actin nucleation by the Arp2/3 complex. On the basis of these findings, we have been able to reconstitute Cdc42-mediated actin assembly *in vitro* using purified components (Ho *et al.*, 2004; Rohatgi *et al.*, 1999), a system that is the topic of the following chapter.

In this chapter, we describe how to prepare *Xenopus* egg extracts optimized for actin assembly experiments and how to prepare and use the agonists Cdc42 and PI(4,5)P$_2$ to stimulate actin polymerization. We also describe two assays to measure actin assembly in extracts: a fluorometric pyrene actin assay for quantitative kinetic analysis of polymerization reactions and a microscopic rhodamine actin assay for quick measurement of actin assembly. Finally, we provide a protocol for immunodepletion of proteins and discuss the use of immunodepletion and rescue experiments for functional analysis of components in the extract.

Preparation of *Xenopus* Egg Extracts for Actin Assembly Experiments

The procedure for preparing cycling and cytostatic factor (CSF)-arrested extracts used in cell cycle studies has been described in great detail (Murray, 1991). Here we will describe a modified protocol for making CSF-arrested extracts optimized for use in actin assembly experiments.

We typically obtain between 1 and 2 ml of concentrated high speed supernatant (HSS) per frog. Because we only use 20 μl of HSS per kinetic assay reaction and 4 μl of HSS per microscopic assay reaction, preparations starting with five frogs allow for relative ease of handling and yield enough material for a few hundred reactions. However, we have successfully prepared extracts starting with as few as one frog and as many as 200. We only use high-quality eggs (described later) to prepare extracts, and depending on the season, the age of the frogs, the water temperature, and a number of other factors, the quality of the eggs can vary significantly. Summer is usually the worst season, older (larger) frogs tend to do poorer than young ones, and water temperatures above 20° seem to be detrimental. Because of this variability, it is advisable to start preparations with a few extra frogs, given that generally some will lay poor-quality eggs that are useless. Except during the summer months, typically ~70% of the frogs lay good-quality eggs. Because the quality of the extracts decreases with prolonged handling time, careful planning and advanced preparation of reagents and instruments is also crucial.

Reagents

Pregnant mare serum gonadotropin (PMSG, Calbiochem 367222): 100 U/ml prepared in sterile water, stored in aliquots at −20° and warmed up to room temperature before use. Make 0.5 ml per frog.

Human chorionic gonadotropin (HCG, Sigma CG-10): 1000 U/ml prepared in sterile water, stored in aliquots at −20° and warmed up to room temperature before use. Make 0.5 ml per frog.

Marc's modified Ringers' (MMR): 100 mM NaCl, 2 mM KCl, 1 mM MgCl$_2$, 2 mM CaCl$_2$, 0.1 mM EDTA, 5 mM Na-HEPES, pH 7.8, at 16°. Prepared from a 25× stock. Make 2 l per frog for egg laying, plus 1 l per frog for washes.

Dejellying solution: MMR + 2% w/v cysteine, pH 7.8, at 16°. Prepared within 1 h of use from 25× MMR and solid L-cysteine (Sigma C-7352), and titrated to pH 7.8 with NaOH. Make 240 ml per frog.

Extract buffer (XB): 100 mM KCl, 0.1 mM CaCl$_2$, 1 mM MgCl$_2$, 10 mM K-HEPES, pH 7.7, at 16°, 50 mM sucrose. Make 400 ml per frog.

Extract buffer for CSF extracts (CSF-XB): 100 mM KCl, 2 mM MgCl$_2$, 10 mM K-HEPES, pH 7.7, at 16°, 50 mM sucrose, 5 mM EGTA. Make 200 ml per frog for washes, plus 20 ml per frog for dilution of extract.

Protease Inhibitors (PIs): 10 mg/ml each leupeptin (Sigma L-2884), pepstatin (Sigma P-5318), and chymostatin (Sigma C-7268) prepared in DMSO, stored in aliquots at −20° and diluted into the buffer immediately before use. Make 32 μl per frog.

Energy mix for CSF extracts (20× CSF-Energy mix): 150 mM phosphocreatine (Sigma P-7936), 20 mM ATP (adjusted to pH 7.0 with Tris base), 20 mM MgCl$_2$, stored in aliquots at −20°. Make 200 μl per frog.

Priming Frogs and Inducing Ovulation

Frogs are kept at 16° in dechlorinated water (preferably treated by reverse osmosis [RO], but tap water can also be used) made 15 mM in NaCl (preferably supplemented with 2% w/w Instant Ocean synthetic sea salts [Aquarium Systems, Inc.], which provide trace elements). On day 1, female frogs are primed for ovulation by injecting 50 units of pregnant mare serum gonadotropin (PMSG) into the dorsal lymph sac using a 27-gauge ½-inch sterile needle. After priming with PMSG, frogs are not fed until after the eggs have been collected to prevent excrement and regurgitated food from ruining the eggs. Ovulation is induced on any day from day 5 to at least day 12 by injecting 500 units of human chorionic gonadotropin (HCG) as described before. Before they begin to lay eggs, frogs are individually transferred to separate containers containing 2 l of Marc's modified Ringers' (MMR), and they are covered with perforated lids to allow air flow. The time from HCG injection to the beginning of egg laying is somewhat variable, but averages 12–14 h at 16°. Therefore, we usually inject the frogs around noon and transfer them in the evening approximately 8 h after injection, so that all the eggs are laid in MMR, but the period during which the frogs are kept in the same container as the eggs is minimized. On the day that we induce ovulation, we also prepare and refrigerate all buffers required for making the extract on the following day (except for dejellying solution, which must be prepared within 1 h of use).

Collecting Eggs

Eggs are collected ∼20 h after HCG injection (ideally no later than 10 h after they have been laid). Primed frogs typically lay packed egg volumes between 20 and 40 ml (measured after the eggs have settled). The frogs are removed from the containers, most of the MMR is poured off, and the eggs are inspected visually to assess their quality. Good-quality eggs are round, uniform in appearance and size, and are detached from one another (Fig. 1A). Poor-quality eggs are often attached together, forming strings or clumps, may have irregular shapes, may appear large and pale, or may exhibit extensive pigment mottling or variegation (Fig. 1B). Any batch in which more than 10% of the eggs are of poor quality is discarded. A few poor-quality eggs can be removed individually from good batches by picking them out with a glass Pasteur pipette whose tip has been broken off to give an orifice about 4 mm in diameter and fire polished. All batches of good-quality eggs are pooled together and washed five times with MMR (∼5 packed egg volumes each wash) to remove shed frog skin, excrement, and other detritus. Because the eggs settle quickly to the bottom of the container, washes are carried out by pouring solution into the container and decanting carefully, but swiftly, while eggs are resettling at the bottom, but the debris is still floating. For a five-frog prep, collection, sorting, and washing should take no longer than 45 min.

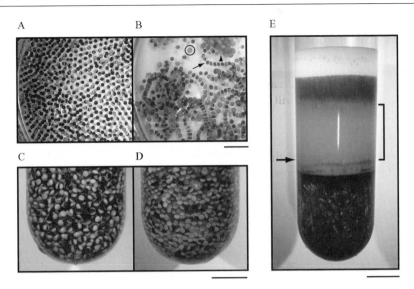

Fig. 1. *Xenopus laevis* eggs. (A) Good-quality eggs are round and uniform. (B) Poor-quality eggs are often attached together forming strings (arrow) or clusters (arrowhead) or may be large and pale (circled). (C) Before dejellying, eggs are separated by their transparent jelly coats. (D) Dejellied eggs pack as tight spheres without any visible separation. (E) After centrifugation, three major layers result: lipid, cytoplasm, and yolk going from top to bottom. Cytoplasmic extract (marked by the bracket) is collected by puncturing the side of the tube near the bottom of the middle layer (at the site indicated by the arrow). Scale bar = 10 mm. (See color insert.)

Making Cytoplasmic Extract

As much MMR as possible is poured off. Eggs are dejellied by adding ~4 volumes of dejellying solution and swirling gently at short intervals for 5–10 min, preferably keeping the temperature as close to 16° as possible. The supernatant is decanted carefully, and the eggs are rinsed for another 2–5 min with ~half the original volume of dejellying solution. Dejellying should not be allowed to proceed longer than necessary, so during this step eggs should be monitored closely. Dejellying is complete when eggs pack as tight spheres without any visible separation caused by their jelly coats (compare Fig. 1C and 1D), and the packed egg volume is reduced to 20–25% of the original volume.

Dejellied eggs are washed four times in extract buffer (XB, ~10 packed dejelied egg volumes each wash) and twice in extract buffer for CSF extracts (CSF-XB, ~10 volumes each wash). Because dejellied eggs are very fragile and prone to lysing spontaneously, these washes should be carried out as quickly as possible. Damaged eggs turn pale and tend to float above intact eggs, so they can be gently poured off after each wash or manually discarded using a wide-mouthed polished Pasteur pipette.

For the subsequent centrifugation steps, any soft-walled centrifuge tube allowing perforation with a needle and suitable for a swinging-bucket rotor able to reach 25,000g can be used. We routinely use Beckman Thinwall Polyallomer or Ultra-Clear 13- × 51-mm tubes for spinning dejellied egg volumes up to 5 ml in SW 50.1 or MLS 50 rotors, 14- × 95-mm tubes for spinning volumes up to 12 ml in an SW 40 Ti rotor, and 25- × 89-mm tubes for spinning volumes up to 36 ml in an SW 28 rotor. Before transferring the eggs, each centrifuge tube is loaded with a volume of CSF-XB plus 100 μg/ml (each) protease inhibitors (PIs) equivalent to $\sim$10% of the volume of eggs that will be transferred. After the final wash of the eggs with CSF-XB, as much buffer as possible is poured off, and the eggs are carefully transferred until each centrifuge tube is full. To avoid damaging the eggs, a wide-mouthed fire polished Pasteur pipette can be used to transfer small volumes, or a Falcon 100 ml serological pipette (BD Biosciences 357600) can be used to transfer large volumes (this pipette has a wide opening and smooth internal edges). After transferring the eggs, excess CSF-XB is removed, eggs are spun at 300g for 1 min to displace interstitial buffer trapped between them, and the excess buffer from the top of the tube is removed again, thereby decreasing the dilution of the cyto-plasm in the final extract. Eggs are then crushed by spinning at 25,000g for 20 min at 4°. After this step, all handling is done at 4°, and extracts are kept on ice.

After centrifugation, eggs will have ruptured and separated into three major layers of lipid, cytoplasm, and yolk going from top to bottom, with other minor layers at the interfaces (Fig. 1E). Cytoplasmic extract (see bracket in Fig. 1E) is collected using a syringe fitted with a 16-gauge beveled needle by puncturing the side of the tube at the bottom of the middle layer (see arrow in Fig. 1E) and very slowly drawing the cytoplasm while avoiding the interfaces. Carefully rotating the needle while punctur-ing the tube can facilitate perforation and avoid clogging the needle with plastic bits, but because the cytoplasm is turbid, it is important to note the beveled side of the needle to orient it upwards before starting to draw the cytoplasm. We typically recover $\sim$20% of the original volume of dejellied eggs as cytoplasmic extract. Protease inhibitors are added to a final con-centration of 10 μg/ml (each), energy mix for CSF extracts (20× CSF-energy mix) is added to a final concentration of 1×, and the extract is mixed gently by flicking or inverting.

Contamination of the cytoplasmic extract with material drawn from the interfaces and adjacent layers is not uncommon, and if we suspect this to be the case, we typically do a second spin as follows. The extract is transferred to new centrifuge tubes of the same kind and appropriate size and spun again at 25,000g for 20 min at 4°. After this spin, the top and bottom layers formed by contaminating material will probably be small, but cytoplasm from the middle layer is still recovered by side puncture as described before. The resulting extract can be used immediately for assays requiring membranes, it can be used to prepare high speed supernatant, or it can be supplemented with sucrose to a final concentration of 200 mM, snap-frozen in liquid nitrogen and stored at −80°. For a five-frog prep, making cytoplasmic extract should take no longer than 2 h.

Making High-Speed Supernatant

Crude cytoplasmic extracts are very viscous and, therefore, must be diluted for efficient sedimentation during high-speed ultracentrifugation. The high-speed supernatant (HSS) must then be concentrated back to the original volume for use in actin assembly assays. The crude extract is diluted 10-fold in CSF-XB plus 10 μg/ml (each) PIs and 1 mM DTT. The dilute extract is then transferred to prechilled ultracentrifuge bottles (Beckman 355618), 0.5 ml of mineral oil (Sigma M-5904) is layered on top, and the extract is centrifuged at 500,000g for 1 h at 4° in a fixed-angle rotor (Beckman Type 70 Ti). Light membranes that do not pellet during high-speed centrifugation will partition into the mineral oil layer and are removed after centrifugation by aspirating the top layer carefully (the rotor and tubes should be handled with care to avoid mixing). A small amount of contamination with light membranes is unavoidable and does not adversely affect the quality of the HSS. The HSS is carefully recovered by pipetting without disturbing the pellet and filtered through 0.22-μm syringe filter units (Millipore SLGV 033) to avoid clogging the concentrator's filter during the subsequent step. The HSS is reconcentrated to the original cytoplasmic extract volume using Centriprep YM-10 Centrifugal Filter Units (Millipore 4304) according to the manufacturer's instructions. When more than one centrifugation step is required to achieve the desired volume, the retentate should be mixed between spins to avoid high local protein concentration. The HSS is recovered from the concentrator, centrifuged at 20,000g for 10 min at 4° to remove particulates, and 20× CSF-energy mix is added to a final concentration of 1×. The final protein concentration in the HSS should be ~25 mg/ml. The HSS can be used immediately for actin polymerization assays or supplement with sucrose to a final concentration of 200 mM, snap-frozen in 100- to 500-μl aliquots in liquid nitrogen, and stored at −80°. Making and concentrating HSS can take several hours.

Preparation of Prenylated Cdc42

Bacterially expressed Cdc42, which lacks the geranylgeranyl modification, fails to induce actin assembly in *Xenopus* egg HSS. We, therefore, use isoprenylated, tagged Cdc42 purified from the membranes of baculovirus infected *Spodoptera frugiperda* Sf9 cells by glutathione affinity chromatography. A protocol for purification of baculovirus-expressed untagged Cdc42 has been described in a previous volume of *Methods in Enzymology* Cerione *et al.*, 1995), but because the protocol we routinely use is significantly different, we will describe it in detail here. For smaller scale purifications, the same basic procedure can also be done starting with transient transfections of a GST-Cdc42 construct into a number of different cell lines (293T, 293F, COS).

Reagents

Phosphate-buffered saline (PBS): 10 mM phosphate, pH 7.4, at 25°, 137 mM NaCl, 2.7 mM KCl.

Sonication buffer: 100 mM NaCl, 5 mM MgCl$_2$, 20 mM Na-HEPES, pH 7.4, at 4°, 1 mM EDTA, 1 mM DTT, 0.1 mM GTP, 1 mM Phenylmethylsulfonyl Fluoride (PMSF, MP Biomedicals 800263, prepared as a 1 M stock in DMSO), 1× Complete EDTA-free protease inhibitor cocktail tablets (Roche Molecular Biochemicals 1873580), 10 μg/ml each leupeptin, pepstatin, and chymostatin.

Extraction buffer: 100 mM NaCl, 5 mM MgCl$_2$, 20 mM Na-HEPES, pH 7.8, at 4°, 1 mM EDTA, 1 mM DTT, 0.1 mM GTP, 1 % Cholate, 1 mM PMSF, 10 μg/ml each leupeptin, pepstatin, and chymostatin.

Wash buffer 1: 100 mM NaCl, 5 mM MgCl$_2$, 20 mM Na-HEPES, pH 7.8, at 4°, 0.1 mM EDTA, 1 mM DTT, 1% cholate.

Wash buffer 2: same as wash buffer 1 but 0.1 % cholate.

Loading buffer: 100 mM NaCl, 50 mM Na-HEPES, pH 7.8, at 4°, 20 mM EDTA, 1 mM DTT, 0.1% cholate.

Storage buffer: 100 mM NaCl, 5 mM MgCl$_2$, 20 mM Na-HEPES, pH 7.8, at 4°, 0.1 mM EDTA, 10% glycerol, 1 mM DTT, 0.1% cholate.

Elution buffer: Storage buffer + 10 mM reduced glutathione, added from a 0.5 M stock of pH 8.0.

Procedure

Three liters of Sf9 cells are infected during log phase growth (density ∼1 million cells/ml) with recombinant baculovirus bearing the sequence for human Cdc42 N-terminally fused to glutathione S-transferase (GST). Cells are harvested 72 h after infection by centrifugation and washed once with phosphate-buffered saline (PBS). The cell pellet (∼30 ml) is resuspended thoroughly in 5 volumes of cold sonication buffer. All subsequent steps are carried out at 4° or on ice unless indicated otherwise. Cells are lysed by sonication (7 × 10-sec pulses on output control setting 5 and duty cycle setting 50% for an analog Branson Sonifier 450 with a 19 mm diameter solid horn). An additional 1 mM PMSF is added to the lysate.

The lysate is centrifuged at 2000g for 20 min in conical tubes using a swinging bucket rotor. This low-speed spin separates nuclei and unbroken cells in the pellet (P1) from cytoplasm and membranes in the supernatant (S1). The pellet (P1) is very loose, so the supernatant (S1) should be recovered carefully by pipetting as opposed to decanting. For increased yield, the pellet (P1) can be resuspended in two volumes of sonication buffer, homogenized in a glass/glass homogenizer (Kontes Dounce Tissue Grinder 40 ml) using the large clearance pestle (pestle A), centrifuged as before, and the supernatant pooled with the rest of the supernatant (S1).

The supernatant (S1) is centrifuged at 300,000g for 1 h in a fixed-angle rotor (Beckman Type 70 Ti). This high-speed spin separates membranes containing prenylated Cdc42 in the pellet (P2) from cytosol in the supernatant (S2). Prenylated Cdc42 is extracted from the membranes as follows. The pellet (P2) is resuspended in 40 ml of extraction buffer and homogenized in a glass/glass homogenizer using the small clearance pestle (pestle B). The homogenate is transferred to a 50-ml conical tube, and extraction is allowed to proceed for 90 min with gentle rotation. The homogenate is then centrifuged at 400,000g for 35 min in a fixed-angle rotor. This clarifying spin separates membranes in the pellet (P3) from extracted prenylated Cdc42 in the supernatant (S3).

The supernatant (S3) is transferred to a 50-ml conical tube and is incubated with 1 ml of glutathione sepharose 4B beads (Amersham Biosciences 17-0756-01, pre-equilibrated in extraction buffer) for 3 h with gentle rotation. The beads are transferred to a 20-ml column (Biorad Econo-Pac 732-1010) and washed by gravity flow with 25 ml of wash buffer 1 followed by 25 ml of wash buffer 2.

We prefer to load prenylated Cdc42 with GDP or GTPγS while the protein is still bound to the beads, because this strategy provides an easy way to remove all unbound nucleotide, which can affect actin assembly assays if added to *Xenopus* egg extracts. The beads are washed by gravity flow with 25 ml of loading buffer, they are resuspended in 3 ml of loading buffer to make 4 ml of a 25% slurry, and half of the slurry is transferred to each of two 2-ml columns (Biorad Bio-Spin 732-6008) capped at the bottom. GDP or GTPγS (Roche Molecular Biochemicals) is added to a final concentration of 1.2 mM, the columns are capped tightly at both ends and incubated at 30° for 15 min, mixing by inversion every 2 min. MgCl$_2$ is added to a final concentration of 30 mM, and the columns are incubated at 4° for 15 min mixing as before. The beads are drained and washed by gravity flow in the same column with 5 ml of storage buffer. A 50% slurry of the beads in storage buffer can be used directly in the microscopic rhodamine actin assay.

To elute bound Cdc42, the beads are drained and resuspended in the same column with 1.5 ml of elution buffer. The columns are capped tightly, and the elution reaction is allowed to proceed for 4 h at 4° with gentle rotation. The eluate is subsequently collected by gravity flow, and a second elution is carried out as before with another 1.5 ml of elution buffer. If the concentration of protein in the first elution is low (<0.1 mg/ml), the second elution can be done under harsher conditions by adjusting the pH of the elution buffer to 8.0, incubating at 4° for 1.5 h, heating to 30° for 10 min, and collecting the eluate immediately. The eluted protein is used directly in actin assembly assays, or it is snap frozen in aliquots and stored at −80°. Our typical yield is ~1 mg of eluted GST-Cdc42 protein, with a concentration range between 0.2–1 mg/ml. Because the GST tag does not affect the activity of Cdc42 in actin assembly assays, we generally do not cleave it. Furthermore, because the concentration of glutathione resulting after dilution of the eluted protein into the assays does not interfere with actin assembly, it is not necessary to dialyze it.

Preparation of Synthetic PI(4,5)P$_2$ Lipid Vesicles

Reagents

All lipids are purchased from Avanti Polar Lipids, Inc. and are stored at −80° in air-tight glass vials closed with Teflon-lined caps. (National Scientific Company B7800-1A).

Phosphatidylinositol-4,5-bisphosphate (PI[4,5]P$_2$): Cat. # 840046X, 5mg/ml in chloroform/methanol/water, MW = 1098.19.

Phosphatidylcholine (PC): Cat. #840051C, 25 mg/ml in chloroform, MW = 760.09.

Phosphatidylinositol (PI): Cat. # 840042C, 10 mg/ml in chloroform, MW =909.12.

Lipid buffer: 20 mM K-HEPES, pH 7.7, at 4°, 1 mM EDTA.

Procedure

To prepare 250 μl of PI(4,5)P$_2$-containing vesicles (PI[4,5]P$_2$:PC:PI, 4:48:48 molar ratio, 1 mM total lipid), solutions of PI(4,5)P$_2$ (2.2 μl), PC (3.65 μl) and PI (10.9 μl) are mixed in a glass test tube and dried under a stream of dry nitrogen. A thin film of lipid should be visible at the bottom of the tube. The lipid is dried further under vacuum for 1 h to remove residual solvent. The dried lipid mixture is resuspended thoroughly in 250 μl of lipid buffer and then extruded 15 passes through a 0.1 μm pore polycarbonate membrane using the Mini-Extruder (Avanti Polar Lipids, Inc.). The resuspended lipid mixture should appear cloudy before extrusion and clear thereafter. Lipid vesicles are stored in the dark at 4° for up to 1 month.

Actin Polymerization Assays

We use two different assays to monitor Cdc42 and PI(4,5)P$_2$-induced actin polymerization in *Xenopus* egg extracts. In the first assay, we supplement the extracts with pyrene-labeled actin and follow the reaction kinetics in a fluorescence spectrophotometer. The fluorescence intensity of pyrene actin increases ~25-fold on polymerization (Kouyama and Mihashi, 1981), making this assay extremely useful for quantitative analysis of the kinetics of actin assembly (Fig. 2A). We have also adapted this assay for high-throughput screening of small molecule inhibitors of PI(4,5)P$_2$-mediated actin polymerization (Peterson *et al.*, 2001). In the second assay, we supplement the extracts with rhodamine-labeled actin and monitor the reaction visually under a fluorescence microscope. Unlike pyrene actin, the

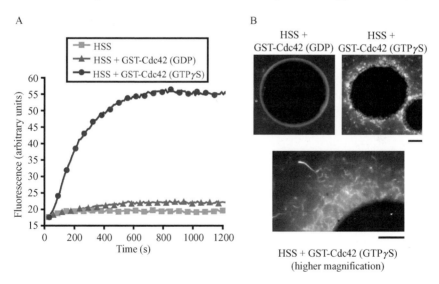

FIG. 2. Kinetic analysis of actin assembly in extracts supplemented with pyrene actin and microscopic analysis of actin assembly in extracts supplemented with rhodamine actin. (A) Prenylated GST-Cdc42 (200 n*M*) charged with GTPγS, but not GDP, induces actin polymerization. (B) Prenylated GST-Cdc42 on glutathione sepharose beads induces formation of actin foci and comet tails in a nucleotide-dependent manner. Scale bar = 25 μm.

fluorescence intensity of rhodamine actin does not increase on polymerization, and instead the observed variation in fluorescent signal is purely due to changes in local rhodamine actin concentration. However, in *Xenopus* egg extracts, Cdc42-induced actin polymerization results in the formation of large actin clusters shaped like foci and comet tails. These clusters are readily observable on the surface of GST-Cdc42 coated glutathione sepharose beads (Fig. 2B) or diffusing freely in reactions supplemented with soluble GST-Cdc42. Because the rhodamine actin assay is quick and requires only a few microliters of material, it is the assay of choice to follow actin polymerization activity during biochemical fractionation experiments.

Reagents

Pyrene actin: Prepared according to an established protocol (Kouyama and Mihashi, 1981) using rabbit skeletal muscle actin and N-(1-pyrene) iodoacetamide (Molecular Probes, Inc., P-29). A detailed protocol for preparing and characterizing pyrene actin has been described in an earlier volume of *Methods in Enzymology* (Zigmond, 2000). Pyrene actin can also be purchased from Cytoskeleton, Inc.

Rhodamine actin: Prepared according to an established protocol (Symons and Mitchison, 1991) using purified rabbit skeletal muscle actin and tetramethylrhodamine-5-iodoacetamide (Molecular Probes, Inc., T-6006). Rhodamine actin is also commercially available from Cytoskeleton, Inc.

$20\times$ CSF-energy mix: 150 mM creatine phosphate, 20 mM ATP, pH 7.0, 20 mM MgCl$_2$.

$10\times$ Assay buffer: 200 mM Hepes, pH 7.7, at 20°, 1 M KCl, 10 mM MgCl$_2$, 1 mM EDTA, 10 mM DTT.

Kinetic Analysis of Actin Assembly Using Pyrene Actin

A typical actin polymerization reaction is assembled as follows:

Volume added	Final concentration
20 μl *Xenopus* egg HSS (25 mg/ml total protein)	1:3 dilution of HSS
5 μl pyrene actin (1 mg/ml = 24 μM)	2 μM
3 μl 20$\times$ CSF-energy mix	1$\times$
4 μl 10$\times$ assay buffer	1$\times$ (HSS is already in a buffer of similar composition)
+	
26 μl ddH$_2$O	
2 μl GST-Cdc42 (0.3 mg/ml eluted protein = 6 μM)	200 nM
or	
27.4 μl ddH$_2$O	
0.6 μl PI(4,5)P$_2$-containing vesicles (1 mM total lipid)	10 μM (total lipid)
60 μl total	

All components except for Cdc42 or PI(4,5)P$_2$-containing vesicles are added in the order listed, adjusting the indicated volumes based on actual protein concentrations. The reaction is mixed and incubated for 5 min at room temperature to allow the basal steady-state level of filamentous actin to stabilize (dilution of the extract and addition of exogenous pyrene actin can cause the F-actin basal level to shift). Cdc42 or PI(4,5)P$_2$ is added to initiate actin polymerization, the reaction is mixed, transferred immediately to a quartz cuvette (Cary Sub-Micro Fluorometer Cell, 40 μl, Varian 66-100216-00) and up to four reactions are monitored simultaneously in a fluorescence spectrophotometer (Varian Cary Eclipse) equipped with a Peltier Multicell Holder, using the provided kinetics software with the temperature control set to 20°. Pyrene fluorescence is measured at 407 nm with excitation at 365 nm.

Some dilution of the extract is necessary to reduce the level of autofluorescence, but diluting the extract too much may result in spontaneous polymerization. A 1:3 dilution of concentrated HSS generally works well, but because there is variability between different batches of extract, the optimal dilution must be determined empirically. The rate and extent of Cdc42 or PI (4,5)P$_2$-mediated actin polymerization also varies according to the quality of the extract, so the optimal concentration of agonist should be established empirically. A control reaction using GST-Cdc42 loaded with GDP should be conducted in parallel to account for nonspecific effects on actin polymerization. In this sample reaction, pyrene actin is added to a final concentration of 2 μM, but depending on the extent of pyrene labeling, between 1 and 3 μM can be added to obtain a good fluorescence signal. Because of the variability inherent in the preparation of reagents, reactions should only be compared when conducted the same day using the same batch of HSS, pyrene actin, and agonist. A representative assay for Cdc42-induced actin assembly is shown in Fig. 2A.

Microscopic Analysis of Actin Assembly Using Rhodamine Actin

A typical actin polymerization reaction is assembled as follows:

4 ul *Xenopus* egg HSS (25 mg/μl total protein)
0.5 ul rhodamine actin (1 mg/μl)
0.35 μl 20× CSF-energy mix
1.65 μl 1× assay buffer
0.5 μl GST-Cdc42 (50% slurry of glutathione sepharose beads or 0.3 mg/ml eluted protein)

_____

7 μl total

All components except for Cdc42 are added in the order listed, mixed, and incubated for 5 min at room temperature to allow the basal steady-state level of F-actin to stabilize. Cdc42 is added to initiate polymerization, and 5 μl of the reaction mixture are squashed between a slide and a coverslip. The sample is immediately mounted on the stage of a fluorescence microscope and visualized or imaged using an appropriate filter set for rhodamine. Depending on the quality of the HSS, many foci and occasional comet tails should begin to appear 15 sec to 2 min after addition of GTPγS loaded GST-Cdc42. Although in our experience the formation of actin foci in response to Cdc42 is highly reproducible, background rhodamine fluorescence can vary depending on a number of factors, including the quality of the extract and proteins used. Control experiments to test the nucleotide dependence of the response to GST-Cdc42 and to test the cytochalasin D sensitivity of the fluorescent clusters should, therefore, be performed in parallel. Representative images of an assay in which actin assembly was induced by GST-Cdc42 on glutathione sepharose beads are shown in Fig. 2B.

Functional Analysis of Actin Assembly by Immunodepletion and Rescue

Immunodepletion is a powerful technique that allows the functional importance of a molecule of interest to be assessed in *Xenopus* egg extracts. The specificity of immunodepletion can be confirmed by adding back recombinant protein to "rescue" the biological activity under investigation. Furthermore, structure–function analysis can be conducted by testing the activities of recombinant proteins added back to the immunodepleted extracts. During the course of our study of the Cdc42-actin signaling pathway, we used immunodepletion to demonstrate the functional requirement of both N-WASP and Toca-1 for Cdc42-mediated actin nucleation in *Xenopus* egg extracts (Rohatgi *et al.*, 1999; Ho *et al.*, 2004) (see Fig. 3 for Toca-1 immunodepletion experiments).

The quality of the antibody used for depletion is the single most important factor determining the ability to deplete a specific protein or protein complex from the extract. In our experience, antibodies that work best for immunodepletion are prepared by immunizing rabbits with recombinant full-length proteins. We commonly use antibodies that have been affinity purified. Crude serum can be used directly if the titer of the specific antibody is sufficiently high. The use of crude serum is preferable in some cases, given that polyclonal antibodies with the highest affinity are sometimes poorly recovered after affinity purification.

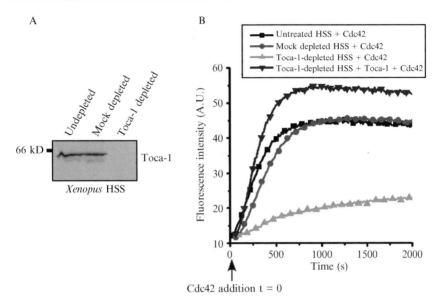

FIG. 3. Immunodepletion of Toca-1 from *Xenopus* egg extract. (A) Western blot analysis showing depletion of Toca-1 from *Xenopus* egg HSS. (B) Pyrene actin assay demonstrating the functional requirement of Toca-1 for Cdc42-mediated actin nucleation in *Xenopus* egg extracts. Addition of recombinant Toca-1 to the Toca-1 depleted extract "rescues" the ability of Cdc42 to induce actin polymerization. (Reprinted with permission from Ho *et al.*, 2004.

Immunodepletion of Xenopus *Egg Extracts*

Seven micrograms of antibody, 25 μl of protein A–coupled magnetic Dynabeads (Dynal Biotech 100.01), and 100 μl of PBS containing 0.1% Triton X-100 are mixed in a 0.5-ml Eppendorf tube by gentle rotation for 1 h at room temperature. We use a magnetic particle concentrator (Dynal Biotech MPC) to retrieve the beads during the following steps. The beads are washed twice with 100 μl of PBS containing 0.1% Triton X-100 and three times with 100 μl of 1× assay buffer. These washes are necessary to remove all unbound antibody and detergent, which can affect subsequent assays using the extracts. Because the beads tend to stick to the tube in the absence of detergent, one should proceed promptly after the washes. After the final wash, all buffer is removed, and 100 μl of *Xenopus* egg HSS is added to the beads. The mixture is incubated for 90 min at 4°, mixing every 20 min. Depending on the antibody, it is sometimes necessary to perform a second round of immunodepletion using fresh antibody-coupled beads.

After the beads are retrieved using the magnetic particle concentrator, the HSS is transferred to a clean tube and centrifuged briefly to remove any residual beads. The supernatant is transferred once more and stored on ice until use.

For actin assembly assays, we use the depleted extracts within 6 h after immunodepletion. Because the immunodepletion procedure can affect the extract, a mock depletion using preimmune IgG or random IgG from the same species should be performed in parallel and taken into consideration when interpreting immunodepletion results. Depletion of the protein of interest can be assessed by Western blot (Fig. 3A). An experiment in which endogenous Toca-1 was immunodepleted from *Xenopus* egg HSS and recombinant Toca-1 protein was added back to rescue Cdc42-mediated actin assembly is shown in Fig. 3B.

Concluding Remarks

In this chapter, we have described a set of tools for studying signal-dependent actin assembly in a biochemically amenable physiological system. We hope we have conveyed how *Xenopus* egg extracts can serve as a powerful system to study regulated actin assembly, both as an engine for the discovery of novel pathway components and as a versatile biochemical tool to investigate pathway mechanisms. Reconstitution of other signal transduction pathways in these extracts should allow detailed biochemical investigations of the kind we have conducted on the $PI(4,5)P_2$ and Cdc42 pathway.

Acknowledgments

We thank Andrew Horwitz, Jeffrey Peterson, Rajat Rohatgi, and Orion Weiner for comments on the manuscript. The studies described in this chapter were supported by grants from The National Institutes of Health to M. W. K.

References

Cerione, R. A., Leonard, D., and Zheng, Y. (1995). Purification of baculovirus-expressed Cdc42Hs. *Methods Enzymol.* **256,** 11–15.

Ho, H. Y., Rohatgi, R., Lebensohn, A. M., Ma, L., Li, J., Gygi, S. P., and Kirschner, M. W. (2004). Toca-1 mediates Cdc42-dependent actin nucleation by activating the N-WASP-WIP complex. *Cell* **118,** 203–216.

Hutchison, C., and Kill, I. (1989). Changes in the nuclear distribution of DNA polymerase alpha and PCNA/cyclin during the progress of the cell cycle, in a cell-free extract of Xenopus eggs. *J. Cell Sci.* **93**(Pt 4), 605–613.

Kouyama, T., and Mihashi, K. (1981). Fluorimetry study of N-(1-pyrenyl)iodoacetamide-labelled F-actin. Local structural change of actin protomer both on polymerization and on binding of heavy meromyosin. *Eur. J. Biochem.* **114,** 33–38.

Ma, L., Cantley, L. C., Janmey, P. A., and Kirschner, M. W. (1998a). Corequirement of specific phosphoinositides and small GTP-binding protein Cdc42 in inducing actin assembly in *Xenopus* egg extracts. *J. Cell Biol.* **140,** 1125–1136.

Ma, L., Rohatgi, R., and Kirschner, M. W. (1998b). The Arp2/3 complex mediates actin polymerization induced by the small GTP-binding protein Cdc42. *Proc. Natl. Acad. Sci. USA* **95,** 15362–15367.

Marchand, J. B., Moreau, P., Paoletti, A., Cossart, P., Carlier, M. F., and Pantaloni, D. (1995). Actin-based movement of Listeria monocytogenes: Actin assembly results from the local maintenance of uncapped filament barbed ends at the bacterium surface. *J. Cell Biol.* **130,** 331–343.

Murray, A. W. (1991). Cell cycle extracts. *Methods Cell Biol.* **36,** 581–605.

Murray, A. W., and Kirschner, M. W. (1989). Cyclin synthesis drives the early embryonic cell cycle. *Nature* **339,** 275–280.

Newport, J., and Spann, T. (1987). Disassembly of the nucleus in mitotic extracts: Membrane vesicularization, lamin disassembly, and chromosome condensation are independent processes. *Cell* **48,** 219–230.

Ohsumi, K., Katagiri, C., and Kishimoto, T. (1993). Chromosome condensation in *Xenopus* mitotic extracts without histone H1. *Science* **262,** 2033–2035.

Peterson, J. R., Lokey, R. S., Mitchison, T. J., and Kirschner, M. W. (2001). A chemical inhibitor of N-WASP reveals a new mechanism for targeting protein interactions. *Proc. Natl. Acad. Sci. USA* **98,** 10624–10629.

Rohatgi, R., Ma, L., Miki, H., Lopez, M., Kirchhausen, T., Takenawa, T., and Kirschner, M. W. (1999). The interaction between N-WASP and the Arp2/3 complex links Cdc42-dependent signals to actin assembly. *Cell* **97,** 221–231.

Sawin, K. E., Le Guellec, K., Philippe, M., and Mitchison, T. J. (1992). Mitotic spindle organization by a plus-end-directed microtubule motor. *Nature* **359,** 540–543.

Symons, M. H., and Mitchison, T. J. (1991). Control of actin polymerization in live and permeabilized fibroblasts. *J. Cell Biol.* **114,** 503–513.

Theriot, J. A., Rosenblatt, J., Portnoy, D. A., Goldschmidt-Clermont, P. J., and Mitchison, T. J. (1994). Involvement of profilin in the actin-based motility of L. monocytogenes in cells and in cell-free extracts. *Cell* **76,** 505–517.

Zigmond, S. H. (2000). *In vitro* actin polymerization using polymorphonuclear leukocyte extracts. *Methods Enzymol.* **325,** 237–254.

[14] *In Vitro* Reconstitution of Cdc42-Mediated Actin Assembly Using Purified Components

By HSIN-YI HENRY HO, RAJAT ROHATGI,
ANDRES M. LEBENSOHN, and MARC W. KIRSCHNER

Abstract

In the accompanying chapter, we describe an *in vitro* system that uses *Xenopus* egg extracts to study actin assembly induced by phosphatidylinositol (4,5)bisphosphate (PIP$_2$) and Cdc42. Biochemical fractionation and candidate screening experiments conducted in the extract system have identified the Arp2/3 complex, the N-WASP-WIP (or N-WASP-CR16) complex, and the Cdc42-binding protein Toca-1 as important mediators of PIP$_2$- and Cdc42-actin signaling. Toward our ultimate goal of reconstituting an *in vitro* system that recapitulates the signaling properties observed *in vivo*, we then developed a purified actin assembly assay system consisting of the regulatory components that we discovered from extracts. In these assays, the stereotypical sigmoidal kinetics of actin polymerization are monitored by pyrene-actin fluorescence in the presence of defined recombinant or purified proteins, enabling the detailed study of mechanism and protein function. In this chapter, we describe the preparation of the components used in these purified actin assembly reactions, as well as the assay conditions under which we monitor actin polymerization kinetics *in vitro*.

Introduction

In the previous chapter, we described an *in vitro* system that uses total cytoplasmic extracts made from *Xenopus* eggs to monitor actin assembly induced by phosphatidylinositol (4,5) bisphosphate (PIP$_2$) and Cdc42. The extract systems are very useful for determining *requirements* for particular components, using either biochemical fractionation or immunodepletion/ add-back of candidates. However, to establish *sufficiency* and to explain the direct interactions between components, we have developed purified actin assembly systems based on the well-established pyrene-actin assay (Cooper *et al.*, 1983; Kouyama and Mihashi, 1980). In these assays, the stereotypical sigmoidal kinetics of pyrene-labeled pure actin is followed in the presence of defined (either recombinant or purified) proteins. Our overall strategy in studying this signaling pathway has been to first identify required proteins using the extract system and then to move to the purified

METHODS IN ENZYMOLOGY, VOL. 406
0076-6879/06 $35.00
DOI: 10.1016/S0076-6879(06)06014-9

system to determine the mechanism and protein function, with the ultimate goal of building a reconstituted system that reflects (as closely as possible) the signaling properties *in vivo*.

We describe here two purified signaling modules that link Cdc42 and PIP$_2$ to the actin nucleating Arp2/3 complex. In the first system (called Reaction #1 henceforth), Cdc42 and PIP$_2$ synergistically activate N-WASP, which in turn stimulates the actin nucleating activity of the Arp2/3 complex. This is a remarkably parsimonious module that shows how the activation properties of N-WASP can dictate actin polymerization precisely at the temporal and spatial intersection of two signals. Since its original report in 1999 (Rohatgi *et al.*, 1999), Reaction #1 has been widely used to study mechanistic aspects of N-WASP regulation, including the autoinhibition of N-WASP and the cooperative mechanism of N-WASP activation by Cdc42 and PIP$_2$ (Higgs and Pollard, 2000; Rohatgi *et al.*, 2000). In addition, it has served as an assay to identify additional regulators of N-WASP activity, such as the adaptor proteins Nck/Grb2 and tyrosine kinases (Carlier *et al.*, 2000; Cory *et al.*, 2002; Rohatgi *et al.*, 2001; Suetsugu *et al.*, 2002; Torres and Rosen, 2003).

Despite the insights gained from studying Reaction #1, detailed analyses of the extracts system have suggested that the Cdc42 signaling pathway is significantly more complicated and involves additional levels of regulation. To understand the molecular logic of this complexity, we have recently identified (through biochemical purification from extracts) two additional components of the Cdc42 pathway, an effort that has allowed us to describe a more complex purified assay system (Reaction #2) (Ho *et al.*, 2004). These components include the protein WIP or the closely related CR16, which exists in a tight complex with N-WASP in these extracts and the novel Cdc42 binding protein Toca-1 (Ho *et al.*, 2001, 2004; Ramesh *et al.*, 1997). We believe that Reaction #2 represents a more physiological recapitulation of the Cdc42 pathway *in vivo*, because it exhibits a number of features that are characteristic of the complete extract system but are lacking in our original Reaction #1.

In this chapter, we describe the preparation of the components used in Reactions #1 and #2, as well as the assay system used to monitor *in vitro* the actin assembly kinetics in these reactions. Finally, some properties of these reactions are discussed.

Preparation of Reaction Components

Preparation of Recombinant N-WASP

We have used untagged recombinant full-length N-WASP (rat origin) in our purified actin polymerization system (Reaction #1). Because a major use of these assays is to study N-WASP activation, it is critical to use protein preparations that show low levels of basal activity toward the Arp2/3 complex. Several affinity-tagged versions of N-WASP have been made in our laboratory, but the activities of these fusion proteins are quite variable depending on the epitope used. Myc(x6)-N-WASP and His(x6)-N-WASP exhibit significant levels of basal activation even in the absence of activators and are, therefore, less useful for most applications (unpublished observations). GST-N-WASP behaves more similarly to untagged N-WASP in that it has a relatively low basal activity that can be further stimulated by Cdc42 and PIP$_2$; however, this protein has the caveat that it is dimerized through the GST moiety (untagged N-WASP is a monomer [Rohatgi et al., 2000]). For these reasons, we have used the untagged protein in most of our experiments that involve recombinant N-WASP.

Untagged N-WASP is expressed in *Spodoptera frugiperda* (Sf9) cells using the Bac-to-Bac baculoviral protein expression system (Invitrogen) and purified using a three-step conventional purification scheme modified from Miki *et al.* (1996) and Rohatgi *et al.* (1999) (Fig. 1). We have not been able to produce recombinant full-length N-WASP in *Escherichia coli*. All chromatographic steps are performed using an FPLC liquid handling system (Amersham Biosciences) equipped with columns purchased from the same supplier. The following procedure is based on a typical preparation that begins with 3 l of Sf9 culture, yielding approximately 10 mg of highly purified N-WASP. With proper preparation, the entire purification can be completed in 1 long day.

Sf9 cells are infected with a recombinant baculovirus overexpressing the N-WASP protein at a density of 1 million cells/ml. Cells are harvested 72 h after infection by centrifugation, washed once with PBS (20 mM Na$_2$HPO$_4$, 2.8 mM KH$_2$PO$_4$, pH 7.3, 140 mM NaCl, 2.7 mM KCl) and resuspended in 200 ml lysis buffer (20 mM Na-Tris, pH 7.5, 20 mM NaCl, 2 mM EDTA, 1 mM PMSF, 1 mM DTT, 10 μg/ml each leupeptin/pepstatin/chymostatin, 1× complete protease inhibitor cocktail [tablets with EDTA, Roche]). Cells are lysed by sonication, and the lysate is cleared by centrifugation at ~160,000g (45,000 rpm in a Beckman Type 45 Ti rotor) for 1 h. The supernatant is filtered through a 0.45-μm syringe filter (Millipore), loaded on two 5-ml HiTrap Heparin columns connected in series, and N-WASP is eluted with a linear gradient of 0–700 mM NaCl developed over 10 column volumes. N-WASP elutes at approximately 600 mM NaCl and can be identified on a Gelcode Blue (Pierce)–stained gel as a prominent protein band migrating at ~66 kDa (see Fig. 1).

The peak N-WASP–containing fractions are pooled and dialyzed against 2 l of 20 mM Na-Tris, pH 8.0, 50 mM NaCl, 1 mM DTT for 1 h (some precipitation may occur). After the dialysis, the protein sample is diluted with an equal volume of buffer Q-A (20 mM Na-Tris, pH 8.3,

50 mM NaCl, 1 mM DTT) and centrifuged at ~180,000g (50,000 rpm in a Beckman Type 70 Ti rotor) for 15 min. The supernatant is filtered through a 0.45-μm syringe filter (Millipore) and loaded onto a 5-ml HiTrap Q column, and N-WASP is eluted with a linear gradient of 50–700 mM NaCl over 10 column volumes. N-WASP elutes at 200–250 mM NaCl, and can be identified on a Gelcode Blue–stained gel as the predominant protein (Fig. 1).

The peak N-WASP–containing fractions are pooled and loaded onto a 120-ml HiLoad Superdex 200 16/60 gel filtration column pre-equilibrated with gel filtration buffer (20 mM HEPES, pH 7.7, 100 mM KCl, 1 mM DTT). The column is run at a flow rate of 0.5 ml/min, and N-WASP elutes as a single symmetric peak at approximately 64 ml. The peak N-WASP fraction has a concentration of ~1 mg/ml determined by the Bradford assay (BioRad), with a purity of >95% as judged by SDS PAGE and Gelcode Blue staining (Fig. 1). Fractions with protein concentrations lower than 1 mg/ml are supplemented with glycerol to a final concentration of 10% (v/v). N-WASP fractions are divided into single-use aliquots, snap frozen in liquid nitrogen, and stored at −80°.

Purification of Native N-WASP-WIP Complex from Xenopus Egg Extracts

Most of the N-WASP in extracts is present as a stable complex with WIP or CR16 (depending on the tissue source), and analysis of the Cdc42-Arp2/3-actin pathway required production of purified complex (Ho *et al.*, 2001, 2004; Ramesh *et al.*, 1997). Despite efforts using insect cells and cultured mammalian cells, we have not been able to produce a complex that exhibits proper regulation using recombinant techniques. Thus, native N-WASP–WIP complex is purified conventionally from the high-speed supernatant (HSS) made from *Xenopus* eggs (Fig. 2). We have previously described the isolation of native N-WASP-CR16 complex from bovine brain extracts (Ho *et al.*, 2001), but it requires a more difficult purification scheme because of the tissue heterogeneity of this starting material.

Throughout the purification, the N-WASP–WIP complex is followed by Western blotting with a rabbit polyclonal α-N-WASP antibody (Rohatgi *et al.*, 1999). All chromatographic media are purchased from Amersham Biosciences. *Xenopus* egg HSS is prepared from 150 frogs as described in the previous chapter, except that the crude cytoplasmic extract (~ 250 ml) is diluted only fivefold before high-speed ultracentrifugation and is not reconcentrated. Given the poor recovery of the complex, this purification scale is necessary to produce sufficient final protein yield (~0.5 mg) and purity (>95%). The HSS is first fractionated by performing a 20–35%

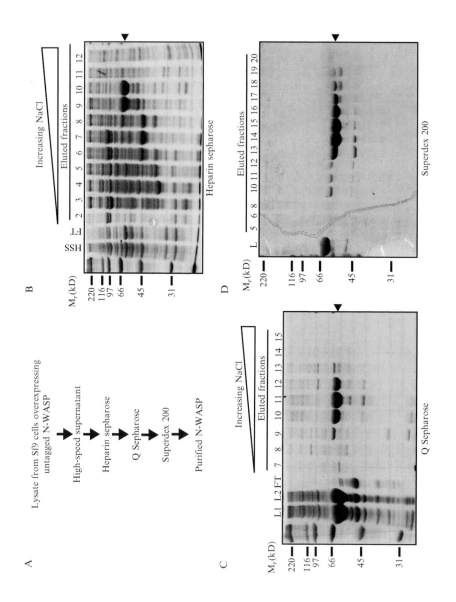

ammonium sulfate cut. The precipitate is collected by centrifugation (21,000g or 15,000 rpm in a Beckman Type 19 Ti rotor for 1 h), resuspended in 140 ml of buffer CSF-XB (100 mM KCl, 2 mM MgCl$_2$, 10 mM K-HEPES, pH 7.7, 50 mM sucrose, 5 mM EGTA), and dialyzed into buffer QA (30 mM Na-Tris, pH 7.8, 100 mM KCl, 1 mM EDTA, 1 mM DTT). The dialyzed protein sample is centrifuged at 265,000g (60,000 rpm in a Beckman Type 70 Ti rotor) for 1 h at 4°, and the supernatant is filtered through 0.22-μm syringe filters (Millipore). The fraction is then loaded onto a 66-ml Q Sepharose HP column (poured in a XK 26/40 column) and eluted with a linear gradient of 100–400 mM KCl developed over 10 column volumes. Peak N-WASP–containing fractions (eluting at ~200 mM NaCl) are pooled and dialyzed against buffer HA (20 mM Na-Tris, pH 7.5, 100 mM NaCl, 1 mM EDTA, 1 mM DTT) and loaded onto two 5-ml HiTrap Heparin HP columns connected in series. N-WASP–WIP is eluted from the heparin column with a linear gradient of 100–700 mM NaCl developed over 10 column volumes. Peak N-WASP–containing fractions (eluting at ~400 mM NaCl) are pooled and dialyzed into buffer SA (20 mM Na-PIPES, pH 6.8, 80 mM NaCl, 10% glycerol, 5 mM β-octylglucoside, 1 mM DTT) and centrifuged at 16,700g (20,000 rpm in a Beckman TLA 100.4 rotor) for 10 min to remove particulates. The material is loaded onto a 1-ml Mono S column, eluted with a linear gradient of 80–400 mM NaCl developed over 20 column volumes (N-WASP elutes at ~200 mM NaCl, see Fig. 2), and the fractions are mixed immediately with the appropriate volume of 1.5 M Tris base to bring to a final pH of 7.2. Before proceeding to the final sucrose gradient step, the peak N-WASP–containing fractions are pooled and concentrated so that all of the material can be loaded on a single 12-ml sucrose gradient. This is performed by binding the protein to a 0.1-ml Mono S column (Amersham Biosciences, Mono S PC 1.6/5 for SMART system; the sample is first diluted fourfold in buffer SA before applying to the column) and eluting it with a step gradient of 400 mM NaCl. The concentrated material is then loaded onto a 12-ml 5–20% (w/v) sucrose gradient poured in buffer XB (30 mM K-HEPES, pH 7.7, 100 mM KCl, 1 mM MgCl$_2$, 0.1 mM EDTA) plus 5 mM β-octylglucoside and 1 mM DTT. The sucrose gradient is centrifuged at 40,000 rpm for 17 h in a Beckman SW40 swinging-bucket rotor. The two subunits of the complex,

FIG. 1. Purification of recombinant N-WASP from Sf9 cells. (A) The fractionation scheme used to purify untagged N-WASP expressed recombinantly in Sf9 cells. (B–D) The protein composition of fractions from the various stages of the purification analyzed by SDS-PAGE and Gelcode Blue staining. The position of N-WASP is indicated by the arrowhead on the right side of each gel. (B) HSS, high-speed supernatant; FT, flow-through. (C) L1, load material before dialysis; L2, load material after dialysis; FT, flow-through. (D) L, load.

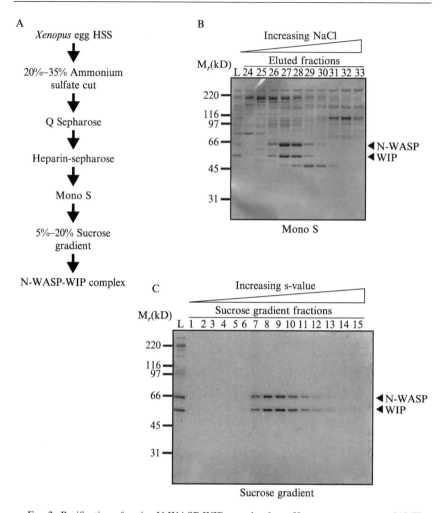

FIG. 2. Purification of native N-WASP-WIP complex from *Xenopus* egg extracts. (A) The fractionation scheme used to purify N-WASP-WIP complex from *Xenopus* egg extracts. (B) The protein composition of fractions from the Mono S step of the purification analyzed by SDS-PAGE and Gelcode Blue staining. The positions of N-WASP and WIP are indicated by the arrowhead on the right. (C) Fractions from the final sucrose gradient step of the purification analyzed by SDS-PAGE and Gelcode Blue staining. Again, the positions of N-WASP and WIP are indicated by the arrowhead on the right. L, load.

N-WASP and WIP, are identified on a Gelcode Blue–stained SDS poly-acrylamide gel as the only detectable protein bands (Fig. 2). Purified N-WASP–WIP complex is stored on ice and used in actin polymerization assays within 48 h. The sucrose gradient fractions can also be divided into single-use aliquots, snap frozen in liquid nitrogen, and stored at $-80°$. We have found that one freeze–thaw cycle does not significantly affect the activity of the complex in actin polymerization assays.

Preparation of Recombinant Toca-1

The wild-type cDNA encoding human Toca-1 is lethal to *E. coli.* Although the nature of this toxicity has remained unknown, we have generated a nontoxic form of human Toca-1 DNA (H-form) by introducing silent wobble-position mutations to codons 126 and 127 (Ho *et al.,* 2004) (Fig. 3). The H-form of human Toca-1 is tagged on its N-terminus with a hexahistadine tag and expressed in insect (Sf9) cells using the Bac-to-Bac baculoviral protein expression system (Invitrogen) and purified in one step over a nickel column. Despite considerable efforts, we have not been able to express recombinant human Toca-1 in *E. coli.*

Three hundred milliliters of Sf9 cells at a density of 1 million cells/ml are infected with the baculovirus overexpressing Toca-1. Cells are harvested 72 h after infection by centrifugation and washed once in PBS. The cell pellet is resuspended in 40 ml lysis buffer (50 mM NaH$_2$PO$_4$, pH 8.0, 300 mM NaCl, 10 mM imidazole, 1% Triton X-100, 5 mM β-mercaptoethanol, 1 mM PMSF, 1× protease inhibitor cocktail [tablets without EDTA, Roche]), homogenized 10 times using a glass/glass homogenizer with a small clearance pestle (Kontes Dounce Tissue Grinder, 40 ml, pestle B), and centrifuged at $\sim$100,000g (36,000 rpm in a Beckman Type 45 Ti rotor) for 1 h. The supernatant is collected and incubated in batch for 2 h to overnight at $4°$ with 0.5 ml of nickel agarose (Qiagen) pre-equilibrated with the lysis buffer. The beads are collected by passing the binding mixture through a 10-ml disposable column (BioRad) by gravity. The beads are washed in the column with 20-ml wash A (50 mM NaH$_2$PO$_4$, pH 8.0, 500 mM NaCl, 5 mM imidazole, 0.5% Triton X-100, 5 mM β-mercaptoethanol), followed by 20 ml of wash B (50 mM NaH$_2$PO$_4$, pH 8.0, 300 mM NaCl, 5 mM imidazole, 0.5% Triton X-100, 5 mM β-mercaptoethanol). Toca-1 is then eluted with 3–5 ml of elution buffer (50 mM NaH$_2$PO$_4$, pH 8.0, 300 mM NaCl, 255 mM imidazole, 5 mM β-mercaptoethanol) in batch. Our typical yield is 3–5 mg of protein from 300 ml of Sf9 culture with a purity of >95% as judged by Gelcode Blue staining (Fig. 3). Fractions with protein concentrations lower than 1 mg/ml are supplemented with glycerol to a final concentration of 10% (v/v). Purified Toca-1 is aliquoted, snap frozen in liquid nitrogen, and stored at $-80°$.

During preparative purifications of recombinant Toca-1, it is important to have at least 300 mM NaCl or KCl in all of the buffers, because Toca-1

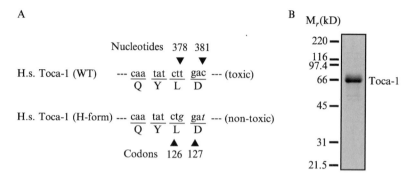

FIG. 3. Preparation of recombinant Toca-1. (A) The wild-type cDNA encoding Toca-1 is lethal to *E. coli*. Although the nature of this toxicity has remained unclear, we have generated a nontoxic form of human Toca-1 DNA (H-form) by introducing silent wobble-position mutations to codons 126 and 127. The H-form of human Toca-1 is tagged on its N-terminus with a hexahistadine tag, expressed in Sf9 cells and purified in one step over a nickel column. (B) A purified preparation of recombinant Toca-1 analyzed by SDS-PAGE and Gelcode Blue staining. (Reprinted with permission from Ho *et al.*, 2004.)

(at a protein concentration between 0.1–1 mg/ml) has a very low solubility under physiological salt concentrations. At a protein concentration of >1 mg/ml, even higher salt is needed to maintain the solubility of Toca-1. For actin polymerization reactions, we dilute Toca-1 directly from stock solutions in 300 mM NaCl into the assay mixtures (which typically contain 100 mM KCl). Given that the final Toca-1 concentration in these reactions is quite low (0.001–0.01 mg/ml), it seems to remain in solution under these conditions.

Purification of the Arp2/3 Complex from Bovine Brain Extracts

Several different purifications of the Arp2/3 complex have been described in the literature (Machesky *et al.*, 1994; Ma *et al.*, 1998; Welch and Mitchison, 1998). For all of the assays described in this chapter, we have used Arp2/3 complex purified from bovine brain extracts by conventional column chromatography (Fig. 4). Perhaps the simplest way to purify the Arp2/3 complex has been published by Egile *et al.* (1999), who isolated the complex from bovine brain based on its affinity for the C-terminus of N-WASP. However, we have found that such preparations have higher basal level of activity in the purified assay system and are less specific than conventional purified complex.

The following procedure is based on a preparation that begins with 20 calf bovine brains, yielding approximately 10 mg of purified Arp2/3

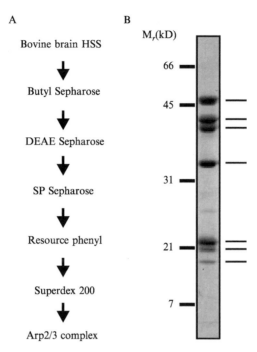

FIG. 4. Purification of the Arp2/3 complex from bovine brain extracts. (A) The fractionation scheme used to purify the Arp2/3 complex from bovine brain extracts. (B) The composition of the purified Arp2/3 complex shown on a 12% SDS polyacrylamide gel stained with Gelcode Blue. The thin, unlabeled lines on the right indicate the seven subunits of the complex. Listed in descending order of molecular weight, the subunits are Arp3, Arp2, p41-ARC, p34-ARC, p21-ARC, p20-ARC, and p16-ARC. (Reprinted with permission from Rohatgi, 1999.)

complex. However, purification from 5 calf brains can be handled more easily and should provide enough material for a few hundred assays. Throughout the purification, the Arp2/3 complex is followed by Western blotting with either α-Arp3 or α-p34ARC IgGs (provided by Dr. Matt Welch of University of California, Berkeley). α-Arp2 and α-Arp3 antibodies that are commercially available from Santa Cruz Biotech can also be used for this purpose. All chromatography media described here are purchased from Amersham Biosciences. First, freshly harvested calf brains obtained from a local slaughterhouse are stripped of meninges and adherent clots, coarsely chopped, and as much fibrous white matter is removed as possible; it is then quickly washed in ice-cold water. The brain pieces are mixed with 1:1 w/v ice-cold lysis buffer (10 mM K-HEPES, pH 7.3, 100 mM

KCl, 0.4 mM EDTA, 2 mM MgCl$_2$, 1 mM PMSF, 10 μg/ml each leupeptin/pepstatin/chymostatin, and 1 mM DTT) and blended in a Waring blender (setting 3) for 1 min. The blended material is then fed through a motorized continuous-flow overhead Teflon homogenizer (Yamato). The homogenate is spun at 12,000g for 30 min to remove particulates. The supernatant from this low-speed spin is collected and spun at 100,000g for 1 h to generate the clear, membrane-free bovine brain high-speed supernatant (bovine brain HSS).

Ammonium sulfate is added to 0.8 M and the bovine brain HSS is incubated in batch with butyl Sepharose. The flow through, containing most of the Arp2/3, is precipitated by adding solid ammonium sulfate to bring to a final concentration of 80% saturation. This precipitation procedure is repeated in several subsequent steps to reduce the volumes of the fractions (see later). The precipitate is collected by centrifugation and dialyzed into DEAE buffer (50 mM Tris-HCl, pH 7.8, 1 mM MgCl$_2$, 1 mM DTT, 0.2 mM ATP) and applied to a DEAE Sepharose column. The flow through contains the Arp2/3 complex and is precipitated again with ammonium sulfate (80%), and the pellet is dialyzed into S buffer (20 mM MES, pH6.1, 1 mM MgCl$_2$, 0.2 mM ATP, 1 mM DTT). The dialyzed material is loaded onto a SP-Sepharose column and eluted with a linear gradient of 0–350 mM KCl. Fractions containing Arp2/3 are neutralized to ~pH 7.0 with 1.5 M Tris base, pooled, and precipitated with ammonium sulfate (80%). The pellet is redissolved in buffer P (50 mM phosphate, pH 7.3, 1.2 M ammonium sulfate, 1 mM MgCl$_2$, 0.2 mM ATP, 1 mM DTT, 0.5 mM EDTA), applied to a Resource Phenyl column, and eluted with a linear gradient of 1.2–0 M ammonium sulfate. Again, fractions containing Arp2/3 complex are precipitated with 80% ammonium sulfate, resuspended in buffer F (20 mM phosphate, pH 7.3, 50 mM KCl, 1 mM MgCl$_2$, 0.2 mM ATP, 1 mM DTT), and fractionated on a Superdex 200 gel filtration column. The Arp2/3 complex elutes from the column at ~200 kDa and is ~90% pure as judged by Gelcode Blue staining (Fig. 4). The protein was supplemented with 10% (w/v) sucrose, snap frozen, and stored at $-80°$.

Cdc42 and PIP$_2$ Vesicles

Preparations of prenylated Cdc42- and PIP$_2$-containing vesicles are described in the previous chapter, except that the vesicle stock is made at 10 mM total lipid. In the purified system (both Reaction #1 and #2), we have found that bacterially produced Cdc42 is also effective at higher concentrations.

Actin and Pyrene Actin

The purification of actin from rabbit skeletal muscle and the procedure for pyrene labeling have been described by Zigmond (2000).

On the day before the assay, frozen aliquots of actin and pyrene actin (stock concentrations are typically 5–10 mg/ml) are thawed, diluted in G buffer (5 mM Tris-HCl, pH 8.0, 0.2 mM CaCl$_2$, 0.2 mM ATP, 0.2 mM DTT) to a final protein concentration of 0.5–1 mg/ml and incubated for ~10 h (usually done overnight) at 4° to allow depolymerization. Residual F-actin is removed by centrifugation at 400,000g (100,000 rpm in a Beckman TLA-100 rotor) for 1 h. Care must be taken when collecting the supernatant to avoid disrupting the pellet, which appears transparent and glossy. Actin concentrations are determined after centrifugation by the Bradford assay. Our typical recovery is 50–80%.

The final concentration of actin in the polymerization reactions is typically 1–2 μM (~35% pyrene actin). However, because of the variability in the quality of actin and the efficiency of pyrene labeling, the actual amount of actin, as well as the ratio of unlabeled/labeled actin used in the reaction, should be adjusted for each batch of actin/pyrene actin to achieve the optimal signal-to-background ratio.

Sample Reactions

> Reaction #1: Synergistic activation of recombinant N-WASP with Cdc42 and PIP$_2$ (Fig. 5)
>
> 39.1 μl H$_2$O
>
> 5.8 μl 10× Assay buffer (10×: 200 mM HEPES, pH 7.7, 1 M KCl, 10 mM MgCl$_2$, 1 mM EDTA, 10 mM DTT)
>
> 1.2 μl ATP (10 mM stock; 0.2 mM final)
>
> 0.2 μl Arp2/3 complex (9 μM stock; 30 nM final)
>
> 0.6 μl recombinant N-WASP (10 μM stock; 100 nM final)
>
> 0.5 μl Cdc42-GTPγS (30 μM stock; 250 nM final)
>
> 0.6 μl PIP$_2$ vesicles (10 mM stock; 100 μM final)
>
> 12.0 μl actin/pyrene actin mixture (10 μM stock, 65% unlabeled, 35% pyrene labeled; 2 μM final)
>
> ─────────────────────
>
> 60-μl reaction
>
> Reaction #2: Stimulation of native N-WASP–WIP complex with Toca-1 and Cdc42 (Fig. 6)
>
> 39.4 μl H$_2$O
>
> 5.8 μl 10× Assay buffer (10×: 200 mM HEPES, pH 7.7, 1 M KCl, 10 mM MgCl$_2$, 1 mM EDTA, 10 mM DTT)
>
> 1.2 μl ATP (10 mM stock; 0.2 mM final)

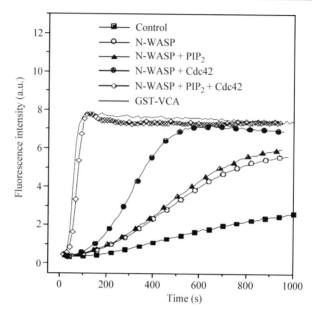

FIG. 5. Synergistic activation of recombinant N-WASP with Cdc42 and PIP$_2$ (Reaction #1). The pyrene actin assay used to monitor the effect of Cdc42-GTPγS (500 nM), PI(4,5)P$_2$-containing vesicles (100 μM, PC:PI:PI(4,5)P2, 48:48:4), or both on actin polymerization (2.5 μM total actin, 40% pyrene labeled, and 60 nM Arp2/3 complex) stimulated by N-WASP (200 nM). The solid line shows stimulation of actin polymerization under the same conditions by a GST fusion of VCA (200 nM), a C-terminal fragment of N-WASP that is constitutively active in stimulating the Arp2/3 complex. Note that the concentrations of reaction components shown in this figure are slightly different from the sample reaction given in the text, but the general properties of the reactions remain similar. (Reprinted with permission from Rohatgi, 1999.)

0.2 μl Arp2/3 complex (9 μM stock; 30 nM final)
0.6 μl native N-WASP–WIP complex (0.6 μM stock; 6 nM final)
0.3 μl Toca-1 (2 μM stock; 10 nM final)
0.5 μl Cdc42-GTPγS (30 μM stock; 250 nM final)
12.0 μl actin/pyrene actin mixture (10 μM stock, 65% unlabeled, 35% pyrene labeled; 2 μM final)

60-μl reaction

Protein stocks are typically prepared in 1$\times$ assay buffer (20 mM HEPES, pH 7.7, 100 mM KCl, 1 mM MgCl$_2$, 0.1 mM EDTA, 1 mM DTT) or in other buffers with similar compositions by dialysis or dilution. The volumes of the reactions components, as well as the amount of 10$\times$ assay buffer needed per reaction to make the final buffer concentration 1$\times$

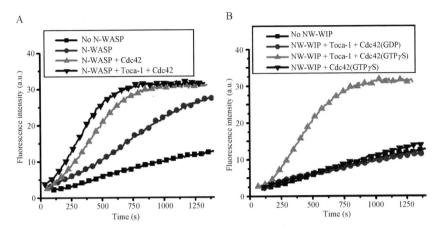

FIG. 6. Stimulation of native N-WASP-WIP complex with Toca-1 and Cdc42 (Reaction #2). (A) The pyrene actin assay used to determine the effect of Toca-1 (10 nM) on actin polymerization (2 μM total G-actin; 35% pyrene labeled) in the presence Arp2/3 complex (30 nM), recombinant N-WASP (100 nM), and Cdc42-GTPγS (250 nM). (B) The pyene actin assay used to compare the activation of the N-WASP-WIP complex (NW-WIP, 6 nM) by Cdc42-GTPγS (250 nM) in the presence or absence of Toca-1 (10 nM). Cdc42-GDP is completely inactive in stimulating actin polymerization under the same conditions. Note the different behaviors of recombinant N-WASP and native N-WASP-WIP in response to Cdc42 and Toca-1 stimulation. (Reprinted with permission from Ho *et al.*, 2004.)

assay buffer, need to be adjusted according to the actual concentrations the protein stocks. All reaction components except actin/pyrene actin are first mixed together in the order listed above and incubated for 5 min at room temperature. The reaction is initiated by adding a mixture of actin and pyrene actin to the preincubated reaction, which is then immediately transferred into a quartz cuvette (Varian Sub-Micro Fluorometer Cell, 40 μl) and monitored in a fluorescence spectrophotometer (Varian Cary Eclipse) using the provided kinetics software with the temperature control set to 20°. Pyrene fluorescence is measured at 407 nm with excitation at 365 nm. With the Varian Cary Eclipse equipped with a Peltier Multicell Holder, four independent actin polymerization reactions can be monitored simultaneously.

In contrast to the extract system, which possesses a high actin-buffering capacity because of the presence of a large number of actin-binding proteins, the purified system lacks such capacity and is prone to a variety of experimental perturbations. For example, the baseline level of actin polymerization in the absence of any stimulatory factor can vary significantly, depending the preparation of the actin, and even within the same actin preparation, it can vary from day to day. On the basis of our experience, at least part of this variability seems to originate from the procedures of the overnight depolymerization and the subsequent removal of residual actin by centrifugation. Given these considerations, proper controls, including the actin alone baseline reaction, must be conducted at the beginning of each day of the assay, and curve-to-curve comparisons should only be made with results obtained using the same actin stock within the same day. Other important controls to account for baseline levels of N-WASP and Arp2/3 activity include the actin + Arp2/3 reaction, the actin + Arp2/3 + N-WASP (or N-WASP–WIP complex) reaction, and the actin + Arp2/3 + agonists (Cdc42) without N-WASP reaction (Figs. 5 and 6).

Kinetic analyses of actin polymerization reactions can be performed using the software provided by the fluorescence spectrophotometer or by using other graphing programs such as Origin (MicroCal Software) or Excel (Microsoft). Polymerization curves can be quantified by the following methods: (1) measuring the slope of the linear, elongating phase of the actin assembly curve; (2) calculating the number of barbed ends from the rate of elongation; and (3) calculating the half-time to steady state. A detailed discussion of these analytical methods can be found in an earlier volume of *Methods in Enzymology* (Mullins and Machesky, 2000).

Properties of the Purified System

The ultimate success and utility of a purified and reconstituted signaling system is its ability to recapitulate the behavior of the pathway in complex extracts (or better yet, cells). Although the components connecting Cdc42 to actin nucleation through the Arp2/3 complex described in Reaction #1 and #2 have not been systematically analyzed *in vivo*, we believe that Reaction #2 represents a more complete reconstitution of the pathway in the extracts. Reaction #1 captures a cardinal feature of pure N-WASP, its ability to be synergistically activated by two upstream activators, Cdc42 and PIP$_2$. However, there are critical differences between this system and the extract system. Most importantly, nearly all the N-WASP in extracts is present as a complex with WIP/CR16, and unlike recombinant N-WASP, this complex exhibits zero basal activity in the absence of upstream stimuli (as one would expect for a central regulator of actin polymerization). Reaction #2 addresses many of these inconsistencies by showing that the purified native N-WASP–WIP complex has no detectable level of basal activation and is also completely insensitive to stimulation by Cdc42 alone. However, it can be activated in a switchlike fashion by Cdc42 and Toca-1, a novel factor that we recently purified from extracts as an essential component of the pathway (Ho *et al.*, 2004).

A number of issues regarding the Cdc42-actin signaling pathway remain to be explained. For example, the precise biochemical mechanism by which the N-WASP–WIP complex is activated by Toca-1 and Cdc42, as well the regulation of Toca-1 itself by other signals remain open questions. We have demonstrated that Toca-1 is a limiting component in extracts, and we expect the levels/activity of Toca-1 to be an important regulator of flux through this pathway (Ho *et al.*, 2004). Finally, systematic testing of these components in *in vivo* systems will be crucial for understanding the cellular function of the Cdc42 pathway. It is our hope that the extract and purified assay systems described in these two chapters will continue to serve as powerful tools for studies of signal-dependent actin regulation, because they provide a way to biochemically dissect signal transduction pathways at a resolution that is difficult to achieve in cell-based assays.

Acknowledgment

We thank Dr. Le Ma for his comments on the manuscript. The studies described in this chapter were supported by grants from National Institutes of Health to M. W. K.

References

Carlier, M. F., Nioche, P., Broutin-L'Hermite, I., Boujemaa, R., Le Clainche, C., Egile, C., Garbay, C., Ducruix, A., Sansonetti, P., and Pantaloni, D. (2000). GRB2 links signaling to actin assembly by enhancing interaction of neural Wiskott-Aldrich syndrome protein (N-WASp) with actin-related protein (ARP2/3) complex. *J. Biol. Chem.* **275,** 21946–21952.

Cooper, J. A., Walker, S. B., and Pollard, T. D. (1983). Pyrene actin: Documentation of the validity of a sensitive assay for actin polymerization. *J. Muscle Res. Cell Motil.* **4,** 253–262.

Cory, G. O., Garg, R., Cramer, R., and Ridley, A. J. (2002). Phosphorylation of tyrosine 291 enhances the ability of WASp to stimulate actin polymerization and filopodium formation. Wiskott-Aldrich Syndrome protein. *J. Biol. Chem.* **277,** 45115–45121.

Egile, C., Loisel, T. P., Laurent, V., Li, R., Pantaloni, D., Sansonetti, P. J., and Carlier, M. F. (1999). Activation of the CDC42 effector N-WASP by the Shigella flexneri IcsA protein promotes actin nucleation by Arp2/3 complex and bacterial actin-based motility. *J. Cell Biol.* **146,** 1319–1332.

Higgs, H. N., and Pollard, T. D. (2000). Activation by Cdc42 and PIP(2) of Wiskott-Aldrich syndrome protein (WASp) stimulates actin nucleation by Arp2/3 complex. *J. Cell Biol.* **150,** 1311–1320.

Ho, H. Y., Rohatgi, R., Lebensohn, A. M., Ma, L., Li, J., Gygi, S. P., and Kirschner, M. W. (2004). Toca-1 mediates Cdc42-dependent actin nucleation by activating the N-WASP-WIP complex. *Cell* **118,** 203–216.

Ho, H. Y., Rohatgi, R., Ma, L., and Kirschner, M. W. (2001). CR16 forms a complex with N-WASP in brain and is a novel member of a conserved proline-rich actin-binding protein family. *Proc. Natl. Acad. Sci. USA* **98,** 11306–11311.

Kouyama, T., and Mihashi, K. (1980). Pulse-fluorometry study on actin and heavy meromyosin using F-actin labelled with N-(1-pyrene)maleimide. *Eur. J. Biochem.* **105,** 279–287.

Ma, L., Rohatgi, R., and Kirschner, M. W. (1998). The Arp2/3 complex mediates actin polymerization induced by the small GTP-binding protein Cdc42. *Proc. Natl. Acad. Sci. USA* **95**, 15362–15367.

Machesky, L. M., Atkinson, S. J., Ampe, C., Vandekerckhove, J., and Pollard, T. D. (1994). Purification of a cortical complex containing two unconventional actins from Acanthamoeba by affinity chromatography on profilin-agarose. *J. Cell Biol.* **127**, 107–115.

Miki, H., Miura, K., and Takenawa, T. (1996). N-WASP, a novel actin-depolymerizing protein, regulates the cortical cytoskeletal rearrangement in a PIP2-dependent manner downstream of tyrosine kinases. *EMBO J.* **15**, 5326–5335.

Mullins, R. D., and Machesky, L. M. (2000). Actin assembly mediated by Arp2/3 complex and WASP family proteins. *Methods Enzymol.* **325**, 214–237.

Ramesh, N., Anton, I. M., Hartwig, J. H., and Geha, R. S. (1997). WIP, a protein associated with Wiskott-Aldrich syndrome protein, induces actin polymerization and redistribution in lymphoid cells. *Proc. Natl. Acad. Sci. USA* **94**, 14671–14676.

Rohatgi, R., Ho, H. Y., and Kirschner, M. W. (2000). Mechanism of N-WASP activation by CDC42 and phosphatidylinositol 4, 5-bisphosphate. *J. Cell Biol.* **150**, 1299–1310.

Rohatgi, R., Ma, L., Miki, H., Lopez, M., Kirchhausen, T., Takenawa, T., and Kirschner, M. W. (1999). The interaction between N-WASP and the Arp2/3 complex links Cdc42-dependent signals to actin assembly. *Cell* **97**, 221–231.

Rohatgi, R., Nollau, P., Ho, H. Y., Kirschner, M. W., and Mayer, B. J. (2001). Nck and phosphatidylinositol 4,5-bisphosphate synergistically activate actin polymerization through the N-WASP-Arp2/3 pathway. *J. Biol. Chem.* **276**, 26448–26452.

Suetsugu, S., Hattori, M., Miki, H., Tezuka, T., Yamamoto, T., Mikoshiba, K., and Takenawa, T. (2002). Sustained activation of N-WASP through phosphorylation is essential for neurite extension. *Dev. Cell* **3**, 645–658.

Torres, E., and Rosen, M. K. (2003). Contingent phosphorylation/dephosphorylation provides a mechanism of molecular memory in WASP. *Mol. Cell* **11**, 1215–1227.

Welch, M. D., and Mitchison, T. J. (1998). Purification and assay of the platelet Arp2/3 complex. *Methods Enzymol.* **298**, 52–61.

Zigmond, S. H. (2000). *In vitro* actin polymerization using polymorphonuclear leukocyte extracts. *Methods Enzymol.* **325**, 237–254.

[15] Biochemical Analysis of Mammalian Formin Effects on Actin Dynamics

By ELIZABETH S. HARRIS and HENRY N. HIGGS

Abstract

Formins are members of a conserved family of proteins, present in all eukaryotes, that regulate actin dynamics. Mammals have 15 distinct formin genes. From studies to date, surprising variability between these isoforms has been uncovered. All formins examined have several common effects on actin dynamics in that they: (1) accelerate nucleation rate; (2) alter filament barbed end elongation/depolymerization rates; and (3) antagonize capping

0076-6879/06 $35.00
DOI: 10.1016/S0076-6879(06)06015-0

protein. However, the potency of each effect can vary greatly between formins. In addition, a subset of formins binds tightly to filament sides and bundle filaments. Even isoforms that are closely related phylogenetically can display marked differences in their effects on actin. This chapter discusses several methods for examining formin function *in vitro*. We also discuss pitfalls associated with these assays. As one example, the effect of profilin on formin function is difficult to interpret by "pyrene-actin" polymerization assays commonly used in the field and requires assays that can distinguish between filament nucleation and filament elongation. The regulatory mechanisms for formins are not clear and certainly vary between isoforms. A subset of formins is regulated by Rho GTPases, and the assays described in this chapter have been used for characterization of this regulation.

Introduction

Formins are widely expressed proteins that are emerging as key regulators of actin filament assembly and elongation. The formin homology 2 (FH2) domain, approximately 400 amino acids in length, defines the formin protein family and is the most conserved domain between family members. For all formins studied, the full FH2 domain is dimeric (Harris *et al.*, 2004; Li and Higgs, 2005; Moseley *et al.*, 2004; Xu *et al.*, 2004). Mammals possess 15 FH2-domain–containing genes that segregate into seven phylogenetic groups (Higgs and Peterson, 2005) (Table I).

The FH2 domain is sufficient for most of the effects formins have on actin dynamics in biochemical assays (reviewed in Higgs [2005]; Wallar and Alberts [2003]; and Zigmond [2004]), including: (1) accelerating nucleation rate; (2) altering elongation/depolymerization rates; and (3) antagonizing barbed end capping by capping proteins (Fig. 1, bottom panel). The relative potencies of each effect vary from formin to formin (reviewed in Higgs [2005]). Current data suggest that all FH2 domain activities are mediated by their ability to bind the filament barbed end and to remain processively attached as the filament elongates (Harris *et al.*, 2004; Higashida *et al.*, 2004; Kovar and Pollard, 2004; Kovar *et al.*, 2003; Moseley *et al.*, 2004; Pruyne *et al.*, 2002; Romero *et al.*, 2004; Zigmond *et al.*, 2003).

Formins also contain a proline-rich FH1 domain just N-terminal to the FH2 domain (Fig. 2). The FH1 domain is a binding site for the actin monomer–binding protein profilin. Profilin binding to FH1-FH2 domain–containing constructs increases the elongation rate of formin-bound filaments (Kovar and Pollard, 2004; Kovar *et al.*, 2003; Romero *et al.*, 2004). Barbed end elongation rates in the absence and presence of profilin vary greatly from formin to formin (Kovar and Pollard, manuscript in

TABLE I
MAMMALIAN FORMINS

Group	Proteins	GTPase binding
Dia	mDia1, mDia2, mDia3	RhoA, B, C, D, Cdc42[a]
DAAM	DAAM1, DAAM2	RhoA
FRL	FRL1, FRL2, FRL3	Rac
FHOD	FHOD1, FHOD2	Rac
FMN	FMN1, FMN2	Unknown
Delphilin	Delphilin	Unknown
INF	INF1, INF2	Unknown

[a] All members of the Dia group do not bind all of these GTPases.

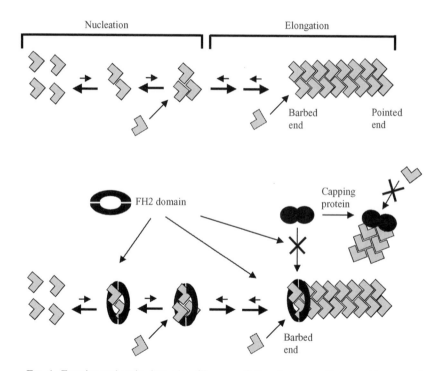

FIG. 1. Formins and actin dynamics. (Top panel) Spontaneous actin assembly occurs in two phases, nucleation and elongation. The nucleation phase refers to the formation of actin trimers, whereas elongation refers to subsequent incorporation of monomers into the filament. In most cells, elongation occurs exclusively at the barbed end. (Bottom panel) Formin FH2 domains have three main effects on actin dynamics: (1) accelerating nucleation, (2) altering elongation/depolymerization rates, and (3) inhibiting barbed end capping by capping proteins.

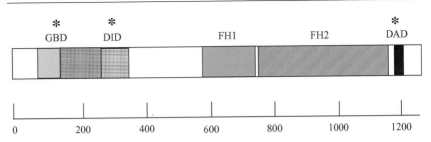

FIG. 2. Bar Diagram of mDia1. Mammals contain 15 formin genes. The domain organization of mDia1 is shown here to illustrate a representative formin. FH1 and FH2 (formin homology) domains are present in all mammalian formins, whereas the DAD (diaphanous auto-regulatory domain), GDB (GTPase binding domain), and DID (diaphanous inhibitory domain) are not (denoted by asterisk). GBD (gray box) and DID (checked box) partially overlap. Below the bar diagram is a ruler indicating amino acids.

preparation), which might have significant implications for isoform-specific actin-based structures in cells.

Regulatory mechanisms have only been studied in detail for the Dia group of mammalian formins, including mDia1, mDia2, and mDia3. mDia1 is regulated by autoinhibition, in which the N-terminal Diaphanous Inhibitory Domain (DID) binds the C-terminal Diaphanous Auto-regulatory Domain (DAD), inhibiting effects of the FH2 domain on actin (Fig. 2). Binding of RhoA to the GTPase binding domain (GBD), which overlaps DID, partially relieves autoinhibition (Alberts, 2001; Li and Higgs, 2003, 2005). The other Dia formins are probably regulated similarly, but with some variations. For instance, mDia2 binds Cdc42 as well as Rho, and Cdc42 binding is mediated by a CRIB motif distinct from the Rho-binding GBD (Peng et al., 2003). In addition, mDia3 might bind Cdc42 (Yasuda et al., 2004), and a splice variant binds RhoD but not RhoA (Gasman et al., 2003). On the basis of DAD and DID sequence similarities, DAAM and FRL group formins are probably autoinhibited as well, with RhoA or Rac, respectively, as possible activators (Table I). Regulation of FHOD, delphilin, FMN, and INF formins is much less well understood (reviewed in Higgs [2005]). Furthermore, Dia formins may require a second activator for full activation (Li and Higgs, 2003, 2005).

We have studied biochemical activities for six mammalian formins: mDia1, mDia2, INF1, INF2, FRL1, and FRL2. The qualitative effects of these formins on actin dynamics are similar, but the quantitative differences are large. For instance, mDia1, mDia2, INF1, and INF2 are potent nucleators, whereas FRL1 and FRL2 are not (Chabra and Higgs, unpublished data; Harris and Higgs, unpublished data; Harris et al., 2004; Li and

Higgs, 2003, 2005). In addition, some formins seem to have unique effects on actin. For example, FRL1 binds filament sides tightly and bundles filaments, whereas mDia1 does not (Harris and Higgs, unpublished data; Harris *et al.*, 2004). The FH2 domain seems sufficient for all of these effects, suggesting that differences in FH2 domain structure alter its function significantly. Formin biochemical variability might serve to mediate formin-specific actin-based structures in cells. Combined with likely regulatory variability between formins, the differences between isoforms are substantial.

Because of multiple effects on actin, results from one assay can be misleading when studying formins biochemically. We propose a combination of techniques to examine formin mechanisms *in vitro*. These assays can be adapted further to examine regulation, as described for mDia1 (see "Extension of the Assays"). An important consideration is that "bulk" polymerization assays, like the "pyrene-actin" assay, (see "Actin Polymerization by Fluorescence Spectroscopy") do not distinguish between nucleation and elongation clearly, which has lead to erroneous initial conclusions for some formin properties (see "Extension of the Assays" for details).

Basic Actin Polymerization Kinetics Description

This section provides information useful to all of the methods described. Actin is an ATPase and binds ATP and an accompanying divalent cation (Mg^{2+} or Ca^{2+}) tightly. Under appropriate conditions (see below), purified actin monomers spontaneously polymerize into actin filaments. Polymerization can be divided into two general phases: nucleation and elongation (Fig. 1, top panel). Nucleation, the assembly of actin monomers into a stable trimeric nucleus, is highly unfavorable, with equilibrium dissociation constants in the millimolar range (Pollard and Cooper, 1986). Elongation, the addition of monomers to the trimeric nucleus, is much more favorable and rapid. Actin filaments are polar, with "barbed" and "pointed" ends. Elongation from the barbed end, the only end that elongates in most cells, is approximately 10 times faster than elongation from the pointed end (Kuhn and Pollard, 2005). The "critical concentration" is the concentration of actin that remains monomeric at polymerization equilibrium. For actin in polymerizing conditions, the critical concentration at the barbed end is 0.1 μM. If the barbed end is "capped," the critical concentration is that of the pointed end, 0.5–0.7 μM. More details can be found in Pollard *et al.* (2000).

Actin is stored in G buffer (composition given in "Materials Used for all Assays"), which has three features favoring the monomeric form:

(1) low ionic strength, (2) high pH, (3) calcium ion. The critical concentration of actin in G buffer is >100 μM, so actin remains monomeric. To polymerize actin, $1 \times$KMEI buffer is added (composition given in "Materials Used for all Assays"), which increases ionic strength and lowers the pH. Both of these effects serve to reduce electrostatic barriers between monomers that inhibit polymerization. In addition, magnesium ion replaces calcium ion in the ATP-binding cleft, resulting in a conformational change of the actin monomer that favors polymerization. Exchange of Ca^{2+} for Mg^{2+} can be a rate-limiting step for nucleation, which is artifactual, because actin monomers are Mg^{2+} bound in cells. Thus, for rigorous polymerization kinetics (see "Actin Polymerization by Fluorescence Spectroscopy"), actin is preincubated briefly with 1 mM EGTA and 0.1 mM MgCl$_2$ before the addition of $1 \times$KMEI. This short pretreatment converts monomers to the Mg^{2+} bound form but does not induce polymerization significantly.

Materials Used for All Assays

Below is a list of common reagents required for all subsequent assays. Other reagents required for specific assays will be defined in their corresponding sections.

1. G buffer: 2 mM Tris, pH 8.0, 0.5 mM DTT, 0.2 mM ATP, 0.1 mM CaCl$_2$, and 0.01% sodium azide (to inhibit bacterial growth during storage).
2. G-Mg buffer: G buffer with 0.1 mM MgCl$_2$ instead of 0.1 mM CaCl$_2$.
3. $10 \times$KMEI: 100 mM imidazole, pH 7.0, 500 mM KCl, 10 mM MgCl$_2$, 10 mM EGTA.
4. KMEI/G-Mg: KMEI diluted to a certain concentration in G-Mg. We dilute in G-Mg to maintain ATP and DTT in the system, both important for actin stability (Straub and Feuer, 1950).
5. 10E/1M: 10 mM EGTA, 1 mM MgCl$_2$.
6. Monomeric actin in G buffer (see purification protocol in "Actin Preparation").
7. Formin protein (see purification protocol in "Expression and Purification of Mammalian Formin FH2 Domains").

Actin Polymerization by Fluorescence Spectroscopy

A widely used biochemical assay for actin assembly/disassembly, known as the "pyrene-actin" assay, uses actin monomers labeled with the fluorophore, pyrene-iodoacetamide. Pyrene-iodoacetamide labels cysteine

374 on actin monomers, and its fluorescence increases 20-fold on monomer incorporation into a filament. Only a small percentage of pyrene-actin is needed as a tracer for these assays (we use 5%). Pyrene-actin polymerizes with similar kinetics to unlabeled actin (Cooper *et al.*, 1983).

This assay is sometimes referred to as the pyrene-actin "nucleation" assay, which is erroneous, because the assay does not distinguish between nucleation and elongation. A more appropriate term is the pyrene-actin "polymerization" assay.

Special Materials

1. Pyrene-labeled actin monomers in G buffer. Prepared as in Cooper *et al.* (1983). Pyrene-iodoacetamide can be purchased from Invitrogent Molecular Probes (P-29).
2. Spectrofluorometer with excitation and emission wavelengths of 365 nm and 407 nm.
3. Quartz cuvette suitable for spectrofluorometer.

Method

This method is adapted for our cuvette volume. See the section on special considerations for discussion of cuvettes.

1. Prepare pyrene-actin monomer stock by mixing pyrene-actin and unlabeled actin monomers to a final concentration of 20 μM actin (5% pyrene) in G buffer on ice.
2. Convert 20 μl of pyrene-actin monomer stock to Mg^{2+}-bound form immediately before polymerization by a 2-min incubation at 23° (room temperature) with 2 μl 10E/1M.
3. During (or before) actin monomer conversion, set up "polymerization mix" as follows: 11 μl 10×KMEI, "X" μl formin or other protein, bring up to 85 μl total volume with G-Mg.
4. After 2 min conversion, start polymerization reaction by adding 78 μl polymerization mix to actin. *This induces polymerization, so subsequent steps must be performed quickly.*
5. Mix sample by pipetting up and down once.
6. Place 78 μl of sample into the cuvette and start fluorescence reading.
7. Record "dead time." See "Special Considerations" section.
8. Record fluorescence until polymerization equilibrium is reached. In our typical assays containing 4 μM actin monomers alone, the time required is 2400–3600 sec (40–60 min) (Fig. 3).

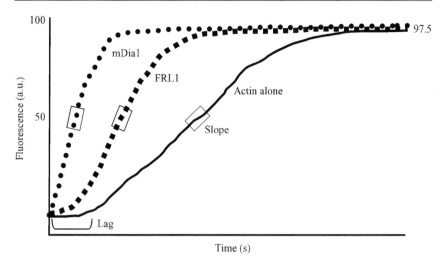

Time (s)

Fig. 3. Pyrene-actin polymerization assay. Schematic of polymerization time course for 4 μM actin monomers alone or with either FRL1 or mDia1. Note the lag in the trace for actin alone, which is due to the unfavorable nucleation phase. A short lag persists in the presence of FRL1, for reasons currently unknown. mDia1 initiates polymerization with no apparent lag. Boxed regions indicated 50% polymerization point, where slopes should be taken for Eq. (2). All three curves reach the same steady-state fluorescence, which is 97.5% of the total monomer because 0.1 μM actin remains monomeric at polymerization equilibrium.

Calculations

The concentration of filaments produced at any point in the polymerization process can be calculated from the slope, as long as the elongation rate is known. We determine slope at the 50% point of polymerization (Fig. 3), using Kaleidagraph (Synergy Software), but any graphing software can be used. Slopes are converted initially to filament concentration under the assumption of unrestricted ATP-actin monomer addition to barbed ends with a rate constant (K_+) of 7.4 $\mu M^{-1}sec^{-1}$ (Kuhn and Pollard, 2005) according to the following equation:

(1) $F = S'/(M_{0.5} \times K_+)$, where F is filament concentration in mM, S' is slope converted to μM/sec, and $M_{0.5}$ is μM monomer concentration at 50% polymerization.

S' is calculated by the equation:

(2) $S' = (S \times M_t)/(f_{max}-f_{min})$, where S is raw slope in a.u./sec (a.u. = arbitrary units), M_t is concentration of total polymerizable monomer in mM, and f_{max} and f_{min} are fluorescence of fully polymerized and unpolymerized actin, respectively, in a.u.

For the assay described above, μ_t is 3.9 mM. The reaction contains 4 uM total actin, leaving 3.9 μM as the total polymerizable fraction, if we subtract the barbed-end critical concentration of 0.1 mM. Thus, $M_{0.5}$ in Eq. (1) is 1.95 μM.

Because most formins change the barbed end elongation rate, this rate must be determined independently for the filament calculation to be valid. The method for determining elongate rate is in "Fluorescence Microscopy (Single Filament Analysis)."

Special Considerations

1. Cuvettes: We find that small-volume quartz cuvettes (Hellma Cells [105.51-QS]; 3-mm light path; center height, 15 mm) provide highly reproducible results and require low volumes (we use 78 μl). Most fluorimeters have the viewing window at either 8 or 15 mm from the bottom of the cuvette holder, so one must purchase cuvettes with their viewing windows at similar heights.

2. Temperatures: We conduct polymerization reactions at "room temperature," which is a fairly constant 23°. Most fluorimeters have jacketed cuvette holders, so that the cuvette can be warmed or cooled if desired. In these cases, care must be taken that the polymerization mix is at or near this target temperature at time of mixing.

3. Ionic strength: Actin polymerization kinetics is strongly altered by ionic strength (Drenckhahn and Pollard, 1986). If adding proteins (formins, others) to the polymerization mix, the ionic strength contributed by the protein's buffer must be accounted for. One solution is to include a constant volume of the protein's buffer in all assays. Another option is to change the volume of 10×KMEI added to compensate.

4. Diluting proteins: Often, the protein added (formin or other) requires substantial dilution for the polymerization assay. We typically dilute in 1×KMEI made up in G-Mg. In addition, we add 0.5 mM of the non-ionic detergent, thesit (Polidocanol, Sigma P-9641). The reason for this addition is that many proteins we study lose activity over time when diluted <1 μM, and thesit preserves activity. We observe no effects of thesit on polymerization kinetics.

5. Dead time: For this assay, as well as any kinetic assay, the reaction must be initiated at a defined time, and the "dead time" between component mixing and the first measured data point must be recorded.

6. Excitation/emission: We recommend using a narrow excitation slit width (we use 0.25 mm) and a wider emission slit width (4 mm). Keeping the excitation slits narrow reduces photobleaching. Although less common, an emission wavelength of 386 nm can also be used. We

prefer to use a 407-nm emission, because the larger Stokes shift allows for wider emission slits.

7. Data collection: It is important to monitor the reactions until steady state has been reached, which can take 40–60 min for 4 μM actin alone. In the presence of a potent actin nucleator, the reaction can reach steady state considerably more quickly. The maximum fluorescence value is required for Eq. (2). Alternately, a second reaction mix can be prepared and saved for several hours (or overnight) in the dark, at which point a single fluorescence reading can be taken.

Uses and Limitations

Fluorescence spectroscopy assays have advanced the field of actin dynamics considerably. Large amounts of data can be collected rapidly, and the required sample volumes are small. The major disadvantage to these bulk assays is that they analyze *populations* of filaments rather than individual filaments. Thus, the contributions of nucleation versus elongation on the observed polymerization rates cannot be distinguished easily. In addition, some proteins have reduced binding to pyrene-actin (ex. profilin [Malm, 1984]), but the use of tracer amounts of pyrene-actin minimizes this problem in most cases.

Variation

A pyrene-actin elongation assay has been used by our laboratory and others to examine the effects of FH2 domains on barbed end elongation. In this assay, a stock of unlabeled actin filaments is made, to which pyrene-labeled monomers are added. Because filaments are already nucleated, fluorescence increases linearly as they elongate. By measuring the initial slope, one can calculate the concentration of filaments present or examine the effects of added proteins on elongation from a fixed concentration of filaments. However, we urge caution when using this assay for formins, because additional activities make interpretation of the results difficult. For example, mDia2 FH2 domain slows barbed end elongation dramatically (Kovar *et al.*, manuscript in preparation), but its potent nucleation ability masks its elongation effect in this assay (Harris and Higgs, unpublished data). We now use a dual filament microscopy assay (see "Fluorescence Microscopy [Single Filament Analysis]") for examining elongation rates. Pyrene-actin elongation assays, however, can be extremely useful for measuring actin polymerization in cell extracts (Zigmond *et al.*, 1998).

Filament Binding

High-speed pelleting assays provide a quick and simple means to determine a protein's affinity for the *sides* of actin filaments. We have found some formins bind tightly to filaments, in 1:1 stoichiometry with actin subunits, whereas others have much lower affinity. We conduct two types of experiments: (1) constant formin at 0.2 μM, with actin concentration varied from 0.2–4 μM; and (2) constant actin at 4 μM, with formin concentration varied from 50 nM to 4 μM. The first experiment is most suitable to determine K_d of formin for filaments.

Special Materials

1. Beckman Optima Tabletop ultracentrifuge, capable of handling TLA-100.1 rotor.
2. Beckman TLA-100.1 rotor.
3. Beckman 7 × 20 mm polycarbonate centrifuge tubes (Beckman 343775).
4. 1xNaMEI: Like KMEI, but with 50 mM NaCl instead of 50 mM KCl because of K^+ precipitation in SDS buffer.
5. Phalloidin: (Sigma P-2141) 1 mM stock in ethanol, store at −20°.
6. Speedvac: we use Savant SC100 speedvac with an oil diffusion pump.

Method

1. Polymerize a 5 μM actin stock for 2 h at 23° in 1×NaMEI/G-Mg.
2. Add phalloidin to 5 μM. Mix by inverting gently and give a brief microfuge at max speed.
3. Dilute formin in 1×NaMEI/G-Mg plus 0.5 mM thesit.
4. Mix formin, filaments, and 1×NaMEI/G-Mg in centrifuge tubes to a final volume of 200 μl.
5. After 10 min incubation at 23°, centrifuge at 80,000 rpm for 20 min at 4° in TLA-100.1 rotor.
6. Remove and transfer 160 μl of supernatant to 1.5-ml Eppendorf tube. Process supernatants as follows (after processing pellets):
   a. Dry in speedvac.
   b. Resuspend in 16 μl 1×SDS-PAGE sample buffer (in water) by pipetting up and down.
   c. Boil 5 min.
   d. Microfuge 1 min at max speed.
7. Remove remaining supernatant and process pellets:
   a. Wash briefly with 200 μl 1×NaMEI.

    b. Resuspend in 20 μl SDS-PAGE sample buffer (in 1×NaMEI) by pipetting up and down.

    c. Place centrifuge tubes in 1.5-ml Eppendorf tubes.

    d. Boil 5 min.

    e. Microfuge 1 min at max speed.

  8. Analyze supernatants and pellets by Coomassie-stained SDS-PAGE. Under the conditions described above, almost all polymerized actin pellets, along with any bound formin.

  9. Scan gels and quantify supernatant and pellets by densitometry.

 10. Plot data using any graphing software, and determine K_d.

Special Considerations

1. Polymerized actin stock: Up to 10 μM actin can be polymerized if desired. Above this concentration, filaments are quite viscous and difficult to pipette accurately.

2. Phalloidin: We stabilize filaments by adding phalloidin. This essentially drives the critical concentration to zero, so that filaments remain polymerized at all dilutions, simplifying analysis. We have not found phalloidin to influence filament binding by any formin, although initial experiments without phalloidin are advised for uncharacterized formins.

3. Mixing components: We mix components directly in centrifuge tubes, adding 1×NaMEI/G-Mg first, formin second, and filaments third. Filaments are pipetted using "cut" tips to minimize filament shearing. We cut the narrow end off of tips using a razor blade. Mix gently up and down once with cut pipette tip.

4. Formin concentration: We use 0.2 μM formin typically. When concentrated 10-fold for SDS-PAGE, this concentration gives a workable detection range for Coomassie stain.

5. Incubation time: A 10-min incubation time is at least 9 min 50 sec longer than required, because the on-rate should be fast!

6. Drying supernatants: For our speedvac, 60–80 min is generally sufficient. We have the heat on "medium" but are careful not to over dry samples.

7. Loading SDS-PAGE: Because of the different salt concentrations, the supernatants and pellets should be run on separate gels. When the dried supernatant is resuspended, the effective buffer concentration is 10×NaMEI. For supernatants, we make up mock loading buffer of 10×NaMEI in 1×SDS-PAGE sample buffer. Load mock buffer in lanes bracketing the outer sample lanes, which avoids distortion of the outer lanes. Do the same for gels of pellets, except use 1×NaMEI. Also on the gel, several

concentrations of the constant component (actin or formin) should be loaded to provide a standard curve. Generally, we use 15-well Bio-Rad minigels, and load five standard concentrations (diluted in either 10×NaMEI or 1×NaMEI, depending on whether they are for supernatants or pellets), one blank lane, eight sample lanes, and one more blank lane.

8. Washing pellet: This wash amounts to carefully adding 200 μl of 1×NaMEI, then taking it off immediately. Its purpose is to remove actin and formins from the tube sides. The actin pellet is glassy, barely visible only at higher concentrations, and found at the base of the outer wall. Hint: mark the outside of the centrifuge tube, and put your pipette tip down the inner side when washing.

9. Plotting data: For the actin concentration curve, plot density of pelleted formins versus actin concentration. We also analyze the loss of formin from the supernatant, which may be especially important if working with formin constructs containing only FH2 domains. FH2 domains are approximately 400 amino acids long and run very near actin (43 kDa) on SDS-PAGE. Using slightly longer formin constructs eliminates this problem.

Uses and Limitations

This is a very robust assay for examining protein binding to filaments. Phalloidin is a variable. In the absence of phalloidin, filaments re-equilibrate to their critical concentration on dilution. For instance, 5 μM polymerized actin actually contains 4.9 μM in filaments and 0.1 μM monomers. On dilution to 0.5 μM, net depolymerization occurs to attain a new equilibrium of 0.4 μM in filaments and 0.1 μM monomers. Full equilibration requires >10 min, so this condition is nonequilibrium. FH2 domains generally slow down depolymerization from barbed ends (Kobielak *et al.*, 2004; Kovar *et al.*, 2003; Li and Higgs, 2003; Pring *et al.*, 2003), so differences in pelleted actin might be observed in the presence of formin.

Variation

A variation of this assay is used to test the ability of formins to crosslink and/or bundle actin filaments. In this "low-speed" pelleting assay, components are mixed in 1.5-ml Eppendorf tubes, and the samples centrifuged in a Microfuge at max speed (13,000–16,000*g*) for 5 min. Supernatants and pellets are analyzed as for high-speed pelleting assays. Under these conditions, only crosslinked and/or bundled filaments pellet, along with their bound formin. Single filaments (bound to formin or unbound) remain in the supernatant.

Fluorescence Microscopy (Single Filament Analysis)

Fluorescence microscopy allows visualization of individual actin filaments. The current "gold standard" of single filament analysis is TIRF (total internal reflection fluorescence) microscopy, using fluorescently labeled monomers. TIRF allows observation of individual filament dynamics in real time (for details of this assay see Amann and Pollard [2001]; Kovar and Pollard [2004]; and Kuhn and Pollard [2005]) (Fig. 4, top panel). Background fluorescence is minimal, because TIRF illuminates only those monomers close to the coverslip surface. Normal epifluorescence microscopy does not work in this setup, because the free monomer background fluorescence is much too high. TIRF allows elongation rates of individual actin filaments to be measured directly, in a manner uncoupled from nucleation. This property alleviates a major problem with pyrene-actin polymerization assays. One disadvantage to TIRF microscopy is it

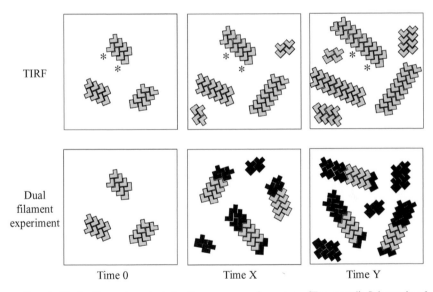

TIRF

Dual
filament
experiment

Time 0   Time X   Time Y

FIG. 4. Single filament analysis by fluorescence microscopy. (Top panel) Schematic of TIRF microscopy experiment. The same population of filaments is observed over all time points. Asterisk marks the barbed and pointed end of one filament. (Bottom panel) Schematic of dual filament experiment using epifluorescence. Each time point represents a different population of filaments. At time zero, only A488P-labeled filament seeds are visible (gray symbols). At subsequent time points, after addition of RhP and actin monomers, elongating segments of RhP-label (black symbols) can be seen at both the barbed and pointed ends of A488P seeds. Newly nucleated filaments, labeled with only RhP, can also be observed. These new filaments also elongate, but barbed and pointed ends cannot be distinguished.

requires more extensive (and expensive) equipment than available to many laboratories.

The experiments described in this section provide a less expensive alternative, requiring only a standard epifluorescence microscope. In this method, fluorescently conjugated phalloidin is used to label actin filaments, which reduces background. A disadvantage to this method is it allows only visualization of filaments at single time points, with each time point representing a different population of filaments taken from the original stock sample. However, we obtain very consistent elongation rates using this method.

Special Materials

1. Fluorescence microscope: TRITC and FITC filter set, 60× or 100× objective (1.4 NA), reasonable digital camera, and image processing software. We have a Nikon inverted TE2000-E microscope, a Roper Cool Snap SE camera, and MetaView software (Universal Imaging Corp.).

2. Fluorescence buffer: 25 mM imidazole, pH 7.0, 25 mM KCl, 4 mM MgCl$_2$, 1 mM EGTA, 100 mM DTT, 0.5% methylcellulose, 3 mg/ml glucose, 18 μg/ml catalase (Sigma C-0515), 100 μg/ml glucose oxidase (Sigma G-6641).

3. Rhodamine-phalloidin (RhP): (Sigma P-1951) Fluoresces red (TRITC channel). Stock at 500 μM in DMSO, store at $-20°$.

4. Alexa-488-phalloidin (A488P): (Invitrogent Molecular Probes A-12379) Fluoresces green (FITC channel). Stock at 100 μM in methanol (corresponds to addition of 100 μl to a vial of 300 units), store at $-20°$.

5. Poly-L-lysine (PLL): (Sigma P-1524) >300,000 MW. Stock at 0.1% in dH$_2$O, store at $-20°$.

6. Compressed air: We use Stoner Premium Air Duster.

7. 12-mm round glass coverslips: To prepare coverslips they are first acid-washed (Fisher Chromerge or equilivant), then rinsed several times with dH$_2$O, before being placed in 0.02% sodium azide for extended storage. Prepare with PLL immediately before use as follows:

    a. Dip in dH$_2$O, then place on laboratory tissue to air dry.
    b. Apply 100 μl of 0.01% PLL.
    c. After 1–2 min, dip once in dH$_2$O, and spray dry with compressed air.
    d. Coverslips are directly applied to 2 μl of sample on a glass slide.

Method

In this experiment, the formin is incubated with prepolymerized filaments labeled with A488P. RhP and actin monomers are then added, and monomers allowed to incorporate onto the A488P-labeled filaments for varying times. Elongation is terminated by dilution with fluorescence buffer. Thus, the A488P-labeled "seed" filaments are green, and the newly elongated RhP-labeled portions of these filaments (as well as newly nucleated filaments) are red (Fig. 4, bottom panel). By measuring the length of the RhP-labeled segments, one can determine barbed and pointed end elongation rates. In the absence of factors that affect elongation, the barbed end will be about 10 times longer than the pointed end.

1. Polymerize actin at 4 μM for 2 h at 23° in 1×KMEI/G-Mg.
2. Add A488P to 4 μM.
3. Make up the following two stock solutions:
    a. 0.375 μM RhP in 2×KMEI/G-Mg
    b. 2.2 μM actin monomers in G buffer
4. Pipette 2 μl of A488P-labeled actin filaments into 1.5-ml Eppendorf tube using a cut pipette tip.
5. Add 2 μl of formin (diluted in 1×KMEI/G-Mg plus 0.5 mM thesit).
6. Mix sample by pipetting up and down once with cut pipette tip and incubate at 23° for 2 min.
7. Convert 18 μl of monomers to Mg^{2+}-bound form with 1.8 μl 10E/ 1 M for 2 min at 23°C.
8. During the 2-min incubation, aliquot 395 μl fluorescence buffer into several 1.5-ml Eppendorf tubes.
9. To the filament/formin mix, add 18 μl of 0.375 μM RhP stock, followed immediately by addition of 18 μl actin monomer stock. Flick tube gently to mix.
10. At various time points after addition of monomers (typically 1–10 min), remove 5 μl and dilute into one tube of 395 μl fluorescence buffer. Mix by gentle inversion.
11. Adsorb samples (2 μl) onto PLL coated coverslips with cut pipette tip.
12. Image using 60× or 100× objective.
13. Record images using both TRITC and FITC filter sets.
14. Measure lengths of newly elongated segments off of A488P-labeled seeds. We measure between 50 and 100 filaments for each condition, take the median lengths for each time point, and plot them versus time, which should be linear over 10 min. Slopes can be converted from mm/min to μM/sec by the estimation that one actin monomer adds about 3 nm to the filament.

Special Considerations

1. Pipetting: *Always* use cut pipette tips when pipetting filaments to reduce mechanical shearing.

2. Fluorophores: We use a low amount of RhP compared with monomer concentration, because this reduces background fluorescence.

We label seed filaments with A488P, because too much background occurs with RhP-labeled seeds.

3. Dilution: Dilution of the sample into fluorescence buffer prevents further elongation, because the samples have been diluted below the critical concentration for the barbed and pointed end. Filaments are stabilized by phalloidin, preventing depolymerization. However, we recommend imaging samples shortly after dilution to minimize mixing of red and green labels.

4. Imaging: Filaments can be difficult to find, so focus up and down carefully.

Variation

The effect of formins on either filament length (i.e., severing ability) or filament organization (i.e., crosslinking and/or bundling ability) can be determined using a simplified microscopy assay. This assay may be important depending on the results obtained from high-speed and low-speed pelleting assays *(see "Filament Binding"),* because binding to filament sides is a requirement for either severing or crosslinking/bundling.

1. Polymerize actin at 4 μM for 2 h at 23° in 1×KMEI/G-Mg.
2. Pipette 10-μl aliquot from this stock into 1.5-ml Eppendorf tube using a cut pipette tip.
3. Add 10 μl of formin (diluted in 1×KMEI/G-Mg plus 0.5 mM thesit) and mix by gently flicking.
4. Add 20 μl rhodamine-phalloidin, diluted to 2 μM in dilution buffer (fluorescence buffer minus glucose, glucose oxidase, and catalase). Mix by gently flicking.
5. Dilute samples between 25- and 200-fold with fluorescence buffer. Mix by gentle inversion.
6. Adsorb 2 μl onto PLL-coated coverslips with cut pipette tip.
7. Image using 60× or 100× objective, and record images.

Special Considerations

1. Dilution: The extent of dilution is varied to obtain the desired level of filament density on coverslips. In addition, diluting the samples extensively may dissociate crosslinked or bundled filaments if the interactions are weak. For filament severing assays, we dilute samples 200-fold with fluorescence buffer. For filament crosslinking/bundling assays we dilute 25-fold.

2. Quantification: We generally take 10–20 images for each sample. For severing assays we perform extensive quantification. All filaments with both ends discernible are measured for each image, often resulting in more than 1000 individual filaments being measured for each condition. Median

filament lengths are obtained for each condition and graphed. In addition, we examine the distribution of filament lengths by making histograms. For crosslinking/bundling experiments, we use this assay only as a visual confirmation of the phenomenon, and quantify by low-speed pelleting assays (see "Filament Binding").

Extensions of the Assays

One can adapt the preceding assays to examine other facets of formin-mediated actin assembly. We briefly discuss two such extensions in this section: (1) regulation of mDia1; and (2) the effects of profilin and capping protein. With the addition of other proteins, care must be taken to ensure that ionic strength and pH remain unchanged.

mDia1 Regulation

mDia1 is autoinhibited by an intramolecular interaction between DAD and DID (Fig. 2). The modular nature of mDia1 allows examination of regulatory mechanisms *in vitro*. In pyrene-actin polymerization assays, an N-terminal mDia1 construct, containing the DID, inhibits nucleation mediated by C-terminal constructs containing the DAD (Li and Higgs, 2003, 2005). For these assays, N- and C-terminal fragments are preincubated with each other (*step 3* in "Actin Polymerization by Fluorescence Spectroscopy") before addition to the actin monomer stock (*step 4*). Using this assay, one can map regions that mediate autoinhibition (Li and Higgs, 2005). This assay can be extended further to examine relief of this auto-inhibition by GTPases. When RhoA is included in the assays, inhibition is partially relieved (Li and Higgs, 2003). Removal of sequence N-terminal to the DID abolishes RhoA relief, suggesting that the GBD extends into this region (Li and Higgs, 2005).

In addition, in high-speed pelleting assays, N-terminal mDia1 constructs inhibit filament binding by C-terminal constructs. In this assay, the N- and C-terminal fragments are preincubated with 1×NaMEI/G-Mg before filaments are added (*step 3* in "Filament Binding").

Sequence alignments indicate DAD and DID sequences are also present in the FRL and DAAM formin groups (Higgs and Peterson, 2005), implying they are likely to be autoinhibited by similar mechanisms. However, the factors that relieve autoinhibition (Rho GTPases or other molecules) are not clear in these cases. Whether the other four metazoan groups (FHOD, INF, FMN, and Delphilin) are autoinhibited remains to be determined, although there is suggestive evidence for such regulation of FHOD and FMN1 (Gasteier *et al.*, 2003; Kobielak *et al.*, 2004; Westendorf, 2001).

Effects of Profilin and Capping Protein

A myriad of actin-binding proteins exist, but here we discuss only profilin and capping protein for the following reasons: (1) their central roles in actin dynamics (Pollard *et al.*, 2000); (2) profilin is a known FH1 domain–binding protein that results in enhanced FH1–FH2 mediated barbed end elongation rate (Kovar and Pollard, 2004; Kovar *et al.*, 2003; Kovar and Pollard, manuscript in preparation; Romero *et al.*, 2004); and (3) all formins examined have antagonistic effects on barbed end capping by capping proteins (Harris *et al.*, 2004; Kovar *et al.*, 2005; Moseley *et al.*, 2004; Zigmond *et al.*, 2003).

Profilin. Profilin's effect on formin-mediated actin assembly serves as an excellent example of why multiple assays are required to understand formin mechanisms. Initial pyrene-actin polymerization experiments showed that profilin either slightly accelerated (Sagot *et al.*, 2002) or slightly reduced (Li and Higgs, 2003; Pring *et al.*, 2003) polymerization rate of FH1–FH2 domain-containing constructs. These results were interpreted as follows: as long as the FH1 domain is present, profilin does not alter nucleation by formins. The problem with these assays was that they could not distinguish nucleation from elongation. In subsequent single filament analysis, profilin clearly had opposing effects on nucleation (inhibited) and elongation (accelerated) (Kovar and Pollard, 2004; Kovar and Pollard, manuscript in preparation; Romero *et al.*, 2004). Using the dual filament assay ("Fluorescent Microscopy [Single Filament Analysis]"), we find similar results (Chhabra and Higgs, unpublished data). Thus, TIRF or dual filament assays are more appropriate for examining the role of profilin. Profilin (dialyzed in G buffer) can be included with the actin stock (*step 3*) in these assays.

Profilin is a 13-kDa monomeric protein and can be prepared easily by bacterial expression. Because of its affinity for poly-proline-agarose, profilin can be purified without an affinity tag (Janmey, 1991).

Capping Protein. Formin's antagonistic effect on barbed end capping by capping proteins can also be examined using the dual filament microscopy assay (in "Fluorescence Microscopy [Single Filament Analysis]"). Here, capping protein is preincubated with formin before addition to A488P-labeled filaments (*step 5*). Thus, both barbed end-binding proteins have access to filaments at the same time. Because both FH2 domain and capping protein have extremely slow off-rates, the first molecule that binds the barbed end generally remains bound for the course of the experiment (Kovar *et al.*, 2005). Given this fact, preincubation of filaments with one component should completely negate the effect of the other. By measuring lengths of the newly elongated RhP segments off of A488P seeds, one can determine whether the elongation rate in the presence of both capping protein and formin is greater than the elongation rate in the presence of capping protein alone.

Heterodimeric capping protein can be prepared from a bis-cistronic bacterial expression vector, as described in (Palmgren *et al.*, 2001; Soeno *et al.*, 1998).

Actin Preparation

High-quality monomeric actin is *crucial* for reproducible and interpretable assembly/disassembly kinetics. Rabbit skeletal muscle actin is most commonly used because of historical reasons and the relative ease of obtaining large quantities.

Method

Actin is extracted from an acetone powder made from rabbit skeletal muscle and purified by rounds of polymerization/depolymerization, essentially as described in Spudich and Watt (1971). One additional and final step that is necessary for obtaining reproducible results is to gel-filter the actin on Sephracyl S-200 or S-300 (MacLean-Fletcher and Pollard, 1980). Without gel filtration, nucleation lag is very short or nonexistent, probably because of short, stable actin oligomers contaminating the monomers. We use only the later fractions from the gel filtration peak (monitored by OD290), which avoids this contamination. To prove that gel filtration matters, one can conduct pyrene actin polymerization assays with 4 μM monomers from individual peak fractions and should find that the lag increases (but maximum fluorescence does not) later in the peak.

Special Considerations

1. Storage: We store purified monomeric actin in G buffer at 4° for up to 3 weeks. Actin used for kinetics should not be frozen. Ideally, actin should be kept in continuous dialysis against G buffer, exchanging the dialysis buffer daily. This procedure ensures that fresh DTT is available continuously, because actin is highly susceptible to cysteine oxidation.

2. Quantifying actin: The extinction coefficient of actin at 290 nm is 26,000 $M^{-1}cm^{-1}$. In other words, an absorbance of 1.0 in a 1-cm cuvette indicates 38.5 μM actin; 290 nm is used due to the 0.2 mM ATP in G buffer. Always blank against G buffer. For pyrene-actin quantification see Cooper *et al.* (1983).

3. Additional sources of actin: Rabbit skeletal muscle actin can be purchased from Cytoskeleton, Inc., but we find this requires additional gel filtration for kinetic assays (no or little lag is observed when prepared

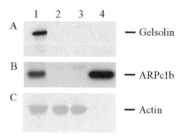

FIG. 5. Western blot analysis of muscle and platelet actin. (A) Anti-gelsolin blot; 2 μg of the following proteins: (1) platelet actin from Cytoskeleton, Inc., prepared according to manufacturer's instructions, (2) platelet actin after purification on SourceQ 5/5 column, (3) muscle actin after gel filtration on S-200 column, (4) empty lane. (B) Anti-ARPC1b (pYQArc) blot. Lanes 1–3 same as above; lane 4 shows 50 ng bovine thymus Arp2/3 complex. (C) Amido black staining of blot in (B) to show actin. Blots of capping protein α_2 subunit and WASp are not shown, because we did not detect either of these two proteins in platelet actin when prepared according to the manufacturer's instructions.

following manufacturer's instructions). Non-muscle actin (β and/or γ isoforms) is closer to "physiological" conditions, although the primary sequence of muscle actin only differs from non-muscle actin by less than 7%. The β isoform can be purified from red blood cells (Sheetz et al., 1976), whereas a mix of β/γ can be purified from other sources, most commonly platelets (Gordon et al., 1977). One source of platelets is from outdated blood bank stocks. Lyophilized platelet actin is available from Cytoskeleton, Inc. (mix of ~85% β/15% γ isoforms). The high concentrations of actin-binding proteins in platelets, however, necessitate extra care. When prepared according to manufacturer's instructions, platelet actin from Cytoskeleton, Inc. contains a significant amount of gelsolin (reported by the manufacturer, see Cytoskeleton catalog). In addition, we detect Arp2/3 complex subunits in this preparation by Western blot analysis (Fig. 5). We are able to remove these two contaminants by further purification steps, either: (1) a round of polymerization/depolymerization followed by gel filtration; or (2) anion exchange chromatography followed by a round of polymerization/depolymerization (Fig. 5). Capping protein (α_2 subunit) and WASp are not detected in the manufacturer's preparation by Western blot.

Expression and Purification of Mammalian Formin FH2 Domains

We have successfully expressed and purified FH2-containing constructs of five different mammalian formins, often several different fragments for each protein, as glutathione S-transferase fusions in E. coli (Harris et al., 2004; Li and Higgs, 2003, 2005). We use the pGEX-KT vector (Hakes and Dixon, 1992), which has a hyperlabile thrombin protease cleavage site

C-terminal to the GST. Removal of the GST tag is especially important, because FH2 domains exist functionally as dimers, and the additional dimeric GST adds potential artifacts. Final yield is 2–6 mg/l of culture, depending on the formin. This procedure is also effective for purification of N-terminal regions of mDia1 (Li and Higgs, 2003, 2005).

Special Materials

1. Rosetta 2, non-DE3 competent cells: (Novagen 71402).
2. Terrific Broth (TB): 12 g/l tryptone, 24 g/l yeast extract, 4.5 ml/l glycerol, 14 g/l dibasic potassium phosphate, and 2.6 g/l monobasic potassium phosphate. Tryptone and yeast extract solution (in 0.9 volumes) is autoclaved separately from glycerol and phosphate solution (in 0.1 volumes). Mix the two solutions together on day of use.
3. Ampicillin: (Fisher BP1760–25) Stock at 100 mg/ml in dH_2O, store at $-20°$.
4. Chloramphenicol: (Sigma C-0378) Stock at 34 mg/ml in ethanol, store at $-20°$.
5. Isopropyl-1-thio-b-D-galactopyranoside (IPTG): (Roche 11 411 446 001) Stock solution at 1 M in dH_2O, store at $-20°$.
6. Complete protease inhibitors: (Roche 11 873 580 001).
7. Extraction buffer (EB): 50 mM Tris-HCl, pH 8.0, 500 mM NaCl, 5 mM EDTA, 1 mM DTT, 1 pill/50 ml complete protease inhibitors.
8. Wash buffer (WB): EB without protease inhibitors but with 0.05% thesit.
9. Glutathione-sepharose 4B (Amersham 27–4574–01): Store in 20% ethanol at $4°$. Before use, wash several times with dH_2O, and then equilibrate with WB.
10. Thrombin (Sigma T-4265): Stock solution at 1 U/ml in PBS. Freeze aliquots at $-70°$.
11. Diisopropyl fluorophosphate (DFP): (Calbiochem 30967).
12. Phenylmethylsulfonyl fluoride (PMSF): (Calbiochem 52332) Stock in ethanol made fresh on the day of use.

Method

1. Transform Rosetta 2 competent cells with the expression construct and grow to OD_{600} of 0.8–1.0 in TB with 100 μg/ml ampicillin and 34 μg/ml chloramphenicol at $37°$.
2. Reduce temperature to $16°$.
3. Add 0.5 mM IPTG and an additional 100 μg/ml ampicillin.
4. Shake cultures overnight at $16°$. *All subsequent purification steps are performed at $4°$ or on ice.*
5. Pellet bacteria and resuspend in 25 ml EB per liter of culture.

6. Extract by sonication. We use probe sonication at 50% duty cycle, 70% maximal power, for 4 × 20 seconds. Cells are in 50-ml plastic tubes, in an ice-water bath.

7. Centrifuge >200,000g for 30 min at 4°. We use a Ti45 rotor at 40,000 rpm.

8. Load supernatant onto glutathione-sepharose 4B column (5 ml beads per liter of original culture), and wash with WB.

9. Add an equal volume of WB to beads.

10. Add thrombin to 10 U/ml, and rotate the suspension for varying times (1 h to overnight) at 4°, depending on protein.

11. Wash cleaved protein from the column with WB and inactivate thrombin with 5 mM DFP/1 mM PMSF for 15 min, after which add DTT to a concentration of 10 mM.

12. Purify further using anion or cation exchange chromatography as necessary.

Special Considerations

1. Thrombin digestion: Digestion time varies depending on the stability of the formin. One hour is sufficient for most proteins. For convenience, mDia1 and mDia2 FH2 domains can tolerate overnight digestions, without additional cleavage products being created. However, a substantial portion of FRL1 gets cleaved further by thrombin when extended digestion times are used.

2. Additional purification: After elution from glutathione-sepharose 4B, mDia1 and mDia2 can be further enriched using Q Sepharose Fast Flow (Li and Higgs, 2003, 2005). FRL1 (Harris *et al.*, 2004), for example, requires more extensive purification by cation exchange chromatography on FPLC.

3. Storage conditions: FRL1 and FRL2 are the only formins we have found to maintain all activities when frozen at −70°. All others loose substantial activity upon freezing (Li and Higgs, 2005), and should be kept at 4° or stored as 50% glycerol stocks at −20°. Typical storage buffer components include NaCl (50–150 mM), 5 mM NaPO$_4$, pH 7.0, 0.5 mM EGTA, 0.5 mM DTT, 0.1 mM MgCl$_2$.

References

Alberts, A. S. (2001). Identification of a carboxyl-terminal diaphanous-related formin homology protein autoregulatory domain. *J. Biol. Chem.* **276,** 2824–2830.

Amann, K. J., and Pollard, T. D. (2001). Direct real-time observation of actin filament branching mediated by Arp2/3 complex using total internal reflection fluorescence microscopy. *Proc. Natl. Acad. Sci. USA* **98,** 15009–15013.

Cooper, J. A., Walker, S. B., and Pollard, T. D. (1983). Pyrene actin: Documentation of the validity of a sensitive assay for actin polymerization. *J. Muscle Res. Cell Motil.* **4,** 253–262.

Drenckhahn, D., and Pollard, T. D. (1986). Elongation of actin filaments is a diffusion-limited reaction at the barbed end and is accelerated by inert macromolecules. *J. Biol. Chem.* **261,** 12754–12758.

Gasman, S., Kalaidzidis, Y., and Zerial, M. (2003). RhoD regulates endosome dynamics through Diaphanous-related Formin and Src tyrosine kinase. *Nat. Cell Biol.* **5,** 195–204.

Gasteier, J. E., Madrid, R., Krautkramer, E., Schroder, S., Muranyi, W., Benichou, S., and Fackler, O. T. (2003). Activation of the Rac-binding partner FHOD1 induces actin stress fibers via a ROCK-dependent mechanism. *J. Biol. Chem.* **278,** 38902–38912.

Gordon, D. J., Boyer, J. L., and Korn, E. D. (1977). Comparative biochemistry of non-muscle actins. *J. Biol. Chem.* **252,** 8300–8309.

Harris, E. S., Li, F., and Higgs, H. N. (2004). The mouse formin, FRLa, slows actin filament barbed end elongation, competes with capping protein, accelerates polymerization from monomers, and severs filaments. *J. Biol. Chem.* **279,** 20076–20087.

Higashida, C., Miyoshi, T., Fujita, A., Oceguera-Yanez, F., Monypenny, J., Andou, Y., Narumiya, S., and Watanabe, N. (2004). Actin polymerization-driven molecular movement of mDia1 in living cells. *Science* **303,** 2007–2010.

Higgs, H. N. (2005). Formin proteins: A Domain-based approach. *Trends Biochem. Sci.* **30,** 342–353.

Higgs, H. N., and Peterson, K. J. (2005). Phylogenetic analysis of the Formin Homology 2 (FH2) domain. *Mol. Biol. Cell* **16,** 1–13.

Janmey, P. A. (1991). Polyproline affinity method for purification of platelet profilin and modification with pyrene-maleimide. *Methods Enzymol.* **196,** 92–99.

Kobielak, A., Pasolli, H. A., and Fuchs, E. (2004). Mammalian formin-1 participates in adherens junctions and polymerization of linear actin cables. *Nature Cell Biol.* **6,** 21–30.

Kovar, D. R., Kuhn, J. R., Tichy, A. L., and Pollard, T. D. (2003). The fission yeast cytokinesis formin Cdc12p is a barbed end actin filament capping protein gated by profilin. *J. Cell Biol.* **161,** 875–887.

Kovar, D. R., and Pollard, T. D. (2004). Insertional assembly of actin filament barbed ends in association with formins produces piconewton forces. *Proc. Natl. Acad. Sci. USA* **101,** 14725–14730.

Kovar, D. R., Wu, J. Q., and Pollard, T. D. (2005). Profilin-mediated competition between capping protein and formin Cdc12 during cytokinesis in fission yeast. *Mol. Biol. Cell.* **16,** 2313–2324.

Kuhn, J. R., and Pollard, T. D. (2005). Real-time measurements of actin filament polymerization by total internal reflection fluorescence microscopy. *Biophys. J.* **88,** 1387–1402.

Li, F., and Higgs, H. N. (2003). The mouse formin mDia1 is a potent actin nucleation factor regulated by autoinhibition. *Curr. Biol.* **13,** 1335–1340.

Li, F., and Higgs, H. N. (2005). Dissecting requirements for auto-inhibition of actin nucleation by the formin, mDia1. *J. Biol. Chem.* **280,** 6986–6992.

MacLean-Fletcher, S., and Pollard, T. D. (1980). Mechanisms of action of cytochalasin B on actin. *Cell* **20,** 329–341.

Malm, B. (1984). Chemical modification of Cys-374 of actin interferes with the formation of the profilactin complex. *FEBS Lett.* **173,** 399–402.

Moseley, J. B., Sagot, I., Manning, A. L., Xu, Y., Eck, M. J., Pellman, D., and Goode, B. L. (2004). A conserved mechanism for Bni1- and mDia1-induced actin assembly and dual regulation of Bni1 by Bud6 and profilin. *Mol. Biol. Cell* **15,** 896–907.

Palmgren, S., Ojala, P. J., Wear, M. A., Cooper, J. A., and Lappalainen, P. (2001). Interactions with PIP2, ADP-actin monomers, and capping protein regulate the activity and localization of yeast twinfilin. *J. Cell Biol.* **155,** 251–260.

Peng, J., Wallar, B. J., Flanders, A., Swiatek, P. J., and Alberts, A. S. (2003). Disruption of the Diaphanous-related formin Drf1 gene encoding mDia1 reveals a role for Drf3 as an effector for Cdc42. *Curr. Biol.* **13**, 534–545.

Pollard, T. D., Blanchoin, L., and Mullins, R. D. (2000). Molecular mechanisms controlling actin filament dynamics in nonmuscle cells. *Annu. Rev. Biophys. Biomol. Struct.* **29**, 545–576.

Pollard, T. D., and Cooper, J. A. (1986). Actin and actin-binding proteins. A critical evaluation of mechanisms and functions. *Annu. Rev. Biochem.* **55**, 987–1035.

Pring, M., Evangelista, M., Boone, C., Yang, C., and Zigmond, S. H. (2003). Mechanisms of formin-induced nucleation of actin filaments. *Biochemistry* **42**, 486–496.

Pruyne, D., Evangelista, M., Yang, C., Bi, E., Zigmond, S., Bretscher, A., and Boone, C. (2002). Role of formins in actin assembly: Nucleation and barbed-end association. *Science* **297**, 612–615.

Romero, S., Le Clainche, C., Didry, D., Egile, C., Pantaloni, D., and Carlier, M. F. (2004). Formin is a processive motor that requires profilin to accelerate actin assembly and associated ATP hydrolysis. *Cell* **119**, 419–429.

Sagot, I., Rodal, A. A., Moseley, J., Goode, B. L., and Pellman, D. (2002). An actin nucleation mechanism mediated by Bni1 and profilin. *Nat. Cell Biol.* **8**, 626–631.

Sheetz, M. P., Painter, R. G., and Singer, S. J. (1976). Relationships of the spectrin complex of human erythrocyte membranes to the actomyosins of muscle cells. *Biochemistry* **15**, 4486–4492.

Soeno, Y., Abe, H., Kimura, S., Maruyama, K., and Obinata, T. (1998). General of functional b-actinin (CapZ) in an *E. coli* expression system. *J. Muscle Res. Cell Motil.* **19**, 639–646.

Spudich, J. A., and Watt, S. (1971). The regulation of rabbit skeletal muscle contraction. I. Biochemical studies of the interaction of the tropomyosin-troponin complex with actin and the proteolytic fragments of myosin. *J. Biol. Chem.* **246**, 4866–4871.

Straub, F. B., and Feuer, G. (1950). Adenosine triphosphate, the functional group of actin. *Biochim. Biophys. Acta* **4**, 180–194.

Wallar, B. J., and Alberts, A. S. (2003). The formins: Active scaffolds that remodel the cytoskeleton. *Trends Cell Biol.* **13**, 435–446.

Westendorf, J. J. (2001). The Formin/diaphanous-related protein, FHOS, interacts with Rac1 and activates transcription from the serum response element. *J. Biol. Chem.* **276**, 46453–46459.

Xu, Y., Moseley, J., Sagot, I., Poy, F., Pellman, D., Goode, B. L., and Eck, M. J. (2004). Crystal Structures of a formin homology-2 domain reveal a tethered dimer architecture. *Cell* **116**, 711–723.

Yasuda, S., Oceguera-Yanez, F., Kato, T., Okamoto, M., Yonemura, S., Terada, Y., Ishizaki, T., and Narumiya, S. (2004). Cdc42 and mDia3 regulate microtubule attachment to kinetochores. *Nature* **428**, 767–771.

Zigmond, S. H. (2004). Formin-induced nucleation of actin filaments. *Curr. Opin. Cell Biol.* **16**, 99–105.

Zigmond, S. H., Evangelista, M., Boone, C., Yang, C., Dar, A. C., Sicheri, F., Forkey, J., and Pring, M. (2003). Formin leaky cap allows elongation in the presence of tight capping proteins. *Curr. Biol.* **13**, 1820–1823.

Zigmond, S. H., Joyce, M., Yang, C., Brown, K., Huang, M., and Pring, M. (1998). Mechanism of Cdc42-induced actin polymerization in neutrophil extracts. *J. Cell Biol.* **142**, 1001–1012.

[16]   Formin Proteins: Purification and Measurement of Effects on Actin Assembly

By James B. Moseley, Sankar Maiti, and Bruce L. Goode

Abstract

We describe methods for expressing and isolating formin proteins from a wide range of species and comparing quantitatively their effects on actin assembly. We first developed these procedures for purification of *S. cerevisiae* formins Bni1 and Bnr1 but have extended them to mammalian formins, including mouse mDia1 and mDia2 and human Daam1. Thus, the approach we describe should be universally applicable to the purification and analysis of formins from any eukaryote. Formins expressed in yeast rather than bacteria usually have improved solubility, yield, and actin assembly activity. Yields are 200–500 μg purified formin per liter of yeast culture. For some applications bacterial expression and purification is preferable, and these methods are also described. For expression of most formins, in either yeast or bacteria, we recommend using an amino terminal 6xHis affinity tag. Active FH1-FH2 containing fragments of the formins Bni1, Bnr1, mDia1, mDia2, and Daam1 are all digomeric. However, they nucleate actin filaments with variable efficiencies, as high as one actin filament per formin complex. In the last section, we outline fluorometric methods for measuring and quantitatively analyzing the *in vitro* activities of formins on actin nucleation and processive capping of actin filaments.

Introduction

Formin proteins are ubiquitous actin nucleation factors with critical roles in assembling cytokinetic rings, stress fibers, filopodia, and actin cables (Wallar and Alberts, 2003). Formins are large multidomain proteins that are thought to be autoinhibited until association with activated Rho proteins. The amino terminus of most formins contains a Rho-binding domain (RBD). Binding of Rho proteins to the RBD is thought to relieve intramolecular (autoinhibitory) interactions between the amino and carboxyl termini, exposing the active carboxyl terminal half, which contains the actin-nucleating formin homology 1 and 2 (FH1 and FH2) domains.

The FH2 is the signature domain of formins and the most highly conserved region across species (Higgs and Peterson, 2005). FH2 is required for actin nucleation *in vivo* and sufficient *in vitro* for nucleating

METHODS IN ENZYMOLOGY, VOL. 406   0076-6879/06 $35.00
             DOI: 10.1016/S0076-6879(06)06016-2

purified actin monomers. Direct actin nucleation by purified FH2-containing fragments of formins was first demonstrated for *S. cerevisiae* formin Bni1 (Pruyne *et al.*, 2002; Sagot *et al.*, 2002) and subsequently for FH2-containing formins from *S. pombe* and mammals (reviewed in Wallar and Alberts, 2003). All FH2-containing formin fragments isolated to date show two activities, albeit with variable potency: (1) actin nucleation and (2) processive capping. Processive capping refers to the ability of FH2 domains to stay persistently associated with the rapidly growing barbed ends of filaments, simultaneously allowing insertional growth and protecting filament ends from capping proteins (Higashida *et al.*, 2004; Kovar and Pollard, 2004; Moseley *et al.*, 2004; Zigmond *et al.*, 2003). All FH2 domains isolated to date display nucleation and processive capping activities but permit variable rates of elongation (Zigmond, 2004).

The FH1 domain, adjacent to the FH2, contains proline-rich sequences that interact directly with profilin. This enables FH1-FH2 fragments to assemble actin filaments from profilin-bound actin monomers, which are considered to be the predominant physiological substrate for actin assembly. FH2 alone cannot nucleate actin assembly from profilin-bound monomers. Some FH1 domains, when coupled to an FH2 domain, accelerate filament elongation of profilin-bound monomers above the rate of elongation at free barbed ends of filaments. This increase in elongation rate correlates roughly with increased number of profilin binding sites in the FH1 (Romero *et al.*, 2004; D. Kovar and T. Pollard, personal communication). Together, these observations provide a working model for formin-mediated actin assembly (Fig. 1). Activated Rho proteins bind the RBD and release formins from an autoinhibited state. FH2 directly nucleates actin assembly and "rides" the fast-growing end of the filament, allowing insertional growth. FH1 interacts with profilin-bound actin monomers, positioning actin subunits for addition at the rapidly growing end of the filament and modulating elongation rate. Sequences carboxyl terminal to the FH2 (the COOH domain) interact with specific formin ligands, such as *S. cerevisiae* Bud6, to stimulate formin-mediated actin assembly (Moseley *et al.*, 2004).

Difficulties with the expression and solubility of full-length formins so far have limited biochemical analysis of formin activities to carboxyl terminal fragments. The most commonly studied fragments extend from FH1 and include FH2 and COOH, referred to as FH1-COOH (Fig. 3A). Here, we describe two systems for the expression and purification of 6xHis-tagged FH1-COOH constructs, one from *E. coli* and one from the budding yeast *S. cerevisiae*. We have used the yeast system to purify highly active constructs (e.g., FH2, FH1-FH2, FH1-COOH, FH2-COOH) of formins from budding yeast (Bni1 and Bnr1), mouse (mDia1 and mDia2), and

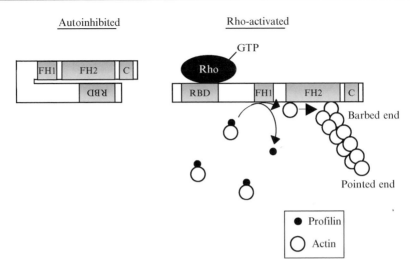

FIG. 1. Schematic of proposed formin molecular regulation. In this model, ligand-free formin is autoinhibited through interactions between its amino and carboxyl termini (left). Binding of an activated Rho protein to the amino terminus relieves autoinhibition (right), liberating the FH1-FH2–containing carboxyl terminus to nucleate actin filament assembly (see "Introduction" for details on mechanism). RBD, Rho-binding domain; FH1, formin homology 1 domain; FH2, formin homology 2 domain; C, carboxyl terminal domain.

human (Daam-1). We also outline techniques for quantifying the activities of purified formins on actin.

Purification Considerations

Choosing the Tag and Expression System

When approaching formin purification, an important first consideration is what epitope tag to use. We have isolated many different formins with different affinity tags (Flag, GST, and 6xHis), and in most cases, optimal solubility, yield, and activity are achieved using an amino-terminal 6xHis tag. The 6xHis tag is advantageous because it is small in size (6 residues), which reduces its chances of interfering with protein folding and/or interactions with actin. For bacterial expression, we use 6xHis-fusion vector pQE-9 (Qiagen, Valencia, CA), and for yeast expression, we built a galactose-inducible 2μ *URA3*-marked 6xHis-fusion overexpression vector (available on request).

A second consideration is which expression system to use. We have compared the efficiencies of expression and isolation of many formins in

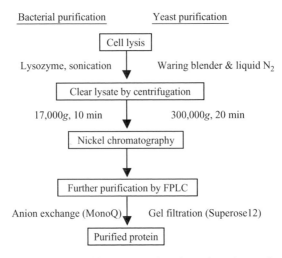

Fig. 2. Flowchart of formin purification steps from bacteria and yeast. See text for details. Reproduced with permission from Kitegawa, M., Mukai, H., Shibata, H., and Ono, Y. (1995). *Biochemical Journal*, **310**, 657–644. © The Biochemical Society.

bacteria and yeast. The two purification schemes are similar (Fig. 2), and both procedures yield highly purified polypeptides (Fig. 3B,C). In initial studies, we used bacterial expression, which has the advantage of more rapid cell growth and the absence of eukaryotic formin ligands that could potentially contaminate preparations. Typical yields of 6xHis-Bni1(FH1-COOH) from bacteria are 300 μg/l of culture. The use of different bacterial strains can greatly affect yield. We recommend *E. coli* strain BL21-Codon Plus (DE3)-RP (Stratagene, La Jolla, CA), which carries extra copies of Arg and Pro tRNAs, important for high expression of proline-rich FH1 domains.

Many formins expressed in *E. coli* are not highly soluble or fully active, and, therefore, we developed methods for overexpression and purification of formins from *S. cerevisiae*. In most cases, expression in yeast has solved the problems with yield, solubility, and activity. For example, Bnr1 FH1-COOH is nearly inactive when expressed in bacteria, but the same construct expressed in yeast shows orders of magnitude higher activity in actin nucleation assays. For most formins (Fig. 4A), yields are about 500 μg/l of yeast culture. Furthermore, posttranslational modification of formins by phosphorylation does not seem to affect activity. Of the formins just mentioned, only Bni1 is phosphorylated (detected by phospho-specific in-gel stains and mobility shift). However, dephosphorylation of Bni1 with lambda phosphatase treatment does not affect its actin assembly activity

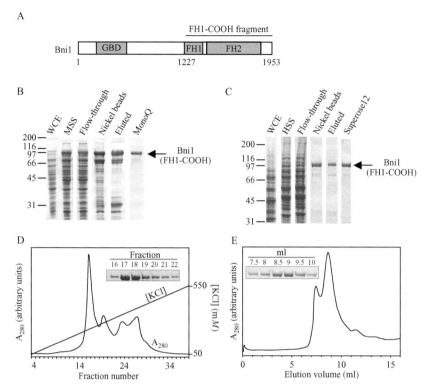

FIG. 3. Purification of the *S. cerevisiae* formin 6xHis-Bni1(FH1-COOH) from both bacterial and yeast expression systems. (A) Schematic of Bni1 domains, highlighting the FH1-COOH fragment (a.a. 1227–1953) purified. (B and C) Coomassie-stained SDS-PAGE gel of fractions through formin purification from bacteria (B) and yeast (C) (see text for details). WCE, whole cell extract; MSS, medium-speed supernatant. (D) Elution profile from monoQ 5/5 column of 6xHis-Bni1(FH1-COOH) purified from bacteria. Sample loading, column washes, and KCl gradient are described in text. SDS-PAGE and Coomassie staining of peak fractions (0.5 ml) revealed highly purified 6xHis-Bni1(FH1-COOH) in fractions 17–18 (inset), eluted with ∼220 m*M* KCl. (E) Superose12 (10/30) elution profile for 6xHis-Bni1(FH1-COOH) purified from yeast. SDS-PAGE and Coomassie staining of fractions (0.5 ml) revealed a minor peak of 6xHis-Bni1(FH1-COOH) in the void (7.7 ml) and a major peak eluting at 8.5–9.0 ml, consistent with formin oligomerization.

(our unpublished data). A final advantage of the yeast expression system is that cells are lysed while still frozen and can be stored as a lysed yeast powder at −80° indefinitely without degradation. This provides added convenience, allowing small-scale or large-scale purifications from the same −80° stocks.

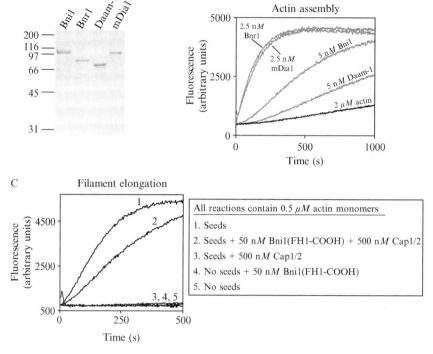

FIG. 4. Fluorometric actin assembly activity assays of purified formins. (A) Coomassie-stained SDS-PAGE gel showing purified FH1-COOH fragments of yeast Bni1 and Bnr1, human Daam-1, and mouse mDia1, all isolated using the yeast expression system. (B) Actin assembly reactions, in which 2 μM monomeric rabbit muscle actin (5% pyrene-labeled) was assembled alone (curve labeled "2 μM actin") or in the presence of the indicated concentrations of different formin proteins (all FH1-COOH fragments). (C) Actin filament elongation reactions, in which different combinations of F-actin "seeds," 50 nM Bni1(FH1-COOH), and/or 500 nM Cap1/2 (*S. cerevisiae* capping protein) were mixed with 0.5 μM monomeric rabbit muscle actin (10% pyrene-labeled) at time = 0. Actin polymerization was monitored by increase in pyrene fluorescence as described in the text.

Monitoring Expression

When first expressing a new formin construct in bacteria or yeast, expression levels and solubility should be assessed. A peroxidase-conjugated anti-polyHis monoclonal antibody from Sigma-Aldrich (St. Louis, MO) gives minimal background on blots of total bacterial and yeast proteins. It is important to optimize time of induction of expression, which can vary for different constructs. Expression levels should be compared at multiple time points after IPTG induction (0, 8, 16, 24 h) in bacteria and GAL induction (0, 2, 4, 6, 8, and 16 h) in yeast, both grown at 25°. The optimal induction time is usually ~16 h for bacteria (IPTG) and ~8 h for yeast (GAL). Solubility should be monitored by comparing formin levels in equivalent amounts of whole cell extracts and clarified extracts (using different centrifugation conditions for bacteria and yeast; see later). If solubility problems are encountered, detergents should be varied (see later), and lower growth temperatures considered for bacteria (e.g., 16°).

Buffers and Detergents

Imidazole buffer is used in the isolation of 6xHis tagged proteins, because it minimizes binding of contaminants to nickel resin; PBS is included as a buffering agent. Purified formins are stored in $HEKG_{10}$ buffer (20 mM HEPES, pH 7.4, 1 mM EDTA, 50 mM KCl, 10% glycerol) with 1 mM DTT. The lack of free amines in this buffer allows formins to be used for chemical cross-linking, conjugation to resins, or labeling with reactive dyes. Inclusion of glycerol in the storage buffer minimizes denaturation during freeze–thaw transitions; 0.5–1.0% NP-40 is included during early stages of 6xHis-tagged purification for optimal solubility and minimizing nonspecific interactions with contaminants; for GST-tagged formins, substitute 0.5–1.0% Triton X-100. If stability and activity are problematic, add Thesit detergent (0.2% final; Sigma-Aldrich, St. Louis, MO) during initial purification and washes (from bacteria and yeast), which improves activity for many formins, most notably mDia1 (Li and Higgs, 2003, and our unpublished observations). Detergents should not be included in later washes or elution, because they can present problems during FPLC analysis.

To Freeze, or Not To Freeze?

An important concern that should be assessed is the possible detrimental effects from freeze–thaw transitions of purified formins. After purification, keep two samples on ice, flash-freeze and thaw two others, then compare their actin nucleation activities (later). Some purified formins maintain full nucleation activity after freeze–thawing (e.g., Bnr1), whereas others lose significant activity and stability (e.g., mDia1). If a formin falls into the latter category, it should always be prepared freshly before activity measurements (and never frozen); formins stored on ice usually maintain full activity for ~2 weeks (Li and Higgs, 2004). If a formin permits freezing, then inclusion of glycerol in the freezing buffers can help minimize loss of activity. In addition, some formins (e.g., Bni1(FH1-COOH)) can be frozen and retain full activity, but only if stored in a stock concentration >2 μM.

Thus, the properties of each formin can vary greatly and must be defined first to optimize handling and storage.

Multimerization—Dimers or Tetramers?

All FH2-containing formin fragments examined multimerize. The FH2 alone forms an extremely stable dimer (Xu *et al.*, 2004), whereas most FH1-COOH fragments are reported to dimerize or tetramerize (Zigmond *et al.*, 2003; our unpublished data), but it is not yet clear whether tetramers are stable or dynamic. When new formin constructs are isolated, multimerization state can be determined most rigorously by analytical ultracentrifugation and/or dynamic light scattering.

Purification Methods

Expression and Purification of FH1-COOH Constructs from E. coli

Solutions. Lysis buffer: 30 mM imidazole (pH 8.0), 1X PBS (20 mM sodium phosphate buffer (pH 7.4), 150 mM NaCl), 0.5 mM DTT, 1% NP-40, 1.0 μg/ml Lysozyme (Sigma-Aldrich), protease inhibitors (final 1.0 μg/μl antipain, leupeptin, pepstatin A, chymostatin, and aprotinin)

> High salt wash buffer: 20 mM imidazole (pH 8.0), 1× PBS, 0.5 mM DTT, 350 mM NaCl
> Low salt wash buffer: 20 mM imidazole (pH 8.0), 1× PBS, 0.5 mM DTT
> Elution buffer: 350 mM imidazole (pH 8.0), 1× PBS, 0.5 mM DTT
> monoQ buffer A: 20 mM HEPES (pH 7.4), 1 mM EDTA, 50 mM KCl, 1 mM DTT
> monoQ buffer B: 20 mM HEPES (pH 7.4), 1 mM EDTA, 1 M KCl, 1 mM DTT

Special Equipment. AKTA FPLC (AP Biotech, Piscataway, NJ) or equivalent chromatography device.

Expression in Bacteria

1. Transform *E. coli* strain BL21-Codon Plus (DE3)-RP (Stratagene, La Jolla, CA) with inducible 6xHis-FH1-COOH expression plasmid, and select for antibiotic resistance.

2. Pick 5–20 freshly transformed colonies and inoculate a 5-ml culture and grow cells at 37°. After 4–5 h growth, use 5ml culture to inoculate 1 l culture in a 4-l flask. Grow at 37°, shaking at ~200 rpm to an OD$_{600}$ 0.3–0.4. Shift culture to 25° shaking, grow for 30 min, then induce

expression by addition of IPTG to 0.4 mM. Continue growth at 25° for 8–24 h, depending on optimization (see previously).

3. Harvest cells by centrifugation at 4000g (5000 rpm in Sorvall GSA rotor) for 10 min at 4°. Resuspend cells in 20 ml ice-cold PBS, transfer to Oak Ridge centrifugation tubes, and pellet cells by centrifugation at 7000g (7000 rpm in Sorvall SA-600 rotor), 4°, 10 min. Discard supernatant, resuspend pellet in 10 ml ice-cold PBS, and store (indefinitely) at −80°.

Purification from Bacterial Pellet

1. Thaw 10-ml pellet by swirling Oak Ridge tube in warm water bath (∼50°). Immediately after pellet thaws, transfer tube to ice. Add 10 ml lysis buffer, mix, and quickly supplement with PMSF (add drop wise to 1 mM final). Incubate on ice 15–30 min; lysate should become viscous from DNA content. Sonicate lysate using a 550 Sonic Dismembranator (Fisher Scientific, Hampton, NH) or equivalent, as per manufacturer's instructions. The viscosity of the lysate should diminish after sonication.

2. Clarify lysate by centrifugation at 17,000g (12,000 rpm, Sorvall SS-34 rotor), 4°, 10 min. Collect the medium speed supernatant (MSS, ∼15 ml) in a 15-ml conical tube; remove ∼50 μl for gel samples.

3. Add 500 μl Ni-NTA beads (Qiagen) or equivalent nickel-agarose resin (prewashed in 10 mM imidazole (pH 8.0), 1X PBS, 0.5 mM DTT) to MSS, and rotate, 4°, 2 h. Pellet beads by centrifugation at 5000g. Remove supernatant (flow-through, FT); save ∼50 μl for gel samples and discard the rest. Wash beads three times in 15 ml high-salt wash, followed by three washes in 15 ml low-salt wash. After the final wash, leave the 500 μl Ni-NTA beads in ∼300 μl low-salt wash buffer. Save 5 μl of slurry for gel sample.

4. To elute protein from nickel beads, remove as much remaining low-salt buffer as possible. Add 500 μl elution buffer rapidly to avoid drying. Mix by pipetting four to five times, and avoid introducing bubbles. Centrifuge briefly to pellet beads (5000 rpm, 5 sec). Collect the supernatant (eluted fraction) containing 6xHis-FH1-COOH. Repeat the elution one to two more times to harvest the remaining protein from beads. A total of two to three elutions removes >95% of 6xHis-FH1-COOH. Levels of protein remaining on the beads should be checked on gels. We find that eluting with an imidazole gradient does not significantly reduce contamination.

5. Analyze samples from the purification steps by SDS-PAGE: initial whole cell extract (WCE), MSS, FT, bead fraction, and eluted protein (see Fig. 3B); 5 μl of the nickel bead fraction and 5–10 μl of the eluted fraction are sufficient to readily detect formin bands (0.2–1.0 μg) on Coomassie-stained gels. Fractions contain prominent contaminants at MW 25, 30, 65,

and 70 kDa, which are removed by subsequent anion-exchange chromatography.

6. Salt levels must be reduced by dialysis or by concentrating the eluted fraction to ~200 μl in a Microcon-10 device (Millipore, Billerica, MA) and diluting with 1.8 ml low-salt wash buffer lacking imidazole.

7. Load eluted protein on a monoQ (5/5) column (AP Biotech, Piscataway, NJ), equilibrated with 10 ml monoQ buffer A. Wash with 5 ml monoQ buffer A (flow rate 0.5 ml/min). Elute with a 20 column volume linear salt gradient (50–500 mM KCl) (0–50% monoQ Buffer B), collecting 0.5 ml fractions. 6xHis-FH1-COOH proteins generally elute at ~220 mM KCl (Fig. 3B,D). Analyze peak fractions by SDS-PAGE and Coomassie staining to estimate concentration and purity. Pool peak fractions and dialyze overnight against storage buffer (HEKG$_{10}$ + 1 mM DTT). Determine final stock concentration by OD$_{280}$ (extinction coefficient for 6xHis-Bni1-FH1-COOH, 43,360 M^{-1} cm^{-1}) and verify on Coomassie-stained gels by comparing band intensity to standards. Flash-freeze small aliquots in liquid N$_2$ and store at $-80°$.

Expression and Purification of 6xHis-FH1-COOH from Yeast

These procedures for expression and purification from *S. cerevisiae* are adapted and modified from previously described methods (Goode, 2002; Rodal *et al.*, 2002). The purification scheme from yeast is similar to bacteria. The main difference is that after elution from nickel resin, proteins are purified by gel filtration (yeast) or anion-exchange (bacteria) chromatography, optimized for removing contaminants from yeast and bacteria, respectively.

Solutions. Resuspension buffer: 30 mM imidazole (pH 8.0), 2× PBS, 0.5 mM DTT, 1% NP-40, protease inhibitors (final 2 mM PMSF and 1.0 μg/μl antipain, leupeptin, pepstatin A, chymostatin, and aprotinin)

High-salt wash, low-salt wash, and elution buffers: same as bacterial purification.

Special Equipment. AKTA FPLC (AP Biotech) or equivalent chromatography device.

Expression in Yeast

1. Transform galactose-inducible 6xHis-FH1-COOH expression plasmid (*URA3*-marked) into yeast strain BJ2168 (Jones, 1991), which is protease deficient and minimizes degradation of proteins during purification. From 2–5 yeast colonies, inoculate 25 ml culture and grow to low saturation (OD$_{600}$ ~1.0) in synthetic complete media lacking uracil and containing 2% w/v raffinose (SC-URA + Raff).

2. Use 25 ml culture to inoculate 1 l SC-URA + Raff in a 4 l baffled flask. Using baffled flasks increases expression of some formins by ~fourfold. Grow yeast cells to OD_{600} 0.4–0.8 at 25°, shaking (~200 rpm). Under these conditions, growth reaches the target OD after about 24 h.

3. To induce protein expression, add 5 g yeast extract and 10 g bacto-peptone, and 100 ml 20% galactose per 1 l culture, and continue shaking at 25° for 8–16 h. These reagents do not need to be sterile.

4. Harvest cells by centrifugation at 4000g for 5 min at 4° (5000 rpm in Sorvall GSA rotor). Resuspend cells in 100 ml ice-cold H_2O and centrifuge as above. Resuspend in 0.2–0.3 volumes cold H_2O, and freeze by adding drop wise into liquid nitrogen. After removal of liquid nitrogen, frozen cells can be stored for years at –80° in plastic containers.

5. Large volumes of frozen yeast should be lysed using liquid nitrogen and a Waring blender as described (Goode, 2002). Smaller quantities of cells (4–20 g) should be lysed using a coffee grinder and liquid nitrogen as described (http://www.bio.brandeis.edu/goodelab). The resulting pow-dered frozen yeast lysate can be stored in plastic containers for years at −80°.

Purification from Frozen Yeast Powder

1. Working quickly, weigh 5 g frozen yeast powder from −80° stock into a disposable 50 ml conical tube. Then add 5 ml ice-cold resuspension buffer, followed by PMSF drop wise to 1 mM final. Mix at room temperature by inverting tube repeatedly, until all frozen clumps of are visibly gone, then immediately proceed to next step.

2. Generate a high-speed supernatant (HSS) by centrifugation at 300,000g (80,000 rpm in Beckman TLA100.3 rotor), 4°, 20 min. Harvest HSS (~8 ml) into a 15 ml conical tube; remove 50 μl HSS for a gel sample. To the remaining HSS, add 500 μl nickel-agarose beads, prewashed as in bacterial method above. Top off the tube with 10 mM imidazole (pH 8.0), 1× PBS, 0.5 mM DTT, 0.5% NP-40, and rotate at 4° for 2 h.

3. Pellet and wash beads as in bacterial method. Save 50 μl FT fraction and 5 μl nickel bead fraction for gel samples.

4. Elute protein as in bacterial method, except with smaller elution volume (facilitates subsequent gel filtration). For maximal yield, perform two 250 μl elutions. In addition, harvest residual eluate surrounding beads by one of two methods: (1) while holding pipette plunger down, insert pipette with beveled tip into bottom of tube. Release plunger while maintaining pressure between pipette tip and bottom of tube. If pipette tip remains flush against bottom of tube, eluate will enter pipette but beads will be excluded. (2) Alternately, use a 20 gauge needle to poke a small hole in the bottom of the Microfuge tube containing the beads. Place this

tube in larger tube and centrifuge to harvest the eluate; beads will remain in smaller tube. If necessary, eluate can be concentrated to 400 μl in a Microcon-10 device (Millipore). Analyze HSS, FT, nickel bead, and eluted fractions on Coomassie-stained gel (Fig. 3C).

5. Load 6xHis-FH1-COOH (total volume <500 μl) on Superose12 HR (10/30) gel filtration column (AP Biotech) equilibrated in storage buffer (HEKG$_{10}$ + 1 mM DTT). Run column at 0.5 ml/min, collecting 0.5-ml fractions. 6xHis-Bni1(FH1-COOH) migrates as a multimer, eluting at ~9 ml (Fig. 3E). Analyze fractions on Coomassie-stained gels to identify peak fractions. Pool fractions, determine protein concentration, and flash-freeze in aliquots as for bacterial method.

Activity Assay Considerations

Selection and Gel Filtration of Actin Monomers

The use of rabbit muscle actin (RMA) is much more convenient, cost-effective, and reliable than isolating actin from other cell types or tissues. RMA can be readily isolated, or alternatively purchased, which is recommended for getting started (Cytoskeleton, Denver, CO). In either case, for nucleation and elongation assays, the RMA must be gel filtered to remove contaminating nuclei. High-speed centrifugation clears actin aggregates, but not smaller nuclei that must be removed for accurate measurement of nucleation and elongation activities. In most cases, use of RMA as a source of actin is not a concern; for example, we find that *S. cerevisiae* Bni1 (FH1-COOH) has equivalent activity on RMA and yeast actin (our unpublished data).

If a sizeable study is undertaken, we recommend purification of RMA as described (Spudich and Watt, 1971) rather than purchasing it. This saves money, and the freshly prepared, gel-filtered RMA gives highly reproducible results. RMA is gel-filtered on an S200 (16/60) column (AKTA FPLC as above). Actin monomer fractions are stable at 4° for >1 month if stored in G buffer made with Tris (pH 8.0) and "refreshed" biweekly by addition of 0.1 mM DTT and 0.1 mM ATP.

Preparation of Pyrene Actin

Pyrenyliodoacetamide-labeled actin monomers (pyrene-RMA) provide a fluorescent readout of actin filament polymerization because of the 30-fold increase in fluorescence that occurs on incorporation of a labeled actin subunit into polymer. Only low levels (5–10%) of pyrene-labeled RMA are required for a strong signal. As with unlabeled RMA, pyrene-RMA can be commercially purchased, which introduces the caveats of high cost, contaminating nuclei, and low labeling efficiency (typically 10% label). Therefore, we recommend pyrene-labeling your own purified RMA as described (Cooper et al., 1983), which yields 100% labeling efficiency, reduces cost, and produces large quantities of pyrene-RMA.

For use in actin nucleation assays, pyrene-RMA is ideally gel-filtered. However, because pyrene-RMA represents a small fraction of the actin in the reaction (typically 5%), low levels of contaminating nuclei do not contribute significantly to the kinetics. Therefore, we store precleared (below), flash-frozen 10 μl pyrene-RMA aliquots at $-80°$. Before each experiment, aliquots are slow-thawed on ice overnight, diluted to 5 μM in G buffer (below), and precleared by centrifugation at 300,000g (90,000 rpm in Beckman TLA100 rotor), 4°, 1 h. The supernatant is removed immediately after centrifugation (for a 50 μl sample, we remove only the upper 40 μl supernatant), which contains primarily pyrene-RMA monomers. By storing small aliquots at $-80°$, one large preparation of pyrene-RMA can be used for years of experiments.

Preparation of Stock Solutions

At the beginning of each day of fluorometry experiments, a fresh actin solution must be prepared. Combine gel-filtered (unlabeled) RMA with freshly precleared pyrene-RMA. For actin assembly reactions (2 μM G-actin final) make a 12 μM actin stock (5% pyrene-RMA). For barbed end actin filament elongation reactions (0.5 μM G-actin final), make a 10 μM actin stock (10% pyrene-RMA). Because each individual actin stock yields slightly different actin assembly kinetics, it is imperative to be generous in estimating the number of reactions to be performed. All reactions that are compared with each other must be performed using the same actin solution. Over the course of the day, actin solutions are stored on ice with a lid and provide reproducible results for $\sim$20 h.

Important Notes for Actin Assembly Reactions

Mg^{2+}-bound actin readily polymerizes, whereas Ca^{2+}-actin does not. This is why G-actin is stored in Ca^{2+} (see G buffer). Thus, the first step in an actin assembly reaction is pre-exchange of Ca^{2+} for Mg^{2+} on actin by incubation for 2 min with 10$\times$ exchange buffer, containing EGTA and $MgCl_2$. If this step is omitted, exchange will occur during the actin assembly reaction (because Mg^{2+} is a component of the actin assembly initiation mix). However, the exchange will generate an artificial lag phase of $\sim$50 sec before polymerization is allowed. Thus, it is important to include the pre-exchange step.

For actin assembly reactions, we typically use a final concentration of 2 μM actin. Higher concentrations can be used (e.g., 3–4 μM), but will promote significant filament assembly in the absence of a nucleating protein (e.g., formins or Arp2/3 complex), making them less optimal for testing activity. At 2 μM actin (in the absence of other proteins) filament assembly should be complete, reaching steady-state polymer equilibrium at $\sim$45 min. Faster polymerization rates suggest contaminating nuclei and that the actin solution should be recentrifuged. Because of these and other considerations, we perform reactions in a specific order for each experiment. The first assembly reaction is 2 μM actin alone. Next, include a high concentration of formin that should give strong nucleation (e.g., 50 nM Bni1(FH1-COOH)). This serves as a checkpoint for the experiment, because these two curves should differ significantly in both the lag before assembly and the slope of assembly kinetics. After $\sim$6–8 h of fluorometry, retest 2 μM actin alone to ensure that baseline assembly kinetics are similar. Also note that different formins have different potencies for actin nucleation (Fig. 4A,B). This does not necessarily suggest that the formin preparation is compromised; differences can result from variation in actin binding affinities and permitted rates of filament elongation.

Tips for Obtaining Consistent Results Between Reactions

To obtain reproducible, consistent results in assembly and elongation assays, it is important to maintain the same buffer and salt conditions in every reaction, because small changes can greatly affect actin polymerization kinetics. Formins are stored in HEKG$_{10}$, and the final volume of HEKG$_{10}$ in every reaction must be constant, regardless of the concentration of formin used.

For instance, in 60 μl assembly reactions, we maintain but never exceed 10 μl of HEKG$_{10}$. A second key to obtaining reproducible results is consistency in pipetting. Use the same pipette for each specific step in all reactions. Small variations in pipette calibrations can drastically alter results. Furthermore, one person must perform all of the reactions in an experiment, because differences in pipetting techniques can generate variability.

Tips for Filament Elongation Reactions

Actin filament elongation reactions are used to test the processive capping activity of formins. Elongation reactions measure the kinetics of actin monomer addition at the barbed ends of preformed actin filaments

(F-actin). When pipetting F-actin, be aware that the narrow opening of most pipette tips (particularly beveled) shears filaments, generating new ends. Therefore, use only "cut" pipette tips ($\sim$5 mm removed with razor blade) for all manipulations of F-actin to minimize variability in reactions. Methods for preparing F-actin stocks and shearing to create "seeds" are described later. The requirements for elongation reactions are that all polymerization being measured occurs at the ends of preformed filaments ("seeds"), and spontaneous actin nucleation is prohibited. For this reason, specific controls must be performed for each set of elongation reactions. In most cases, use of a low concentration of G-actin (0.5 μM) prohibits spontaneous nucleation during the time frame of reactions ($\sim$500 sec) and restricts actin addition to barbed ends (because the critical concentration for growth at pointed ends >0.5 μM). Recommended reaction order:

1. *Seeds + 0.5 μM G-actin.* Actin polymerization should be rapid (and without lag) and reach steady-state by $\sim$500 sec.

2. *No seeds, 0.5 μM G-actin.* With no seeds, actin should not assemble.

The assay for processive capping by formins requires addition of a conventional capping protein to reactions. We use purified heterodimeric *S. cerevisiae* capping protein (Cap1/2), which blocks barbed end RMA filament elongation (half maximal capping at 20 nM, >95% capping at 500 nM), although other capping proteins can be substituted (e.g., gelsolin and CapZ; Zigmond *et al.*, 2003). Cap1/2 is purified as described (Amatruda and Cooper, 1992), and stocks (20 μM) are stored in HEK buffer (20 mM HEPES (pH 7.4), 1 mM EDTA, 50 mM KCl). Whatever capping protein is used, its concentration required to inhibit RMA filament elongation must be determined empirically. In testing processive capping, it is optimal to use a concentration of capping protein that abolishes $\sim$95% elongation (not more). The next reaction:

3. *Seeds + 500 nM Cap1/2 + 0.5 μM G-actin.* Almost no assembly should occur. If assembly is observed, adjust concentration of Cap1/2 (e.g., 1 μM). To measure processive capping activity, formins are added to reactions containing seeds, 500 nM Cap1/2, and G-actin. Higher formin concentrations yield faster rates of filament elongation, a measure of "protection" of filament ends from capping protein (Fig. 4C). The formin concentration supporting half maximal (50%) protection from capping protein provides an index (K_{app}) of formin affinity for filament ends (Moseley *et al.*, 2004). Several reactions must be included to control for possible effects of the formin on: (1) nucleation of new filaments and (2) rate of elongation in the absence of Cap1/2:

4. *Formin(FH1-COOH)* + *seeds* + *0.5 μM G-actin*. This will define what effect formin alone has on rate of elongation at barbed ends (highly variable among formins).

5. *Formin(FH1-COOH)* + *0.5 μM G-actin*. Within this time frame (500 sec), most formins do *not* nucleate 0.5 μM actin, but if assembly occurs, reduce formin concentration.

6. *Formin(FH1-COOH)* + *500 nM Cap1/2* + *seeds* + *0.5 μM G-actin*. Most formins processively cap barbed ends under these conditions, protecting ends from the inhibitory effects of Cap1/2. Keep in mind that *S. pombe* formin Cdc12 requires profilin for processive capping (Kovar *et al.*, 2003), so it is advisable to test elongation in the presence and absence of profilin (3–10 μM).

Data Analysis and Formatting

Kinetic data for each assembly or elongation reaction can be presented in graphical form (AU of fluorescence versus time) (Fig. 4B). To compare rates of actin polymerization between reactions, the slope of each curve is measured using software included with most fluorometers (e.g., later). Concentration of filament ends generated at any time point in a reaction can be calculated from the slope at that point and Eq. (1).

$$\text{Rate}(\mu M sec^{-1}) = k_+[N][\text{G-actin}] - k_-[N] \qquad (1)$$

where, at pH 7.5, $k_+ = 11.6 \ \mu M^{-1} sec$, $k_- = 1.4 \ sec^{-1}$, [G-actin] = actin monomer concentration at time rate measured, [N] = filament concentration (μM). To compare reactions, the slope should be determined at the same point of assembly in each reaction (optimal window: 25–50% polymerization). This allows determination of the molar ratio of formins to filaments nucleated in assembly reactions, and in elongation assays, it provides the number of filament ends [N] undergoing elongation (i.e., seed concentration). Under our conditions, 330 nM F-actin seeds corresponds to ~0.5 nM ends (average filament length ~660 actin subunits, or ~2 μm).

After actin assembly reactions reach steady state, the pyrene fluorescence plateaus. If reactions do not contain factors that change the critical concentration for actin assembly (ActinCc), the plateaus should be equivalent in all reactions (assuming they contain equal concentrations of actin). Capping proteins increase ActinCc significantly, to that of the pointed end of the filament. Formins have variable effects on ActinCc (ranging from no effect to effects similar to capping proteins).

Minor changes in the plateau can result from variations in pyrene labeling between preparations and differences in photobleaching during handling of reactions in an experiments. Although curves can be normalized to have the same plateau, we recommend not doing this, because it may hide important differences between reactions. Tiny air bubbles introduced during reactions also can displace curves along the y-axis; if this occurs, repeat those reactions. For data formatting, the data points can be exported and graphed in Microsoft Excel or an equivalent program.

Actin Assembly and Elongation Assays

Actin Assembly Assay

Solutions

G-buffer (store $-20°$): 10 mM Tris (pH 7.5), 0.2 mM ATP, 0.2 mM CaCl$_2$, and 0.2 mM DTT.

20× initiation mix (IM) (store $-20°$): 40 mM MgCl$_2$, 10 mM ATP, 1 M KCl.

Exchange buffer (store room temp): 10 mM EGTA, 1 mM MgCl$_2$

Special Equipment. QuantaMaster™ QM-5/2005 Spectrofluorometer with PMT detector and pulsed xenon lamp (Photon Technology International, Lawrenceville, NJ) or equivalent time-based spectrofluorometry device

Quartz Ultra-Micro Fluorescence Cell 105.251, with 3 × 3 mm light path and 15 mm center height (Hellma, Plainview, NY).

Software. FeliX32™ Advanced Fluorescence Analysis Software (Photon Technology International) or equivalent spectrofluorometer software.

Procedure

1. Turn on spectrofluorometer, allow xenon lamp to warm up for ~10 min.

2. Open software, and select time-based acquisition. Set fluorescence spectrofluorometer to monitor pyrene fluorescence (excitation 365 nm and emission 407 nm) in time-based mode for 2000 sec, taking readings at 3-sec time points. Use circulating water bath to maintain cuvette (reaction) chamber of $25°$.

3. Prepare 12 μM stock solution of monomeric RMA (5% pyrene-labeled) in G–buffer, and keep on ice. Final reaction volume is 60 μl, containing 2 μM actin. Thus, 10 μl of stock actin solution is required per reaction.

4. Allow aliquots of G–buffer, exchange buffer, and 20× IM to come to room temperature.

5. Transfer 3 μl 20× IM to the bottom of the cuvette.

6. Prepare the following "solution A" (47 μl total) in a 1.5 ml Eppendorf tube. Add ingredients in this order: 32.3 μl G buffer, 4.7 μl exchange buffer, 10 μl G-actin stock solution (12 μM). Mix twice by gentle pipetting. Incubate at room temperature 2 min before initiating the assembly reaction–facilitates Mg^{2+} exchange (see preceding).

7. During this time, prepare the following "solution B" (10 μl total), which contains: 10 μl $HEKG_{10}$ (formin storage buffer) or desired amount of 6xHis-Bni1(FH1-COOH) in 10 μl of $HEKG_{10}$.

8. After 2 min Mg^{2+} exchange, add 47 μl solution A to 10 μl solution B and mix. Then, immediately transfer this mixture to the 3 μl 20× IM in the cuvette. Mix by pipetting three times; be careful to avoid bubbles. Place cuvette in spectrofluorometer and initiate time-based fluorescence acquisition.

Actin Filament Elongation Methods

Solutions

G–buffer (as above)

10× initiation mix (IM): 20 mM $MgCl_2$, 5 mM ATP, 500 mM KCl.

F buffer: 10 mM Tris (pH 8.0), 0.7 mM ATP, 0.2 mM $CaCl_2$, 0.2 mM DTT, 2 mM $MgCl_2$, 50 mM KCl.

Procedure

1. Prepare spectrofluorometer for time-based acquisition as earlier.

2. To assemble actin filaments (F-actin), add initiation mix (1× final) to 10 μM monomeric RMA (no pyrene-label) and incubate 40 min at room temperature; 5 μl of this F-actin stock is required per reaction (remember to overestimate use).

3. Prepare 10 μM monomeric RMA stock (10% pyrene-labeled) in G–buffer as above, store on ice. Reaction volume is 60 μl, containing 0.5 μM G-actin; thus, 3 μl of G-actin stock is required per reaction.

4. Allow G-buffer and 10× IM stock solutions to come to room temperature.

5. Transfer 4 μl 10× IM to the bottom of cuvette.

6. Prepare the following "solution A" (36 μl total) in one tube: 33 μl G buffer + 3 μl 10 μM G-actin (10% pyrene-labeled).

7. Prepare the following "solution B" (50 μl total) in another tube. Combine the following ingredients in this order: 31.25 μl F buffer, 10 μl $HEKG_{10}$ (formin storage buffer) or formin in $HEKG_{10}$, 3.75 μl HEK buffer or Cap1/2 (20 μM) in HEK buffer, and 5 μl of 10 μM F-actin.

8. Shear solution B with five passages through a 27-gauge needle using a 1 ml syringe.

9. Immediately add 20 μl of solution B to the 36 μl of solution A, then transfer this mixture (56 μl) to the cuvette. Mix with $10\times$ IM by pipetting three times; be careful to avoid bubbles. Place cuvette in spectrofluorometer and initiate time-based fluorescence acquisition.

Acknowledgments

We are grateful to I. Sagot, A. Rodal, and members of the Goode laboratory for valued comments on the manuscript. This work was supported a grant to B. G. from the NIH (GM63691).

References

Amatruda, J. F., and Cooper, J. A. (1992). Purification, characterization, and immunofluorescence localization of *Saccharomyces cerevisiae* capping protein. *J. Cell Biol.* **117,** 1067–1076.

Cooper, J. A., Walker, S. B., and Pollard, T. D. (1983). Pyrene actin: Documentation of the validity of a sensitive assay for actin polymerization. *J. Musc. Res. Cell Motil.* **4,** 253–262.

Goode, B. L. (2002). Purification of yeast actin and actin-associated proteins. *Methods Enzymol.* **351,** 433–441.

Higashida, C., Miyoshi, T., Fujita, A., Oceguera-Yanez, F., Monypenny, J., Andou, Y., Narumiya, S., and Watanabe, N. (2004). Actin polymerization-driven molecular movement of mDia1 in living cells. *Science* **303,** 2007–2010.

Higgs, H. N., and Peterson, K. J. (2005). Phylogenetic analysis of the formin homology 2 domain. *Mol. Biol. Cell* **16,** 1–13.

Jones, E. W. (1991). Tackling the protease problem in *Saccharomyces cerevisiae*. *Methods Enzymol.* **194,** 428–453.

Kovar, D. R., Kuhn, J. R., Tichy, A. L., and Pollard, T. D. (2003). The fission yeast cytokinesis formin Cdc12p is a barbed end actin filament capping protein gated by profilin. *J. Cell Biol.* **161,** 875–887.

Kovar, D. R., and Pollard, T. D. (2004). Insertional assembly of actin filament barbed ends in association with formins produces piconewton forces. *Proc. Natl. Acad. Sci. USA* **101,** 14725–14730.

Li, F., and Higgs, H. N. (2003). The mouse formin mDia1 is a potent actin nucleation factor regulated by autoinhibition. *Curr. Biol.* **13,** 1335–1340.

Li, F., and Higgs, H. N. (2004). Dissecting requirements for auto-inhibition of actin nucleation by the formin, mDia1. *J. Biol. Chem.* **280,** 6986–6992.

Moseley, J. B., Sagot, I., Manning, A. L., Xu, Y., Eck, M. J., Pellman, D., and Goode, B. L. (2004). A conserved mechanism for Bni1- and mDia1-induced actin assembly and dual regulation of Bni1 by Bud6 and profilin. *Mol. Biol. Cell* **15,** 896–907.

Pruyne, D., Evangelista, M., Yang, C., Bi, E., Zigmond, S., Bretscher, A., and Boone, C. (2002). Role of formins in actin assembly: Nucleation and barbed end association. *Science* **297,** 612–615.

Rodal, A. A., Duncan, M., and Drubin, D. (2002). Purification of glutathione S-transferase fusion proteins from yeast. *Methods Enzymol.* **351,** 168–172.

Romero, S., Le Clainche, C., Didry, D., Egile, C., Pantaloni, D., and Carlier, M. F. (2004). Formin is a processive motor that requires profilin to accelerate actin assembly and associated ATP hydrolysis. *Cell* **119,** 419–429.

Sagot, I., Rodal, A. A., Moseley, J., Goode, B. L., and Pellman, D. (2002). An actin nucleation mechanism mediated by Bni1 and profilin. *Nat. Cell Biol.* **4,** 626–631.

Spudich, J. A., and Watt, S. (1971). The regulation of rabbit skeletal muscle contraction. I. Biochemical studies of the interaction of the tropomyosin-troponin complex with actin and the proteolytic fragments of myosin. *J. Biol. Chem.* **246,** 4866–4871.

Wallar, B. J., and Alberts, A. S. (2003). The formins: Active scaffolds that remodel the cytoskeleton. *Trends Cell Biol.* **13,** 435–446.

Xu, Y., Moseley, J. B., Sagot, I., Poy, F., Pellman, D., Goode, B. L., and Eck, M. J. (2004). Crystal structures of a formin homology-2 domain reveal a tethered dimer architecture. *Cell* **116,** 711–723.

Zigmond, S. H. (2004). Formin-induced nucleation of actin filaments. *Curr. Opin. Cell Biol.* **16,** 99–105.

Zigmond, S. H., Evangelista, M., Boone, C., Yang, C., Dar, A. C., Sicheri, F., Forkey, J., and Pring, M. (2003). Formin leaky cap allows elongation in the presence of tight capping proteins. *Curr. Biol.* **13,** 1820–1823.

[17] Purification and Kinase Assay of PKN

By HIDEYUKI MUKAI and YOSHITAKA ONO

Abstract

PKN is a serine/threonine protein kinase, which has a catalytic domain highly homologous to that of protein kinase C (PKC) in the carboxyl-terminal region and three repeats of the antiparallel coiled coil (ACC) domain in the amino-terminal region. Mammalian PKN has three isoforms each derived from different genes, PKN1 (PKNα/PRK1/PAK1), PKN2 (PRK2/PAK2/PKNγ), and PKN3 (PKNβ). PKN isoforms show different enzymatic properties and tissue distributions and have been implicated in various distinct cellular processes (reviewed in Mukai [2003]). This chapter discusses methods to prepare purified enzymes and to assay substrate phosphorylation activities.

Introduction

PKN cDNA was first identified in 1994 by a low-stringency hybridization using the catalytic domain of PKC βII as a probe (Mukai and Ono, 1994), and PKN protein was later purified to homogeneity from rat testes

METHODS IN ENZYMOLOGY, VOL. 406   0076-6879/06 $35.00
DOI: 10.1016/S0076-6879(06)06017-4

using a specific antibody raised against recombinant PKN as a guide (Kitagawa *et al.*, 1995). This enzyme was first described as a fatty acid– or phospholipid-activated protein kinase and also as a protease-activated protein kinase (Kitagawa *et al.*, 1995; Mukai *et al.*, 1994; Peng *et al.*, 1996). On the other hand, PKN was elucidated as a target molecule of the small GTPase RhoA using GTP-RhoA affinity chromatography (Amano *et al.*, 1996) and a gel overlay method (Watanabe *et al.*, 1996). Subsequent molecular cloning and protein analysis revealed that PKN is a family consisting of multiple isoforms, and hence the original PKN was redesignated as "PKN1." Because these studies have been carried out independently in several laboratories, different nomenclatures are proposed for the PKN isoforms derived from different genes. The nomenclature adopted herein uses PKN1 (instead of PKNα [Mukai and Ono, 1994]/ PRK1 [Palmer *et al.*, 1995b]/PAK1 [Peng *et al.*, 1996]), PKN2 (instead of PRK2 [Palmer *et al.*, 1995b]/PAK2 [Yu *et al.*, 1997]/PKNγ), and PKN3 (instead of PKNβ [Oishi *et al.*, 1999]) according to the Human Kinome (Manning *et al.*, 2002). Sequence comparison reveals that PKN isoforms have conserved domains: ACC domains and the C2-like domain in the amino-terminal region and the catalytic serine/threonine kinase domain in the carboxyl-terminal region. However, PKN isoforms show different enzymatic properties and tissue distributions and have been implicated in various distinct cellular processes (reviewed in Mukai, 2003).

This chapter describes methods to purify endogenous PKN from mammalian tissues and recombinant enzyme from insect cells and to assay for kinase activity. Modifiers of PKN kinase activity are also summarized.

Purification of PKN

This section describes the purification procedures of PKN from various sources. The first two methods enable purification of the endogenous PKN1 to homogeneity from mammalian tissues as determined by silver staining after SDS-polyacrylamide gel electrophoresis (PAGE), and the third method is to purify recombinant PKN isoforms from insect cells. (The apparent molecular weights of PKN1, PKN2, and PKN3 are different from one another, and these enzymes are easily distinguished on SDS-PAGE.)

PKN1 is expressed in every tissue and is especially enriched in the testes and spleen. In so far as cell lines, leukemic cell lines such as Jurkat and U937 cells are also abundant in PKN1. It should be noted at this point that PKN1 actually consists of at least two types of alternatively spliced isoforms with almost the same molecular weight (manuscript in preparation), and the relative amounts of these isoforms differ among tissues. PKN2 is also widely expressed in adult tissues, with the highest levels being in the liver and lung (Vincent and Settleman, 1997). PKN3 is almost undetectable in normal adult tissues. However, culture cells such as HeLa, chronic myelogenous leukemia K-562, and colorectal adenocarcinoma SW480 cells contain significant amounts of this enzyme (Oishi *et al.*, 1999). Subcellular fractionation experiments revealed that about half of all endogenous PKN1 resides in the soluble cytosolic fraction (Kitagawa *et al.*, 1995) of brain tissue. It was recently suggested that a part of PKN1 exists as the integral membrane form and has modifications different from the cytosolic/peripheral membrane form of the enzyme (Zhu *et al.*, 2004). The source of the endogenous enzyme should be determined in view of the preceding information.

Purification of PKN1 from the Soluble Cytosolic Fraction of Rat Testes

The purification protocol for endogenous PKN1 from rat testes was published in detail (Kitagawa *et al.*, 1995). An outline of the method and previously undescribed information will be provided herein. The flowchart of the procedure is depicted in Fig. 1A. All procedures are carried out at $0-4°$. Rat testes (60 g fresh weight) are added to 4 volumes of the buffer Te-A (50 mM Tris-HCl at pH 7.5, 5 mM EDTA, 5 mM EGTA, 0.5 mM dithiothreitol, 10 μg/ml leupeptin, 0.02% NaN$_3$, and 1 mM phenylmethanesulfonyl fluoride) and homogenized with a Polytron homogenizer (KINEMATICA) or a Teflon homogenizer. The resulting crude homogenate is then centrifuged at 24000g for 1 h, and the supernatant is loaded onto a 180-ml DE52 (Whatman) column equilibrated in buffer Te-B (25 mM Tris-HCl at pH 7.5, 1 mM EDTA, 1 mM EGTA, 0.5 mM dithiothreitol, 100 ng/ml leupeptin, and 0.02% NaN$_3$). After washing, PKN1 is eluted with buffer Te-B containing 200 mM NaCl and immediately subjected to a 40% ammonium sulfate (AS) fractionation. The precipitate is resuspended in buffer Te-B containing 20% AS and loaded onto a 20-ml Butyl-Sepharose column (Amersham) equilibrated with the same buffer. The eluate from Butyl-Sepharose with buffer, Te-B containing 10% AS is dialyzed against buffer Te-B and is loaded onto a 5-ml HiTrap Heparin Sepharose column (Amersham) equilibrated with buffer Te-B, which is connected to a Pharmacia FPLC system. PKN1 is eluted from the column by application of a linear concentration gradient of NaCl (0–0.6 M) in buffer Te-B. The major peak of PKN1 detected by anti-PKN1 antibody is dialyzed against buffer Te-B and then loaded onto a Mono Q 5/5HR column (Amersham) equilibrated with the same buffer using an FPLC system. PKN1 is eluted from the column by application of a 40-ml linear concentration gradient of NaCl (0–0.4 M) in buffer Te-B, and the major peak of PKN1 is loaded onto a 4-ml protamine-CH-Sepharose column.

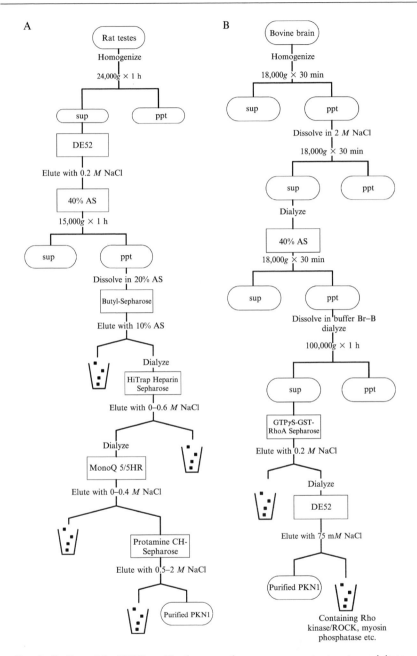

FIG. 1. Outline of the PKN1 purification procedure. sup, supernatant; ppt, precipitate.

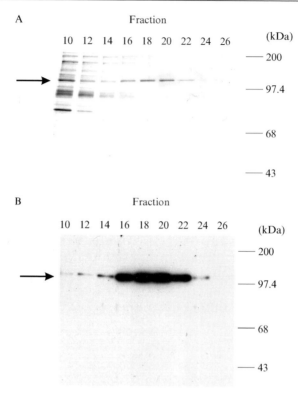

FIG. 2. PKN1 obtained from rat testes after protamine-CH-Sepharose chromatography. (A) Silver-stained SDS/polyacrylamide gel. (B) Western blotting using polyclonal antiserum against PKN1. The arrows indicate the position of PKN1. Reproduced with permission from Kitagawa *et al.*, 1995. © The Biochemical Society.

(This column is prepared by coupling protamine sulfate to the activated CH-Sepharose 4B according to the methods of Amersham.) Purified PKN1 tightly bound to the column is eluted by 2 M NaCl (Fig. 2). Approximately 12 μg of PKN1 is purified about 8000-fold to an apparent homogeneity from 60 g of starting material. Purified enzyme should be dialyzed against Tris/dithiothreitol buffer for enzyme assays, because the salt at high concentrations inhibits the activity of PKN1 (refer to the "Modifiers of PKN" section). The purified enzyme is stable for more than 6 months in 50% glycerol at −80°. (Glycerol does not inhibit and rather mildly increases the kinase activity of PKN1.) Even in the absence of glycerol, PKN1 can be preserved for about 1 month at −80° without a significant loss of enzyme activity, although freeze–thaw cycles should be avoided.

A Thr-Sepharose column with a salt gradient is useful for partial purification of PKN1, although we did not include this step in the preceding protocol. We previously loaded the sample from the MonoQ 5/5HR in Fig. 1A onto the Thr-Sepharose column and applied the eluate to a TSK G3000SW gel filtration column in the presence of 0.3 M NaCl. PKN1 was separated in roughly two peaks with high (>200 kDa) and low (monomer) molecular weight fractions, suggesting the presence of PKN1 that has oligomerized or has interacted with other proteins under these conditions.

Rapid Purification of PKN1 from the Membrane Fraction of Bovine Brain

PKN1 binds to the small GTPase RhoA in a GTP-dependent manner. So the GTPγS-loaded glutathione S-transferase (GST)-RhoA affinity column is applicable for the purification of PKN1 (Amano *et al.*, 1996). The GTPγS-GST-RhoA affinity column is prepared as described previously (Amano *et al.*, 2000). The flowchart of the purification procedure is depicted in Fig 1B. All procedures are carried out at 0–4°. Bovine brain is cut into small pieces. A 30-g aliquot is homogenized with 90 ml of buffer Br-A (25 mM Tris-HCl at pH 7.5, 10% sucrose, 5 mM EDTA, 1 mM dithiothreitol, 10 mM MgCl$_2$, and 0.1 μg/ml leupeptin) and 0.1 mM phenylmethanesulfonyl fluoride in a Teflon glass homogenizer and filtered through three layers of gauze. The crude homogenate from 200 g of brain in total is put together and is centrifuged at 18,000g for 30 min. The precipitate is resuspended in 400 ml of buffer Br-A, and the protein from the membrane fraction is extracted by mixing with 400 ml of buffer Br-A containing 4 M NaCl, being stirred for 1 h. After the solution is centrifuged at 18,000g for 30 min, the supernatant is filtered through three layers of gauze and is dialyzed against buffer Br-B (20 mM Tris-HCl at pH 7.5, 1 mM EDTA, 1 mM dithiothreitol, and 5 mM MgCl$_2$]. After centrifugation at 18,000g for 30 min, solid ammonium sulfate is added to the supernatant to a final concentration of 40% saturation. The precipitate is dissolved in 16 ml of buffer Br-B, dialyzed against buffer Br-B, and centrifuged at 100,000g for 1 h followed by filtrations through 0.80-μm and 0.45-μm filters sequentially. After being passed over a 1-ml glutathione-Sepharose column, the flow through fraction is loaded onto a 1-ml glutathione-Sepharose column carrying GTP-γS-GST-RhoA. The column is washed extensively (more than 20 volumes) of buffer Br-B, and proteins are eluted by addition of buffer Br-B containing 0.2 M NaCl. One-milliliter fractions are collected, and the amount of PKN1 purified is examined by silver staining and/or immunoblotting with anti-PKN1 antibody. (Typically, fractions #3–#10 contain relatively rich amounts of PKN1.) Fractions containing PKN1 are pooled together and dialyzed against buffer Br-B. The dialyzed sample is subjected to a 0.3-ml DE52 column, and the column is washed with buffer Br-B containing 50 mM NaCl. PKN1 is eluted by loading buffer Br-B containing 75 mM NaCl, and 0.3-ml fractions are collected. (Typically, fractions #2–#4 contain purified PKN1.)

Expression and Purification of Recombinant PKNs from Insect Cells

It is difficult to obtain active PKN1 from *Escherichia coli* expression systems. PKN2 can be expressed as an active enzyme in *E. coli,* but its specific activity is low (Yoshinaga *et al.,* 1999), and it is difficult to obtain rich amounts of the full-length enzyme from the bacteria. The eukaryotic insect cell expression system is useful to obtain isotype-pure and sufficient amount of functional PKNs. The GST/glutathione Sepharose or His$_6$ tag/Ni-NTA one-step purification system is useful for preparing active enzymes quickly. Further purification of the enzyme can be carried out by combining these two affinity chromatographies sequentially or by adding some of the purification steps described in the previous section. The GST and His$_6$ tag may potentially induce steric hindrance or conformational changes of the enzyme itself and affect substrate recognition because of dimer formation of GST (Kaplan *et al.,* 1997). However, the wild-type PKN1 expressed as a fusion protein with the GST tag or His$_6$ tag in Sf9 cells has almost the same properties as endogenous PKN1 purified from the soluble cytosolic fraction of rat testes with regard to the substrate specificity and response of effectors (Yoshinaga *et al.,* 1999). The insect cell expression system is useful for obtaining each isoform as described previously, although we should keep in mind that the purified PKN still might be composed of heterogeneous molecules with different modifications in insect cells. Construction of a baculovirus for expression of PKN1 and the purification protocols have been described elsewhere in detail (Mukai and Ono, 2003), so the outline of the procedure with previously undescribed information will be commented on herein.

pBlueBacHisGST (pR538) vector, a transfer vector for the expression of His$_6$-GST–tagged fusion proteins, is made by subcloning the coding region for GST and a thrombin recognition site in frame to the multiple cloning site of pBlueBacHis-B (Invitrogen). The cDNA for PKN is subcloned into the multicloning site of this vector in frame. For the protein product, the His$_6$ tag can be cleaved by enterokinase if so desired, yielding a GST fusion protein. Both the His$_6$ tag and the GST tag can be cleaved by thrombin. The pAcGHLT vector for expression of GST-His$_6$–tagged fusion proteins is commercially available (Pharmingen). However, it should be kept in mind that the PKA phosphorylation site is included in the product enzyme without cleavage tags. PKN efficiently phosphorylates this

PKA phosphorylation site. Information about construction of the catalytically active fragment or introduction of the kinase-negative mutation has been described (Yoshinaga *et al.*, 1999). Sf9 cells are seeded at 1×10^7 cells on a 10-cm culture dish. About 2–20 dishes are used for purification of each enzyme depending on the expression levels of the recombinant proteins. Cells are incubated at 28° for 1 h so that they attach to the dish. After removing the medium, recombinant baculovirus is added at an appropriate multiplicity (usually in a range of 5–10 pfu/cell) to the cells. After incubating the virus/cells for 1 h at room temperature, 10 ml of medium is added to the dish and incubated at 28° for 2–3 days until a cytopathic effect is observed. The cells are harvested by scraping them from the dish with a rubber policeman. The cell pellets are collected by centrifugation at 1000g for 5 min at 4° and stored at −80°. It is basically easier to get rich amounts of PKN1 in the soluble fraction than PKN2 and PKN3, although the amounts of the soluble recombinant enzymes vary among the deleted and mutated constructs. To increase the recovery of PKN2 and PKN3, higher multiplicities of infection and longer incubation times after infection with virus are necessary, although too high of a concentration of virus and long incubation times will sometimes result in a decrease of recovery of the enzyme in the supernatant fraction.

The sequential affinity purification of His$_6$- and GST-tagged PKN is as follows. All procedures are carried out at 0–4°. The cell pellet is resuspended in 4 ml of buffer Sf-A (50 mM Tris-HCl at pH 8.0, 1 μg/ml of leupeptin, 1 mM MgCl$_2$, 0.3 M NaCl, and 10 mM β-ME) containing 1 mM phenylmethanesulfonyl fluoride and 10 mM imidazole and is homogenized with a Dounce homogenizer with 30 strokes. Triton X-100 is added at 1% and incubated for 10 min. After centrifuging at 100,000g for 30 min, the supernatant is added to 400 μl of 50% Ni-NTA agarose equilibrated with buffer Sf-A and is rotated for 1 h. The lysate-resin mixture is loaded into an empty column and washed with 30 column volumes or more of buffer Sf-A containing 10 mM imidazole. The recombinant protein is eluted with 800 μl of buffer Sf-A containing 250 mM imidazole; 200 μl of 50% glutathione Sepharose (Amersham) equilibrated with buffer Sf-B (50 mM Tris-HCl at pH 7.5, 1 μg/ml leupeptin, 1 mM EGTA, 1 mM EDTA, 3 mM MgCl$_2$, and 1 mM dithiothreitol) is added to the eluate and rotated for 1 h. The sample is loaded onto an empty column and washed with 30 column volumes or more of buffer Sf-B. The recombinant protein is eluted four times with 100 μl buffer Sf-C (50 mM Tris-HCl at pH 8.0, 1 μg/ml leupeptin, 1 mM EDTA, 3 mM MgCl$_2$, 1 mM dithiothreitol, and 10 mM glutathione) in each tube. The advantage of this sequential chromatography method is to provide highly pure enzyme quickly and to remove imidazole, which is inhibitory to PKN, during the glutathione Sepharose column step.

Kinase Assay

To assess the protein kinase activity of PKN isoforms, purified enzyme ($\sim$10 ng) is incubated for 5 min at 30° in a reaction mixture (final volume 25 μl) containing 20 mM Tris-HCl at pH 7.5, 4 mM MgCl$_2$, 40 μM ATP, 18.5 kBq of [γ-^{32}P]ATP, 40 μM oligopeptide phosphate acceptor as described later, 0.1 mg/ml BSA or recombinant GST as a stabilizer, and in the presence or absence of modifiers such as 40 μM arachidonic acid (Table I). (PKN exhibits a characteristic Mg^{2+} dependence for a protein kinase, with an optimal Mg^{2+} ion concentration in the range of 2–5 mM [Morrice *et al.*, 1994]). Reactions are terminated by spotting the mixture onto P81 phosphocellulose paper (Whatman) and submersing them in 75 mM phosphate and then washing them three times for 10 min. The incorporation of ^{32}P phosphate into the oligopeptide phosphate acceptor is assessed by the Cherenkov counting.

As speculated from the structural resemblance among catalytic domains of PKNs and PKCs, the consensus phosphorylation motif sites for PKNs are very similar to each other and also to those of PKCs (manuscript in preparation). This is also supported by the report that PKN1 phosphorylates the same sites of MARCKS (Palmer *et al.*, 1996) and vimentin (Matsuzawa *et al.*, 1997) as PKCs *in vitro*. Oligopeptide substrates synthesized based on the pseudosubstrate sites of PKCs can be used as good phosphate acceptors for PKN (Mukai *et al.*, 1994). For example, AMFPTMNRRGSIKQAKI is an efficient substrate for PKN and corresponds to amino acids 137–153 of PKCδ, substituting Ser for Ala$_{147}$. Thus, PKN seems to phosphorylate PKC substrates unscrupulously. However, relatively optimized peptide substrates for either PKN or PKC are available for kinase assays (manuscript in preparation). Protein substrates such as myelin basic protein (MBP), protamine sulfate, and histone H1 can be used instead of oligopeptide phosphate acceptors. In this case, the radiolabeled bands can be visualized by autoradiography after resolving the reacted mixture by SDS-PAGE. In the absence of modifiers such as arachidonic acid, these protein substrates are efficient substrates. However, this test does not reflect well the conformational change of PKN by the modifier (Kitagawa *et al.*, 1995). PKN does not phosphorylate GST alone, so GST fused to an oligopeptide optimized for the PKN assay can be used as a good stoichiometric protein substrate instead of those mentioned previously. (It is noteworthy that PKN weakly phosphorylates the Ser residue coded by the *Bam*HI [GGA-$_{Gly}$:TTC-$_{Ser}$] site of some pGEX vectors, especially when Met follows this Ser residue.) This method is more sensitive than using the P81 phosphocellulose paper method as described previously.

TABLE I
EFFECTS OF VARIOUS ACTIVATORS ON PKN1 ACTIVITY

Modifier	Activity (%)	Enzyme	Substrate	Reference
10 μg/ml cardiolipin	380	Purified PKN1 from rat liver	S6-peptide (AKRRRLSSLRA)	(Morrice et al., 1994a)
40 μM arachidonic acid	850	purified PKN1 from rat testes	PKC δ peptide (AMFPTMNRRG-SIKQAKI)	(Kitagawa et al., 1995)
40 μM linoleic acid	900			
8μg/ml lyso-PA	430			
60 μg/ml PtdIns-4,5-P$_2$, 60 μg/ml PtdIns-3,4,5-P$_3$	robust activation	Recombinant hyman PKN1 purified from COS7 cells	MBP	(Palmer et al., 1995)
0.04% DOC	730	purified PKN1 from rat testes	PKC δ peptide (AMFPTMNRRG-SIKQAKI)	(Kitagawa et al., 1995)
0.004% SDS	690			

Phosphate incorporation in the absence of modifiers was taken as 100% activity. Each value of "activity (%)" represents the mean of the observations.

Lipid activators are freshly prepared as follows. The required amounts of lipids in organic solvent are dried in a nitrogen stream and then sonicated on ice in distilled water for 5 sec to make sonicated micelles.

Palmer et al. (1995a) used the detergent mixed-micellar assay, first developed by Hannun et al. (1985) for the study of PKC activity, to the kinase assay for PKN1. In the case of PKC, phosphatidylserine (PtdSer) is a potent activator when presented alone. However, in detergent mixed micelles, activation of PKC becomes dependent on diacylglycerol or phorbol esters (Hannun et al., 1985), suggesting that detergent mixed micelles, although not physiological, support an in vitro behavior of PKC consistent with that observed in vivo. It was reported that PKN1 is activated by both phosphatidylinositol 4,5-bisphosphate (PtdIns 4,5-P2) and phosphatidylinositol 3,4,5-trisphosphate (PtdIns-3,4,5-P3) with similar potency either as pure sonicated lipids or in detergent mixed micelles. However, sonicated phosphatidylinositol (PtdIns) and PtdSer are less effective in detergent mixed micelles than in pure sonicated ones. The detergent mixed micellar assay might be useful, especially for the specific modulation of PKN by lipids under defined conditions.

The kinase assay can be performed conventionally using immunoprecipitation of PKN from cell extracts. However, especially in this case, the following points should be taken into account: (1) Various detergents such as non-ionic Triton X-100 and NP-40 affect the kinase activity of PKN in a biphasic manner (Kitagawa et al., 1995). PKN is also sensitive to various ions. For example, $CaCl_2$, $MnCl_2$, and NaCl inhibit PKN activity in a dose-dependent manner with IC_{50} values of $\sim$600 μM, 40 μM, and 160 mM, respectively for the basal kinase activity (Table II). Thus, the final precipitate should be washed before application to the enzyme assay to remove these reagents commonly used for the immunoprecipitation. (2) The Kd value for PKN1 with GTPγS-RhoA is $\sim$0.4 μM (Amano et al., 1996), and the in vivo binding between these proteins is not strong enough to be kept in the extensive washing step. When the recombinant PKN is immunoprecipitated and subjected to the kinase assay, the following additional points should be noted. The carboxyl-terminally tagged catalytic domain of PKN1 bound to anti-tag (such as FLAG and HA) antibody does not show a significant kinase activity. The non-tagged catalytic domain of PKN1 also does not have kinase activity when bound to the anti-carboxyl-terminal region antibody such as αC6 (Mukai et al., 1996). The kinase activity can be restored once the enzyme is dissociated from the antibody by low pH treatment after immunoprecipitation (Takahashi et al., 1998). Therefore, the antibody bound to the carboxyl-terminal region of the catalytic domain of PKN probably induces steric hindrance or a conformational change in the enzyme. This is not the case for the full-length PKN1. Leenders et al. (2004)

TABLE II

$A_{0.5}$ AND IC_{50} VALUES FOR THE VARIOUS AGENTS ON PKN1 ACTIVITY

Modifier	IC_{50}	Enzyme	Substrate	Reference
sodium glycerophosphate	54 mM	PKN1 purified from rat liver, activated by trypsin	S6-peptide (AKRRRLSSLRA)	(Morrice et al., 1994)
NaF	30 mM			
staurosporine	0.03 μM			
bisindolylmaleimide I (Gö6850)	0.2 μM	recombinant full length GST-PKN1 from Sf9 cells, activated by 40 μM arachidonic acid,	PKC δ peptide (AMFPTMNRRGSIKQAKI)	
H-7	1 μM			
HA-1077	1.7 μM	GST-catalytic domain of PKN1 from Sf9 cells,	PKC substrate peptide (RFARKGSLRQKNVHEVK)	(Amano et al., 1999)
Y-32885	0.4 μM			
Y-27632	0.5 μM	GST-catalytic domain of PKN1 from Sf9 cells,	PKC substrate peptide (RFARKGSLRQKNVHEVK)	(Amano et al., 2000)
NaCl	160 mM	purified PKN1 from rat testes	PKC δ peptide (AMFPTMNRRGSIKQAKI)	(Kitagawa et al., 1995)
MnCl₂	40 μM			
CaCl₂	610 μM			
His6-1 alpha fragment (human PKN1 455–511 peptide)	0.05 μM	GST-catalytic domain of PKN1	PKC δ peptide (AMFPTMNRRGSIKQAKI)	(Yoshinaga et al., 1999)
Modifier	$A_{0.5}$	Enzyme	Substrate	Reference
arachidonic acid	7 μM	purified PKN1 from rat testes	PKC δ peptide (AMFPTMNRRGSIKQAKI)	(Kitagawa et al., 1995)
lyso-PA	5 μg/ml			
DOC	0.005%			
SDS	0.0004%			
cardiolipin	1.7 μM	purified PKN1 from rat liver	S6-peptide (AKRRRLSSLRA)	(Morrice et al., 1994)

For agents that inhibit kinase activivty, the concentration of the agent which causes 50% maximal inhibition of kinase activity (IC_{50}) is listed. For agents that activate kinase activity, the concentration of the agent which causes 50% maximal activation of kinase activity ($A_{0.5}$) is listed. Each value represents the mean of the observations.

reported that the catalytic region fragment has just slight or no kinase activity compared with the full-length fragment of PKN3. This controversial result might be obtained with the immunoprecipitation kinase assay when using an antibody against the carboxyl-terminal tag or a part of PKN.

Because PKN is sensitive to detergents and salts *in vitro* and may change its activity drastically within a few minutes after stimulation *in vivo*, a quick and accurate method of PKN in the crude extract is longed for studying the physiological significance of this enzyme. However, the development of this type of assay has been hampered by the lack of a highly specific and efficient substrate. From another point of view, the previous substrates that are supposed to be specific for PKC are not always specific for PKC because the finding of the related kinase, PKN. Although some commercial peptide substrates have been available for the *in vitro* assay of PKC in crude extracts, these same peptides also serve as substrates for PKN and might partly reflect PKN activation.

PKN1 is phosphorylated at several sites *in vivo* and also by autophosphorylation *in vitro*. Phosphorylation-dependent activation is suggested (Peng *et al.*, 1996; Yoshinaga *et al.*, 1999), and the activation of PKN seems to be correlated with the appearance of hyperphosphorylated bands migrating more slowly (Mellor *et al.*, 1998). Therefore, it may be helpful for the assessment of PKN kinase activity to measure the electrophoretic mobility shift of the kinase. However, it should be noted that the activation loop phosphorylation corresponding to the Thr 774 site of human PKN1, which is thought to be critical for activity, is not directly responsible for the apparent band-shift of PKN1 (manuscript in preparation), and that the extent of autophosphorylation is diminished especially in the presence of good substrate peptides *in vitro*. Anti-phospho-Thr (in the activation loop) antibody is now commercially available, and several groups have suggested the *in vivo* signal-induced activation of PKN by detecting the phosphorylation of this site (Dong *et al.*, 2000; Leenders *et al.*, 2004; Torbett *et al.*, 2003; Yau *et al.*, 2003). However, care should be taken about the interpretation of activation loop phosphorylation. At least for PKN2, activation loop phosphorylation may not be a critical point of acute regulation of the enzyme (discussed in detail in Mukai [2003]).

Modifiers of Kinase Activity

The effects of various activators on the kinase activity of PKN1 are summarized in Table I. The effective doses of modifiers of PKN1 are listed in Table II. Some of the well-known potent inhibitors for PKC, such as bisindolylmaleimide I and H-7, and for Rho kinase/ROCK, such as Y27632, also inhibit PKN1 kinase activity with a similar efficiency. Despite the many similarities among the isoforms of PKN, responsiveness to some modifiers is different for each isoform (Oishi *et al.*, 1999; Yu *et al.*, 1997). For example, responsiveness to unsaturated fatty acid, one of the potent activators of PKN1, is clearly different among isoforms, with the order of PKN1>PKN2>PKN3 when a PKCδ peptide is used as a substrate.

Comment

As discussed in the "Kinase Assay" section, there is no standard appropriate measure to monitor PKN activity *in vivo* thus far. PKN-specific *in vivo* substrates have not been identified yet, and the correlation of phosphorylation of PKN itself with its substrate phosphorylation activity has not been established. Biochemical assays of PKN activation have limited predictive value for the understanding of structure–activity relationships because of extensive modulation of PKN behavior by the cellular environment, which includes contributions from lipids, ions, and cellular binding proteins, including Rho family members. A number of studies have shown that fluorescence resonance energy transfer (FRET) reporters are useful tools with which to probe kinase-mediated phosphorylation in living cells. Much information can be obtained by these kinds of reporters, provided that such reporters do not significantly perturb cellular function (for example, by the buffering of cell signals resulting from reporter overexpression) and provided that reporter specificity is maintained in cells. Some of these reporters consist of portions of a kinase substrate fused to a fluorescent protein in an arrangement such that a conformational change occurs on phosphorylation, resulting in a change in FRET signal (Kunkel *et al.*, 2005; Nagai *et al.*, 2000; Sato *et al.*, 2002; Ting *et al.*, 2001; Violin *et al.*, 2003). Among these is included a C kinase activity reporter (CKAR), which reports phosphorylation driven by PKC (Violin *et al.*, 2003). With the fact that PKN1 phosphorylates the original CKAR *in vitro*, this reporter might also be affected by PKN activity. Further studies are expected to develop an efficient PKN-specific reporter, "NKAR."

Acknowledgments

We thank Dr. Kozo Kaibuchi for collaborative study and Dr. Ushio Kikkawa for critical reading of the manuscript.

References

Amano, M., Chihara, K., Nakamura, N., Kaneko, T., Matsuura, Y., and Kaibuchi, K. (1999). The COOH terminus of Rho-kinase negatively regulates rho-kinase activity. *J. Biol. Chem.* **274**, 32418–32424.

Amano, M., Fukata, Y., Shimokawa, H., and Kaibuchi, K. (2000). Purification and *in vitro* activity of Rho-associated kinase. *Methods Enzymol.* **325,** 149–155.

Amano, M., Mukai, H., Ono, Y., Chihara, K., Matsui, T., Hamajima, Y., Okawa, K., Iwamatsu, A., and Kaibuchi, K. (1996). Identification of a putative target for Rho as the serine-threonine kinase protein kinase. N. *Science* **271,** 648–650.

Dong, L. Q., Landa, L. R., Wick, M. J., Zhu, L., Mukai, H., Ono, Y., and Liu, F. (2000). Phosphorylation of protein kinase N by phosphoinositide-dependent protein kinase-1 mediates insulin signals to the actin cytoskeleton. *Proc. Natl. Acad. Sci. USA* **97,** 5089–5094.

Hannun, Y. A., Loomis, C. R., and Bell, R. M. (1985). Activation of protein kinase C by Triton X-100 mixed micelles containing diacylglycerol and phosphatidylserine. *J. Biol. Chem.* **260,** 10039–10043.

Kaplan, W., Husler, P., Klump, H., Erhardt, J., Sluis-Cremer, N., and Dirr, H. (1997). Conformational stability of pGEX-expressed *Schistosoma japonicum* glutathione S-transferase: a detoxification enzyme and fusion-protein affinity tag. *Protein Sci.* **6,** 399–406.

Kitagawa, M., Mukai, H., Shibata, H., and Ono, Y. (1995). Purification and characterization of a fatty acid-activated protein kinase (PKN) from rat testis. *Biochem. J.* **310**(Pt 2), 657–664.

Kunkel, M. T., Ni, Q., Tsien, R. Y., Zhang, J., and Newton, A. C. (2005). Spatio-temporal dynamics of protein kinase B/Akt signaling revealed by a genetically encoded fluorescent reporter. *J. Biol. Chem.* **280,** 5581–5587.

Leenders, F., Mopert, K., Schmiedeknecht, A., Santel, A., Czauderna, F., Aleku, M., Penschuck, S., Dames, S., Sternberger, M., Rohl, T., Wellmann, A., Arnold, W., Giese, K., Kaufmann, J., and Klippel, A. (2004). PKN3 is required for malignant prostate cell growth downstream of activated PI 3-kinase. *EMBO J.* **23,** 3303–3313.

Manning, G., Whyte, D. B., Martinez, R., Hunter, T., and Sudarsanam, S. (2002). The protein kinase complement of the human genome. *Science* **298,** 1912–1934.

Matsuzawa, K., Kosako, H., Inagaki, N., Shibata, H., Mukai, H., Ono, Y., Amano, M., Kaibuchi, K., Matsuura, Y., Azuma, I., and Inagaki, M. (1997). Domain-specific phosphorylation of vimentin and glial fibrillary acidic protein by PKN. *Biochem. Biophys. Res. Commun.* **234,** 621–625.

Mellor, H., Flynn, P., Nobes, C. D., Hall, A., and Parker, P. J. (1998). PRK1 is targeted to endosomes by the small GTPase, RhoB. *J. Biol. Chem.* **273,** 4811–4814.

Morrice, N. A., Fecondo, J., and Wettenhall, R. E. (1994a). Differential effects of fatty acid and phospholipid activators on the catalytic activities of a structurally novel protein kinase from rat liver. *FEBS Lett.* **351,** 171–175.

Morrice, N. A., Gabrielli, B., Kemp, B. E., and Wettenhall, R. E. (1994). A cardiolipin-activated protein kinase from rat liver structurally distinct from the protein kinases C. *J. Biol. Chem.* **269,** 20040–20046.

Mukai, H., and Ono, Y. (1994). A novel protein kinase with leucine zipper-like sequences: Its catalytic domain is highly homologous to that of protein kinase C. *Biochem. Biophys. Res. Commun.* **199,** 897–904.

Mukai, H., Miyahara, M., Sunakawa, H., Shibata, H., Toshimori, M., Kitagawa, M., Shimakawa, M., Takanaga, H., and Ono, Y. (1996). Translocation of PKN from the cytosol to the nucleus induced by stresses. *Proc. Natl. Acad. Sci. USA* **93,** 10195–10199.

Mukai, H. (2003). The structure and function of PKN, a protein kinase having a catalytic domain homologous to that of PKC. *J. Biochem. (Tokyo)* **133,** 17–27.

Mukai, H., Kitagawa, M., Shibata, H., Takanaga, H., Mori, K., Shimakawa, M., Miyahara, M., Hirao, K., and Ono, Y. (1994). Activation of PKN, a novel 120-kDa protein kinase with leucine zipper-like sequences, by unsaturated fatty acids and by limited proteolysis. *Biochem. Biophys. Res. Commun.* **204,** 348–356.

Mukai, H., and Ono, Y. (2003). Expression and purification of protein kinase C from insect cells. *Methods Mol. Biol.* **233,** 21–34.

Nagai, Y., Miyazaki, M., Aoki, R., Zama, T., Inouye, S., Hirose, K., Iino, M., and Hagiwara, M. (2000). A fluorescent indicator for visualizing cAMP-induced phosphorylation *in vivo.* *Nat. Biotechnol.* **18,** 313–316.

Oishi, K., Mukai, H., Shibata, H., Takahashi, M., and Ono, Y. (1999). Identification and characterization of PKNbeta, a novel isoform of protein kinase PKN: Expression and arachidonic acid dependency are different from those of PKNalpha. *Biochem. Biophys. Res. Commun.* **261,** 808–814.

Palmer, R. H., Dekker, L. V., Woscholski, R., Le Good, J. A., Gigg, R., and Parker, P. J. (1995a). Activation of PRK1 by phosphatidylinositol 4,5-bisphosphate and phosphatidylinositol 3,4,5-trisphosphate. A comparison with protein kinase C isotypes. *J. Biol. Chem.* **270,** 22412–22416.

Palmer, R. H., Ridden, J., and Parker, P. J. (1995b). Cloning and expression patterns of two members of a novel protein-kinase-C-related kinase family. *Eur. J. Biochem.* **227,** 344–351.

Palmer, R. H., Schonwasser, D. C., Rahman, D., Pappin, D. J., Herget, T., and Parker, P. J. (1996). PRK1 phosphorylates MARCKS at the PKC sites: Serine 152, serine 156 and serine 163. *FEBS Lett.* **378,** 281–285.

Peng, B., Morrice, N. A., Groenen, L. C., and Wettenhall, R. E. (1996). Phosphorylation events associated with different states of activation of a hepatic cardiolipin/protease-activated protein kinase. Structural identity to the protein kinase N-type protein kinases. *J. Biol. Chem.* **271,** 32233–32240.

Sato, M., Ozawa, T., Inukai, K., Asano, T., and Umezawa, Y. (2002). Fluorescent indicators for imaging protein phosphorylation in single living cells. *Nat. Biotechnol.* **20,** 287–294.

Takahashi, M., Mukai, H., Toshimori, M., Miyamoto, M., and Ono, Y. (1998). Proteolytic activation of PKN by caspase-3 or related protease during apoptosis. *Proc. Natl. Acad. Sci. USA* **95,** 11566–11571.

Ting, A. Y., Kain, K. H., Klemke, R. L., and Tsien, R. Y. (2001). Genetically encoded fluorescent reporters of protein tyrosine kinase activities in living cells. *Proc. Natl. Acad. Sci. USA* **98,** 15003–15008.

Torbett, N. E., Casamassima, A., and Parker, P. J. (2003). Hyperosmotic-induced protein kinase N 1 activation in a vesicular compartment is dependent upon Rac1 and 3-phosphoinositide-dependent kinase 1. *J. Biol. Chem.* **278,** 32344–32351.

Vincent, S., and Settleman, J. (1997). The PRK2 kinase is a potential effector target of both Rho and Rac GTPases and regulates actin cytoskeletal organization. *Mol. Cell Biol.* **17,** 2247–2256.

Violin, J. D., Zhang, J., Tsien, R. Y., and Newton, A. C. (2003). A genetically encoded fluorescent reporter reveals oscillatory phosphorylation by protein kinase C. *J. Cell Biol.* **161,** 899–909.

Watanabe, G., Saito, Y., Madaule, P., Ishizaki, T., Fujisawa, K., Morii, N., Mukai, H., Ono, Y., Kakizuka, A., and Narumiya, S. (1996). Protein kinase N (PKN) and PKN-related protein rhophilin as targets of small GTPase Rho. *Science* **271,** 645–648.

Yau, L., Litchie, B., Thomas, S., Storie, B., Yurkova, N., and Zahradka, P. (2003). Endogenous mono-ADP-ribosylation mediates smooth muscle cell proliferation and migration via protein kinase N-dependent induction of c-fos expression. *Eur. J. Biochem.* **270,** 101–110.

Yoshinaga, C., Mukai, H., Toshimori, M., Miyamoto, M., and Ono, Y. (1999). Mutational analysis of the regulatory mechanism of PKN: the regulatory region of PKN contains an arachidonic acid-sensitive autoinhibitory domain. *J. Biochem. (Tokyo)* **126,** 475–484.

Yu, W., Liu, J., Morrice, N. A., and Wettenhall, R. E. (1997). Isolation and characterization of a structural homologue of human PRK2 from rat liver. Distinguishing substrate and lipid activator specificities. *J. Biol. Chem.* **272,** 10030–10034.

Zhu, Y., Stolz, D. B., Guo, F., Ross, M. A., Watkins, S. C., Tan, B. J., Qi, R. Z., Manser, E., Li, Q. T., Bay, B. H., Teo, T. S., and Duan, W. (2004). Signaling via a novel integral plasma membrane pool of a serine/threonine protein kinase PRK1 in mammalian cells. *FASEB J.* **18,** 1722–1724.

[18] Purification and Enzyme Activity of ACK1

By NORIKO YOKOYAMA and W. TODD MILLER

Abstract

The activated Cdc42 associated kinases (ACKs) are nonreceptor tyrosine kinases that are specific targets of Cdc42. To study the biochemical properties of ACK1, we expressed and purified the enzyme using the baculovirus/Sf9 cell system. This ACK1 construct contains (from N- to C-terminus) the kinase catalytic domain, SH3 domain, and Cdc42-binding CRIB domain. We describe enzyme activity assays based on synthetic peptide substrates. The best such substrate is a peptide derived from the site of ACK1-catalyzed phosphorylation of the Wiskott-Aldrich syndrome protein (WASP). Although the SH3 and CRIB domains of purified ACK1 are able to bind ligands (a polyproline peptide and Cdc42, respectively), the ligands did not stimulate *in vitro* tyrosine kinase activity. Purified ACK1 undergoes autophosphorylation at Tyr284, and autophosphorylation increases kinase activity.

Introduction

The activated Cdc42-associated kinases (ACKs), ACK1 and ACK2, are a family of nonreceptor tyrosine kinases that associate specifically with Cdc42 (not Rac or Rho) and act as Cdc42 effectors in several signaling pathways (Manser *et al.*, 1993; Yang and Cerione, 1997). ACK1 and ACK2 contain similar functional domains but differ at the amino- and carboxy termini. The domain structure of ACK kinases is unique among nonreceptor tyrosine kinases: it consists of an N-terminal tyrosine kinase catalytic domain followed by an SH3 domain, a Cdc42/Rac interactive binding (CRIB) domain, and a proline-rich region (Manser *et al.*, 1993). The position of the SH3 domain C-terminal to the catalytic domain contrasts

METHODS IN ENZYMOLOGY, VOL. 406
Copyright 2006, Elsevier Inc. All rights reserved.

0076-6879/06 $35.00
DOI: 10.1016/S0076-6879(06)06018-6

with the N-terminal SH3 domain observed in the Src, Csk, Abl, Frk, and Tec families. The C-terminal proline-rich domain of ACK1 interacts with the adaptor proteins Nck (Teo *et al.*, 2001), Grb2 (Satoh *et al.*, 1996), sorting nexin protein 9 (SH3PX1) (Lin *et al.*, 2002; Worby *et al.*, 2002) and Hck (Yokoyama and Miller, 2003).

In this chapter, we describe procedures for the expression and purification of active ACK1. We have used purified ACK1 to study how enzyme activity is regulated. We have also developed a convenient synthetic peptide substrate to be used in ACK1 activity assays.

Peptide Synthesis and Characterization

The following synthetic peptides were used in this study: Abl substrate (EAIYAAPFAKKKG), Src substrate (AEEEIYGEFEAKKKKG), EGFR substrate (AEEEEYFELVAKKKG), IR substrate (KKEEEEY MMMMG), IRS-1 Y987 peptide (KKSRGDY-Nle-TMQIG), SH3-binding substrate (KKAEEEIYGEFGGGGGGRPLPSPPKFG), WASP peptide (KVIYDFIEKKKG), and a proline-rich SH3 binding peptide (DFPLG PPPPLPPRATPSR). The synthetic peptides were purified by preparative reversed-phase high-performance liquid chromatography (HPLC) and characterized by matrix-assisted laser description/ionization time of flight mass spectrometry.

Baculovirus Expression Vector for the Kinase-SH3-CRIB Construct of ACK1

We expressed and purified active ACK1 using the baculovirus/Sf9 cell system. This ACK1 construct contains (from N- to C-terminus) the kinase catalytic domain, SH3 domain, and Cdc42-binding CRIB domain (residues 110–476). The ACK1 construct possesses CBD- and His-tags at the C-terminus.

In our initial attempts to produce active ACK1, we expressed a number of other ACK1 constructs in Sf9 cells. These included full-length ACK1, kinase domain alone (residues 110–383), N-terminus plus kinase domain (residues 1–383), and N-terminus plus kinase, SH3, and CRIB domains (residues 1–476). These constructs were expressed with a variety of different affinity tags at the N- and C-termini. Although the various constructs were expressed to reasonable levels and were active in Sf9 cells, we found that only the kinase–SH3-CRIB construct was stable and active throughout the multiple chromatographic steps necessary for purification.

The kinase–SH3-CRIB construct was generated by the polymerase chain reaction (PCR). The 5′ PCR primer had the sequence GGGATC

CGGGGGAGGGGCCCCTGCA G, and the 3' primer was GGAATT-
CAAGTCCCGCAGGGCCACAAAC. These primers had 27 nucleotides
(5'-primer) and 28 nucleotides (3'-primer) of complementarily with the
template and encoded unique restriction sites (*Bam*HI at the 5'-end and
*Eco*R1 at the 3'-end). The PCR product was ligated into plasmid pCR-
BluntII-TOPO (Invitrogen). The resulting plasmid was digested with
*Bam*H1/*Eco*R1, and the ACK1 insert was purified on an agarose gel. The
ACK1 fragment was subcloned into plasmid pBACgus-9 (N-terminal T7
tag and C-terminal CBD-tag and poly-Histidine-tag, Novagen), and expressed in Sf9 cells using the BacVector-3000 DNA transfection kit (Novagen). The virus was amplified twice to generate high-titer virus, and the
final titer was determined by plaque assay. Sf9 cells were grown in a 1-1
spinner flask in Ex-cell-401 media with L-glutamine (JRH Biosciences) and
2.5% fetal calf serum. For protein production, 0.6 l of Sf9 cells (1.8×10^6
cells/ ml) were infected with recombinant ACK1 baculovirus at a multiplicity of infection (MOI) of 7.5. After 4 days of infection, cells were
harvested and washed with phosphate-buffered saline two times.

Purification of ACK1 Kinase–SH3-CRIB Construct

We purified the kinase–SH3-CRIB construct to homogeneity by chromatography on Source-Q, Ni-NTA, and Mono-Q columns. All chromatography steps were carried out at 4°. Gel filtration was also effective as the
third step in the purification and could be used in place of Mono-Q anion
exchange. For experiments not requiring homogeneous protein, ACK1 can
be used after the Ni-NTA step, at which point it is substantially pure.

Step 1. Preparation of cell lysates. Sf9 Cells were lysed in a French
pressure cell two times in 20 mM Tris-HCl buffer (pH 8.0) containing
2 mM Na$_3$VO$_4$, 5 mM 2-mercaptoethanol, 10 μg/ml leupeptin, 10 μg/ml
aprotinin, and 1 mM phenylmethylsulfonyl fluoride (PMSF). Homogenates
were centrifuged at 40,000g for 30 min. The supernatant was filtered
(0.8 μm).

Step 2. Source Q FPLC column chromatography. A 1.6×10 cm
Source-Q FPLC column (Amersham Biosciences, Piscataway, NJ) was
pre-equilibrated with loading buffer (same as lysis buffer, but without
leupeptin, aprotinin, and PMSF). The ACK1 filtrate was applied to the
column at 2 ml/min. After washing extensively with loading buffer, ACK1
kinase was eluted with a linear gradient of NaCl (0–0.6 M). Fractions of
1.3 ml were collected and 10-μl aliquots were assayed for ACK1 kinase
activity using the phosphocellulose paper–binding assay (Casnellie, 1991).
The peptide EAIYAAPFAKKKG was used as a substrate in the assays at
a concentration of 0.5 mM. ACK1 activity eluted at 0.2–0.4 M NaCl.

The peaks that showed high ACK1 activity were pooled, and 0.3 M NaCl was added.

Step. 3. Ni^{2+}-affinity column chromatography. Pooled fractions were applied to a 3 ml Ni-NTA column (QIAGEN), which was pre-equilibrated with 20 mM Tris-HCl (pH 8.0), 0.5 M NaCl, 2 mM Na$_3$VO$_4$, 10% glycerol, 5 mM 2-mercaptoethanol. The column was washed with 125 ml of buffer containing 15 mM imidazole, 0.3 M NaCl, 2 mM Na$_3$VO$_4$, 10% glycerol, 5 mM 2-mercaptoethanol, 20 mM Tris-HCl (pH. 8.0), and further washed with 25 ml of buffer containing 1M NaCl, 2 mM Na$_3$VO$_4$, 5 mM 2-mercaptoethanol, 20 mM Tris-HCl (pH. 8.0). ACK1 was eluted with buffer containing 150 mM imidazole, 0.15 M NaCl, 5 mM 2-mercaptoethanol, 2 mM Na$_3$VO$_4$, 10% glycerol, and 20 mM Tris-HCl (pH. 8.0). Fractions were tested for ACK1 activity using the phosphocellulose paper–binding assay described previously.

Step. 4. Mono Q FPLC chromatography. Peaks of activity were pooled, concentrated in an Amicon Ultra 30,000 molecular weight cutoff device (Millipore), and applied to a Mono Q HR 5/5 FPLC column (Amersham Biosciences, Piscataway, NJ) at a rate of 1 ml/min. The column was washed with 5–10 column volumes of buffer containing 20 mM Tris (pH 8.0) and 0.05 M NaCl, and proteins were eluted with a linear gradient of 0.05–0.6 M NaCl in the same buffer. ACK1 eluted at 0.2–0.25 M NaCl. The protein was identified by SDS-PAGE, Western blotting, and by activity assay. The purified ACK1 was stored in buffer containing 20 mM DTT, 10% glycerol, 2 mM Na$_3$VO$_4$, and 20 mM Tris-HCl (pH 8.0) at –20°. The CBD- and His-tags at the C-terminus of ACK1 could be removed by thrombin treatment. ACK1 (300 μg) was treated with 0.4 units of thrombin at 4° overnight. After digestion, the tags were separated from ACK1 by gel filtration or Mono-Q chromatography.

Fig. 1 shows an SDS-PAGE analysis of the purification steps. Purified ACK1 migrates with the expected molecular weight ($\approx$60.4 kDa). ACK1 reacted with a rabbit polyclonal antibody raised against a peptide sequence in the ACK1 catalytic domain (KPDVLSQPEAMDDFI). ACK1 also reacted with anti-phosphotyrosine antibody, suggesting that purified ACK1 is autophosphorylated or phosphorylated by an endogenous Sf9 cell kinase (Fig. 1).

Substrate Specificity of ACK1 Kinase Using Synthetic Peptide Substrates

Because relatively few protein substrates have been identified for ACK1, we used synthetic peptides as an initial screen of substrate specificity. We used substrates that were derived from peptide library screens of

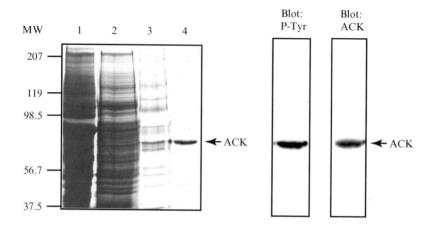

FIG. 1. Purification of ACK1 kinase-SH3-CRIB construct. (Left) Proteins from each step of the purification were separated by SDS-PAGE and visualized with Coomassie brilliant blue. Lane 1, crude Sf9 cell lysate. Lane 2, Pooled fractions after Source-Q chromatography. Lane 3, Pooled fractions after Ni-NTA chromatography. Lane 4, Pooled fractions after Mono-Q chromatography. (Right) Purified ACK1 was analyzed by Western blotting using anti-phosphotyrosine and anti-ACK antibodies. (From Yokoyama and Miller, 2003).

four prototypical tyrosine kinases: Src, Abl, EGF receptor, and insulin receptor (Songyang, 1995). We included a substrate that is derived from the sequence of IRS-1 that has been shown to be an excellent substrate for insulin receptor kinase (Shoelson et al., 1992), as well as for Jak2 (Yokoyama et al., 2003). Finally, we included a substrate that possesses an SH3 domain–binding polyproline sequence. For Src-family kinases, addition of an SH3 domain ligand to a substrate increases its phosphorylation by activating down-regulated kinase and by decreasing substrate K_m (Scott and Miller, 2000).

ACK1 kinase activity was determined using the p81 phosphocellulose paper assay (Casnellie, 1991). Reaction mixtures (20 μl) contained 20 mM Tris-HCl (pH 7.4), 10 mM MgCl$_2$, 0.1 mM Na$_3$VO$_4$, 0.5 mM DTT, 0.25 mM ATP, 5 mg/ml BSA, varying concentrations of peptide substrate, and $[\gamma\text{-}^{32}\text{P}]$-ATP (200–400 cpm/pmol). Reactions were initiated by addition of $[\gamma\text{-}^{32}\text{P}]$-ATP. Reactions were terminated by addition of 50% acetic acid, and samples were spotted on P81 phosphocellulose paper (Whatman). Incorporation of ^{32}P into peptide was determined by Cerenkov counting. The values of K_m for peptide were determined using a range of peptide concentrations (0.05–2.0 mM) and 0.25 mM $[\gamma\text{-}^{32}\text{P}]$-ATP. Kinetic parameters were calculated by fitting data to the Michaelis–Menten equation.

Initial studies on several peptides (Src, EGFR, IRS-1, SH3, and IR peptides) established that their Michaelis constants were in the millimolar range ($\gg 2$ mM). Thus, it was not possible to determine accurate Michaelis constants by using initial rate kinetics. The complete time course for the phosphorylation of each these peptides was measured by using peptide concentrations less than the Michaelis constant. For these peptides, we analyzed the data graphically as described (Thomas *et al.*, 1987) to determine k_{cat}/K_m.

Table I shows a summary of the ACK1 kinetic parameters using different peptide substrates. Most of the peptides from this original group were poor substrates for ACK1, with the exception of the Abl peptide. We carried out kinetic measurements with saturating concentrations of ATP and varying concentrations of Abl peptide. These experiments yielded a Km for the Abl peptide of 507 μM and a k_{cat} of 0.249 min^{-1} ($k_{cat}/K_m = 4.8 \times 10^{-4}$ min^{-1} μM^{-1}) (Table I). By comparison, the Src-specific peptide was phosphorylated with a k_{cat}/K_m of 2.2×10^{-5} min^{-1} μM^{-1}, approximately 20 times lower than the Abl substrate (Table I). Phosphorylation of the EGFR- and insulin receptor–specific peptides was barely detectable above background. The SH3-binding substrate was not phosphorylated appreciably by ACK1. SH3-binding peptides containing either a longer spacer length between the substrate sequence and the SH3 ligand, or with these sequences reversed, were also poorly phosphorylated (data not shown). This is in contrast to results from Src-family kinases and suggests that the SH3 domain of ACK1 may not play a direct role in substrate targeting. The substrate specificity of ACK1 seems to be narrower than other tyrosine kinases. These results are consistent with a recent report in which only four of 160 diverse peptides containing known tyrosine kinase phosphorylation sites showed a low level of phosphorylation (Lougheed *et al.*, 2004).

We recently identified WASP as a protein substrate for ACK1 (N. Y. and W. T. M., unpublished observations). We mapped the site of phosphorylation as Tyr291 of WASP, in the sequence VIYDFIE. To determine whether this sequence would be phosphorylated in the context of a synthetic peptide, we prepared a peptide with the sequence KVIYD-FIEKKKG. As shown in Table I, this peptide is by far the best substrate yet reported for ACK1. This WASP peptide had a k_{cat}/K_m of 0.136 min^{-1} μM^{-1}, approximately 450 times higher the Abl peptide. The kcat value of 107.6 min^{-1} is comparable to turnover numbers for other tyrosine kinases in their maximum-activity states (e.g., Hck with $k_{cat} = 60$ min^{-1}; LaFevre-Bernt *et al.*, 1998). Thus, the WASP peptide is a convenient substrate for *in vitro* assays of ACK1 kinase activity, and it should be useful in future studies of ACK1 function.

TABLE I
KINETIC PARAMETERS FOR ACK1

Peptide	Sequence	k_{cat}/K_m $(\text{min}^{-1}\ \mu M^{-1})$	k_{cat} (min^{-1})	K_m (μM)
Abl substrate	EAIYAAPFAKKKG	4.8×10^{-4}	0.24	507
Src substrate	AEEEIYGEFEAKKKKG	2.2×10^{-5}	*	*
EGFR substrate	AEEEEYFELVAKKKG	$< 3 \times 10^{-6}$	*	*
IR substrate	KKEEEEYMMMMG	$< 3 \times 10^{-6}$	*	*
sequence from IRS-1 Y987	KKSRGDY-Nle-TMQIG	1.2×10^{-5}	*	*
SH3-binding substrate	KKAEEEIYGEFGGGG-GGRPLPSPPKFG	5.8×10^{-6}	*	*
WASP substrate	KVIYDFIEKKKKG	1.36×10^{-1}	107.6	789

Kinetic parameters were determined by the phosphocellulose binding assay, as described in experimental procedures.

*It was not possible to determine accurate Michaelis constants for these peptides using initial rate kinetics. We analyzed the data graphically (Thomas *et al.*, 1987) to determine k_{cat}/K_m, using ATP and peptide concentrations of 1 mM and 0.5 mM, respectively.

We have carried out initial experiments aimed at understanding how ACK1 activity is regulated. Addition of activated Cdc42 to purified ACK1 *in vitro* did not stimulate ACK1 activity, although the CRIB domain of this construct is able to bind Cdc42 (Yokoyama and Miller, 2003). In contrast, both ACK1 and ACK2 have been shown to be activated by Cdc42 in intact mammalian cells (Yang *et al.*, 1999; Yokoyama and Miller, 2003). We were also unable to activate ACK1 *in vitro* by addition of a polyproline-containing peptide (DFPLGPPPPLPPRATPSR). Thus, it is not clear whether the SH3 domains of ACK kinases play a role in regulating enzyme activity. For Src- and Abl-family nonreceptor tyrosine kinases, the SH3 domains are involved in autoinhibition; they bind to polyproline type II helices in the linker region between the SH2 domain and the catalytic domain. Protein or peptide ligands for the SH3 domains of Src- or Abl-family kinases disrupt these intramolecular interactions and increase enzymatic activity (Alexandropoulos and Baltimore, 1996; Briggs *et al.*, 1997; Moarefi *et al.*, 1997). The SH3 domains of ACK kinases, which are on the C-terminal side of the catalytic domain, may not play an analogous role.

Autophosphorylation of ACK1

We carried out experiments to measure the ability of purified ACK1 to undergo autophosphorylation. Purified ACK1 was incubated with 0.25 mM $[\gamma\text{-}^{32}\text{P}]$-ATP (400–700 cpm/pmol) in kinase buffer containing

20 mM Tris-HCl (pH 7.5), 10 mM MgCl$_2$, 0.5 mM DTT and 0.1 mM Na$_3$VO$_4$ at 30°. Reactions were stopped by the addition of SDS-sample buffer and analyzed by SDS-PAGE and autoradiography. ACK1 was autophosphorylated in a time-dependent manner (Fig. 2A). Preincubation of ACK1 with 0.25 mM unlabeled ATP in kinase buffer for 40 min, followed by incubation with [γ-^{32}P]-ATP for 40 min at 30°, led to a decrease in the level of ^{32}P incorporation (Fig 2A). To measure changes in ACK1 activity after autophosphorylation, ACK1 was incubated with 0.5 mM unlabeled ATP in kinase reaction buffer for 40 min at 30°. Activity was then measured using the phosphocellulose paper assay. Autophosphorylation activated ACK1 kinase threefold compared with unphosphorylated ACK1 (Fig. 2B). This increase in ACK1 activity is low

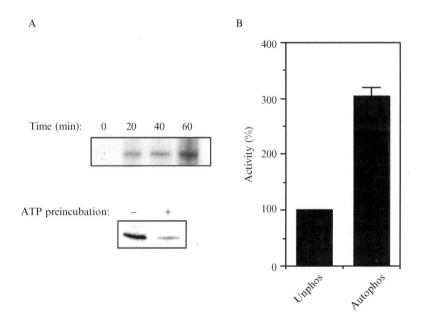

FIG. 2. ACK1 autophosphorylation. (A) Top, the ACK1 kinase-SH3-CRIB construct was incubated with 0.25 mM [γ-^{32}P]-ATP in kinase assay buffer for 0, 20, 40, or 60 min. Reactions were terminated by addition of SDS-sample buffer and analyzed by SDS-PAGE and autoradiography. Bottom, ACK1 was preincubated with 0.25 mM ATP for 40 min, followed by an incubation with [γ-^{32}P]-ATP for 40 min as described above. Reactions were followed by SDS-PAGE and autoradiography. (B) The ACK1 kinase-SH3-CRIB construct was preincubated in the presence or absence of 0.5 mM ATP ("autophos." and "unphos.", respectively) in kinase reaction buffer for 1 h at 37°. ACK1 activity was then determined using the phosphocellulose binding assay. Kinase assays contained 0.1 mM [γ-^{32}P]-ATP and 0.25 mM Abl peptide, and proceeded for 30 min. (From Yokoyama and Miller, 2003).

compared with other kinases that are autoinhibited by their unphosphory-lated activation loop (Nolen *et al.*, 2004). Interestingly, the recently re-ported structure of ACK1 shows that the activation loop of unphosphorylated ACK1 has a partially "active" conformation that is generally observed only in phosphorylated protein kinases (Lougheed *et al.*, 2004). Thus, it is not surprising that autophosphorylation triggers only a modest increase in ACK1 activity.

We have identified Tyr 284 as the major autophosphorylation site of ACK1 by mass spectrometry (Yokoyama and Miller, 2003). After autopho-sphorylation and trypsin digestion, one tryptic fragment of ACK1 was shifted by the mass of a phosphate (+80 daltons) relative to the unphosphorylated sample. This tryptic fragment had the sequence ^{276}ALPQNDDHYVMQEHR^{290} and contains a single tyrosine residue, Tyr284. We confirmed the identity of this autophosphorylated peptide by sequence analysis by LC/MS/MS. On the basis of a sequence alignment of the ACK kinase catalytic domain with the Src catalytic domain, Tyr284 of ACK1 is predicted to be in the kinase activation loop. The sequence surrounding Tyr284 of ACK1 is similar to the sequence surrounding Tyr416 of Src, the major autophosphorylation site. Expression of a full-length Y284F mutant form of ACK1 in COS-7 cells showed a dramatically re-duced level of tyrosine phosphorylation relative to wild-type ACK1 (Yo-koyama and Miller, 2003). We purified a Y284F mutant version of ACK1 using the Sf9/baculovirus system. The Y284F mutant is capable of additional autophosphorylation after incubation with $[\gamma\text{-}^{32}\text{P}]$-ATP, sug-gesting that there are additional, unidentified autophosphorylation sites on ACK1.

Concluding Remarks

We report here the first purification of active ACK1 from eucaryotic cells, using the Sf9/baculovirus expression system. Of four peptides tested that are specific for different families of tyrosine kinases, we show that the specific activity of ACK1 most closely resembles that of Abl. However, the peptide derived from the phosphorylation site of WASP is a superior peptide substrate, and it is the most potent *in vitro* substrate yet identified for this enzyme. We also identified Tyr284 as the major autophosphoryla-tion site of ACK1; autophosphorylation enhances ACK1 activity. Neither an SH3 ligand nor Cdc42 activates ACK kinase *in vitro*, although both ligands bind ACK1. At present, the roles of the SH3 domain and polypro-line region of ACK1 in enzyme regulation have not been explained. A full understanding of the intramolecular interactions of ACK1 will require additional structure–function studies on full-length ACK1.

Acknowledgments

This work was supported by National Institutes of Health Grant CA 28146 to W. T. M. We are grateful to Dr. Edward Manser (Institute of Molecular and Cell Biology, Singapore) for supplying ACK1 plasmids. We thank Dr. Nicolas Nassar (SUNY Stony Brook) for bacterial expression vectors for His-tagged Cdc42 (wild-type and G12V), as well as for anti-Cdc42 antibody.

References

Alexandropoulos, K., and Baltimore, D. (1996). Coordinate activation of c-Src by SH3- and SH2-binding sites on a novel p130Cas-related protein, Sin. *Genes Dev.* **10,** 1341–1355.

Briggs, S. D., Sharkey, M., Stevenson, M., and Smithgall, T. E. (1997). SH3-mediated Hck tyrosine kinase activation and fibroblast transformation by the Nef protein of HIV-1. *J. Biol. Chem.* **272,** 17899–17902.

Casnellie, J. E. (1991). Assay of protein kinases using peptides with basic residues for phosphocellulose binding. *Methods Enzymol.* **200,** 115–120.

LaFevre-Bernt, M., Sicheri, F., Pico, A., Porter, M., Kuriyan, J., and Miller, W. T. (1998). Intramolecular regulatory interactions in the Src family kinase Hck probed by mutagenesis of a conserved tryptophan residue. *J. Biol. Chem.* **273,** 32129–32134.

Lin, Q., Lo, C. G., Cerione, R. A., and Yang, W. (2002). The Cdc42 target ACK2 interacts with sorting nexin 9 (SH3PX1) to regulate epidermal growth factor receptor degradation. *J. Biol. Chem.* **277,** 10134–10138.

Lougheed, J. C., Chen, R. H., Mak, P., and Stout, T. J. (2004). Crystal structures of the phosphorylated and unphosphorylated kinase domains of the Cdc42-associated tyrosine kinase ACK1. *J. Biol. Chem.* **279,** 44039–44045.

Manser, E., Leung, T., Salihuddin, H., Tan, L., and Lim, L. (1993). A non-receptor tyrosine kinase that inhibits the GTPase activity of p21cdc42. *Nature* **363,** 364–367.

Moarefi, I., LaFevre-Bernt, M., Sicheri, F., Huse, M., Lee, C. H., Kuriyan, J., and Miller, W. T. (1997). Activation of the Src-family tyrosine kinase Hck by SH3 domain displacement. *Nature* **385,** 650–653.

Nolen, B., Taylor, S., and Ghosh, G. (2004). Regulation of protein kinases; controlling activity through activation segment conformation. *Mol. Cell* **15,** 661–675.

Satoh, T., Kato, J., Nishida, K., and Kaziro, Y. (1996). Tyrosine phosphorylation of ACK in response to temperature shift-down, hyperosmotic shock, and epidermal growth factor stimulation. *FEBS Lett.* **386,** 230–234.

Scott, M. P., and Miller, W. T. (2000). A peptide model system for processive phosphorylation by Src family kinases. *Biochemistry* **39,** 14531–14537.

Shoelson, S. E., Chatterjee, S., Chaudhuri, M., and White, M. F. (1992). YMXM motifs of IRS-1 define substrate specificity of the insulin receptor kinase. *Proc. Natl. Acad. Sci. USA* **89,** 2027–2031.

Songyang, Z., Carraway, K. L., III, Eck, M. J., Harrison, S. C., Feldman, R. A., Mohammadi, M., Schlessinger, J., Hubbard, S. R., Smith, D. P., Eng, C., Lorenzo, M. J., Poner, B. A. J., Mayer, B. J., and Cantley, L. C. (1995). Catalytic specificity of protein-tyrosine kinases is critical for selective signalling. *Nature* **373,** 536–539.

Teo, M., Tan, L., Lim, L., and Manser, E. (2001). The tyrosine kinase ACK1 associates with clathrin-coated vesicles through a binding motif shared by arrestin and other adaptors. *J. Biol. Chem.* **276,** 18392–18398.

Thomas, N. E., Bramson, H. N., Miller, W. T., and Kaiser, E. T. (1987). Role of enzyme-peptide substrate backbone hydrogen bonding in determining protein kinase substrate specificities. *Biochemistry* **26,** 4461–4466.

Worby, C. A., Simonson-Leff, N., Clemens, J. C., Huddler, D., Jr., Muda, M., and Dixon, J. E. (2002). *Drosophila* Ack targets its substrate, the sorting nexin DSH3PX1, to a protein complex involved in axonal guidance. *J. Biol. Chem.* **277,** 9422–9428.

Yang, W., and Cerione, R. A. (1997). Cloning and characterization of a novel Cdc42-associated tyrosine kinase, ACK-2, from bovine brain. *J. Biol. Chem.* **272,** 24819–24824.

Yang, W., Lin, Q., Guan, J. L., and Cerione, R. A. (1999). Activation of the Cdc42-associated tyrosine kinase-2 (ACK-2) by cell adhesion via integrin beta1. *J. Biol. Chem.* **274,** 8524–8530.

Yokoyama, N., and Miller, W. T. (2003). Biochemical properties of the Cdc42-associated tyrosine kinase ACK1. Substrate specificity, autophosphorylation, and interaction with Hck. *J. Biol. Chem.* **278,** 47713–47723.

Yokoyama, N., Reich, N. C., and Miller, W. T. (2003). Determinants for the interaction between Janus kinase 2 and protein phosphatase 2A. *Arch. Biochem. Biophys.* **417,** 87–95.

[19]   Direct Activation of Purified Phospholipase C Epsilon by RhoA Studied in Reconstituted Phospholipid Vesicles

By JASON P. SEIFERT, JASON T. SNYDER,
JOHN SONDEK, and T. KENDALL HARDEN

Abstract

Phospholipase C-ε (PLC-ε) was shown recently to be a downstream effector of Rho GTPases, and we have used an *in vitro* phospholipid vesicle reconstitution system with purified proteins to show this regulation to be direct. This chapter describes high-level expression of a hexahistidine-tagged fragment of PLC-ε encompassing the catalytic core of the enzyme through the tandem RA domains by use of a recombinant baculovirus and High Five insect cells. The recombinant protein is purified to homogeneity using metal chelate affinity and size exclusion chromatography. The small GTPase RhoA also is expressed to high levels in a lipidated form after baculovirus expression in High Five cells and is purified to near homogeneity after detergent extraction and metal chelate affinity chromatography. The capacity of GTPγS-bound RhoA to stimulate the phospholipase activity of PLC-ε is assessed by reconstitution of the RhoA in mixed-detergent phospholipid micelles containing PtdIns(4,5)P$_2$ substrate.

METHODS IN ENZYMOLOGY, VOL. 406   0076-6879/06 $35.00
Copyright 2006, Elsevier Inc. All rights reserved.   DOI: 10.1016/S0076-6879(06)06019-8

Introduction

An extensive array of extracellular signaling molecules including hormones, neurotransmitters, and growth factors elicit their physiological effects through receptor-mediated stimulation of phospholipase C (PLC) (Berridge and Irvine, 1989; Rhee, 2001). PLC enzymes respond by hydrolyzing the minor membrane phospholipid phosphatidylinositol (4,5)-bisphosphate [PtdIns(4,5)P$_2$] into the second messenger molecules inositol (1,4,5)-trisphosphate [Ins(1,4,5)P$_3$] and diacylglycerol, which are responsible for the release of Ca^{2+} from intracellular stores and the activation of protein kinase C, respectively.

The six PLC isoforms (PLC-β, -γ, -δ, -ε, -ζ, and -η; Fig. 1) are characterized by the presence of conserved X- and Y-boxes, which fold together to form a triose-phosphate isomerase (TIM) barrel and the catalytic core of the enzyme (Rhee, 2001). These isozymes are differentially elaborated with structural/regulatory domains outside of the catalytic region. For example, the two PLC-γ isozymes contain SH2 and SH3 domains and are subject to regulation by tyrosine kinases (Meisenhelder *et al.*, 1989; Wahl *et al.*, 1989), and the four PLC-β isozymes contain a long carboxy terminal domain that confers capacity to bind and be activated by Gα subunits of Gq family of heterotrimeric G proteins (Smrcka *et al.*, 1991; Taylor *et al.*, 1991; Waldo *et al.*, 1991).

Kataoka and colleagues first identified PLC-ε as a Ras-binding protein in *C. elegans* (Shibatohge *et al.*, 1998), and subsequent work from the laboratories of Kataoka (Song *et al.*, 2001), Lomasney (Lopez *et al.*, 2001), and Kelley and Smrcka (Kelley *et al.*, 2001) revealed the presence of PLC-ε in mammalian tissues. Two Ras-associating (RA) domains in the carboxyl terminus of PLC-ε underlie direct regulation of this signaling protein by Ras and Rap (Evellin *et al.*, 2002; Lopez *et al.*, 2001; Song *et al.*, 2001) and a conserved CDC25 guanine nucleotide exchange (GEF) domain in the amino terminus provides a poorly understood mechanism whereby PLC-ε acts as an upstream regulator of Ras, Rap, and potentially other Ras superfamily GTPases (Jin *et al.*, 2001; Lopez *et al.*, 2001). Although heterotrimeric G proteins of the Gq family do not regulate the activity of PLC-ε, this isozyme is activated by Gα_{12} (Lopez *et al.*, 2001), Gα_{13} (Wing *et al.*, 2001), and G$\beta\gamma$-subunits (Wing *et al.*, 2001). Regulation by these heterotrimeric G protein subunits may not be direct, and the fact that several RhoGEFs are effectors of G$\alpha_{12/13}$ (Fukuhara *et al*, 1999; Hart *et al.*, 1998) suggested to us that Rho might be responsible for activation of PLC-ε by G$\alpha_{12/13}$. Indeed, RhoA, RhoB, and RhoC, but not Rac or Cdc42, markedly stimulate the lipase activity of PLC-ε after coexpression in COS-7 cells (Wing *et al.*, 2003). Whereas truncation of the carboxy

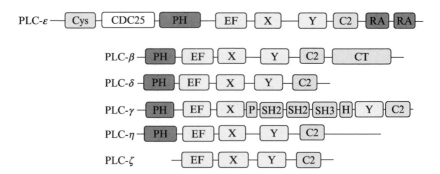

FIG. 1. Domain organization of the phospholipase C (PLC) isozymes. The structural motifs present within all PLC family members are outlined. All isozymes share common domains, including a pleckstrin homology (PH) domain (excluding PLC-ζ), EF hand domain, catalytic TIM-barrel (X, Y), and C2 domain. *Cys*, cysteine rich region; *CDC25*, Ras superfamily guanine nucleotide exchange factor domain; *RA*, Ras-associating domain; *CT*, region at carboxyl terminus of PLC-β responsible for specific binding and activation by G-protein $\alpha_{q/11}$ subunits; *SH2* and *SH3*, an array of Src-homology-2 and -3 domain embedded within the catalytic regions of PLC-γ.

terminal RA domains of PLC-ε results in loss of regulation by Ras, the capacity of both G$\alpha_{12/13}$ and Rho to activate PLC-ε is retained. Expression of C3 botulinum toxin, which ADP-ribosylates and inactivates Rho, results in concomitant loss of capacity of Rho, Gα_{12}, Gα_{13}, or LPA or thrombin receptor agonists (which activate G$\alpha_{12/13}$ signaling pathways) to activate PLC-ε (Hains *et al.*, 2005). These data are consistent with the idea that heterotrimeric G protein–coupled receptors that activate G$\alpha_{12/13}$ regulate PLC-ε through activation of a RhoGEF and conversion of Rho to its GTP-bound form. Pull-down experiments illustrating GTP-dependent interaction of RhoA with PLC-ε also were consistent with this possibility (Wing *et al.*, 2003).

To address whether PLC-ε is directly activated by Rho, we have developed an *in vitro* system that allows quantification of the enzymatic activity of purified PLC-ε and the capacity of PLC-ε to function as a direct effector for RhoA (Seifert *et al.*, 2004). This chapter describes in detail methods for purifying to homogeneity a catalytically active fragment of PLC-ε, as well as methods for the purification of lipidated RhoA. In both cases, we have generated a recombinant baculovirus and expressed the PLC isozyme or RhoA to high levels using an insect cell expression system. We also present methods for quantification of the activity of this Rho effector enzyme using PtdIns(4,5)P$_2$ substrate-containing mixed-detergent phospholipid vesicles. Similar methods for study of regulation of PLC-β2 by Rac have been described previously (Illenberger et al., 2000).

Protein Purification

Virus Preparation

Baculovirus encoding a fragment of PLC-ε encompassing the EF hand domains through tandem RA domains (PLC-ε EF-RA2: amino acids 1258–2215) was prepared by PCR amplification of the desired regions of rat PLC-ε and subsequently cloned into pFastBacHTb (Invitrogen), which incorporates a hexahistidine affinity tag at the amino terminus of the coding sequence. The human monomeric GTPase RhoA also was cloned into pFastBacHTb for viral propagation. Viruses are amplified according to the manufacturer's instructions, and the titer of the working stock viruses was approximately 7×10^7 pfu/ml. Working stock viruses are stored at 4° in the dark. Additional aliquots of P1 and P2 viral stocks are frozen at $-80°$ for long-term storage.

Cell Culture

High Five cells (Invitrogen) are maintained in shaker flasks at 27° with Express Five SFM supplemented with 10 mM L-glutamine. Cell density is monitored daily and maintained between 0.8 and 4×10^6 cells/ml. One day before infection, cells are diluted to a density of 0.8×10^6 cells/ml in a final volume of 1 l per 4-l shaker flask. Cells are grown overnight in shaker flasks at 150 rpm and 27° to reach a target density of $\sim 2.0 \times 10^6$ cells/ml the following day.

PLC-ε EF-RA2 Purification

Cell Infection and Harvest of Cytosolic Fraction

Overexpression of recombinant protein is somewhat variable using a baculovirus expression system, and optimal expression conditions are determined empirically. In our experience, the PLC-ε EF-RA2 fragment is expressed to high levels after 48 h of infection using a multiplicity of infection (MOI, number of virus particles per cell) of 1.

Four liters of High Five cells ($\sim 2.0 \times 10^6$ cells/ml) are infected with working stock virus encoding hexahistidine-tagged PLC-ε EF-RA2. Forty-eight hours after infection, cells are harvested by centrifugation at 1000g for 10 min in a J6 Beckman swinging bucket centrifuge at 4°. All buffers and subsequent steps during the purification should be at 4°.

Cells are resuspended in 100 ml of lysis buffer (lysis buffer: 20 mM HEPES, pH 8, 300 mM NaCl, 15 mM imidazole, 1 mM CaCl$_2$, 1 mM β-mercaptoethanol (β-ME), 10% glycerol, and 2 EDTA-free complete protease inhibitor tablets [Roche Applied Sciences]) and lysed using an EmulsiFlex C5 cell homogenizer (Avestin). Cell lysate is cleared of intact cells and nuclei by low-speed centrifugation (500g, 15 min, J6 Beckman centrifuge). The supernatant from the low-speed spin is centrifuged at 150,000g for 45 min in a Type 60Ti rotor in an ultracentrifuge (Beckman). Although most recombinant protein exists in the pellet after the ultracentrifugation step, sufficient PLC-ε EF-RA2 for subsequent purification is recovered in the soluble fraction. The supernatant is diluted to a final volume of 150 ml with lysis buffer and passed through a 5-μm syringe filter to remove any residual debris.

Metal Chelate Affinity Chromatography

Wash a 5-ml HiTrap Chelating HP column (GE Healthcare) with 5 column volumes of deionized water using a disposable syringe with Luer fittings to remove the ethanol in which the column is stored. The column is then charged with Ni^{2+} by injecting a single column volume (5 ml) of 0.1 M NiSO$_4$ with a disposable syringe. Connect the column to the FPLC and equilibrate with the following washes: 5 column volumes of starting buffer (starting buffer: lysis buffer minus protease inhibitors), 5 column volumes of elution buffer (elution buffer: 20 mM HEPES, pH 8, 300 mM NaCl, 1 M imidazole, 1 mM CaCl$_2$, 1 mM β-ME, and 10% glycerol), and 5 column volumes of starting buffer.

The soluble fraction of the whole cell lysate containing approximately 500 μg of PLC-ε EF-RA2 per liter of cells is loaded onto the chelating column at a rate of 1 ml/min, maintaining a column pressure of <0.5 mPa. The column is washed with 10 column volumes of starting buffer, followed by 10 column volumes of 50 mM imidazole (5% elution buffer) to remove proteins nonspecifically bound to the Ni^{2+}–charged resin. Ultraviolet absorbance levels measured at A$_{280}$ should return to near baseline during the 5% elution buffer wash. Recombinant protein is eluted using a 50 mM–1 M imidazole gradient (5–100% elution buffer) over 10 column volumes, collecting 2.5-ml fractions. A peak of absorbance measured at A$_{280}$ will indicate fractions containing the recombinant protein, which is confirmed using Coomassie-stained 7.5 % SDS-PAGE gels (Fig. 2).

Fractions containing recombinant protein are pooled, and hexahistidine-tagged tobacco etch virus (TEV) protease is added to the sample to cleave the hexahistidine tag from the PLC-ε fragment. This mixture is dialyzed overnight against 4 l of dialysis buffer (dialysis buffer: 20 mM HEPES, pH 8, 300 mM NaCl, 1 mM CaCl$_2$, and 10% glycerol) at 4° to dilute the

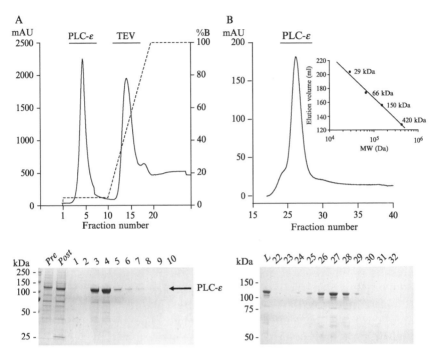

FIG. 2. Purification of phospholipase C-ε (PLC-ε) EF-RA2 with metal chelate and size exclusion chromatography. (A) Above, the left-hand y-axis and solid line indicate mAU (A_{280}), and the right-hand y-axis and hatched line indicate the percentage of elution buffer (1 M imidazole). The absorbance peak at 5% elution buffer corresponds to the cleaved PLC-ε EF-RA2 fragment, and the second absorbance peak indicates the elution of hexahistidine-tagged TEV protease and other nonspecific proteins. Below, a Coomassie-stained SDS-PAGE gel showing the uncleaved, pooled fractions from the initial metal chelate elution (*Pre*), TEV protease cleaved PLC-ε EF-RA2 (*Post*) at a lower molecular weight, and elution of 1-ml fractions of cleaved PLC-ε from the second metal chelate column at 50 mM imidazole (5% elution buffer) over a range of 10 ml. Note the presence of TEV protease as a band at ~25 kDa in the postcleavage lane. (B) Above, elution profile of PLC-ε EF-RA2 after injection onto an S-300–size exclusion column measuring ultraviolet absorbance (A_{280}). Protein standards were used to calibrate the S-300 column (inset). Below, a load sample (*L*) and fractions corresponding to a peak in the ultraviolet absorbance are shown on a Coomassie-stained SDS-PAGE gel. Most PLC-ε EF-RA2 was recovered in fraction 27 (160–165 ml), suggesting the elution of a monomeric protein of ~117 kDa corresponding to the correct size for PLC-ε EF-RA2 compared with protein standards used to calibrate the S-300 column (inset).

imidazole used to elute the recombinant protein from the metal chelate column. The amount of TEV protease used should be sufficient to completely cleave the hexahistidine tag from the PLC-ε fragment. Because individual TEV protease preparations have varying specific activities, the amount of enzyme required for complete cleavage is determined empirically. Purification of hexahistidine-tagged TEV protease was described previously (Kapust et al., 2001).

A small amount of precipitation may appear in the sample after overnight dialysis. This material is removed by centrifuging the sample for 10 min at 2000g. The sample is then loaded onto a 1-ml HiTrap Chelating HP column using the FPLC. The 1-ml column is prepared in the same manner as the 5-ml HiTrap Chelating HP column. A flow rate of 0.5 ml/min is used to load the column, keeping the column pressure <0.5 mPa. Although the hexahistidine tag is cleaved from PLC-ε EF-RA2 by TEV protease, this PLC-ε fragment possesses a low affinity for the Ni^{2+}-charged resin and is retained on the column during the loading procedure. The loaded column is washed with 10 column volumes of dialysis buffer. The cleaved PLC-ε fragment is eluted from the column using a 50 mM imidazole (5% elution buffer) wash over 10 column volumes, collecting 1-ml fractions. The presence of a noncleavable hexahistidine tag on TEV protease results in retention of this enzyme by the column during the elution of PLC-ε. However, the TEV can be eluted with a 5-column volume, 50 mM–1 M imidazole (5–100% elution buffer) gradient. Other nonspecific proteins also will be eluted over this range. The presence of the recombinant proteins is confirmed using Coomassie-stained SDS-PAGE gels (Fig. 2).

Sephacryl S-300 Gel-Filtration Chromatography. Fractions from the second metal chelate affinity column purification step containing recombinant PLC-ε EF-RA2 are pooled and applied in a volume of 10 ml to a HiPrep 26/60 Sephacryl S-300 High Resolution column (26 × 600 mm, 320 ml bed volume; GE Healthcare) equilibrated with buffer A (buffer A: 20 mM HEPES, pH 8, 150 mM NaCl, 1 mM $CaCl_2$, 1 mM DTT, and 5% glycerol). This step is performed to ensure the recombinant PLC-ε EF-RA2 is monodispersed and not aggregated; however, this step does not significantly increase the purity of the sample. A flow rate of 1 ml/min is used to elute proteins, and 5-ml fractions are collected starting at injection of the sample. The elution volume (V_e) of the purified PLC is approximately 162 ml as determined by a peak of protein absorbance measured at A_{280}, which is verified with Coomassie-stained SDS-PAGE gels (Fig. 2). These results indicate that PLC-ε EF-RA2 exists as a 117-kDa monomer relative to protein standards used to calibrate the S-300 column. Fractions containing the recombinant protein are pooled and concentrated using a 50,000 MWCO PES centrifugal filtering device (Vivascience) to a concentration of 1–2 mg/ml. Aliquots of 25–50 μl are snap frozen and stored at −80°. Repeated freeze-thaw or freezing of diluted samples should be avoided, because the protein may lose enzyme activity. Under optimum conditions,

TABLE I
Properties of Purified PLC-ε EF-RA2

Property	Comments
SDS-PAGE M_r	$\sim$117,000 Da
Purity	>95%
Substrate preference	
PtdIns	V_{max} <1 μmol/min/mg
PtdIns(4)P	V_{max} $\sim$3 μmol/min/mg
PtdIns(4,5)P$_2$	V_{max} $\sim$10 μmol/min/mg, K_m = 6 μM
Ca^{2+}-dependence	EC$_{50}$ $\sim$65 nM, max $\sim$1 μM
RhoA-dependent activation	EC$_{50}$ $\sim$58 nm

See Seifert *et al.* (2004) for further details.

the purification process should take no more than 72 h between cell lysis and the freezing of aliquots. Typical yields of near homogeneous, monodispersed 117-kDa PLC-ε EF-RA2 are 1–2 mg per 4 l of baculovirus-infected High Five cells. The properties of the purified enzyme are summarized in Table I.

Prenylated RhoA Purification

Cell Infection and Membrane Protein Extraction

One liter of High Five cells ($\sim$2.0 $\times$ 10^6 cells/ml) is infected with baculovirus encoding hexahistidine-tagged RhoA at an MOI of 1. Cells are harvested by low-speed centrifugation (1000g for 15 min in J6 swinging bucket centrifuge, Beckman) approximately 48 h after infection. The cell pellet is resuspended in 25 ml of ice-cold lysis buffer (lysis buffer: 20 mM HEPES, pH 8, 150 mM NaCl, 5 mM MgCl$_2$, 1 mM β-ME, 5% glycerol, plus one EDTA-free complete protease inhibitor tablet [Roche]) and lysed using an EmulsiFlex C5 cell homogenizer (Avestin). All subsequent steps are carried out at 4°. The cell lysate is cleared of intact cells and nuclei by low-speed centrifugation (500g, 15 min, J6 Beckman centrifuge). Membranes containing prenylated RhoA are harvest by centrifugation of the slow-speed supernatant at 150,000g for 45 min.

The pelleted membranes are resuspended in 20 ml of extraction buffer (extraction buffer: 20 mM HEPES, pH 8, 150 mM NaCl, 5 mM MgCl$_2$, 1 mM β-ME, 5% glycerol, 1% sodium cholate, and one EDTA-free complete protease inhibitor tablet) and homogenized using approximately 20 strokes with a glass/glass Dounce homogenizer. The protein concentration of the sample is determined using a Bradford assay, and the sample is diluted to a final protein concentration of 5 mg of protein/ml in extraction buffer. Membrane proteins are extracted by continuous stirring of the resuspended membrane/detergent mixture at 4° for 1 h. Solubilized membrane proteins are recovered by ultracentrifugation of the detergent extracted membranes at 100,000g in a Type 60Ti rotor for 1 h. The clarified supernatant is diluted threefold in lysis buffer to achieve a final sodium cholate concentration of approximately 0.3%. Dilution of the detergent sample is important, because sodium cholate is an ionic detergent and may interfere with protein binding to the Ni^{2+}–charged resin.

Metal Chelate Affinity Chromatography

A 1-ml HiTrap Chelating HP column is prepared as previously described. Using an FPLC, the diluted supernatant containing recombinant RhoA is loaded onto the column at a rate of 0.5 ml/min, maintaining a column pressure of <0.5 mPa. The column is washed with subsequent steps of 10 column volumes of buffer A (buffer A: 20 mM HEPES, pH 8, 150 mM NaCl, 5 mM $MgCl_2$, 1 mM β-ME, 0.3% sodium cholate, and 5% glycerol), and 10 column volumes of 5% buffer B (buffer B: buffer A + 1 M imidazole). Recombinant RhoA is eluted using a 50–750 mM imidazole gradient (5–75% buffer B) over 20 column volumes, collecting 1-ml fractions over the gradient. Absorbance measured at A_{280} is used to identify fractions containing recombinant protein, and the presence of RhoA is verified using 12.5% SDS-PAGE Coomassie-stained gels. Fractions containing the recombinant RhoA are pooled and concentrated using a 10,000 MWCO PES centrifugal filtering device (Vivascience).

The concentration of RhoA is determined by replicating the reconstitution assay conditions described later in the presence of $[^{35}S]$GTPγS. Approximately 50 pmol (estimated from Coomassie-stained SDS-PAGE gels) was reconstituted with PE:PtdIns(4,5)P$_2$-containing vesicles in a final volume of 60 μl (see following for assay conditions). The assay included a 10-fold molar excess of unlabeled GTPγS and approximately 100,000 cpm/assay of $[^{35}S]$GTPγS. The reaction is incubated at 25° for 30 min, allowing the intrinsic exchange of GTPγS for GDP on RhoA, and terminated by the addition of 4 ml of stop buffer (stop buffer: 20 mM HEPES, pH 8, 120 mM NaCl, and 25 mM $MgCl_2$). The amount of $[^{35}S]$GTPγS bound RhoA is quantified by filtering samples over 0.45-μm nitrocellulose filters and liquid scintillation counting of the nitrocellulose membranes.

PLC Assay

Prepare phospholipid vesicles by combining 30 nmol/assay L-α-phosphatidylethanolamine (PE), 3 nmol/assay PtdIns(4,5)P$_2$, and ~10,000 cpm/assay $[^3H]$PtdIns(4,5)P$_2$, in a 12 × 75 mm glass borosilicate tube and

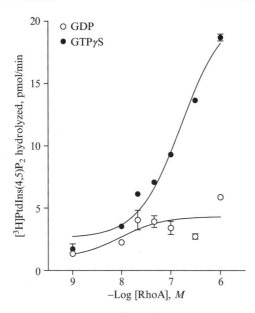

FIG. 3. Activation of phospholipase C-ε (PLC-ε) by RhoA in phospholipid vesicles. A representative concentration-effect curve showing the GTPγS-dependent activation of PLC-ε by RhoA. RhoA-containing mixed-detergent phospholipid vesicles were preincubated for 20 min at 25° in the presence of either 10 μM GDP (open circles) or 10 μM GTPγS (filled circles). Assays were initiated by the addition of 2 ng of PLC-ε and incubated for 10 min at 25°. Basal PLC-ε activity in the absence of RhoA (7 pmol/min) was subtracted to give the final values.

drying the lipids under a constant stream of nitrogen. [³H]PtdIns(4,5)P₂ is isolated from [³H]-inositol labeled turkey erythrocytes as previously described (Waldo *et al.*, 1991) or obtained from commercial sources. The dried lipids are resuspended in 20 μl of 20 mM HEPES, pH 7.4, per assay, and the mixture is probe-sonicated with approximately five 10-sec pulses. Once a homogeneous mixture of lipids is observed, the solution is vortexed and placed on ice.

Assays are performed in a final volume of 60 μl in 12 × 75 mm conical polypropylene tubes. Reaction mixtures are prepared in an ice-water bath and each assay includes: (1) 10 μl of 6× assay buffer (6× assay buffer: 120 mM HEPES, pH 7.4, 420 mM KCl, 12 mM DTT, 18 mM EGTA, and 17.1 mM CaCl₂ [this concentration of EGTA and CaCl₂ will yield a final free calcium concentration of 1 μM]); (2) 20 μl of phospholipid vesicles giving 30 nmol/assay (500 μM) of PE and 3 nmol/assay (50 μM) PtdIns(4,5)P₂; (3) 10 μl of 60 μM GDP or GTPγS; (4) 9 μl of 1 mg/ml fatty acid free bovine serum albumin (FAF-BSA); and (5) 10 μl of purified RhoA diluted in buffer A from the FPLC purification. The tubes are vortexed and incubated at 25° for 20 min to allow GDP or GTPγS to bind to RhoA. Reactions are placed on ice and cooled to 4° before the addition of PLC enzyme.

The amount of PLC enzyme used is determined empirically, because different purified fractions potentially exhibit varying specific activities. If 10,000 cpm of PtdIns(4,5)P$_2$ is used per assay, the amount of hydrolysis observed by PLC-ε in the absence of G protein should be approximately 100–200 cpm above background (usually $\sim$50 cpm). Under any given condition, the total substrate hydrolyzed should not exceed 30% of the starting [^{3}H]PtdIns(4,5)P$_2$. Given the specific activity of the purified fragment, 1–2 ng of purified protein usually is sufficient to achieve the desired levels of enzyme activity. PLC-ε is diluted to 1 ng/μl in 1 mg/ml FAF-BSA, and the assay is initiated by the addition of 1 μl of enzyme to each tube, which are then vortexed, placed in a 25° water bath, and incubated for 10 min. Final assay conditions are 500 μM PE, 50 μM PtdIns(4,5)P$_2$, 20 mM HEPES, pH 7.4, 70 mM KCl, 2 mM DTT, 3 mM EGTA, 2.85 mM CaCl$_2$, 0.16 mg/ml FAF-BSA, and 0.05% sodium cholate (final sodium cholate concentration is below the critical micelle concentration for sodium cholate).

The reaction is terminated by the addition of 200 μl of ice-cold 10% trichloroacetic acid and 100 μl of 10 mg/ml FAF-BSA, which will precipitate proteins and nonhydrolyzed lipids. The tubes are vortexed and centrifuged for 10 min at 3000g in a J6 swinging bucket centrifuge at 4°. This step separates the soluble [^{3}H]Ins(1,4,5)P$_3$ cleavage product from uncleaved [^{3}H]PtdIns(4,5)P$_2$. [^{3}H]Ins(1,4,5)P$_2$ is quantified by liquid scintillation counting of the reaction supernatant.

The specific activity of the phospholipid substrate in cpm/mol is determined by dividing the total cpm of [^{3}H]PtdIns(4,5)P$_2$ used per assay by the total number of moles of nonradioactive PtdIns(4,5)P$_2$ used per assay (labeled substrate is not included in the total pmol of PtdIns[4,5]P$_2$, because there is a substantial molar excess of unlabeled substrate). A concentration of 50 μM PtdIns(4,5)P$_2$ in a final assay volume of 60 μl gives 3000 pmol of substrate per assay. The specific activity is approximately 3.3 cpm/pmol if 10,000 cpm of [^{3}H]PtdIns(4,5)P$_2$ is used per assay. Dividing this number by the assay duration in minutes will give a final value of pmol/min.

Acknowledgments

    Supported by National Institutes of Health grants GM29536, GM57391, and GM65533. J. S. recognized support from the Pew Charitable Trusts.

References

Berridge, M. J., and Irvine, R. F. (1989). Inositol phosphates and cell signalling. *Nature* **341**, 197–205.

Evellin, S., Nolte, J., Tysack, K., Vom, D. F., Thiel, M., Weernink, P. A., Jakobs, K. H., Webb, E. J., Lomasney, J. W., and Schmidt, M. (2002). Stimulation of phospholipase C-ε by the M3 muscarinic acetylcholine receptor mediated by cyclic AMP and the GTPase Rap2B. *J. Biol. Chem.* **277**, 16805–16813.

Hains, M. D., Wing, M. R., Rogan, S. C., Siderovski, D. P., and Harden, T. K. (2005). Galpha12/13- and Rho-dependent activation of phospholipase C-ε by lysophosphatidic acid and thrombin receptors. *Submitted manuscript*.

Illenberger, D., Stephan, I., Gierschik, P., and Schwald, F. (2000). Stimulation of phospholipase C-β 2 by Rho GTPases. *Methods Enzymol.* **325**, 167–177.

Kapust, R. B., Tozser, J., Fox, J. D., Anderson, D. E., Cherry, S., Copeland, T. D., and Waugh, D. S. (2001). Tobacco etch virus protease: Mechanism of autolysis and rational design of stable mutants with wild-type catalytic proficiency. *Protein Eng.* **14**, 993–1000.

Kelley, G. G., Reks, S. E., Ondrako, J. M., and Smrcka, A. V. (2001). Phospholipase C(ε): A novel Ras effector. *EMBO J.* **20**, 743–754.

Lopez, I., Mak, E. C., Ding, J., Hamm, H. E., and Lomasney, J. W. (2001). A novel bifunctional phospholipase C that is regulated by Gα 12 and stimulates the Ras/mitogen-activated protein kinase pathway. *J. Biol. Chem.* **276**, 2758–2765.

Meisenhelder, J., Suh, P. G., Rhee, S. G., and Hunter, T. (1989). Phospholipase C-gamma is a substrate for the PDGF and EGF receptor protein-tyrosine kinases *in vivo* and *in vitro*. *Cell* **57**, 1109–1122.

Rhee, S. G. (2001). Regulation of phosphoinositide-specific phospholipase C. *Annu. Rev. Biochem.* **70**, 281–312.

Seifert, J. P., Wing, M. R., Snyder, J. T., Gershburg, S., Sondek, J., and Harden, T. K. (2004). RhoA activates purified phospholipase C-ε by a guanine nucleotide-dependent mechanism. *J. Biol. Chem.* **279**, 47992–47997.

Shibatohge, M., Kariya, K., Liao, Y., Hu, C. D., Watari, Y., Goshima, M., Shima, F., and Kataoka, T. (1998). Identification of PLC210, a Caenorhabditis elegans phospholipase C, as a putative effector of Ras. *J. Biol. Chem.* **273**, 6218–6222.

Smrcka, A. V., Hepler, J. R., Brown, K. O., and Sternweis, P. C. (1991). Regulation of poly-phosphoinositide-specific phospholipase C activity by purified Gq. *Science* **251**, 804–807.

Song, C., Hu, C. D., Masago, M., Kariyai, K., Yamawaki-Kataoka, Y., Shibatohge, M., Wu, D., Satoh, T., and Kataoka, T. (2001). Regulation of a novel human phospholipase C, PLCε, through membrane targeting by Ras. *J. Biol. Chem.* **276**, 2752–2757.

Taylor, S. J., Chae, H. Z., Rhee, S. G., and Exton, J. H. (1991). Activation of the β 1 isozyme of phospholipase C by α subunits of the Gq class of G proteins. *Nature* **350**, 516–518.

Wahl, M. I., Nishibe, S., Suh, P. G., Rhee, S. G., and Carpenter, G. (1989). Epidermal growth factor stimulates tyrosine phosphorylation of phospholipase C-II independently of receptor internalization and extracellular calcium. *Proc. Natl. Acad. Sci. USA* **86**, 1568–1572.

Waldo, G. L., Boyer, J. L., Morris, A. J., and Harden, T. K. (1991). Purification of an AlF4- and G-protein β γ-subunit-regulated phospholipase C-activating protein. *J. Biol. Chem.* **266**, 14217–14225.

Wing, M. R., Houston, D., Kelley, G. G., Der, C. J., Siderovski, D. P., and Harden, T. K. (2001). Activation of phospholipase C-ε by heterotrimeric G protein β γ-subunits. *J. Biol. Chem.* **276**, 48257–48261.

Wing, M. R., Snyder, J. T., Sondek, J., and Harden, T. K. (2003). Direct activation of phospholipase C-ε by Rho. *J. Biol. Chem.* **278**, 41253–41258.

[20]   Regulation of PLCβ Isoforms by Rac

By JASON T. SNYDER, MARK R. JEZYK, SVETLANA GERSHBURG,
T. KENDALL HARDEN, and JOHN SONDEK

Abstract

Small GTPases function as molecular switches, which transduce cellular signals from upstream regulators to downstream effectors in a guanine nucleotide–dependent manner. Direct binding partners of small GTPases fall into four classes of both regulators and effectors that can be differentiated on the basis of the state of nucleotide required for binding. Here we describe a procedure for the rapid screening and quantitative assessment of direct interactions of the Rho family of small GTPases with effector molecules of the phospholipase Cβ class of enzymes using surface plasmon resonance technology. The experimental format described is also readily adaptable toward characterizing guanine nucleotide–dependent binding events of both small and heterotrimeric G proteins with various classes of GTPase regulatory proteins.

Introduction

Rho-family GTPase members influence critical cellular events by their inherent ability to recognize various classes of regulatory and effector proteins depending on their bound guanine nucleotide (Hall, 1998). These four classes of direct binding partners can be delineated as guanine nucleotide exchange factors (GEFs), guanine nucleotide dissociation inhibitors (GDIs), GTPase accelerating proteins (GAPs), or downstream effectors. GEFs stimulate the release of GDP from a GTPase and promote subsequent binding of GTP, thereby activating the GTPase. GDIs directly bind the GDP-bound form of small GTPases, but in contrast to GEFs, they prevent dissociation of GDP, thereby sequestering the GTPase in its inactive state. Once activation of a GTPase occurs, it can then be inactivated by a GAP, which stabilizes the transition state of guanine nucleotide hydrolysis and acts to accelerate the enzymatic removal of the terminal phosphate of bound GTP. The state of a guanine nucleotide bound to a GTPase is kept under such strict control by the aforementioned classes of proteins to tightly regulate their ability to recognize downstream effectors. These effectors are bound to and stimulated by the activated GTP-bound form of a small GTPase and control a wide variety of cellular functions.

METHODS IN ENZYMOLOGY, VOL. 406   0076-6879/06 $35.00

The identification of the class of a prospective direct GTPase regulator can be accomplished by analysis of the nucleotide dependence of the GTPase-regulator interaction. We have found surface plasmon resonance (SPR) technology to be an excellent format for the rapid examination of a large set of GTPase-regulator interactions. Specifically, this chapter highlights the design of a protein–protein interaction screen to identify the Rac isoforms as unique among Rho GTPase family members in their ability to bind phospholipase Cβ (PLCβ) enzymes and stimulate their capacity to hydrolyze phosphatidylinositol 4,5-bisphosphate. Much of this work has been reported elsewhere (Kimple and Snyder, 2004; Snyder *et al.*, 2003).

Protein Production

Many G proteins receive posttranslational lipid modifications at either their amino or carboxy terminus to allow these molecules to partition to various cellular membranes (Chen and Manning, 2001). For the SPR assay described in the following, all recombinant proteins were produced lacking their known sites of lipidation. These subtle truncations allow for high-yield expression of soluble proteins and obviate the need for lipid vesicles or other membrane mimetics during assessment of direct protein–protein interactions using biophysical techniques.

Expression Constructs

The coding sequences for human PLCβ1, PLCβ2, residues 1–798 of PLCβ2, PLCβ3, Gαq (residues 6–350), and Gβ1 are amplified by PCR and inserted into pFastBacHTb (Invitrogen) by restriction-independent directional subcloning (Zeng, 1998) using the *Nco*1 and *Xho*1 restriction sites within the vector. Human Gγ1 cDNA representing residues 1–67 is amplified exactly as described previously but inserted into pFastBac1 between the *Bam*H1 and *Xho*1 restriction sites. DH10Bac *Escherichia coli* cells (Invitrogen) are used to generate bacmid DNA harboring each open reading frame exactly as described within the Bac-To-Bac Baculovirus Expression Systems manual (Invitrogen). Bacmid DNA is isolated according to the manufacturer's protocol. The construction of bacterial expression constructs engineered to produce hexahistidine-tagged pleckstrin homology (PH) domain (residues 1–144) and C-terminus (residues 871–1151) of PLCβ2, and Rho GTPase proteins from pProExHTb (Invitrogen) has been described (Singer *et al.*, 2002; Snyder *et al.*, 2003). For the Rho-family member BTB2, only the GTPase portion of the protein corresponding to amino acids 1–127 was created. Each Rho GTPase construct, with the exception of BTB2, is truncated at the site of C-terminal isoprenylation with a terminal cysteine-to-serine substitution.

Baculovirus Generation

Spodoptera frugiperda (Sf9) cells are cultured in spinner flasks at 27° using Graces Media (Gibco) supplemented with 10% fetal bovine serum (Gemini), 125 ng/ml Fungizone, 50 U/ ml penicillin, and 50 μg/ml streptomycin. In 6-well plates, 1×10^6 Sf9 cells from a stock of cells at 1.5×10^6 cells/ml are seeded in 2-ml Graces media. The wells are washed with 2 ml of Sf-900 II SFM (Gibco), followed by removal of the media, and overlaid with a mixture of bacmid DNA and CELLFECTIN reagent (Invitrogen) exactly as described by the manufacturer. After 5 h, the transfection mixtures are removed by aspiration, and the wells are overlaid with 2.5 ml supplemented Graces media. At 5 days after transfection, the media are collected and used to infect spinner flasks containing Sf9 cells growing in log-phase at a ratio of 1 ml transfection liquid to 100 ml cells. At 4 days after infection, the baculovirus supernatant is collected by removing the cell pellet by centrifugation and stored at 4°.

Protein Expression

A High Five insect cell stock is maintained in Express Five SFM (Gibco) media supplemented with 10 mM L-glutamine (Gibco) in a 100-ml spinner flask at 27°. For recombinant protein expression, High Five cells are gradually expanded to 1 l in a 4-l shaker flask, shaking at 140 rpm, by splitting the cells to 1×10^6 cells/ml. One liter of High Five cells at a density of 2×10^6 cells/ml is infected with 10 ml baculovirus stock at an approximate multiplicity of infection (MOI) of 2:1, and the cells are harvested 2 days after infection by centrifugation. For production of recombinant Gβ1γ1 dimer, High Five cells are coinfected with 10 ml of each baculovirus stock maintaining the MOI of 2:1 for each construct.

The PH domain and C-terminus of PLCβ2 and Rho GTPase pProExHT expression constructs are used to transform BL21(DE3) *E. coli*, and cells are grown at 37° for 12 h in Luria broth (LB) with 100 μg/ml ampicillin. These starter cultures are used to inoculate 1-l cultures of LB in the presence of the selective antibiotic, ampicillin, in 4-l baffled flasks (Bellco), shaken at 230 rpm, at 37°. When an optical density of $\lambda600 = 1.0$ is reached, cultures are cooled to 27°, and IPTG is added to 1 mM final concentration. Cells are harvested 10 h after induction by centrifugation.

Protein Purification

All protein purification steps are conducted at 4°. The cell pellet from 1 l of culture is resuspended in 40 ml of N1 buffer: 20 mM Tris (pH = 8), 200 mM NaCl, 10% glycerol, and 10 mM imidazole. The cell slurry is passed through an Emulsi-Flex C5 (Avestin) at a pressure of 15,000 psi, and the resulting cell lysate is fractionated by centrifugation using a Ti70 rotor (Beckman Coulter) in a Beckman L8-M ultracentrifuge at 55,000 rpm for 30 min. The PH domain of the PLCβ2 sample is extracted from the pellet fraction of *E. coli* cells, renatured, and purified as previously described in detail (Snyder *et al.*, 2003). For all other preparations, the supernatant is decanted and applied to a 5-ml HiTrap Chelating Sepharose column (Amersham) that is pre-equilibrated with 100 mM NiCl$_2$. After loading 20 bed volumes of the N1 buffer, a gradient of solution N1 supplemented with 1 M imidazole is applied to the column, and the recombinant hex-ahistidine-tagged protein is collected from the 400 mM imidazole fraction. For the proteins expressed in insect cells, the fractions rich in the desired recombinant protein are next applied to a HiPrep 26/60 Sephacryl-200 (Amersham) gel filtration column preequilibrated in 20 mM Tris (pH = 8), 200 mM NaCl, 10% glycerol, and 1 mM DTT. Fractions containing monomeric (or heterodimeric Gβ1γ1) proteins are pooled and analyzed by SDS-PAGE (Fig. 1.). All purified proteins are concentrated to >1 mg/ml and stored at $-80°$.

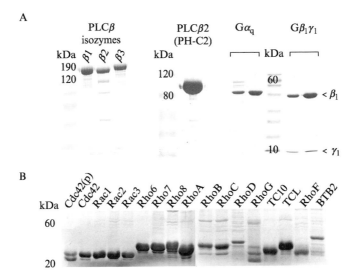

FIG. 1. Purified recombinant PLCβ isoforms, heterotrimeric G proteins, and Rho GTPases. The recombinant proteins used for the surface plasmon resonance assay were expressed in either insect (A) or bacteria (B) host cells, purified from the soluble supernatant fraction of the cell lysates, analyzed by SDS-PAGE, followed by Coomassie blue staining. Molecular weight standards are indicated. Cdc42(p) is the placental isoform of Cdc42. Portions of this figure are reproduced with permission (Snyder *et al.*, 2003).

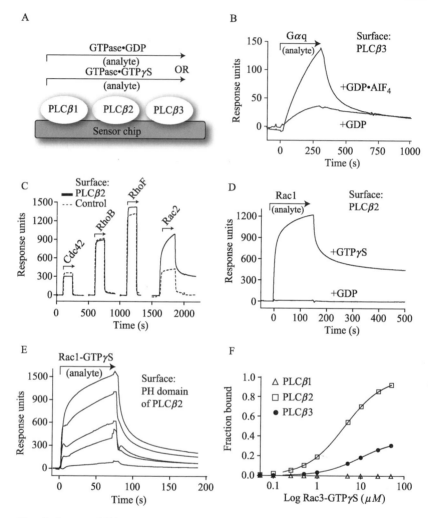

FIG. 2. Screen of Rho GTPase-PLCβ interactions using surface plasmon resonance. (A) Schematic of the surface plasmon resonance protein–protein interaction assay. PLCβ ligands are immobilized on individual flow cells of a CM5 sensor chip (Biacore), and GTPase analytes, loaded with GDP, GDP + AlF4, or GTPγS, are injected over the sensor chip. (B) Sensorgram displaying the signal in response units from injections of Gαq-GDP, in the presence or absence of AlCl3 and NaF, over the PLCβ3 flow cell as a function of time. The injection of analyte protein (association phase) is labeled with an arrow and is followed by injection of running buffer (dissociation phase). Sensorgrams are normalized to the signal achieved by injection of the analyte over a blank, control flow cell. (C) Raw sensorgrams of a representative set of injections of the GTPγS-loaded Rho GTPases over both the PLCβ2 and control flow cells. (D) Sensorgrams of data normalized as described for (B) from injections of

Surface Plasmon Resonance Screen

We have found that effective measurement of GTPase affinity toward macromolecules is readily accomplished using Biacore technology. Potential GTPase binding ligands are immobilized to individual flow cells on a Biacore sensor chip, and GTPase analytes are individually injected over the sensor chip surfaces (Fig. 2A). This format permits assessment of binding affinity for each ligand under conditions that mimic various guanine nucleotide states: GDP (inactive state), GDP + AlF4 (transition state), or GTPγS (active state). In addition, the running buffer can be easily supplemented with Mg^{2+} to stabilize bound nucleotide or, conversely, EDTA to promote nucleotide release and recognition of the GTPase by guanine nucleotide exchange factors (Rossman *et al.*, 2002).

Preparation of Analytes

Rho GTPase, Gαq, and Gβ1γ1 samples are diluted to 100 μM in 1 ml, and dialyzed against 20 mM Tris (pH = 8), 150 mM NaCl for 12 h at 4°. For the Rho GTPases, after dialysis each sample receives 2 mM EDTA and is then split into two 500-μl aliquots at room temperature, when 1 mM GDP and 1 mM GTPγS (Sigma) are added to the aliquots, respectively, for 30 min. Each Rho GTPase sample is then dialyzed against running buffer: 20 mM Tris (pH = 8), 150 mM NaCl, 10 mM MgCl$_2$ for 12 h at 4°. Gαq exhibits hindered capacity to load free guanine nucleotide (Chidiac *et al.*, 1999). Therefore, to mimic the activated GTP-bound form, each Gαq dilution used for injection into the Biacore is first supplemented with 1 mM NaF, 50 μM AlCl3, and 10 mM MgCl$_2$, whereas the inactive GDP-bound sample is only given 10 mM MgCl$_2$.

Immobilization of PLCβ Ligands on Sensor Surfaces

Several techniques for immobilizing GTPase regulatory proteins to Biacore sensor chips exist. We have successfully characterized GTPase binding events with ligands attached by direct covalent modification (CM5 chip) (Snyder *et al.*, 2003), GST-tagged protein bound to covalently

---

Rac1 loaded with either GDP or GTPγS over the PLCβ2 flow cell. (E) A titration experiment of a range of injections (0.1–50 μM) of GTPγS-loaded Rac1 over the PH domain of PLCβ2 surface. (F) Plot of the steady-state equilibrium-binding signal from the GTPγS-loaded Rac3 titrations over the PLCβ1, PLCβ2, and PLCβ3 surfaces. These binding isotherms were used to estimate dissociation constants (Kd) for each titration. This figure is reproduced with permission (Snyder *et al.*, 2003).

immobilized anti-GST antibody (CM5 chip) (Kimple and Snyder, 2004), hexahistidine-tagged protein bound to immobilized Ni^{2+} (NTA chip) (Rossman et al., 2002), and biotinylated protein bound to immobilized streptavidin (SA chip) (unpublished data). Specifically, for analyzing Rac GTPase-PLCβ interactions, each tested PLCβ protein retains binding functions when directly immobilized to a CM5 sensor chip.

Before creating new surfaces, the unclogging and desorb maintenance routines are executed exactly as described by Biacore to ensure high-quality data collection. After cleaning, the covalent amine coupling chemistry approach of ligand immobilization onto flow cells within a CM5 sensor chip (research grade) (Biacore) using a Biacore 3000 is performed exactly as described by the manufacturer. Purified PLCβ samples are diluted to 100 μg/ml in running buffer. All surfaces are activated with a multichannel injection of the N-hydroxysuccinimide/1-ethyl-3-(3-dimethylaminopropyl) carbodiimide solution. PLCβ samples are injected at 5 μl/min through individual channels until the desired surfaces are created. For screening potential GTPase binding targets, high surface levels >1000 response units (RU) are created, whereas for titration experiments of detected binding events, lower levels of ligand are preferred (50–200 RU). A single control flow cell is activated but does not receive any ligand. After immobilization, all surfaces are then deactivated with a single injection of 1 M ethanolamine.

Injection of Analytes

For screening a large set of potential GTPase–GTPase regulator binding events, all analytes are diluted in running buffer to 10 μM in 150-μl samples in 7-mm plastic vials and placed in the Thermo-rack of the Biacore. A script delineating the injection format of each GTPase sample is next created using the Biacore 3000 methods language. Kinetic injections (Kinject) of 25 μl with a dissociation time of 200 sec are created for each analyte overflow path 1–2–3–4 at a rate of 25 μl/ min. Each analyte is injected in duplicate.

Regeneration of Sensor Surfaces

For the Rac GTPase–PLCβ binding experiments, no regeneration injections are used between Kinject steps. Both GDP, and GTPγS–loaded GTPase samples do not present significant accumulation of response units throughout the set of injections. However, in general, regeneration injections of 50 mM EDTA can destabilize lingering GTPases from the flow cell surfaces immobilized with GTPase effectors. As the PLCβ proteins are covalently attached to the CM5 chip, harsh regeneration treatments such as

NaOH and SDS are avoided to maintain the integrity of the immobilized proteins.

Data Analysis

Resulting data from the GTPase injections are analyzed using BIAevaluation software. Raw data are normalized to the signal achieved from injections over the blank, control surface. As expected for a known stimulator of PLCβ isoforms, Gαq displays nucleotide-dependent binding to the PLCβ surfaces (Fig. 2B). Importantly, of the 17 tested Rho-family GTPases, only the Rac isoforms (Rac1, Rac2, and Rac3) show affinity for the PLCβ surfaces over the control surfaces (Fig. 2C). This binding is only observed with the GTPγS-loaded Rac GTPases, because the GDP-bound samples display no significant affinity for any PLCβ fragment (Fig. 2D). In addition, these activated Rac GTPase proteins bind to the PH-C2 and PH domain fragments of PLCβ2, but not the C-terminal PLCβ2 fragment (Snyder *et al.*, 2003) (Fig. 2E).

To quantify these interactions, a titration series of the Rac-GTPγS samples ranging in concentrations of 100 nM–50 μM are injected over the PLCβ surfaces. Data are aligned and normalized to the control surface sensorgrams. The set of titrations from each binding event are aligned, and thermodynamic parameters are estimated by use of curve fitting with global analysis (Roden and Myszka, 1996). The steady-state equilibrium signal achieved for each injection is plotted as a function of Rac GTPase concentration, and dissociation constants (Kd) are estimated for each binding isotherm using a 1:1 binding model. The measured interactions display a saturable binding signal, with estimated Kd values in the low micromolar range (Fig. 2F) (Snyder *et al.*, 2003). In general, strong binding interactions with Kd values in the nanomolar range or tighter are more suited for kinetic analysis, whereas data generated for weaker interactions such as the Rac GTPase–PLCβ binding events are more difficult to interpret from the ratio of estimated association and dissociation rate constants.

Discussion

Biacore technology using surface plasmon resonance is very useful for evaluating the specificity of GTPase targets and selectivity of the guanine nucleotide state of a potential GTPase-binding protein. A large collection of purified GTPases can be rapidly auditioned for interactions with several ligands simultaneously. We have discovered that only the Rac GTPase members of the Rho-family of GTPases recognize PLCβ isoforms, and these interactions are specific to the activated form of Rac. Furthermore, we have delineated the binding surface for Rac GTPases on the multi-domain PLCβ proteins to the N-terminal PH domain. Detecting other GTPase-regulatory protein interactions may require exploration of various immobilization strategies and buffer conditions to achieve acceptable binding signal (Van Regenmortel, 2003). We note that although this set of assays focused on the Rho family of small GTPases and one class of effectors, the assay format can be readily expanded and tailored to the identification of binding partners such as GEFs, GDIs, and GAPs of any guanine nucleotide–binding protein including those of the Ras and heterotrimeric G protein families.

References

Chen, C. A., and Manning, D. R. (2001). Regulation of G proteins by covalent modification. *Oncogene* **20**, 1643–1652.

Chidiac, P., Markin, V. S., and Ross, E. M. (1999). Kinetic control of guanine nucleotide binding to soluble G alpha (q). *Biochem. Pharmacol.* **58**, 39–48.

Hall, A. (1998). Rho GTPases and the actin cytoskeleton. *Science* **279**, 509–514.

Kimple, R. J., and Snyder, J. T. (2004). Screening direct interactions with G proteins by surface plasmon resonance. *Biacore J.* **4**, 20–23.

Roden, L. D., and Myszka, D. G. (1996). Global analysis of a macromolecular interaction measured on BIAcore. *Biochem. Biophys. Res. Commun.* **225**, 1073–1077.

Rossman, K. L., Worthylake, D. K., Snyder, J. T., Cheng, L., Whitehead, I. P., and Sondek, J. (2002). Functional analysis of cdc42 residues required for guanine nucleotide exchange. *J. Biol. Chem.* **277**, 50893–50898.

Singer, A. U., Waldo, G. L., Harden, T. K., and Sondek, J. (2002). A unique fold of phospholipase C-β mediates dimerization and interaction with G alpha-q. *Nat. Struct. Biol.* **9**, 32–36.

Snyder, J. T., Singer, A. U., Wing, M. R., Harden, T. K., and Sondek, J. (2003). The pleckstrin homology domain of phospholipase C-β2 as an effector site for Rac. *J. Biol. Chem.* **278**, 21099–21104.

Van Regenmortel, M. H. (2003). Improving the quality of BIACORE-based affinity measurements. *Dev. Biol. (Basel)* **112**, 141–151.

Zeng, G. (1998). Sticky-end PCR: New method for subcloning. *Biotechniques* **25**, 206–208.

[21]   Biochemical Properties and Inhibitors of (N-)WASP

By Daisy W. Leung, David M. Morgan, and Michael K. Rosen

Abstract

The Wiskott–Aldrich syndrome protein (WASP) is an effector of the Rho GTPase Cdc42 and a key component of signaling pathways that regulate the actin cytoskeleton. WASP is regulated by a number of ligands, and the mechanisms by which these act are beginning to be understood through detailed biochemical analyses. Here we describe the protocols we use to study WASP proteins, including the methods we use to purify signaling components and the assays we use to quantitatively characterize the biochemical and biophysical properties of WASP, its activation by Cdc42, and its inhibition by the small molecule wiskostatin. These methods have broad use within the WASP-related cytoskeletal-signaling pathway but are also applicable to investigations of other intramolecular and intermolecular interactions.

Introduction

The Wiskott–Aldrich syndrome protein (WASP) and its homolog N-WASP are autoinhibited effectors of the Rho family GTPase Cdc42. WASP regulates cytoskeletal architecture and dynamics through stimulating the activity of the actin-nucleating machine, Arp2/3 complex. The activity of WASP toward Arp2/3 complex can be activated by a variety of upstream ligands, including Cdc42, PIP2, Nck, Grb2, Lck, and Toca-1 (Benesch *et al.*, 2002; Higgs and Pollard, 2000; Ho *et al.*, 2004; Prehoda *et al.*, 2000; Rohatgi *et al.*, 1999, 2001; Scaplehorn *et al.*, 2002; Torres and Rosen, 2003; Yarar *et al.*, 1999). Many of these binding partners can act cooperatively to relieve WASP autoinhibition. Biophysical and structural studies indicate that Cdc42 and PIP2 act through biasing a two-state autoinhibitory equilibrium in WASP and N-WASP toward the active state (Buck *et al.*, 2004; Leung and Rosen, 2005; Prehoda *et al.*, 2000). The mechanisms of other activators are not well understood. WIP and the small molecule wiskostatin can inhibit the activity of N-WASP (Peterson *et al.*, 2004; Ramesh *et al.*, 1997; Volkman *et al.*, 2002). WIP has been shown to bind to the N-terminus of N-WASP, although the mechanism by which it controls activity is unknown. Wiskostatin selectively stabilizes the autoinhibited N-WASP conformation and prevents its allosteric activation. Quantitative measurements of the intramolecular and intermolecular

METHODS IN ENZYMOLOGY, VOL. 406   0076-6879/06 $35.00
Copyright 2006, Elsevier Inc. All rights reserved.   DOI: 10.1016/S0076-6879(06)06021-6

interactions that control WASP activity are critical to understanding how WASP is regulated and can be differentially and cooperatively activated by various signals. Understanding of the biochemical properties of WASP *in vitro* can provide a valuable segue toward the understanding of WASP function *in vivo*.

Purification of Proteins

The assays we describe in the following sections require highly purified protein reagents; we and others have found that some, particularly the pyrene-actin polymerization assay, are quite sensitive to the presence of contaminants, which adversely affect the reproducibility of results between trials and accuracy of any quantitative measurement. The buffers listed in the following have also been optimized for purification and, in some cases, for increased sensitivity in the respective assays. The chromatography columns we use are purchased from Amersham/GE Healthcare, unless stated otherwise. For spectroscopic assays, buffers are degassed to decrease noise and artifacts that arise from outgassing-induced light scattering. In all cases, multiple proteins used in the same assay are dialyzed together in the same buffer the night before to maximally normalize buffer conditions.

Buffers

A. 20 mM Tris, pH 8, 50 mM NaCl, 1 mM EDTA, 2 mM DTT, 1 mM benzamidine, 1 μg/ml pepstatin, 1 μg/ml leupeptin, 1 μg/ml antipain

B. 20 mM Tris, pH 8, 50 mM NaCl, 1 mM EDTA, and 2 mM DTT

C. 20 mM Tris, pH 8, 1 M NaCl, 1 mM EDTA, and 2 mM DTT

D. 25 mM sodium phosphate, pH 7, 150 mM NaCl, 1 mM EDTA, and 2 mM DTT

E. 20 mM Tris, pH 8, 1 mM EDTA, 2 mM DTT, 1mM benzamidine, 1 μg/ml pepstatin, 1 μg/ml leupeptin, 1 μg/ml antipain

F. 50 mM HEPES, pH 7.25, 500 mM KCl, 5 mM β-mercaptoethanol, 0.01% IGEPAL, 5% glycerol, 1 mM benzamidine, 1 μg/ml pepstatin, 1 μg/ml leupeptin, 1 μg/ml antipain

G. 500 mM imidazole, 500 mM KCl, 25 mM sodium phosphate, and 5 mM β-mercaptoethanol

H. 25 mM sodium phosphate, pH 6, 50 mM NaCl, 1 mM EDTA, 2 mM DTT

I. 25 mM sodium phosphate, pH 6, 1 M NaCl, 1 mM EDTA, 2 mM DTT

J. 20 mM Tris, pH 8.5, 1 mM EDTA, 2 mM DTT

K. 20 mM Tris, pH 8.5, 1 M NaCl, 1 mM EDTA, 2 mM DTT

L. 20 mM Tris, pH 8, 1 M NaCl, 2 mM DTT, 1 mM EDTA

M. 20 mM Tris, pH 7.5, 100 mM NaCl, 2 mM DTT, and 1 mM EDTA

N. 25 mM sodium phosphate, pH 7, 150 mM NaCl, 2 mM MgCl$_2$, 2 mM DTT

Purification of Human WASP Proteins

We have purified a series of human WASP proteins of varying lengths that contain different domains and mutations. Full-length WASP contains an N-terminal Ena/VASP homology domain 1 (EVH1), a basic region (B), a GTPase binding domain (GBD), a polyproline linker region (P), and a verprolin homology/central domain/acidic region (collectively termed VCA) at the C-terminus (Fig. 1). The most stable of these WASP proteins, GBD-C, is composed of a truncated GBD tethered to the C region by a GGSGGS linker and forms the minimum unit required for autoinhibition (Kim *et al.*, 2000). Expression and purification of GBD-C is relatively straightforward. However, we have found that addition of other domains that are primarily unstructured, including the polyproline linker region, the verprolin homology domain (V), and acidic region (A), results in more difficult bacterial expression and/or purification, because these domains are generally prone to proteolysis and degradation.

Purification of GBD-C

Buffers with: A, B, C, D

OVERVIEW: DEAE, MONO Q,SD75. The GBD-C (residues 242–310, GGSGGS linker, residues 461–492) protein is expressed from the Novagen pET11a vector in *E. coli* BL21(DE3) cells. Transformed cells are cultured in shaker flasks at 37° until OD$_{600}$ = 1.0 and induced with 1 mM IPTG for 3 h. Cells are harvested, resuspended in 20 ml of cold buffer A per 1 l of culture, and stored at −80°. Cells from 1 l growth are lysed by sonication

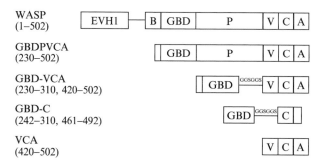

FIG. 1. Domain organization of WASP constructs.

and clarified by centrifugation at 39,000g for 30 min at 4°. The supernatant is loaded onto a DEAE column (50 ml, XK 26/40 column) (6 column volumes [CV] buffer B to 30% buffer C). Peak fractions as monitored by A_{280} and confirmed by SDS-PAGE are pooled and diluted to ~2 mS/cm with 20 mM Tris, pH 8. The protein is then loaded onto a Mono Q 10/100 column (12 CV buffer B to 30% buffer C). Peak fractions are loaded onto an SD75 26/60 gel filtration column (buffer D) as a final purification step. GBD-C is flash frozen in liquid N_2 and stored at −80°. Yield is ~20 mg/l of culture.

Purification of VCA

Buffers: E, C, B, D

OVERVIEW: DEAE, GLUTATHIONE SEPHAROSE, THROMBIN CLEAVAGE, MONO Q, SD75. The isolated VCA domain (residues 420–502) is difficult to purify, because it degrades easily, especially in the acidic region. Furthermore, the high acidic content carries many impurities that are difficult to separate. To aid expression and purification, we express VCA as a GST fusion protein from a pGEX-2T vector in *E. coli* BL21(DE3) cells. Cells are cultured at 37° and induced at OD_{600} 0.6-0.8 with 1 mM IPTG for 3 h. Cultures are harvested, resuspended in 20 ml cold buffer E per liter of culture, and stored at −80°. Cells from 1 l culture are lysed by sonication and clarified by centrifugation at 39,000g for 30 min at 4°. Lysate supernatant is first passed through a DEAE column (50 ml, in XK 26/40 column; 8 CV buffer E to 40% buffer C, 5 CV to 40% buffer C) before peak fractions are collected and incubated in batch with 4 ml of glutathione sepharose beads for 30 min at 4°. We have found that adding a DEAE step before glutathione sepharose beads eliminates impurities that are not visible by SDS-PAGE but that cause VCA degradation over time and inconsistencies in quantitation and assays. The bead suspension is then transferred to a 20-ml disposable column (Bio-Rad), and the beads are washed with 3 CV of buffer B. The beads are resuspended in 20 ml buffer B, and 200 units of thrombin protease (Amersham/GE Healthcare) is added. The suspension is nutated at 25° for at least 2 h until cleavage is complete as monitored by SDS-PAGE. VCA is eluted from the beads with 12 ml of 30 mM glutathione, pH 8, diluted to conductivity ~2 mS/cm with buffer E, and loaded onto Mono Q 10/100 (0.5 CV buffer E to 35% buffer C, 20 CV to 50% buffer C). We have observed that most degradation of VCA occurs at its C terminus. However, the Mono Q column efficiently separates full-length VCA from its degradation products, which elute earlier, because they have lost acidic C-terminal residues. This is a critical step, because it eliminates any degradation products that are lacking a full acidic region, including residue Trp500, which are inactive. Finally, the peak fractions from Mono Q are loaded onto an SD75 26/60 gel filtration column (buffer D). VCA is flash frozen and stored at $-80°$. Yields are ~1 mg of VCA per liter of culture.

Purification of GBD-VCA

Buffers: F, G, H, I, J, K, D

OVERVIEW: NI-NTA, DESALTING, MONO S, TEV CLEAVAGE, Mono S, Mono Q, SD75. Expression and purification of GBD-VCA (residues 230–310, GGSGGS linker, 420–502) has the same issue of C-terminal degradation as VCA. To facilitate expression and purification, GBD-VCA was expressed as a $(His)_6$-fusion from a Novagen pET-15b vector in *E. coli* BL21(DE3) that was modified to include a Tev protease site C-terminal to the $(His)_6$ tag. We have found that Tev protease produces fewer nonspecific cleavage products compared with thrombin and that a $(His)_6$-tag is easier to separate from GBD-VCA than a GST-tag after cleavage. Transformed cells are cultured at 37° and induced at OD_{600} 0.6–0.8 with 1 mM IPTG for 3 h. Cells are harvested, resuspended in 20 ml of cold buffer F per liter of cells, and stored at $-80°$. Cells from a 6 l culture are lysed by sonication and clarified by centrifugation at $39,000g$ for 30 min at 4°. Lysate supernatant is applied to Ni-NTA beads (8 ml bead volume/6 l cell culture) and incubated in batch at 4° for 30 min. Ni-NTA bead suspension is transferred to a disposable column and washed with additional lysis buffer before elution with 7.5 ml of buffer G. Eluate is desalted into buffer H using a High Prep 26/10 desalting column before loading onto a Mono S 10/10 column (15 CV buffer H to 35% buffer I). Peak fractions are pooled, and concentrated to ~2 ml using Amicon Ultra 15 (10,000 MWCO). The pH is adjusted to 8 with Na_2HPO_4 before addition of His-Tev protease (Invitrogen) for incubation at 25° until the cleavage reaction is complete as monitored by SDS-PAGE. The reaction mixture is loaded onto a Mono S column as previously to eliminate any uncleaved protein, His-tag, and His-Tev protease. Flow-through fractions containing GBD-VCA are collected and applied onto a Mono Q 10/100, where full-length GBD-VCA is resolved from proteins that are degraded at the C terminus (15 CV buffer J from 20–60% buffer K). Peak fractions are pooled and loaded onto a SD75 26/60 gel filtration column as a final purification and buffer exchange step (buffer D). GBD-VCA is flash frozen and stored at $-80°$. Yields are ~7 mg of GBD-VCA per 6 l of culture.

Purification of GBD-P-VCA

Buffers: F, G, L, M, N

OVERVIEW: NI-NTA, MONO Q, SD75. The WASP protein we use that most closely represents full-length WASP is GBD-P-VCA (residues 230–502), which lacks only the N-terminal EVH1 domain (Torres and

Rosen, 2003). GBD-P-VCA expresses very poorly in *E. coli* and is easily degraded during purification. We have found that the expression protocol and the lysis step we have optimized below are extremely important to obtaining full-length protein in practical yields. GBD-P-VCA is expressed from the same vector as GBD-VCA (see earlier) in BL21(DE3) Codon$^+$ cells (Novagen). Transformed cells are cultured at 37° to $OD_{600} = 0.6$, cooled on ice to ~20°, and induced with 0.5 mM IPTG at 20° for 16 h. Harvested cells are resuspended in 20 ml of cold buffer F per liter of culture and stored at −80°. Cells from 12 l of culture are lysed by cycles of flash freezing and thawing (repeated three times); lysis by sonication produces significantly lower yields. Lysate is clarified by centrifugation at 39,000g for 30 min at 4°, and supernatant is incubated with Ni-NTA beads at 4° for 20 min. Beads are washed with additional buffer J before elution with buffer G. Eluate is diluted 10-fold with buffer J and loaded onto a Mono Q 10/100 column with buffer J. GBD-P-VCA typically elutes between 30 and 40% buffer L. Fractions containing GBD-P-VCA are pooled, concentrated, and loaded onto a SD75 26/60 gel filtration column (buffer M). GBD-P-VCA purified in this fashion can be stored on ice for several weeks without degradation. Yields are ~2 mg/ 12 l of culture.

Purification of Cdc42t (-GDP, -GMPPNP, -mantGMPPNP)

Studies in our laboratory, including fluorescence binding and actin polymerization assays, are conducted using a Cdc42 construct lacking the C-terminal CAAX box (Cdc42t) (residues 1–179). The purification and loading of Cdc42t with nucleotide have been discussed in detail previously (Manor, 2000).

Purification of Arp2/3 Complex, Actin, and Pyrene-Labeled Actin

Although a number of purification protocols for Arp2/3 complex exist, we find that we consistently obtain high quality and quantity of the assembly from bovine thymus using the protocol outlined by Higgs *et al.* (1999). Actin and pyrene-labeled actin from rabbit skeletal muscle are obtained using protocols as described previously (Cooper and Pollard, 1982).

Handling of Wiskostatin

Wiskostatin (Fig. 2) was discovered using a forward chemical genetic screen developed to identify inhibitors of actin assembly (Peterson *et al.*, 2001, 2004; Specht and Shokat, 2002). It was found in the DiverSet E library supplied by Chembridge (San Diego, CA) and is now commercially available through EMD Biosciences under the Calbiochem brand name.

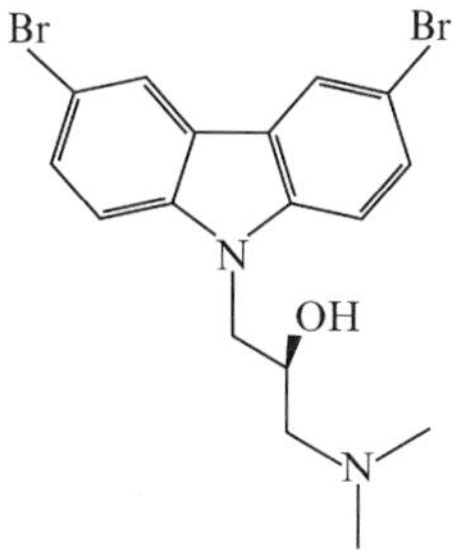

FIG. 2. (S-) Wiskostatin, (2s)-1-(3,6-dibromo-9h-carbazol-9-yl)-3-(dimethylamino) propan-2-ol.

Wiskostatin is a derivative of carbazole. The *Merck Index* describes the parent compound of wiskostatin as "an extremely weak base" that is "insoluble in water" and that "exhibits strong fluorescence and long phosphorescence on exposure to ultraviolet light." Wiskostatin shares these photophysical properties with its parent compound but is more complex and more soluble. Its carbazole backbone is symmetrically brominated at the 3- and 6-positions, has an N-propyl side chain with a hydroxyl group at the 2-position, and has a dimethylamino function at the 3-position. Initial biochemical studies largely used racemic wiskostatin, but biochemical analyses of the pure (S-) enantiomer have suggested that its stereochemistry is not important to its inhibitory function (Peterson *et al.*, 2004). However, structure–function analyses have revealed other critical features. Halogenation of the ring system seems to be a requirement: the dichlorinated derivative is only slightly less active than the dibrominated derivative, but the unhalogenated analog is inactive. The tertiary amino function is also important for activity: an analog containing a hydroxyl substituent in place of the tertiary amino function is also inactive, although a variety of amines are fully active.

The solubility of wiskostatin in aqueous buffers was not exhaustively tested, but it exceeds that of its parent compound. Although its propyl side chain is polar, only its tertiary amino group may carry charge. Therefore, the compound was found to be reasonably soluble in the pH range of 5–8. Wiskostatin degrades over time in 0.1% trifluoroacetic acid solution (pH ∼2).

To facilitate biochemical and biophysical experiments with wiskostatin, a cosolvent is introduced, and solution pH is maintained in the range 5.5–7.5. Although wiskostatin is soluble in a variety of organic solvents, including acetone and ethanol, our studies use DMSO, in which wiskostatin can be stably dissolved at concentrations in excess of 100 m*M*. Wiskostatin stocks are prepared in neat DMSO at concentrations slightly greater than 100 mM ($\varepsilon_{348} = 3501\ M^{-1}\ cm^{-1}$ as determined gravimetrically). For experimental solutions, wiskostatin is added to pH 5.5–pH 7.5 buffer already containing DMSO, such that the final concentration of DMSO on addition of wiskostatin is 5% (v/v). Control experiments indicate that neither WASP structure, nor WASP-mediated actin assembly kinetics are affected by the presence of this quantity of DMSO. In our standard NMR buffer conditions (5% DMSO, 20 mM MES buffer, pH 5.5, 50 mM NaCl, 2 mM MgCl$_2$, 1 mM DTT) and 100 μM WASP GBD (residues 230–310), wiskostatin is soluble at concentrations up to $\sim$ 7 mM. This seems to be the maximum solubility of the compound; at higher pH values, wiskostatin cannot be maintained in solution at these concentrations with or without protein.

Fluorescence studies using Cdc42 loaded with the mant nucleotide derivative, mantGMPPNP, in the presence of wiskostatin are complicated by the fact that wiskostatin and mant have overlapping excitation and emission bands. In the buffer conditions described previously, fluorescence spectra of wiskostatin exhibit excitation maxima at 298 and 360 nm and an emission maximum at 395 nm. Under these conditions, mantGMPPNP bound to Cdc42 exhibits excitation bands at 260, 280, and 365 nm and an emission band centered at 444 nm. The fluorescence of wiskostatin is much weaker than that of mant nucleotides, but its phosphorescence can pose severe complications. At appropriately rapid scan rates (>1 nm/sec), it is possible to take absorbance and fluorescence spectra in the presence of wiskostatin that are relatively uncomplicated by its phosphorescence. However, the phosphorescence quickly overwhelms other signals during single wavelength measurements, in which signal averaging may occur over periods of many seconds. Optical measurements in the presence of wiskostatin are facilitated by the use of fluorophores with distinct spectral properties, including tetramethylrhodamine, pyrene, and coumarin derivatives (Molecular Probes).

Assays

Guanidine Hydrochloride or Urea Denaturation of WASP Proteins

The stability of WASP against unfolding can be determined by titration of urea or guanidine hydrochloride (GdnHCl), which can be followed by CD spectroscopy. Both denaturants give similar estimates of WASP stability (Leung and Rosen, 2005).

We typically perform chemical denaturant titrations at 25° on a 10 μM WASP protein sample. The titrant is either 8 M urea or 6 M GdnHCl. To maintain a constant WASP concentration during titration, 10 μM WASP is added to the titrant. Both sample and titrant are allowed to equilibrate at room temperature for 30 min before the experiment is begun. The CD signal of the protein is then monitored on each addition of denaturant at 222 nm or 225 nm for urea and guanidine hydrochloride, respectively (guanidine hydrochloride absorbs at 222 nm), making sure that the sample equilibrates between additions.

We fit the titration data with the six-parameter equation described by C. N. Pace, which takes into account errors in fitting in the pretransition and posttransition baselines (Pace, 1990):

$$y = \frac{[(y_f + m_f[D]) + (y_u + m_u[D])(\exp - (\Delta G/RT - m[D]/RT))]}{[1 + \exp - (\Delta G/RT - m[D]/RT)]} \quad (1)$$

where y_f and y_u are the slopes of the pretransition and posttransition baselines, m_f and m_u are the intercepts of the pretransition and posttransition baselines, ΔG is the free energy of unfolding, R is the ideal gas constant, T is temperature, and m describes the dependence of ΔG on the concentration of denaturant.

Fluorescence Binding Experiments

The binding affinity of Cdc42t for WASP can be measured using a number of techniques. We have focused on two methods that use a fluorescent molecule to report on the binding interaction. The first method measures affinity by direct binding between WASP and Cdc42t loaded with a GTP analog conjugated to a mant fluorophore. The second method measures affinity by WASP-mediated displacement of Cdc42t prebound to a fluorescent sensor. Although the first method is the simpler of the two, we and others have found that the affinities measured in this manner are 7-fold to 10-fold weaker than those measured for non-mant–labeled nucleotides using alternative assays (Buchwald et al., 2001; Buck et al., 2001; Leung and Rosen, 2005). This may not be a factor when comparing differences in binding affinity among similar effector constructs using the same assay. However, the effect of the mant moiety on binding affinity becomes a greater issue when comparing values obtained from two different assays using mant- or non-mant-nucleotide–loaded Cdc42t. An alternative measure of binding affinity in this instance can be found in the second method.

Mant Fluorescence Experiments. The binding affinity of GTP-bound Cdc42t for WASP can be measured using GTPase loaded with a fluorescent GTP analog, mantGMPPNP ($\lambda_{exc} = 360$ nm; $\lambda_{em} = 440$ nm). This method and considerations using the mant fluorophore have been described in detail previously (Manor, 2000) and will thus be described here only briefly. Binding of Cdc42t-mantGMPPNP to WASP proteins is performed in buffer N by monitoring the decrease of mant fluorescence at 466 nm during titration. Cdc42t is used at concentrations approximately equal to its K_D, and WASP is added from $\sim 0.1 K_D$ to $\sim 10 K_D$. K_D is obtained by fitting to the quadratic equation describing a single-site binding isotherm:

$$y = F_f + (F_b - F_f) \frac{\left((K_D + [L] + [R]) - \sqrt{(K_D + [L] + [R])^2 - (4[L][R])} \right)}{2[R]}$$

(2)

where $[L]$ is the total ligand WASP concentration, $[R]$ is the total receptor Cdc42t-mantGMPPNP concentration, F_f and F_b are the fluorescence of Cdc42t-mantGMPPNP in the free and WASP-bound states, and K_D is the binding affinity (Marchand *et al.*, 2001).

Fluorescence Competition Experiments. The binding affinities of WASP proteins for Cdc42t-GMPPNP and Cdc42t-GDP can be determined using fluorescence competition experiments. The basic strategy involves measuring the binding affinity of a receptor for a fluorescent ligand and then determining the affinity of a nonfluorescent ligand for receptor by displacement of the fluorescent ligand. Affinities obtained from fluorescence competition experiments are not as accurate as those obtained from direct binding because of errors in the K_D of the fluorescent ligand. However, by taking effort to obtain a precise measurement of K_D of the fluorescent ligand and by dialyzing all proteins into the same buffer, we can minimize the error propagated into the measurement of the K_D of the nonfluorescent ligand.

To measure WASP affinity for Cdc42t by competition, we used a previously described fluorescence resonance energy transfer (FRET)–based sensor for the GTPase (Seth *et al.*, 2003). This reagent consists of the WASP GBD (residues 230–310) flanked by the fluorophores enhanced cyan fluorescent protein (eCFP) and enhanced yellow fluorescent protein (eYFP) at the N– and C–termini, respectively. The isolated sensor has a high FRET value (ratio of YFP emission, 526 nm, to CFP emission, 476 nm) upon CFP excitation ($\lambda_{ex} = 433$ nm) of ~ 2.4 in buffer containing 20 mM HEPES, pH 7.5, 150 mM KCl, 2 mM MgCl$_2$, and 2 mM DTT. Binding of Cdc42t changes the FRET value to ~ 1.6. Using this sensor, the K_D for Cdc42t can be determined by monitoring the FRET change during titration with Cdc42t. For example, up to 1 μM of Cdc42t-GMPPNP is titrated into a sample containing 15 nM of FRET sensor. The FRET values are plotted against Cdc42t concentration and fit to a single site-binding isotherm (Equation 2, with sensor and Cdc42t as R and L, respectively) to obtain a K_D (29 nM). The K_D of WASP for Cdc42t can subsequently be determined by monitoring the FRET change during titration of WASP, because Cdc42t is displaced from the sensor. For example, up to 50 μM of WASP mutant GBD-VCA I290Q is titrated into a sample containing 15 nM of FRET sensor and a saturating 1 μM of Cdc42t-GMPPNP. Under conditions in which the concentration of FRET sensor is significantly smaller than both ligands, such that the free and total ligand concentrations are approximately equal, the competition data obtained are fit according to:

$$y = F_f + (F_b - F_f)/(K_{D,1}([C] + K_{D,2})/(K_{D,2}[L] + 1)) \qquad (3)$$

where F_f and F_b are the FRET ratios of the free and Cdc42t-bound sensor, $[L]$ is the total ligand Cdc42t concentration, $[C]$ is the total WASP concentration, $K_{D,1}$ is the equilibrium dissociation constant of Cdc42t-GMPPNP for sensor, and $K_{D,2}$ is the equilibrium dissociation constant of the competing ligand for Cdc42t-GMPPNP (Vinson *et al.*, 1998). In cases in which these experimental conditions cannot be achieved, the data must be analyzed with the general fourth-order equation describing competitive binding, which must be solved numerically (Panchal *et al.*, 2003). A detailed description of this fitting procedure is available from the authors.

Pyrene Actin Polymerization Assays

The ability of WASP proteins to activate Arp2/3 complex can be measured using the pyrene–actin polymerization assay that has been described by many laboratories (Higgs *et al.*, 1999; Mullins and Machesky, 2000; Rohatgi *et al.*, 1999). This assay is based on the increase in fluorescence of pyrene–actin when it incorporates into a filament. Thus, actin polymerization can be measured by after the change in pyrene fluorescence over time in the presence of Arp2/3 complex and activators.

Several experimental parameters should be considered when preparing the assay. Photobleaching of pyrene should be avoided by decreasing the excitation monochromator slit width. The resultant decrease in sensitivity can be offset by increasing the emission monochromator slit width and/or the percentage of pyrene actin (although pyrene–actin doping to >10% can inhibit polymerization). Temperature also affects the reproducibility of polymerization assays. For this reason, all buffers should be equilibrated at room temperature for 30–60 min before the start of experiments. Protein components that are added in very small volumes are kept on ice until the start of a reaction.

Buffers

KMEI: 10 mM imidazole, pH 7, 50 mM KCl, 1 mM MgCl$_2$, 1 mM EGTA

Actin-Ca: 2 mM Tris, pH 8, 25°, 0.2 mM ATP, 0.5 mM DTT, 0.1 mM CaCl$_2$, 1 mM NaN3

Actin-Mg: 2 mM Tris, pH 8, 25°, 0.2 mM ATP, 0.5 mM DTT, 0.1 mM MgCl$_2$, 1 mM NaN3

10E/1M: 10 mM EGTA, pH 8, and 1 mM MgCl$_2$

A typical polymerization reaction is performed as follows:

1. 32 μl of 25 μM Ca^{2+}-actin (5% pyrene, stored in buffer Actin-Ca) is added to an Eppendorf tube. Mg^{2+} is exchanged for Ca^{2+} by addition of 3.2 μl of buffer 10E/1M. 64.8 μl of buffer Actin-Mg is added to bring the volume to 100 μl, and the mixture is incubated for 2 min at room temperature.

2. In a separate Eppendorf tube, 2 μl of 1 μM bovine Arp2/3, 2 μl of 50 μM WASP, 10 μl of 10× buffer KMEI, and 78 μl of 1× buffer KMEI are mixed together and incubated for 1 min at room temperature. In reactions with wiskostatin, the small molecule is added to this mixture in DMSO (to give final DMSO <5% v/v).

3. The contents of both Eppendorf tubes are quickly combined to 200 μl volume, and the sample is transferred to a quartz cuvette and into the fluorimeter.

The polymerization reaction here is performed with final concentrations of 4 μM actin (5% pyrene), 10 nM bArp2/3 complex, and 500 nM WASP in buffer KMEI. The time it takes to mix the components and start the fluorimeter is noted and must be consistent among experiments to obtain reproducible results. Pyrene–actin fluorescence (λ_{ex} = 365 nm, λ_{em} = 407 nm) is measured every second until it plateaus (~50–1000 sec, depending on the Arp2/3 complex activator).

Several metrics have been described in the literature to quantitatively analyze Arp2/3 complex-mediated actin polymerization, including maximum polymerization rate (Higgs *et al.*, 1999; Prehoda *et al.*, 2000), polymerization rate at 50% polymerization (Dueber *et al.*, 2003), time to 50% polymerization (Zalevsky *et al.*, 2001), and Arp2/3 activation rate constants for nucleation-promoting factors (Zalevsky *et al.*, 2001). Our laboratory favors measurement of maximum polymerization rate, which is converted to concentration of filament barbed ends based on the known rate of actin addition at barbed ends (Higgs *et al.*, 1999; Mullins and Machesky, 2000). Assuming that each Arp2/3 complex nucleates only a single filament during the assay and that filament growth occurs only at the barbed ends, the maximum concentration of barbed ends is equal to the concentration of active Arp2/3 complex.

The normalized fluorescence values from each actin polymerization curve are converted to actin filament concentrations, and then the rate of filament change with time (elongation rate) is converted to the barbed end concentration using the following equation:

$$[\text{ends}] = \frac{\text{elongation rate}}{(k_+[actin] - k_-)}$$

$$= \left(\frac{[actin]_0 - cc}{F_{\max} - F_0}\right)\left(\frac{\Delta F}{\Delta t}\right) * \frac{1}{k_+\left[\left(\left(\frac{F_{\max} - F_x}{F_{\max} - F_0}\right)([actin]_0 - cc)\right) + cc\right] - k_-}$$

$$(4)$$

where $[actin]_0$ is the initial actin concentration, cc is the filament barbed end critical concentration (0.1 μM), $F_{\max}$ is the maximum fluorescence of the polymerization curve, F_x is the instantaneous fluorescence, $\Delta F/\Delta t$ is the instantaneous slope of the polymerization curve, k_+ is the barbed end Mg^{2+}–actin association rate constant (11.6 μM^{-1} s^{-1}), and k_- is the barbed end Mg^{2+}–actin dissociation rate constant (1.3 s^{-1}). The calculation of barbed ends is greatly affected by noise in the polymerization curve. Thus, we average over a window of points when determining the instantaneous actin concentration and slope ($\Delta F/\Delta t$) at each point of the polymerization curve. Values for F_x and $F_{\max}$ are each an average of more than 10 data points, and the slope is calculated from a linear fit of at least 20 data points. Furthermore, although spontaneous actin nucleation is negligible when Arp2/3 complex is stimulated by strong activators (e.g., free VCA peptides), this is not the case for weaker activators (e.g., GBD-VCA). Thus, to achieve accurate quantitation across a range of activators, it is necessary to appropriately subtract the concentration of barbed ends generated by actin alone from the Arp2/3-mediated concentration.

NMR Analyses of Wiskostatin and its Interactions with WASP

To investigate how wiskostatin affects the allosteric equilibrium of WASP, we examined the binding of wiskostatin to the WASP GBD and the structural effects of that interaction. We use NMR spectroscopy, because it is suitable for studying binding events in the μM range, where wiskostatin and the WASP GBD interact well, and also because it provides structural information. NMR samples for titration experiments are prepared in the following way: a receptor solution is prepared from a buffer composed of 23.4 mM MES, pH 5.5, containing 58.5 mM NaCl, 2.34 mM

$MgCl_2$, and 1.17 mM DTT. Protein is previously exchanged into this buffer and further diluted to 1.17-fold its desired concentration. d_6-DMSO and D_2O are added such that the final buffer contains 20 mM MES, pH 5.5, 50 mM NaCl, 2 mM $MgCl_2$, 1 mM DTT, 5% d_6-DMSO, 10% D_2O, and the protein at its desired concentration. A titrant solution is prepared similarly, except that it also includes wiskostatin diluted from a stock previously prepared in neat d_6-DMSO. Typical protein concentrations are in the 100 μM range; typical wiskostatin concentrations are in the 1–3 mM range. Titrant and receptor solutions are mixed in appropriate ratios to afford a range of protein/wiskostatin ratios. For each point in the titration, a one-dimensional proton spectrum is acquired on a 600 MHz instrument, with 1152 complex points and a sweep width of 9000 Hz, with the carrier centered on the water resonance. Because wiskostatin induces folding of the WASP GBD, the progress of the titration is followed using chemical shift changes of two peaks characteristic of the folded GBD: the δ-methyl resonance of Leu281, which shifts from ~0.5 ppm to ~0 ppm on folding, and the indole NH resonance of Trp252, which shifts from ~10 ppm to ~9.7 ppm on folding. Water suppression is accomplished with a flip-back Watergate sequence (Piotto et al., 1992).

Summary

Understanding how WASP is regulated by different signals to stimulate changes in the actin cytoskeleton requires quantitative analysis of the protein's biochemical properties and the elements that contribute to its inhibition or activation. The methods described here have been used recently to study the interactions among WASP, Cdc42, and wiskostatin. These strategies can also be used to quantitatively characterize the interactions of WASP with other known regulators, including Nck, Toca-1, Grb2, and WIP. Furthermore, analogous approaches should be widely applicable to other signaling systems to examine intramolecular and intermolecular regulatory interactions.

References

Benesch, S., Lommel, S., Steffen, A., Stradal, T. E., Scaplehorn, N., Way, M., Wehland, J., and Rottner, K. (2002). Phosphatidylinositol 4,5-biphosphate (PIP2)-induced vesicle movement depends on N-WASP and involves Nck, WIP, and Grb2. J. Biol. Chem. 277, 37771–37776.

Buchwald, G., Hostinova, E., Rudolph, M. G., Kraemer, A., Sickmann, A., Meyer, H. E., Scheffzek, K., and Wittinghofer, A. (2001). Conformational switch and role of phosphorylation in PAK activation. Mol. Cell Biol. 21, 5179–5189.

Buck, M., Xu, W., and Rosen, M. K. (2001). Global disruption of the WASP autoinhibited structure on Cdc42 binding. Ligand displacement as a novel method for monitoring amide hydrogen exchange. *Biochemistry* **40,** 14115–14122.

Buck, M., Xu, W., and Rosen, M. K. (2004). A two-state allosteric model for autoinhibition rationalizes WASP signal integration and targeting. *J. Mol. Biol.* **338,** 271–285.

Cooper, J. A., and Pollard, T. D. (1982). Methods to measure actin polymerization. *Methods Enzymol.* **85**(Pt B), 182–210.

Dueber, J. E., Yeh, B. J., Chak, K., and Lim, W. A. (2003). Reprogramming control of an allosteric signaling switch through modular recombination. *Science* **301,** 1904–1908.

Higgs, H. N., Blanchoin, L., and Pollard, T. D. (1999). Influence of the C terminus of Wiskott-Aldrich syndrome protein (WASp) and the Arp2/3 complex on actin polymerization. *Biochemistry* **38,** 15212–15222.

Higgs, H. N., and Pollard, T. D. (2000). Activation by Cdc42 and PIP(2) of Wiskott-Aldrich syndrome protein (WASp) stimulates actin nucleation by Arp2/3 complex. *J. Cell Biol.* **150,** 1311–1320.

Ho, H. Y., Rohatgi, R., Lebensohn, A. M., Le, M., Li, J., Gygi, S. P., and Kirschner, M. W. (2004). Toca-1 mediates Cdc42-dependent actin nucleation by activating the N-WASP-WIP complex. *Cell* **118,** 203–216.

Kim, A. S., Kakalis, L. T., Abdul-Manan, N., Liu, G. A., and Rosen, M. K. (2000). Autoinhibition and activation mechanisms of the Wiskott-Aldrich syndrome protein. *Nature* **404,** 151–158.

Leung, D. W., and Rosen, M. K. (2005). The nucleotide switch in Cdc42 modulates coupling between the GTPase-binding and allosteric equilibria of Wiskott-Aldrich syndrome protein. *Proc. Natl. Acad. Sci. USA* **102,** 5685–5690.

Manor, D. (2000). Measurement of GTPase.effector affinities. *Methods Enzymol.* **325,** 139–149.

Marchand, J. B., Kaiser, D. A., Pollard, T. D., and Higgs, H. N. (2001). Interaction of WASP/Scar proteins with actin and vertebrate Arp2/3 complex. *Nat. Cell Biol.* **3,** 76–82.

Mullins, R. D., and Machesky, L. M. (2000). Actin assembly mediated by Arp2/3 complex and WASP family proteins. *Methods Enzymol.* **325,** 214–237.

Pace, C. N. (1990). Measuring and increasing protein stability. *Trends Biotechnol.* **8,** 93–98.

Panchal, S. C., Kaiser, D. A., Torres, E., Pollard, T. D., and Rosen, M. K. (2003). A conserved amphipathic helix in WASP/Scar proteins is essential for activation of Arp2/3 complex. *Nat. Struct. Biol.* **10,** 591–598.

Peterson, J. R., Bickford, L. C., Morgan, D., Kim, A. S., Ouerfelli, O., Kirschner, M. W., and Rosen, M. K. (2004). Chemical inhibition of N-WASP by stabilization of a native autoinhibited conformation. *Nat. Struct. Mol. Biol.* **11,** 747–755.

Peterson, J. R., Lokey, R. S., Mitchison, T. J., and Kirschner, M. W. (2001). A chemical inhibitor of N-WASP reveals a new mechanism for targeting protein interactions. *Proc. Natl. Acad. Sci. USA* **98,** 10624–10629.

Piotto, M., Saudek, V., and Sklenar, V. (1992). Gradient-tailored excitation for single-quantum NMR spectroscopy of aqueous solutions. *J. Biomol. NMR* **2,** 661–665.

Prehoda, K. E., Scott, J. A., Mullins, R. D., and Lim, W. A. (2000). Integration of multiple signals through cooperative regulation of the N-WASP-Arp2/3 complex. *Science* **290,** 801–806.

Ramesh, N., Anton, I. M., Hartwig, J. H., and Geha, R. S. (1997). WIP, a protein associated with Wiskott-Aldrich syndrome protein, induces actin polymerization and redistribution in lymphoid cells. *Proc. Natl. Acad. Sci. USA* **94,** 14671–14676.

Rohatgi, R., Ma, L., Miki, H., Lopez, M., Kirchhausen, T., Takenawa, T., and Kirschner, M. W. (1999). The interaction between N-WASP and the Arp2/3 complex links Cdc42-dependent signals to actin assembly. *Cell* **97,** 221–231.

Rohatgi, R., Nollau, P., Ho, H. Y., Kirschner, M. W., and Mayer, B. J. (2001). Nck and phosphatidylinositol 4,5-bisphosphate synergistically activate actin polymerization through the N-WASP-Arp2/3 pathway. *J. Biol. Chem.* **276,** 26448–26452.

Scaplehorn, N., Holmstrom, A., Moreau, V., Frischknecht, F., Reckmann, I., and Way, M. (2002). Grb2 and Nck act cooperatively to promote actin-based motility of vaccinia virus. *Curr. Biol.* **12,** 740–745.

Seth, A., Otomo, T., Yin, H. L., and Rosen, M. K. (2003). Rational design of genetically encoded fluorescence resonance energy transfer-based sensors of cellular Cdc42 signaling. *Biochemistry* **42,** 3997–4008.

Specht, K. M., and Shokat, K. M. (2002). The emerging power of chemical genetics. *Curr. Opin. Cell Biol.* **14,** 155–159.

Torres, E., and Rosen, M. K. (2003). Contingent phosphorylation/dephosphorylation provides a mechanism of molecular memory in WASP. *Mol. Cell* **11,** 1215–1227.

Vinson, V. K., De La Cruz, E. M., Higgs, H. N., and Pollard, T. D. (1998). Interactions of Acanthamoeba profilin with actin and nucleotides bound to actin. *Biochemistry* **37,** 10871–10880.

Volkman, B. F., Prehoda, K. E., Scott, J. A., Peterson, F. C., and Lim, W. A. (2002). Structure of the N-WASP EVH1 domain-WIP complex: Insight into the molecular basis of Wiskott-Aldrich syndrome. *Cell* **111,** 565–576.

Yarar, D., To, W., Abo, A., and Welch, M. D. (1999). The Wiskott-Aldrich syndrome protein directs actin-based motility by stimulating actin nucleation with the Arp2/3 complex. *Curr. Biol.* **9,** 555–558.

Zalevsky, J., Lempert, L., Kranitz, H., and Mullins, R. D. (2001). Different WASP family proteins stimulate different Arp2/3 complex-dependent actin-nucleating activities. *Curr. Biol.* **11,** 1903–1913.

Further Reading

"The Merck Index." CambridgeSoft, Cambridge, Mass.

[22] The Use of GFP to Localize Rho GTPases in Living Cells

By DAVID MICHAELSON and MARK PHILIPS

Abstract

The green fluorescent protein (GFP) of the jellyfish *Aequorea victoria* has revolutionized the study of protein localization and dynamics. GFP fusions permit analysis of proteins in living cells and offer distinct advantages over conventional immunofluorescence. Among these are lower background, higher resolution, robust dual color colocalization, and avoidance of fixation artifacts. In the case of Ras and Rho family proteins, GFP fusions have allowed breakthroughs in the understanding of how CAAX

METHODS IN ENZYMOLOGY, VOL. 406
Copyright 2006, Elsevier Inc. All rights reserved.

0076-6879/06 $35.00
DOI: 10.1016/S0076-6879(06)06022-8

proteins are targeted to specific cell membranes and how signaling at different membranes can result in different cellular responses. GFP-tagged Rho proteins have also been informative in analyzing the interactions with the cytosolic chaperone, RhoGDI. The major disadvantages of studying GFP fusion proteins is that they are generally overexpressed relative to endogenous proteins, and the GFP tag can, in principle, affect protein function. Fortunately, in the case of Ras and Rho family proteins, a GFP tag at the N terminus seems to have little effect on protein targeting and function. Nevertheless, it is prudent to confirm GFP fusion protein data with the study of the endogenous protein. This chapter describes the tagging of Rho proteins with GFP and the analysis of GFP–Rho protein localization by epifluorescence and confocal microscopy. It further describes methods of analyzing endogenous Rho proteins as confirmation of data acquired using GFP–Rho fusion proteins. These techniques will be useful for anyone studying Rho protein function and are widely applicable to many cell types and signal transduction systems.

Introduction

Rho proteins are Ras-related GTPases that regulate a wide variety of cellular processes. More than 15 mammalian Rho proteins have been described, including RhoA–E, G, and H; Rac1-3; two isoforms of Cdc42hs; and TC10. The diversity of functions among Rho proteins belies the extensive sequence homology found among them (Fig. 1). The Rho protein amino acid sequences are highly conserved, including residues found in the GTP binding domains and the effector loops, giving an overall amino acid identity among family members of 50% or greater with further sequence similarities. Many of the amino acid differences among the Rho proteins occur within the C-terminal 20 amino acids, the so-called hypervariable region. In fact, some Rho proteins are virtually identical except for their hypervariable regions (e.g., Rac 1, 2, and 3; Rho A, B and C; the two isoforms of Cdc42hs). This region is not involved in the GTPase activity of Rho proteins or in the interaction of Rho GTPases with effectors, but it plays a critical role in localization. Of all the various protein–protein interactions in which Rho proteins engage, only the interaction with GDP dissociation inhibitors (GDIs) has been found to involve amino acids in the hypervariable region (Hoffman *et al.*, 2000). Some of the diverse functions of Rho family proteins are due to differences in affinities for GEFs, GAPs, and effectors because of subtle differences in the amino acid sequences of the GTPases. However, the considerable sequence divergence within the hypervariable region suggests that some of the diversity

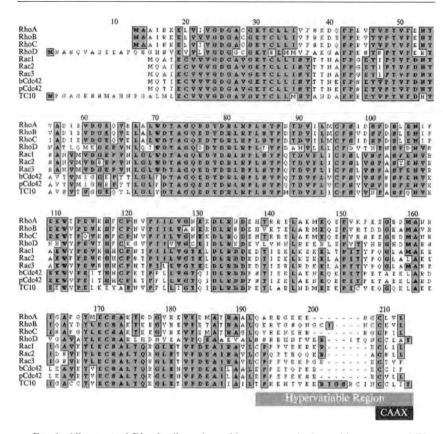

FIG. 1. Alignment of Rho family amino acid sequences. Amino acid sequences of Rho proteins were aligned using the CLUSTAL method. Amino acids boxed in gray show regions of high sequence identity. The C-terminal hypervariable region with the lowest sequence similarity is underscored with a gray box. The CAAX motif is underscored with a black box. All of the membrane-targeting information is contained in this region.

in function will be due to diversity of sequence within this region, leading to localization differences.

Recent analyses of the localization of Ras proteins using GFP technology have forced a reevaluation of long-held assumptions about Ras trafficking (Choy *et al.*, 1999). Whereas it was previously assumed that Ras translocated directly from the cytosol where it is synthesized to the plasma membrane by virtue of CAAX processing, analysis of GFP–Ras fusion proteins revealed endomembrane trafficking en route to the plasma membrane. This endomembrane localization has been shown to be functional

(Chiu *et al.*, 2002) and regulated (Rocks *et al.*, 2005). Prenylation targets CAAX proteins specifically to the endomembrane where they are further processed (proteolyzed and methylated) and then sent to the plasma membrane only if they possess a secondary targeting sequence (palmitoylation sites or a polybasic domain). Moreover, the pathway to the plasma membrane used by N-Ras and H-Ras (palmitoylated) seems to differ from that used by K-Ras (polybasic). This difference can be observed using GFP-tagged Ras protein in that, whereas GFP-N-Ras and GFP-H-Ras are visualized on Golgi membranes, GFP-K-Ras is not. More recently, the same GFP technology has extended these observations to heterotrimeric GTPases (Michaelson *et al.*, 2002) and Rho family proteins (Michaelson *et al.*, 2001). This chapter will detail the methods used for tagging Rho proteins with GFP, imaging the resultant fusion proteins in live cells to observe subcellular localization and membrane trafficking, and how these results compare with immunofluorescence and cell fractionation analyses of endogenous Rho proteins.

Expression Constructs

The original clone of the green fluorescent protein (GFP) of the jelly-fish *Aequorea victoria* isolated in 1992 (Prasher *et al.*, 1992) has been widely disseminated and extensively mutated. GFP variants are commercially available from several venders. The GFP-derived expression vectors we have used are from BD Biosciences Clontech's LivingColors line of products. This line includes enhanced GFP (EGFP), cyan (CFP) and yellow (YFP) mutants of the original GFP, and a red fluorescent protein (DsRed) from an IndoPacific sea anemone relative, *Discosoma* sp. (Matz *et al.*, 1999). Each of these fluorescent tags can be used individually to tag proteins for observation in live cells. Selected pairs of these tags can be used for two-color colocalization studies of two proteins simultaneously. By use of standard confocal microscopy, CFP can be paired with YFP, and DsRed can be paired with any of the others. For our single color experiments, we have used pEGFP (BD Biosciences Clontech) for our Rho constructs. This plasmid directs the expression of enhanced GFP under the control of a CMV immediate early promoter and enhancer. EGFP has been optimized for brighter fluorescence and higher expression in mammalian cells and has an excitation maximum at 488 nm and emission maximum at 507 nm. pEGFP variants are available with a multiple cloning site (MCS) 5' or 3' of the EGFP sequence, each in three reading frames. To tag Rho proteins or fragments thereof we have used almost exclusively pEGFP-C3, which allows for insertion of Rho protein coding sequences 3' of the EGFP sequence. The EGFP tag must be N terminal to the Rho protein sequence to ensure a free C-terminal CAAX box for processing. Our inserts are derived from PCR amplification allowing for adjustment of the reading frame to the C3 variant and introduction of convenient 5' and 3' restriction sites compatible with the pEGFP MCS (usually 5' *Hind*III and 3' *Eco*RI) by primer design.

By use of the cDNA of choice (e.g., wild type, constitutively active or dominant negative Rac, Rho, or Cdc42 GTPases), the coding sequence can be PCR amplified with appropriate restriction sites designed into the primers using a polymerase of moderate fidelity (e.g., Taq). The PCR product is then double digested and inserted into the MCS of pEGFP using standard molecular biology techniques.

Primer Design

In general, PCR primers should be least 15 bases long with at least six contiguous bases complementary to the target sequence at the 3' end. Ideally, the most 3' nucleotide of each primer (the anchor nucleotide from which the polymerase will extend) should be a G or C residue. The melting temperature of the primers should be similar and be approximately 60°. A simple equation for determining the melting temperature of a primer is:

$$Tm = 4(G + C) + 2(A + T)°C$$

In designing the 5' primer, the following should be considered:

1. A 5' terminal extension of two to four nucleotides to facilitate restriction enzyme activity.
2. A restriction site compatible with the MCS.
3. Ten to 15 nucleotides of the coding sequence of the Rho insert in-frame with the EGFP coding sequence.
4. If the initiation codon of the Rho protein insert is retained, then eliminate, if possible, any Kozak sequence.

In designing the 3' primer, the following should be considered:

1. A 5' terminal extension of two to four nucleotides to facilitate restriction enzyme activity.
2. A restriction site found in the MCS 3' of that used for the 5' primer.
3. A stop anti-codon in-frame with the coding sequence of the Rho protein being subcloned.
4. Ten to 15 nucleotides of the anticoding sequence of the Rho protein being subcloned.

Cloning Methods

Full-length Rho Protein Constructs. We have used the following PCR conditions to amplify inserts for EGFP-Rho protein constructs using a Perkin Elmer thermal cycler.
REACTANTS AND BUFFER
20 μM 5′ primer 5 μl
20 μM 3′ primer 5 μl
1 μg/μl target cDNA plasmid 1 μl
10 mM dNTP 2.5 μl
Taq polymerase 0.5 μl
10× polymerase buffer 5 μl
ddH$_2$O 31 μl.
PCR CYCLES

Cycle#	Melting	Annealing	Extension
#1	5 min at 98°	1 min at 55°	1 min at 72°
#2–#30	1 min at 98°	1 min at 55°	1 min at 72°
#31			10 min at 72°
#32 Hold at 4°			

Annealing temperature should be adjusted on the basis of the primer melting temperatures. The ideal annealing temperature should be 5° below the lower melting temperature of the two primers. Optimal primer and PCR conditions can be determined for each primer/template combination. A good resource for optimizing primer design and PCR conditions is *PCR Protocols* (Innis *et al.*, 1989).

Rho protein amplification products are purified by 1% agarose gel electrophoresis and, for the full length cDNAs, should migrate at ∼0.6 kb (visualized by ethidium bromide). The amplification product can be excised from the gel with a sterile razor or spatula and extracted by column purification (e.g., Qiagen gel extraction kit) and stored at 1 μg/μl in 10 mM Tris-HCl, pH 7.4.

RESTRICTION DIGESTS (MUST BE APPLIED TO BOTH PCR PRODUCT AND GFP EXPRESSION VECTOR)
1 mg/ml PCR product or pEGFP DNA 1.0 μl
Restriction enzyme A (to cut 5′ end) 0.5 μl
Restriction enzyme B (to cut 3′ end) 0.5 μl
10× restriction buffer (as per manufacturer's suggestion) 2.0 μl
ddH$_2$O 16 μl.

Incubate restriction digest at 37° for 2 h (or as per manufacturer's suggestion). In cases in which the two restriction enzymes are not functional in compatible buffers, sequential digestion with each enzyme is necessary with either purification between each digestion or heat inactivation of the first enzyme followed by adjustment of the buffer's salt conditions to be compatible with the second enzyme. Digested DNA can be column purified (e.g., Qiagen plasmid purification kit) as per manufacturer's instructions. The PCR product should be brought to 50 ng/μl and the vector at 1 μg/μl each in 10 mM Tris-HCl, pH 7.4.

LIGATION REACTIONS. To optimize conditions and to troubleshoot the digestion, ligation, and transformation steps, six ligation reactions are performed in 0.5-ml Eppendorf tubes as specified in Table I. Incubate the ligation reactions at 14° for >16 h. Ligation reactions are extremely sensitive to the age of the buffer (because of decay of DTT and ATP) and enzyme and to impurities left from DNA purification. Failed ligations are usually due either to impure DNA, which can be cleaned up using kits like the Qiagen plasmid purification kit, or because of old buffer or enzyme. Fresh enzyme should be used and T4 DNA ligase must be kept below 4° at all times. Old buffer can be supplemented with fresh DTT or ATP to boost activity.

BACTERIAL TRANSFORMATION AND DNA ISOLATION. Competent bacterial cells (e.g., DH5α or XL1blue) are prepared by growing bacteria in standard Luria Broth (LB) to mid log phase (Optical density at 550 nm wavelength of 0.45–0.55). Cells are iced 10 min followed by centrifugation at 2500 rpm for 5 min, 4° to pellet bacteria. Bacterial cells are gently resuspended in 10 mM CaCl$_2$, 15% glycerol to 1/12.5 of the original culture volume, and iced 45 min. Cells can either be used immediately or frozen at −80° for future use; 100 μl of competent bacteria is mixed with 10 μl of each of the ligation reactions, and the mixture is iced for 45 min followed by heat shock at 42° for 90 sec followed by incubation over ice for 3 min. LB is added to a final volume of 500 μl, and cells are incubated for 45 min at 37° with shaking to allow expression of the antibiotic resistance gene on the GFP vector. Cells are spread on LB plates containing selective antibiotics (e.g., 50 μg/ml kanamycin for pEGFP). Incubate at 37° for 18 h. In general, this technique works well but is not an optimized protocol. An excellent discussion of the parameters to be considered when optimizing bacterial transformations can be found in Huff et al. (1990).

Expected results are as follows: Condition 1 should give >500 colonies per plate if the bacteria are competent. Condition 2 should give 50–200 colonies if the ligase is active. Condition 3 should give <5 colonies per plate if the vector is completely digested by both restriction enzymes. Conditions 4–6 should give at least twofold the number of colonies

TABLE I
LIGATION CONDITIONS

Condition	Uncut vector (1 mg/ml)	Single cut vector (1 mg/ml)	Double cut vector (1 mg/ml)	Double cut PCR prod. (50 ng/ml)	T4 DNA ligase	10× Buffer with ATP	ddH$_2$O	Purpose
1.	1 μl	0 μl	0 μl	0 μl	0 μl	2 μl	17 μl	Bacterial competence
2.	0 μl	1 μl	0 μl	0 μl	1 μl	2 μl	16 μl	Ligase activity
3.	0 μl	0 μl	1 μl	0 μl	1 μl	2 μl.	16 μl	Vector digestion
4.	0 μl	0 μl	1 μl	1 μl	1 μl	2 μl	15 μl	Cloning
5.	0 μl	0 μl	1 μl	2 μl	1 μl	2 μl	14 μl	Cloning
6.	0 μl	0 μl	1 μl	4 μl	1 μl	2 μl	12 μl	Cloning

observed in condition 3 if insertion of the PCR product is successful. Choose the cloning condition (condition 4–6) that gives the most colonies per plate and pick 5–10 colonies for analysis. Prepare small-scale plasmid DNA (e.g., using Qiagen Miniprep kit) from 2-ml overnight cultures of each of these colonies. Double digest 1 μg of plasmid DNA from each clone and analyze the insert size by 1% agarose gel electrophoresis to confirm that the plasmid contains an insert of the correct size. Clones that contain an appropriate-sized insert should be confirmed by DNA sequencing. Prepare large-scale plasmid DNA of desired clone from 100–500 ml LB cultures (e.g., using Qiagen Maxi Prep kit).

C-terminal Constructs. Many Rho proteins bind to a cytosolic chaperone called RhoGDI, which sequesters the inactive Rho GTPase in the cytosol (see following). We have found that the hypervariable C-terminal membrane targeting domains of Rho proteins, when fused to the C terminus of GFP, target the fusion protein in a manner independent of RhoGDI, revealing the underlying membrane targeting function of the hypervariable domain (Michaelson *et al.*, 2001). To prepare these short constructs, we have used synthetic complementary oligonucleotides in place of the PCR products discussed previously. DNA inserts of 45 bp or less used to produce C-terminal constructs are generated by synthesizing complementary positive and negative oligonucleotides (with appropriate restriction site overhangs) that have a 5′ *Hind*III, a 3′ stop codon, and a 3′ *Eco*RI site in their sequence. These single-stranded oligonucleotides are purified by polyacrylamide gel electrophoresis and resuspended at 10 mM Tris-HCl, pH 7.4. Twenty nmol (2 μl) of each complementary strand are mixed in 0.5 ml Eppendorf tubes with 14 μl ddH$_2$O and heated to 100° for 5 min. The tubes are allowed to slowly cool down to allow the two DNA strands to properly anneal, then placed on ice; 2 μl of 10× ligation buffer, 1 U of T4 DNA ligase, and 1 μg of pEGFP-C3 vector linearized by double digestion with *Hind*III and *Eco*RI are added to the tube and incubated at 14° for >16 h. Transformation of competent bacteria with these constructs is accomplished as described previously.

Transfection of Mammalian Cells

GFP-tagged proteins expressed by transfection into cultured mammalian cells have several advantages over traditional immunofluorescence. The primary advantage is the ability to observe the subcellular localization of tagged proteins in living cells in real time. This approach has lower background and higher resolution than immunofluorescence and avoids fixation artifacts. Currently available high-efficiency transfection methods (e.g., LipofectaAMINE and SuperFect) permit a wide variety of cell lines to be studied. We have successfully transfected COS-1, CHO, MDCK, ECV309, NIH3T3, PAE, and HeLa cells for studying the subcellular localization of Rho proteins. The various morphologies of the different cell types can be exploited to study specific subcellular compartments. For example, whereas sessile, well-spread cells like COS-1 and ECV309 are optimal for visualizing the ER, confluent epithelial cells such as MDCK or HeLa that assume a semicolumnar morphology are best suited for scoring plasma membrane.

Cell Culture and Transfection Techniques

COS-1, CHO, MDCK, ECV309, PAE, and HeLa cells can be maintained in 5% CO_2 at 37° in Dulbecco's modified of Eagle's medium (DMEM) containing 10% fetal bovine serum (FBS). NIH3T3 fibroblasts are best maintained in the same medium containing 10% calf serum instead of FBS. In general, we maintain cell culture stocks in 10-cm tissue culture dishes. To visualize live cells on an inverted microscope, we have found that the most practical plates are 35-mm culture dishes that have a 14-mm cutout at the bottom that forms a miniculture well sealed by a No.1.5 glass coverslip (MaTTek Corporation, Ashland, MA). Cells can be grown, transfected and observed all in the same plate.

Transfection Method

1. One day before transfection, seed 1.5×10^5 cells into 35-mm MaTTek dishes.
2. Twenty four hours after seeding, cells should be approximately 30–50% confluent and ready to be transfected using one of several transfection techniques. Although traditional calcium phosphate or DEAE–dextran methods will suffice, we have found either LipofectA-MINE (Invitrogen) or SuperFect (Qiagen) when used according to the manufacturer's protocols to be efficient and reliable. For Rho proteins, the amount of GFP–Rho protein DNA used for transfection depends on the desired experiment (see Fig. 2). When 0.1–0.3 μg of DNA is used in most cell types, the level of expression of the GFP-Rho protein will be low enough that endogenous RhoGDI suffices to sequester the GFP-Rho protein in the cytosol. When >0.5 μg of DNA is used, the expression of the GFP–Rho protein overwhelms the endogenous RhoGDI, and so the membrane localization of the GFP-Rho protein is revealed. When a RhoGDI expression plasmid (e.g., pcDNA-RhoGDI) is cotransfected to compensate, cytosolic localization is restored. To accomplish this, 0.5–1 μg of GFP–Rho protein and 1 μg of pcDNA–RhoGDI DNA were generally

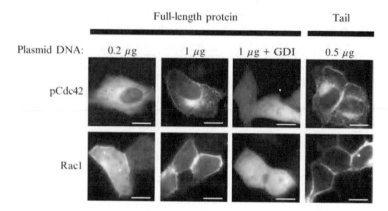

FIG. 2. Localization of GFP-tagged full-length Rho proteins and Rho protein hypervariable domains (Tail) in live MDCK cells. The indicated quantity of plasmid DNA directing expression of GFP-Rac1 or GFP-pCdc42 was transfected into MDCK cells with or without cotransfection of an equal amount of pcDNA–RhoGDI. Live cells were imaged by epifluorescence 24 h after transfection. Bars indicate 10 μm.

used because higher levels of overexpression of RhoGDI are detrimental to the cell.

3. Fluorescence can be observed as early as 5 h after transfection, although full expression revealing steady-state localization of the GFP–Rho protein is best observed after 16–48 h.

Establishment of Stable Cell Lines. Many of the commercially available GFP vectors, including pEGFP, incorporate selectable markers and are, therefore, useful in establishing cell lines that stably express GFP-tagged proteins. We have found that it is relatively easy to establish cell lines that stably express GFP-tagged Rho proteins using pEGFP and combining aminoglycoside (G418) selection and fluorescence activated cell sorting (FACS). A G418 killing curve (0.1–3 mg/ml) must be performed for each cell type before selection to establish the appropriate dose for killing off all nonexpressing cells within 5–7 days. In general, a concentration between 0.5 and 1 mg/ml is sufficient for most cell lines. In addition, a nontransfected control should be treated in parallel with transfected cells to verify selection of resistant clones. We have used the following method:

1. We transfect cells, with a parallel mock-transfected control, as outlined for transient transfections previously.

2. Twenty-four hours after transfection, both control and transfected plates grown in appropriate media containing the concentration of G418

established by the killing curve. Medium should be changed (maintaining selection) every few days to remove dead, floating cells.

3. After 3–9 days, when all cells in the nontransfected control plate have died, the plates with transfected cells should have numerous colonies of fluorescent cells.

4. Surviving transfected cells should be harvested by trypsinization and sorted by FACS, selecting for moderately green cells. Sorted cells should be plated at an average of 0.5 cells per well in the wells of a 96-well plate to establish clonal lines of transfected cells. These clones should be grown in selective medium and scaled up to larger plates as the cells proliferate. Once sufficient numbers of cells are achieved, samples of each surviving clonal line should be properly frozen on liquid nitrogen for long-term storage.

Imaging

EGFP-tagged proteins can be imaged with conventional FITC filter sets, although custom sets optimized for GFP are also available (Chroma Technology Corp., www.chroma.com). Whether using epifluorescence or confocal imaging, combining an inverted microscope with the MaTTek culture dishes described previously allows repeated imaging of the same culture plate without contamination and, when combined with a heated stage or a microincubator (PDMI-2, Harvard Apparatus, www.harvardapparatus.com or Zeiss incubator S, www.zeiss.de), allows kinetic analysis and directly observed pharmacological manipulation in real time or by time-lapse photography. We have found that a laser scanning confocal microscopy, although indispensable for certain applications, and useful for most applications, is not required to obtain membrane-targeting information for GFP–Rho proteins expressed in live cultured cells. Although epifluorescence microscopes equipped with film cameras will suffice, digital image acquisition and manipulation offers tremendous advantages. These include very high light sensitivity (with cooled CCD cameras), instantaneous acquisition setting adjustment for optimized exposure, time-lapse capability, digital image enhancement and analysis, and digital storage of data. Such systems are available from a variety of vendors (e.g., Leica, Nikon, and Zeiss). We have used a Zeiss Axiovert 100 microscope equipped with a Princeton Instruments (www.prinst.com) cooled CCD camera optimized for GFP (model RTE/CCD-1300Y/HS). The halogen and mercury lamps, filter wheels, shutters, and camera are all controlled by MetaMorph software (Universal Imaging, www.image1.com), which also offers state-of-the-art image enhancement and analysis. This system is also ideal for capturing time-lapse image sequences that can be converted to QuickTime

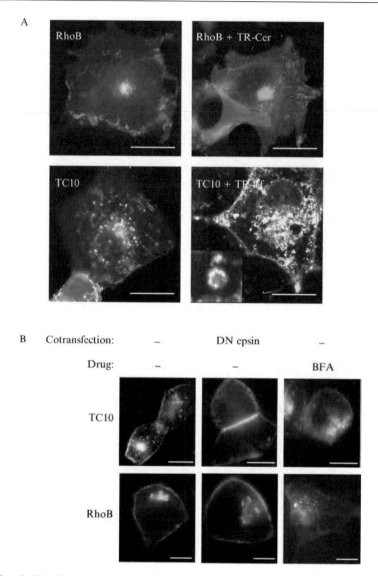

FIG. 3. Identification of subcellular compartments in live cells. (A) Cos-1 cells were transfected as described with GFP-RhoB or GFP-TC10. The Golgi apparatus and endosomes were marked by incubating the cells for 30 min before imaging with, respectively, Texas Red–conjugated BODIPY ceramide (10 μg/ml, Molecular Probes) or transferrin (100 μg/ml, Molecular Probes). Excess label was washed away immediately before imaging. The inset on the lower right panel shows an enlargement of an endosome that reveals that the

or avi movies. For confocal imaging, we use a standard Zeiss 510 laser scanning confocal microscope that incorporates an inverted Zeiss Axiovert 200 microscope and 40-mW argon 458/488/514 and dual HeNe 543/633 lasers.

Unlike observations of GFP–Ras proteins in live cells, observation of GFP Rho proteins must take into account the cytosolic chaperone Rho-GDI. Fig. 2 shows this phenomenon for the Rho family proteins Rac1 and Cdc42 (placental isoform). When 0.2 μg of GFP-Rac1 or GFP-pCdc42 DNA is transiently transfected into MDCK cells, the low level of expression is still within the capability of endogenous RhoGDI to bind and sequester the exogenous protein in the cytosol. However, the protein expressed by transfection of 1 μg of GFP-Rac1 or GFP-pCdc42 DNA into MDCK cells overwhelms the endogenous RhoGDI, and the exogenous protein is seen in the membrane compartment specified by its hyper-variable domain (plasma membrane for GFP-Rac1, endomembrane, and plasma membrane for GFP-pCdc42). Co-overexpression of 1 μg each of pcDNA-RhoGDI and of GFP-Rac1 or GFP-pCdc42 DNA reestablishes the balance between the GTPase and the chaperone and restores a cyto-solic phenotype. It should be noted that >1 μg of pcDNA-RhoGDI trans-fected into most cell types we have tried leads to cell death. High levels of some GFP–Rho protein expression also leads to cell death and morpho-logical artifacts. One way to observe the membrane localization of a Rho protein without excessive overexpression artifacts is to use the last 10–20 amino acids of the Rho protein, the hypervariable domain. When the hypervariable domain of Rac1 or pCdc42 alone is attached to GFP, the resulting fusion protein is targeted to the same membrane compartments as the full-length fusion proteins when overexpressed at high levels. Some Rho proteins do not bind RhoGDI. Two such proteins, RhoB and TC10, are shown in Figs. 3 and 4.

Identification of Subcellular Compartments in Live Cells

The identification of subcellular membranes where GFP–Rho proteins are targeted can be made based on morphology and colocalization with compartment markers. For example, RhoB and TC10 are both targeted to

---

colocalization is on the limiting membrane. Bars indicate 10 μm. (B) MDCK cells were transfected with GFP-TC10 or GFP-RhoB with or without dominant negative epsin and imaged alive 16 h later. Cells transfected with GFP-TC10 or GFP-RhoB were treated with brefeldin A (BFA) for 5 h before imaging, causing dispersal of the Golgi. Bars indicate 10 μm. (See color insert.)

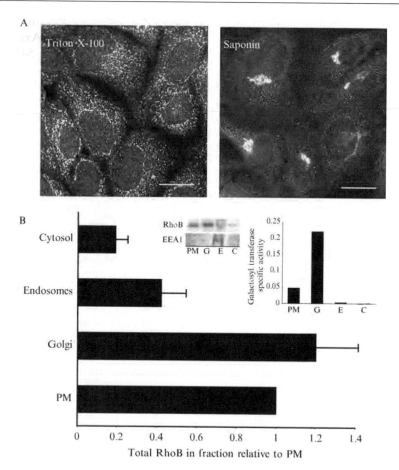

FIG. 4. Localization of endogenous RhoB on the Golgi apparatus. (A) Indirect immunofluorescence to detect endogenous RhoB was performed as described on paraformaldehyde-fixed MDCK cells using either Triton X-100 or saponin to permeabilize the cells before staining. Whereas a vesicular pattern was observed after Triton treatment, a Golgi pattern was visible after saponin. Bars indicate 10 μm. (B) Subcellular fractionation of MDCK cells was performed as described. The main graph shows the distribution of RhoB detected by immunoblot among the various fractions, normalized to the PM. Results shown are mean $\pm$ SEM ($n = 3$). Sample immunoblots for the endosomal marker EEA1 and RhoB are shown as an inset. The inset graph shows the distribution of activity of the Golgi marker galactosyl transferase among the various fractions (mean, $n = 2$).

the PM and internal membranes. Although the RhoB- and TC10-containing compartments can, at times, be difficult to distinguish, the experienced observer has little trouble making distinctions. Unlike GFP-RhoB, GFP-TC10 is visible on numerous distinct, motile vesicles that extend throughout the cytoplasm but coalesce in a perinuclear structure. It is the perinuclear structure that presents the biggest challenge. However, on close inspection the morphologies of the perinuclear compartments differ; GFP-RhoB localizes to a compact, Golgi-like compartment, whereas GFP-TC10 localizes to vesicles that coalesce in the same region but appear as a more diffuse structure (Fig. 3). Identification of these compartments in live cells can be accomplished using fluorescent compartment markers. These include CFP/YFP-tagged proteins available from Clontech that can be coexpressed with Rho proteins tagged with the spectrally distinct fluorescent protein, membrane-permeable vital dyes (e.g., Texas Red conjugated BODIPY Ceramide), and fluorescent markers that are taken into the cell by endocytosis or pinocytosis (e.g., Texas Red–conjugated transferrin). Texas Red–conjugated BODIPY Ceramide (10 μg/ml, Molecular Probes) and Texas Red–conjugated transferrin (100 μg/mL, Molecular Probes) can be added directly to culture medium 20–30 min before imaging, incubated at 37°, then rinsed out with fresh medium. The ceramide reagent concentrates in the Golgi. Fluorescent transferrin is actively taken up by endocytosis and can be observed in endocytic vesicles throughout the cell after 5–10 min. After 30 min, transferrin will collect in the perinuclear region in a structure known as the endosomal recycling compartment, which is difficult to distinguish from the Golgi apparatus. As seen in Fig. 3A, RhoB colocalizes with Texas Red–conjugated BODIPY ceramide, whereas TC10 colocalizes with transferrin, identifying the respective compartments as Golgi and endosomes. A close-up of the TC10/transferrin colocalization shows that the GFP-TC10 often forms a ring around a transferrin-containing vesicle (see inset in Fig. 3A), indicating that TC10 associates with the outer membrane of these organelles.

Confirmation of subcellular localization can be performed by disrupting the compartment in question. Brefeldin A (BFA, Sigma) is commonly used to disrupt Golgi structure. BFA can be added 5 h after transfection of the GFP-Rho proteins to a final concentration of 5 μg/ml. Epsin and eps15 (two alpha adaptin-interacting proteins [Chen *et al.*, 1998]) are regulators of the endocytic pathway. cDNAs encoding dominant negative forms of epsin and eps15 can be cotransfected along with GFP–Rho proteins to block formation of endosomes. Fig. 3B shows that the compact perinuclear localization of RhoB is dispersed by BFA treatment but is unaffected by cotransfection with either dominant negative epsin or eps15. Conversely, TC10's vesicular localization is unaffected by BFA treatment but disappears with cotransfection of dominant negative epsin or eps15. The efficacy of the dominant negative epsin and eps15 can be tested by analyzing uptake of Texas Red–conjugated transferrin as described previously.

Comparison of GFP-Tagged Proteins with Endogenous Protein

The main disadvantages of using GFP-tagged Rho proteins to determine the subcellular localization of Rho proteins is that the proteins observed are both overexpressed and tagged with a relatively large, albeit inert, protein. It is, therefore, often necessary to validate results seen with GFP-tagged proteins using methods that detect endogenous protein. The two most common methods of observing endogenous proteins are immunofluorescence microscopy and subcellular fractionation. The disadvantage of immunofluorescence is that it requires suitable antibodies that recognize with high specificity Rho proteins in fixed cells (not widely available), and it necessitates fixation and cell permeabilization with attendant artifacts. The advantage of subcellular fractionation is that the readout is by immunoblot for which there are numerous excellent commercially available antibodies. The disadvantage is that the methods are cumbersome, time-consuming, and a clean separation of membranes from various compartments is almost never achieved. The methods for immunofluorescence and subcellular fractionation that we have used to localize Rho proteins are briefly described in the following.

Indirect Immunofluorescence

1. Circular #1 coverslips were washed with 70% ethanol.
2. Cos-1 cells were plated at a density of 2×10^5 cells per well into 6-well plates containing four round glass coverslips per well.
3. Twenty-four hours after plating, cells were rinsed twice with PBS and fixed with 4% paraformaldehyde at room temperature for 10 min.
4. Cells were washed three times with PBS followed by one of the following permeabilization treatments:

   a. 0.2% Triton X-100 at room temperature for 2 min.
   b. 0.5% saponin at room temperature for 2 min.

5. Fixed and permeabilized cells were then washed twice with PBS and blocked with 1% milk in PBS for 1 h.
6. Antibodies (e.g., anti-RhoB antiserum, Santa Cruz Biotechnology, Inc.) was added at dilutions ranging from 1:200 to 1:2000 for 1 h followed, after extensive washing, by a Texas Red–conjugated secondary antiserum (1:10,000 dilution, Jackson Immunoresearch Laboratories) for 45 min.

7. Coverslips were rinsed three times with PBS and once with dH$_2$O, dried, mounted on slides with a bleach retardant (e.g., Mowiol 4–88, Calbiochem), and imaged by epifluorescence or confocal microscopy.

Subcellular Fractionation

Fractionation methods are cell type specific. We have used the following protocol for MDCK cells (Choy *et al.*, 1999):

1. MDCK cells were grown to 90% confluence and suspended in homogenization buffer (0.3 *M* sucrose, 10 m*M* Tris-HCl [pH 7.5], 10 m*M* KCl, 1 m*M* DTT and protease inhibitors) and broken open using a ball-bearing homogenizer.

2. Unbroken cells and nuclei were removed by centrifugation at 600*g*, 5 min.

3. Membranes and cytosol were separated by centrifugation at 160,000*g*, 120 min.

4. The membrane pellet was resuspended in 1.35 *M* sucrose and separated by flotation through a discontinuous density gradient (0.25, 0.9, 1.35 *M* sucrose) by centrifugation at 350,000*g*, 120 min. Purified Golgi membranes were isolated from the interface between the 0.25 and 0.9 *M* sucrose layers. Galactosyl transferase activity was used as a marker of Golgi membranes (Choy *et al.*, 1999). Endosomes were isolated from the 1.35-*M* sucrose layer and verified by detection by immunoblot of early endosomal antigen-1 (EEA-1). A mixed fraction of PM and smooth ER was isolated from the interface between the 0.9 and 1.35 *M* sucrose layers.

5. PM was separated from other light membranes by loading the PM/SER fraction on top of a linear gradient of 5–20% Optiprep (Nycomed) and centrifuging at 95,000*g*, 18 h. PM was isolated from fractions near the top of the gradient and validated by immunoblot for the Na$^+$/K$^+$ ATPase.

Membrane fractions were analyzed by immunoblot for Rho proteins. To observe total rather than specific activity of each protein, it is important to load the immunoblots not with equal amounts of protein but rather with equal cell equivalents. Using [^{125}I]-labeled Protein A (Amersham) as the secondary reagent in immunoblots greatly facilitates quantification as a PhosphorImager can be used affording a very large dynamic range.

RhoB is an example of a Rho protein for which assignment to a specific subcellular compartment has not been straightforward. Unlike most Rho proteins, RhoB does not bind RhoGDI and, therefore, its endogenous, resting localization is not the cytosol but rather membrane compartments. We have consistently found GFP-RhoB targeted to the plasma membrane and to a perinuclear structure that colocalizes with Golgi markers (e.g., BODIPY ceramide and CFP-tagged galactosyltransferase) (Fig. 3 and Michaelson *et al.* [2001]). This localization parallels exactly that of GFP-H-Ras, an observation that makes sense, because the C-terminal targeting motifs of these proteins are very similar (a CAAX sequence preceded by dual palmitoylation sites). However, others have reported RhoB localized entirely to endosomes (Ellis and Mellor, 2000). It is, therefore, particularly important to examine the localization of endogenous RhoB using immunofluorescence and subcellular fractionation.

Using the most commonly used anti-RhoB antibody (Santa Cruz Biotechnologies), we observed that the apparent localization of endogenous RhoB in MDCK cells depended on the method of permeabilization. In Triton X-100 permeabilized cells, we observed a vesicular pattern consistent with published reports of an endosomal localization (Ellis and Mellor, 2000). In contrast, gentle permeabilization with saponin revealed a Golgi pattern identical to that observed with GFP-RhoB (Fig. 4A). Subcellular fractionation revealed that whereas 42% of total cellular RhoB was in the Golgi-enriched fraction and 36% co-purified with the PM, only 14% was found in fractions enriched for endosomes (Fig. 4B). Thus, although some ambiguity persists, the evidence for Golgi localization of endogenous RhoB is strong.

Acknowledgments

This work was supported by grants from the National Institutes of Health and the Burroughs Wellcome Fund.

References

Chen, H., Fre, S., Slepnev, V. I., Capua, M. R., Takei, K., Butler, M. H., Di Fiore, P. P., and De Camilli, P. (1998). Epsin is an EH-domain-binding protein implicated in clathrin-mediated endocytosis. *Nature* **394,** 793–797.

Chiu, V. K., Bivona, T., Hach, A., Sajous, J. B., Silletti, J., Wiener, H., Johnson, R. L., Cox, A. D., and Philips, M. R. (2002). Ras signalling on the endoplasmic reticulum and the Golgi. *Nat. Cell Biol.* **4,** 343–350.

Choy, E., Chiu, V. K., Silletti, J., Feoktistov, M., Morimoto, T., Michaelson, D., Ivanov, I. E., and Philips, M. R. (1999). Endomembrane trafficking of ras: The CAAX motif targets proteins to the ER and Golgi. *Cell* **98,** 69–80.

Ellis, S., and Mellor, H. (2000). Regulation of endocytic traffic by Rho family GTPases. *Trends Cell Biol.* **10,** 85–88.

Hoffman, G. R., Nassar, N., and Cerione, R. A. (2000). Structure of the Rho family GTP-binding protein Cdc42 in complex with the multifunctional regulator RhoGDI. *Cell* **100,** 345–356.

Huff, J. P., Grant, B. J., Penning, C. A., and Sullivan, K. F. (1990). Optimization of routine transformation of *Escherichia coli* with plasmid DNA. *Biotechniques* **9,** 570–577.

Innis, M., Gelfand, D., Sninsky, J., and White, T. (1989). "PCR Protocols." Elsevier, New York.

Matz, M. V., Fradkov, A. F., Labas, Y. A., Savitsky, A. P., Zaraisky, A. G. M., and Lukyanov, S. A. (1999). Fluorescent proteins from nonbioluminescent Anthozoa species. *Nature Biotechnology* **17**, 969–973.

Michaelson, D., Ahearn, I., Bergo, M., Young, S., and Philips, M. (2002). Membrane trafficking of heterotrimeric G proteins via the endoplasmic reticulum and Golgi. *Mol. Biol. Cell* **13**, 3294–3302.

Michaelson, D., Silletti, J., Murphy, G., D'Eustachio, P., Rush, M., and Philips, M. R. (2001). Differential localization of Rho GTPases in live cells. Regulation by hypervariable regions and rhogdi binding. *J. Cell Biol.* **152**, 111–126.

Prasher, D. C., Eckenrode, V. K., Ward, W. W., Prendergast, F. G., and Cormier, J. J. (1992). Primary structure of the *Aequorea victoria* green-fluorescent protein. *Gene* **111**, 229–233.

Rocks, O., Peyker, A., Kahms, M., Verveer, P. J., Koerner, C., Lumbierres, M., Kuhlmann, J., Waldmann, H., Wittinghofer, A., and Bastiaens, P. I. (2005). An acylation cycle regulates localization and activity of palmitoylated Ras isoforms. *Science* **307**, 1746–1752.

[23] Analysis of the Spatiotemporal Activation of Rho GTPases Using Raichu Probes

By Takeshi Nakamura, Kazuo Kurokawa, Etsuko Kiyokawa, and Michiyuki Matsuda

Abstract

GFP-based FRET probes that can visualize local activity changes in Rho GTPases in living cells are now available for examining the spatiotemporal regulation of these proteins. We previously developed FRET probes for Rho (and Ras) GTPases and collectively designated them "Ras and interacting protein chimeric unit" (Raichu) probes. In this chapter, we describe the principles and strategies used to develop Raichu-type FRET probes for Rho-family GTPases. The procedures for characterizing candidate probes, setting up the imaging system, and image acquisition/ processing are also explained. An optimal FRET probe should: (1) have a wide dynamic range (i.e., a high sensitivity); (2) demonstrate high fluorescence intensity (i.e., a high signal-to-noise ratio); (3) show target specificity; and (4) cause minimal perturbation of endogenous signaling cascades. Although improvements of FRET probes should be executed in a trial-and-error manner, we provide practical tips for their optimization. In addition, some experimental results are presented to illustrate the expanding number of fields for the application of Raichu-RhoA/Rac1/Cdc42, and the advantages and disadvantages of Raichu probes are discussed.

METHODS IN ENZYMOLOGY, VOL. 406   0076-6879/06 $35.00

Introduction

Rho-family GTPases are molecular switches that perceive changes in the extracellular or intracellular environment and transduce these changes into downstream effectors that regulate a wide spectrum of cellular functions, such as remodeling of the actin cytoskeleton, membrane trafficking, transcriptional activation, and cell cycle progression (Hall, 1998; Van Aelst and D'Souza-Schorey, 1997). In other words, these GTPases have roles in specifying when and where downstream events should happen in response to upstream cues. Therefore, to understand how Rho GTPases control various cellular functions, there is a clear need for knowledge regarding the spatiotemporal changes in the activities of these GTPases. This requirement has led us and others to develop fluorescent probes that can directly and nondestructively monitor the activation state of Rho GTPases in living cells (del Pozo *et al.*, 2002; Itoh *et al.*, 2002; Kraynov *et al.*, 2000; Nalbant *et al.*, 2004; Yoshizaki *et al.*, 2003). These probes are based on the principle of fluorescence resonance energy transfer (FRET). FRET is a process by which a fluorophore (donor) in an excited state transfers its energy to a neighboring fluorophore (acceptor) non-radioactively (Pollok and Heim, 1999; Tsien and Miyawaki, 1998), thereby causing the acceptor to emit fluorescence at its characteristic wavelength. FRET depends on a proper spectral overlap between the donor and the acceptor, the distance between the two fluorophores, and their relative orientation. The physical principles underlying FRET have been extensively reviewed elsewhere (Jares-Erijman and Jovin, 2003; Periasamy and Day, 1999).

We previously developed GFP-based FRET probes for Rho (and Ras) GTPases and collectively designated them "Ras and interacting protein chimeric unit" (Raichu) probes. The main purpose of this chapter is to provide a practical guide for how to make and use Raichu-type FRET probes. The advantages and disadvantages of Raichu probes are also discussed. To date, most researchers, including ourselves, have preferred to use a cyan-emitting mutant of GFP (CFP) and a yellow-emitting mutant of GFP (YFP) rather than the original pair of a blue-emitting mutant of GFP (BFP) and GFP (Zhang *et al.*, 2002). This is partly because the use of near-UV light for BFP excitation is harmful to cells and partly because the quantum efficiency of BFP is lower than that of CFP. Another disadvantage of BFP is its vulnerability to photobleaching. Excited CFP emanates cyan fluorescence at 475 nm. If YFP is placed in close proximity to CFP, FRET occurs and YFP emanates yellow fluorescence at 530 nm. It should be noted that, in the case of GFP and its mutants, the donor and acceptor fluorophores must be placed within 5 nm of each other for detectable FRET. Because GFP and its variants have a radial diameter of 6 nm, the presence of FRET between CFP and YFP indicates nearly direct contact between the two molecules.

Development of Raichu Probes for Rho GTPases

Basic Structure of Raichu Probes

The basic structure of Raichu probes is very similar to that of chameleon (Miyawaki *et al.*, 1997), a FRET probe used for monitoring the Ca^{2+} concentration. The Raichu probes are composed of four modules: the donor (CFP), the acceptor (YFP), a GTPase, and the GTPase-binding domain of its binding partner (Fig. 1A). In the Raichu probes for Rho GTPases, the YFP, GTPase-binding domain, GTPase, and CFP are sequentially connected from the N-terminus by spacers (Itoh *et al.*, 2002; Yoshizaki *et al.*, 2003). In the inactive GDP-bound form, the CFP and YFP in the probe are located at a distance from each other, mostly resulting in emission from CFP (Fig. 1A, top). On stimulation, the GDP on the GTPase

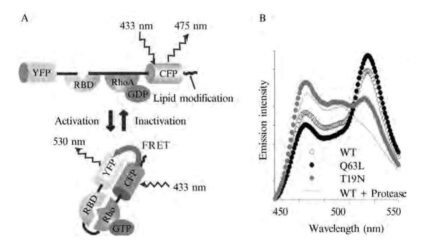

FIG. 1. GFP-based FRET probe for Rho GTPases (Raichu-RhoA). (A) Basic structure of Raichu-RhoA. The FRET probe for RhoA activation comprises the Rho-binding domain (RBD) of PKN and RhoA sandwiched between the GFP mutants YFP and CFP. For membrane localization, this probe is fused to the carboxyl-terminal region of RhoA (amino acids 173–193) or K-Ras4B (amino acids 169–188). On stimulation, the GDP on RhoA is exchanged for GTP, and FRET subsequently occurs. (B) Fluorescence profiles of Raichu-RhoA. Emission spectra of Raichu-RhoA expressed in 293T cells at an excitation wavelength of 433 nm are shown. WT, wild type; Q63L, constitutively active mutant; T19N, dominant-negative mutant. In the protease-treated sample, Raichu-RhoA-WT was cleaved with trypsin and proteinase K before analysis.

is exchanged for GTP, which induces interaction between the GTP-bound GTPase and the GTPase-binding domain of the effector protein. This intramolecular binding brings CFP into close proximity to YFP, thereby causing FRET (Fig. 1A, bottom). FRET is simultaneously manifested by quenching of the CFP fluorescence and emanation of the YFP fluorescence, and, therefore, the YFP/CFP ratio of Raichu probes is conveniently used as a representation of the FRET efficiency. Because the FRET efficiency of a Raichu probe is correlated with the GTP/GDP ratio (Itoh *et al.*, 2002; Yoshizaki *et al.*, 2003), the activities of Rho GTPases can be estimated from measurements of the YFP/CFP ratio.

Design Considerations

The mechanisms of activation/inactivation are essentially the same among different members of the Rho-family GTPase. Thus, the most straightforward way to generate a prototype FRET probe for any Rho GTPase is to adopt the basic Raichu probe structure shown in Fig. 1A. However, some specific issues should be considered as beneficial for cell imaging. An optimal probe should: (1) have a wide dynamic range (i.e., a high sensitivity), (2) demonstrate high fluorescence intensity (i.e., a high signal-to-noise ratio), (3) show target specificity, and (4) cause minimal perturbation to endogenous signaling. Among these criteria, prototype probes usually show room for improvement in the dynamic range. Under our criteria, Raichu-type probes are only recognized as good when the maximum increase (%) in the YFP/CFP ratio exceeds 30%.

To improve the dynamic range of a probe, it is first desirable to search for a GTPase-binding domain that has a moderate affinity for the GTPase (Yoshizaki *et al.*, 2003). A plausible explanation for this is that the GTPase-binding domain competes with GTPase activating proteins (GAPs) in the cells (Kurokawa *et al.*, 2005; Pertz and Hahn, 2004). This inhibition of GAPs leads to a relatively high GTP level in the probe, even in the un-stimulated state, and may therefore cause a narrowing of the dynamic range. For example, in our efforts to develop a probe for RhoA, we tested the RhoA-binding domains of four effector proteins (mDia1, Rhotekin, ROCK, and PKN) and found that only the RhoA-binding domain of PKN could be used for cell imaging (Yoshizaki *et al.*, 2003).

Crystallographic data for the GTPase and GTPase-binding domain can help to determine the minimum regions to be incorporated into the probe. Unfortunately, there are currently insufficient crystallographic data for optimal design of a Raichu probe in most cases. Therefore, trying various lengths of the GTPase and GTPase-binding domain is highly recommended. In addition, various order combinations of the four modules (i.e., YFP, CFP, GTPase, and GTPase-binding domain) should be tested.

The proper choice of GFP variants is important for the development of a bright and sensitive probe. Although we can easily make a probe brighter by using brighter YFP variants, such as Citrine and Venus (Griesbeck et al., 2001; Nagai et al., 2002), such a replacement does not always extend the dynamic range of the probe in our experience. The use of circularly permutated YFPs dramatically improved the dynamic range of the Ca^{2+} indicator chameleon (Nagai et al., 2004). Raichu probes have the same basic structure as chameleon probes, and it may therefore be possible to improve Raichu probes by using circularly permutated GFPs. Monomeric GFP variants (Zacharias et al., 2002) may be other candidates for such improvement. In fact, there do not seem to be any fixed rules that can predict which variant will be optimal for a given probe, and the choice of GFP variants should be executed in a trial-and-error manner. In addition, 11 amino acids at the carboxyl terminus of GFP can be truncated without affecting its fluorescence profile. In most of our Raichu probes, we have removed the 11 carboxyl-terminal residues of YFP, hoping to reduce the flexibility between YFP and the subsequent module.

The length and sequence of the spacers are also critical. If the FRET efficiency of a prototype probe varies to some extent on activation, the possibility of further improvement by changing the spacer between the GTPase and the GTPase-binding domain should be considered. As spacers, we usually use one to six repeats of the sequence Gly-Gly-Ser-Gly-Gly. It is considered that Gly provides flexibility, whereas Ser prevents aggregation of peptide chains. Misfolding of CFP, a C-terminally located fluorophore, occasionally occurs, and this can sometimes be rectified by modifying the spacer before the CFP.

The ideal location for a probe in a cell has been a matter of debate (Kurokawa et al., 2005; Pertz et al., 2004; Yoshizaki et al., 2003). One idea is that the probe should be located diffusely throughout the cell so that any signal changes can be detected at any region. However, if the probe is diffusible freely, the activated probe is diluted by the large amount of inactive cytosolic probe in the thick parts of cells. To avoid this dilution effect, we add either of two types of modification to the C terminus of a Raichu probe (Fig. 1A). One is the GTPase's own authentic CAAX-box, and the other is the CAAX-box of K-Ras4B. The former is expected to enable the probe to colocalize with the endogenous protein (Yoshizaki et al., 2003). In some cases, this strategy works well. However, this approach does not always yield a high signal-to-noise ratio, especially when only a limited fraction of the GTPase is activated on stimulation. In this case, choosing to locate the probes at the site of interest is a possibility, and the addition of the CAAX-box of K-Ras4B enables the probe to monitor activity changes at the plasma membrane. As an example, we adduce

Cdc42, which is located at the plasma membrane, at the endomembrane as well as in the cytoplasm. On stimulation by growth factors, Cdc42 is activated primarily at the plasma membrane (Kurokawa *et al.*, 2004). When the probe is localized at the same place as the endogenous Cdc42 and viewed from the top of the cell using epifluorescence microscopy, Cdc42 activation seems to occur mostly in the peripheral regions of the cell. This is because most of the probe at the endomembrane does not respond to the stimulation. Therefore, fusing the CAAX-box of K-Ras4B and only placing the Raichu-Cdc42 probe at the plasma membrane can dramatically improve the signal-to-noise ratio, such that diffuse activation of Cdc42 throughout the plasma membrane can be observed (Kurokawa *et al.*, 2004).

Characterization of Candidate Probes

In the past, the most demanding task during the development of a novel probe was functional screening using spectrometric characterization. This situation has now changed, and preparation of probes for spectral analysis using 293T cells, rather than *Escherichia coli*, is the key to rapid characterization. Most Raichu-type probes become insoluble when prepared in bacteria, whereas 293T cells produce sufficient amounts of soluble probes for use in fluorescence analysis. Note that candidate probes without lipid modification at the C terminus is often preferable for spectral analysis, because this lipid modification generally reduces the level of probe expression.

The method for fluorescence spectral analysis using 293T cells is as follows.

1. Transfect 10 μg of a probe-encoding plasmid into 293T cells grown on 100 mm collagen-coated dishes using a calcium phosphate coprecipitation method.

2. At 6 h after the transfection, change the culture medium to a phenol red-free medium.

3. At 48 h after the transfection, harvest the cells in lysis buffer (20 mM Tris-HCl [pH 7.5], 100 mM NaCl, 5 mM MgCl2, 0.5% Triton X-100). Centrifuge the lysates at 15,000 rpm for 15 min at 4°. Alternatively, ultracentrifugation at 50,000 rpm is recommended for the complete removal of cell debris, which increases the background fluorescence.

4. Transfer the cleared lysates into 3-ml cuvettes and place the cuvettes in an FP-750 spectrophotometer (JASCO, Tokyo, Japan). Next, illuminate the lysates with an excitation wavelength of 433 nm and obtain a fluorescence spectrum from 450–550 nm. The background is subtracted by using the spectrum of cell lysates prepared without transfection.

For the characterization of a candidate probe, we introduce a constitutively active or inactive mutation into the GTPase in the probe for comparison with the same probe containing the wild-type GTPase. An example of the results obtained is shown in Fig. 1B. Alternatively, we cotransfect them with guanine nucleotide exchange factor (GEF) and a GAP toward the GTPase and compare the new spectrum with that of samples transfected with the probe alone. In addition, the presence of FRET can easily be confirmed by protease treatment. GFP proteins are exceptionally resistant to proteases, and, therefore, lysates containing Raichu probes produce free YFP and CFP after treatment with trypsin and proteinase K. The disappearance of YFP fluorescence in the spectrum after protease treatment indicates the presence of FRET (Fig. 1B). Fluorescence spectral analysis becomes even easier when the FreeStyle 293-F cell line (Invitrogen), a variant of the 293 cell line adapted for suspension growth, is used. This cell line is very easy to culture, transfect, and harvest, and only 1.5 ml of cell suspension is sufficient to obtain a fluorescence spectrum.

Practically, further evaluation of a probe is recommended before use in a wide range of applications to check the following: (1) Does the probe show a linear correlation between its GTP loading and FRET efficiency when cotransfected with various quantities of GEF or GAP? (2) do the probe and its endogenous counterpart show comparable responses to physiological stimulations when examined by biochemical methods?

Time-Lapse Imaging Using Raichu Probes

Imaging System

Our basic instrument consists of a normal inverted fluorescence microscope, filter changers for dual-emission imaging, and a cooled CCD camera (Fig. 2). The filter changers and the camera are under the control of a computer. The stage of the microscope is covered by an aryl chamber with a thermocontroller. The components of our system are:

Microscope (Fig. 2-1): an IX70, IX71 or IX81 inverted epifluorescence microscope (Olympus, Tokyo, Japan). For long-term and multipoint image acquisition, IX81 can be equipped with a focus-drift compensation system (IX2-ZDC; Olympus) and an automatic programmable XY stage (MD-XY30100T-META, Sigma Koki, Tokyo, Japan).

Objective lens (Fig. 2-2): UPlanApo 10× and 40× oil, PlanApo 60× oil (Olympus) and others if necessary.

Filter changers (Fig. 2-3A,B): MAC 5000 (Ludl Electronic Products Ltd., Hawthorne, NY).

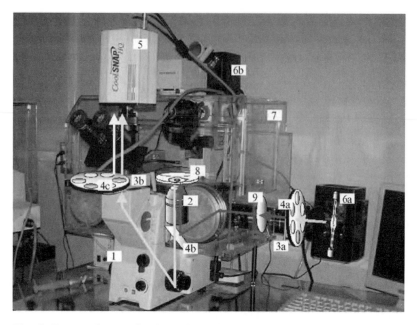

Fig. 2. Recent photograph of our basic instrument for FRET imaging. Elements numbered 1–9 are explained in the text.

Filter set (Fig. 2-4A-C): the following two requirements should be fulfilled based on the fluorescence profiles of CFP and YFP: (1) maximization of the signal intensity captured by the CCD camera, and (2) minimization of donor emission bleed-through and direct acceptor excitation (Hailey *et al.*, 2002). For routine dual-emission imaging studies, we use an XF1071 (440AF21) excitation filter, an XF2034 (455DRLP) dichroic mirror, and two emission filters (XF3075 [480AF30] for CFP and XF3079 [535AF26] for YFP) (Miyawaki and Tsien, 2000) (all from Omega Optical, Brattleboro, VT). Equivalent filters from other companies are also used.

Cooled CCD camera (Fig. 2-5): Cool SNAP-HQ (Roper Scientific) or its equivalent.

Light sources (Fig. 2-6A,B): 75-W Xenon lamp (Olympus) for fluorescence imaging and a halogen lamp (Olympus) for DIC image acquisition.

Transparent image acquisition system: to record the cell morphology, the microscope must be equipped with either a phase-contrast or DIC system. However, the use of phase-contrast is not recommended, because 10–20% of the fluorescence is lost with a phase-contrast lens. To use phase-contrast, it is preferable to place the phase-contrast unit outside the objective lens. Note that, when a DIC system is used, the analyzer for the DIC images must be set in the emitter wheel changer, not in the body of the microscope.

Acryl chamber (Fig. 2-7): custom-made or Olympus IBMU. A chamber is very desirable, because the temperature, pH, and osmotic pressure of the culture medium should ideally be kept constant to maintain the cells in a healthy state. Furthermore, a stable microscope temperature helps to keep a sharp focus during time-lapse experiments. To further stabilize the pH and osmotic pressure of the medium, we also use a small custom-made chamber that just covers the dish. A CO_2 supply to this small chamber is available, although we prefer to add 15–25 mM HEPES buffer (final concentration) to maintain the medium at a neutral pH.

Software to control the camera and filter changers: MetaMorph (Universal Imaging, West Chester, PA).

Cells are cultured in 35-mm glass-bottom dishes (Fig. 2-8; Asahi Techno Glass, Tokyo, Japan), because plastic is unsuitable for fluorescence and DIC observation. If necessary, the dishes are coated with collagen or other substrates to increase cell adhesiveness. The culture medium should be as transparent as possible, and, therefore, phenol red and serum should either be removed entirely or at least lowered. The cells are illuminated with a 75 W-Xenon lamp through bandpass and ND filters (Fig. 2-9).

Image Acquisition

We strongly recommend starting with the easiest experiment using established probes to become familiar with the technique. For example, we recommend monitoring agonist-induced calcium oscillation with a recently developed calcium sensor chameleon YC3.60, which has an extraordinary wide dynamic range (Nagai *et al.*, 2004). Imaging of Ras activation in EGF-stimulated COS cells with Raichu-Ras probes (Mochizuki *et al.*, 2001) is another fairly easy example. The following is a standard procedure for image acquisition:

1. Prepare cells expressing Raichu probes in 35-mm glass-bottom dishes. The volume of the imaging medium is usually 2 ml.

2. Turn on the thermocontroller, Xenon lamp, and other electrical devices 1 h before observation to eliminate focus drift during time-lapse experiments.

3. If cells should be treated with factors or drugs, dilute them beforehand in a small amount (300–500 μl) of conditioned medium.

4. Search for an appropriate cell to image. Cells transfected with Raichu probes show various shapes and fluorescence intensities, and it is, therefore, critical to find cells that give clear and reproducible results.

In general, cells expressing higher amounts of Raichu probes and showing no damage from overexpression are good candidates. In addition, each Raichu probe has a characteristic range for the YFP/CFP ratio. Therefore, cells with ratio values outside the proper range should be avoided, because misfolding of either the YFP or the CFP in the probe is possible.

5. Set up the imaging conditions on the multidimensional acquisition module of the MetaMorph software and start image acquisition. The main parameters are: (1) the number of wavelengths (usually three, i.e., YFP, CFP and DIC); (2) the camera binning factor (usually 4×4 in our system); (3) the expose time at each wavelength; and (4) the time interval between frames.

Note: The cells should be minimally illuminated. Never forget to put an appropriate ND filter into the light path. Usually, we use a 6–12% transmission filter. There are two types of GFP photoconversions: reversible photochromism and irreversible photobleaching (Prendergast, 1999). Among the various GFP mutants, YFP is particularly labile. Therefore, by searching for and focusing on a cell, YFP fluorescence is more or less decayed. If cells should be treated with factors or drugs (prepared in step 3), wait until the fluorescence intensity reaches a stationary value. Usually, we need to wait for 5–10 min before the YFP fluorescence recovers. In addition, to minimize irreversible photobleaching, the exposure time during time-lapse observations should be as short as possible, while still obtaining a sufficient level of signals.

FRET imaging using Raichu probes is a quantitative method for measuring the activities of small GTPases, and, therefore, improvement of the signal-to-noise ratio is more crucial than for usual fluorescent imaging. The signal-to-noise ratio can be improved by decreasing the background noise, which consists of camera noise and background fluorescence. In most cases, the camera noise is a fixed value. Background fluorescence can be reduced by optimizing the ingredients of the imaging medium. Another way to improve the signal-to-noise ratio is to increase the net amount of signals. As stated previously, one easy way to achieve this is to use brighter YFP variants, such as Citrine and Venus. In addition, reducing the resolution of the image will increase the amount of signals per pixel. Alternatively, a higher number of signals per pixel can be achieved by increasing the concentration of the probe or the exposure time for image acquisition. In general, a long exposure time or reduced resolution is unfavorable, and, therefore, increasing the probe concentration in the cells seems to be the easiest way to obtain brighter images. However, we must keep the probe concentration at the lowest level possible to minimize interference with endogenous signaling (Miyawaki, 2003b). All considered, FRET imaging is confronted with conflicting requirements, and thus researchers must resort to trial-and-error to identify the optimal conditions.

Image Analysis

If researchers are familiar with cell culture, gene transfer, and microscopic observation, there is no need for specific skills to obtain fluorescence images using Raichu probes. However, to obtain correct FRET images by image processing requires considerable skill, and expert training is desirable.

From the raw images of YFP and CFP fluorescence, ratio images are constructed using the image-processing functions of the MetaMorph software. The following is the minimum procedure for obtaining ratio images:

1. Save the images as a stack file ("Review Multi Dimensional Data" command in MetaMorph).
2. Set a region containing no cells as the background region. Next, subtract the average intensity of the background region from the intensity of the stack file made in step 1 ("Use Region For Background" command). Note that the same region should be used for the background subtractions of the stack files of the YFP and CFP images.
3. Make ratio images of YFP/CFP ("Ratio images" command). We present FRET images in the intensity-modulated display (IMD) mode (for an example, see Fig. 3). This mode associates the color hue with the ratio value and the intensity of each hue with the source image brightness. Thus, the color represents the GTPase activity (e.g., red, high activity; blue, low activity), and the intensity represents the probe localization. Furthermore, by using the MetaMorph software, we can calculate and plot time-dependent changes of the YFP/CFP ratio within selected regions or cells.

Applications

Rho GTPases control rearrangements of the actin cytoskeleton in a wide range of cellular functions, such as cell migration, cytokinesis, phagocytosis, and neurite outgrowth (Hall, 1998; Van Aelst and D'Souza-Schorey, 1997). Thus, the fields for application of Raichu-RhoA/Rac1/Cdc42 have been expanding. Some examples are presented in the following.

Cell Migration

In randomly migrating HT1080 cells, the activities of Rac1 and Cdc42 show increasing gradients from the tail to the leading edge (Itoh *et al.*, 2002). FRET imaging also delineated a difference between the localizations of

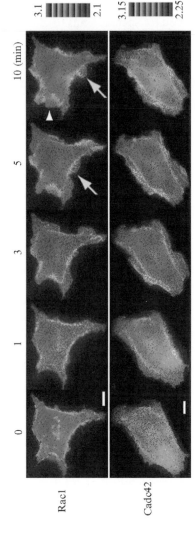

FIG. 3. Time-lapse experiments of Rac1 and Cdc42 activations upon EGF stimulation in Cos1 cells. Serum-starved Cos1 cells expressing Raichu-Rac1 or Raichu-Cdc42 were treated with EGF and imaged every 30 sec. FRET images at the indicated time points after EGF addition are shown in the IMD mode (see text). The yellow arrows and white arrowheads indicate nascent and retracting lamellipodia, respectively. The upper and lower limits of the ratio range are shown on the right. Bars indicate 10 μm. Reproduced with permission from Kurokawa *et al.* (2004). (See color insert.)

activated Rac1 and Cdc42 within nascent lamellipodia, because the Rac activity was highest immediately behind the leading edge, whereas the Cdc42 activity was most prominent at the tip of the leading edge. Similar distributions of active Rac1 and Cdc42 were observed in directionally migrating MDCK cells during the wound healing process (Takaya *et al.*, 2004).

Growth Factor–Stimulated Morphological Changes

In EGF-stimulated Cos1 and A431 cells, both Rac1 and Cdc42 were activated diffusely at the plasma membrane, followed by lamellipodia protrusion (Fig. 3; Kurokawa *et al.*, 2004). Although the Rac1 activity subsided rapidly, the Cdc42 activity was sustained at the lamellipodia. These FRET data and other functional studies argue that Rac1 and Cdc42 synergistically induce lamellipodia in EGF-stimulated Cos1 and A431 cells.

We also examined the spatiotemporal regulations of Rac1 and Cdc42 activities during NGF-induced neurite outgrowth in PC12 cells (Aoki *et al.*, 2004). Immediately after NGF addition, Rac1 and Cdc42 were transiently activated in broad areas of the cell periphery, and a repetitive activation-and-inactivation cycle was then observed at the motile tips of protrusions. These localized activations, which were more evident in PC12 cells treated with NGF for more than 24 h, may be required for neurite extension, because the expressions of constitutively active mutants of Rac1 and Cdc42 throughout the cells abrogated NGF-induced neurite outgrowth.

Cytokinesis

FRET imaging during cell division revealed an unexpected regulation of Rho GTPases (Yoshizaki *et al.*, 2003). In HeLa cells, the activities of RhoA, Rac1, and Cdc42 were high at the plasma membrane during interphase and then decreased rapidly on entry into M phase. After anaphase, the RhoA activity increased at the plasma membrane including the cleavage furrow. Rac1 activity was suppressed at the spindle midzone and then increased on the polar side of the plasma membranes after telophase. Cdc42 activity was suppressed at the plasma membrane and high in the intracellular membrane compartments.

Discussion

Advantages and Disadvantages of Raichu Probes

In general, GFP-based FRET probes are classified into two types: intermolecular (or bimolecular) and intramolecular (or unimolecular) (Miyawaki, 2003b). In an intermolecular probe, YFP and CFP are fused

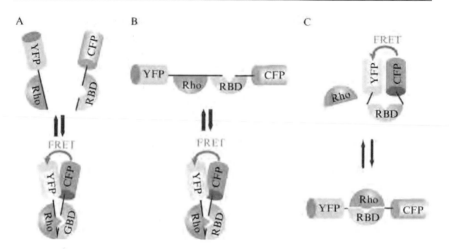

FIG. 4. Intermolecular and intramolecular FRET probes for Rho GTPases. (A) Intermolecular probe, in which YFP and CFP are fused to a Rho GTPase and a Rho GTPase-binding domain (RBD), respectively. On stimulation, association of the Rho GTPase and the RBD brings YFP into close proximity to CFP, and FRET subsequently occurs. (B) Intramolecular probe (complete-type) in which a Rho GTPase and an RBD are sandwiched between YFP and CFP. (C) Intramolecular probe (incomplete-type) in which only an RBD is sandwiched between YFP and CFP. In this probe, FRET occurs in the unstimulated state. On stimulation, endogenous active Rho GTPases bind to the RBD in the probe, thereby dissociating YFP and CFP and decreasing the FRET efficiency.

to a GTPase and a GTPase-binding domain, respectively, and interact with one another on stimulation (Fig. 4A). The Raichu probes are intramolecular probes, in which all four modules are combined into a single peptide chain (Fig. 4B). In our opinion, using an intramolecular probe is a better way to start, provided good probes are available, because the biggest hurdle of an intramolecular probe is the large amount of labor required to identify its optimal design. There are two reasons why we chose to use an intramolecular probe as the basic design for generating Raichu probes. First, in an intramolecular probe, a GTPase and a GTPase-binding domain are placed in close proximity, such that the GTPase-binding domain can easily find the activated GTPase on stimulation. This will increase the percentage of real FRET signals versus undesired signals arising from donor emission bleed-through and direct acceptor excitation (Hailey et al., 2002; Kurokawa et al., 2005). Concurrently, perturbation of the endogenous signaling will be reduced (Miyawaki, 2003b). Second, the molar ratio of CFP and YFP is a fixed constant at every pixel in an intramolecular probe, and, therefore, a simple ratio of the CFP and YFP fluorescence intensities suffices to represent the FRET efficiency.

From a general point of view, it should be noted that the applications suitable for intermolecular and intramolecular probes differ. Thus, in practice, the type of probe is chosen depending on the aim of the experiment. In the case of monitoring an interaction between proteins A and B, it is natural to select an intermolecular probe. However, correction of the FRET signals obtained with an intermolecular probe is elaborate but executable (Kraynov et al., 2000; Sekar and Periasamy, 2003). In contrast, to visualize changes in protein activity, pH, Ca^{2+} concentration, etc., an intramolecular probe is preferable, because in this case, researchers can take advantage of the merits described previously.

Insensitivity to RhoGDI activity has been pointed out as a drawback of the Raichu probes for Rho GTPases (Pertz et al., 2004). In addition to GEFs and GAPs, Rho-family GTPases are regulated by another class of regulators, RhoGDIs. The interactions between RhoGDIs and Rho GTPases involve both the isoprenoid moiety and the switch I and II domains (Hoffman et al., 2000). Therefore, even a Raichu probe with an authentic CAAX-box in its C terminus escapes RhoGDI regulation. In other words, Raichu probes for RhoA/Rac1/Cdc42 can monitor the balance between the GEF and GAP activities at membranes. Thus, we have developed another type of FRET probe for Rho-family GTPases (i.e., an incomplete-type intramolecular probe), which can monitor the local concentration of the endogenous active GTPase and thus enable RhoGDI activity to be considered (Itoh et al., 2002; Yoshizaki et al., 2003). In this type of probe, the GTPase-binding domain of an effector protein is sandwiched by YFP and CFP (Fig. 4C). Interaction of the endogenous GTPase with the GTPase-binding domain in the probe dissociates YFP and CFP, thereby decreasing the FRET efficiency. It should be noted that the signal-to-noise ratio of the incomplete-type probe is significantly lower than that of the corresponding intramolecular FRET probe in most cases.

Besides the Raichu probes, two other fluorescent probes for monitoring the activity of Rho GTPases in living cells have been reported (Kraynov et al., 2000; Srinivasan et al., 2003). The first is an intermolecular FRET probe for Rac activation that consists of a GFP-Rac fusion protein (donor) and the CRIB domain of PAK labeled with an Alexa-546 acceptor dye (Kraynov et al., 2000). The CRIB domain was produced in bacteria, and the labeled CRIB domain was microinjected into GFP-Rac–expressing cells. In the case of this intermolecular probe, the level of Rac activation cannot be obtained from the simple YFP/CFP ratio, and an elaborate correction considering bleed-through and direct acceptor excitation is needed (Kraynov et al., 2000; Sekar et al., 2003). Furthermore, the amount of the labeled CRIB domain needs to be tightly controlled, because the expression of the CRIB domain alone can exert dominant-negative effects on the endogenous Rac protein. Although this Rac monitor can deduce useful information (del Pozo *et al.*, 2002), the drawbacks noted have prevented widespread application of this type of intermolecular FRET probe. The second is a fluorescent indicator for Rac activation and involves YFP fusion to the CRIB domain of PAK (Srinivasan *et al.*, 2003). Pertz and Hahn (Pertz *et al.*, 2004) pointed out the lower sensitivity of this type of indicator compared with the corresponding FRET probe, because its localization has to be discerned over a background of randomly distributed indicators that are fluorescing at the same wavelength. Furthermore, the expression level of this Rac indicator needs to be tightly controlled because of its potential dominant-negative effect on endogenous signaling. Taking into consideration all the facts noted previously, we believe that the Raichu probes currently represent the most reliable and easiest tools for monitoring the activities of Rho GTPases in living cells.

Future Perspectives

GFP-based FRET probes, including Raichu probes, hold promise for facilitating better understanding of the spatiotemporal dynamics of the activation of Rho GTPases in living cells. The next step will be their application to more physiological systems. Using specific promoters and targeting signals, FRET probes are introduced into an intact organism and directed to specific tissue regions (Miyawaki, 2003a). Reconstruction of three-dimensional images using a two-photon microscope (Nikolenko *et al.*, 2003) and a confocal microscope equipped with a Nipkow disc (Nakano, 2002) will further increase the range of applications for these probes.

Acknowledgments

This work was supported by a grant from the Ministry of Education, Science, Sports and Culture of Japan and a grant from the Health Science Foundation, Japan. We thank N. Yoshida, N. Fujimoto, and K. Fukuhara for their technical assistance and members of the Matsuda laboratory for technical advice and helpful input.

References

Aoki, K., Nakamura, T., and Matsuda, M. (2004). Spatio-temporal regulation of Rac1 and Cdc42 activity during nerve growth factor-induced neurite outgrowth in PC12 cells. *J. Biol. Chem.* **279,** 713–719.
Del Pozo, M. A., Kiosses, W. B., Alderson, N. B., Meller, N., Hahn, K. M., and Schwartz, M. A. (2002). Integrins regulate GTP-Rac localized effector interactions through dissociation of Rho-GDI. *Nat. Cell Biol.* **4,** 232–239.

Griesbeck, O., Baird, G. S., Campbell, R. E., Zacharias, D. A., and Tsien, R. Y. (2001). Reducing the environmental sensitivity of yellow fluorescent protein. Mechanism and applications. *J. Biol. Chem.* **276**, 29188–29194.

Hailey, D. W., Davis, T. N., and Muller, E. G. (2002). Fluorescence resonance energy transfer using color variants of green fluorescent protein. *Methods Enzymol.* **351**, 34–49.

Hall, A. (1998). Rho GTPases and the actin cytoskeleton. *Science* **279**, 509–514.

Hoffman, G. R., Nassar, N., and Cerione, R. A. (2000). Structure of the Rho family GTP-binding protein Cdc42 in complex with the multifunctional regulator RhoGDI. *Cell* **100**, 345–356.

Itoh, R. E., Kurokawa, K., Ohba, Y., Yoshizaki, H., Mochizuki, N., and Matsuda, M. (2002). Activation of rac and cdc42 video imaged by fluorescent resonance energy transfer-based single-molecule probes in the membrane of living cells. *Mol. Cell Biol.* **22**, 6582–6591.

Jares-Erijman, E. A., and Jovin, T. M. (2003). FRET imaging. *Nat. Biotechnol.* **21**, 1387–1395.

Kraynov, V. S., Chamberlain, C., Bokoch, G. M., Schwartz, M. A., Slabaugh, S., and Hahn, K. M. (2000). Localized Rac activation dynamics visualized in living cells. *Science* **290**, 333–337.

Kurokawa, K., Itoh, R. E., Yoshizaki, H., Nakamura, T., Ohba, Y., and Matsuda, M. (2004). Coactivation of Rac1 and Cdc42 at lamellipodia and membrane ruffles induced by epidermal growth factor. *Mol. Biol. Cell* **15**, 1003–1010.

Kurokawa, K., Takaya, A., Fujioka, A., Terai, K., and Matsuda, M. (2004). Visualizing the signal transduction pathways in living cells with GFP-based, FRET probes. *Acta Histochem. Cytochem.* **34**, 347–355.

Miyawaki, A. (2003a). Fluorescence imaging of physiological activity in complex systems using GFP-based probes. *Curr. Opin. Neurobiol.* **13**, 591–596.

Miyawaki, A. (2003b). Visualization of the spatial and temporal dynamics of intracellular signaling. *Dev. Cell* **4**, 295–305.

Miyawaki, A., Llopis, J., Heim, R., McCaffery, J. M., Adams, J. A., Ikura, M., and Tsien, R. Y. (1997). Fluorescent indicators for Ca^{2+} based on green fluorescent proteins and calmodulin. *Nature* **388**, 882–887.

Miyawaki, A., and Tsien, R. Y. (2000). Monitoring protein conformations and interactions by fluorescence resonance energy transfer between mutants of green fluorescent protein. *Methods Enzymol.* **327**, 472–500.

Mochizuki, N., Yamashita, S., Kurokawa, K., Ohba, Y., Nagai, T., Miyawaki, A., and Matsuda, M. (2001). Spatio-temporal images of growth-factor-induced activation of Ras and Rap1. *Nature* **411**, 1065–1068.

Nagai, T., Ibata, K., Park, E. S., Kubota, M., Mikoshiba, K., and Miyawaki, A. (2002). A variant of yellow fluorescent protein with fast and efficient maturation for cell-biological applications. *Nat. Biotechnol.* **20**, 87–90.

Nagai, T., Yamada, S., Tominaga, T., Ichikawa, M., and Miyawaki, A. (2004). Expanded dynamic range of fluorescent indicators for Ca^{2+} by circularly permuted yellow fluorescent proteins. *Proc. Natl. Acad. Sci. USA* **101**, 10554–10559.

Nakano, A. (2002). Spinning-disk confocal microscopy—a cutting-edge tool for imaging of membrane traffic. *Cell Struct. Funct.* **27**, 349–355.

Nalbant, P., Hodgson, L., Kraynov, V., Toutchkine, A., and Hahn, K. M. (2004). Activation of Endogenous Cdc42 Visualized in Living Cells. *Science* **305**, 1615–1619.

Nikolenko, V., Nemet, B., and Yuste, R. (2003). A two-photon and second-harmonic microscope. *Methods* **30**, 3–15.

Periasamy, A., and Day, R. N. (1999). Visualizing protein interactions in living cells using digitized GFP imaging and FRET microscopy. *Methods Cell Biol.* **58**, 293–314.

Pertz, O., and Hahn, K. M. (2004). Designing biosensors for Rho family proteins-deciphering the dynamics of Rho family GTPase activation in living cells. *J. Cell Sci.* **117,** 1313–1318.

Pollok, B. A., and Heim, R. (1999). Using GFP in FRET-based applications. *Trends Cell Biol.* **9,** 57–60.

Prendergast, F. G. (1999). Biophysics of the green fluorescent protein. *Methods Cell Biol.* **58,** 1–18.

Sekar, R. B., and Periasamy, A. (2003). Fluorescence resonance energy transfer (FRET) microscopy imaging of live cell protein localizations. *J. Cell Biol.* **160,** 629–633.

Srinivasan, S., Wang, F., Glavas, S., Ott, A., Hofmann, F., Aktories, K., Kalman, D., and Bourne, H. R. (2003). Rac and Cdc42 play distinct roles in regulating PI(3,4,5)P3 and polarity during neutrophil chemotaxis. *J. Cell Biol.* **160,** 375–385.

Takaya, A., Ohba, Y., Kurokawa, K., and Matsuda, M. (2004). RalA activation at nascent lamellipodia of epidermal growth factor-stimulated Cos7 cells and migrating Madin-Darby canine kidney cells. *Mol. Biol. Cell* **15,** 2549–2557.

Tsien, R. Y., and Miyawaki, A. (1998). Seeing the machinery of live cells. *Science* **280,** 1954–1955.

Van Aelst, L., and D'Souza-Schorey, C. (1997). Rho GTPases and signaling networks. *Genes Dev.* **11,** 2295–2322.

Yoshizaki, H., Ohba, Y., Kurokawa, K., Itoh, R. E., Nakamura, T., Mochizuki, N., Nagashima, K., and Matsuda, M. (2003). Activity of Rho-family G proteins during cell division as visualized with FRET-based probes. *J. Cell Biol.* **162,** 223–232.

Zacharias, D. A., Violin, J. D., Newton, A. C., and Tsien, R. Y. (2002). Partitioning of lipid-modified monomeric GFPs into membrane Microdomains of live cells. *Science* **296,** 913–916.

Zhang, J., Campbell, R. E., Ting, A. Y., and Tsien, R. Y. (2002). Creating new fluorescent probes for cell biology. *Nat. Rev. Mol. Cell Biol.* **3,** 906–918.

[24] Measurement of Activity of Rho GTPases During Mitosis

By FABIAN OCEGUERA-YANEZ and SHUH NARUMIYA

Abstract

The members of Rho-family GTPases regulate progression through mitosis. Rho induces the contractile ring at the equatorial cortex of the dividing cell, and thus works as a molecular switch between nuclear division and cytokinesis. Cdc42 regulates the progression from prometaphase to metaphase by stabilizing microtubule attachment to the kinetochore. These results suggest that Rho GTPases are activated at specific points in mitosis and regulate each step. Here we describe the methods to analyze the activity and regulation of Rho GTPases during mitosis.

METHODS IN ENZYMOLOGY, VOL. 406   0076-6879/06 $35.00
DOI: 10.1016/S0076-6879(06)06024-1

Introduction

The members of Rho-family GTPases regulate cell morphogenesis by controlling specific types of actin cytoskeleton and by local regulation of microtubule (MT) dynamics (Etienne-Manneville and Hall, 2002). Through these actions, they also regulate progression through mitosis. It is long known that Rho works as a molecular switch between nuclear division and cytokinesis by inducing the contractile ring at the equatorial cortex of dividing cells (Mabuchi et al., 1993). Recently it has been shown that Cdc42 regulates the progression from prometaphase to metaphase by stabilizing microtubule attachment to the kinetochore (Yasuda et al., 2004). These results suggest that Rho GTPases are activated at specific points in mitosis and regulate each step. We have applied the pull-down assays for the GTP-bound form of Rho, Rac, and Cdc42 and found out that Rho is, indeed, activated from anaphase to telophase, whereas Cdc42 is activated in metaphase (Kimura et al., 2000; Oceguera-Yanez et al., 2005). Using this method, we have also examined the identity of a Rho guanine nucleotide exchange factor (GEF) involved in these processes (Kimura et al., 2000; Oceguera-Yanez et al., 2005). In this chapter, we describe the methods we used in these studies.

Activity of Cdc42 or Rac During Mitosis

We measure changes in the levels of GTP-bound Cdc42 or Rac during mitosis by applying a pull-down assay using a glutathione S-transferase (GST) fusion of the Cdc42-Rac-interactive–binding (CRIB) domain of Pak1 to cell cycle–synchronized HeLa cells. The CRIB-Pak domain we use has been characterized as a useful tool to precipitate GTP-bound Cdc42 and Rac in a variety of systems and conditions (Azim et al., 2000; Matsuo et al., 2002). We first enrich HeLa S3 cells in early S phase by the use of the double thymidine block and then use nocodazole to collect cells in prometaphase. Cells harvested at different times after removal of nocodazole are snap-frozen and used for preparation of cell lysates for the pull-down assay.

Cell Cycle Synchronization of HeLa S3 Cells

HeLa S3 cells are maintained in Dulbecco's modified Eagle medium (DMEM) supplemented with 10% FCS and antibiotics at 37° in an atmosphere containing 5% CO_2. The cells are synchronized at early S phase by the double thymidine block (Bostock et al., 1971). Cells grown in one 10-cm dish at 90% confluency, are trypsinized, and seeded onto 20 10-cm dishes. After overnight culture, the medium is replaced by DMEM containing

Microtubules   Chromosomes

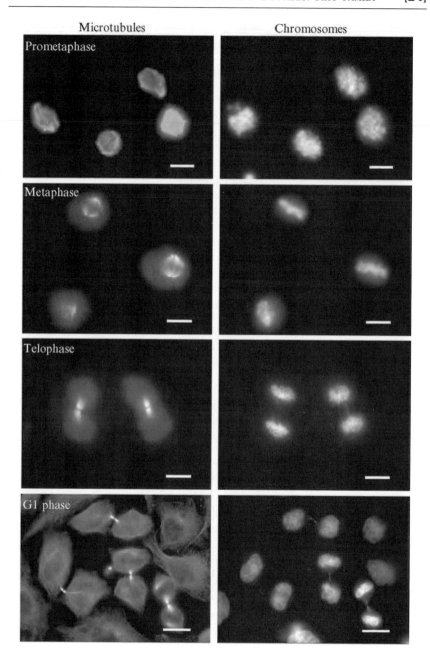

10 mM thymidine and 5% FCS, and the cells are incubated for 12–14 h. The thymidine-containing medium is then removed, and the cells are washed twice with PBS without divalent cations (PBS(–)). The cells are cultured further in DMEM containing 10% FCS for 10 h and then subjected to the second thymidine block for 12 h. The cells are washed twice with PBS and placed in fresh DMEM containing 10% FCS. After 8 h of incubation, four 10-cm dishes of the culture is collected as the cells in G2 phase, and the rest of the culture is incubated with 40 ng/ml nocodazole for 6 h. Round cells arrested in prometaphase are collected by the shake-off method, washed twice with PBS(–), reseeded, and further incubated in DMEM containing 10% FCS. The cells are collected at 25, 45, 75, 90, and 180 min as populations in prometaphase, metaphase, anaphase, telophase, and G1 phase, respectively (see Fig. 1 for example). The mitotic stage of each cell population is corroborated by staining MTs and chromosomes with an antibody to β-tubulin and DAPI, respectively. The cells recovered from the dishes are washed quickly with PBS(–) and immediately frozen at −80°. In overexpression experiments, we prolong the time of the second thymidine block to 15 h and perform transfection from 8–11 h after the addition of thymidine (see later).

Purification of GST CRIB–Pak Fusion Protein

Competent *Escherichia coli* (DH5α strain) are transformed with pGEX4T-CRIB-Pak (Matsuo *et al.*, 2002), and plated on LB agar containing 100 μg/ml ampicillin. A single colony of the bacteria is picked and cultured overnight in 10 ml of 2×YT medium containing 100 μg/ml ampicillin at 37° with agitation. The culture is added to two Erlenmeyer flasks, each containing 200 ml of 2×YT medium, which are shaken at 37° for 1 h until the OD$_{600}$ of the culture reaches 0.6. Isopropyl β-D-thiogalactoside (IPTG; Wako chemicals) is then added at 0.1 mM to induce the expression of the recombinant protein, and the culture is continued for another 3 h at 37°. The bacteria are pelleted by centrifugation and frozen at −20° until use. For protein extraction, the bacteria pellets are equilibrated to 4° and

---

FIG. 1. Cell cycle synchronization. HeLa S3 cells synchronized in G1-S phase with thymidine were arrested in prometaphase with nocodazole. After removal of nocodazole, the cells were seeded on poly-L-lysine–coated coverslips and further incubated for 25 min for prometaphase, 45 min for metaphase, 90 min for telophase, and 180 min for G1 phase. The cells were fixed and stained with anti-β-tubulin antibodies (microtubules) and with DAPI (chromosomes). Images were acquired with 63× N.A. 1.3 lenses on an inverted microscope equipped with a CCD camera.

suspended in 20 ml of PBS containing 1 mM dithiothreitol and supplemented with 100 μg/ml PMSF, 10 μg/ml leupeptin, and 10 μg/ml pepstatin. The suspension was sonicated on ice for 10 sec five times followed by the addition of Triton X-100 at 1% and incubated for 30 min at 4° with continuous agitation. The lysates are clarified by centrifugation at 10,000g for 15 min. The supernatant is recovered and incubated with 500 μl of glutathione-Sepharose beads (Amersham Pharmacia, Inc.) for 1 h at 4° with continuous agitation. The beads are recovered by centrifugation at 1000g for 5 min at 4° and washed three times with the preceding PBS solution. The purity and amount of the GST-CRIB-Pak protein bound to the Sepharose beads is determined by analyzing an aliquot by SDS-PAGE in a 10% gel and comparing the amount of the protein with standard amounts of BSA on Coomassie blue staining. The fusion protein is frozen in small aliquots at −80° until use for the pull down assay.

Precipitation of GTP-Bound Cdc42 or Rac in HeLa S3 Cells in Mitosis

Frozen pellets of synchronized HeLa cells are suspended by pipetting up and down in 50 mM Tris-HCl, pH 7.5, containing 100 mM NaCl, 2 mM MgCl$_2$, 10% glycerol, and 1% Nonidet P-40 supplemented with 10 mM NaF, 1 mM Na$_3$VO$_4$, 100 μg/ml PMSF, 10 μg/ml leupeptin, and 10 μg/ml pepstatin (the lysis buffer). The suspension is incubated for 5 min on ice and further for 15 min at 4° with continuous agitation, sonicated on ice for 1 sec 10 times (level 5, Handy Sonic UR-20P, Tomy Seiko Co., LTD), and clarified by centrifugation at 18,000g for 15 min at 4°. The supernatants are saved, and the protein amount is determined by the Lowry method (Dulley and Grieve, 1975). An aliquot containing 800 μg of protein is then mixed with 40 μg of CRIB-Pak-beads, and the suspension is incubated for 1 h at 4° with continuous agitation. The beads are collected by centrifugation, washed three times with the lysis buffer, and boiled in 40 μl of 2× Laemmli sample buffer.

Equal fractions of the precipitates are applied to SDS-PAGE in separate gels in parallel to determine the amounts of GTP-Rac and GTP-Cdc42, respectively, and a fraction of the supernatant is used to determine the total amounts of Rac and Cdc42. Proteins resolved by SDS-PAGE are transferred onto nitrocellulose membranes (Schleicher & Schuell, Protran BA 83) using a transfer buffer containing 3.03 g Trizma (base), 14.41 g glycine, 200 ml methanol, and H$_2$O in a total volume of 1 l. The membranes are blotted with specific antibodies to Cdc42 (rabbit polyclonal P-1 antibody, Santa Cruz Biotechnology) or Rac1 (mouse monoclonal antibody clone 23A8, Upstate Biotechnology Inc.), further incubated with the respective secondary antibodies coupled to HRP, and developed by the ECL method (Amersham Biosciences, Inc.). The intensity of each band is measured with a densitometer (BIO-RAD GS-700 Imaging Densitometer), and the amount of each GTPase is calculated by the following equation:

$$[(X - Z) \times Pn/(Y - Z) \times Sn] \times 100 = \text{GTP-bound GTPase (\% of total)} \tag{1}$$

Where X is the intensity of a band in the precipitate fraction; Y is the intensity of a band in the supernatant fraction; Z is the background of the film; Pn is a ratio of total precipitates to a loaded fraction; and Sn is a ratio of total supernatant to a loaded fraction; Fig. 2 illustrates an example of this analysis. GTP-bound Cdc42 is present at about 10% of the total amount during G2 phase and remains at this level in nocodazole-arrested cells. The level of GTP-Cdc42 decreases shortly after nocodazole removal in prometaphase, then increases transiently in metaphase, and then decreases again in telophase. On the other hand, the level of GTP-bound Rac1 does not change significantly during mitosis.

Activity of Rho During Mitosis

To measure the activity of Rho during mitosis, we use the Rho binding domain (RBD) of mDia1 that binds to GTP-RhoA with high affinity. We have constructed a GST fusion of the RBD located at the N-terminal region of mDia1 (mDia1-N1; hereafter named mDia1-RBD) that prevents the formation of stress fibers when expressed in HeLa cells (Watanabe *et al.*, 1999). We used GST-mDia1-RBD to trap GTP-bound RhoA from cell lysates of cells synchronized in their cell cycles.

Solutions Used To Prepare Cell Lysates

Lysis buffer A is used for lysis of bacteria; 50 mM Tris-HCl, pH 7.5, containing 100 mM NaCl, 5 mM MgCl$_2$, 1 mM EDTA, 10 μg/ml leupeptin, 10 μg/ml pepstatin, 100 μg/ml PMSF.

Lysis buffer B is used to prepare lysates of HeLa cells in the pull-down assay; 50 mM Tris-HCl, pH 7.5, containing 100 mM NaCl, 5 mM MgCl$_2$, 1 mM EDTA, 10% glycerol, 0.1% NP-40, 50 mM NaF, 1 mM Na$_3$VO$_4$, 1 mM DTT, 10 μg/ml leupeptin, 10 μg/ml pepstatin, 100 μg/ml PMSF.

Preparation of GST-mDia1-RBD

E. coli BL21 (DE3) pTrx are transformed with pGEX4T-mDia1-RBD (aa–2–304) (Kimura *et al.*, 2000; Watanabe *et al.*, 1999) and plated on

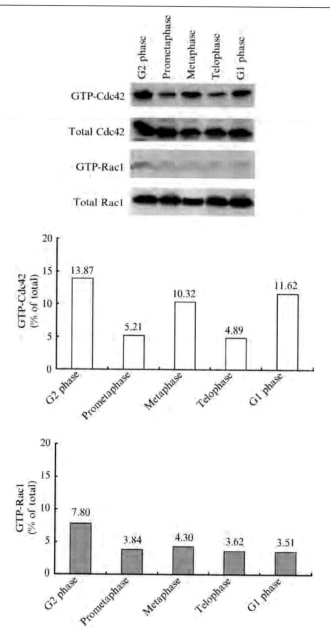

100 μg/ml ampicillin supplemented with 25 μg/ml chloramphenicol LB-agar medium. A single colony of bacteria is picked and grown overnight on 10 ml of 100 μg/ml ampicillin supplemented with 25 μg/ml chloramphenicol LB liquid medium at 37° with agitation. This culture is used to inoculate two 200-ml LB cultures containing 100 μg/ml ampicillin and 25 μg/ml chloramphenicol in 1-l Erlenmeyer flasks, which are agitated at 37° for 1 h. IPTG is added at 0.1 mM when the OD_{600} of the culture reaches 0.6. After agitation for another 15 h at 20°, the bacteria are harvested by centrifugation at 1000g for 10 min at 4° and suspended in 20 ml of ice-cold lysis buffer A. GST-mDia1 RBD is extracted and bound to glutathione sepharose-4B beads essentially as described for GST-CRIB-Pak. The beads were collected after centrifugation at 1000g and washed two times with lysis buffer A, once with lysis buffer A containing 1 M NaCl instead of 0.1 M NaCl, and two times with lysis buffer A. After the protein concentration of GST-mDia1 RBD bound to the beads is determined by SDS-PAGE as described previously for GST-CRIB-Pak, the mDia1-RBD beads are suspended in 1 volume of lysis buffer B (50% slurry). The beads containing 30 μg protein of mDia1-RBD are used in the pull-down assay described later. The binding affinity of the fusion protein to GTP-Rho decreases significantly after freezing and thawing. We, therefore, always use freshly prepared GST-mDia1-RBD for the assay.

Precipitation of GTP-Bound Rho in HeLa S3 Cells in Mitosis

Frozen cells recovered from four 10-cm dishes of G1-S phase and of each stage of mitosis are suspended by pipetting up and down in 200 μl of lysis buffer B and incubated on ice for 15 min. The cell lysates are sonicated for 1 sec 10 times. After centrifugation at 18,000g for 15 min at 4°, the supernatant is saved, sonicated again for 1 sec 10 times, and centrifuged again at 18,000g for 10 min at 4°. The final supernatant is recovered in an Eppendorf tube, and the protein content is determined by a modified Lowry method (Dulley and Grieve, 1975). In parallel, freshly

FIG. 2. Changes in the activity of Cdc42 in HeLa S3 cells during mitosis. HeLa S3 cells were synchronized and collected in G2 phase, prometaphase, metaphase, telophase, and G1 phase. Lysates of these cells were subjected to the pull-down assay with CRIB-Pak and the GTPases bound to the sepharose-beads were analyzed by SDS-PAGE and Western blotting with anti-Cdc42 or anti-Rac1 antibodies (upper panel). The signals of Cdc42 or Rac1 were measured from the films and calculated the proportions of each GTPase in the different phases of mitosis. One eighth or one fourth of the precipitates, and one twenty-fifth or one fiftieth of the supernatant were used to calculate GTP-Cdc42 or GTP-Rac1 from the total protein, respectively (lower panels).

prepared GST-mDia1-RBD Sepharose-beads are washed two times with lysis buffer B. An aliquot of the cell supernatant containing 800 μg of protein is incubated with 30 μg of mDia1-RBD beads for 1 h at 4° with agitation. The suspension is centrifuged at 5000g, and the supernatant saved is mixed with one volume of 2× Laemmli sample buffer. The precipitated beads are washed three times with 10 volumes of lysis buffer B and finally suspended in 40 μl of 2× Laemmli sample buffer. Fractions of the supernatant and the precipitates are subjected to SDS-PAGE and Western blot analysis using an antibody to RhoA (mouse monoclonal antibody 26C4; Santa Cruz Biotechnology) as described for Rac and Cdc42. Figure 3 shows an example of the analysis. GTP-bound RhoA begins to accumulate in anaphase and reaches the maximum accumulation at telophase and then decreases on the G1 phase entry.

Effect of Ect2 Overexpression and Depletion on the Levels of GTP-RhoA and GTP-Cdc42 During Mitosis

Rho GTPases shuttle between the GDP-bound inactive state and the GTP-bound active state by the catalysis of guanine nucleotide exchange factors (GEFs) and GTPase activating proteins (GAPs). Ect2 is a GEF for

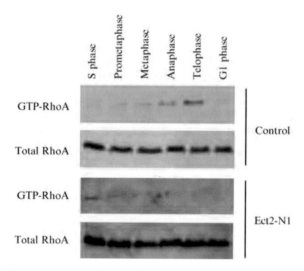

FIG. 3. Ect2 regulates accumulation of GTP-RhoA in telophase. HeLa S3 cells were synchronized in G1-S phase and transfected with pCEV32-F-Ect2-N1 (Ect2-N1) or not (control) in the middle of the second thymidine block. Cells were collected in prometaphase, metaphase, anaphase, telophase, and G1 phase in addition to S phase as reported by Kimura (2000), and the GTP-RhoA level was measured.

Rho-family GTPases containing the hallmark of a Dbl homology (DH) domain followed by a pleckstrin homology (PH) domain in tandem (Miki *et al.*, 1993). Prokopenko *et al.* (1999) showed in *Drosophila melanogaster* that Pebble, the Ect2 ortholog, interacts genetically with RhoA and is required for the formation of the contractile ring and initiation of cytokinesis. Tatsumoto *et al.* (1999) found in mammalian cells that inhibition of Ect2 function either by overexpression of deletion mutants lacking the DH and PH domains or by injection of anti-Ect2 antibodies generated multinucleated cells and indicated that Ect2 is involved in cytokinesis. Although these findings suggest that Ect2 is a GEF for Rho in cytokinesis, Ect2 is known to catalyze GDP-GTP exchange toward Rac1 and Cdc42 as potently as Rho. We have combined overexpression of the Ect2 mutant or depletion of Ect2 by RNAi with the pull-down assays for GTP-Rho GTPases described previously and found that interference of Ect2 by either method suppresses activation of not only Rho in telophase but also that of Cdc42 in metaphase (Kimura *et al.*, 2000; Oceguera-Yanez *et al.*, 2005). We describe in the following the methods we used in these studies.

Transfection of Ect2-N1 During Cell Cycle Synchronization of HeLa S3 Cells

Two 10 cm-dishes of HeLa S3 cells at semiconfluence are trypsinized and seeded onto 40 10-cm dishes in DMEM supplemented with 10% FCS and antibiotics at $37°$ with an atmosphere containing 5% CO_2. After overnight culture, the cells are subjected to the double thymidine block as described previously. Eight hours after the onset of the second block, the cells are transfected with pCEV32-F-Ect2-N1 using Lipofectamine Plus essentially as described by the manufacturer (Invitrogen). For each dish, 2 μg of the plasmid DNA is used and dissolved in 200 μl of Opti-MEM (Gibco). The plasmid solution is mixed with 20 μl of Plus reagent in 200 μl of Opti-MEM medium immediately. Separately, 20 μl of Lipofectamine reagent is diluted in 400 μl of Opti-MEM. After incubation for 15 min at room temperature, the two solutions are mixed and incubated for another 15 min at room temperature. The plasmid-Lipofectamine mixture is then added to the cells that have been washed twice with PBS(–) and replenished with Opti-MEM. After incubation for 3 h at $37°$ with 5% CO_2, the medium is replaced with the thymidine-containing medium, and the cells are cultured for another 4 h to complete the second thymidine block. Cells in different phases of the cell cycle are then collected as described in "*Cell Cycle Synchronization of HeLa S3 Cells.*" Using this protocol, we can achieve a transfection efficiency of 70% as determined by staining for the Flag epitope. Expression of flag-tagged Ect2-N1 can also be confirmed by Western blotting in lysates of HeLa S3 cells collected 10 h after transfection. Therefore, the preceding protocol allows substantial expression of Ect2-N1 protein by the time HeLa cells reach mitosis. Typical results are shown in Figs. 3 and 4. Expression of Ect2-N1 results in poor accumulation of GTP-bound Rho in telophase (Fig. 3). The Ect2-N1 expression also suppresses accumulation of GTP-Cdc42 in metaphase (Fig. 4). As a consequence, the level of GTP-Cdc42 is maintained low from prometaphase to telophase in the Ect2-N1 overexpressing cells. On the other hand, Ect2-N1 expression had no effect on the level of GTP-bound Rac1 (data not shown).

Depletion of Ect2 by RNA Interference

siRNA for Ect2 is generated by the BLOCK-iT RNAi-TOPO Transcription and the BLOCK-iT Dicer RNAi kits (Invitrogen). A 1000 nucleotide-long cDNA for the N-terminal part of human Ect2 is amplified by PCR using the following primers:

ATGGCTGAAAATAGTGTATTAACATCCACT
ATTTCTTGAGCTCAGGAGTATTTGCCTTTT.

The PCR fragment obtained is ligated to the T7 promoter. Using a combination of T7 primer and the gene-specific forward or reverse primer, T7-linked PCR products are amplified to produce sense and antisense

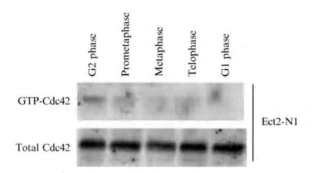

FIG. 4. Ect2 regulates the accumulation of GTP-Cdc42 in metaphase. HeLa S3 cells were synchronized in G1-S phase and transfected with pCEV32-F-Ect2-N1 (Ect2-N1) in the middle of the second thymidine block. Cells were collected in G2 phase, prometaphase, metaphase, telophase, and G1 phase, and the lysates of these cells were subjected to the pull-down assay with CRIB-Pak. The fractions of GTP-Cdc42 bound to the sepharose-beads were analyzed by SDS-PAGE and Western blotting with anti-Cdc42 antibodies.

DNA templates, which are used in an *in vitro* transcription reaction to produce antisense and sense RNA transcripts, respectively. Sense and antisense RNA transcripts are purified and annealed to form double-strand RNA. The double-strand RNA obtained are cleaved into 21–23 nucleotide fragments with Dicer enzyme and purified by using the RNA purification reagent supplied in the kit. As a control, we use siRNA for the *E. coli* LacZ gene as provided by the manufacturer.

One 10-cm dish of HeLa cells of 90% confluency is trypsinized and seeded onto 14 10-cm dishes. After overnight culture, the medium is replaced with DMEM containing 10 m*M* thymidine and 5% FCS, and the cells are cultured for 14 h. The cells are transfected with siRNA for Ect2 immediately after the first thymidine block using Lipofectamine 2000

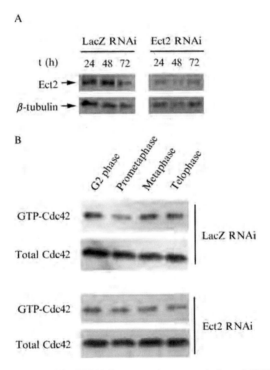

FIG. 5. Depletion of Ect2 by RNAi decreases the accumulation of GTP-bound Cdc42 in metaphase. (A) Ect2 depletion. HeLa S3 cells transfected with siRNA for LacZ or Ect2 were harvested at the indicated times and subjected to SDS-PAGE and Western blot analysis with rabbit polyclonal anti-Ect2 antibody (Santa Cruz Biotechnology) or anti-β-tubulin antibodies. (B) Impaired accumulation of GTP-Cdc42 in metaphase in Ect2-depleted cells. HeLa S3 cells subjected to RNAi for Ect2 were analyzed for the level of GTP-Cdc42 during mitosis.

(Invitrogen). For each dish, we mix 1.8 μg of either control or Ect2 siRNA with 0.5 ml of Opti-MEM (Gibco). In parallel, we dilute 36 μl of Lipofectamine 2000 in 0.5 ml of Opti-MEM and incubate the mixture for 5 min at room temperature. The siRNA solution is mixed with the diluted Lipofectamine 2000 suspensions and incubated for 20 min at room temperature. The medium containing thymidine is then aspirated, and the cells are washed twice with PBS(–) and replenished with 3 ml of Opti-MEM. The siRNA-Lipofectamine 2000 mixture is added to the culture and mixed carefully by rocking the plates. Transfection is allowed to occur for 2 h at 37° with 5% CO_2. The cells are then cultured in DME containing 10% FCS for 10 h and subjected to the second thymidine block for 12 h. The cells in different phases of the cell cycle are collected as described in the *"Cell Cycle Synchronization of HeLa S3 Cells."* Typically, transfection of siRNA for Ect2 by the preceding protocol results in depletion of endogenous Ect2 protein at 24, 48, and 72 h after the transfection (Fig. 5A). In these cells, elevation of GTP-Cdc42 in metaphase is significantly suppressed (Fig. 5B).

References

Azim, A. C., Barkalow, K. L., and Hartwig, J. H. (2000). Determination of GTP loading on Rac and Cdc42 in platelets and fibroblasts. *Methods Enzymol.* **325,** 257–263.

Bostock, C. J., Prescott, D. M., and Kirkpatrick, J. B. (1971). An evaluation of the double thymidine block for synchronizing mammalian cells at the G1-S border. *Exp. Cell Res.* **68,** 163–168.

Dulley, J. R., and Grieve, P. A. (1975). A simple technique for eliminating interference by detergents in the Lowry method of protein determination. *Anal. Biochem.* **64,** 136–141.

Etienne-Manneville, S., and Hall, A. (2002). Rho GTPases in cell biology. *Nature* **420,** 629–635.

Kimura, K., Tsuji, T., Takada, Y., Miki, T., and Narumiya, S. (2000). Accumulation of GTP-bound RhoA during cytokinesis and a critical role of ECT2 in this accumulation. *J. Biol. Chem.* **275,** 17233–17236.

Mabuchi, I., Hamaguchi, Y., Fujimoto, H., Morii, N., Mishima, M., and Narumiya, S. (1993). A rho-like protein is involved in the organisation of the contractile ring in dividing sand dollar eggs. *Zygote* **1,** 325–331.

Matsuo, N., Hoshino, M., Yoshizawa, M., and Nabeshima, Y. (2002). Characterization of STEF, a guanine nucleotide exchange factor for Rac1, required for neurite growth. *J. Biol. Chem.* **227,** 2860–2868.

Miki, T., Smith, C. L., Long, J. E., Eva, A., and Fleming, T. P. (1993). Oncogene *ect2* is related to regulators of small GTP-binding proteins. *Nature* **362,** 462–465.

Oceguera-Yanez, F., Kimura, K., Yasuda, S., Higashida, C., Kitamura, T., Hiraoka, Y., Haraguchi, T., and Narumiya, S. (2005). Ect2 and MgcRacGAP regulate the activation and function of Cdc42 in mitosis. *J. Cell Biol.* **168,** 221–232.

Prokopenko, S. N., Brumby, A., O'Keefe, L., Prior, L., He, Y., Saint, R., and Bellen, H. J. (1999). A putative exchange factor for Rho1 GTPase is required for initiation of cytokinesis in *Drosophila. Genes Dev.* **13**, 2301–2314.

Tatsumoto, T., Xie, X., Blumenthal, R., Okamoto, I., and Miki, T. (1999). Human ECT2 is an exchange factor for Rho GTPases, phosphorylated in G2/M phases, and involved in cytokinesis. *J. Cell Biol.* **147**, 921–928.

Watanabe, N., Kato, T., Fujita, A., Ishizaki, T., and Narumiya, S. (1999). Cooperation between mDia1 and ROCK in Rho-induced actin reorganization. *Nat. Cell Biol.* **1**, 136–143.

Yasuda, S., Oceguera-Yanez, F., Kato, T., Okamoto, M., Yonemura, S., Terada, Y., Ishizaki, T., and Narumiya, S. (2004). Cdc42 and mDia3 regulate microtubule attachment to kinetochores. *Nature* **428**, 767–771.

[25]   Inhibition of Rho GTPases by RNA Interference

By Yukinori Endo,* Sharona Even-Ram,* Roumen Pankov, Kazue Matsumoto, and Kenneth M. Yamada

Abstract

Selective down-modulation or silencing of individual members of the Rho-GTPase family is now practical using RNA interference. Transfection of mammalian cells with an individual siRNA duplex or siRNA pools can suppress expression of a specific isoform to understand its function. By adjusting the dose of siRNA, intermediate levels of suppression can be attained to test the biological role of different levels of a GTPase such as Rac. Nevertheless, there are significant potential pitfalls, including "off-target" effects of the siRNA on other genes. Besides demonstrating successful, noncytotoxic suppression of protein and activity levels of a specific GTPase, controls are essential to establish specificity. In this chapter, we provide methods for selective knockdown of expression by siRNA and confirmation of the effectiveness of Rho GTPase silencing, as well as descriptions and some examples of controls for specificity that include evaluations of dose-response, negative and positive controls, GTPase specificity, confirmation by using more than one siRNA for the same gene, rescue by a mutated siRNA-resistant cDNA encoding the target gene, and complementary supporting evidence. Selective silencing of specific Rho family GTPases should provide increasing insight into the regulatory and functional roles of each isoform in a wide variety of biological processes.

*Y. Endo and S. Even-Ram contributed equally to this work.

METHODS IN ENZYMOLOGY, VOL. 406   0076-6879/06 $35.00
                                                                DOI: 10.1016/S0076-6879(06)06025-3

Introduction

Rho-family GTPases have been implicated in such a wide variety of adhesive, migratory, and other biologically important cellular processes that it is crucial to have selective and specific approaches to testing the functions of each GTPase. Classically, the specific functions of small GTPases have been characterized by experimental manipulations that have included: (1) increasing activity, usually by overexpressing a constitutively activated mutant, (2) decreasing activity using inhibitors such as C3 or overexpressing dominant-interfering mutants, (3) ablating the gene by homologous recombination to completely abolish activity, or (4) modulating activity by inhibition of upstream or downstream regulatory or signaling molecules such as GEFs, GAPs, and ROCK and other effectors. Although these approaches are powerful and have provided important functional insights, concerns about specificity have been raised about each approach. For example, constitutively active GTPases are unable to cycle or respond to natural regulators, mutant proteins or other inhibitors can be rather broad in their effects, cells with gene ablations often attempt to compensate by altering expression of other genes, and regulatory or signaling molecules are often functionally associated with multiple Rho GTPases.

Recent breakthroughs in the experimental application of RNA interference (RNAi) have permitted selective suppression ("knockdown" or silencing) of individual genes in mammalian and other genomes (e.g., see recent review by Hannon and Rossi [2004]). A theoretical advantage of RNAi is its specificity for only a single gene if it can be designed sufficiently selectively, although "off-target" effects on other genes can occur. Thus, an accompanying disadvantage is the need to perform a number of controls to rule out nonspecific effects, although one can argue that similar care should be applied to the use of other experimental approaches using stimulation or inhibition of GTPases. Another advantage is that it is possible by using dose-response curves to use RNAi to induce partial suppression of a gene to test the effects of different levels of total activity of the molecule on biological functions.

The phenomenon of RNAi and discussions of the use of small inhibitory RNA (siRNA) have been reviewed extensively elsewhere (Hannon and Rossi, 2004; Huppi et al., 2005; Juliano et al., 2005; Meister and Tuschl, 2004). Basically, providing cells with RNA duplexes 21–23 base pairs in length in the form of either synthetic siRNA molecules or short hairpin RNA (shRNA) targets the endogenous cellular machinery of the RISC (RNA-induced silencing complex). In the RISC, the antisense strand of an exogenous RNA duplex corresponding to a coding sequence of the gene binds to a target mRNA. If the match is sufficiently close in mammalian cells, the enzyme Argonaute2 cleaves the RNA complex 10 nucleotides from the 5' end of the siRNA, thereby destroying the mRNA (Liu *et al.*, 2004; Meister *et al.*, 2004). Alternately, the RISC can sometimes use siRNAs to suppress translation of a protein. The result in both cases is suppression of synthesis of a particular protein and experimental silencing of the gene.

The experimental knockdown of specific Rho-family GTPases has been achieved in a variety of recent articles, which has revealed roles in a number of important biological processes. Some arbitrarily chosen recent examples include siRNA- or shRNA-mediated knockdown of Rac to demonstrate roles in integrin function and chemotaxis (Deroanne *et al.*, 2003; Weiss-Haljiti *et al.*, 2004), of Rho to identify roles in neurite outgrowth and tumor invasion (Ahmed *et al.*, 2005; Simpson *et al.*, 2004), and of Cdc42 to show roles in motility and protease regulation (Deroanne *et al.*, 2005; Wilkinson *et al.*, 2005). However, it is also possible to take advantage of the finding (see example later) that a partial decrease in protein levels of a molecule such as Rac results in a partial decrease or knockdown of total activity (Pankov *et al.*, 2005). Thus, it is also possible to test the roles of specific overall levels of activity on biological functions.

The following sections provide protocols for transfection of cells with a specific siRNA. Using primary human fibroblasts and Rac as the example, we show dose-response effects of transfected siRNA on activity and protein concentrations. With this approach, direct determinations of activity are needed for each experiment because of day-to-day biological variability. In subsequent sections, we describe important controls and caveats, along with an example application of the methods to examine biological functions. The following protocol was developed for primary human fibroblasts but has also been applied successfully to the nontransformed human breast epithelial cell line MCF-10A and the human glioblastoma line U87-MG and to suppress the activities of Rac1, Cdc42, and RhoA. The subsequent protocol shows the use of a different transfection reagent for a different cell line.

Stepwise Knockdown of Activity

Materials for siRNA Transfection

Purified siRNA duplexes can be purchased from a variety of sources. In the protocols we describe, they are from the following sources:

   Rac1 siRNAs, the control nontargeting siRNA pool, and siGLO RISC-Free siRNA are from Dharmacon.

The individual sequences (sense strand) for each Rac1 siRNA duplex are:

#5: AGACGGAGCUGUAGGUAAAUU

#7: UAAGGAGAUUGGUGCUGUAUU

#8: UAAAGACACGAUCGAGAAAUU

#9 (with ON TARGET modification for enhanced specificity): CGGCACCACUGUCCCAACAUU

#9 scrambled sequence: AGCACACGACCGCUCUCCAUU

The preceding four siRNAs are used either as a premixed pool of all four together from the manufacturer or as the individual siRNA duplexes to compare specificity.

siRNA dilution buffer: 20 mM KCl, 0.2 mM $MgCl_2$, and 6 mM HEPES, pH 7.5, in RNase-free water.

Cdc42 siRNA catalog number sc29256 and RhoA siRNA catalog number sc29471 are from Santa Cruz Biotechnology.

Lipofectamine 2000 Transfection Reagent is from Invitrogen.

RNAiFect Transfection Reagent is from Qiagen.

Opti-MEM I Reduced Serum Medium is from Invitrogen. Divide into small aliquots (e.g., 5–10 ml in tightly capped 15-ml tubes) to avoid exposing the main stock to pH changes.

Regular culture medium: For human primary fibroblasts and glioblastoma cells, we use Dulbecco's modified Eagle medium (DMEM) containing 10% fetal bovine serum, 100 U/ml penicillin, and 100 μg/ml streptomycin, and cells are maintained in a humidified 10% CO_2 tissue culture incubator.

Culture medium without antibiotics: Identical cell culture medium lacking all antibiotics but containing serum.

Sterile, RNase-free, polypropylene microcentrifuge tubes.

Sterile, RNase-free 10-μl, 100-μl, and 1000-μl pipet tips with filter plugs and appropriate pipettes.

Procedure for siRNA Transfection

1. One day before the transfection, passage the cells as usual, but then plate in 6-well culture plates so that they will reach 60–80% confluence 20 h later (e.g., 1.5–2 × 10^5 primary human fibroblasts per well) in regular culture medium *without* antibiotics. Note that omitting antibiotics at this step is important. Change the antibiotic-free medium approximately 3–4 h before transfection to remove any dead cells and to provide the cell monolayer with fresh medium.

2. Suspend siRNA using siRNA dilution buffer to 20–40 μM concentration. Store small aliquots at $-20°$. Calculate concentrations of siRNA to be used and dilute to make 200× stock solutions in at least 6 μl siRNA dilution buffer.

3. In a sterile tube, dilute 5 μl siRNA stock solution into 250 μl Opti-MEM without serum to make a 4× siRNA solution and mix gently by pipetting up and down five times using a 1000-μl pipet tip.

4. Mix Lipofectamine 2000 before aliquoting into a separate sterile tube: add 5 μl Lipofectamine 2000 to 250 μl Opti-MEM without serum, mix gently by pipetting, and incubate for 5 min at room temperature. The Lipofectamine and the siRNA stocks can be scaled up in volume to transfect multiple wells.

5. Pipet the Lipofectamine 2000 solution into the siRNA solution, mix gently by pipetting, and incubate for 30 min at room temperature to generate transfection complexes.

6. Aspirate the medium from the cells and replace with 500 μl regular culture medium without antibiotics but with serum.

7. Add the siRNA-Lipofectamine 2000 mixture drop wise (500 μl), mix by gently rocking the plate manually forward-and-back and then side-to-side (twice each), and culture for at least 4 h (4–16 h).

8. Add 1 ml additional antibiotic-free regular culture medium 4 h after transfection or replace the medium with fresh antibiotic-free medium after overnight incubation.

9. Detach the transfected cells 50–55 h after transfection and plate in a dish for GTPase pull-down assays (e.g., 10^6 human fibroblasts from 3–4 wells of a 6-well plate in a 100-mm tissue culture dish) and for any biological assays needed, plus a small quantity in culture dishes or on coverslips for examining morphology. Use the cells 72 h after transfection. Before using, examine for morphological changes and evidence of cytotoxicity. Particularly striking changes in cell morphology are seen after knockdown of Rac activity, with a loss of peripheral lamellae (Fig. 1).

Alternative Procedure: Transfection of GD25 Cells with siRNA Using RNAiFect

1. One day before the transfection, passage and plate 8×10^4 GD25 cells per well in a 6-well plate in regular culture medium containing serum and antibiotics.

2. Dilute siRNA stock (e.g., 7.5 μl of a 50-μM siRNA stock) into 92.5 μl media and mix gently by vortexing.

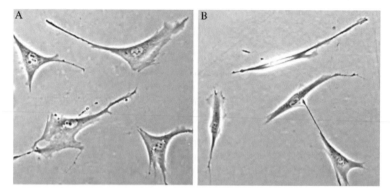

FIG. 1. Morphology of human fibroblasts after knockdown of Rac1 activity. Primary human foreskin fibroblasts were transfected as described in this chapter using (A) 1 nM control nonspecific pool siRNA or (B) 1 nM Rac1 siRNA pool and examined by phase contrast microscopy. Note that cells after Rac1 knockdown are often more spindle-shaped and have fewer leading edges (lamellae) associated with cell migration. As a result, they tend to migrate in a directionally persistent manner compared with the relatively random migration of untreated or control cells (Pankov et al., 2005).

3. Add 30 μl RNAiFect reagent to the diluted siRNA stock, mix by vortexing 10 sec, and incubate 10–15 min at room temperature to generate transfection complexes.

4. Wash the GD25 cells with Dulbecco's phosphate-buffered saline without divalent cations.

5. Remove excess extracellular matrix from these cells by incubating 1 min in 0.5 ml trypsin solution, then add 1.5 ml regular serum-containing medium to stop proteolysis. The purpose of the trypsin treatment is only to remove extracellular matrix, and the cells should not detach. Remove the trypsin/medium mixture and replace with 2 ml regular culture medium containing serum and antibiotics.

6. Add the siRNA mixture to the cells and mix by rocking manually.

7. Detach the cells 60 h after transfection for replating and use the cells 72 h after transfection; the medium may require changing before use if the cells are dense.

GTPase Assays and Immunoblotting

The effectiveness and specificity of the knockdown of Rho GTPase activity is evaluated using standard assays. The following protocol is derived from Sander et al. (1998) with minor modifications.

Materials for GTPase Assays

Lysis buffer: 300 mM NaCl, 10 mM MgCl$_2$, 50 mM Tris-Cl, pH 7.5, 2% Igepal, and protease inhibitor cocktail from Cytoskeleton.

Wash buffer: 25 mM Tris-Cl, pH 7.5, 30 mM MgCl$_2$, and 40 mM NaCl.

2× gel electrophoresis sample buffer is from Invitrogen.

Immunoblot blocking buffer: 10% nonfat dry milk in 150 mM NaCl, 50 mM Tris-Cl, pH 7.4, and 0.1% Tween 20.

PAK1 PBD conjugated to agarose and Rhotekin-Rho binding domain bound to glutathione agarose are from Upstate or Cytoskeleton.

8–16% gradient gels and 0.2-μm pore-size nitrocellulose membranes are from Novex.

Anti-Rac antibody is from Upstate.

Anti-Cdc42 antibody is from BD Transduction.

Anti-RhoA antibody is from Cytoskeleton.

Anti-actin clone AC-40 antibody is from Sigma.

Secondary horseradish peroxidase–conjugated anti-mouse antibodies are from Amersham.

ECL system and Hyperfilm X-ray film are from Amersham Biosciences.

Procedure for GTPase Assays

Steps 1–3 below are performed in a 4° cold room:

1. Wash cells with PBS without Ca^{2+} and Mg^{2+} once at room temperature. Immediately scrape the cells into ice-cold lysis buffer and then centrifuge the lysates for 5 min at 6000g. Save aliquots of the cleared lysates for total protein determination and immunoblotting for specific proteins. If you need to store lysates, snap freeze in liquid nitrogen and store at −80°.

2. Incubate the cleared lysate with 10 μg or 20 μg PAK-1 PBD agarose (to assay Rac or Cdc42 activity) or 40 μg GST-tagged Rhotekin Rho-binding domain bound to glutathione agarose (to assay Rho activity) for 60 min at 4° with continual mixing on a rotator. Quantities and incubation lengths should be optimized for each laboratory.

3. Wash the beads two times with washing buffer by suspension and centrifugation at 1500g for 3 min. Remove all of the supernatant solution and resuspend the beads in 20 μl 2× electrophoresis sample buffer with 10% β-mercaptoethanol.

4. Incubate 5 min in a boiling water bath.

5. Separate the released proteins on 8–16% gradient gels in parallel with aliquots of the lysates to estimate both active and total protein levels

of the GTPase. Keep track of all volumes of lysates, eluates, and samples loaded onto the gels for future calculations.

6. After electrotransfer to nitrocellulose membranes, block the filters with immunoblot blocking buffer and probe with the appropriate antibodies, followed by appropriate secondary horseradish peroxidase–conjugated antibodies.

7. Immunoblots are visualized using the ECL system and Hyperfilm X-ray film.

For quantification, the X-ray film can be scanned if care is taken to remain in the linear range (tested by showing linearity of twofold dilutions of samples). Alternately, a chemiluminescent imager can be used. If greater sensitivity is required, we recommend the Super Signal West Femto Maximum Sensitivity Substrate from Pierce. Figure 2A shows stepwise decreases in Rac activity as increasing concentrations of Rac1 siRNA are used for transfection. Figure 2B shows the relatively parallel decreases in total Rac protein with increasing concentrations of Rac1 siRNA.

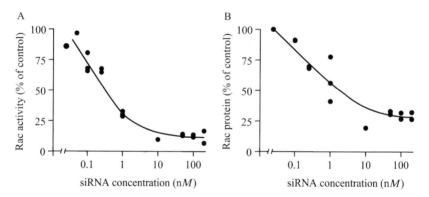

FIG. 2. Dose-response curves of Rac1 siRNA transfection showing knockdown of Rac activity and total protein levels. Primary human fibroblasts were transfected with control pool siRNA or the indicated range of concentrations of Rac1 siRNA pool and assayed for Rac protein and activity 72 h later; the concentration of the control siRNA was equal to the highest concentration of Rac1 siRNA being tested. Results from five independent experiments were combined. (A) Rac activity measured by pull-down assay normalized to the control as 100%. Note the dose-response relationship with substantial suppression of Rac activity at even 1 nM siRNA. (B) Rac protein levels in lysates of cells treated with control or increasing concentrations of Rac1 siRNA. Although there is rough proportionality between Rac protein and activity levels, the correlation is not exact. It is, therefore, important to assay directly the activity of Rac relative to control directly in each experiment for comparison with biological effects.

Maximizing Efficiency of siRNA Transfection

The effectiveness of transfection of siRNA duplexes seems to depend strongly on choosing the most effective transfection agent or protocol for each specific cell type. Unless an efficient procedure is known for the particular cell line being used, it is best to test a variety of agents or methods to identify the conditions that permit maximal transfection with minimal cytotoxicity. Efficiency of transfection should be tested using a fluorophore-labeled siRNA probe, for example using siGLO RISC-free siRNA (Dharmacon). The latter probe is modified so that it does not compete with functional siRNAs in the RISC, because the RISC can potentially become saturated with exogenous siRNA molecules.

The procedure is to use concentrations of labeled siRNA identical to those to be used for transfection of functional siRNAs, then to examine cells plated on a glass substrate by fluorescence microscopy. Alternatives are to combine the probe with functional siRNAs or to use a functional siRNA labeled with a fluorophore. As determined by fluorescence microscopy, the proportion of cells that is labeled and the intensity of the fluorescence will depend on the efficiency of transfection and the concentration of siRNA tested. If the transfection conditions are ideal, the transfection agent will permit 100% transfection efficiency (Fig. 3). Testing a range of different siRNA concentrations will indicate whether the same high efficiency is retained at progressively lower concentrations, although the intensity of fluorescence should decrease at low siRNA concentrations and require increased sensitivity of detection (Fig. 3B–D).

Evaluation of Specificity

Although RNAi-mediated knockdowns of gene expression using siRNA can be extremely specific, it is crucial to test specificity. There is a general consensus that as many of the following tests as practical should ideally be performed ("Whither RNAi?" *Nat. Cell Biol.*, 2003; Hannon and Rossi, 2004; Huppi *et al.*, 2005; Juliano *et al.*, 2005; Medema, 2004).

1. Negative siRNA control: Synthetic peptide studies often use a control peptide with a scrambled sequence of the same residues, and an siRNA control using a scrambled sequence can demonstrate that the specific sequence order is essential (e.g., Fig. 4A). However, this approach can be misleading as a control for functional siRNA activity, because random scrambling can produce sequence combinations that can target other genes by chance even after using NCBI BLAST searches to avoid matches with any known gene of the species being targeted. The most common type of functional control is an siRNA (or a pool of several

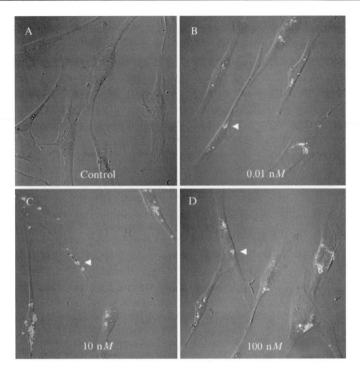

FIG. 3. Evaluation of efficiency of siRNA transfection using a fluorescent siRNA probe. Primary human fibroblasts were transfected with the indicated concentrations of fluorophore-labeled siGLO RISC-Free siRNA (Dharmacon), and living cells were examined by fluorescence microscopy 72 h after transfection without fixation. The images show superimposed fluorescence (white) and transmitted light microscopy using differential interference contrast microscopy to visualize the cells (gray). The arrowheads show examples in each panel of fluorescence uptake of the siGLO siRNA. Note that all cells in (B–D) are labeled over a 10,000-fold range.

siRNAs if the siRNA being used is a pool) that cannot target genes in the cell type being studied yet can enter the RISC. The control is transfected in parallel under identical conditions as the active siRNA (Fig. 4). If there are side effects because of the transfection procedure, the incubation of cells with siRNA, or saturation of the RISC complex and resultant defects in endogenous RNA-mediated regulation, this control should show effects compared with nontreated controls.

2. Target specificity control: In studies of small GTPases, it is important to demonstrate that only the targeted molecule is decreased. For example, if Rac1 is targeted, then one should show by quantification not only that Rac activity and protein levels are decreased but that other

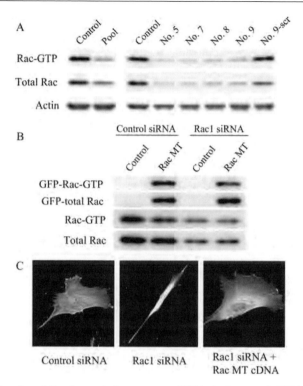

FIG. 4. Silencing of Rac by pooled or separate siRNA duplexes and cDNA rescue from knockdown. (A) This example compares Rac knockdown after transfection of primary human fibroblasts with 1 nM control siRNA versus Rac1 siRNA pools (left) or by 1 nM of four different individual Rac1 siRNA duplexes. The siRNA pool or each duplex produces a substantial reduction in Rac activity (Rac-GTP) in pull-down assays and total Rac protein by immunoblotting of whole-cell lysates with anti-Rac antibody. In contrast, the scrambled version of number 9 has little effect. Actin levels in the lysates serve as the loading control. (B) Rescue of Rac siRNA knockdown as performed by cotransfection of 1 nM Rac1 siRNA no. 9 with a GFP-tagged mutated Rac1 cDNA containing silent mutations. Note that the mutated Rac1 (Rac MT) has normal Rac activity (GFP-Rac-GTP) and that the sum of the activities of knocked-down endogenous Rac and exogenous GFP-Rac1 with silent mutations is similar to the original activity of Rac in control cells. (C) Rescue of cell morphology by expression of siRNA-resistant Rac1 cDNA. Cells transfected with scrambled control or Rac1 siRNA and visualized using cotransfected EGFP show the characteristic spindle-shaped morphological change after reduction of total Rac levels, whereas the mutated Rac1 cDNA (Rac MT) was not sensitive to RNA interference and was thus able to rescue normal morphology in Rac1 siRNA–treated cells.

Rho-family GTPases are not affected (e.g., no changes in protein or activity levels of Cdc42 and Rho). Isoform specificity should ideally also be evaluated (e.g., by quantitative realtime RT-PCR). Theoretically, performing microarray analysis of general gene expression could provide a broad control, but it may be difficult to separate out the specific downstream effects of pleiotropic regulators such as Rho-GTPase family members from nonspecific changes in gene expression.

3. Multiple active siRNAs: Effective evidence for specificity can be obtained if several individual siRNA sequences targeting different sites on mRNA from the same gene produce the same effects (i.e., knockdown of activity; Fig. 4A) and the same biological changes.

4. Low siRNA concentration. Whenever possible, the lowest possible concentrations of siRNA duplexes should be used. Nonspecific "off-target" effects are more likely when using higher concentrations. We have found that effective knockdown of Rac activity can be obtained with only 1 nM siRNA (Fig. 4). As the concentration of siRNA is raised, there can be additional biological effects that can be attributed to nearly complete silencing (Fig. 5) or potentially to off-target effects that can prove difficult to exclude. Tests of specificity for just the gene of interest can help resolve questions about specificity, but there may always be lingering doubts.

5. Rescue: Another test of specificity is to reverse the knockdown by expressing a cDNA for the protein of interest that has a sequence that is not susceptible to the siRNA knockdown. Introduction of several nucleotide mismatches in the cDNA sequence should produce resistance to siRNA-mediated effects. Alternately, an orthologous cDNA from

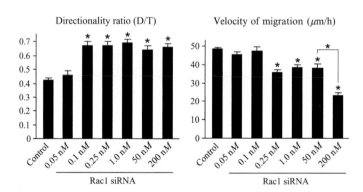

FIG. 5. Dose-response effects of Rac1 knockdown. This example shows the value of dose-response curves in evaluating two types of biological activity assayed after Rac1 knockdown. Directionality (persistence of migration in one direction) and the velocity of migration of primary human fibroblasts were measured as described (Pankov *et al.*, 2005). Note that even though directional persistence of migration switches to more directional between 0.05 and 0.1 nM siRNA, there seems to be a two-step decrease in cell velocity of migration: the first change occurs between 0.1 nM and 0.25 nM and then a further drop occurs at 200 nM siRNA. All asterisks indicate $p < 0.001$ by ANOVA; error bars indicate standard error.

another species with such mismatches can also be used. A complication with this approach can occur if double transfection of siRNA and an expression plasmid induce cytotoxicity, necessitating a search for alternative transfection methods. Another caveat is that the level of expression should ideally be in a physiological range rather than overexpression. The protocol in the next section and Fig. 4B–C show examples of this type of rescue control for siRNA specificity.

6. Complementary supporting evidence: If possible, alternative mechanisms for reducing activity should be compared. For example, an integrin mutant selectively targets Rac activity compared with Cdc42 and Rho, and similar biological effects are observed by this mechanism of lowering Rac activity (Pankov et al., 2005). Although the widely used inhibitor C3 exotoxin is not specific for only Rac, an inhibitor now exists: the Rac inhibitor NSC23766 was recently reported to be a Rac-specific small-molecule inhibitor that targets Rac activation by GEF (Gao et al., 2004). The compound was designed, synthesized, and kindly provided by the Drug Synthesis and Chemistry Branch, Developmental Therapeutics Program, Division of Cancer Treatment and Diagnosis, National Cancer Institute.

Rescue of Function by Expressing siRNA-Resistant cDNA

This excellent type of control for the specificity of siRNA silencing involves reversing the knockdown of protein expression by expressing a cDNA for the protein of interest. The cDNA needs to contain silent mutations that do not change the amino acid sequence but that create mismatches with the siRNA to resist RNAi. Because double transfections of primary human fibroblasts with siRNA plus plasmids using a cationic lipid can anomalously produce cytotoxicity, the following protocol uses electroporation with cell-cycle synchronization to enhanced transfection efficiency. It involves coelectroporation of a mixture of siRNA with a plasmid containing a cDNA mutated to resist siRNA silencing. As a positive control, it is also useful to test in parallel the corresponding nonmutated cDNA to confirm knockdown of exogenous transfected wild-type, but not mutated, cDNA by the siRNA.

Materials for Cotransfection Rescue

Purified siRNA duplex for the gene to be silenced. In the example for Fig. 4, Rac1 siRNA #9 duplex and the control scrambled #9 duplex are from Dharmacon.

Vector expressing EGFP for monitoring transfection by fluorescence microscopy. The appropriate cDNA is cloned into an EGFP expression vector by standard molecular biology techniques, and then three silent mutations are introduced into the cDNA segment corresponding to the siRNA by site-directed mutagenesis.

In this example, six different silent mutations are introduced into sites corresponding to duplexes #5 and #9 of Rac1 cDNA using the Stratagene QuikChange Multi Site-Directed Mutagenesis kit. The original sequences GTAGGTAAA and TGTCCCAAC are replaced with **GTGGGAAAG** and **TGCCCGAAT**, respectively, where the nucleotide substitutions are indicated by bold letters. The mutated Rac1 cDNA is excised and inserted into appropriate sites of a pEGFP expression vector from Clontech. The mutations are verified by sequencing the entire cDNA.

Electroporation buffer: 20 mM HEPES, pH 7.05, 137 mM NaCl, 5 mM KCl, 0.7 mM Na$_2$HPO$_4$, and 6 mM dextrose. Make a 10× stock solution and sterilize by filtration, then store at 4°. The 10× stock solution is diluted with RNase-free water to make a 1× solution.

> 0.5 M solution of sodium butyrate in water (100× concentration)
> Sterile, RNase-free polypropylene microcentrifuge tubes
> Sterile, RNase-free pipette tips with filter plugs and appropriate pipettes
> Sterile pipettes: 1 ml, 10 ml, and 25 ml
> 15-ml sterile tubes
> 0.4-cm gap Gene Pulsar cuvettes from BioRad
> BioRad Gene Pulser with Capacitance Extender or other appropriate electroporator

Procedure for Cotransfection Rescue Experiment

1. Two days before the transfection, passage the cells as usual into 180 cm^2 flasks at a concentration that will achieve a density of approximately 70% confluence by the next day.

2. On the evening before the day of transfection, aspirate the medium and add 20 ml fresh medium containing 270 μl 0.2 M thymidine (warm the thymidine stock in a 37° waterbath to dissolve). Mix by swirling and culture the cells overnight.

3. On the morning of the day of transfection, wash the cells with Hanks balanced salt solution and add 25 ml fresh medium to release the cells from the thymidine block.

4. In the late afternoon (after about 6 h), check for synchronized mitosis (ideally >4 round cells per field).

5. Harvest the cells using trypsin-EDTA and inhibit proteolysis by adding regular culture medium containing serum. Collect the cells by centrifugation, resuspend in medium, and count the cells with a hemocytometer or electronic cell counter.

6. Centrifuge and resuspend the cells in EP buffer to a concentration of $3–4 \times 10^6$ cells/ml.

7. Warm up the electroporator (at least 5 min), adjust the settings to 170 V and 960 μFd using external capacitance, and set it to monitor the time constant.

8. Transfer 525 μl of the cell suspension to a sterile microcentrifuge tube. Add 15–30 μg/500 μl plasmid DNA and siRNA to 200 nM. Mix gently by pipetting up and down.

9. Transfer 500 μl cell suspension into a 0.4-cm gap cuvette (avoid air bubbles).

10. Insert the cuvette into the holder with the bump facing forward and press both red pulse buttons until the buzzer sounds. The time constant should be 18–22 msec.

11. Incubate for 5 min at room temperature, then transfer cells using a 1-ml pipette into a 15-ml tube containing 7 ml regular cell culture medium containing 5 mM sodium butyrate. Rinse residual cells from the cuvette with medium and add to the same tube.

12. Centrifuge and resuspend the cells in 10 ml regular medium containing 5 mM butyrate and plate in 10-cm plastic tissue culture dishes. In parallel, place a small amount of the cell suspension on coverslips in culture dishes or into tissue culture plates for examination of morphology, as well as plating other aliquots for appropriate biological assays. Incubate at 37° overnight in a humidified CO_2 incubator.

13. Wash the cells with Hanks balanced salt solution and incubate in butyrate-free regular culture medium, or detach and count the cells, then replate an appropriate number of cells for biological or biochemical assays.

14. Incubate the cells an additional 24 h, then perform GTPase pull-down assays, immunological staining, and/or any other biological assays as necessary.

The efficiency of plasmid transfection using this method is roughly 50–60% rather than the 100% efficiency of single siRNA transfection using Lipofectamine 2000. Consequently, we recommend using transfection markers (e.g., GFP-tagged plasmids and Cy3-labeled siRNA) to identify the transfected cells. An alternative approach is to use a fluorescence-activated cell sorter (FACS) to isolate GFP-expressing cells to enrich for transfected cells before performing biological assays.

Troubleshooting

The efficiency of transfection and biological effects of siRNA treatment on cells can be affected by the condition of the cells. Maintain cells in healthy condition in regular culture medium by regular passage and avoidance of over-confluence; some cells such as primary fibroblasts are also sensitive to being cultured too sparsely. It is important to limit the number of passages of primary cells used.

Another source of poor transfection or cell death is the presence of antibiotics in media during transfection with Lipofectamine 2000. The cells should be switched to antibiotic-free medium the day before transfection.

A particularly important source of variation involves the type of cell being transfected. It is crucial to adapt the transfection protocol to the particular cell type being used. Surprising differences in effectiveness of different transfection agents occur depending on the cell system, which may necessitate laborious comparisons of different reagents and transfection protocols. A well-established positive control for transfection efficiency is the knockdown of lamin A/C (Elbashir *et al.*, 2001). For further discussions of transfection optimization strategies, see Brazas and Hagstrom (2005) and web sites of companies that sell siRNA duplexes.

Summary

RNA interference provides a powerful new tool for selectively reducing specific Rho-family GTPases. By the use of dose-response analyses, it is even possible to identify distinct roles for different levels of a GTPase such as Rac. Nevertheless, there are potential pitfalls, particularly those associated with potential "off-target" effects of the siRNA on other genes. Application of a series of controls, as well as using complementary approaches to modulate levels of a particular GTPase, will increase confidence in the specificity of the knockdown. This approach should contribute to the identification and characterization of the functions of specific small GTPases, as well as helping to sort out the roles of different isoforms.

Acknowledgments

This research was supported by the Intramural Research Program of the NIH, NIDCR. Y. E. is a JSPS Research Fellow in Biomedical and Behavioral Research at NIH.

References

Ahmed, Z., Dent, R. G., Suggate, E. L., Barrett, L. B., Seabright, R. J., Berry, M., and Logan, A. (2005). Disinhibition of neurotrophin-induced dorsal root ganglion cell neurite outgrowth on CNS myelin by siRNA-mediated knockdown of NgR, p75NTR and Rho-A. *Mol. Cell Neurosci.* **28,** 509–523.

Brazas, R. M., and Hagstrom, J. E. (2005). Delivery of small interfering RNA to mammalian cells in culture by using cationic lipid/polymer-based transfection reagents. *Methods Enzymol.* **392,** 112–124.

Deroanne, C., Vouret-Craviari, V., Wang, B., and Pouyssegur, J. (2003). EphrinA1 inactivates integrin-mediated vascular smooth muscle cell spreading via the Rac/PAK pathway. *J. Cell Sci.* **116,** 1367–1376.

Deroanne, C. F., Hamelryckx, D., Ho, T. T., Lambert, C. A., Catroux, P., Lapiere, C. M., and Nusgens, B. V. (2005). Cdc42 downregulates MMP-1 expression by inhibiting the ERK1/2 pathway. *J. Cell Sci.* **118,** 1173–1183.

Elbashir, S. M., Harborth, J., Lendeckel, W., Yalcin, A., Weber, K., and Tuschl, T. (2001). Duplexes of 21-nucleotide RNAs mediate RNA interference in cultured mammalian cells. *Nature* **411,** 494–498.

Gao, Y., Dickerson, J. B., Guo, F., Zheng, J., and Zheng, Y. (2004). Rational design and characterization of a Rac GTPase-specific small molecule inhibitor. *Proc. Natl. Acad. Sci. USA* **101,** 7618–7623.

Hannon, G. J., and Rossi, J. J. (2004). Unlocking the potential of the human genome with RNA interference. *Nature* **431,** 371–378.

Huppi, K., Martin, S. E., and Caplen, N. J. (2005). Defining and assaying RNAi in mammalian cells. *Mol. Cell* **17,** 1–10.

Juliano, R. L., Dixit, V. R., Kang, H., Kim, T. Y., Miyamoto, Y., and Xu, D. (2005). Epigenetic manipulation of gene expression: A toolkit for cell biologists. *J. Cell Biol.* **169,** 847–857.

Liu, J., Carmell, M. A., Rivas, F. V., Marsden, C. G., Thomson, J. M., Song, J. J., Hammond, S. M., Joshua-Tor, L., and Hannon, G. J. (2004). Argonaute2 is the catalytic engine of mammalian RNAi. *Science* **305,** 1437–1441.

Medema, R. H. (2004). Optimizing RNA interference for application in mammalian cells. *Biochem. J.* **380,** 593–603.

Meister, G., Landthaler, M., Patkaniowska, A., Dorsett, Y., Teng, G., and Tuschl, T. (2004). Human argonaute2 mediates RNA cleavage targeted by miRNAs and siRNAs. *Mol. Cell* **15,** 185–197.

Meister, G., and Tuschl, T. (2004). Mechanisms of gene silencing by double-stranded RNA. *Nature* **431,** 343–349.

No authors listed. (2003). Whither RNAi? *Nat. Cell Biol.* **5,** 489–490.

Pankov, R., Endo, Y., Even-Ram, S., Araki, M., Clark, K., Cukierman, E., Matsumoto, K., and Yamada, K. M. (2005). A Rac switch regulates random versus directionally persistent cell migration. *J. Cell Biol.* **170,** 783–802.

Sander, E. E., van Delft, S., ten Klooster, J. P., Reid, T., van der Kammen, R. A., Michiels, F., and Collard, J. G. (1998). Matrix-dependent Tiam1/Rac signaling in epithelial cells promotes either cell-cell adhesion or cell migration and is regulated by phosphatidylinositol 3-kinase. *J. Cell Biol.* **143,** 1385–1398.

Simpson, K. J., Dugan, A. S., and Mercurio, A. M. (2004). Functional analysis of the contribution of RhoA and RhoC GTPases to invasive breast carcinoma. *Cancer Res.* **64,** 8694–8701.

Weiss-Haljiti, C., Pasquali, C., Ji, H., Gillieron, C., Chabert, C., Curchod, M. L., Hirsch, E., Ridley, A. J., van Huijsduijnen, R. H., Camps, M., and Rommel, C. (2004). Involvement of phosphoinositide 3-kinase gamma, Rac, and PAK signaling in chemokine-induced macrophage migration. *J. Biol. Chem.* **279,** 43273–43284.

Wilkinson, S., Paterson, H. F., and Marshall, C. J. (2005). Cdc42-MRCK and Rho-ROCK signalling cooperate in myosin phosphorylation and cell invasion. *Nat. Cell Biol.* **7,** 255–261.

[26] RNA Interference Techniques to Study Epithelial Cell Adhesion and Polarity

By XINYU CHEN and IAN G. MACARA

Abstract

Polarized epithelial cells are characterized by distinct plasma membrane domains and asymmetrical distribution of cell surface proteins and lipids. In vertebrates, tight junctions act as a fence between the apical and basolateral domains. Although many of the key components of the polarity machinery have been identified, their functions in cell polarization and junction formation remain to be determined. With the rapid improvement of the RNA interference (RNAi) technique, it is now possible to silence the expression of these polarity proteins in mammalian cells and to systematically analyze their distinct roles in orchestrating the polarization program. Here we describe approaches to achieve specific gene suppression in MDCK cells, a well-established cell culture model of canine kidney cells. We discuss the potential challenges and problems associated with the RNAi technique and describe basic protocols for suppressing gene expression using a vector-based short hairpin RNA (shRNA) expression system coupled with nucleofection.

Introduction

The establishment and maintenance of epithelial polarity are crucial for normal development and for the functions of various internal organs. Polarized epithelial cells have two distinct plasma membrane domains, an apical domain facing the lumen or external environment and the basolateral domain that is in contact with adjacent cells and the extracellular matrix. Different proteins are targeted to and retained within these two domains, creating a surface asymmetry that underlies many of the key functions performed by epithelial cells, such as vectorial transport and the controlled exchange of molecules across the epithelial layer (Rodriguez-Boulan and Nelson, 1989). The formation of cell–cell junctions plays a critical role during epithelial polarization (Knust and Bossinger, 2002; Yeaman *et al.*, 1999). Tight junctions (TJs) form at the apex of the lateral domain, acting as a barrier that prevents the free diffusion of proteins and lipids between the membrane domains (D'Atri and Citi, 2002; Matter and Balda, 2003b). Adherens junctions (AJs) are located

METHODS IN ENZYMOLOGY, VOL. 406
0076-6879/06 $35.00
DOI: 10.1016/S0076-6879(06)06026-5

below TJs along the lateral membrane and are assembled from cadherins and their associated proteins (Perez-Moreno *et al.*, 2003; Tepass, 2002). They provide dynamic and adhesive connections between adjacent cells. Desmosomal junctions between the lateral membranes contribute to the mechanical strength of the epithelial layer and are essential in maintaining tissue integrity (Getsios *et al.*, 2004).

Three evolutionarily conserved protein complexes have been identified to be key regulators of cell polarity. These include the Crumbs complex, the Par-3/Par-6/atypical protein kinase C (aPKC) complex, and the Scribble complex (Macara, 2004; Ohno, 2001; Roh and Margolis, 2003). Extensive genetic studies in *Drosophila* and *C. elegans* support their essential roles in polarity establishment. Recent studies in mammalian systems have linked the Par-3/Par-6/aPKC complex to various polarization processes, including epithelial polarity, polarized migration, and neuronal polarity.

Rapid advances in RNA interference (RNAi) have made it possible to efficiently and specifically suppress the expression of a target protein in mammalian systems, and the technology provides a powerful tool to investigate mechanisms of epithelial polarization and the biological functions of polarity proteins. Several groups have demonstrated the successful use of RNAi in functional studies of polarity proteins in mammalian cell culture (Chen and Macara, 2005; Parkinson *et al.*, 2004; Shin *et al.*, 2005). This chapter describes protocols for gene silencing in Madin–Darby Canine Kidney (MDCK) cells and discusses procedures for the analysis of defects in cell adhesion and polarity.

Technical Background

Two different methods of RNA interference are commonly used in mammalian systems. These involve the introduction of either chemically synthesized short interfering RNAs (siRNAs) or short hairpin RNAs (shRNAs) produced from shRNA expression vectors (Elbashir *et al.*, 2001; Paddison *et al.*, 2004). Both methods can result in highly specific gene suppression, but the effects of siRNAs are usually transient, whereas long-term suppression or inducible suppression can be achieved in stable cell lines expressing shRNA constructs.

The first step for successful RNAi-based experiments is to choose effective and specific target sequences. The accuracy of predicting effective targets has been greatly improved by recent studies (Reynolds *et al.*, 2004), and several algorithms are available on public or commercial web sites that identify target sequences based on a set of criteria known to be associated with efficient suppression (for example, http://web.mit.edu/mmcmanus/www/home1.2files/siRNAs; http://hydra1.wistar.upenn.edu/Projects/siRNA/

siRNAindex.htm; http://design.dharmacon.com). Care should be taken to minimize homology between target sequences and other genes in the genome to reduce the chance of nonspecific suppression. Another known issue of RNAi is its off-target effects (Jackson *et al.*, 2003). Such effects include the inadvertent silencing of other unknown genes and induction of the interferon response. Therefore, it is crucial to test several different target sequences to determine whether they give the same phenotype and whether the severity of the phenotype correlates with the level of suppression. It is also important to include a negative control siRNA or shRNA to confirm that the phenotype is only produced with the gene-specific RNAi. Good negative controls include siRNAs or shRNAs targeting luciferase or green fluorescent protein, which are not expected to affect endogenous protein expression in mammalian cells.

To best demonstrate that the observed phenotype of RNAi is due to suppression of the target protein, rescue experiments should be performed. The goal is to revert the phenotype by coexpressing a version of the targeted gene not recognized by the interfering RNA. This can be accomplished by introducing silent mutations into the cDNA that encodes the target protein or by expression of a homolog from another species with a different sequence at the RNAi target region. In either case, it is wise to ensure that there are at least two nucleotide differences between the rescue sequence and the target sequence. We have on occasion achieved equally efficient suppression using a single shRNA in both human and murine cells, where the target sequence differed by a single nucleotide.

MDCK I and MDCK II cells are commonly used for the study of epithelial cell adhesion and polarity. They are canine kidney epithelial cells, and the two clones differ in the tightness of their junctions. However, both cell types are difficult to transfect. In our hands, Lipofectamine 2000 (Invitrogen, Carlsbad, CA) and similar reagents can achieve efficiencies of approximately 20% in MDCK II cells, which is too low for many types of experiments. MDCK cells can be efficiently transduced using retroviruses or adenoviruses (Schuck *et al.*, 2004; Shinomiya *et al.*, 2004), but these reagents can be time-consuming to generate. Schuck and colleagues described a retroviral vector for gene silencing by RNAi, called pRVH1-puro, in which shRNA expression is driven from an H1 promoter. The vector also contains a puromycin resistance gene for selection of transduced cells. Retroviruses are pseudotyped with the VSV-G envelope protein to enhance stability and infection of MDCK cells. After selection, the efficiency of target gene silencing was in the range of 70–95% (Schuck *et al.*, 2004).

As a more rapid alternative approach, we have used the Nucleofection procedure developed by Amaxa Inc. (Gaithersburg, MD) to introduce shRNA constructs or cDNA expression plasmids into MDCK cells with more than 80% efficiency and can achieve gene silencing with similar efficiencies to the retroviral method.

Generation of shRNA-Expressing Constructs with the pSUPER Vector

We use the pSUPER vector provided by Reuven Agami (Netherlands Cancer Institute) to drive the expression of shRNA in mammalian cells. pSUPER is based on the pBlueScript-KS vector and contains a polymerase-III H1-RNA promoter that directs the synthesis of shRNAs (Brummelkamp et al., 2002). The BglII and HindIII sites downstream of the H1-RNA promoter are used to clone the gene-specific insert (Fig. 1A). For each target gene, several 19-nucleotide (nt) long, gene-specific target sequences are selected. Most canine gene sequences are now available from the NCBI dog genome database (http://www.ncbi.nlm.nih.gov/projects/genome/guide/dog/). The genomic sequences are fragmented but can be easily assembled using the tBLASTn search engine to identify exons from the cosmids.

The canine genes can then be used to find suitable shRNA target sequences. This process can be simplified by the use of algorithms available on the Web, which consider various parameters to select for likely targets. The target sequence is preferably within the coding region of the mRNA, with at least a 100-bp distance from the start and termination codons. The GC content of the 19-nt target sequence should be at least 30%, and it should not contain a stretch of four or more adenines or thymidines that will cause premature termination of the transcript (Elbashir et al., 2002; Reynolds et al., 2004). Potential targets are then BLASTed against genome-specific databases to ensure that they do not share high homology to other genes.

A pair of complementary 64-nt DNA oligonucleotides (oligos) for each target sequence are annealed and cloned into the pSUPER vector (Fig. 1B). The annealed oligos have 5' overhangs complementary to the overhangs on pSUPER that has been digested with BglII and HindIII.

Annealing of Oligos

Oligos can be ordered or synthesized at the 50-nmol scale with salt-free purification. They are dissolved in H_2O to a final concentration of about 3 $\mu g/\mu l$, then 1 μl each of forward and reverse oligo is mixed with 48 μl of annealing buffer (100 mM potassium acetate, 30 mM HEPES-KOH, pH 7.4, 2 mM magnesium acetate). Incubate the oligo mixture at 95° for 4 min, then incubate at 70° for 10 min. In the meantime, prepare several hundred

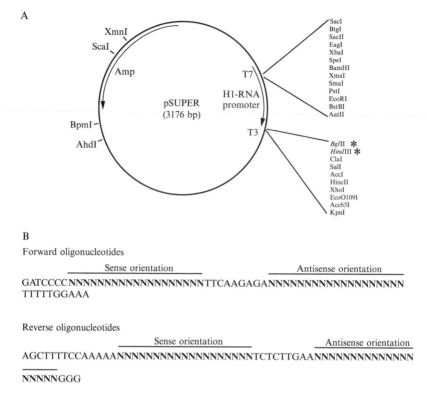

FIG. 1. Schematics of the pSUPER vector and design of oligos. (A) The BglII and HindIII sites used to clone the gene-specific insert are indicated by asterisks. (B) Design of the forward and reverse oligos. Each oligo is 64-nt long. The 19-nt target sequence in its sense or antisense orientation is represented by bold Ns. Information on pSUPER was provided by Reuven Agami (Netherlands Cancer Institute).

milliliters of 70° water in a large beaker and immediately transfer the tubes containing the oligo mixture into the 70° water after the end of the 70° incubation. Let the water bath and oligo mixture slowly cool at room temperature for about 1 h, then transfer them to 4° until annealed oligos cool to 4°. This step of slow cooling is critical for the successful annealing of these oligos. Annealed oligos can be stored at −20° if they are not used immediately.

Phosphorylation of Annealed Oligos

Take 2 μl of the annealed oligos, add 1 μl of 10 × T4 polynucleotide kinase buffer (or 10 × T4 DNA ligase buffer), add 1 μl of 10 m*M* ATP (not necessary if using T4 DNA ligase buffer), add DTT to a final concentration of 5 mM, then add 1 μl of T4 kinase, adjust the total volume to 10 μl with H$_2$O, and incubate the mixture at 37° for 30 min. T4 kinase is then heat-inactivated by incubating the reaction mixture at 70° for 10 min, followed by slow cooling in large volume of water to 4°.

Digestion of the pSUPER Vector

The cloning sites on pSUPER that we typically use are *Hind*III and *Bgl*II. The pSUPER is digested first with *Hind*III for 4 h at 37°, then with *Bgl*II at 30° overnight. The digested vector is dephosphorylated with calf intestinal alkaline phosphatase (CIP) and purified with the QIAquick PCR purification kit from Qiagen (Valencia, CA). The digested vector can be frozen at −20° for future use.

Ligation of Annealed Oligos into the pSUPER Vector

Ligate 2 μl of annealed oligos with 1 μl of digested pSUPER vector for 1 h at room temperature. A vector alone ligation should be included as a negative control, and 3–5 μl of ligation products is used to transform competent bacteria. We screen three to five colonies for positive clones by restriction digests with *Eco*RI and *Hind*III. Empty pSUPER vector produces an insert of ~230 bp, whereas positive clones have inserts of ~290 bp. Alternately, the colonies can be screened directly by PCR using T7 and T3 promoter primers. Positive clones give products of ~440 bp, whereas negative clones give products of ~380 bp. The positive clones are then sequenced with the T7 primer to confirm that they contain the desired 64-nt insert at the correct location.

Sometimes the sequencing result cannot proceed through the insert region, probably because of the high stability of the hairpin structure. This problem can usually be resolved by adding 5% DMSO to the sequencing reaction. In our experience, mutations occur in approximately 20% of all pSUPER inserts. These mutations range from large deletions of the front or rear of the insert to single base changes at random locations within the insert. The mutations probably arise from the formation of secondary structures within the insert oligonucleotides.

Transient Transfection of RNAi Constructs by Nucleofection

Another key component of successful RNAi experiments is to intro-duce siRNA or shRNA constructs into a large fraction of cells so that any morphological or biochemical changes caused by the gene silencing can be easily detected. We have used the nucleofection technique developed by

Amaxa Inc. to transiently transfect MDCK cells with more than 80% efficiency. Nucleofection is a nonviral, electroporation-based transfection method that uses the Nucleofector device, which delivers unique electrical parameters, and a proprietary solution. Among the different RNAi targets that we tested, suppression of protein expression is usually maximal after 48–72 h, and can last 5–7 days. The following is the optimized protocol we routinely use for transfecting MDCK cells with shRNA constructs and cDNA expression constructs.

1. Passage MDCK cells 1–2 days before transfection, so that cells are less than 80% confluent and in their logarithmic growth phase on the day of transfection. Each transfection uses 2×10^6 cells. A confluent 100-mm dish contains approximately 1×10^7 MDCK cells.

2. Prepare 2 μg of pSUPER construct or pSUPER-luciferase (negative control) DNA in individual microcentrifuge tubes. When performing cotransfection with shRNA construct and cDNA expression construct, use 2 μg of each construct. We also routinely cotransfect 0.5 μg of plasmid DNA expressing superenhanced YFP as a transfection marker.

3. Prewarm the Nucleofector solution T and supplement to room temperature. Prewarm RPMI medium (GIBCO-BRL, Gaithersburg, MD) containing 10% fetal calf serum at 37°. Prepare 6-well plates by filling wells with normal MDCK culture medium containing serum and antibiotics and preincubate the plates in a 37° incubator.

4. Wash MDCK cells once with PBS and trypsinize cells with trypsin/ EDTA for a few minutes at 37°. Trypsinization time should be kept as short as possible to reduce cell death after nucleofection. Add culture medium to stop trypsinization and resuspend cells into a conical tube. Determine cell density using a hemocytometer, and transfer the required number of cells (2×10^6 cells per transfection) into a conical tube.

5. Centrifuge cells at 1000–2000 rpm for 2 min and gently remove the supernatant so that virtually no residual medium covers the cell pellet. Add 85 μl Nucleofection solution T and 19 μl supplement 1 per transfection into the cell pellet and resuspend the cells to a final concentration of approximately 2×10^6 cells/100 μl. Avoid storing the cell suspension longer than 15–20 min in the Nucleofector solution, because this reduces cell viability and transfection efficiency. This time is typically sufficient for only four to six transfections, so we recommend repeating the procedure from step 2 if more transfections need to be performed.

6. Transfer 100 μl of cell suspension into each tube containing DNA, then perform steps 7–9 for each transfection separately.

7. Mix the cells with DNA a few times with a pipette, then transfer the mixture into the electroporation cuvette. Make sure that the sample covers

the bottom of the cuvette and avoid air bubbles. Close the cuvette with the blue cap.

8. Insert the cuvette into the cuvette holder on the Nucleofector device and rotate the wheel clockwise to the final position. Choose the program T-23 on the screen, and press the "X" key to start the program. When the screen displays "OK," immediately take out the cuvette and add 500 μl of prewarmed RPMI medium with serum. Use plastic pipettes to transfer cells into a microcentrifuge tube.

9. Plate desired number of cells into each well of a 6-well plate for Western blot analysis; 150–200 μl cells are usually enough to achieve confluency in each well after 2–3 days.

To confirm transfection efficiency, YFP expression can be monitored after several hours or overnight incubation by fluorescence microscopy. There will be some floating dead cells, but excessive cell death might indicate a problem. We found that minimizing the trypsinization time is critical for keeping cells viable, and cells with old passage numbers might also be more susceptible to damage during transfection.

We have generated several pSUPER constructs that express shRNAs targeting different regions of canine Par-3 mRNA (Chen and Macara, 2005). Each of the constructs can suppress the endogenous Par-3 level in MDCK cells to a different degree. Western blot analysis of many other junctional and polarity proteins did not detect any obvious changes in their total level, suggesting that the RNAi effect is specific for Par-3 (Fig. 2A).

Analysis of Defects in Cell Adhesion and Polarity After Gene Suppression

Confluent MDCK cells form a polarized monolayer when grown on plastic or filter support, and each cell has distinct apical and basolateral domains that are enriched with different transmembrane or membrane associated proteins. By performing immunofluorescence of these specific markers, the effects of RNAi on the polarized distribution of proteins can be analyzed. In addition to two-dimensional (2-D) culture systems, MDCK cells can also form polarized cysts when embedded in extracellular matrix (Zegers *et al.*, 2003). Because of the complex stages required for the complete formation of polarized cysts in 3-D cultures, this system can reveal defects in epithelial polarity and morphogenesis that are not evident in 2-D cultures. It usually takes 7–10 days for MDCK cells to grow into cysts lined with a monolayer of polarized cells; therefore, it might be necessary to establish stable cell lines expressing specific shRNA constructs to study the consequences of suppressing a certain gene on cysts formation.

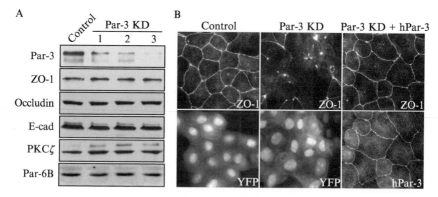

FIG. 2. Suppression of Par-3 expression in MDCK cells. (A) MDCK cells were transiently transfected with either control pSUPER vector (control) or three pSUPER constructs targeting different sequences of canine Par-3 mRNA (Par-3 KD). Equal amounts of total proteins were analyzed by SDS-PAGE and immunoblot. (B) Suppression of Par-3 causes a tight junction assembly defect. MDCK cells were transfected with control or pSUPER-Par-3 (Par-3 KD) constructs. Yellow fluorescent protein (YFP) was coexpressed as a marker for transfected cells. To rescue the RNAi phenotype, a construct encoding HA-tagged human Par-3 (hPar-3) was cotransfected with pSUPER-Par-3. Cells were fixed and immunostained for ZO-1 and ectopic HA-tagged hPar-3. Adapted from Chen and Macara (2005).

Another good indicator of effects on cell polarity is the assembly of TJs. TJs form at the boundary between the apical and basolateral domains, where both the Crumbs complex and the Par-3/Par-6/aPKC complex are concentrated in polarized MDCK cells. TJs act as a fence between the two membrane domains to prevent the intermixing of membrane components (fence functions) and restrict the paracellular diffusion of ions and small molecules (gate function) (Schneeberger and Lynch, 2004). Therefore, defects in TJ assembly can contribute to compromised cell polarization.

In a confluent monolayer, TJ proteins such as occludin and ZO-1 mostly localize along cell borders in a continuous fashion. If a protein is important for the proper assembly of TJs, reduction in its level by RNAi could result in mislocalization of TJ components as determined by indirect immunofluorescence.

As one example, when endogenous Par-3 expression was suppressed by RNAi, ZO-1 and occludin staining were fragmented or completely absent from cell borders, indicating a critical role of Par-3 in TJ formation (Fig. 2B). The defects in TJ assembly can be more carefully examined by the calcium switch procedure. When confluent monolayer of MDCK cells are incubated in low calcium medium, their TJs and AJs disassemble, and the cells pull apart from one another and round up. To make low calcium medium, minimum essential medium (S-MEM, GIBCO-BRL, Gaithersburg, MD) is supplemented with 2% dialyzed fetal calf serum (FCS). Serum is dialyzed at 4° against 0.15 M NaCl, then 0.15 M NaCl with 0.2 mM EDTA, and then two changes with 0.15 M NaCl. Each change is against 100-fold volume of buffer for 12–24 h (Gumbiner and Simons, 1986). After incubation in low-calcium medium overnight, normal calcium medium (DMEM with 10% FCS) is added back for different periods of time before cells are fixed and subjected to immunofluorescence. This procedure follows the assembly of new cell junctions, and it is able to reveal defects of TJ assembly that may not be visible at steady state. For example, suppression of Par-3 caused a profound delay in TJ assembly after calcium addition, as evidenced by the severely disrupted localization of occludin (Fig. 3A).

Several functional assays are also commonly used in combination with immunofluorescence studies to assess the integrity of TJs. These include the measurement of transepithelial electrical resistance (TER), which monitors the paracellular gate functions of TJs; and the fluorescent lipid probe diffusion assay, which monitors the fence functions (Matter and Balda, 2003a). Disrupted localization of TJ components often results in compromised gate and fence functions. For example, MDCK cells with reduced level of Par-3 also exhibited a much delayed increase of TER after calcium addition, and the peak value of TER was significantly lower than that of control cells (Fig. 3B).

To measure the TER, 6.6×10^5 MDCK cells are plated into poly-carbonate Transwell filters (0.4-μm pore size, 12-mm diameter; Corning Costar, Corning, NY). After 40–44 h, cells are washed twice with PBS and incubated in low calcium medium for 16–20 h. Cells are then incubated in normal calcium medium, and TER is measured at various times with an Epithelial Voltohmmeter (EVOM, World Precision Instruments, Sarasota, FL). TER values are averaged from three parallel filters and are expressed in ohm.cm^2 after subtracting the blank value from a filter without cells.

Rescue the RNAi Phenotype with Ectopic Protein Expression

To confirm that the phenotype observed in an RNAi experiment is specifically due to the suppression of the target protein rather than an off-target effect, it is useful to perform rescue experiments to test whether restoring the protein levels can reverse the phenotype. This is usually achieved by introducing silent mutations into the cDNA of the target protein to enable its ectopic expression in the presence of RNAi constructs. Alternately, expression of a homolog from another species can also be used to functionally replace the endogenous protein. It should be noted that ectopic expression does not always fully revert the phenotype induced by

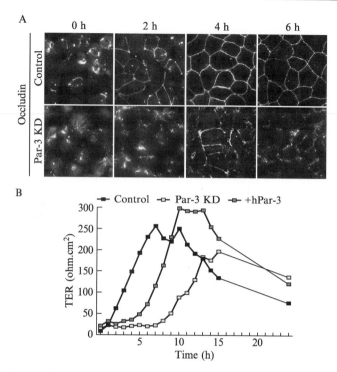

FIG. 3. Suppression of Par-3 severely delays tight junction assembly after calcium switch. (A) Control or Par-3 knockdown (Par-3 KD) MDCK cells were subjected to calcium switch and stained for occludin at indicated times after addition of normal calcium medium. (B) Kinetics of transepithelial electrical resistance (TER) establishment of MDCK cells was monitored for 24 h after addition of normal calcium medium. Expression of human Par-3 in Par-3 KD cells (+hPar-3) rescued TER establishment. Each value is the mean of triplicate measurement. Adapted from Chen and Macara (2005).

RNAi, probably because it is difficult to achieve the same level of homogenous expression as the endogenous protein in the entire cell population. If the ectopic expression level is too high, it might sequester other endogenous proteins and produce adverse effects. To rescue the TJ assembly defect in Par-3 knockdown cells, we cotransfected a construct that expresses human Par-3 together with the shRNA construct targeting canine Par-3. After 2–3 days, the level of ectopic Par-3 was very similar to the level of endogenous Par-3, as determined by Western blot, and ectopic Par-3 localized to cell–cell borders (Fig. 2B). Calcium switch experiments showed that ectopic Par-3 expression efficiently rescued the speed of TJ assembly and also partially restored the kinetics of TER development (Fig. 3B).

The relatively low level expression of ectopic protein after nucleofection minimizes potential problems associated with overexpression.

Using the approach of RNAi coupled with rescue experiments, it is feasible to dissect the functional domains of a protein required for certain cellular processes. This is advantageous over simple overexpression experiments, because the readout is the rescue of certain phenotypic defects in the absence of most of the endogenous protein. Different mutants or deletion fragments of the protein can be expressed to examine the functional significance of specific regions of the protein to a certain phenotype, and in the case in which other proteins are known to bind to a certain region, a deletion fragment lacking this region could help clarify the role of the binding partner in its physiological activities. We have used this rescue approach to dissect the involvement of various domains of Par-3 in supporting TJ formation and demonstrated that Par-3 plays important roles in regulating TJs independent of its binding to Par-6 or aPKC. Instead, the C terminus of Par-3 seems to be critical for stable TJ assembly in MDCK cells (Chen and Macara, 2005).

References

Brummelkamp, T. R., Bernards, R., and Agami, R. (2002). A system for stable expression of short interfering RNAs in mammalian cells. *Science* **296,** 550–553.

Chen, X., and Macara, I. G. (2005). Par-3 controls tight junction assembly through the Rac exchange factor Tiam1. *Nat. Cell Biol.* **7,** 262–269.

D'Atri, F., and Citi, S. (2002). Molecular complexity of vertebrate tight junctions (Review). *Mol. Membr. Biol.* **19,** 103–112.

Elbashir, S. M., Harborth, J., Lendeckel, W., Yalcin, A., Weber, K., and Tuschl, T. (2001). Duplexes of 21-nucleotide RNAs mediate RNA interference in cultured mammalian cells. *Nature* **411,** 494–498.

Elbashir, S. M., Harborth, J., Weber, K., and Tuschl, T. (2002). Analysis of gene function in somatic mammalian cells using small interfering RNAs. *Methods Enzymol.* **26,** 199–213.

Getsios, S., Huen, A. C., and Green, K. J. (2004). Working out the strength and flexibility of desmosomes. *Nat. Rev. Mol. Cell. Biol.* **5,** 271–281.

Gumbiner, B., and Simons, K. (1986). A functional assay for proteins involved in establishing an epithelial occluding barrier: Identification of a uvomorulin-like polypeptide. *J. Cell Biol.* **102,** 457–468.

Jackson, A. L., Bartz, S. R., Schelter, J., Kobayashi, S. V., Burchard, J., Mao, M., Li, B., Cavet, G., and Linsley, P. S. (2003). Expression profiling reveals off-target gene regulation by RNAi. *Nat. Biotechnol.* **21,** 635–637.

Knust, E., and Bossinger, O. (2002). Composition and formation of intercellular junctions in epithelial cells. *Science* **298,** 1955–1959.

Macara, I. G. (2004). Parsing the polarity code. *Nat. Rev. Mol. Cell. Biol.* **5,** 220–231.

Matter, K., and Balda, M. S. (2003a). Functional analysis of tight junctions. *Methods Enzymol.* **30,** 228–234.

Matter, K., and Balda, M. S. (2003b). Signalling to and from tight junctions. *Nat. Rev. Mol. Cell. Biol.* **4,** 225–236.

Ohno, S. (2001). Intercellular junctions and cellular polarity: The PAR-aPKC complex, a conserved core cassette playing fundamental roles in cell polarity. *Curr. Opin. Cell Biol.* **13**, 641–648.

Paddison, P. J., Caudy, A. A., Sachidanandam, R., and Hannon, G. J. (2004). Short hairpin activated gene silencing in mammalian cells. *Methods Mol. Biol.* **265**, 85–100.

Parkinson, S. J., Le Good, J. A., Whelan, R. D., Whitehead, P., and Parker, P. J. (2004). Identification of PKCzetaII: An endogenous inhibitor of cell polarity. *EMBO J.* **23**, 77–88.

Perez-Moreno, M., Jamora, C., and Fuchs, E. (2003). Sticky business: Orchestrating cellular signals at adherens junctions. *Cell* **112**, 535–548.

Reynolds, A., Leake, D., Boese, Q., Scaringe, S., Marshall, W. S., and Khvorova, A. (2004). Rational siRNA design for RNA interference. *Nat. Biotechnol.* **22**, 326–330.

Rodriguez-Boulan, E., and Nelson, W. J. (1989). Morphogenesis of the polarized epithelial cell phenotype. *Science* **245**, 718–725.

Roh, M. H., and Margolis, B. (2003). Composition and function of PDZ protein complexes during cell polarization. *Am. J. Physiol. Renal Physiol.* **285**, F377–F387.

Schneeberger, E. E., and Lynch, R. D. (2004). The tight junction: A multifunctional complex. *Am. J. Physiol. Cell Physiol.* **286**, C1213–C12128.

Schuck, S., Manninen, A., Honsho, M., Fullekrug, J., and Simons, K. (2004). Generation of single and double knockdowns in polarized epithelial cells by retrovirus-mediated RNA interference. *Proc. Natl. Acad. Sci. USA* **101**, 4912–4917.

Shin, K., Straight, S., and Margolis, B. (2005). PATJ regulates tight junction formation and polarity in mammalian epithelial cells. *J. Cell Biol.* **168**, 705–711.

Shinomiya, N., Gao, C. F., Xie, Q., Gustafson, M., Waters, D. J., Zhang, Y. W., and Vande Woude, G. F. (2004). RNA interference reveals that ligand-independent met activity is required for tumor cell signaling and survival. *Cancer Res.* **64**, 7962–7970.

Tepass, U. (2002). Adherens junctions: New insight into assembly, modulation and function. *Bioessays* **24**, 690–695.

Yeaman, C., Grindstaff, K. K., and Nelson, W. J. (1999). New perspectives on mechanisms involved in generating epithelial cell polarity. *Physiol. Rev.* **79**, 73–98.

Zegers, M. M., O'Brien, L. E., Yu, W., Datta, A., and Mostov, K. E. (2003). Epithelial polarity and tubulogenesis *in vitro. Trends Cell Biol.* **13**, 169–176.

[27] Nucleofection of Primary Neurons

By Annette Gärtner, Ludovic Collin, and Giovanna Lalli

Abstract

Efficient gene transfer is an important tool for the study of neuronal function and biology. This has proved difficult and inefficient with traditional transfection strategies, which can also be fairly toxic, whereas viral-mediated gene transfer, although highly efficient, is often time-consuming. The recently developed Amaxa Nucleofector technology, based on electroporation with preset parameters in a cell-type–specific solution, enables direct delivery of DNA, small interfering (si)RNA oligonucleotides and

METHODS IN ENZYMOLOGY, VOL. 406
Copyright 2006, Elsevier Inc. All rights reserved.

0076-6879/06 $35.00
DOI: 10.1016/S0076-6879(06)06027-7

siRNA vectors into the cell nucleus. This strategy results in reproducible, rapid, and efficient transfection of a broad range of cells, including primary neurons. Nucleofected neurons survive for up to 3 weeks and remain functional. We are currently using this transfection method to examine the contribution of Rho GTPase signaling pathways in the establishment of neuronal polarity, neuronal migration, and neurite outgrowth. Here, we describe three protocols to efficiently nucleofect rat cerebellar granule, cortical, and hippocampal neurons.

Principle

The nucleofection technology is based on an improved method of cell electroporation. Cell-type–specific electrical current and electroporating solutions allow the transfer of polyanionic macromolecules directly into the cell nucleus, thus permitting transfection of slowly dividing or mitotically inactive cells, such as embryonic and adult mammalian neurons (Dityateva *et al.*, 2003). So far, this technique has been successfully applied to a variety of both mammalian and avian neuronal cell types, including hippocampal, cerebellar granule, cortical, sympathetic, sensory neurons, and also to neural stem cells (for an updated reference list, consult the Amaxa scientific citation database http://www.amaxa.com/citations.html).

Advantages

- Easy, fast, and safe to use.
- Transfection efficiency is reproducible, and, depending on the neuronal cell type, allows biochemical assays to be performed.
- Transfection is performed before cell plating, and protein expression is usually observed 1–2 h after nucleofection. This makes the technique suitable for the study of gene function during early developmental events, such as neuronal migration and neurite formation and outgrowth (Solecki *et al.*, 2004; Zhou *et al.*, 2004).
- Long-term expression allows the study of protein function in later developmental processes such as protein/receptor trafficking, synaptogenesis, and synaptic plasticity (Bezzerides *et al.*, 2004; Govek *et al.*, 2004; Viard *et al.*, 2004).

Drawbacks

- Large amounts of cells (minimum 1×10^6) are needed for each nucleofection.
- Transfection efficiency depends on the neuronal cell type (varies from 10–15% to 70–80%).

- Success dependent on the type of construct used.
- Decrease of expression can be observed in long-term experiments.
- Early expression of some transgenes may interfere with cell attachment or with other developmental events preceding the process of interest (in these cases inducible promoters could be used).

General Considerations

1. Good DNA quality: This is essential for a successful nucleofection. In general, the purity of each plasmid, checked by measurement of the A260/A280 ratio, should be at least 1.8. We observed good transfection efficiency of different plasmids purified with QiaFilter Plasmid Maxi kits (Qiagen cat. No. 12263) and EndoFree Plasmid kits (Qiagen cat. No. 12362).

2. Number of neurons per nucleofection: In our experience, optimal results have been achieved using between 1 and 6×10^6 cells per nucleofection, depending on the experimental conditions and assays. A minimum of 1×10^6 of neurons is generally recommended.

3. Gentle handling of neurons: Before nucleofection, cells must be resuspended in a small amount (100 μl per transfection) of the nucleofector solution provided with the Amaxa kit. The nucleofector solution should be prewarmed at room temperature. We recommend gently resuspending the neurons by pipetting up and down a few times with a P1000 Gilson pipette and avoiding air bubbles when transferring the cell suspension in the bottom of the nucleofection cuvette. For transfection of multiple samples, the entire nucleofection procedure should not take longer than 5–10 min, because cell viability and transfection efficiency decrease if neurons are kept in the nucleofector solution for longer times.

4. Speed during the nucleofection procedure: After nucleofection, it is very important to quickly resuspend and plate neurons in pre-warmed Dulbecco's modified Eagle medium (DMEM) supplemented with 10% fetal calf serum (FCS), previously equilibrated in the culture incubator for at least a couple of hours. This medium helps to improve neuronal survival during the first 2–4 h after nucleofection and is then replaced by the standard culture medium according to the neuronal cell type used.

Preparation and Nucleofection of Cerebellar Granule Neurons for Use in Migration Assays

Cerebellar granule neurons (CGNs) are the most abundant neurons of the cerebellum. Progenitors of these neurons actively proliferate during the first 2 postnatal weeks of life in the most superficial layer of the developing cerebellar cortex, the external granular layer, before migrating to reach their final destination, the internal granular layer. CGNs can be purified in large quantity from postnatal cerebellum (the usual yield is 1–2 × 10^6 cells per rat pup aged P5–P6) and are commonly used in various cell-based assays to study neuronal migration, such as glial/neuronal co-cultures (Hatten *et al.*, 1998) or neuronal reaggregates (Bix and Clark, 1998; Kholmanskikh *et al.*, 2003; Tanaka *et al.*, 2004). Here, we describe a protocol for nucleofection of CGNs to study neuronal migration with a neuronal reaggregate assay.

Reagents and Materials

Medium A: Dulbecco's modified Eagle's medium–Ham's F12 medium (1:1) (Invitrogen), 6 m*M* D-glucose, 10 μg/ml transferrin, 10 μg/ml insulin, 0.2 m*M* glutamine, 10 ng/ml selenium, 100 U/ml penicillin, 100 μg/ml streptomycin.

Medium B: Medium A containing 10% FCS and 10% horse serum
Percoll gradient (Amersham Pharmacia) (10 ml):

- 35%: 1 ml PBS (10×), 3.5 ml Percoll, 5.5 ml ddH$_2$O.
- 60%: 1 ml PBS (10×), 6 ml Percoll, 3 ml ddH$_2$O, 30 μl 0.4% Trypan blue.

Add 5 ml 35% Percoll to a 15-ml Falcon tube and draw 5 ml 60% Percoll up into a 5-ml pipette. By slow expulsion, carefully form an underlying 60% Percoll layer beneath the 35% layer. Prepare six gradients for one CGN preparation.

Medium C: Hank's balanced salt solution without Ca^{2+} and Mg^{2+} (HBSS–/–, Invitrogen), 20% FCS.

Medium D: HBSS–/–, 0.1% DNAse I.
Medium E: Medium D containing 12.5 mg/ml trypsin.
30-mm bacterial Petri dishes (Bibby Sterilin).
40-μm cell strainer (BD Falcon).
Amaxa Nucleofector rat primary neuron kit.

Coating solution for glass coverslips: 25 μg/ml poly-D-lysine in ddH$_2$O. Coverslips are incubated for 45 min in a humidified 37°/5% CO$_2$ incubator and then overnight with 25 μg/ml laminin in PBS.

Dissociation Procedure

The cerebella from seven P6 rat brains are dissected, placed in ice-cold HBSS–/–, and carefully minced. The tissue fragments are transferred into a sterile 15-ml Falcon tube and washed three times with 12 ml of ice-cold

HBSS–/–. Trypsinization is performed by adding 5 ml of pre-warmed medium E and incubating for 10 min in a water bath at 37° (shaking every 2–3 min); 10 ml of medium C is then immediately added to the trypsin solution. The medium is aspirated, and the tissue fragments are washed three times with 12 ml of ice-cold HBSS–/–. After the last wash, the cerebellar fragments are triturated in 5 ml of medium D using a 5-ml plastic pipette. The resulting cell suspension is filtered through a 40-μm cell strainer to eliminate cell debris. The filter is then washed with 10 ml of medium C, and the cell suspension is centrifuged for 10 min at 1000 rpm. The pellet is gently resuspended in 3 ml of medium C; 500 μl of cell suspension is carefully layered underneath each of the six 35% Percoll gradients using a P1000 Gilson pipette and centrifuged for 10 min at 3000 rpm with the brake off. A cooled centrifuge is recommended, but a conventional benchtop swing rotor centrifuge also works well. The CGNs, which are located at the 35–60% interface, are carefully removed using a glass Pasteur pipette (the volume removed should be approximately 2 ml). The CGN fractions are equally distributed in three 15-ml Falcon tubes (approximately 4–5 ml/tube), which are then filled with medium C, inverted three to four times and centrifuged for 5 min at 1000 rpm. The pellets are resuspended in a total volume of 3 ml of medium C and incubated for 30 min at room temperature in a plastic cell culture dish coated with 25 μg/ml of poly-D-lysine. This step increases the purity of the culture by eliminating glial cells that attach quickly to the dish. After 30 min, the medium is gently pipetted several times over the surface of the dish to resuspend the CGNs, transferred to a 15-ml Falcon tube, and centrifuged for 5 min at 1000 rpm. Cells are resuspended in 2 ml of medium C and counted. With this protocol, the CGN purity of the preparation obtained is approximately 95–99%.

Nucleofection

Thirty-millimeter bacterial Petri dishes containing 2 ml of medium B are preequilibrated in a humidified 37°/5% CO_2 incubator for 3–4 h before nucleofection; 6–7 $\times$ 10^6 CGNs per nucleofection are resuspended in medium C and centrifuged for 5 min at 1000 rpm. The pellet is gently resuspended using a P1000 Gilson pipette in 100 μl of nucleofection solution and mixed with 3 μg of the supplied pmaxGFP plasmid. The cell suspension is transferred to the nucleofector cuvette and nucleofected using program O-03. Immediately after nucleofection, 1 ml of pre-warmed medium B is added to the cuvette. Cells are transferred to the Petri dish containing the remaining 1 ml of medium B using a plastic pipette provided in the Amaxa kit. Two fractions of 3–3.5 $\times$ 10^6 granule neurons in a final volume of 2 ml of medium B are cultured in suspension in bacterial Petri dishes for 90 min in a humidified 37°/5% CO_2 incubator. The cell suspensions are then removed using a P1000 Gilson pipette and centrifuged for 5 min at 1000 rpm. Each CGN fraction is gently resuspended in 2 ml of prewarmed medium B and incubated in a new 30-mm bacterial Petri dish in a humidified 37°/5% CO_2 incubator to allow formation of neuronal reaggregates (Fig. 1 A,B). After 18 h, neuronal reaggregates from each fraction are centrifuged for 5 min at 1000 rpm and resuspended in 4 ml of medium B. Reaggregates are separated by gently pipetting up and down. Twenty to 30 reaggregates containing GFP-positive cells (Fig. 1 A,B) are seeded on 13-mm coverslips. After 45 min, medium B is carefully removed, and migration is initiated by addition of 400 μl/well of medium A. Cells are usually fixed 24 h after the start of migration, 48 h after nucleofection. In these conditions, nucleofection efficiency is approximately 15–25%. By use of a different experimental procedure, it has been reported that the nucleofection efficiency of CGNs can reach 38% (Kovacs et al., 2004). In our assay, CGNs migrate along neurites formed by other neurons; the migration is robust, quantifiable, and reproducible (Kholmanskikh et al., 2003; Tanaka et al., 2004). Pmax-GFP is not harmful for neurons, because migration of nucleofected CGNs is not affected, and the morphology of nucleofected CGN neurons is similar to that of non-nucleofected cells (Fig. 1 C,D and inset) (Hatten, 2002).

Preparation and Nucleofection of Rat Cortical Neurons

Reagents and Materials

> Dissection medium: HBSS (Invitrogen) with 10 mM HEPES.
> Plating medium: DMEM (Invitrogen) with 10% FCS.
> Culture medium: Neurobasal medium (Invitrogen), 50 × B27 supplement (Invitrogen, diluted 1:50), 2 mM L-glutamine, 0.6% D-glucose.
> Trypsin (0.05%)–EDTA (0.53 mM) solution (Invitrogen).
> Coating solution for glass coverslips or plastic dishes: 0.5 mg/ml poly-DL-ornithine in PBS.

Coverslips and plates are incubated with the coating solution overnight at 37°. The following day, before starting the dissection, they are washed three times with sterile ddH_2O, covered with an adequate amount of plating medium and kept in a humidified 37°/5% CO_2 incubator.

> 40-μm cell strainer (BD Falcon).
> Amaxa Nucleofector rat primary neuron kit.

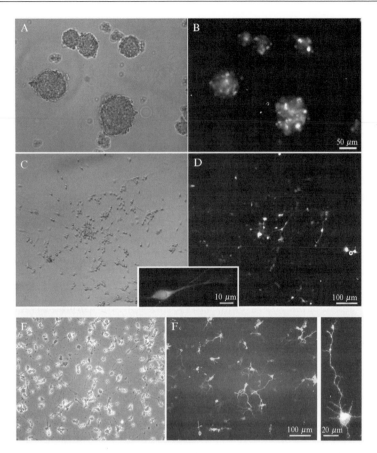

FIG. 1. Nucleofection of cerebellar granule and cortical neurons. Phase-contrast (A) and fluorescent (B) images of cerebellar granule neurons nucleofected with pmaxGFP and cultured in suspension for 18 h to allow the formation of neuronal reaggregates. After plating of the reaggregates, neurons are able to migrate on laminin-coated coverslips (C–D and inset). (E–F) Rat cortical neurons nucleofected with a plasmid coding for farnesylated EGFP (EGFP-F) to visualize neuronal morphology. Phase-contrast (E) and fluorescent (F) images of a cortical culture transfected using the O-03 nucleofection program and maintained *in vitro* for 24 h. The morphology of the nucleofected neurons appears normal, with multiple dendritic processes and an axon, as also shown at higher magnification (right inset).

Dissociation Procedure

Cortices are dissected from the brain of four E18 rat embryos (Banker and Goslin, 1998) and transferred to a sterile 50-ml Falcon tube medium full of dissection medium and placed in ice. At the end of the dissection, the medium is aspirated, and the cortices are washed twice with 6–7 ml of trypsin–EDTA solution using a 10-ml plastic pipette. Washing is performed by swirling the tube each time and by carefully aspirating the trypsin solution after cortices settle at the bottom of the tube. Cortices are incubated with 8 ml of 0.05% trypsin–EDTA for 20 min at $37°$ (shaking every 2–3 min). The trypsin solution is then aspirated, and the cortices are washed five times with 5 ml of plating medium. After aspirating the last wash, the tissues are triturated in 5 ml of plating medium by pipetting up and down 10 times using a 10-ml plastic pipette. The tissue fragments are left to settle at the bottom of the tube, and the supernatant containing the cells in suspension is transferred to another 50-ml Falcon tube. Three milliliters of plating medium is added to the fragments, which are gently triturated with 15 passages through a glass Pasteur pipette (the borders of the pipette bore is usually slightly narrowed and smoothened with a flame). The tissue fragments are left to settle, and the supernatant is pooled with the cell suspension obtained in the previous trituration. The dissociation in 3 ml of plating medium is then repeated another couple of times, always pooling all the supernatants. At the end, the pooled supernatants (approximately 15 ml) are passed through a 40-μm cell strainer to eliminate bigger fragments and cell aggregates. After cell counting, an adequate amount of cells for nucleofection is placed in a 15-ml Falcon tube. We usually use 2–2.5 × 10^6 cells per nucleofection, and perform up to 12 nucleofections per preparation. Cells are spun at 1000 rpm for 5 min at room temperature. During the centrifugation, an Eppendorf tube for each nucleofection is prepared with the DNA to transfect (between 3 and 6 μg per nucleofection).

Nucleofection

The supernatant is aspirated, and cells are resuspended using a P1000 tip in the appropriate amount of pre-warmed Amaxa rat neuron nucleofector solution, with 100 μl of solution per nucleofection; 100 μl of cell suspension is mixed with DNA in each Eppendorf tube by gently pipetting up and down a few times. The cell suspension is then immediately placed in the bottom of a nucleofection cuvette and nucleofected using program O-03 or O-05. Immediately after each nucleofection, 1 ml of plating medium (prewarmed in a $37°/5\%$ CO_2 incubator) is added to the cuvette, using one of the plastic pipettes provided in the nucleofection kit. The procedure is repeated as quickly as possible with the other samples. Cells are then plated on coverslips or plastic dishes and kept at $37°/5\%$ CO_2. To remove cellular debris and the nucleofector solution, the plating medium is replaced 2–3 h after nucleofection with culture medium, previously equilibrated in a $37°/5\%$ CO_2 incubator. Protein expression is usually already detectable a few hours after nucleofection. This protocol allows neurons to be maintained for up to 1 week in culture and gives a transfection efficiency of between 50 and 70% (Fig. 1E, F), sufficient to allow biochemical experiments.

Preparation and Nucleofection of Hippocampal Neurons

Hippocampal neurons are extensively used to study neuronal development differentiation and synaptic properties in culture, because cultures
derived from E18 or 19 embryonic dissociated hippocampi provide a very
homogenous population of about 90% pyramidal neurons and only very
few interneurons and astrocytes (Banker and Goslin, 1998). Pyramidal
neurons have one axon and several dendrites and can establish functional
synaptic contacts with each other.

Reagents and Materials

HBSS–HEPES: HBSS (Invitrogen) with 10 mM HEPES (Invitrogen).
Plating medium: DMEM (Invitrogen) with 10% FCS or horse serum.
Complete medium (modified from Brewer and Cotman, 1989): for
100 ml DMEM

Component	Sigma order #	Amount in 100 ml
BSA	A-9418	250 mg
Transferrin	T-5391	500 μg
Insulin	I-5500	400 μg
Water-soluble stocks:		
L-Alanine	A-7627	200 μg
Biotin	B-4501	10 μg
L-Carnitine	C-0283	200 μg
Ethanolamine	E-9508	100 μg
D(+)-Galactose	G-0625	1,5 mg
L-Proline	P-5607	776 μg
Putrescine	P-7505	1,61 mg
Sodium pyruvate	P-5280	2,5 mg
Sodium selenite	S-1382	1,6 μg
Vitamin B$_{12}$	V-2876	34 μg
Zinc sulfate	Z-4750	19,4 μg
Catalase	C-40	1,6 mg
L-Glutathione	G-6013	100 μg
Superoxide dismutase	S-2515	250 μg
Ethanol-soluble stocks:		
Linoleic acid	L-1376	100 μg
Linolenic acid	L-2376	100 μg
($\pm$)α-Tocopherol	T-3251	100 μg
($\pm$)α-Tocopherol acetate	T-3001	100 μg
All-trans-retinol	R-7632	10 μg
Retinyl acetate	R-7882	10 μg
Progesterone	P-8783	630 ng
3,3′,5-Triiodo-L-thyronine	T-6397	500 ng

Ethanol and water stocks are prepared as $1:1 \times 10^6$ and $1:1 \times 10^3$ dilutions, respectively, and kept in aliquots at $-20°$. After mixing all components, the medium is filtered through a $0.22\text{-}\mu m$ filter. Alternatively, Neurobasal medium with B27 supplement (Invitrogen) as described previously for cortical neurons can be used for culturing hippocampal neurons without changing nucleofection efficiency.

Paraffin pastilles (Merck 1-07158).
Trypsin (0.05%)–EDTA (0.53 mM) (Invitrogen).
Amaxa Nucleofector rat primary neuron kit.

Preparation of coverslips: 15-mm coverslips are washed overnight in 70% HNO_3, washed in ddH_2O, dried, and autoclaved. With a Pasteur pipette, three paraffin dots are applied to the edge of each coverslip placed in a Petri dish. Coverslips are covered with 0.5 mg/ml (for plastic dishes a concentration of 0.1 mg/ml is sufficient) poly-DL-ornithine solution in borate buffer (1.24 g boric acid and 1.9 g sodium tetraborate in 400 ml ddH_2O, pH 8.5), incubated for 15–24 h at $37°$ and washed three times 30 min with ddH_2O. After the last wash, plating medium (DMEM-FCS or DMEM-HS) is added.

Preparation of Astrocyte Feeder Layers

Astrocytes are isolated from embryonic (E18 or E19) or postnatal (P0–P1) cortices or hippocampi. Tissues are dissected, cut in small pieces, and incubated for 15 min with trypsin–EDTA at $37°$. After washing with HBSS–HEPES, tissues are triturated with a fire-polished Pasteur pipette and placed in DMEM–15% FCS in uncoated flasks in a 10% CO_2 incubator. After 4–6 h, neuronal cells that do not adhere are removed by changing the culture medium. In the following 3–5 days, the culture flasks are vigorously shaken each day, and detached cells are removed by changing the medium until the cultures contain only astrocytes. Astrocytes are passaged in a ratio of 1:5 in 12-well plates or 6-cm dishes 3–5 days before starting the hippocampal cultures to gain 50% confluent dishes; 6–20 h before the hippocampal preparation, the medium is replaced by complete medium.

Preparation of Hippocampal Neurons

E18–E19 pregnant Wistar or Sprague–Dawley rats are killed by cervical dislocation, and embryos are removed. Hippocampi are dissected out from brains placed in HBSS–HEPES and incubated in trypsin–EDTA for 15 min at $37°$, washed five times with HBSS–HEPES, and triturated with a fire-polished Pasteur pipette with only a slightly reduced bore to limit mechanical stress. Tissues should dissociate after 10–20 triturations. The expected yield is approximately 0.3×10^6/hippocampus. Neurons are immediately plated in preequilibrated dishes at a density of 2500 cells/cm^2. Instead of plating cells at this step, neurons can be nucleofected before plating as described in the following section. Neurons should adhere by 2–4 h after plating, and coverslips can then be inverted (paraffin dots down facing the astrocytes) onto the astrocyte feeder layers. For biochemical experiments, neurons are usually plated on plastic dishes at a density higher than 50,000 cells/cm^2, and the plating medium is replaced by preequilibrated complete medium 2–4 h after plating. If cells are cultured for more than 1 week, Ara-C is added to the cultures to a final concentration of 1 μM to decrease growth of contaminating astrocytes.

Nucleofection

To achieve best survival rates, nucleofection should be carried out immediately (and no later than 20–25 min) after tissue dissociation. Fewer than 1×10^6 neurons per nucleofection can be successfully transfected, but with lower efficiency and higher toxicity. The required number of neurons is placed in one 15-ml Falcon tube. Pooling an amount of cells for more than three nucleofections should be avoided, because prolonged incubation in the nucleofector solution is detrimental to the neurons. Cells are collected by centrifugation at 110g for 5 min, resuspended in 100 μl (per nucleofection) of nucleofector solution using a P1000 Gilson pipette, pipetted and mixed once with the DNA or siRNA, and nucleofected using program O-03 or O-05. Immediately after nucleofection, cells are resuspended in prewarmed plating medium and plated. The onset of expression, using the N1-EGFP vector (Invitrogen), is as early as 1 h. After 4 h, a clear GFP expression can be detected in approximately 80% of neurons (Fig. 2A). Cells remain healthy and extend neurites (Fig. 2A), which are similar in length to non-nucleofected neurons (Table I). A normal morphology with one axon and several dendrites is established by 48 h in culture (Table I). The amount of DNA required depends very much on the type of construct used. Transgenes under the control of a CMV promoter are, in general, very well expressed. Depending on expression levels and possible deleterious effects of the transgene, the amount of DNA should be carefully titrated (in the range of 0.5–8 μg, usually starting with 2–3 μg). If the experiment requires expression of the transgene before plating, it is possible to incubate cells after nucleofection in uncoated dishes in a large amount of medium without agitation to prevent cell adhesion and clumping. Neurons will, therefore, start to express the transgene in suspension. Cells are subsequently harvested by gentle

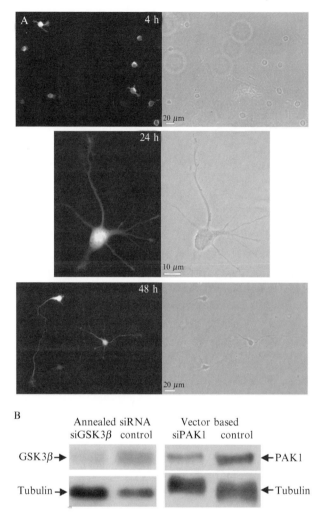

FIG. 2. Nucleofected hippocampal neurons expressing EGFP under the control of the CMV promoter. (A) Phase-contrast and fluorescent pictures taken at the indicated times after nucleofection. By 4 h, approximately 80% of the neurons express GFP. Nucleofected neurons maintain normal cell morphology. (B) Downregulation of GSK3β (left panel) or PAK1 (right panel) can be observed 48 h after nucleofecting hippocampal neurons with either double-stranded RNA oligomers or plasmid-based introduction of siRNA.

TABLE I

MORPHOLOGICAL COMPARISON OF CONTROL AND NUCLEOFECTED HIPPOCAMPAL NEURONS[a]

Time after plating		Control neurons length (μm) (mean $\pm$ SD)	EGFP nucleofected length (μm) (mean $\pm$ SD)
7 h	minor neurites	11.2 $\pm$ 4.6	11.5 $\pm$ 5.4
	axons	no axons	no axons
	soma diameter	11.9 $\pm$ 2.0	12.1 $\pm$ 2.3
	n	12	12
21 h	minor neurites	19.1 $\pm$ 8.8	22.0 $\pm$ 14.0
	axons	85 $\pm$ 46.5	67.6 $\pm$ 36.7
	soma diameter	14.1 $\pm$ 1.7	14.8 $\pm$ 2.2
	n	12	14
45 h	minor neurites	18.5 $\pm$ 15.0	22.5 $\pm$ 19.6
	axons	177.6 $\pm$ 94.7	167 $\pm$ 74.6
	soma diameter	14.4 $\pm$ 2.3	14.5 $\pm$ 3.5
	n	10	11
	% of neurons with 1 axon	75	76
	% of neurons with no axon	16	19

[a] Neurons nucleofected with the N1-EGFP vector grow neurites of the same length as nonnucleofected control neurons. In addition, the onset of axon formation (monitored by using an anti-tau-1 antibody), and soma diameter are similar in control and nucleofected cells.

centrifugation (100g for 3 min), carefully resuspended using a P1000 Gilson pipette, and plated. Because the onset of expression after nucleofection is 1 h, this incubation period can be limited to 1–2 h, but we also successfully tested incubation times up to 24 h for RNAi experiments.

RNAi Using Nucleofection

Nucleofection can also be used to introduce siRNAs into neurons either by using vector-based systems (Brummelkamp et al., 2002) or annealed standard purified dsRNAi oligomers (Ambion). In both cases, the procedure for preparing neurons and nucleofection is similar to what is described previously, and transfection efficiency is comparable to the one observed using GFP under the control of the CMV promoter. We usually use 4 μg of DNA or 8 μg of ds-RNA. We could detect by Western blot a decrease in protein levels of targeted proteins after 48 h (Fig. 2B). The use of a vector-based system has the advantage that transfected cells can be identified by using GFP expressed under an independent promoter, as a reporter gene (Li *et al.*, 2003). For downregulation of GSK3β, we used dsRNA oligonucleotides and could detect by immunofluorescence significantly lower protein levels in most (>90%) of the neurons. For single-cell observation, the vector-based system is preferable, because transfected cells can be identified, whereas oligo-based siRNA can be suitable for biochemical experiments because of their highly efficient delivery.

Acknowledgments

We thank the Hall laboratory for valuable discussions, and T. Ferraro, J. Moneypenny, and A. Riccio for useful suggestions about neuronal cultures. This work was supported by a Deutsche Forschungsgemeinschaft fellowship (A.G.), an EMBO postdoctoral fellowship (L.C.), and a Wellcome Trust Project Grant (G.L.).

References

Banker, G., and Goslin, K. (1998). "Culturing Nerve Cells." MIT Press, Cambridge, MA.

Bezzerides, V. J., Ramsey, I. S., Kotecha, S., Greka, A., and Clapham, D. E. (2004). Rapid vesicular translocation and insertion of TRP channels. *Nat. Cell Biol.* **6**, 709–720.

Bix, G. J., and Clark, G. D. (1998). Platelet-activating factor receptor stimulation disrupts neuronal migration *in vitro*. *J. Neurosci.* **18**, 307–318.

Brewer, G. J., and Cotman, C. W. (1989). Survival and growth of hippocampal neurons in defined medium at low density: Advantages of a sandwich culture technique or low oxygen. *Brain Res.* **494**, 65–74.

Brummelkamp, T. R., Bernards, R., and Agami, R. (2002). A system for stable expression of short interfering RNAs in mammalian cells. *Science* **296**, 550–553.

Dityateva, G., Hammond, M., Thiel, C., Ruonala, M. O., Delling, M., Siebenkotten, G., Nix, M., and Dityatev, A. (2003). Rapid and efficient electroporation-based gene transfer into primary dissociated neurons. *J. Neurosci. Methods* **130**, 65–73.

Govek, E. E., Newey, S. E., Akerman, C. J., Cross, J. R., Van der Veken, L., and Van Aelst, L. (2004). The X-linked mental retardation protein oligophrenin-1 is required for dendritic spine morphogenesis. *Nat. Neurosci.* **7**, 364–372.

Hatten, M. B., Gao, W. Q., Morrison, M. E., and Mason, C. A. (1998). The cerebellum: Purification and coculture of identified cell populations. *In* "Culturing Nerve Cells" (B. A. Goslin, ed.). MIT Press, Cambridge, MA.

Hatten, M. E. (2002). New directions in neuronal migration. *Science* **297**, 1660–1663.

Kholmanskikh, S. S., Dobrin, J. S., Wynshaw-Boris, A., Letourneau, P. C., and Ross, M. E. (2003). Disregulated RhoGTPases and actin cytoskeleton contribute to the migration defect in Lis1-deficient neurons. *J. Neurosci.* **23**, 8673–8681.

Kovacs, A. D., Chakraborty-Sett, S., Ramirez, S. H., Sniderhan, L. F., Williamson, A. L., and Maggirwar, S. B. (2004). Mechanism of NF-kappaB inactivation induced by survival signal withdrawal in cerebellar granule neurons. *Eur. J. Neurosci.* **20**, 345–352.

Li, Z., Hannigan, M., Mo, Z., Liu, B., Lu, W., Wu, Y., Smrcka, A. V., Wu, G., Li, L., Liu, M., Huang, C. K., and Wu, D. (2003). Directional sensing requires G beta gamma-mediated PAK1 and PIX alpha-dependent activation of Cdc42. *Cell* **114**, 215–227.

Solecki, D. J., Model, L., Gaetz, J., Kapoor, T. M., and Hatten, M. E. (2004). Par6alpha signaling controls glial-guided neuronal migration. *Nat. Neurosci.* **7,** 1195–1203.

Tanaka, T., Serneo, F. F., Higgins, C., Gambello, M. J., Wynshaw-Boris, A., and Gleeson, J. G. (2004). Lis1 and doublecortin function with dynein to mediate coupling of the nucleus to the centrosome in neuronal migration. *J. Cell Biol.* **165,** 709–721.

Viard, P., Butcher, A. J., Halet, G., Davies, A., Nurnberg, B., Heblich, F., and Dolphin, A. C. (2004). PI3K promotes voltage-dependent calcium channel trafficking to the plasma membrane. *Nat. Neurosci.* **7,** 939–946.

Zhou, F. Q., Zhou, J., Dedhar, S., Wu, Y. H., and Snider, W. D. (2004). NGF-induced axon growth is mediated by localized inactivation of GSK-3beta and functions of the microtubule plus end binding protein APC. *Neuron* **42,** 897–912.

[28] Dock180–ELMO Cooperation in Rac Activation

By MINGJIAN LU and KODI S. RAVICHANDRAN

Abstract

Dock180 superfamily of proteins has been recently identified as novel, unconventional guanine nucleotide exchange factors (GEF) for Rho-family GTPases. Unlike most other GEFs for Rho-family GTPases, Dock180 family members do not contain the characteristic Dbl homology (DH) domain. Instead, they use a conserved "Docker" or "CZH2" domain to mediate the nucleotide exchange on Rho-family GTPases. The Dock180 family members are evolutionarily conserved from worms to mammals. They play critical roles in a number of biological processes essential for the normal development of entire organisms, as well as for the physiological responses of these organisms, including removal of apoptotic cells and directed cell migration in *C. elegans*; myoblast fusion, and dorsal closure in *Drosophila*; lymphocyte migration, T-cell activation, tumor metastasis, HIV infection, and development of neuronal degenerative diseases in mammals. All these biological activities of the Dock180 family members have been linked to their ability to activate their specific GTPase substrate. At least four members of the Dock180 family bind to another evolution-arily conserved protein ELMO to optimally activate the Rac GTPase. The best characterized is the Rac activation by the Dock180–ELMO complex. ELMO modulates the Rac activation by Dock180 by means of at least three distinct mechanisms: helping Dock180 stabilize Rac in its nucleotide-free transition state; relieving a self-inhibition of Dock180; and targeting Dock180 to the plasma membrane to gain access to Rac. Thus, Dock180 and ELMO function together as a bipartite GEF to optimally activate Rac

METHODS IN ENZYMOLOGY, VOL. 406
0076-6879/06 $35.00
DOI: 10.1016/S0076-6879(06)06028-9

on upstream stimulation to mediate the engulfment of apoptotic cells and cell migration.

Introduction

Rac is a Rho family GTPase that alternates between the GDP-bound inactive state and the GTP-bound active state. In its GTP-bound form, Rac interacts with its specific effectors to elicit downstream signaling events, including actin cytoskeletal reorganization, gene transcription, cell proliferation, and differentiation (Bishop and Hall, 2000; Rossman et al., 2005). Because of its intrinsic GTP hydrolysis activity, which is further enhanced by the GTPase activating protein (GAP), Rac is associated with GDP at its dormant state. On stimulation by upstream signaling molecules, Rac becomes activated through replacing the bound GDP with GTP, a process referred as guanine nucleotide exchange, which is catalyzed by guanine nucleotide exchange factors (GEF).

The conventional GEFs for Rho family GTPases share common structural features (Hoffman and Cerione, 2002; Schmidt and Hall, 2002; Rossman et al., 2005). With a few exceptions, they contain a Dbl homology (DH) domain and a tandemly associated pleckstrin homology (PH) domain. Although the DH domain is responsible for the catalytic activity, the functions of the PH domains vary in different GEFs (Rossman et al., 2003): some PH domains help target the GEFs to the plasma membrane by means of binding to the phospholipids, whereas other PH domains facilitate the conformational change of the GEFs or make direct contact with GTPase to subtly affect the rate of nucleotide exchange; still other PH domains interact with the catalytic DH domain and inhibit the basal activity of the GEF (Schmidt and Hall, 2002).

Recently, the Dock180 family of proteins has been identified as novel GEFs for Rho-family GTPases (Brugnera et al., 2002; Cote and Vuori, 2002; Meller et al., 2002). Unlike the previously known GEFs for Rho-family GTPases, they lack the DH domain characteristic of the conventional GEFs. Interestingly, many of the family members also lack a PH domain.

Dock180 Family of Proteins

Dock180 is a prototype member of this superfamily and was the first to be identified as an upstream activator for Rac (Brugnera et al., 2002; Kiyokawa et al., 1998). Overexpression of membrane-targeted Dock180 in cells induced membrane ruffling (Hasegawa et al., 1996) and also resulted in Rac activation in cells (Kiyokawa et al., 1998). However, because

Dock180 does not possess the typical DH domain, how it led to Rac activation remained a mystery for several years. Recently, several groups independently identified a conserved domain within the Dock180 super-family members, termed as "Docker" domain (Brugnera et al., 2002), "DHR-2" domain (Cote and Vuori, 2002), or "CZH-2" domain (Meller et al., 2002). This domain directly and specifically binds to the nucleotide-free Rho family GTPase and is necessary and sometimes sufficient to mediate the nucleotide exchange on specific Rho family GTPase.

Bioinformatics analysis revealed that the characteristic "Docker," "DHR-2," or "CZH-2" domain is conserved within the eukaryotic king-dom, ranging from yeast, plant, to mammals (Cote and Vuori, 2002; Meller et al., 2002). Eleven human proteins have been identified. On the basis of their sequence similarity, these 11 members and their homologs from other organisms are divided into four subfamilies. With Dock180, Dock2, and Dock5 in the DOCK A subfamily; Dock3 and Dock4 in the DOCK B subfamily; Dock6, Dock7, and Dock8 in the DOCK C subfamily; and Dock9, Dock10, and Dock11 in the DOCK D subfamily (Cote and Vuori, 2002; Meller et al., 2002). Except for Dock5, whose substrate specificity is unknown, the other mammalian members in the DOCK A and DOCK B subfamilies have been characterized as GEFs for Rac (Brugnera et al., 2002; Cote and Vuori, 2002; Namekata et al., 2004; Sanui et al., 2003b; Yajnik et al., 2003). These mammalian DOCK A and DOCK B members share an overall domain organization (Fig. 1), which is conserved in the C. elegans CED-5 protein (Wu and Horvitz, 1998) and the Drosophila Myoblast city (Mbc) protein (Erickson et al., 1997). Thus, they are collec-tively referred to as CDM proteins (CED-5, Dock180, and Myoblast city). Proteins in the DOCK D subfamily, Dock9/Zizimin1 (Meller et al., 2002) and Dock11/Zizimin2 (Nishikimi et al., 2005), have been shown to bind and

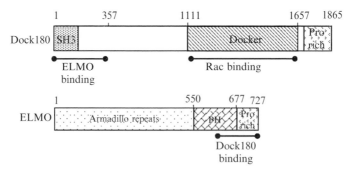

FIG. 1. Schematic domain organization of Dock180 and ELMO proteins (not drawn to scale).

activate Cdc42. The GTPase substrate specificity for other Dock180 superfamily members remains to be defined.

CDM proteins play a key role in a wide variety of biological processes conserved from worm to mammals. CED-5 is essential for the clearance of apoptotic bodies and distal tip cell (DTC) migration in *C. elegans* (Wu and Horvitz, 1998). Mutations within Mbc are associated with multiple developmental defects in *Drosophila*, such as myoblast fusion, dorsal closure (Erickson *et al.*, 1997), and thoracic closure (Ishimaru *et al.*, 2004). In mammals, Dock180 is important for the engulfment of apoptotic cells and cell migration (Albert *et al.*, 2000; Brugnera *et al.*, 2002; Grimsley *et al.*, 2004; Gumienny *et al.*, 2001). Moreover, gene-targeting experiments in mice demonstrated that Dock2 is essential for the lymphocytes to migrate to the peripheral lymphoid compartments (Fukui *et al.*, 2001), as well as T-cell activation (Sanui *et al.*, 2003a). In addition, a gain of function mutation in Dock4 promotes tumor metastasis (Yajnik *et al.*, 2003). Given the critical roles the CDM proteins play in the various biological contexts, it is not surprising that the activities of these proteins are under tight regulation.

CED-12/ELMO Proteins, Master Regulators of the CDM Proteins

C. elegans CED-12 and mammalian ELMO proteins were identified as conserved players in the signaling pathways involved in the engulfment of apoptotic cells and DTC migration. It has been demonstrated that CED-2, CED-5, and CED-12 (and their mammalian homologs CrkII, Dock180, and ELMO, respectively) function upstream of CED-10, the worm homolog of mammalian Rac, to reorganize the actin cytoskeleton during engulfment and cell migration (Gumienny *et al.*, 2001; Reddien and Horvitz, 2000; Wu and Horvitz, 1998; Wu *et al.*, 2001; Zhou *et al.*, 2001).

CED-12 is an evolutionarily conserved protein with homologs in fly, plants, and mammals, but not in yeast. There are three known ELMO proteins in human and mice: ELMO1, ELMO2, and ELMO3 (Gumienny *et al.*, 2001). Among these, ELMO1 is the best studied to date. Biochemical and functional studies in mammalian cells have demonstrated that Dock180 requires ELMO1 to function as an effective Rac activator and for Rac activation–dependent biological responses. Dock180–ELMO complex binds better to the nucleotide-depleted Rac than Dock180 alone, which correlates with the higher *in vitro* nucleotide-exchange rate by means of the Dock180–ELMO complex on Rac (Fig. 2) (Brugnera *et al.*, 2002; Lu *et al.*, 2004). In the cellular context, whereas overexpression of Dock180 alone does not induce obvious effects, coexpression of Dock180 and ELMO strongly promotes the formation of membrane

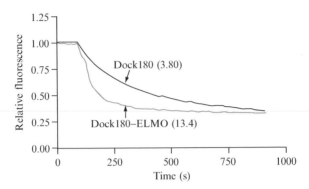

FIG. 2. Dock180–ELMO complex possesses higher guanine nucleotide exchange activity compared with Dock180 alone. The Flag-tagged Dock180 protein or Dock180–ELMO complex was affinity purified from transfected 293T cells and eluted with the Flag peptide. The purified proteins with equal amounts of Dock180 were tested in a fluorescence-based GEF assay. In this experiment, the Rac was preloaded with Mant-GDP, and the decrease in fluorescence reflected the dissociation of Mant-GDP from Rac. The observed Mant-GDP dissociation rate constants $\kappa_{obs}(s^{-1} \times 10^{-3})$ were indicated in parentheses.

ruffles, phagocytosis, and cell migration in transfected cells (Brugnera *et al.*, 2002; Grimsley *et al.*, 2004; Gumienny *et al.*, 2001).

CED-12/ELMO proteins contain a proline-rich tail at their very C termini, preceded by a PH domain (Gumienny *et al.*, 2001; Wu *et al.*, 2001; Zhou *et al.*, 2001). The N terminal two thirds comprises several Armadillo (ARM) repeats (deBakker *et al.*, 2004) (Fig. 1). CED-12/ELMO proteins form a complex with CDM proteins, with at least one of the interactions mediated through their proline-rich region binding to the SH3 domain of the CDM proteins (Gumienny *et al.*, 2001; Lu *et al.*, 2005; Sanui *et al.*, 2003b; Wu *et al.*, 2001; Zhou *et al.*, 2001). For Dock180–ELMO complex, there is a second interaction site in the region adjacent to the SH3 domain, which can be disrupted by a G171E mutation within Dock180 (Lu *et al.*, 2005).

Characterization of the Rac activation by the Dock180–ELMO complex revealed at least three distinct mechanisms by which ELMO directly or indirectly modulates the Rac activation by Dock180.

ELMO PH Domain-dependent Stabilization of the Complex Formation between Dock180 and Nucleotide-free Rac. The PH domain of ELMO forms a ternary complex with Dock180 and nucleotide-free Rac and seems to directly affect the nucleotide exchange rate on Rac (Lu *et al.*, 2004). The formation of the tight ternary complex between Dock180, ELMO, and nucleotide-free Rac is independent of the interactions between Dock180 and ELMO. Since nucleotide-free Rac does not bind ELMO under the conditions tested to date, the working model is that the Dock180–nucleotide-free Rac complex creates an additional contact site for ELMO, and the nucleotide-free Rac–Dock180–ELMO ternary complex formation further stabilizes Rac in its nucleotide-free transition state, and thereby facilitates the nucleotide exchange (Fig. 3). This effect of ELMO depends on its PH domain, and the ELMO PH domain–mediated contribution to the Rac activation is functionally relevant as demonstrated through a combination of studies using transfected cell lines and genetic rescues in *C. elegans*. The genetic rescue studies in *C. elegans* using analogous PH domain mutants of CED-12 also revealed that this CED-12/ELMO PH domain–mediated regulation of nucleotide exchange by means of CED-5/Dock180 is biologically relevant at the whole-organism level and is evolutionarily conserved.

Binding of the PxxP Motif of ELMO to the SH3 Domain of Dock180 Relieves a Steric Inhibition of Dock180. Recent biochemical, functional, and genetic rescue studies in *C. elegans* suggested that the N-terminal region of Dock180 regulates the Docker domain, which is separated by ~1000 amino acids in the primary sequence (Lu *et al.*, 2005). Functional studies revealed that the SH3 domain of Dock180 negatively regulates the activity of Dock180. Further biochemical characterization demonstrated that the SH3 domain interacts with the catalytic Docker domain and blocks the access of Rac to the Docker domain (Lu *et al.*,

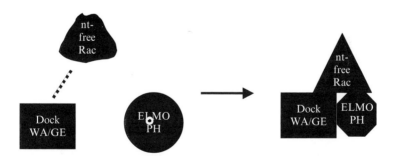

Fig. 3. ELMO PH domain helps Dock180 stabilize Rac in its nucleotide-free transition state. Dock WA/GE is a Dock180 mutant that does not bind to ELMO. Like the wild-type Dock180, the Dock WA/GE mutant binds to nucleotide-free Rac with relatively low affinity (compared with Dock180–ELMO complex). However, when ELMO is present, the association of nucleotide-free Rac with Dock WA/GE is substantially enhanced, and ELMO (by means of its PH domain) becomes part of this complex, despite the fact ELMO itself does not interact with either Dock WA/GE or nucleotide-free Rac (Lu *et al.*, 2004).

2005). These observations suggest that, at basal state, the Docker domain may not be sufficiently accessible to Rac. The binding of ELMO, by means of its PxxP motif, to the Dock180 SH3 domain displaces the SH3 domain from the Docker domain and thereby relieves a self-inhibition of Dock180 (Fig. 4). These studies also revealed that the ELMO-mediated relief of self-inhibition might be applicable to several other CDM family proteins (Dock2, Dock4, and possibly other members of the CDM family) (Lu *et al.*, 2005).

ELMO Facilitates Translocation of the Dock180–ELMO Complex to the Plasma Membrane. Dock180 is cytoplasmic protein that needs to be targeted to the membrane to activate Rac. In fact, cytoskeletal changes induced by a membrane-targeted Dock180 was the first indication that Dock180 may regulate the actin cytoskeleton, possibly through activating Rac (Hasegawa *et al.*, 1996). Several recent studies have suggested that the N terminal two thirds of the ELMO proteins may regulate the localization of Dock180–ELMO complex to the membrane. Grimsley *et al.* observed a requirement for the N-terminal 550 amino acids of ELMO in targeting Dock180 during cell migration (Grimsley *et al.*, 2004), and Katoh *et al.* elegantly demonstrated that the targeting via ELMO requires an interaction of ELMO with a GTPase RhoG (Katoh and Negishi, 2003). More

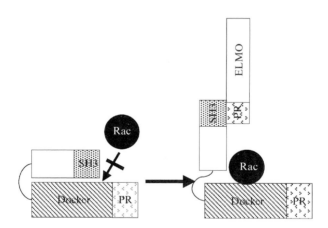

FIG. 4. ELMO relieves a self-inhibition of Dock180. At its basal state, the interaction between the SH3 domain and the catalytic Docker domain of Dock180 blocks the access of Rac to the Docker domain and inhibits the GEF activity of Dock180. This self-inhibition of Dock180 is relived by ELMO through its PxxP motif, which binds to the SH3 domain of Dock180 and disrupts the inhibitory SH3–Docker interaction. This allows better access of Rac to the Docker domain (Lu *et al.*, 2005).

recent studies by deBakker *et al.* (2004) have identified several Armadillo (ARM) repeats within the N-terminus of ELMO and that mutations of specific ARM repeats affect the interaction between ELMO and RhoG with functional implications (Fig. 5). The biochemical observations are consistent with those from genetic studies in *C. elegans*. Mig-2, the worm homolog of RhoG, binds similarly to CED-12 and is involved in phagocytosis. It is important to note that CED-12/ELMO bind to the active GTP–bound form of Mig-2/RhoG, which targets the ELMO–Dock180 complex to the membrane, and thereby promotes Rac activation. Thus, the N-terminal ARM repeats of ELMO seem to play a key role in linking two distinct GTPases during phagocytosis and cell migration (deBakker *et al.*, 2004; Katoh and Negishi, 2003).

   In summary, the Dock180 family proteins are novel, unconventional GEFs for Rho-family GTPases. ELMO proteins and their homologs in other organisms are key regulators for the GEF activity of the CDM proteins. By forming a protein complex, the CDM proteins and ELMO proteins optimally activate Rac, which is essential for a number of important biological responses.

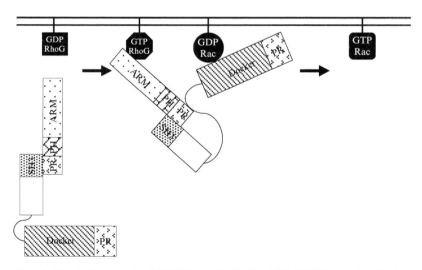

FIG. 5. The ARM repeats of ELMO targets the Dock180–ELMO complex to plasma membrane through interaction with active RhoG. When stimulated by upstream signaling, the plasma membrane–associated RhoG is activated and interacts with the Armadillo repeats of ELMO, which brings the Dock180–ELMO complex to the plasma membrane to gain access to the membrane-associated Rac. This, in turn, leads to Rac-GTP loading by means of the Dock180–ELMO complex (deBakker *et al.*, 2004; Katoh and Negishi, 2003).

Experimental Procedures

Transfection of 293T Cells

293T cells were maintained in DMEM medium supplemented with 10% FBS and 1% penicillin/streptomycin/glutamine. One day before transfection, cells were plated on 100-mm tissue culture dishes, allowing the cells to grow to 40–50% confluency before transfection. Calcium phosphate was used to transfect 293T cells. Two hours before transfection, the culture medium was replaced with fresh growth medium. In a sterile tube, DNA was added as needed. After mixing with water to make total volume of 375 μl, 125 μl of $CaCl_2$ solution was added and mixed by vortex; 500 μl of 2× phosphate buffer was added, vortexed to mix, and immediately added to the cell culture plate. The cells were allowed to grow for 2–3 days.

Transfection of LR73 Cells

This protocol for transfecting LR73 cells in 24-well plates can be scaled up. LR73 cells are CHO cell–derived fibroblasts (Gumienny *et al.*, 2001). They were maintained in Alpha MEM medium supplemented with 10% FBS and 1% penicillin/streptomycin/glutamine. One day before transfection, in a 24-well plate, 1×10^5 cells per well were seeded. For each transfection, an appropriate amount of DNA was added into 50 μl of serum-free medium. In a separate tube, 1.5 μl of Lipofectamine 2000 was added into 50 μl of serum-free medium, mixed, and incubated for 5 min. The DNA-containing solution was mixed with the lipid-containing solution and incubated for 20 min. Meanwhile, the culture medium was replaced with 500 μl of fresh growth medium. At the end of the 20 min, the transfection mixture was added to the cell culture; and 6 h later, the growth medium was replaced with fresh medium, and the cells were allowed to grow for 1–2 days.

Binding of Nucleotide-free Rac to the Dock180–ELMO

Purification of GTS-Rac from E. coli. GST-Rac was purified from *E. coli* strain BL21 transformed with pGEX4T2-Rac vector. Bacteria were grown in 5 ml culture overnight at 37°. The bacteria were diluted in 250 ml culture, incubated for 1 h at 37°, and induced with 1 mM IPTG. After additional 4–6 h of growth, the bacteria were harvested by centrifugation at 8000 rpm for 10 min. Te bacteria pellet could be frozen at −80° at this stage. The bacteria were resuspended in 8 ml lysis buffer: 100 mM Tris, pH 7.6, 100 mM NaCl, 1 mM EDTA, 1 mg/ml lysozyme, 1 mM DTT, 10 μg/ml pepstatin, aprotinin, leupeptin. After lysis for 20 min on ice, the bacteria were sonicated three times for 15 secs each with 2-min intervals on ice; 80 μl of 10% deoxycholate was added and incubated at room temperature for 20 min; 25 μl of 10 mg/ml DNAse was added, incubated at room temperature for 20 min, and centrifuged at 14,000 rpm for 30 min. The supernatants were collected and incubated with 300 μl of glutathione agarose beads (pre-washed with lysis buffer). After incubation on a rotator for 2 h at 4°, the beads were washed eight times with wash buffer: 50 mM Tris, pH 7.6, 150 mM NaCl, 1% NP-40, 1 mM DTT, 10 μg/ml pepstatin, aprotinin, leupeptin, 10% glycerol. After the final wash, the beads were resuspended to a 50% solution in the preceding wash buffer with 20% glycerol and frozen in aliquots at $-80°$. Protein concentration was estimated by running sample of beads along with BSA standards followed by Coomassie staining.

GST-Rac Pull-down of the Dock180–ELMO Complex. 239T cells were transfected on 100-mm plates with 10 μg of pCXN2-Flag-Dock180 plasmid (Brugnera *et al.*, 2002) and 3 μg of pEBB-ELMO-Flag (Lu *et al.*, 2004) plasmid. Dock180 alone and ELMO alone were used as controls; 48 h after transfection, the cells were washed with ice-cold PBS and lysed with 1 ml of lysis buffer: 50 mM Tris, pH 7.6, 150 mM NaCl, 10 mM EDTA, 10 mM sodium pyrophosphate, 10 mM NaF, 1 mM sodium orthovanadate, 10 μg/ml of pepstatin, aprotinin, leupeptin, and AEBSF each, 1% NP-40. After incubation for 20 min at 4°, the cell lysates were cleared through centrifugation at maximal speed for 10 min. The supernatant was collected and incubated with pre-washed beads with 1$\sim$2 μg of GST-Rac for 2 h at 4°. The beads were washed four times with wash buffer: 20 mM Tris, pH 7.6, 150 mM NaCl, 10 mM EDTA, 10% glycerol, 5 mM NaF, 1 mM sodium orthovanadate, 1 μg/ml of pepstatin, aprotinin, leupeptin, and AEBSF each, 0.1% of NP-40. The beads were boiled with 100 μl of 1 × SDS-PAGE loading buffer, and the proteins were resolved by SDS-PAGE and analyzed by Western blotting.

In Vivo *Rac Activation by the Dock180–ELMO Complex*

Purification of the GST-CRIB Protein. CRIB is the Cdc42 and Rac binding domain of p21 activated kinase (PAK1). GTS-CRIB was purified from *E. coli* strain BL21 transformed with pGEX4T2-CRIB vector. The bacteria were grown in 200 ml culture overnight at 37°. The bacteria were diluted in 2000 ml culture for 1$\sim$1.5 h at 37° until OD_{600} reached 0.8. After induction with 0.5 mM IPTG, the cultures were grown for additional 2 h. The bacteria were harvested by centrifugation at 8000 rpm for 10 min. If necessary, the bacteria pellet can be frozen at $-80°$ at this stage. The bacteria were resuspended in 20 ml lysis buffer: 50 mM Tris, pH 7.6, 1% Triton X-100, 150 mM NaCl, 5 mM $MgCl_2$, 1 mM DTT, 10 μg/ml of aprotinin, leupeptin, and 1 mM PMSF. The bacteria were then sonicated six times for 15 sec each with 2 min intervals on ice. After centrifugation at 14,000 rpm for 30 min, the supernatants were collected and incubated with 300 μl of glutathione agarose beads (pre-washed with lysis buffer). After incubating on a rotator for 2 h at 4°, the beads were washed eight times with wash buffer: 50 mM Tris, pH 7.6, 150 mM NaCl, 1% NP-40, 1 mM DTT, 10 μg/ml pepstatin, aprotinin, leupeptin, 10% glycerol. After the final wash, the beads were resuspended to a 50% solution in preceding wash buffer with 20% glycerol and frozen in aliquots at −80°. Protein concentration was estimated by running a sample of beads along with BSA standards followed by Coomassie staining.

GTP-Rac Pull-down. LR73 cells were transfected on 100-mm plates with 15 μg of pCXN2-Flag-Dock180 plasmid and 12 μg of pEBB-ELMO-Flag plasmids using Lipofectamine 2000 (Invitrogen). Dock180 alone and ELMO alone were used as controls; 20 h after transfection, cells were washed with ice-cold PBS and 1 ml of lysis buffer added: 50 mM Tris, pH 7.2, 1% Triton X-100, 0.5% sodium deoxycholate, 0.1% SDS, 500 mM NaCl, 10 mM MgCl$_2$, 10 μg/ml of aprotinin, leupeptin each, 1 mM PMSF. Cells were scraped off plates and incubated on ice for 10 min. Cell lysates were cleared by centrifugation at maximal speed for 10 min. The supernatant was collected and incubated with 10 μl of pre-washed GST-CRIB beads for 2 h at 4°. The beads were washed four times with wash buffer: 50 mM Tris, pH 7.2, 1% Triton X-100, 150 mM NaCl, 10 mM MgCl$_2$, 10 μg/ml of aprotinin, leupeptin each, 0.1 mM PMSF. The beads were boiled with 100 μl of 1 × SDS-PAGE loading buffer, and the bound proteins were resolved by SDS-PAGE and analyzed by Western blotting.

In Vitro *GEF Assay*

Purification of the Dock180 and ELMO Proteins from Transfected 293 T Cells. 293T cells were transfected on 100-mm plates as described previously. To purify Flag-Dock180 or ELMO-Flag protein individually, cells were transfected with pCXN2-Flag-Dock180 and pEBB-ELMO-Flag, respectively. To purify Dock180–ELMO complex, cells were cotransfected with pCXN2-Flag-Dock180 and pEBB-ELMO-GFP (Gumienny *et al.*, 2001); 48 h after transfection, cells were washed with ice-cold PBS and lysed with 1 ml of lysis buffer containing: 50 mM Tris, pH 7.6, 150 mM NaCl, 10 mM sodium pyrophosphate, 10 mM NaF, 1 mM sodium orthovanadate, 10 μg/ml of pepstatin, aprotinin, leupeptin, and AEBSF each, 1% Triton X-100. After incubation for 20 min at 4°, the cell lysates were cleared by centrifugation at maximal speed for 10 min. The supernatant was removed and incubated with 30 μl of prewashed M2 anti-Flag antibody–conjugated agarose beads (Sigma) for 2 h at 4°. The beads were washed four times with wash buffer: 20 mM Tris, pH 7.6, 150 mM NaCl, 10% glycerol, 5 mM NaF, 1 mM sodium orthovanadate, 1 μg/ml of pepstatin, aprotinin, leupeptin and AEBSF each, 0.1% of Triton X-100. The beads were then washed twice with elution buffer (10 mM Tris, pH 7.6, 150 mM NaCl, 10% glycerol, 0.5 mg/ml BSA) to get rid of detergent. Flag-tagged proteins were eluted by incubating beads with 60 μl of elution buffer containing 100 μg/ml of Flag peptide (Sigma) 30 min on ice. This was repeated a second time. 20 μl of the eluted proteins were loaded on to an SDS-PAGE gel and the amount of Dock180 and ELMO proteins were quantitated after Western blotting and densitometry.

Radioactivity-Based GEF Assay. The following buffers were prepared before the start of the assay.

QUENCH BUFFER. 0.9 g of sodium phosphate (monobasic), 1.05 g of sodium phosphate (dibasic), 2.03 g of MgCl$_2$, 0.55 g of ATP, 0.2 g of sodium azide, add water to 1 ;, pH was adjusted to 6.8 with 5 M NaOH.

MOPS/EDTA. 125 μl of 1 M MOPS (pH 7.1), 10 μl of 500 mM EDTA, add water to 5 ml. This was made fresh.

REACTION BUFFER. 200 μl of 1 M MOPS (pH 7.1), 50 μl of 1 M MgCl$_2$, 100 μl of 100 mM GTP, 10 μl of 500 mM sodium phosphate (monobasic), 400 μl of 10 mg/ml BSA, 160 μl of 5 M NaCl, add water to 8 ml. This was made fresh.

Loading of Rac with [α-^{32}P]GDP. The following reaction was assembled on ice: 198 μl of MOPS/EDTA, 17 μl of 10 mg/ml BSA, 5 μl of [α-^{32}P] GDP ($\sim$0.05 mCi), 7 μl of 10 μM GDP, 5 μg of recombinant Rac. After keeping on ice for 20 min, MgCl$_2$ was added to a final concentration of 10 mM to stop the reaction. This was kept on ice for 20 min before use.

The Nucleotide Exchange Reaction. The following reaction was assembled on ice: 100 μl of reaction buffer, 3~5 μl of eluted Flag-Dock180 protein (keeping an equal amount of eluted Dock180 in each reaction). The reaction solution was warmed in a 30° water bath for approximately 90 sec; 5 μl of [α-^{32}P]GDP-loaded Rac was added, vortexed briefly, and incubated at 30° for 5–15 min; 50 μl of this reaction solution was added to a nitrocellulose filter prewashed with quench buffer. The filter was washed twice with quench buffer. The filter was removed, and the radioactivity retained on the filter was counted with a scintillation counter.

Phagocytosis Assay

LR73 cells were transfected in 24-well plates with Dock180 plasmids with either GFP or ELMO-GFP; 20 h after transfection, 1.5 μl of 2-μm carboxylate-modified fluorescent latex beads (Molecular Probes) was added into 300 μl of serum-free Alpha MEM medium. The cell culture medium was replaced with this mixture. The cells were incubated for 2 h. After taking off the medium, the cells were detached with trypsin/EDTA, and 300 μl of cold growth medium containing 1% sodium azide was added; the cells were analyzed by two-color flow cytometry. The phagocytic activity of the transfected cells was determined by measuring the percentage of GFP-positive cells with engulfed particles.

Transwell Cell Migration Assay

LR73 cells were seeded in a 6-well plate to allow cells to grow to approximately 70% confluency before transfection. LR73 cells were transiently transfected with pGL3-CMV-luciferase as a reporter construct in addition to the appropriate Dock180 and ELMO plasmids as needed. After 20 h, the cells were harvested using trypsin/EDTA and resuspended in Opti-MEM medium supplemented with 2% FBS; 10^5 cells were added in duplicates to the upper chambers of an untreated polycarbonate Transwell filter with 8-μm pores (Costar). Opti-MEM medium supplemented with 2% FBS was added to the lower chamber; 10^5 cells were separately plated in parallel wells without Transwell filters, based on which the percent migration was estimated for each transfection condition. After 6 h of incubation, Transwell filters were removed. The number of cells migrating completely through the filter to the lower chamber was quantitated by the luciferase activity (Promega). The percentage of cell migration was determined by dividing the luciferase activity from the cells migrated to the lower chamber by that from the cells plated in the wells without a Transwell filter.

Acknowledgment

We would like to thank Kent Rossman and John Sondek for helping with the GEF assay in Fig. 2. We also thank the members in Ravichandran laboratory for helpful discussion. This work was supported by grants from NIH (GM-064709) to K. S. R.

References

Albert, M., Kim, J., and Birge, R. (2000). avb5 integrin recruits the CrkII–Dock180–Rac1 complex for phagocytosis of apoptotic cells. *Nat. Cell Biol.* **2,** 899–905.

Bishop, A. L., and Hall, A. (2000). Rho GTPases and their effector proteins. *Biochem. J.* **348,** 241–255.

Brugnera, E., Haney, L., Grimsley, C., Lu, M., Walk, S. F., Tosello-Trampont, A. C., Macara, I. G., Madhani, H., Fink, G. R., and Ravichandran, K. S. (2002). Unconventional Rac-GEF activity is mediated through the Dock180-ELMO complex. *Nat. Cell Biol.* **4,** 574–582.

Cote, J. F., and Vuori, K. (2002). Identification of an evolutionarily conserved superfamily of DOCK180-related proteins with guanine nucleotide exchange activity. *J. Cell Sci.* **115,** 4901–4913.

DeBakker, C. D., Haney, L. B., Kinchen, J. M., Grimsley, C., Lu, M., Klingele, D., Hsu, P. K., Chou, B. K., Cheng, L. C., Blangy, A., Sondek, J., Hengartner, M. O., Wu, Y. C., and Ravichandran, K. S. (2004). Phagocytosis of apoptotic cells is regulated by a UNC-73/TRIO-MIG-2/RhoG signaling module and armadillo repeats of CED-12/ELMO. *Curr. Biol.* **14,** 2208–2216.

Erickson, M. R., Galletta, B. J., and Abmayr, S. M. (1997). Drosophila myoblast city encodes a conserved protein that is essential for myoblast fusion, dorsal closure, and cytoskeletal organization. *J. Cell Biol.* **138,** 589–603.

Fukui, Y., Hashimoto, O., Sanui, T., Oono, T., Koga, H., Abe, M., Inayoshi, A., Noda, M., Oike, M., Shirai, T., and Sasazuki, T. (2001). Haematopoietic cell-specific CDM family protein DOCK2 is essential for lymphocyte migration. *Nature* **412,** 826–831.

Grimsley, C. M., Kinchen, J. M., Tosello-Trampont, A. C., Brugnera, E., Haney, L. B., Lu, M., Chen, Q., Klingele, D., Hengartner, M. O., and Ravichandran, K. S. (2004). Dock180 and ELMO1 proteins cooperate to promote evolutionarily conserved Rac-dependent cell migration. *J. Biol. Chem.* **279,** 6087–6097.

Gumienny, T. L., Brugnera, E., Tosello-Trampont, A. C., Kinchen, J. M., Haney, L. B., Nishiwaki, K., Walk, S. F., Nemergut, M. E., Macara, I. G., Francis, R., Schedl, T., Qin, Y., Van Aelst, L., Hengartner, M. O., and Ravichandran, K. S. (2001). CED-12/ELMO, a novel member of the crkII/Dock180/Rac pathway, is required for phagocytosis and cell migration. *Cell* **107,** 27–41.

Hasegawa, H., Kiyokawa, E., Tanaka, S., Nagashima, K., Gotoh, N., Shibuya, M., Kurata, T., and Matsuda, M. (1996). DOCK180, a major CRK-binding protein, alters cell morphology upon translocation to the cell membrane. *Mol. Cell. Biol.* **16,** 1770–1776.

Hoffman, G. R., and Cerione, R. A. (2002). Signaling to the Rho GTPases: Networking with the DH domain. *FEBS Lett.* **513,** 85–91.

Ishimaru, S., Ueda, R., Hinohara, Y., Ohtani, M., and Hanafusa, H. (2004). PVR plays a critical role via JNK activation in thorax closure during Drosophila metamorphosis. *EMBO J.* **23,** 3984–3994.

Katoh, H., and Negishi, M. (2003). RhoG activates Rac1 by direct interaction with the Dock180-binding protein Elmo. *Nature* **424,** 461–464.

Kiyokawa, E., Hashimoto, Y., Kobayashi, S., Sugimura, H., Kurata, T., and Matsuda, M. (1998). Activation of Rac1 by a Crk SH3-binding protein, DOCK180. *Genes Dev.* **12,** 3331–3336.

Lu, M., Kinchen, J. M., Rossman, K. L., Grimsley, C., deBakker, C., Brugnera, E., Tosello-Trampont, A. C., Haney, L. B., Klingele, D., Sondek, J., Hengartner, M. O., and Ravichandran, K. S. (2004). PH domain of ELMO functions *in trans* to regulate Rac activation via Dock180. *Nat. Struct. Mol. Biol.* **11,** 756–762.

Lu, M., Kinchen, J. M., Rossman, K. L., Grimsley, C., Hall, M., Sondek, J., Hengartner, M. O., Yajnik, V., and Ravichandran, K. S. (2005). A steric-inhibition model for regulation of nucleotide exchange via the Dock180 family of GEFs. *Curr. Biol.* **15,** 371–377.

Meller, N., Irani-Tehrani, M., Kiosses, W. B., Del Pozo, M. A., and Schwartz, M. A. (2002). Zizimin1, a novel Cdc42 activator, reveals a new GEF domain for Rho proteins. *Nat. Cell Biol.* **4,** 639–647.

Namekata, K., Enokido, Y., Iwasawa, K., and Kimura, H. (2004). MOCA induces membrane spreading by activating Rac1. *J. Biol. Chem.* **279,** 14331–14337.

Nishikimi, A., Meller, N., Uekawa, N., Isobe, K., Schwartz, M. A., and Maruyama, M. (2005). Zizimin2: A novel, DOCK180-related Cdc42 guanine nucleotide exchange factor expressed predominantly in lymphocytes. *FEBS Lett.* **579,** 1039–1046.

Reddien, P. W., and Horvitz, H. R. (2000). CED-2/CrkII and CED-10/Rac control phagocytosis and cell migration in *Caenorhabditis elegans*. *Nat. Cell Biol.* **2**, 131–136.

Rossman, K. L., Cheng, L., Mahon, G. M., Rojas, R. J., Snyder, J. T., Whitehead, I. P., and Sondek, J. (2003). Multifunctional roles for the PH domain of Dbs in regulating Rho GTPase activation. *J. Biol. Chem.* **278**, 18393–18400.

Rossman, K. L., Der, C. J., and Sondek, J. (2005). GEF means go: Turning on RHO GTPases with guanine nucleotide-exchange factors. *Nat. Rev. Mol. Cell. Biol.* **6**, 167–180.

Sanui, T., Inayoshi, A., Noda, M., Iwata, E., Oike, M., Sasazuki, T., and Fukui, Y. (2003a). DOCK2 is essential for antigen-induced translocation of TCR and lipid rafts, but not PKC-theta and LFA-1, in T cells. *Immunity* **19**, 119–129.

Sanui, T., Inayoshi, A., Noda, M., Iwata, E., Stein, J. V., Sasazuki, T., and Fukui, Y. (2003b). DOCK2 regulates Rac activation and cytoskeletal reorganization through interaction with ELMO1. *Blood* **102**, 2948–2950.

Schmidt, A., and Hall, A. (2002). Guanine nucleotide exchange factors for the Rho GTPases: Turning on the switch. *Genes Dev.* **16**, 1587–1609.

Wu, Y. C., and Horvitz, H. R. (1998). C. elegans phagocytosis and cell-migration protein CED-5 is similar to human DOCK180. *Nature* **392**, 501–504.

Wu, Y. C., Tsai, M. C., Cheng, L. C., Chou, C. J., and Weng, N. Y. (2001). C. elegans CED-12 acts in the conserved crkII/DOCK180/Rac pathway to control cell migration and cell corpse engulfment. *Dev. Cell* **1**, 491–502.

Yajnik, V., Paulding, C., Sordella, R., McClatchey, A. I., Saito, M., Wahrer, D. C., Reynolds, P., Bell, D. W., Lake, R., van den Heuvel, S., Settleman, J., and Haber, D. A. (2003). DOCK4, a GTPase activator, is disrupted during tumorigenesis. *Cell* **112**, 673–684.

Zhou, Z., Caron, E., Hartwieg, E., Hall, A., and Horvitz, H. R. (2001). The C. elegans PH domain protein CED-12 regulates cytoskeletal reorganization via a Rho/Rac GTPase signaling pathway. *Dev. Cell* **1**, 477–489.

[29]   Rho GTPase Activation by Cell–Cell Adhesion

By JENNIFER C. ERASMUS and VANIA M. M. BRAGA

Abstract

Cell–cell adhesion can occur in a calcium-dependent or calcium-independent manner, depending on the type of receptor involved. Establishment of cell contacts by either type of cell–cell adhesion (calcium-dependent or calcium-independent) has been shown to activate Rho GTPases in different cells. In this chapter, we describe the method used to assess the activation of Rho GTPases by cadherins, the prototype calcium-dependent adhesion receptor in epithelial cells. We cover the optimal cell culture conditions and controls to ensure that the activation of the GTPases is specifically triggered by the formation of cadherin-dependent cell–cell contacts. Controls described herein determine the specificity of activation of Rho proteins with respect to cadherin adhesion

METHODS IN ENZYMOLOGY, VOL. 406
Copyright 2006, Elsevier Inc. All rights reserved.
0076-6879/06 $35.00
DOI: 10.1016/S0076-6879(06)06029-0

and exclude the contribution of other adhesive receptors, calcium-signaling, cell spreading, and migration. Although we focus on cadherin receptors and normal human keratinocytes as our model system, the methods described can be easily adapted to other adhesion receptors and different cell types.

Introduction

Adhesion mediated by cadherin receptors has been shown to modulate the activity of Rho proteins in epithelial, endothelial, muscle, fibroblasts, and transformed cells. Of note is that the pattern of activation obtained for Rho, Rac, and Cdc42 varies considerably, depending on the cell type (i.e., epithelial versus muscle cells) or even between different types of epithelial cells (i.e., stratified or simple epithelia; reviewed by Braga, 2002b). Therefore, it is important to bear in mind the cellular context when investigating the activation and signaling downstream of Rho proteins by cadherin adhesion.

When assessing the activation of small GTPases by cadherin-dependent cell–cell contacts, it is important to determine the specificity of activation with respect to (1) junction assembly, (2) calcium signaling, and (3) the receptor itself. It is essential to remove any contribution of cell retraction, spreading, or migration, because these processes are also known to activate Rho GTPases. This can be easily done by using confluent monolayers to induce cell–cell contacts and avoid cell migration before adhesion to neighboring cells. The specificity toward the type of receptors is complicated by the fact that other adhesive receptors accumulate at junctions very shortly after the initial cadherin-dependent cell–cell contact is formed. Thus, when cell–cell contacts are established, distinct adhesive receptors can contribute toward cell–cell adhesion and signaling to different extents. Here we describe techniques to eliminate or minimize the participation of additional receptors in small GTPase activation.

Cell Culture and Induction of Cell–Cell Contacts

Two points should be considered when optimizing activation assays in epithelial cells: the levels of endogenous Rho proteins in a particular cell type and the basal levels of activity. Regarding the former, it is essential to start with enough material, because activation of endogenous small GTPases after a physiological stimulus is usually seen in approximately 0.5–1% of the total pool of Rho proteins. We use approximately $0.7–1 \times 10^7$ normal keratinocytes per tube (i.e., a confluent monolayer from a 9-cm dish).

To detect activation of endogenous Rho proteins, a reduction of the basal levels of GTPase activity must be performed before the assays. This is an important step, because cell–cell adhesion-induced activation of endogenous Rho small GTPases in keratinocytes is usually about twofold to fivefold above background levels. The standard starvation method to reduce basal levels of activity, however, is not appropriate for keratinocytes (incubate cells overnight with reduced amount of serum, but see "Serum Stimulation"). Instead, we leave confluent monolayers without feeding for at least 2 days before testing for GTPase activity. This method is efficient and does not cause changes in cell morphology as seen during starvation.

The assay used for establishment of cadherin-dependent adhesion is known as calcium switch. Normal keratinocytes are grown until postconfluence in medium containing low levels of calcium ions (0.02 mM, for specific growth and medium conditions please see Braga et al. [1997] and Hodivala and Watt [1994]). Analysis of Rho protein activation by cell–cell adhesion in subconfluent cells is not appropriate, because migration and spreading are required to establish cell–cell contacts. Calcium ions (1.8 mM) can then be added to the medium and cells incubated for different periods of time to induce junction formation (calcium switch). It is important to avoid adding new medium containing 1.8 mM calcium ions, because fresh serum activates Rho GTPases.

Some cell types are not adapted to grow in low calcium medium. To obtain cells without calcium-dependent adhesion, calcium can be chelated with EGTA in confluent monolayers to disassemble preformed contacts (Braga et al., 1999). However, a careful time course and titration of EGTA concentration should be performed to avoid perturbing attachment to substratum and retraction of cells. Alternately, briefly wash confluent cells with EGTA (twice) and then place cells in low calcium medium (0.02 mM) to allow complete disruption of cell–cell contacts (be aware that serum contains enough calcium to induce junction assembly; see Braga [2002a] for chelation of calcium ions from serum).

Activation Assays

Activation assays (pull-down) involve the use of GST-PAK-CRIB (Cdc42 and Rac), GST-WASP-CRIB (Cdc42), or GST-Rhotekin (Rho) to precipitate the active, GTP-bound form of Rho proteins. The pull-down method used to detect active small GTPases has been described before (Ren et al., 1999; Sander et al., 1999). For a successful activation assay, it is imperative to obtain good quality fusion proteins with high affinity for the active form of the Rho small GTPase to be tested. Here we describe the optimizations performed in our laboratory to make the method more consistent and to normalize the different batches of fusion proteins produced. Because the yield of fusion protein can vary, each batch of fusion proteins should be titrated in a pull-down assay. We routinely test each batch as described in the following ("Determination of Amount of Beads Required per Pull-down").

Preparation and Storage of Fusion Proteins Used in the Pull-Down Assay

Standard protocols are used for the preparation of GST-PAK-CRIB and Rhotekin beads used in the pull-downs. After purification of the fusion proteins, we keep them on beads (GSH-Sepharose brand) as a 1:1 slur (packed beads/buffer containing 50% glycerol). We store the slur in small aliquots (50–100 μl) at $-80°$ (beads do not freeze in the presence of glycerol). Aliquots are thawed only once; any unused material is discarded. As control to determine the quality and yield of fusion proteins, we run aliquots of the slur (5 μl, 10 μl, 15 μl, and 20 μl) and BSA standard (0.5 μl, 1 μl, and 2 μl of a stock solution at 2 mg/ml). The activity of each preparation is then determined in a pull-down assay as described in the following section.

Determination of Amount of Beads Required per Pull-Down

To determine the amount of beads that gives the best ratio of active over inactive Rho GTPases, different amounts of bead slur are used in a pull-down assay. We use two methods to activate Rho proteins to test the amount of fusion proteins to use: GTP-loading or serum stimulation. The first method results in most endogenous GTPases in your sample loaded with either GTP or GDP and will give a clear result of whether your fusion protein is active or not (Fig. 1A). However, a given stimuli activates only a fraction of Rho GTPases present in the sample. The second method, serum stimulation, is our preferred method and provides a better optimization (Fig. 1B). It gives a more physiological level of activation of GTPases and is, thus, more consistent with the levels observed after cell–cell adhesion.

GTP/GDP Loading

1. Grow cells in 6 × 9-cm plates in low calcium medium until confluent.
2. There is no need for starvation, because lysates will be loaded with different guanosine nucleotides.
3. Chill plates on ice, wash with ice-cold PBS, and extract protein lysates (see "Pull-down").
4. Load the lysates with GTP-γ-S (3 dishes) or GDP (3 dishes) as follows (per ml of lysate):

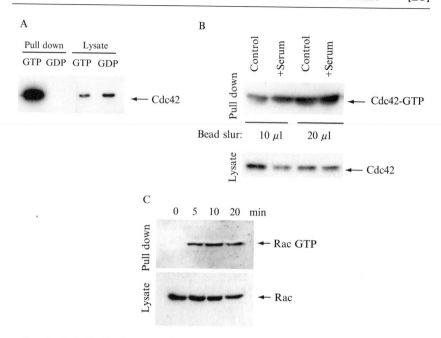

FIG. 1. Test of effectiveness and amount of GST-PAK-CRIB to be used in pull-down experiment. (A) GTP/GDP loading. After protein extraction, lysates were loaded with GTP or GDP to obtain endogenous GTPases activated or inactivated, respectively. After pull-down with GST-PAK-CRIB, a higher proportion of Cdc42 is precipitated from the GTP-loaded lysates. (B) Serum stimulation. Different amounts of GST-PAK-CRIB (10 or 20 μl) were used in a pull-down assay. Increasing the amount of bead slur leads to saturation and masks an increase in the amount of Cdc42-GTP (twofold to fivefold). This control is more relevant and representative to experimental design than GTP/GDP loading, in which most endogenous GTPases are activated. (C) Rac levels are measured during a short time course after induction of cell–cell adhesion by the addition of calcium ions.

    a. 10 mM EDTA (20 μl of 0.5 M stock)
    b. 1 mM GTP-γ-S or GDP (20 μl of 50 mM stock)
5. Mix and incubate for 10 min at room temperature.
6. Stop reaction by adding 20 mM MgCl$_2$ (20 μl of 1 M stock), mix well, and chill on ice. Because lysates are loaded with GTP-γ-S, they can be frozen for later use.
7. Test different amounts of beads (e.g., 10 μl and 20 μl) with the GTP- and GDP-loaded lysates. *Make sure to cut end of tip and mix the bead slur frequently to pipette an accurate amount of beads.*
8. Follow pull-down method as described in the section "Pull-down."
9. Run a SDS-PAGE gel and perform a Western blot (Fig. 1A).

Serum Stimulation

1. Grow 6 × 9 cm dishes of keratinocytes in low calcium medium until confluent.
2. Leave cells for 48 h without feeding before the experiment.
3. Starve cells briefly in low calcium media containing 0.5% serum (for about 4 h).
4. Stimulate three dishes by replacing the starvation medium with fresh medium containing 10% serum for 3 min. (*Stimulation with EGF or PDGF for a few minutes is also effective*).
5. Chill plate on ice, wash in ice-cold PBS, and extract protein lysates (see next section).
6. Test different amount of beads (i.e., 10 μl, 20 μl) with and without serum stimulation. *Make sure to cut end of tip and mix the bead slur frequently to suck up the correct amount of beads.*
7. Follow pull-down method as described in the next section.
8. Run an SDS-PAGE gel and perform a Western blot to determine the amount of beads that gives the highest ratio of Rho.GTP/Rho.GDP (Fig. 1B).

Pull-Down

An example of cell–cell adhesion-induced activation of Rac in human keratinocytes is shown in Fig. 1C. For a successful pull-down, it is essential to observe the low temperature and time involved in each step. Speed and low temperature are necessary to avoid GTP hydrolysis in your sample. For example, lysates should not be frozen at any stage during the experiment nor can protein concentration be measured. We have found that when many samples are processed, more variability is seen between experiments. Herein we describe some tips optimized in our laboratory to minimize the time used in each step.

Equipment and Reagents

- Lysis buffer for PAK-CRIB (50 mM Tris-HCl, pH 7.5, 1% Triton X-100, 0.5% sodium deoxycholate, 0.1% SDS, 150 mM NaCl, 10 mM MgCl$_2$, leupeptin [5 μg/ml], pepstatin [5 μg/ml], Pefabloc [5 μg/ml], PMSF [50 μg/ml])
- Lysis buffer for Rhotekin (50 mM Tris-HCl, pH 7.5, 1% Triton X-100, 0.5% sodium deoxycholate, 0.1% SDS, 500 mM NaCl, 10 mM MgCl$_2$, leupeptin [5 μg/ml], pepstatin [5 μg/ml], Pefabloc [5 μg/ml], PMSF [50 μg/ml])
- Wash buffer (50 mM Tris-HCl, pH 7.5, 1% Triton X-100, 150 mM NaCl, 10 mM MgCl$_2$, leupeptin [5 μg/ml], pepstatin [5 μg/ml], Pefabloc [5 μg/ml], PMSF [50 μg/ml])

- 4 × Sample buffer
- Refrigerated centrifuge
- Heating block
- 15% SDS-PAGE gel
- PONCEAU S (2% w/v Ponceaus, 5% Acetic Acid, water q.s.p.).

Before starting the pull-down, make sure all solutions/PBS are pre-chilled on ice, all tubes are labeled and on ice, refrigerated centrifuge is cold (centrifuges in cold room are not good enough because they warm up considerably during centrifugation), pipettes and scrapers are at hand, and rotating wheel is at 4°. If more than six dishes are needed per experiment, we usually get help with the lysis and cell-scraping step.

1. Culture keratinocytes in low calcium until a postconfluent mono-layer of cells is obtained. Use one 9 cm dish of keratinocytes per pull-down.

2. Place allotted amount of beads (as determined in "Determination of Amount of Beads Required per Pull-down") into numbered 1.5-ml tubes and chill on ice. *Make sure to cut end of tip and mix the bead slur frequently to suck up the correct amount of beads. Check if similar amount of beads have been pipetted into each tube beforehand.*

3. Induce cell–cell contacts by adding calcium to 1.8 mM low calcium medium. (*Do not replace the medium*).

4. Place dishes on ice and wash once with 10 ml of cold PBS.

5. To normalize the amount of time different samples are lysed, we tilt the ice tray containing dishes about 40° and add 800 μl cold buffer at the bottom corner of all samples.

6. Immediately before scraping each sample, swirl the lysis buffer over the entire area of the dish and quickly scrape the cells. Scrape remaining samples as described.

7. Collect the lysates and place in prelabeled, cold 1.5-ml tubes. Spin lysates down 2 min at 14,000 rpm in a refrigerated centrifuge.

8. Remove supernatant and place in labeled tubes containing beads (from step 2).

9. Incubate lysate with beads for 30–45 min at 4° on a rotating wheel.

10. Spin down beads for 1 min at 14,000 rpm at 4°. (*The beads are spun down at a high speed but for a short time to avoid damage*).

11. Remove supernatant and place it in a new fresh chilled tube (lysate). Keep it for later. (*To prevent sucking up the beads, leave about 20 μl of lysate on the beads*).

12. Wash beads three times with 1 ml cold wash buffer. Spin down each time at 14,000 rpm for 1 min. (*Be careful not to loose beads during washes [i.e., leave about 100 μl wash buffer in tube]*).

13. After the last wash, carefully aspirate the final 100 μl of wash buffer without removing any beads; use a Gilson pipette loaded with a long, thin loading tip. Leave about 15–20 μl on beads.

14. Add appropriate amount of 4 × sample buffer and resuspend beads. Boil for 5 min. (*This sample contains activated GTPases. It can be frozen after boiling*).

15. To break down any DNA present, syringe the lysates (from step 11) using 1 ml syringe and a thin needle. Syringe four times and measure the volume of lysate for quantification later. Lysates can now be frozen at −70°.

16. Separate an aliquot of the supernatant (10–20 μl) into a new tube, add SDS-sample buffer, and boil for 5 min. This is necessary to measure the endogenous levels of Rho GTPases that will be needed in the quantification.

17. Run all the volume from the beads sample (step 14) and 10–20 μl of the lysate on a 15% SDS-PAGE gel. Transfer to PVDF membrane.

18. Stain membrane with Ponceau S and scan to check whether the amount of GST-PAK-CRIB or GST-Rhotekin per sample is similar. Scan and keep for your records.

19. Probe with antibody for Rac (Upstate, #05–389, clone23A8), or Cdc42 (BD Transduction Laboratories, #610928), or if using Rhotekin beads, probe with an antibody against RhoA (Santa Cruz, #sc-418).

20. When developing the blot, make multiple exposures to ensure that the intensity obtained is on the linear range of the film (at least three exposures in which doubling the exposure time leads to increased intensity). These times may be different for the activated (pull-down) and the total amount of endogenous GTPase (lysate) samples. Because of the lower levels of GTPases in keratinocytes, we use SuperSignal WestDura Extended Duration Substrate (Pierce, #34075). In addition to being more sensitive, this substrate allows prolonged exposure time, which is required when faint bands for active Rac are detected.

Quantification. The amount of active GTPase (pull-down) is calculated as the percentage of the total amount of endogenous GTPase present in each sample (lysates). Control levels (i.e., activation seen in low calcium medium) are arbitrarily set to 1. All other samples are expressed as relative to controls (Fig. 3B).

Controls for the Specificity of Activation of Rho Small GTPases by Cadherin-Dependent Adhesion

In addition to the use of confluent monolayers as described previously, experiments should be performed to eliminate the contribution of signaling mediated by other receptors and by an increase in intracellular calcium ions during cell–cell contact formation. To assess the contribution of other receptors for activation of Rho proteins during cell–cell contact assembly, the best strategy is to inhibit cadherin function during the calcium switch, which allows any other adhesive receptor to function (including other calcium-dependent receptors because calcium ions will be present). This technique addresses whether cadherin-mediated adhesion is necessary for activation of Rho proteins.

It is known that during the calcium switch, a transient raise in intracellular calcium levels is observed (Price *et al.*, 2003; Sharpe *et al.*, 1993; Tu *et al.*, 2001). To eliminate the contribution of calcium ions to the activation of small GTPases, cadherin receptors must be clustered independently of calcium ions. This can be achieved by artificially clustering cadherins with antibody-coated beads in the absence of calcium ions. Clustering cadherins with antibody-coated beads mimics the initial events observed after cadherin adhesion (actin recruitment, activation of signaling pathways, etc.) (Betson *et al.*, 2002; Braga *et al.*, 1997; Goodwin *et al.*, 2003; Lambert *et al.*, 2002). However, it does not reproduce the full set of events triggered by cell–cell adhesion (remodeling of cytoskeleton, formation of desmosomes, tight junction, etc). Despite this caveat, this technique can be very useful if the signaling event occurs early enough. If, after bead-induced receptor clustering, the level of GTPase activation is similar to the activation observed during the calcium switch, then cadherin clustering and adhesiveness is sufficient to activate Rho GTPases. Otherwise, additional factors other than cadherin-mediated adhesion contribute to the full activation levels of Rho proteins seen after cell–cell contact assembly.

The combined use of the two preceding techniques in our laboratory indicated that cadherin-mediated adhesion is necessary (inhibition assays) and sufficient (clustering assays) for small GTPase activation (Betson *et al.*, 2002). A detailed description of the methods optimized in our laboratory is described herein.

Inhibition of Cadherin Adhesion

Keratinocytes express E-cadherins and P-cadherins. Both receptors should be blocked to fully prevent cell–cell contact assembly and keratinocyte stratification (Hodivala and Watt, 1994; Wheelock and Jensen, 1992). Inhibition of cadherins can be achieved by preincubation with inhibitory antibodies or blocking peptides (Noe *et al.*, 1999) in low calcium medium before induction of junction assembly. Titration of the antibody concentration, incubation times, and effectiveness of blocking cadherin adhesion should be determined beforehand by immunofluorescence (Braga *et al.*, 1997; Hodivala and Watt, 1994). As control, a nonspecific antibody is used at the same final concentration as the blocking antibodies. Antibody binding can induce receptor recycling, and, therefore, an excess of antibody should be used and a time course carried out to determine the minimum incubation necessary for inhibition. Results for inhibitory assays are shown in Fig. 2.

Reagents

- HECD-1 mouse monoclonal antibody (anti-E-cadherin)
- NCC-CAD-299 mouse monoclonal antibody (anti-P-cadherin)
- IgG (goat anti-rabbit)

1. Prepare confluent keratinocytes dishes (9 cm) by removing low calcium medium to a separate tube (*Cells should not be fed for 2 days before the experiment*).
2. Pipette back 2 ml of medium to each dish. *Do not change the media because fresh serum can activate small GTPases.*
3. Dilute antibodies in low calcium medium taken from the cells (step1 above) to desired concentration (10 μg/ml HECD-1 and 4 μg/ml NCC-CAD-299; as control 14 μg/ml IgG).
4. Add antibodies or IgG to keratinocyte dishes and incubate at 37° for 16 min.

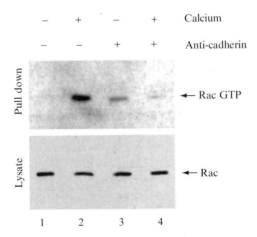

FIG. 2. Activation of Rac does not result from calcium signaling. Establishment of calcium-dependent cell–cell adhesion (30 min calcium switch) results in a twofold to fourfold fold increase in levels of Rac GTP (lane 2, + calcium). Addition of inhibitory antibodies to block cadherin-mediated adhesion during the calcium switch completely prevents Rac activation (lane 4, + calcium, + anti-cadherin).

5. Switch cells to standard calcium by adding calcium ions to a final concentration of 1.8 mM.
6. Perform pull-down (see "Pull-down" section).

Clustering of Cadherin Receptors

Clustering of adhesion receptors has been traditionally performed by primary antibody binding (against relevant receptor) followed by incubation with secondary antibodies (specific to the Fc portion of the primary antibody used). However, when this method was used to cluster cadherins in keratinocytes, Rac activity was increased to a fraction of the activation seen during the calcium switch (Fig. 3A,B; Betson *et al.*, 2002). This result indicates that a minimal threshold of cadherin clustering is required to promote activation of Rho GTPases to similar levels obtained by induction of cell–cell contacts. This technical problem can be circumvented by adding beads coated with anti-cadherin antibodies, which increases the number of clustered receptors per area (Braga *et al.*, 1997).

Before assessing activation of signaling pathways using this method, specific clustering of cadherins around the beads is monitored by the presence of cytoplasmic proteins known to bind to cadherin tail (catenins) and the absence of proteins not found at cell–cell contacts (for example, focal adhesion proteins such as talin (Braga *et al.*, 1997). Controls used are beads coated with heat-denatured BSA (to determine base line of activation) or coated with antibodies against a different receptor (i.e., anti-integrin antibodies) to determine specificity of signaling by means of cadherin clustering (Fig. 3C,D; Braga *et al.*, 1997).

In addition to antibodies, cadherin extracellular domain-coupled to the Fc portion of IgG$_1$ can also be used for coating the beads. This is an established method that specifically clusters cadherin receptors and is useful to determine the contribution of cadherin signaling with respect to other adhesion receptors (Niessen and Gumbiner, 2002; Goodwin *et al.*, 2003). However, one drawback is that the cadherin extracellular domain–coated bead method requires calcium ions for homophilic binding and clustering of surface cadherin receptors. Thus, this method cannot be used to determine the contribution of calcium ions for GTPase activation.

Equipment and Reagents

- Beads for clustering (Polybead Polystyrene 15.0 micron, Polysciences Inc., [9003–53–6] #18328). Count the amount of beads per microliter for each new bottle.
- HECD-1 mouse monoclonal antibody, anti-E-cadherin

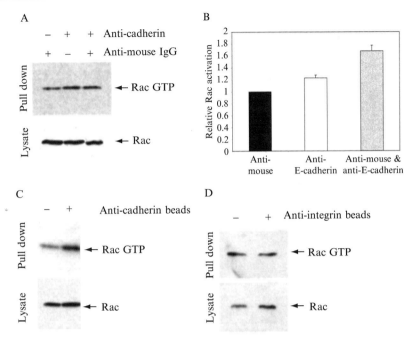

FIG. 3. Demonstration of specificity of Rac activation by cadherin receptor. (A) Clustering of cadherin receptors using soluble E-cadherin and P-cadherin mouse monoclonal antibodies with (+) or without (−) anti-mouse IgG. (B) Quantification of multiple results of experiment shown in (A) reveals approximately 30–60% activation of Rac compared with controls. This level of Rac activation is much lower than the one observed during the calcium switch (twofold to fourfold). (C) Clustering of cadherin receptors using antibody-coated latex beads without addition of calcium ions. Keratinocytes maintained without cell–cell contacts were incubated with anti-E-cadherin–coated beads for 5–10 min in low calcium medium. Upregulation of Rac activity is seen (twofold to fourfold), suggesting that a threshold of cadherin clustering is required. (D) Clustering of integrins using anti-integrin–coated beads (anti-α3 integrins) does not result in Rac activation. This control shows that Rac activation in keratinocytes is specifically triggered by cadherin clustering, not α3-integrins.

- Heat denatured BSA (10 mg/ml) (Braga, 2002a)
- PBS
- Benchtop centrifuge

1. Aliquot 500 μl of bead slur per tube (i.e., 10^7 beads per 9-cm dish).
2. Pulse beads and remove supernatant. Do not spin beads down for too long because this can cause damage.

3. Add HECD-1 antibody to 1 mg/ml final concentration to packed beads and 600 µl BSA to another tube as a control.
4. Resuspend beads and incubate for 1 h at room temperature, mixing frequently by inversion. Do not vortex.
5. Place beads at 4° overnight.
6. Top up volume of beads to 1 ml with PBS.
7. Pulse and remove supernatant.
8. Wash beads with 1 ml PBS.
9. Aspirate supernatant and add 100 µl BSA to each tube. Block for 1 h at room temperature.
10. Wash beads twice with 1 ml PBS. Keep at room temperature.
11. Prepare confluent keratinocytes dishes by removing low calcium medium to a separate tube. Cells should not be fed for 2 days before the experiment. Pipette back 2 ml of medium to each dish.
12. Resuspend beads in 500 µl of low calcium medium reserved as described previously in step 11.
13. Add bead suspension (step 12) to the 2 ml of low calcium medium on keratinocytes; mix well.
14. Incubate beads on cells for 5–30 min at 37°. Incubation time depends on the peak of activation of the GTPase seen during the calcium switch.
15. Stop incubation by placing dishes on ice.
16. Proceed with the pull-down assay ("Pull-down Assay" section).

References

Betson, M., Lozano, E., Zhang, J., and Braga, V. M. M. (2002). Rac activation upon cell-cell contact formation is dependent on signaling from the epidermal growth factor receptor. *J. Biol. Chem.* **277,** 36962–36969.

Braga, V. M. M. (2002a). Cadherin adhesion regulation in keratinocytes. *In* "Cell–cell Interactions—A Practical Approach" (T. Fleming, ed.), p. 268. Oxford University Press, Oxford.

Braga, V. M. M. (2002b). Cell-cell adhesion and signalling. *Cur. Opin. Cell Biol.* **14,** 546–556.

Braga, V. M. M., Del Maschio, A., Machesky, L. M., and Dejana, E. (1999). Regulation of cadherin function by Rho and Rac: Modulation by junction maturation and cellular context. *Mol. Biol. Cell* **10,** 9–22.

Braga, V. M. M., Machesky, L. M., Hall, A., and Hotchin, N. A. (1997). The small GTPases Rho and Rac are required for the establishment of cadherin-dependent cell-cell contacts. *J. Cell Biol.* **137,** 1421–1431.

Goodwin, M., Kovacs, E. M., Thoreson, M. A., Reynolds, A. B., and Yap, A. S. (2003). Minimal mutation of the cytoplasmic tail inhibits the ability of E-cadherin to activate Rac but not phosphatidylinositol 3-kinase: Direct evidence of a role for cadherin-activated Rac signaling in adhesion and contact formation. *J. Biol. Chem.* **278,** 20533–20539.

Hodivala, K. J., and Watt, F. M. (1994). Evidence that cadherins play a role in the down-regulation of integrin expression that occurs during keratinocyte terminal differentiation. *J. Cell Biol.* **124,** 589–600.

Lambert, M., Choquet, D., and Mege, R.-M. (2002). Dynamics of ligand-induced, Rac1-dependent anchoring of cadherins to the actin cytoskeleton. *J. Cell Biol.* **157,** 469–479.

Niessen, C. M., and Gumbiner, B. M. (2002). Cadherin-mediated cell sorting not determined by binding or adhesion specificity. *J. Cell Biol.* **156,** 389–400.

Noe, V., Willems, J., Vandekerckhove, J., Roy, F. V., Bruyneel, E., and Mareel, M. (1999). Inhibition of adhesion and induction of epithelial cell invasion by HAV-containing E-cadherin-specific peptides. *J. Cell Sci.* **112**(Pt. 1), 127–135.

Price, L. S., Langeslag, M., Klooster, J. P.t., Hordijk, P. L., Jalink, K., and Collard, J. G. (2003). Calcium signaling regulates translocation and activation of Rac. *J. Biol. Chem.* **278,** 39413–39421.

Ren, X.-D., Kiosses, W. B., and Schwartz, M. A. (1999). Regulation of the small GTPase-binding protein Rho by cell adhesion and the cytoskeleton. *EMBO J.* **18,** 578–585.

Sander, E. E., ten Klooster, J. P., van Delft, S., van der Kammen, R. A., and Collard, J. G. (1999). Rac downregulates Rho activity: Reciprocal balance between both GTPases determines cellular morphology and migratory behaviour. *J. Cell Biol.* **147,** 1009–1021.

Sharpe, G. R., Fisher, C., Gillespie, J. I., and Greenwell, J. R. (1993). Growth and differentiation stimuli induce different and distinct increases in intracellular free calcium in human keratinocytes. *Arch. Dermatol. Res.* **284,** 445–450.

Tu, C.-L., Chang, W., and Bikle, D. D. (2001). The extracellular calcium-sensing receptor is required for calcium-induced differentiation in human keratinocytes. *J. Biol. Chem.* **276,** 41079–41085.

Wheelock, M. J., and Jensen, P. J. (1992). Regulation of keratinocyte intercellular junction organization and epidermal morphogenesis by E-cadherin. *J. Cell Biol.* **117,** 415–425.

[30] Activation of Rap1, Cdc42, and Rac by Nectin Adhesion System

By Hisakazu Ogita and Yoshimi Takai

Abstract

Nectin is an immunoglobulin-like cell–cell adhesion molecule and forms adherens junctions cooperatively with cadherin. Trans-interaction of nectin induces activation of Rap1, Cdc42, and Rac small G proteins. The activity of these small G proteins can be analyzed by the pull-down assay using GST-Ral-GDS fusion protein for Rap1 and GST-PAK-CRIB for Cdc42 and Rac. The fluorescent resonance energy transfer (FRET) system is also available to spatially and temporally detect the activity of these small G proteins in the living cells. In addition to these assays, the activity

METHODS IN ENZYMOLOGY, VOL. 406
0076-6879/06 $35.00
DOI: 10.1016/S0076-6879(06)06030-7

of Cdc42 and Rac is indirectly, but easily, evaluated by the cell-spreading assay to examine formation of filopodia and lamellipodia, respectively. To clearly explore the effect of trans-interacting nectin on the small G proteins, we use L fibroblasts stably expressing nectin-1 (nectin-1-L cells) and the extracellular domain of nectin-3 fused to human IgG Fc (Nef-3). Treatment of nectin-1-L cells with Nef-3 remarkably increases both the amount of the GTP-bound form and the FRET efficiency of all Rap1, Cdc42, and Rac small G proteins. In the cell-spreading assay, Cdc42 and Rac activated in this way promote the formation of filopodia and lamellipodia, respectively. Here, we focus on how the activity of Cdc42 and Rac induced by trans-interacting nectin is examined by use of the pull-down and the cell-spreading assays.

Introduction

An immunoglobulin-like cell–cell adhesion molecule (CAM) nectin constitutes a family of four members, nectin-1, nectin-2, nectin-3, and nectin-4 (Sakisaka and Takai, 2004; Takai and Nakanishi, 2003; Takai et al., 2003). Nectin first forms homo-cis-dimers, followed by formation of homo- or hetero-trans-dimers (trans-interaction) in a Ca^{2+}-independent manner, causing cell–cell contacts. Regarding the hetero-trans-interaction, nectin-1 interacts with either nectin-3 or nectin-4, and nectin-2 with nectin-3, but not with nectin-1 or nectin-4. These hetero-trans-interactions show stronger cell–cell adhesion activity than the homo-trans-interactions. Nectin recruits cadherin, another CAM at adherens junctions (AJs), to the nectin-based cell–cell contact sites. Thus, nectin forms AJs cooperatively with cadherin. Nectin also associates with the actin cytoskeleton through afadin, a nectin- and actin-filament (F-actin)–binding protein, as cadherin associates with the actin cytoskeleton through many peripheral membrane proteins, including α- and β-catenins. The trans-interaction of nectin induces formation of filopodia and lamellipodia through the respective activation of Cdc42 and Rac small G proteins mediated by the activation of Rap1 small G protein (Fukuhara et al., 2004; Fukuyama et al., 2005; Kawakatsu et al., 2002, 2005). Nectin first recruits and activates c-Src (Fukuhara et al., 2004). Activated c-Src then phosphorylates FRG, a Cdc42-GDP/GTP exchange factor (GEF), and Vav2, a Rac-GEF, and it also activates Rap1 through the Crk-C3G complex (Fukuhara et al., 2004; Fukuyama et al., 2005; Kawakatsu et al., 2005). The activation of both c-Src and Rap1 is necessary for the activation of FRG, followed by the activation of Cdc42 (Fukuhara et al., 2004; Fukuyama et al., 2005). Similarly, the activation of both c-Src and Cdc42 is necessary for the activation of

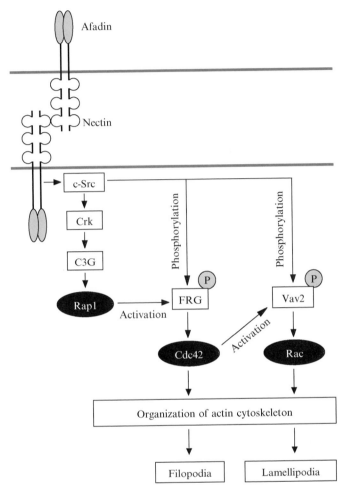

FIG. 1. Nectin-induced signaling toward the activation of Cdc42 and Rac and subsequent formation of filopodia and lamellipodia, respectively. Details are described in the "Introduction" section.

Vav2, eventually causing the activation of Rac (Kawakatsu *et al.*, 2005). The entire relationship in the nectin-induced activation of Cdc42 and Rac and subsequent formation of filopodia and lamellipodia, respectively, is depicted in Fig. 1. Filopodia induced in this way increase the number of cell–cell contact sites at the initial stage of the formation of AJs. On the other hand, lamellipodia induced in this way efficiently expand the cell–cell adhesion between filopodia, acting like a "zipper". Thus, the nectin-induced formation of filopodia and lamellipodia significantly increase the velocity of the formation of AJs.

The activity of Rap1, Cdc42, and Rac is analyzed by the pull-down assay using GST-Ral-GDS fusion protein for Rap1 and GST-PAK-CRIB for Cdc42 and Rac. The fluorescent resonance energy transfer (FRET) system is also available to detect the activity of these small G proteins, especially in the living cells on a spatiotemporal basis, although this system requires exclusive equipment and the skilled techniques. Moreover, the activity of Cdc42 and Rac is indirectly, but easily, examined by the cell-spreading assay to measure the formation of filopodia and lamellipodia. We use here L fibroblasts stably expressing nectin-1 (nectin-1-L cells) and the extracellular domain of nectin-3 fused to human IgG Fc (Nef-3). L cells do not express any cadherins (Nagafuchi et al., 1991). The treatment of nectin-1-L cells with Nef-3, which is in advance preclustered with the antibody against human IgG Fc, induces the activation of Cdc42 and Rac. In this chapter, we describe how to analyze the nectin-induced activation of Cdc42 and Rac by the two assay methods, the pull-down and the cell-spreading assays.

Materials

Mouse L fibroblasts are used (Nagafuchi et al., 1991). The following materials are purchased from commercial sources: pFastBac1 baculovirus vector, BL21 and DH10Bac Escherichia coli competent cells, Sf9 insect cells, High Five insect cells, Grace's medium, fetal calf serum (FCS), and Lipofectamine 2000 transfection reagent are from Invitrogen (Carlsbad, CA), LB medium and Dulbecco's modified Eagle's medium (DMEM) are from Nacalai Tesque (Kyoto, Japan), EX-CELL 400 medium is from JRH Biosciences (Lenexa, KS), sodium deoxycholate, dithiothreitol (DTT), p-amidinophenyl methanesulfonyl fluoride hydrochloride (APMSF), isopropyl-β-D-thiogalactopyranoside (IPTG), leupeptin, aprotinin, ampicillin, and reduced glutathione are from Wako Pure Chemicals (Osaka, Japan), and glutathione-sepharose 4B beads are from Amersham Biosciences (Piscataway, NJ). All other chemicals are of reagent grade. The goat anti-human IgG (Fc specific) polyclonal antibody (pAb) and human IgG are purchased from Sigma (St. Louis, MO). The mouse anti-Cdc42 and anti-Rac monoclonal antibodies (mAbs) are also purchased from Pharmingen (San Diego, CA) and Upstate Biotechnology (Charlottesville, VA), respectively. Rhodamine-phalloidin is obtained from Molecular Probes (Eugene, OR).

Methods

Purification of GST-PAK-CRIB

To obtain an expression vector to produce a glutathione-S-transferase (GST) fusion protein GST-PAK-Cdc42 and Rac interacting and binding domain (CRIB), we amplify DNA of PAK-CRIB (amino acids 68–150) by PCR and insert it into pGEX vector. The GST-PAK-CRIB fusion protein was purified from overexpressing *E. coli* (BL21) by the following steps: (1) cultivation of *E. coli* and induction of GST-PAK-CRIB, (2) preparation of crude supernatant, and (3) affinity purification of GST-PAK-CRIB. Buffers used in the purification of GST-PAK-CRIB are as follows:

Buffer A: phosphate-buffered saline (PBS) containing 10 μM APMSF, 10 μg/ml leupeptin, and 2 μg/ml aprotinin.

Buffer B: PBS containing 10 mM reduced glutathione, 10 μM APMSF, 10 μg/ml leupeptin, 2 μg/ml aprotinin, and 10 mM NaOH. Adjust pH to 8.0 with Tris-HCl solution.

Cultivation of E. coli *and Induction of GST-PAK-CRIB.* BL21 competent *E. coli* transformed with pGEX-PAK-CRIB is cultured at 37° in 1 l of LB medium containing 50 μg/ml ampicillin to an OD_{595} 0.2–0.3. After the addition of IPTG at a final concentration of 0.1 mM, bacteria are further cultured at 26° for 4 h to induce the production of GST-PAK-CRIB. All procedures after this step should be performed at 0–4°. After bacteria are harvested by centrifugation at 8,000 rpm at 4° for 5 min, the supernatant is discarded, and the bacteria pellet is suspended with 20 ml ice-cold PBS. The suspension is centrifuged at 8,000 rpm at 4° for 5 min and then the supernatant is discarded. The bacteria pellet can be frozen at −80° for at least 2 weeks.

Preparation of Crude Supernatant. The pellet is quickly thawed at 37° and is suspended with 20 ml of ice-cold Buffer A, and the suspension is sonicated by Ultrasonic Processor (Taitec, Tokyo, Japan) on ice for 30 sec four times at a 1-min intervals. The homogenate is centrifuged at 100,000g at 4° for 1 h. The supernatant is used for the affinity purification.

Affinity Purification of GST-PAK-CRIB. Glutathione-sepharose 4B beads are packed onto a 10-ml disposable syringe (bed volume, 1 ml). The beads are washed with 20 ml of PBS twice and equilibrated with 20 ml of PBS. Twenty milliliters of crude supernatant prepared as described previously is applied to the syringe column, and the pass fraction is reapplied to the column. After the column is washed with 20 ml of PBS, GST-PAK-CRIB is eluted with 5 ml of Buffer B. The eluate containing GST-PAK-CRIB (5 ml) is dialyzed against 500 ml of PBS three times.

The dialyzed sample can be kept as purified recombinant GST-PAK-CRIB at $-80°$ for at least 6 months without loss of activity.

Preparation of Nef-3

Recombinant Nef-3 is obtained by using a baculovirus system (Satoh-Horikawa *et al.*, 2000). A baculovirus transfer vector for Nef-3 is constructed as follows: pFastBac1-Msp-Fc is first constructed by sub-cloning the inserts encoding the honeybee melittin signal peptide (ATGAAATTCTTAGTCAACGTTGCCCTTGTTTTTATGGTCGTGT ACATTTCTTACA TCTATGCG) (Tessier *et al.*, 1991) and human IgG Fc into pFastBac1. A fragment of an extracellular domain of mouse nectin-3α (aa 51–400) is then inserted into pFastBac1-Msp-Fc to express the chimeric protein fused to the N-terminal signal peptide and the C-terminal IgG Fc (Nef-3). A construct of pFastBac1-Msp-Fc-nectin-3α-ex is transformed into DH10Bac *E. coli* competent cells, which have the parent bacmid, to obtain the bacmid harboring the cDNA of Msp-Fc-nectin-3α-ex by transposition. The recombinant bacmid is purified and transfected to Sf9 insect cells cultured in Grace's medium to gain the baculovirus with the cDNA of Msp-Fc-nectin-3α-ex. High Five insect cells are grown in serum-free EX-CELL 400 medium, infected with the recombinant baculovirus, and cultured at $26°$ for 72 h. Culture supernatants are collected, subjected to a protein A–sepharose column, and then eluted with 0.1 M glycine-HCl at pH 2.5. The eluted proteins are immediately neutralized with 1.5 M Tris-HCl at pH 8.8 and dialyzed against PBS. Twenty-five micrograms of Nef-3 or human IgG are preclustered using 10 μg of the anti-human IgG pAb in 50 μl of PBS for 1 h at room temperature. These preclustered Nef-3 or IgG as a control are used to stimulate nectin-1-L cells in pull-down assay as described in the following.

Generation of Nectin-1-L Cells and Cell Culture

pCAGGS-nectin-1 is constructed by inserting human nectin-1α cDNA into pCAGGS eukaryote expression vector (Niwa *et al.*, 1991). To obtain nectin-1-L cells, L cells are transfected with pCAGGS-nectin-1 using Li-pofectamine 2000. The cells are cultured in DMEM containing 10% FCS for 1 day, replated, and selected by the treatment with 500 μg/ml Geneticin (Takahashi *et al.*, 1999). We usually obtain nectin-1-L cells that express approximately 1×10^6 nectin-1 molecules/cell. Nectin-1-L cells are seeded on a 60-mm dish at the density of 5×10^5 cells/dish 48 h before the assay for the activity of Cdc42 and Rac and are cultured in DMEM with 10% FCS. For 24 h before the beginning of the assay, the cells are cultured in DMEM without serum.

Pull-down Assay for Cdc42 and Rac

The serum-starved nectin-1-L cells are incubated with preclustered Nef-3 or IgG as a control in 1 ml of DMEM for the indicated periods of time. Nectin-1-L cells are treated with Nef-3, because nectin-1 and nectin-3 most strongly trans-interact with each other (Honda *et al.*, 2003). After being washed with ice-cold PBS, the cells are lysed with 0.5 ml of Lysis buffer (50 mM Tris-HCl, pH 7.4, 100 mM NaCl, 10 mM MgCl$_2$, 0.5% sodium deoxycholate, 0.1% SDS, 1 mM DTT, 1% Triton X-100, 10 μg/ml leupeptin, 10 μg/ml aprotinin, 10 μM APMSF) containing 20 μg of recombinant GST-PAK-CRIB and incubated at 2° for 30 min with gentle rotation. After the incubation, the lysates are cleared by centrifugation at 15,000 rpm at 2° for 15 min. The cleared lysates are then incubated with 60 μl of 50% slurry glutathione-sepharose 4B beads, which is washed in advance with 1 ml of PBS twice at 2° for 1 h with gentle rotation. After the centrifugation at 3,000 rpm at 2° for 2 min, the beads are washed with

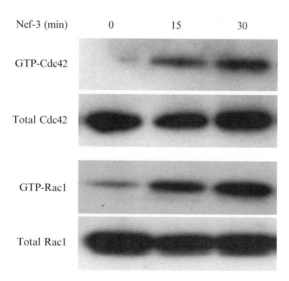

FIG. 2. Nectin-induced activation of Cdc42 and Rac. Nectin-1-L cells were incubated with Nef-3 for the indicated periods of time. The cells were lysed and subjected to the pull-down assay using GST-PAK-CRIB followed by Western blotting with the anti-Cdc42 and anti-Rac1 mAbs.

0.5 ml of Lysis buffer three times. The beads are incubated with Laemmli buffer (Laemmli, 1970) and boiled at 97° for 5 min. The GTP-bound form of Cdc42 and Rac is eluted and separated by SDS-polyacrylamide gel electrophoresis using 15% polyacrylamide gel, followed by Western blotting with the anti-Cdc42 and anti-Rac mAbs. The whole cell lysates are also analyzed for the presence of Cdc42 and Rac for normalization. As shown in Fig. 2, the amounts of the GTP-bound form of Cdc42 and Rac in nectin-1-L cells incubated with Nef-3 for 30 min markedly increase, indicating that the trans-interaction of nectin actually induces the activation of Cdc42 and Rac.

Cell-Spreading Assay

The cell-spreading assay is performed as described (Kawakatsu et al., 2002). Nef-3- or IgG-coated coverslips are prepared by coating coverslips with adequate amounts of Nef-3 or IgG (50 µg/ml) for 15 h and then blocking them with 1% BSA in Hanks' balanced salt solution for 1 h. Nectin-1-L cells are washed with PBS, incubated with 0.2% trypsin and

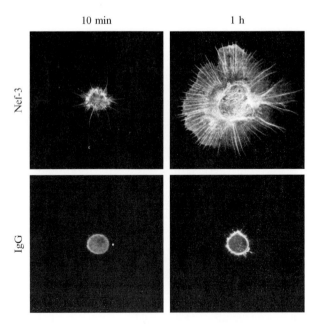

FIG. 3. Nectin-induced formation of filopodia and lamellipodia. Nectin-1-L cells were plated on the Nef-3-or IgG-coated coverslips for the indicated periods of time. The cells were fixed and immunostained for F-actin with rhodamine-phalloidin.

1 mM EDTA at 37° for 5 min, and gently dispersed by pipetting. The dispersed cells were then suspended in DMEM containing 10% FCS. The cells (1×10^4) were placed on the coverslips precoated with Nef-3 or IgG as a control on a 24-well dish and cultured for the indicated periods of time. The cells are then fixed with 4% formaldehyde for 15 min and immunostained with F-actin for rhodamine-phalloidin for 30 min, followed by observation with confocal laser scanning microscopy to measure the formation of filopodia and lamellipodia. Nectin-1-L cells incubated on the Nef-3-coated coverslips for 1 h markedly form filopodia and lamellipodia, whereas the cells on the IgG-coated coverslips hardly form these protrusions (Fig. 3), suggesting that the trans-interaction of nectin induces the activation of Cdc42 and Rac.

Comments

We have introduced here the procedures to examine the nectin-induced activation of Cdc42 and Rac small G proteins by use of nectin-1-L cells and Nef-3. Although we have only demonstrated the model system to clearly explain the nectin-induced activation of these small G proteins, essentially the same results are obtained in other cell lines including NIH3T3 mouse fibroblasts (data not shown). Because NIH3T3 cells express both nectin-1 and nectin-3, treatment of the cells with an extracellular domain of nectin-1 fused to human IgG Fc Nef-1 or Nef-3 increases the GTP-bound form of Cdc42 and Rac. However, because of the low expression level of each nectin in NIH3T3 cells, the increased levels of the GTP-bound form of Cdc42 and Rac in NIH3T3 cells are lower than those in nectin-1-L cells. The pull-down and the cell-spreading assays are also available for other cell lines to assess the nectin-induced activation of Cdc42 and Rac by treatment of the cells with either of Nef-1, Nef-2, or Nef-3, depending on an appropriate counterpart of nectins on the surface of the cells.

References

Fukuhara, T., Shimizu, K., Kawakatsu, T., Fukuyama, T., Minami, Y., Honda, T., Hoshino, T., Yamada, T., Ogita, H., Okada, M., and Takai, Y. (2004). Activation of Cdc42 by trans interactions of the cell adhesion molecules nectins through c-Src and Cdc42-GEF FRG. J. Cell Biol. 166, 393–405.

Fukuyama, T., Ogita, H., Kawakatsu, T., Fukuhara, T., Yamada, T., Sato, T., Shimizu, K., Nakamura, T., Matsuda, M., and Takai, Y. (2005). Involvement of the c-Src-Crk-C3G-Rap1 signaling in the nectin-induced activation of Cdc42 and formation of adherens junctions. J. Biol. Chem. 280, 815–825.

Honda, T., Shimizu, K., Kawakatsu, T., Fukuhara, A., Irie, K., Nakamura, T., Matsuda, M., and Takai, Y. (2003). Cdc42 and Rac small G proteins activated by trans-interactions of nectins are involved in activation of c-Jun N-terminal kinase, but not in association of nectins and cadherin to form adherens junctions, in fibroblasts. *Genes Cells* **8,** 481–491.

Kawakatsu, T., Ogita, H., Fukuhara, T., Fukuyama, T., Minami, Y., Shimizu, K., and Takai, Y. (2005). Vav2 as a Rac-GDP/GTP Exchange Factor Responsible for the Nectin-induced, c-Src-and Cdc42-mediated Activation of Rac. *J. Biol. Chem.* **280,** 4940–4947.

Kawakatsu, T., Shimizu, K., Honda, T., Fukuhara, T., Hoshino, T., and Takai, Y. (2002). *trans*-Interactions of nectins induce formation of filopodia and lamellipodia through the respective activation of Cdc42 and Rac small G proteins. *J. Biol. Chem.* **277,** 50749–50755.

Laemmli, U. K. (1970). Cleavage of structural proteins during the assembly of the head of bacteriophage T4. *Nature* **227,** 680–685.

Nagafuchi, A., Takeichi, M., and Tsukita, S. (1991). The 102 kd cadherin-associated protein: Similarity to vinculin and posttranscriptional regulation of expression. *Cell* **65,** 849–857.

Niwa, H., Yamamura, K., and Miyazaki, J. (1991). Efficient selection for high-expression transfectants with a novel eukaryotic vector. *Gene* **108,** 193–199.

Sakisaka, T., and Takai, Y. (2004). Biology and pathology of nectins and nectin-like molecules. *Curr. Opin. Cell Biol.* **16,** 513–521.

Satoh-Horikawa, K., Nakanishi, H., Takahashi, K., Miyahara, M., Nishimura, M., Tachibana, K., Mizoguchi, A., and Takai, Y. (2000). Nectin-3, a new member of immunoglobulin-like cell adhesion molecules that shows homophilic and heterophilic cell-cell adhesion activities. *J. Biol. Chem.* **275,** 10291–10299.

Takahashi, K., Nakanishi, H., Miyahara, M., Mandai, K., Satoh, K., Satoh, A., Nishioka, H., Aoki, J., Nomoto, A., Mizoguchi, A., and Takai, Y. (1999). Nectin/PRR: An immunoglobulin-like cell adhesion molecule recruited to cadherin-based adherens junctions through interaction with Afadin, a PDZ domain-containing protein. *J. Cell Biol.* **145,** 539–549.

Takai, Y., Irie, K., Shimizu, K., Sakisaka, T., and Ikeda, W. (2003). Nectins and nectin-like molecules: Roles in cell adhesion, migration, and polarization. *Cancer Sci.* **94,** 655–667.

Takai, Y., and Nakanishi, H. (2003). Nectin and afadin: Novel organizers of intercellular junctions. *J. Cell Sci.* **116,** 17–27.

Tessier, D. C., Thomas, D. Y., Khouri, H. E., Laliberte, F., and Vernet, T. (1991). Enhanced secretion from insect cells of a foreign protein fused to the honeybee melittin signal peptide. *Gene* **98,** 177–183.

[31] Analysis of Activated GAPs and GEFs in Cell Lysates

By Rafael García-Mata, Krister Wennerberg, William T. Arthur, Nicole K. Noren, Shawn M. Ellerbroek, and Keith Burridge

Abstract

An assay was developed that allows the precipitation of the active pools of Rho-GEFs, Rho-GAPs, or effectors from cell or tissue lysates. This assay can be used to identify GEFs, GAPs, and effectors involved in specific cellular pathways to determine their GTPase specificity and to monitor the temporal activation of GEFs and GAPs in response to upstream signals.

Introduction

Rho GTPases control many aspects of cellular behavior, including the organization of the cytoskeleton, cell migration, cell adhesion, cell cycle, and gene expression (Burridge and Wennerberg, 2004; Hall, 1998; Van Aelst and D'Souza-Schorey, 1997). Like all GTPases, Rho proteins act as molecular switches by cycling between an active (GTP bound) and an inactive (GDP bound) state. Exchange of GTP for GDP allows the GTPase to interact with downstream effectors to modulate their activity and localization. Cycling between the GDP-bound and GTP-bound state is regulated primarily by two distinct families of proteins: guanine-nucleotide exchange factors (GEFs) activate Rho proteins by catalyzing the exchange of GDP for GTP, the GTPase activating proteins or GAPs negatively regulate GTPase function by stimulating GTP hydrolysis.

Since the development of the Rho pull-down assay in 1999 (Ren *et al.*, 1999), a lot of progress has been made in determining the identity of the Rho-GTPases that are activated in response to certain stimuli or signals. However, much less has been done in terms of identifying the molecules involved in the activation and inactivation of each particular Rho-GTPase. The issue becomes more complicated considering that the human genome contains more than 60 RhoGEFs and approximately 80 RhoGAPs (Moon and Zheng, 2003; Peck *et al.*, 2002; Rossman *et al.*, 2005; Schmidt and Hall, 2002).

We have developed an affinity precipitation assay that allows us to specifically pull down RhoGEFs, RhoGAPs, or effectors that are being

METHODS IN ENZYMOLOGY, VOL. 406
0076-6879/06 $35.00
DOI: 10.1016/S0076-6879(06)06031-9

activated in the cell at any given time or condition. The assay takes advantage of constitutively active and dominant negative Rho-family mutants that bind either to Rho-GAPs and effectors or to RhoGEFs, respectively. Constitutively active Rho mutants were originally designed based on their analogous Ras mutants (G12V and Q61L) (Garrett et al., 1989; Ridley and Hall, 1992; Ridley et al., 1992). These mutants have lost both their intrinsic capacity and their GAP-mediated ability to hydrolyze GTP, and they bind with high affinity to GAPs and effectors (Barbacid, 1987; Trahey and McCormick, 1987). Traditional dominant negative mutants like S17N-Ras and the analogous Rho mutants have been shown to bind GDP with similar affinity to the corresponding wild-type form. However, their affinity for GTP is extremely low, so virtually all of the protein is found in the GDP-bound form (Feig, 1999; Ridley et al., 1992). Another dominant negative Ras mutant (RasG15A) has been previously shown to bind very poorly to both GDP and GTP, existing virtually in a nucleotide free state (Chen et al., 1994). A nucleotide-free GTPase is one of the intermediates of the nucleotide exchange reaction and is able to form a high affinity binary complex with the GEF (Cherfils and Chardin, 1999). This intermediate is rapidly dissociated by GTP and does not accumulate in cells. We took advantage of the properties of these mutants and used them to pull down Rho effectors and GAPs from tissue and cell lysates (Arthur et al., 2002; Noren et al., 2003). We generated mutant versions of various representative GTPases of the Rho subfamily that harbor mutations equivalent to the Q61L and G15A mutations in Ras. We then used GST fusion proteins of these mutants to specifically pull down Rho family GEFs, GAPs, or effectors from cell or tissue lysates. This assay can be used to determine Rho protein specificity on GEFs, GAPs, or effectors (Arthur et al., 2002; Ellerbroek et al., 2004; Noren et al., 2003; Wennerberg et al., 2003). It can also be used as a simple way to find interactors such as GEFs, GAPs, and effectors to GTPases where little is known about upstream and downstream regulation. In addition, many GEFs and GAPs seem to be activated by making the binding site to their target GTPase available. This can be achieved either by unmasking an intramolecular inhibitory domain (Vav1, Dbl, Asef), by association with or dissociation from other proteins (p115-RhoGEF, Sos, Dock180), or by release of the protein from a sequestering cellular compartment (GEF-H1/Lfc, Net1, Ect2) (Schmidt and Hall, 2002; Rossman et al., 2005). Given this type of regulation, it is likely that the activated and nucleotide-free mutants of Rho proteins will have highest affinity toward "activated" GAPs and GEFs, respectively, and that our assay can, indeed, be used to detect activation of GEFs and GAPs by specific stimuli.

   Activation and inactivation of GEFs and GAPs can be studied either by analyzing the precipitation of known candidate components by immunoblot or by determining the identity of the bound proteins by mass spectrometry. In addition, GEF and GAP pull-downs can be an extremely useful tool to follow the pattern of GAP or GEF activation over time or in response to different upstream signals.

Materials

DNA Constructs

Human cDNA for RhoA, Rac1, and Cdc42 were subcloned into pGEX 4T-1 (Amersham) between the *Eco*RI and *Xho*I sites. Empty-nucleotide mutants, G17A (RhoA), G15A (Rac1 and Cdc42), and constitutively active mutants, Q63L (RhoA), and Q61L (Rac1 and Cdc42) were generated by site-directed mutagenesis using the Quick Change Site–directed mutagenesis kit (Stratagene) following the manufacturer's instructions.

Antibodies

Antibodies against p190RhoGAP, ROCK, Sos1, and GFP were from BD Biosciences; Lsc, LARG, and RhoGDI were from Santa Cruz Biotechnology; and anti-myc was from Invitrogen.

Solutions

Lysis Buffer: 20 mM HEPES, pH 7.5; 150 mM NaCl; 5 mM MgCl$_2$; 1% Triton X-100; 1 mM DTT with protease inhibitors (1 mM PMSF, 10 μg/ml aprotinin, 10 μg/ml leupeptin). DTT and protease inhibitors should be added fresh before use. In addition to these, other inhibitors can be added. If pervanadate is added, DTT should be omitted, because DTT will reduce it and inactivate it.

HBS: 20 mM HEPES, pH 7.5; 150 mM NaCl.

Methods

Expression and Purification of GST-Rho Constructs

1. Transform and grow up *Escherichia coli* with the appropriate pGEX construct. For RhoA, Rac1, and Cdc42 proteins, DH5α or any regular strain will work. For some less-soluble Rho proteins (RhoB, RhoC, RhoG), it is good to use a codon optimized strain such as codon plus BL21 (Stratagene).

2. Grow up a 50-ml culture overnight to full density in LB with 50 μg/ml ampicillin (LB-Amp)(O.D. > 1.0).

3. Dilute the culture into 450 ml of LB-Amp and let grow for 30 min at 37°.

4. Induce the protein by adding IPTG to a final concentration of 100 μM and shake the culture at room temperature over night (~16 h).

5. Spin down bacteria and resuspend in 10 ml of lysis buffer. All purification steps are carried out at 4°.

6. Sonicate the resuspended bacteria for 1 min.

7. Spin at 15,000–20,000g for 15 min.

8. Transfer supernatant to a new tube and add 100–500 μl of glutathione-sepharose (Amersham) pre-equilibrated in lysis buffer (amount depends on the solubility of the protein). Rotate tube at 4° for 45–60 min (RhoA ~200 μl of 50% Glutathione sepharose slurry, Rac1, and Cdc42: 500 μl).

9. Spin down sepharose and wash it with 2 × 10 ml lysis buffer and 2 × 10 ml with HBS with 5 mM MgCl$_2$ and 1 mM DTT.

10. Aspirate sufficient wash buffer to get initial sepharose volume (i.e., get approximately 50% slurry). Add 0.5 volumes of glycerol.

11. Estimate protein concentration (see following) directly on beads using Coomassie Plus protein reagent (Pierce).

12. (Optional) If possible, dilute the beads to a final concentration of 1 mg/ml with glutathione-sepharose in 2/3 HBS w/5 mM MgCl$_2$, 1 mM DTT, and 1/3 glycerol. (It is advantageous to have the different fusion proteins at the same concentration so you add equal amounts of sepharose to each sample in the pull-down.)

13. Store the beads at −20°. Typically, the activated mutants are stable for at least a month, whereas the nucleotide-free mutants are stable 1–2 weeks. For longer storage, the beads can be aliquoted, snap frozen and stored at −70°.

Estimation of GST-Rho Protein Concentration on Sepharose Beads

The amount of purified protein can be estimated without the need of eluting the proteins from the beads. We use the Coomassie Plus reagent (Pierce) following the manufacturer's directions and use a small aliquot of the purified GST-beads (2.5 μl is usually enough) to measure protein concentration.

Pull-Down of GEFs or GAPs

1. Treat the cells in the desired way. How many cells you will need for each sample will have to be optimized for each different situation, depending on cell type, level of expression of protein of interest, its affinity

for the Rho protein tested, and also if is going to be used for immunodetection or mass-spectrometry analysis. The quality and detection level of the antibody used to detect the protein is also relevant. A good starting point is a 100-mm dish for Western blot experiments and 1–2 × 150-mm dishes for silver stain/mass-spectrometry analysis.

2. Immediately after treatment, wash the cells twice with ice-cold HBS.

3. Lyse the cells in 400 μl lysis buffer/100-mm dish.

4. Scrape off the cells and spin down debris at 16,000g for 1 min.

5. Measure protein concentration using BioRad DC Protein Assay. It is important to equalize both the total amount of protein and also the volume used for each condition.

6. Preclear the lysate by rotating it with 50 μl of GST bound to glutathione-sepharose (1 mg/ml) for 10 min at 4°. (This step is optional and can be left out if you are looking for conformational states of binding proteins that might be short-lived.)

7. Spin down the GST beads and transfer the cleared lysate to a new tube. Save 10–20 μl of lysate for a loading (mix this with an equal volume of 2× SDS-PAGE sample buffer).

8. Add 10 μg of fusion protein on beads (GST or GST-Rho protein) to each lysate. Rotate for 45–60 min at 4°.

9. Wash the beads three times with lysis buffer, aspirate off all buffer, and dissolve protein complexes by adding 20–40 μl of 1× SDS-PAGE sample buffer and boil. For mass spectrometry analysis, it is recommended to wash more extensively (5–6 times).

10. Run the lysate and the pull-down samples on appropriate SDS-PAGE gels, transfer, and blot for the protein of interest.

11. For the identification of novel proteins, stain the gels with Coomassie blue or silver stain and submit the samples for mass spectrometry analysis.

Results

GAP and GEF Pull-Downs

Wild-type RhoA, nucleotide-free RhoA (G17A-RhoA), and constitutively active Q63L-RhoA were expressed as GST fusion proteins, and affinity precipitations were performed on lysates of CHO cells to assess the ability of these GST fusion proteins to bind to known regulators of Rho proteins. We immunoblotted the precipitated proteins with antibodies against a RhoA-GAP (p190RhoGAP), a RhoA effector ROCK (ROK or Rho kinase), a RhoA-GEF (Lsc/p115 RhoGEF), and RhoGDI. Our results revealed that constitutively active Q63L-RhoA selectively bound both the

RhoA-specific GAP p190RhoGAP and the RhoA effector ROCK but not the RhoGEF Lsc or RhoGDI. In contrast, wtRhoA bound and RhoGDI and G17-Rho bound to Lsc but neither was unable to precipitate p190-RhoGAP and the effector ROCK (Fig. 1A).

As mentioned in the "Introduction," the GEF pull-down assay can also be used to characterize the specificity of a Rho-GEF. We tested the ability of three RhoGEFs of known specificity to bind the empty nucleotide mutants of RhoA, Rac1, and Cdc42. We transiently expressed XPLN (RhoA), an activated mutant of Tiam1 (Rac1), and the DH-PH tandem domains of Intersectin (Cdc42) in NIH 3T3 cells. We found that XPLN coprecipitated only with G17A-RhoA but not with G15A-Rac1 or G15A-Cdc42, Tiam1 coprecipitated only with G15A-Rac1 but not with G17A-RhoA or G15A-Cdc42, and Intersectin coprecipitated only with G15A-Cdc42 but not with G17A-RhoA or G15A-Rac1 (Fig. 1B).

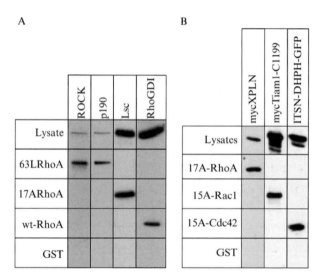

Fig. 1. (A) Selective precipitation of Rho-GEFs, GAPs, and effectors from cell lysates. CHO lysates were subjected to pull-down analysis with GST alone, GST-wt-RhoA, GST-G17A-RhoA, and GST-Q63L-RhoA as described in "Methods." The precipitated material was run on an SDS-PAGE gel and immunoblotted for ROCK, p190RhoGAP, Lsc, and RhoGDI. (B) Analysis of Rho-GEF specificity. NIH 3T3 cells were transiently transfected with mycXPLN, mycTiam1-C1199, or ITSN-DHPH-GFP. After 24 h, cells were lysed, and the lysates were subjected to pull-down analysis with GST alone, GST-G17A-RhoA, and GST-G15A-Rac1 or GST-G15A-Cdc42 as described in "Methods." The precipitated material was run on an SDS-PAGE gel and immunobloted for the transfected proteins with either anti-myc or anti-GFP antibodies.

These results are consistent with previous studies that demonstrated that XPLN, Intersectin, and Tiam1 are specific for RhoA, Cdc42, and Rac1, respectively (Arthur *et al.*, 2002; Hussain *et al.*, 2001; Michiels *et al.*, 1995).

To determine whether the pull-downs were specific enough and whether they could be used to identify true interactors of Rho GTPases, NIH 3T3 lysates were subjected to pull-downs with GST, GST-wtRhoA, GST-G15A-RhoA, and GST-Q63L-RhoA (10 μg fusion protein/pull-down), and bands specifically precipitated with the different variants of RhoA were identified by mass spectrometry (Fig. 2). Of eight proteins identified, three well-known effectors (ROCK2, PKN1, and PKN2) and three well-known RhoA GEFs (Lfc, Lsc, and SmgGDS) were found. Two of the eight identified proteins did not obviously fall in to any Rho-regulatory class of proteins: tubulin in the G17A-RhoA pull-down and myosin heavy chain in the Q63L-RhoA pull-down. It is possible that tubulin coprecipitated with the tubulin-binding RhoGEF, Lfc/GEF-H1 and that myosin heavy chain coprecipitated with one of the Rho effectors. It should be noted that GEFs precipitated not only with the G17A mutant but also to a lesser degree with the Q63L mutant. Because GEFs are able to exchange both GDP and GTP *in vitro* (Cherfils and Chardin, 1999), it is possible that at the concentration used in the assay, some Rho-GEFs can bind and coprecipitate not only with the GDP-bound Rho but also with GTP-bound Rho. *In vivo*, this does not occur, because the concentration of GTP and the interaction of the GTP-bound form with the effectors drive the reaction in the GDP > GTP direction. Overall, these results indicate that the method is suitable for the identification of novel GEFs, GAPs, and effectors of less well-studied GTPases, as well as identification of changes in activity of GEFs or GAPs for well-studied GTPases.

One exciting application for this kind of pull-down assay is to follow the activation of a particular GEF or GAP in response to an upstream signal. We hypothesized that RhoGEFs and RhoGAPs will only be able to bind their target GTPases when they are in an active state and that this pull-down method, therefore, can distinguish between active and inactive GEFs. To test this, we treated CHO cells with PDGF that activates Rac1 and used GST-15A-Rac to pull-down binding partners (i.e., GEFs that had been activated by the PDGF treatment) from the lysate. One GEF that has been implicated in the activation of Rac downstream of PDGF is Sos1 (Nimnual *et al.*, 1998). We blotted for Sos1 in the pull-downs and observed an increase in the amount of Sos1 that was precipitated with G15A-Rac in response to PDGF (Fig. 3). In contrast, Tiam1, another Rac-specific GEF that was present in the lysates, did not bind to G15A-Rac in the presence or absence of PDGF. We have previously shown (Arthur *et al.*, 2002 and Fig. 1B) that a constitutively active mutant of Tiam1 can associate with

Protein used for pulldown

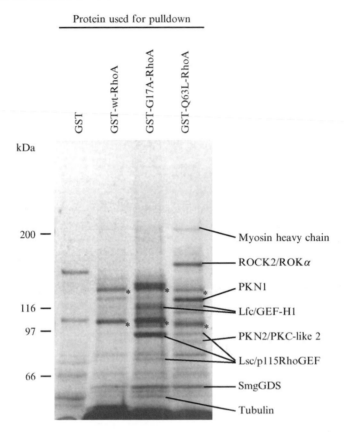

kDa

GST

GST-wt-RhoA

GST-G17A-RhoA

GST-Q63L-RhoA

200 —   Myosin heavy chain

ROCK2/ROKα

PKN1

116 —   Lfc/GEF-H1

97 —   PKN2/PKC-like 2

Lsc/p115RhoGEF

66 —   SmgGDS

Tubulin

FIG. 2. Precipitation of RhoA-interacting proteins from NIH 3T3 cell lysates. NIH 3T3 lysates were subjected to pull-down analysis with GST alone, GST-wt-RhoA, GST-G17A-RhoA, and GST-Q63L-RhoA as described in "Methods." The precipitated material was run on an SDS-PAGE gel and stained with Coomassie blue, and bands that were specifically precipitated with the RhoA variants were identified by mass spectrometry. Eight proteins were identified and are marked on the left side of the gel. The GST-RhoA fusion protein can be seen at the bottom of the gel image, as well as in higher migrating bands marked with asterisks (*). These are likely dimers and trimers of the fusion protein.

G15A-Rac, suggesting that in this case the Tiam1 present in the PDGF-treated cells is inactive and thus unable to bind the GTPase. Supporting these results, we have previously shown that only the active cellular pool of p190-RhoGAP is able to bind to Q63L-RhoA (Noren et al., 2003).

In another experiment, we wanted to examine the RhoA-regulatory signals in nontransformed immortalized human embryonic kidney cells and their Ras-transformed counterparts (Hahn et al., 1999). Several studies

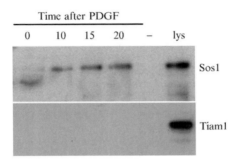

FIG. 3. Sos1 activation in PDGF-treated cells. After PDGF stimulation, CHO cells were lysed, and the lysates incubated with GST-G15A-Rac1 to pull down Rac1-specific GEFs. Samples were analyzed by SDS-PAGE and immunoblotted with the indicated antibodies.

have suggested a role for RhoA regulation in Ras transformation of cells (Chen *et al.*, 2003; Khosravi-Far *et al.*, 1995; Zhong *et al.*, 1997). Interestingly, we did not detect any significant difference in overall RhoA-GTP levels between the two cell types (data not shown). However, in G17A-RhoA pull-downs, we noticed increased pull-down of a band of about 200 kDa from the Ras-transformed cells (Fig. 4A). This protein was identified as LARG, a RhoA-specific GEF. When lysates were blotted for the presence of LARG, we could detect an upregulation of LARG protein levels in the Ras-transformed cells (Fig. 4B). We also noticed a gel shift of the main isoform of LARG, suggesting that either the Ras-transformed cells mainly express a different splice variant of LARG than the nontransformed cells or that the LARG protein in the Ras-transformed cells is posttranslationally modified. Because p190RhoGAP has been suggested to play a role in Ras-mediated regulation of RhoA (Chen *et al.*, 2003), we did pull-downs with Q63L-RhoA to assess its activity in these cells (Noren *et al.*, 2003). We detected increased pull-down of p190RhoGAP without the protein being overexpressed (Fig. 4C), indicating that p190RhoGAP is activated in the Ras-transformed cells. Together, these results suggest that, at least in the immortalized HEK system, Ras transformation leads to both RhoA-activating and inactivating signals without drastically changing the overall levels of RhoA-GTP. This change in regulation could result in increased cycling between the GTP and GDP-bound states of RhoA in the Ras-transformed cells because of increased GEF and GAP activities. Alternatively, this could result in differential subcellular regulation of RhoA in the Ras-transformed cells where RhoA-GTP levels are high in cellular compartments where LARG is located and low in compartments

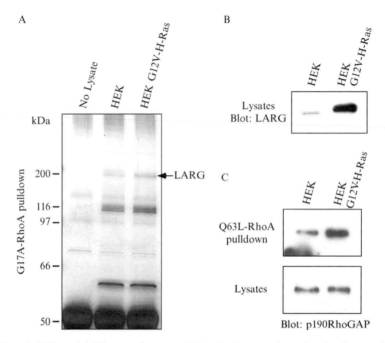

FIG. 4. Differential Rho regulatory activities in Ras-transformed cells. Immortalized human embryonic kidney cells (HEK) and Ras-transformed HEKs (HEK G12V-H-Ras) were tested for activities of RhoGEFs and RhoGAPs by pull downs as described in "Methods." (A) Precipitation of G17A-RhoA–interacting proteins from HEK and HEK G12V-H-Ras cell lysates. The precipitated material was run on an SDS-PAGE gel and Coomassie-stained. As a control, fusion protein alone was run in one lane (No Lysate). A band at approximately 200 kDa was pulled down in a larger amount from the HEK G12V-H-Ras HEK lysates (arrow). This band was identified as LARG by mass spectrometry. (B) Immunoblotting for LARG protein in HEK and HEK G12V-H-Ras lysates reveals that LARG is overexpressed in HEK G12V-H-Ras cells. (C) A pull down with GST-Q63L-RhoA indicates increased p190RhoGAP activity in HEK G12V-H-Ras cells while expression is unchanged.

where p190RhoGAP is located. Future studies will aim at determine whether this change in RhoA regulation contributes to cell transformation.

Troubleshooting

Stability of GST-Rho Proteins

Some of the GST-Rho mutants, in particular the empty nucleotide constructs, are not as stable as their wild-type counterparts when expressed in bacteria. Protein stability decreases even more when proteins are eluted from beads. One way to deal with this problem is to prepare the GST proteins fresh when needed and to keep them at $-20°$ with glycerol (see "Methods") for no longer than 1 week.

Bacterial Contaminants

The method described here to purify the GST fusion proteins allows fast purification. However, depending on the expression level and the stability of each GST fusion protein, a significant amount of bacterial contaminant proteins can be nonspecifically purified with the GST. This is usually not a problem when precipitated proteins are analyzed by Western blot. However, it can be a major inconvenience if the goal of the experiment is to identify novel proteins by mass spectrometry analysis. A control lane should always be run with the GST fusion alone to distinguish contaminating bands from specific bands pulled down from the lysate. Alternatively, and depending on the stability of the protein, additional purification steps can be performed after elution of the GST fusion protein from glutathione-beads. It is not recommended to elute the free nucleotide forms from the beads, because they tend to degrade fast after elution, but it can be tried for the constitutively active mutants.

Even if contaminants are kept to a minimum, the range of molecular weights of GST-Rho proteins can still be a problem for mass spectrometry analysis. One way to overcome this problem is to cross-link the GST fusion proteins to the glutathione beads. There are very few GEFs and GAPs in the molecular weight range of GST-Rho proteins; however, some effectors are small and would be masked by the presence of the GST fusion protein.

Acknowledgments

We thank Dr. Janiel Shields and Dr. Channing Der for the Ras-transformed cell lines. We would also like to thank Adi Dubash for the critical reading of this manuscript. This work was supported by NIH grants GM29860 and HL45100. RGM was supported by a postdoctoral fellowship from the Susan G. Komen Breast Cancer Foundation.

References

Arthur, W. T., Ellerbroek, S. M., Der, C. J., Burridge, K., and Wennerberg, K. (2002). XPLN, a guanine nucleotide exchange factor for RhoA and RhoB, but not RhoC. *J. Biol. Chem.* **277**, 42964–42972.
Barbacid, M. (1987). Ras genes. *Annu. Rev. Biochem.* **56**, 779–827.
Burridge, K., and Wennerberg, K. (2004). Rho and Rac take center stage. *Cell* **116**, 167–179.

Chen, J. C., Zhuang, S., Nguyen, T. H., Boss, G. R., and Pilz, R. B. (2003). Oncogenic Ras leads to Rho activation by activating the mitogen-activated protein kinase pathway and decreasing Rho-GTPase-activating protein activity. *J. Biol. Chem.* **278,** 2807–2818.

Chen, S. Y., Huff, S. Y., Lai, C. C., Der, C. J., and Powers, S. (1994). Ras-15A protein shares highly similar dominant-negative biological properties with Ras-17N and forms a stable, guanine-nucleotide resistant complex with CDC25 exchange factor. *Oncogene* **9,** 2691–2698.

Cherfils, J., and Chardin, P. (1999). GEFs: Structural basis for their activation of small GTP-binding proteins. *Trends Biochem. Sci.* **24,** 306–311.

Ellerbroek, S. M., Wennerberg, K., Arthur, W. T., Dunty, J. M., Bowman, D. R., De Mali, K. A., Der, C., and Burridge, K. (2004). SGEF, a RhoG guanine nucleotide exchange factor that stimulates macropinocytosis. *Mol. Biol. Cell* **15,** 3309–3319.

Feig, L. A. (1999). Tools of the trade: Use of dominant-inhibitory mutants of Ras-family GTPases. *Nat. Cell Biol.* **1,** E25–E27.

Garrett, M. D., Self, A. J., van Oers, C., and Hall, A. (1989). Identification of distinct cytoplasmic targets for ras/R-ras and rho regulatory proteins. *J. Biol. Chem.* **264,** 10–13.

Hahn, W. C., Counter, C. M., Lundberg, A. S., Beijersbergen, R. L., Brooks, M. W., and Weinberg, R. A. (1999). Creation of human tumour cells with defined genetic elements. *Nature* **400,** 464–468.

Hall, A. (1998). Rho GTPases and the actin cytoskeleton. *Science* **279,** 509–514.

Hussain, N. K., Jenna, S., Glogauer, M., Quinn, C. C., Wasiak, S., Guipponi, M., Antonarakis, S. E., Kay, B. K., Stossel, T. P., Lamarche-Vane, N., and McPherson, P. S. (2001). Endocytic protein intersectin-l regulates actin assembly via Cdc42 and N-WASP. *Nat. Cell Biol.* **3,** 927–932.

Khosravi-Far, R., Solski, P. A., Clark, G. J., Kinch, M. S., and Der, C. J. (1995). Activation of Rac1, RhoA, and mitogen-activated protein kinases is required for Ras transformation. *Mol. Cell. Biol.* **15,** 6443–6453.

Michiels, F., Habets, G. G., Stam, J. C., van der Kammen, R. A., and Collard, J. G. (1995). A role for Rac in Tiam1-induced membrane ruffling and invasion. *Nature* **375,** 338–340.

Moon, S. Y., and Zheng, Y. (2003). Rho GTPase-activating proteins in cell regulation. *Trends Cell Biol.* **13,** 13–22.

Nimnual, A. S., Yatsula, B. A., and Bar-Sagi, D. (1998). Coupling of Ras and Rac guanosine triphosphatases through the Ras exchanger Sos. *Science* **279,** 560–563.

Noren, N. K., Arthur, W. T., and Burridge, K. (2003). Cadherin engagement inhibits RhoA via p190RhoGAP. *J. Biol. Chem.* **278,** 13615–13618.

Peck, J., Douglas, G. T., Wu, C. H., and Burbelo, P. D. (2002). Human RhoGAP domain-containing proteins: Structure, function and evolutionary relationships. *FEBS Lett.* **528,** 27–34.

Ren, X. D., Kiosses, W. B., and Schwartz, M. A. (1999). Regulation of the small GTP-binding protein Rho by cell adhesion and the cytoskeleton. *EMBO J.* **18,** 578–585.

Ridley, A. J., and Hall, A. (1992). The small GTP-binding protein rho regulates the assembly of focal adhesions and actin stress fibers in response to growth factors. *Cell* **70,** 389–399.

Ridley, A. J., Paterson, H. F., Johnston, C. L., Diekmann, D., and Hall, A. (1992). The small GTP-binding protein rac regulates growth factor-induced membrane ruffling. *Cell* **70,** 401–410.

Rossman, K. L., Der, C. J., and Sondek, J. (2005). GEF means go: Turning on RHO GTPases with guanine nucleotide-exchange factors. *Nat. Rev. Mol. Cell. Biol.* **6,** 167–180.

Schmidt, A., and Hall, A. (2002). Guanine nucleotide exchange factors for Rho GTPases: Turning on the switch. *Genes Dev.* **16,** 1587–1609.

Trahey, M., and McCormick, F. (1987). A cytoplasmic protein stimulates normal N-ras p21 GTPase, but does not affect oncogenic mutants. *Science* **238**, 542–545.

Van Aelst, L., and D'Souza-Schorey, C. (1997). Rho GTPases and signaling networks. *Genes Dev.* **11**, 2295–2322.

Wennerberg, K., Forget, M. A., Ellerbroek, S. M., Arthur, W. T., Burridge, K., Settleman, J., Der, C. J., and Hansen, S. H. (2003). Rnd proteins function as RhoA antagonists by activating p190 RhoGAP. *Curr. Biol.* **13**, 1106–1115.

Zhong, C., Kinch, M. S., and Burridge, K. (1997). Rho-stimulated contractility contributes to the fibroblastic phenotype of Ras-transformed epithelial cells. *Mol. Biol. Cell* **8**, 2329–2344.

[32] Degradation of RhoA by Smurf1 Ubiquitin Ligase

By HONG-RUI WANG, ABIODUN A. OGUNJIMI, YUE ZHANG, BARISH OZDAMAR, ROHIT BOSE, and JEFFREY L. WRANA

Abstract

The Rho family of small GTPases plays a key role in the dynamic regulation of the actin cytoskeleton that underlies various important cellular functions such as shape changes, migration, and polarity. We found that Smurf1, a HECT domain E3 ubiquitin ligase, could specifically target RhoA but not Cdc42 or Rac1 for degradation. Smurf1 interacts with the dominant inactive form of RhoA, RhoA N19, which binds constitutively to guanine nucleotide exchange factors (GEFs) *in vivo*. Smurf1 also interacts directly with either nucleotide-free or GDP-bound RhoA *in vitro*; however, loading with GTPγS inhibits the interaction. RhoA is ubiquitinated by wild-type Smurf1 but not the catalytic mutant of Smurf1 (C699A) *in vivo* and *in vitro*, indicating that RhoA is a direct substrate of Smurf1. In this chapter, we summarize the systems and methods used in the analyses of Smurf1-regulated RhoA ubiquitination and degradation.

Introduction

The Rho family of small GTPases is a subset of the Ras superfamily and is an important regulator of the cytoskeletal dynamics that control cell shape, motility, and polarity (Bar-Sagi and Hall, 2000; Bishop and Hall, 2000; Etienne-Manneville and Hall, 2002; Hall and Nobes, 2000; Van Aelst and Symons, 2002). The activity of Rho family members is regulated by their nucleotide-bound state, cycling between an active GTP-bound form and an inactive GDP-bound form. This cycling is tightly controlled by

0076-6879/06 $35.00
DOI: 10.1016/S0076-6879(06)06032-0

associated cofactors such as GTPase activating proteins (GAPs), which stimulate the intrinsic GTPase activity to convert bound GTP to GDP and guanine nucleotide exchange factors (GEFs), which stimulate exchange of GDP for GTP. Current evidence suggests that GEF-dependent nucleotide exchange is the key control point for regulating the biological function of GTPases (Bar-Sagi and Hall, 2000; Bishop and Hall, 2000; Hall and Nobes, 2000; Van Aelst and Symons, 2002).

Ubiquitin-dependent proteolysis is a key regulatory mechanism that controls the degradation of intracellular and membrane proteins that have been tagged by ubiquitin for degradation by the proteasome or lysosome, respectively (Hershko and Ciechanover, 1998). Conjugation of ubiquitin to protein targets is mediated by an enzymatic cascade. An E1 ubiquitin–activating enzyme that transfers ubiquitin to an E2 ubiquitin–conjugating enzyme, which can then either conjugate ubiquitin directly onto protein targets using a set of substrate specific adaptors, or transfer ubiquitin to E3 ligases, which in turn transfer ubiquitin to the substrate.

The Smurfs belong to the HECT family ubiquitin ligases, which contain C2 and WW domains as well as a conserved C-terminal HECT domain (Harvey and Kumar, 1999). The C2 domain can mediate interactions with membrane lipids and proteins, whereas the WW domains bind proline-tyrosine (PY) motifs. The Smurf C2-WW-HECT domain ubiquitin ligases have been shown to regulate TGFβ signal transduction by targeting Smad signaling molecules for degradation (Lin et al., 2000; Zhang et al., 2001; Zhu et al., 1999). In addition, Smurfs are recruited by Smads to other non-PY containing substrates including TGFβ ser/thr kinase receptors (Ebisawa et al., 2001; Kavsak et al., 2000), as well as the nuclear oncoprotein SnoN (Bonni et al., 2001). In our recent study, we found that Smurf1 also controls RhoA levels at active cellular protrusions through ubiquitin-mediated degradation (Wang et al., 2003). Our findings highlight an unexpected mechanism regulating RhoA activity by the ubiquitin-dependent pathway.

Methods

In Vitro *Interaction and Ubiquitination Assay of RhoA*

Purification of Bacterially Expressed Smurf1 and RhoA. Smurf1 (WT or C699A) is fused to the C-terminus of the *Schistosoma japonicum* glutathione *S*-transferase gene by cloning into the *SalI/NotI* sites of pGEX-4T-1. *Escherichia coli* (DH5α) is transformed with GST-Smurf1 (WT or C699A). The next day, a single colony is inoculated into 5 ml LB medium containing 100 μg/ml ampicillin and incubated with shaking at 37° overnight. The culture of *E. coli* is then diluted into 100 ml of fresh LB/ampicillin and kept growing at 37° until the OD_{600} reaches 0.5. GST-Smurf1 (WT or C699A) expression is then induced by adding 1 ml of 0.1 M isopropylthio-galactoside (IPTG) stock solution (made in water and stored at $-20°$) to reach a final concentration of 1 mM. The culture is incubated with shaking at room temperature for a further 4 h or overnight. The cells are harvested by centrifugation at 4000 rpm for 10 min at 4°. The cell pellet is then resuspended and sonicated in 15 ml of lysis buffer (0.5% Triton X-100, 50 mM Tris-HCl, pH 7.5, 150 mM NaCl, 1 mM EDTA, 1 mM DTT) with 1 mM phenylmethylsulfonyl fluoride (PMSF), 10 μg/ml pepstatin A, and 10 μg/ml leupeptin. The lysate is cleared by centrifugation at 11,000 rpm for 15 min, and the supernatant is transferred to a 15-ml Falcon tube. Gluta-thione sepharose 4B beads (Amersham Biosciences) are washed twice and subsequently prepared as 50% slurry with lysis buffer. One milliliter of the 50% slurry is added to the supernatant and incubated with constant mixing overnight at 4°. After incubation, the beads are washed five times with TNTE buffer (0.1% Triton X-100, 50 mM Tris-HCl, pH 7.5, 150 mM NaCl, 1 mM EDTA) by centrifugation at 3000 rpm for 5 min. The beads are then resuspended in a roughly equal volume of the TNTE buffer and kept at 4°.

pGEX-4-TEV expression vector is generated by cloning the spacer region and TEV protease cleavage site (GAT TAC GAT ATC CCA ACG ACC GAA AAC CTG TAT TTT CAG GGC) from pProEX-Hta (Invitrogen) into the *Bam*HI/*Sma*I sites of pGEX-4T-1. To express GST-RhoA with the TEV protease cleavage site between GST and RhoA, RhoA is cloned into the *Sal*I/*Not*I sites of pGEX-4-TEV.

GST-RhoA is expressed in *E. coli* (BL21) and purified as previously described (Self and Hall, 1995). To generate nonfused RhoA, 100 μl of glutathione beads containing GST-RhoA is incubated with 10 units His-tagged TEV protease (Invitrogen) in 200 μl of 50 mM Tris-HCl, pH 7.6, 150 mM NaCl, 10 mM DTT 10 mM $MgCl_2$ at 4° overnight. TEV protease can then be removed by adding 25 μl equilibrated Ni-NTA beads (Qiagen) to the supernatant incubating at 4° for 30–60 min. The supernatant is then quickly frozen in liquid nitrogen in 20-μl aliquots and stored at $-70°$.

Nucleotide-free GST-RhoA is prepared by incubating glutathione beads containing GST-RhoA in 50 mM HEPES, pH 7.6, and 20 mM EDTA for 20 min at room temperature (Zhang *et al.*, 2000). After incuba-tion, the beads are washed three times in 50 mM Tris-HCl, pH 7.6, 150 mM NaCl, 1 mM DTT by centrifugation at 3000 rpm for 1 min. Nonfused nucleotide–free RhoA is cleaved as described previously, except $MgCl_2$ is omitted from the buffer.

In Vitro *Interaction between Smurf1 and RhoA.* To generate GDP or GTPγS bound form of RhoA, 0.2 μg of nucleotide free RhoA is incubated with 250 μM GDP or GTPγS in 10 μl 50 mM HEPES, pH 7.5, 100 mM

NaCl, 10 mM MgCl$_2$ at room temperature for 30 min. For *in vitro* binding assay, 0.2 μg of each different RhoA sample is incubated with 15 μl of glutathione beads with GST (control) or GST-Smurf1 in 0.5 ml 50 mM HEPES, pH 7.5, 100 mM NaCl, 10 mM MgCl$_2$, 0.1% Triton X-100 at 4° for 3 h. After incubation, the beads are washed five times with the same solution and boiled in 20 μl 2× SDS loading buffer for 5 min. The samples are run on 11% SDS-PAGE gel and transferred to nitrocellulose membrane (Bio-Rad). RhoA and Smurf1 are detected by immunoblotting with anti-RhoA monoclonal antibody (Santa Cruz) and laboratory-made anti-Smurf1 polyclonal antibody, respectively. GST is detected by Ponceau S staining (Fig. 1A).

In Vitro *Ubiquitination Assay of RhoA.* The *in vitro* RhoA ubiquitination reaction is carried out in 15 μl of reaction volume prepared by using 5× reaction buffer (250 mM Tris-HCl, pH 7.5, 50 mM MgCl$_2$, 50 μM DTT, 20 mM ATP). On ice, 10 μg of ubiquitin, 0.35 μg of rabbit E1 (ubiquitin activating enzyme) (Boston Biochem), 0.4–0.5 μg of UbcH5c (Boston Biochem), and at least 0.35 μg RhoA are added to 3 μl 5× reaction buffer for each reaction. Sterile distilled water is added and mixed thoroughly to increase the volume to 15 μl. The reaction cocktail is then added to the tube containing either GST beads (control) or GST-Smurf1 (WT or C699A) beads and incubated for 1 h at room temperature.

The reaction is terminated by adding 185 μl of TNTE buffer containing 1% SDS and 1% deoxycholate and heating for 20 min at 70°. To avoid taking any beads, only 150 μl of the boiled sample is transferred to a fresh tube and diluted 10 times with TNTE buffer. RhoA is immunoprecipitated overnight at 4° using anti-RhoA monoclonal antibody (Santa Cruz) and protein G sepharose beads. The beads are washed four times with TNTE buffer and then boiled in 25 μl 2× SDS gel loading buffer. The samples are run on 7.5% SDS-PAGE gel and transferred to nitrocellulose membrane. Ubiquitination of RhoA is detected by immunoblotting with anti-ubiquitin antibody (Santa Cruz, P4D1) (Fig. 1B).

Ubiquitination and Degradation of RhoA in Mammalian Cells

Cell Type, Growth Conditions, and Transient Transfection Method. Human embryonic kidney (HEK) 293T transformed cell line is used to study the degradation and ubiquitination of RhoA. The cells are grown in high-glucose Dulbecco's modified Eagle's medium (DMEM; HyClone Lab.) supplemented with 10% (v/v) fetal bovine serum (HyClone Lab.) and 100 units/ml streptomycin and penicillin at 37° in a humidified 5% CO$_2$ incubator. Calcium phosphate precipitation is used to transiently transfect HEK 293T cells.

A

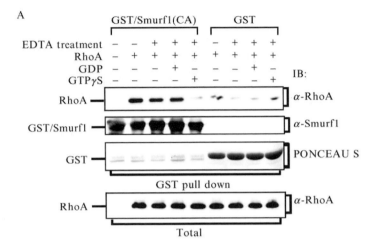

FIG. 1. *In vitro* interaction and ubiquitination assay of RhoA. (A) Interaction between Smurf1 and RhoA *in vitro*. Aliquots of bacterially produced RhoA treated with or without 0.25 mM GDP or GTPγS as indicated were incubated with glutathione beads coupled to GST or GST/Smurf1(CA). Associated RhoA (top panel) and input RhoA (bottom panel) were determined by immunoblotting with the indicated antibody. GST was detected by PONCEAU S staining. (B) Direct ubiquitination of RhoA by Smurf1. RhoA and WT or C699A Smurf1 were purified from bacteria and subjected to an *in vitro* ubiquitination assay (Wang *et al.*, 2003).

REAGENTS FOR TRANSFECTION. 2×HeBs: 16.4 g of NaCl, 11.9 g of HEPES acid, and 0.21 g of Na_2HPO_4 are added to 900 ml of distilled H_2O. The pH is adjusted to 7.05 with NaOH. An exact pH is important for good DNA precipitation formation. The volume is completed to 1000 ml, filter sterilized, aliquoted and stored at $-20°$.

2.5 M CaCl$_2$: Made in distilled water, filter sterilized, and stored at 4°.

PROCEDURE FOR TRANSFECTION. Cells are split the day before transfection to achieve approximately 40–50% confluence when DNA is applied. A 12-well dish is used for degradation assays, and a 100-mm dish is used for the interaction and ubiquitination assays. Approximately 3 h before transfection, prewarmed fresh medium (with antibiotics if desired) is added to the cells. The volume should be 9/10 of the final volume (that is, 9 ml per 100-mm dish or 0.9 ml per 12-well dish).

To transfect cells in a 100-mm dish, a total of 10 μg of DNA (including expression constructs and empty vector carrier) is added to a sterile Eppendorf tube. The DNA amount and the reaction volume should be scaled appropriately for other sized dishes. The reaction volume is completed to 450 μl with autoclaved water. Then 50 μl of 2.5 M CaCl$_2$ is added and mixed well with a pipetteman. In a second tube, 0.5 ml of 2×HeBS is prepared before the DNA/CaCl$_2$ solution is added drop wise and mixed with a pipetteman. The tube is then vortexed for 10 sec before incubation at room temperature for 15–20 min. The reactions are staggered so that each reaction is given the exact same amount of time to form precipitate. The precipitated suspension is then added to the cells and spread evenly over the dish. The dish is shaken gently to distribute the DNA evenly and incubated overnight at 37° in a humidified 5% CO$_2$ incubator. The following day, the old medium is removed by aspiration, and prewarmed fresh medium is added to the cells. The cells are harvested for the assay after an additional 24 h of incubation.

Immunoblotting and Immunoprecipitation. To prepare the cell lysate, the culture medium is aspirated and rinsed gently twice with PBS before adding ice-cold lysis buffer (0.5% Triton X-100, 50 mM Tris-HCl, pH 7.5, 150 mM NaCl, 1 mM EDTA) containing 10 μg/ml pepstatin A, 10 μg/ml leupeptin, and 1 mM PMSF. One milliliter of lysis buffer is used for a 100-mm dish, and the volume of lysis buffer should be scaled appropriately for other sized dishes. The dish is shaken for 15 min at 4°, and the cell lysate is transferred to an Eppendorf tube. The lysate is then cleared by centrifugation at 14,000 rpm for 10 min at 4°. After transferring the supernatant to a new tube (on ice), a 20-μl aliquot of the lysate is taken for immunoblotting assay. The remaining lysate can then be used for immunoprecipitation assays using appropriate antibodies.

Degradation Assay of RhoA In Vivo. RhoA degradation assay is performed in 12-well tissue culture plate using HEK 293T cells. In a single well, 0.1 μg of RhoA expression construct with or without 0.1 μg Smurf1 (WT or C699A) expression construct is transfected using calcium phosphate transfection method as described previously. The protein level is detected by immunoblotting as described previously. Wild-type Smurf1 decreases the steady-state level of RhoA but has no effect on Cdc42 or Rac1. Moreover, neither Smurf2 nor Smurf1 (C699A) targets RhoA for degradation (Fig. 2A). To confirm that RhoA degradation occurs through the proteasome pathway, cells are treated with proteasome inhibitors 4 h (10 μM lactacystin [BostonBiochem] or 40 μM LLnL [BostonBiochem]). Both lactacystin and LLnL inhibit the Smurf1-dependent degradation of RhoA (Fig. 2B). Because transfection of Smurf1 may cause changes in cell morphology and affect cell function, the ubiquitin protein reference (UPR) technique is used to normalize RhoA levels (Levy *et al.*, 1996; Wang *et al.*, 2003). The UPR-RhoA construct is generated as described in Fig. 2C. As in the degradation assay, 0.1 μg UPR-RhoA construct is transfected to HEK 293T cells with or without 0.1 μg Smurf1 (WT or C699A) expression construct in a 12-well tissue culture plate, and levels of RhoA and the reference protein (DHFR) are determined by immunoblotting (Fig. 2D).

In Vivo *Interaction Between Smurf1 and RhoA.* Because binding of Smurf ubiquitin ligases to their substrates can lead to degradation of the substrate, the catalytically inactive Smurf1 mutant (C699A) is used to trap the ligase substrate. To test *in vivo* interaction between Smurf1 and RhoA, 3 μg of DNA encoding N-terminal HA-tagged wild-type RhoA, dominant active RhoA (RhoA V14), or dominant inactive RhoA (RhoA N19) are transfected together with or without 1 μg Flag-tagged Smurf1 (C699A) expression construct to HEK 293T cells in a 100-mm dish using calcium phosphate transfection method as described previously. Interaction between Smurf1 (C699A) and RhoA is detected by immunoprecipitation of Flag-Smurf1 (C699A) using anti-Flag M2 antibody (Sigma) followed by immunoblotting HA-tagged RhoA with rat monoclonal anti-HA antibody (Roche). Use of the rat monoclonal anti-HA antibody from Roche results in a minimal mouse IgG light chain background. As shown in Fig. 3A, Smurf1 (C699A) only interacts with the dominant inactive form of RhoA, RhoA N19, which binds constitutively to GEFs and thus accumulates in a nucleotide-free state in cells (Schmidt and Hall, 2002).

Ubiquitination Assay of RhoA In Vivo. In a 100-mm dish, HEK 293T cells are transfected with 2 μg N-terminal Flag-tagged RhoA expression construct, 2 μg myc-tagged Smurf1 (WT or C699A) expression construct and 3 μg HA-tagged ubiquitin expression construct as desired. Because RhoA levels are strongly decreased by Smurf1, the ubiquitination of RhoA can only be detected in the presence of proteasome inhibitor. After overnight treatment of 20 μM LLnL, the cells are lysed as described previously. An aliquot of 15 μl anti-Flag M2 agarose beads (Sigma) is added to the cell lysate followed by incubation at 4° for 3 h on a rocker. After the incubation, the beads are washed three times with TNTE buffer by centrifugation at 3000 rpm for 1 min, then boiled with 100 μl 1% SDS for 5 min. The

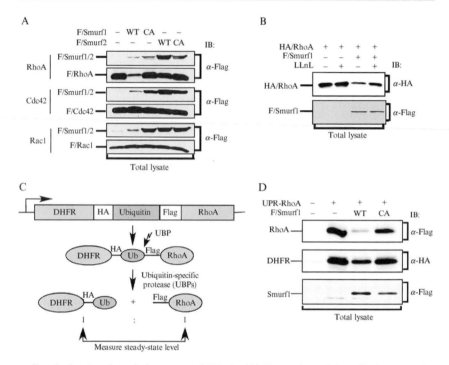

FIG. 2. *In vivo* degradation assay of RhoA. (A) Expression of Smurf1 decreases the steady-state level of RhoA. HEK293T cells were transiently transfected with the indicated combinations of wild-type (WT) or catalytically inactive (CA) Flag-tagged Smurf1 (F/Smurf1) or Flag-tagged Smurf2 (F/Smurf2) and Flag-tagged RhoA (F/RhoA), Cdc42 (F/Cdc42), or Rac1 (F/Rac1). Steady-state protein levels were determined by immunoblotting (IB) total cell lysates with anti-Flag. (B) Proteasome inhibitors decrease RhoA down-regulation by Smurf1. HEK293T cells transfected with HA-tagged RhoA (HA/RhoA) and F/Smurf1 were treated for 4 h with or without 40 μM LLnL, and RhoA steady-state levels determined. (C) A schematic of the UPR-RhoA construct and the UPR assay to produce equimolar amounts of two specific proteins. In this procedure, a reference protein (mouse DHFR) is linked to an ubiquitin (Ub) moiety by means of a 20-residue spacer that also contains the HA epitope. The C-terminus of Ub is in turn linked to Flag-tagged RhoA. This tripartite fusion is cleaved cotranslationally by ubiquitin-specific processing proteases (UBPs) *in vivo* at the Ub-F/RhoA junction to yield equimolar amounts of the F/RhoA and the reference protein DHFR bearing a C-terminal Ub moiety. (D) HEK293T cells were transfected with UPR-RhoA and wild-type (WT) or catalytically inactive (CA) Flag-tagged Smurf1 (F/Smurf1) as indicated. Steady-state protein levels were determined by immunoblotting total cell lysates using anti-HA and anti-Flag antibodies as indicated (Wang *et al.*, 2003).

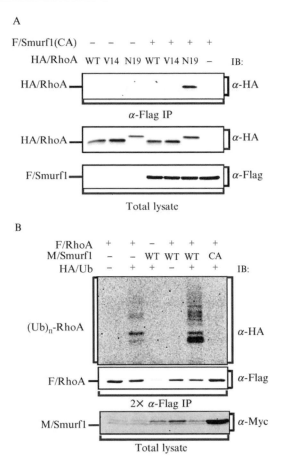

FIG. 3. *In vivo* interaction and ubiquitination assay of RhoA. (A) Interaction of Smurf1 with RhoA. HEK293T cells were transfected with F/Smurf1(CA) and various versions of HA/RhoA (WT, V14, or N19). Cell lysates were subjected to anti-Flag immunoprecipitation (IP) followed by immunoblotting with rat anti-HA antibody to detect associated RhoA. Total protein expression was confirmed by immunoblotting as shown. (B) Ubiquitination of RhoA in HEK293T cells. After overnight treatment with 20 μM LLnL, lysates from cells transfected with HA-tagged ubiquitin (HA/Ub), F/RhoA or Myc-tagged Smurf1 (M/Smurf1) (WT or CA) was subjected to anti-Flag immunoprecipitation, eluted by boiling in 1% SDS, and then reprecipitated with anti-Flag (2× IP). Ubiquitin-conjugated ((Ub)n-RhoA) and free RhoA were detected by immunoblotting with the appropriate antibodies. M/Smurf1 expression was confirmed by anti-Myc immunoblotting samples of total lysates (Wang *et al.*, 2003).

supernatant is transferred to a new tube and diluted 10 times in TNTE buffer. Another 15 μl anti-Flag M2 agarose beads are added to the diluted supernatant and incubated with rocking overnight at 4°. The beads are boiled in 20 μl 2× SDS loading buffer for 5 min. The samples are run on 11% SDS-PAGE gel and transferred to nitrocellulose membrane (Bio-Rad). The ubiquitin-conjugated RhoA or free RhoA is detected by immunoblotting with rat monoclonal anti-HA antibody (Roche) or mouse monoclonal anti-Flag M2 antibody (Sigma). As shown in Fig. 3B, in the absence of Smurf1, some ubiquitin-conjugated RhoA can be detected and coexpression of Smurf1 causes a significant increase in ubiquitinated RhoA. In contrast, expression of Smurf1 (C699A) prevents both the basal and Smurf1-dependent ubiquitination of RhoA, suggesting that basal ubiquitination of RhoA may be due to endogenous Smurf1.

Acknowledgments

This work was supported by grants from the National Cancer Institute of Canada and the Canadian Institutes of Health Research to J. L. W. Y. Z. is supported by a CIHR Postdoctoral Fellowship. B. O. is supported by a CIHR Doctoral Studentship Awards. R. B. is supported by a CIHR MD-PhD Studentship. J. L. W. is a CIHR Senior Investigator and an International Scholar of the Howard Hughes Medical Institute.

References

Bar-Sagi, D., and Hall, A. (2000). Ras and Rho GTPases: A family reunion. *Cell* **103,** 227–238.
Bishop, A. L., and Hall, A. (2000). Rho GTPases and their effector proteins. *Biochem. J.* **348** (Pt 2), 241–255.
Bonni, S., Wang, H. R., Causing, C. G., Kavsak, P., Stroschein, S. L., Luo, K., and Wrana, J. L. (2001). TGF-beta induces assembly of a Smad2-Smurf2 ubiquitin ligase complex that targets SnoN for degradation. *Nat. Cell Biol.* **3,** 587–595.
Ebisawa, T., Fukuchi, M., Murakami, G., Chiba, T., Tanaka, K., Imamura, T., and Miyazono, K. (2001). Smurf1 interacts with transforming growth factor-beta type I receptor through Smad7 and induces receptor degradation. *J. Biol. Chem.* **276,** 12477–12480.
Etienne-Manneville, S., and Hall, A. (2002). Rho GTPases in cell biology. *Nature* **420,** 629–635.
Hall, A., and Nobes, C. D. (2000). Rho GTPases: Molecular switches that control the organization and dynamics of the actin cytoskeleton. *Philos. Trans. R. Soc. Lond. B. Biol. Sci.* **355,** 965–970.
Harvey, K. F., and Kumar, S. (1999). Nedd4-like proteins: An emerging family of ubiquitin-protein ligases implicated in diverse cellular functions. *Trends Cell Biol.* **9,** 166–169.
Hershko, A., and Ciechanover, A. (1998). The ubiquitin system. *Annu. Rev. Biochem.* **67,** 425–479.
Kavsak, P., Rasmussen, R. K., Causing, C. G., Bonni, S., Zhu, H., Thomsen, G. H., and Wrana, J. L. (2000). Smad7 binds to Smurf2 to form an E3 ubiquitin ligase that targets the TGF beta receptor for degradation. *Mol. Cell* **6,** 1365–1375.

Levy, F., Johnsson, N., Rumenapf, T., and Varshavsky, A. (1996). Using ubiquitin to follow the metabolic fate of a protein. *Proc. Natl. Acad. Sci. USA* **93**, 4907–4912.

Lin, X., Liang, M., and Feng, X. H. (2000). Smurf2 is a ubiquitin E3 ligase mediating proteasome-dependent degradation of Smad2 in transforming growth factor-beta signaling. *J. Biol. Chem.* **275**, 36818–36822.

Schmidt, A., and Hall, A. (2002). Guanine nucleotide exchange factors for Rho GTPases: Turning on the switch. *Genes Dev.* **16**, 1587–1609.

Self, A. J., and Hall, A. (1995). Purification of recombinant Rho/Rac/G25K from *Escherichia coli. Methods Enzymol.* **256**, 3–10.

Van Aelst, L., and Symons, M. (2002). Role of Rho family GTPases in epithelial morphogenesis. *Genes Dev.* **16**, 1032–1054.

Wang, H. R., Zhang, Y., Ozdamar, B., Ogunjimi, A. A., Alexandrova, E., Thomsen, G. H., and Wrana, J. L. (2003). Regulation of cell polarity and protrusion formation by targeting RhoA for degradation. *Science* **302**, 1775–1779.

Zhang, B., Zhang, Y., Wang, Z., and Zheng, Y. (2000). The role of Mg2+ cofactor in the guanine nucleotide exchange and GTP hydrolysis reactions of Rho family GTP-binding proteins. *J. Biol. Chem.* **275**, 25299–252307.

Zhang, Y., Chang, C., Gehling, D. J., Hemmati-Brivanlou, A., and Derynck, R. (2001). Regulation of Smad degradation and activity by Smurf2, an E3 ubiquitin ligase. *Proc. Natl. Acad. Sci. USA* **98**, 974–979.

Zhu, H., Kavsak, P., Abdollah, S., Wrana, J. L., and Thomsen, G. H. (1999). A SMAD ubiquitin ligase targets the BMP pathway and affects embryonic pattern formation. *Nature* **400**, 687–693.

[33] Ubiquitin-Mediated Proteasomal Degradation of Rho Proteins by the CNF1 Toxin

By ANNE DOYE, LAURENT BOYER, AMEL METTOUCHI, and EMMANUEL LEMICHEZ

Abstract

The CNF1 toxin is produced by uropathogenic and meningitis-causing *Escherichia coli*. CNF1 catalyzes the constitutive activation of Rho proteins by deamidation. The threshold of activation of Rho proteins by CNF1 is, however, attenuated because of a concomitant decrease of their cellular levels. Depletion of activated-Rac1 is catalyzed by ubiquitin-mediated proteasomal degradation. Consequently, we show by effector-binding pull-down that co-treatment of intoxicated cells with the MG132 proteasome-inhibitor results in a higher level of activation of Rac, as well as RhoA and Cdc42. We show that CNF1 induces the transient recruitment of Rho proteins to cellular membranes. Interestingly, at the difference of Rac and Cdc42, the inhibition of the proteasome during CNF1 treatment does not result in a significant accumulation of RhoA to cellular

METHODS IN ENZYMOLOGY, VOL. 406
0076-6879/06 $35.00
DOI: 10.1016/S0076-6879(06)06033-2

membranes. Using an *in vivo* ubiquitylation assay, we evidence that mutation of the geranylgeranyl acceptor cysteine of Rac1 (Rac1$_{C189G}$) abolished the sensitivity of permanently activated-Rac1 to ubiquitylation, whereas Rac1$_{C189G}$ remained able to bind to the effector-binding domain of p21-PAK. Collectively, these results indicate that association with the cellular membranes is a necessary step for activated-Rac1 ubiquitylation.

Introduction

Uropathogenic strains of *Escherichia coli* (UPECs) account for 70–95% of urinary tract infections (UTIs), a pathology resulting in $1.6 billion in medical expenditures in the United States each year (Foxman, 2002). The cytotoxic necrotizing factor-1 (CNF1) (Falbo *et al.*, 1993) is a classical protein toxin found in 30% of UPECs (Landraud *et al.*, 2004). CNF1 is also associated to pathogenic K1 *E. coli* responsible for meningitis in childhood (Xie *et al.*, 2004). CNF1 binds to host cells through the 67-kDa laminin receptor (Kim *et al.*, 2005). The receptor-toxin complex is internalized into endocytic vesicles up to acidic endosomes (Landraud *et al.*, 2004). Acidic conditions found in endosomes are sufficient to trigger the penetration of CNF1 catalytic domain across endosomal membranes to the host cell cytosol (Landraud *et al.*, 2004). On reaching the host cell cytosol, CNF1 enzymatic domain catalyzes the posttranslational modification of RhoA, Rac1, and Cdc42. CNF1 toxin produces a counterintuitive molecular mechanism of action on Rho GTPases. CNF1 catalyzes the deamidation into a glutamic acid of the glutamine 63 for RhoA and the equivalent glutamine 61 for Rac1 and Cdc42 (Flatau *et al.*, 1997; Lerm *et al.*, 1999; Schmidt *et al.*, 1997). The posttranslational modification of the glutamine 63 of RhoA impairs its GTPase activity, thus conferring RhoA permanent activation properties. Nevertheless, activation of Rac by CNF1 is rapidly counteracted into cells by its sensitization to ubiquitin-mediated proteasomal degradation (Doye *et al.*, 2002; Lerm *et al.*, 2002). As a result of this balance between the activation (toxin-induced deamidation) and the subsequent deactivation (cell-induced degradation), the threshold of activation of Rho proteins by the toxin is lowered (Doye *et al.*, 2002). This dual mechanism of action confers a high efficiency of UPEC's internalization into host cells and the induction of epithelial intercellular junction dynamic, as well as epithelial cell motility (Doye *et al.*, 2002). Activation of Rho proteins triggers a genetic program of production of inflammatory mediators aimed at leukocyte recruitment and activation by intoxicated cells (Munro *et al.*, 2004). The ubiquitin-mediated proteasomal degradation of Rac leads to a reduction of Rac activation, and, consequently, it lowers the threshold of inflammatory mediator production (Munro *et al.*, 2004).

CNF1 Toxin Purification and Activity Normalization

For CNF1 toxin purification, the *E. coli* OneShot strain carrying the pCRIIcnf1 plasmid was grown overnight at 37° in LB medium. Bacteria were lysed in PBS using a French press. Bacterial lysate was centrifuged at 10,000g, and the supernatant was precipitated with an equal volume of saturated ammonium sulfate for 5–8 h at 4°. The precipitate was dialyzed against TN buffer (25 m*M* Tris-HCl, pH 7.4, 50 m*M* NaCl), overnight at 4° and applied on a DEAE fast flow column (11 × 1.5 cm; Pharmacia Biotech) equilibrated with TN buffer. The column was washed 200 min at 1 ml/min flow rate with TN buffer. CNF1 was eluted at 200 m*M* NaCl (gradient of 50–300 m*M* NaCl, 100 min). The peak of eluted protein was determined by measure of the absorbance at 280 nm. The fractions containing CNF1 were pooled, dialyzed against TN buffer overnight, and applied on a Superdex 75 column at 0.3 ml/min (30 × 1 cm, Pharmacia Biotech). The fractions containing CNF1 were pooled, concentrated on a 30-kDa cutoff vivaspin (Vivascience) and applied on a monoQ column (5 × 0.5 cm, Pharmacia Biotech). The column was washed 30 min at 1 ml/min with TN buffer and CNF1 was eluted at 350 m*M* NaCl (gradient of 50–400 m*M* NaCl, 20 min). The level of CNF1 purification is assessed by SDS-PAGE followed by Coomassie brilliant blue staining. CNF1 appears as a single band migrating at 110 kDa (Fig. 1A). The activity of the different batches of CNF1 has to be normalized. This can be achieved by a measure of the dose-dependent multinucleation effect induced by CNF1 on human epithelial larynx cells (HEp-2: CCL-23, ATCC). Cells were seeded at 5×10^3 cells/well in 24-well plates the day before intoxication. They were intoxicated 48 h with

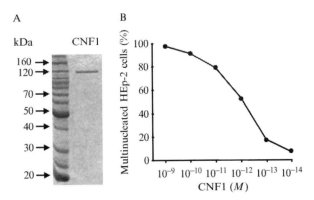

FIG. 1. CNF1 toxin. (A) 1 μg of purified-CNF1 toxin resolved on a 10% SDS-PAGE together with the BenchMark protein ladder (Invitrogen). Proteins were stained with Coomassie brilliant blue R-250. (B) Dose-dependent multinucleation effect induced by CNF1 on HEp-2 cells.

different dilutions of toxin. After intoxication, cells were washed once in PBS and stained with Giemsa. Multinucleated cells were enumerated, and the results were expressed as percentages of the total cell population. The CNF1 toxin purified in the condition described previously gives a multinucleation effect of 50% at 10^{-12} M (Fig. 1B). This assay allows normalization of the CNF1 toxin activity and thus provides reproducibility in kinetics and amplitudes of Rho activation and cellular depletion.

Human Umbilical Vein Endothelial Cell Culture and Transfection

Human umbilical vein endothelial cells (HUVECs) were obtained from PromoCell (Heidelberg, Germany). Cells were grown in human endothelial SFM medium (Invitrogen Co., Scotland), supplemented with defined growth factors (10 ng/ml EGF and 20 ng/ml bFGF, Invitrogen), 1 μg/ml heparin (Sigma-Aldrich), penicillin and streptomycin with 20% (v/v) fetal bovine serum (FBS) (Invitrogen). Cells were grown on 0.2% (w/v) gelatin-coated dishes (Sigma-Aldrich). For ubiquitylation assays, cells were transfected by electroporation. This was performed with 30 μg total DNA for 10^7 cells (in this study: 15 μg of a 6xHis-ubiquitin expression plasmid and 15 μg of Rho expression plasmids) in 4-mm electroporation cuvettes at 300 V, 450 μF, one pulse (Easyject Plus, Equibio) in a volume of 300 μl SFM. Cells were replated in complete SFM, and the medium was replaced 3 h after transfection. Plasmids used in this study pKH3-HA$_3$-Rac1, Rac1$_{Q61L}$, RhoA, RhoA$_{Q63L}$, Cdc42, Cdc42$_{Q61L}$, and pRBG4-His$_6$-ubiquitin expression plasmid (pCW7) were kindly provided by Dr. D. Manor and Dr. R. R. Kopito, respectively. HA-Rac1 double mutants were obtained using the QuickChange Site Directed Mutagenesis Kit (Stratagene Europe) on pXJ-HA-Rac1$_{Q61L}$ (Doye et al., 2002) as follows: TGC to GGC for HA-Rac1$_{Q61L,C189G}$, TAT to TGT for HA-$_{Q61L,Y40C}$, ACC to GCC for HA-Rac1$_{Q61L,T35A}$, and TTT to GCT for HA-Rac1$_{Q61L,F37A}$. The absence of other mutations was verified by sequence analysis.

CNF1 Induces a Transient Activation and Recruitment to Cellular Membranes of Rho Proteins

Recruitment of Rho Proteins to Cellular Membranes

The in vivo activation of Rho GTPases by CNF1 can be assessed either directly by effector-binding pull-down assays or indirectly by measure of Rho GTPases association with cellular membranes, as described hereafter. HUVECs were plated a day before the experiment at the density of 2×10^7 cells/150-mm dish in complete-SFM. One dish was used for each condition. Cell culture medium was replaced by complete-SFM containing 10^{-9} M CNF1 $\pm$ 10 μM MG132 (Biomol), 3 or 6 h before fractionation. After intoxication, cell dishes were chilled on ice and rinsed twice with cold PBS. Further steps are performed at 4°. Plates were scraped in 5 ml of cold PBS, centrifuged 5 min at 1000 rpm, and pellets were homogenized in 0.25 ml cold BSI buffer (3 mM imidazole, pH 7.4, 250 mM sucrose) supplemented extemporaneously with 1 mM phenylmethylsulfonyl fluoride (PMSF). The mixture was transferred into a 1.5-ml tube, and cells were lysed by passing 40 times through a 1-ml syringe equipped with a 25G $\times$ 5/8″-needle (U-100 Insulin, Terumo). Cell lysis can be followed under microscope. Nuclei were removed by centrifugation 10 min at 10,000g at 4°. Protein concentrations of the postnuclear supernatants (PNS) were normalized. PNS were centrifuged 1 h at 100,000g at 4°. Supernatants (cytosolic fractions) were transferred in a new 1.5-ml tube, and pellets (membrane fractions) were homogenized in an equal volume of BSI. Pellets and supernatants were mixed with Laemmli blue buffer and boiled 5 min at 100°. Samples, about 30 μl, were resolved on 12% SDS-PAGE and transferred onto PVDF transfer membrane (Hybond-P, Amersham Biosciences). Immunoblots were performed using monoclonal antibodies incubated overnight at 4° at a dilution 1:250 for anti-RhoA (clone 26C4; Santa Cruz) and anti-Cdc42 (clone 44; Transduction Laboratories), or 1 h at room temperature at a dilution of 1:1000 for anti-Rac1 (clone 102; Transduction Laboratories), and 1:3000 for anti-RhoGDI polyclonal antibodies (A-20; Santa Cruz).

Measure of Rho Protein Activation by Effector-Binding Pull-Down

Methods of activated-Rho pull-down, with either GST-PAK$^{70\text{--}106}$ (for Rac and Cdc42) or GST-Rhotekin RBD (for Rho) were previously established (Manser *et al.*, 1998; Ren *et al.*, 1999). HUVECs, 1×10^7/assay, were intoxicated with 10^{-9} M CNF1 $\pm$ 10 μM MG132. Cells were lysed at 4° in 1 ml of either Rho lysis buffer (50 mM Tris-HCl, pH 7.5, 0.5 M NaCl, 10 mM MgCl$_2$, 1% Triton-X100, 0.5% Na-Deoxycholate, 0.1% SDS) or Rac/Cdc42 lysis buffer (25 mM Tris-HCl, pH 7.5, 0.15 M NaCl, 5 mM MgCl$_2$, 0.5% Triton-X100, 4% glycerol, 10 mM NaF) supplemented extemporaneously with 1 mM PMSF, 2 mM Na$_3$VO$_4$, 2 mM DTT, 20 mM beta-glycerophosphate. Lysates were centrifuged 10 min at 10,000g at 4°. An aliquot of 50 μl was collected (Total Rho protein input) and mixed with Laemmli blue buffer. Lysates were incubated with 30 μg of either GST-PAK$^{70\text{--}106}$ or GST-Rhotekin RBD bound to glutathione-agarose beads (Sigma-Aldrich) 45 min at 4° on a rotating shaker. Beads were

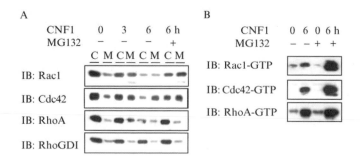

FIG. 2. CNF1-induced transient activation and recruitment of Rho proteins to cell membranes. (A) Immunoblots showing the distribution of Rho proteins between membrane (M) and cytosolic (C) fractions at different periods of HUVECs intoxication by 10^{-9} M CNF1 toxin (representative of two independent experiments). (B) Immunoblots showing the levels of activated-Rho proteins in HUVECs intoxicated with 10^{-9} M CNF1 toxin $\pm$ MG132 proteasome inhibitor (representative of two independent experiments).

washed twice with 1 ml of Rho washing buffer (50 mM Tris-HCl, pH 7.5, 0.15 M NaCl, 10 mM MgCl$_2$, 1% Triton-X100) or Rac lysis buffer. Beads were mixed with 30 μl of Laemmli blue buffer, and proteins were resolved on a 12% SDS-PAGE for western blot analysis.

Comments

These experiments show that CNF1 induces a transient recruitment of Rho proteins to cellular membranes after 3 h of HUVECs intoxication (Fig. 2A). These results are in accordance with the findings of Munro (2004) showing by effector-binding pull-down experiments that CNF1 induced a transient activation of all three Rho proteins in these cells. The addition of the proteasome inhibitor MG132 during the intoxication of cells by CNF1 results in an increase of the levels of permanently activated-Rho, Rac, and Cdc42 (Fig. 2B). Interestingly, at the difference of Rac and Cdc42, RhoA did not accumulate significantly in cellular membranes on addition of MG132 during CNF1 intoxication (Fig. 2A), whereas the level of activated-RhoA increased (Fig. 2B). This suggests that the transient characteristic of the recruitment of RhoA to cellular membranes observed at 3 h of CNF1-intoxication may involve other cellular regulations in addition to RhoA ubiquitin-mediated proteasomal degradation.

Measure of the *In Vivo* Ubiquitylation Sensitivity of Rho Proteins

In Vivo *Ubiquitylation of Rho Proteins*

Sensitivity of activated-Rac1 mutants to ubiquitylation is a function of their strength of activation, independently of the type or position mutated

(Doye *et al.*, 2002). Sensitization of wild-type Rac1 to ubiquitylation can also be triggered by overexpression of the GEF domain of Dbl (Doye *et al.*, 2002). Here we have compared the ubiquitylation sensitivity of permanently activated-Rac1, RhoA and Cdc42 with that of their wild-type counterpart. This question was assessed *in vivo* using a His-tagged ubiquitin pull-down assay. HUVECs, 2×10^7, except for RhoA$_{Q63L}$, 6×10^7, were electroporated 6 h before lysis (refer to "Human Umbilical Vein endothelial Cell Culture and Transfection"). Cells were washed once in PBS and scraped at room temperature in 1 ml of BU buffer (20 mM Tris-HCl, pH 7.5, 200 mM NaCl, 10 mM imidazole, 0.1% (v/v) Triton X-100, 8 M urea). Samples were homogenized using vortex and centrifuged 10 min at 10,000g at room temperature. An aliquot of 50 μl was collected (Total Rho protein input) and mixed with Laemmli blue buffer. In parallel, cobalt beads (Talon, Clontech) were prepared as follows. A volume of 60-μl slurry beads/assay was centrifuged 2 min at 1000 rpm at room temperature. Beads were washed twice in 1 ml of BU and saturated 1 h in 1 ml of BU supplemented with 5.0 μg/ml bovine serum albumin (BSA RIA grade, Sigma-Aldrich) on a rotating shaker. Beads were washed twice in 1 ml of BU, and the final volume was adjusted with BU to 0.1 ml/assay. Volumes of 0.1 ml of beads were added to each 0.9 ml assay supernatants and incubated at room temperature 1 h on a rotating shaker. Beads were washed five times in BU, then resuspended in Laemmli blue buffer and boiled 5 min. A first immunoblot using anti-HA at a dilution 1:3000 (clone 11; BabCo) was performed to quantify the amount of total HA-Rho protein input of the different assays. Usually, variations of HA-Rho protein levels expressed in cells did not exceed fourfold differences. In these conditions, levels of ubiquitylated-Rho protein detected after His$_6$-ubiquitin pull-down were linear, with regard either to levels of Rho protein expressed in cells or to the level of Rho proteins engaged in the pull-down. Volumes of purified His-ubiquitylated proteins and total lysates were normalized with BU buffer according to total HA-Rho protein inputs and were next resolved on 12% SDS-PAGE. Immunoblotting anti-HA was performed to detect ubiquitylated HA-tagged proteins and to verify the levels of total HA-Rho proteins (Fig. 3A).

Comments

In these experimental conditions, permanently-activated Q(61)L mutants of Rac1 and Cdc42 appeared more ubiquitylated than their wild-type counterpart, 7.2 and 4.3-fold, respectively (Fig. 3A). This indicates that, similarly to Rac1, the permanent activation of Cdc42 sensitizes it to ubiquitylation (Fig. 3A). In contrast, this assay did not allow us to observe significant differences in the sensitivity of RhoA$_{Q63L}$ to ubiquitylation

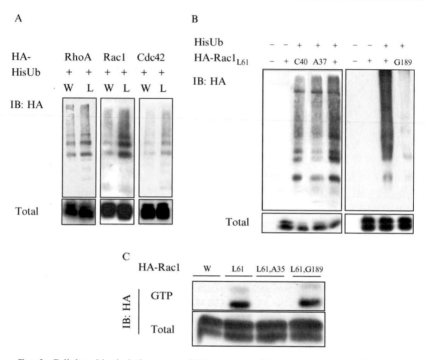

FIG. 3. Cellular ubiquitylation assay of Rho proteins. (A) Immunoblots anti-HA showing the HA-Rho-Ub(n) ubiquitylation profiles, as well as the sensitivity of permanently activated Rho proteins (noted L and corresponding to $RhoA_{Q63L}$, $Rac1_{Q61L}$ and $Cdc42_{Q61L}$) compared with their wild-type counterpart (noted W) (representative of three independent experiments). (B) Immunoblots anti-HA showing the HA-Rac1-Ub(n) ubiquitylation profile, as well as the sensitivity of Rac_{Q61L} compared with that of $Rac_{Q61L,Y40C}$, $Rac1_{Q61L,F37A}$, $Rac_{Q61L,C189G}$ (representative of three independent experiments). (C) Immunoblots anti-HA showing activated Rac proteins associated to GST-PAK[70-106]. This result shows that $Rac1_{Q61L,C189G}$ remains able to bind to the CRIB-domain of p21-PAK at the difference with $Rac1_{Q61L,T35A}$ (representative of two independent experiments).

compared with its wild-type counterpart. This difference might be due to the threefold lower level of HUVECs transfection by $RhoA_{Q63L}$ compared with that of wild-type RhoA (see section "*In Vivo* Ubiquitylation of Rho Proteins"). On the other hand, the difference obtained with RhoA compared with Rac1 and Cdc42 might be linked to the observation that addition of MG132 during HUVECs intoxication by CNF1 does not produce an accumulation of RhoA to cellular membranes, whereas the level of activated RhoA increases (Fig. 2A and B). Collectively, these results reinforce the idea that the transient characteristic of the activation of RhoA by CNF1 might involve several molecular events.

The *in vivo* ubiquitylation assay described here can be used to study the effect of other types of mutations on permanently activated Rac1. Most notably the mutation of the prenyl-acceptor cysteine-189 of Rac1 completely abolished permanently activated-Rac1 sensitivity to ubiquitylation (Fig. 3B), whereas $Rac1_{Q61L,C189G}$ remains able to bind efficiently to the CRIB domain of p21-PAK (Fig. 3C). That Rac is ubiquitylated on membrane association is in accordance with the fact that its sensitivity to ubiquitylation is a function of the strength of its activation and that addition of MG132 during CNF1 treatment increases both the levels of activated Rac and membrane-associated Rac. This is also consistent with the observation that the deletion of the CAAX-box of Rac1 impairs its proteasomal degradation produced by CNF1, as previously described (Pop *et al.*, 2004). Both mutations Y40C and F37A in the switch-I effector-binding domain of Rac1 partially inhibit permanently activated $Rac1_{Q61L}$ ubiquitylation, by 25 and 30%, respectively (Fig. 3B). This result suggests that Rac1 may contact specifically the ubiquitylation machinery outside its switch-I effector-binding domain. Consistently, other determinants outside the switch-I domain corresponding to the carboxyterminal polybasic region of Rac1, as well as amino acids N107, A151, F90, and K147, have previously been shown to cooperate collectively to the turnover of Rac1 (Pop *et al.*, 2004).

References

Doye, A., Mettouchi, A., Bossis, G., Clement, R., Buisson-Touati, C., Flatau, G., Gagnoux, L., Piechaczyk, M., Boquet, P., and Lemichez, E. (2002). CNF1 exploits the ubiquitin-proteasome machinery to restrict Rho GTPase activation for bacterial host cell invasion. *Cell* **111**, 553–564.

Falbo, V., Pace, T., Picci, L., Pizzi, E., and Caprioli, A. (1993). Isolation and nucleotide sequence of the gene encoding cytotoxic necrotizing factor 1 of *Escherichia coli*. *Infect. Immun.* **61**, 4909–4914.

Flatau, G., Lemichez, E., Gauthier, M., Chardin, P., Paris, S., Fiorentini, C., and Boquet, P. (1997). Toxin-induced activation of the G protein p21 Rho by deamidation of glutamine. *Nature* **387**, 729–733.

Foxman, B. (2002). Epidemiology of urinary tract infections: Incidence, morbidity, and economic costs. *Am. J. Med.* **113**(Suppl. 1A), 5S–13S.

Kim, K. J., Chung, J. W., and Kim, K. S. (2005). 67-kDa laminin receptor promotes internalization of cytotoxic necrotizing factor 1-expressing *Escherichia coli* K1 into human brain microvascular endothelial cells. *J. Biol. Chem.* **280**, 1360–1368.

Landraud, L., Pulcini, C., Gounon, P., Flatau, G., Boquet, P., and Lemichez, E. (2004). E. coli CNF1 toxin: A two-in-one system for host-cell invasion. *Int. J. Med. Microbiol.* **293**, 513–518.

Lerm, M., Pop, M., Fritz, G., Aktories, K., and Schmidt, G. (2002). Proteasomal degradation of cytotoxic necrotizing factor 1-activated rac. *Infect. Immun.* **70**, 4053–4058.

Lerm, M., Selzer, J., Hoffmeyer, A., Rapp, U. R., Aktories, K., and Schmidt, G. (1999). Deamidation of Cdc42 and Rac by *Escherichia coli* cytotoxic necrotizing factor 1: Activation of c-Jun N-terminal kinase in HeLa cells. *Infect. Immun.* **67**, 496–503.

Manser, E., Loo, T. H., Koh, C. G., Zhao, Z. S., Chen, X. Q., Tan, L., Tan, I., Leung, T., and Lim, L. (1998). PAK kinases are directly coupled to the PIX family of nucleotide exchange factors. *Mol. Cell* **1**, 183–192.

Munro, P., Flatau, G., Doye, A., Boyer, L., Oregioni, O., Mege, J. L., Landraud, L., and Lemichez, E. (2004). Activation and proteasomal degradation of rho GTPases by cytotoxic necrotizing factor-1 elicit a controlled inflammatory response. *J. Biol. Chem.* **279**, 35849–35857.

Pop, M., Aktories, K., and Schmidt, G. (2004). Isotype-specific degradation of Rac activated by the cytotoxic necrotizing factor 1. *J. Biol. Chem.* **279**, 35840–35848.

Ren, X. D., Kiosses, W. B., and Schwartz, M. A. (1999). Regulation of the small GTP-binding protein Rho by cell adhesion and the cytoskeleton. *EMBO J.* **18**, 578–585.

Schmidt, G., Sehr, P., Wilm, M., Selzer, J., Mann, M., and Aktories, K. (1997). Gln 63 of Rho is deamidated by *Escherichia coli* cytotoxic necrotizing factor-1. *Nature* **387**, 725–729.

Xie, Y., Kim, K. J., and Kim, K. S. (2004). Current concepts on *Escherichia coli* K1 translocation of the blood-brain barrier. *FEMS Immunol. Med. Microbiol.* **42**, 271–279.

[34]   Regulation of Superoxide-Producing NADPH Oxidases in Nonphagocytic Cells

By Ryu Takeya, Noriko Ueno, and Hideki Sumimoto

Abstract

The membrane-integrated protein gp91^{phox} functions as the catalytic center of the superoxide-producing phagocyte NADPH oxidase. Recent studies have identified homologs of gp91^{phox} in nonphagocytic cells, which constitute the NADPH oxidase (Nox) family. Activation of the Nox oxidases leads to production of reactive oxygen species (ROS), thereby participating in a variety of biological events, such as host defense, hormone biosynthesis, and signal transduction. The activity of the Nox enzymes is regulated by various proteins, including the small GTPase Rac; regulatory mechanisms differ dependent on the type of the Nox proteins. For example, an oxidase activator (p47^{phox} or Noxo1) and an oxidase activator (p67^{phox} or Noxa1) are absolutely required for superoxide production by gp91^{phox} and Nox1, but not by Nox3. Rac, albeit probably dispensable to the Nox3 activity, plays an essential role in activation of gp91^{phox}. Thus, functional reconstitution of Nox systems is crucial for the study of Nox regulation. Here we describe a basic method for the reconstitution of Nox systems by expression of oxidase proteins in transfectable cells.

METHODS IN ENZYMOLOGY, VOL. 406
0076-6879/06 $35.00
DOI: 10.1016/S0076-6879(06)06034-4

Introduction: The Phagocyte NADPH Oxidase

It is widely accepted that reactive oxygen species (ROS) participate in a variety of biological events, including host defense, hormone biosynthesis, and signal transduction. ROS are not only generated as by-products in aerobic metabolism but also produced by specialized enzymes such as NADPH oxidases. Among them, the NADPH oxidase in phagocytes (e.g., neutrophils and macrophages) has been studied for several decades, which helps our understanding of the mechanisms of ROS production and their functions (Cross and Segal, 2004; Nauseef, 2004; Quinn and Gauss, 2004). The phagocyte NADPH oxidase (phox), dormant in resting cells, becomes activated during phagocytosis of invading microbes to produce superoxide, which is converted to microbicidal ROS. The importance of this enzyme in host defense is evident from the fact that recurrent and life-threatening infections occur in patients with chronic granulomatous disease (CGD) whose phagocytes lack the superoxide-producing activity.

The catalytic center of the phagocyte oxidase is membrane-spanning flavocytochrome b_{558} composed of the two subunits $p22^{phox}$ and $gp91^{phox}$ (also termed Nox2), the latter of which contains the complete electron-transporting apparatus from NADPH to molecular oxygen for superoxide production (see Fig. 1A). Successful electron transfer in $gp91^{phox}$ (i.e., activation of $gp91^{phox}$) requires stimulus-induced membrane translocation of three proteins present in the cytoplasm of resting cells: the small GTPase Rac and the two specialized adaptor proteins $p67^{phox}$ and $p47^{phox}$, each harboring two SH3 domains (Cross and Segal, 2004; Nauseef, 2004; Quinn and Gauss, 2004). Indeed, the phagocyte NADPH oxidase activity can be reconstituted in a cell-free system with cytochrome b_{558}, $p47^{phox}$, $p67^{phox}$, and Rac (Abo and Segal, 1995; Sarfstein et al., 2004). In the activation process, $p47^{phox}$ directly binds to $p22^{phox}$ via the SH3 domains and recruits $p67^{phox}$ to the oxidase complex. On the other hand, Rac is targeted to the membrane in a manner independent of $p47^{phox}$ and $p67^{phox}$. At the membrane, Rac in the GTP-bound state interacts with $p67^{phox}$ via binding to the N-terminal tetratricopeptide repeat (TPR) domain, leading to superoxide production (Koga et al., 1999; Lapouge et al., 2000). The binding to the $p67^{phox}$ TPR domain is specific to Rac; Cdc42 neither binds to $p67^{phox}$ nor activates $gp91^{phox}$. Another adaptor protein, $p40^{phox}$, which tightly associates with $p67^{phox}$ in phagocytes, positively regulates activation of $gp91^{phox}$ via facilitating membrane translocation of $p47^{phox}$ and $p67^{phox}$, although $p40^{phox}$ is dispensable to the oxidase activation (Kuribayashi et al., 2002).

Recent expansion of information available in genome databases has enabled us to identify several novel homologs of $gp91^{phox}$ (Bokoch and

A   B

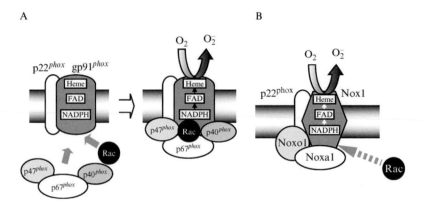

FIG. 1. (A) Activation of gp91^{phox}/Nox2 complexed with p22^{phox} requires stimulus–induced membrane translocation of p47^{phox}, p67^{phox}, and the small GTPase Rac. p40^{phox} enhances gp91^{phox} activation via facilitating membrane translocation of p47^{phox} and p67^{phox}. For details, see text. (B) Activation of Nox1, likely complexed with p22^{phox}, requires both Nox01 and Noxoa1. The small GTPase Rac also seems to participate in Nox1 activation. For details, see text.

Knaus, 2003; Geiszt and Leto, 2004; Lambeth, 2004; Takeya and Sumimoto, 2003) that constitute the NADPH oxidase (Nox) family.

Regulation of Nox1, a Nonphagocytic NADPH Oxidase

Nox1, the first identified homolog of gp91^{phox}, is abundantly expressed in colon epithelial cells and at lower levels in various types of cells including vascular smooth muscle cells (Bánfi et al., 2000; Suh et al., 1999). This nonphagocytic oxidase is considered to be involved in host defence at the colon (Geiszt et al., 2003a; Kawahara et al., 2004) and proposed to participate in signal transduction leading to hypertrophy and angiogenesis (Arnold et al., 2001; Suh et al., 1999).

A regulatory mechanism for the Nox1 activity had remained obscure until the discovery of Nox01 (Nox organizer 1) and Noxa1 (Nox activator 1)—respective homologs of p47^{phox} and p67^{phox} (Bánfi et al., 2003; Cheng and Lambeth, 2004; Geiszt et al., 2003b; Takeya et al., 2003). Nox1 likely forms a complex with p22^{phox} as does gp91^{phox}/Nox2 (Ambasta et al., 2004; Takeya et al., 2003), and superoxide production by Nox1 absolutely requires Nox01 and Noxa1, which can be replaced by p47^{phox} and p67^{phox}, respectively, but to a lesser extent (Takeya et al., 2003) (Fig. 1B). In the presence of Nox01 and Noxa1, Nox1 produces superoxide without cell stimulants such as phorbol 12-myristate 13-acetate (PMA), a potent in vivo activator of the phagocyte NADPH oxidase, albeit PMA treatment of cells enhances the superoxide production (Bánfi et al., 2003; Cheng and Lambeth, 2004; Geiszt et al., 2003b; Takeya et al., 2003). The constitutive activity of Nox1 seems to be at least partially because, in contrast to p47^{phox}, Noxo1 can bind to p22^{phox} even in the resting state (Cheng and Lambeth, 2004; Takeya et al., 2003). It seems likely that Rac participates in the Nox1 activation: Nox1-dependent superoxide production in LPS-stimulated gastric mucosal cells is enhanced by activation of Rac1, but not by that of Cdc42 (Kawahara et al., 2005); GTP-bound Rac, but not GTP-bound Cdc42, binds to the N-terminal TPR domain of the oxidase activator Noxa1 (Takeya et al., 2003).

Other Nonphagocytic NADPH Oxidases

Other nonphagocytic NADPH oxidases include three Nox enzymes (Nox3–5) and two dual oxidases (Duox1 and Duox2) that contain two oxidase modules each: an N-terminal extracellular peroxidase-like domain and a C-terminal gp91^{phox}-homologous oxidase portion.

Nox3, initially identified as an oxidase expressed in the human fetal kidney, exhibits the closest similarity to gp91^{phox} among the Nox-family oxidases (Cheng et al., 2001; Kikuchi et al., 2000). This oxidase is present in the inner ear of mouse and plays a crucial role in formation of otoconia, tiny mineralized structures, which is required for perception of balance and gravity (Paffenholz et al., 2004). When Nox3 is ectopically expressed in cells, it forms a functional complex with p22^{phox} and seems to be constitutively active even without coexpression of an oxidase organizer (p47^{phox} or Noxo1) or an oxidase activator (p67^{phox} or Noxa1) (Ueno et al., 2005). Although the organizers and activators can regulate superoxide production by Nox3 (Bánfi et al., 2004; Cheng et al., 2004; Ueno et al., 2005), Rac does not seem to be involved (Ueno et al., 2005).

Nox4 is highly expressed in the adult and fetal kidney and produces a small but significant amount of superoxide in a constitutive manner (Geiszt et al., 2000; Shiose et al., 2001; Yang et al., 2001). It is also known that Nox4, as well as gp91^{phox} and Nox1, is present in the cardiovasculature (Griendling, 2004). Although Nox4 also seems to be complexed with p22^{phox} (Ambasta et al., 2004), regulatory mechanism for the Nox4 activity is presently unclear at the molecular level.

Nox5, abundant in the testis and spleen, has an N-terminal cytoplasmic extension with four Ca^{2+}-binding EF hand motifs (Bánfi et al., 2001). As can be expected from this structure, cells ectopically expressing Nox5 produce high amounts of superoxide in response to the Ca^{2+}-ionophore ionomycin (Bánfi et al., 2001). The Nox5 activity does not seem to require an oxidase organizer or an oxidase activator.

Duox1 and Duox2 are highly expressed in the thyroid gland (De Deken *et al.*, 2000; Dupuy *et al.*, 1999). Duox2 is essential for thyroid hormone synthesis; mutations in Duox2 lead to congenital hypothyroidism, even in heterozygotes (Moreno *et al.*, 2002). Rac is likely dispensable to thyroid oxidase activity (Fortemaison *et al.*, 2005). Heterologous expression of Duox enzymes in several mammalian cells fails to reconstitute ROS production; unknown oxidase components may be required for Duox activity.

Reconstitution of NADPH Oxidases in Cultured Cells

Because the activities of NADPH oxidases are dependent on the presence of various proteins, as described previously, functional reconstitution of the enzyme systems is essential for investigation of their regulatory mechanism. The phagocyte NADPH oxidase can be reconstituted in a cell-free system using the recombinant cytosolic proteins $p67^{phox}$, $p47^{phox}$, and Rac together with purified cytochrome b_{558} or with neutrophil membranes rich in the cytochrome (Abo and Segal, 1995; Sarfstein *et al.*, 2004). The cell-free assay has delineated the basic properties of the phagocyte oxidase, such as the absolute requirement of Rac for enzymatic activity. There are, however, some differences in the oxidase activation between the *in vitro* system and intact cells (e.g., the cell-free activation does not require $p47^{phox}$ phosphorylation, which is indispensable at a cellular level). For *in vivo* reconstitution of the phagocyte oxidase, whole-cell systems using cell lines of hematopoietic lineage have developed at early stage, which include the undifferentiated multipotent leukemic cell line K562, expressing endogenous $p22^{phox}$ and Rac, cotransfected with cDNAs for $gp91^{phox}$, $p47^{phox}$, and $p67^{phox}$ (Ago *et al.*, 1999; de Mendez and Leto, 1995; Koga *et al.*, 1999). Although expression of the transgenes has provided important information, these cell lines suffer variously from limitations, especially on the ease of transfection and level of superoxide production. On the other hand, monkey kidney COS-7 cells are easily transfectable, and transgenic expression of phagocyte oxidase proteins in these cells confers high-level production of superoxide (Biberstine-Kinkade *et al.*, 2002; Price *et al.*, 2002).

The phagocyte oxidase has been successfully reconstituted in various nonhematopoietic cell lines such as COS-7, HeLa, HEK293, and CHO cells (Cheng *et al.*, 2004; Price *et al.*, 2002; Takeya *et al.*, 2003). The cell types determine proteins that should be ectopically expressed for the reconstitution of the oxidases. HeLa and HEK293 cells are likely suitable for investigation of the role of Rac in NADPH oxidase activation; superoxide production by $gp91^{phox}$ in these cells is largely dependent on ectopic expression of an active form of Rac, whereas the expression is not required for that in COS-7 and CHO cells (Cheng and Lambeth, 2004; Takeya *et al.*, 2003; Ueno *et al.*, 2005). As shown in Fig. 2A, when HeLa cells were transfected with the cDNAs for gp91^{phox}, p47^{phox}, and p67^{phox} and simultaneously with the cDNA encoding a constitutively active form of Rac1, Rac1(Q61L), they produced superoxide in response to PMA; the superoxide production was not observed in cells without Rac1(Q61L). Even in the presence of Rac1(Q61L), the phagocyte oxidase was not activated by coexpression of a mutant p67^{phox} carrying the R102E substitution in the TPR domain, defective in binding to Rac (Koga *et al.*, 1999) (Fig. 2A). Thus, Rac in the GTP-bound form seems to activate gp91^{phox} by interacting with p67^{phox}. This effect is probably specific for Rac, because Cdc42 (Q61L) was much less active than the corresponding form of Rac (Fig. 2B). The requirement of ectopic expression of p22^{phox} for the oxidase reconstitution is also dependent on the cell types used. Since the endogenous message for p22^{phox} is present in a variety of cells including COS-7, HeLa, and HEK-293 cells, transfection of these cells with the p22^{phox} cDNA is not

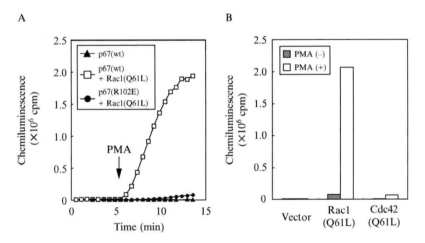

FIG. 2. Role of the small GTPase Rac in activation of the phagocyte NADPH oxidase reconstituted in HeLa cells. (A) Approximately 80% confluent HeLa cells in a 60-mm dish were transfected with 1 μg each of pcDNA3.0–gp91^{phox}, pEF-BOS–p47^{phox}, and pEF-BOS–p67^{phox} (wt) or pEF-BOS–p67^{phox} (R102E) and simultaneously with or without 1 μg of pEF-BOS–Rac1(Q61L) using 12 μl of LipofectAMINE. The transfected cells (1 × 10^6 cells) were incubated for 5 min at 37° and then stimulated with PMA (200 ng/ml). Chemiluminescence change was continuously monitored with DIOGENES. (B) HeLa cells were transfected with pcDNA3.0–gp91^{phox}, pEF-BOS–p47^{phox}, and pEF-BOS–p67^{phox} and simultaneously with pEF-BOS–Rac1(Q61L), pEF-BOS–Cdc42(Q61L), or the pEF-BOS vector. Superoxide production was assayed by chemiluminescence using DIOGENES in the presence or absence of PMA (200 ng/ml).

essential for gp91^{phox} -dependent superoxide production. In the reconstitution in CHO cells, which contain only a small amount of the p22^{phox} mRNA, the gp91^{phox} activity absolutely requires ectopic expression of p22^{phox} (Takeya *et al.*, 2003; Ueno *et al.*, 2005).

The nonphagocytic oxidases Nox1 and Nox3 are also reconstituted in COS-7, HeLa, HEK293, and CHO cells (Bánfi *et al.*, 2003, 2004; Cheng *et al.*, 2004; Takeya *et al.*, 2003; Ueno *et al.*, 2005). As shown in Fig. 3A, when Nox1 was expressed together with Nox1, Noxa1, and p22^{phox} in CHO cells, this oxidase produced a substantial amount of superoxide even in the absence of stimulants added, whereas the superoxide production was increased by treatment of the cells with PMA. This is in contrast with activation of gp91^{phox} that is totally dependent on the addition of PMA to cells (Fig. 2A). The superoxide production by Nox1 is inhibited by diphenylene iodonium (DPI), an inhibitor of Nox family oxidases (Fig. 3B). The Nox1 activity absolutely requires both Noxo1 and Noxa1 and partially depends on ectopic expression of p22^{phox} (Fig. 3). On the other hand,

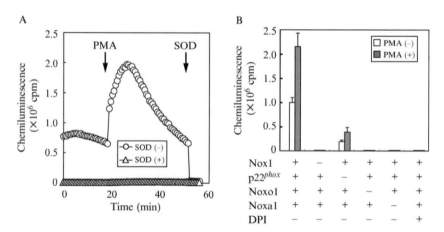

FIG. 3. Reconstitution of Nox1 activity in CHO cells. (A) Approximately 80% confluent CHO cells in a 60-mm dish were transfected simultaneously with 1 μg each of pcDNA3.0–Nox1, pEF-BOS–Noxo1, pEF-BOS–Noxa1, and pEF-BOS–p22^{phox} using 12 μl of FuGENE6. The transfected cells (1 × 10^6 cells) were incubated for 20 min at 37° and then stimulated with PMA (200 ng/ml). Alternatively, cells were treated with SOD (50 μg/ml) before measurement of chemiluminescence. Chemiluminescence change was continuously monitored with DIOGENES, and SOD (50 μg/ml) was added as indicated. (B) CHO cells were transfected with the indicated combination of pcDNA3.0–Nox1, pEF-BOS–Noxo1, pEF-BOS–Noxa1, and pEF-BOS–p22^{phox} using FuGENE6 in the presence or absence of 5 μM DPI. Superoxide production was assayed by chemiluminescence using DIOGENES in the presence or absence of PMA. Each graph represents the mean ± SD of the peak chemiluminescence values obtained from three independent transfections.

Nox3 as well as gp91^{phox} does not produce superoxide in CHO cells untransfected with the p22^{phox} cDNA (Ueno et al., 2005). Thus, Nox1 may bind to p22^{phox} with a higher affinity or might be complexed with a heretofore unidentified homolog of p22^{phox}.

Methods

Transfection of Cultured Cells

Plasmid DNAs for transfection of eukaryotic cells must be clean and free from phenol and sodium chloride as usual: we usually isolate the DNAs using the column purification kit (QIAGEN Plasmid Maxi Kit). The cationic liposome–mediated transfection method (lipofection) is generally used for reconstitution of Nox enzymes in cultured cells. Optimal transfection reagents seem to vary with the cell types used. We usually transfect COS-7 and HeLa cells using LipofectAMINE (Invitrogen), whereas we use FuGENE6 Transfection Reagent (Roche Molecular Biochemicals) for transfection of CHO cells. In our preliminary experiments using HEK293 and COS-7 cells, the use of LipofectAMINE confers a higher oxidase activity than that of FuGENE6. On the other hand, in CHO cells, the use of FuGENE6 results in higher superoxide production than that of LipofectAMINE2000. For transfection of K562 cells, we perform an electroporation method (Ago et al., 1999; Koga et al., 1999).

Procedure

1. Culture CHO or HeLa cells in Ham's F12 medium or Dulbecco's modified Eagle's medium (DMEM), respectively, supplemented with 10% FCS.
2. The day before transfection, trypsinize and count cells, plating them at 1–3×10^5 cells per 60-mm dish so that they are 70–90% confluent on the day of transfection.
3. Transfect CHO cells with the cDNAs using FuGENE6: the mixture of 12 μl of FuGENE6 reagent and 1–8 μg of mixed plasmids DNA is used for a 60-mm dish. In the case of HeLa cells, use LipofectAMINE: the mixture of 10 μl of LipofectAMINE reagent and 1–8 μg of plasmid DNAs is used for 60-mm dish.
4. Incubate the cells at 37° in a CO$_2$ incubator for 18–24 h.

Measurement of Superoxide Production

Cytochrome c reduction assay and chemiluminescence method are used for measuring superoxide production in a whole-cell system using cultured cells. Although cytochrome c reduction is a well-accepted technique for quantification of superoxide, the activity of NADPH oxidases expressed in nonphagocytic cells is often below the detection limit of this assay; the chemiluminescence method is usually used for the detection of superoxide generated in a whole-cell system.

Several substances are available that interact with superoxide and thus emit a photon, which can be detected in a luminometer. The use of a cellular luminescence enhancement system for superoxide detection (DIOGENES; National Diagnostics) enables us to detect a small amount of superoxide specifically and reproducibly. Whereas luminol (5-amino-2,3-dihydro-1,4-phthalazinedione) reacts with hydrogen peroxide (H_2O_2), sensitivity to superoxide in the DIOGENES system is at least 10^6 greater than that to hydrogen peroxide, according to the manufacturers' specifications. Indeed, chemiluminescence signals in a whole-cell system for phagocyte oxidase activation are completely abolished by superoxide dismutase (SOD) (de Mendez et al., 1995). Alternately, we use DIOGENES-luminol solution (the modified DIOGENES solution; see later): dilution of the DIOGENES solution by luminol does not alter specificity and sensitivity to superoxide (unpublished observation). Inhibitors are also useful for confirmation of the origin for chemiluminescence signals: diphenylene iodonium (DPI), a selective but not specific inhibitor of NADPH oxidases, is usually used at the range of 1–20 μM. Chemiluminescence changes can be continuously recorded with a luminometer such as Auto Lumat LB953 (EG&G Berthold). The use of a multiwell plate reader such as Wallac 1420 ARVOsx (PerkinElmer Life Sciences) allows measurement of superoxide in attached cultured cells.

Materials

HEPES-BUFFERED SALINE (HBS): 120 mM NaCl, 5 mM KCl, 5 mM glucose, 1 mM MgCl$_2$, 0.5 mM CaCl$_2$, and 17 mM HEPES, pH 7.4.

DIOGENES SOLUTION (THE ORIGINAL DIOGENES SOLUTION): Dissolve the DIOGENES Reagent and the DIOGENES Activator (National Diagnostics; CL-202) in 10 ml deionized water. The solution can be stored at 4° for up to 60 days.

DIOGENES-LUMINOL SOLUTION (THE MODIFIED DIOGENES SOLUTION). Dissolve the DIOGENES Reagent and the DIOGENES Activator in 12 ml PBS containing 25 mM luminol. Divide the solution into aliquots and freeze at $-20°$.

SUPEROXIDE DISMUTASE (SOD). Dissolve in deionized water to 1 mg/ml and store at $-20°$.

PHORBOL 12-MYRISTATE 13-ACETATE (PMA). Dissolve in dimethyl sulfoxide (DMSO) to 1 mg/ml for stock solution. Divide the solution into aliquots and freeze at $-20°$. On the occasion of use, dilute the stock solution into 5% DMSO in PBS to 10 μg/ml.

DIPHENYLENE IODONIUM (DPI). Dissolve in DMSO to 10 mM and store at $-20°$.

Procedure

1. After cultured for 18–24 h, harvest adherent cells by incubating with trypsin/EDTA for 1 min at $37°$, and washed with HBS.

2. Centrifuge the cells for 3 min at 1500 rpm and washs with HBS.

3. Count the cell numbers and resuspend in HBS containing 0.03% BSA to 1 $\times$ 10^6 cells/ml.

4. Add 1 ml of the cell suspension to polystyrene round tubes containing 100 μl of the original DIOGENES solution or 40 μl of the DIOGENES-luminol solution, and set the tubes into the luminometer.

5. Monitor chemiluminescence continuously at $37°$.

6. After 5–10 min of incubation, add 200 μl of PMA (10 μg/ml).

7. Add 50 μl of SOD (1 mg/ml) to confirm the generation of superoxide.

Acknowledgments

We are grateful to Yohko Kage (Kyushu University and JST), Miki Matsuo (Kyushu University), and Natsuko Yoshiura (Kyushu University) for technical assistance, and Minako Nishino (Kyushu University and JST) for secretarial assistance. This work was supported in part by Grants-in-Aid for Scientific Research and National Project on Protein Structural and Functional Analyses from the Ministry of Education, Culture, Sports, Science and Technology of Japan, and CREST and BIRD projects of JST (Japan Science and Technology Agency).

References

Abo, A., and Segal, A. W. (1995). Reconstitution of cell-free NADPH oxidase activity by purified components. *Methods Enzymol.* **256,** 268–278.

Ago, T., Nunoi, H., Ito, T., and Sumimoto, H. (1999). Mechanism for phosphorylation-induced activation of the phagocyte NADPH oxidase protein p47^{phox}: Triple replacement of serines 303, 304, and 328 with aspartates disrupts the SH3 domain-mediated intramolecular interaction in p47^{phox}, thereby activating the oxidase. *J. Biol. Chem.* **274,** 33644–33653.

Ambasta, R. K., Kumar, P., Griendling, K. K., Schmidt, H. H., Busse, R., and Brandes, R. P. (2004). Direct interaction of the novel Nox proteins with p22phox is required for the formation of a functionally active NADPH oxidase. *J. Biol. Chem.* **279,** 45935–45941.

Arnold, R. S., Shi, J., Murad, E., Whalen, A. M., Sun, C. Q., Polavarapu, R., Parthasarathy, S., Petros, J. A., and Lambeth, J. D. (2001). Hydrogen peroxide mediates the cell growth and transformation caused by the mitogenic oxidase Nox1. *Proc. Natl. Acad. Sci. USA* **98,** 5550–5555.

Bánfi, B., Maturana, A., Jaconi, S., Arnaudeau, S., Laforge, T., Sinha, B., Ligeti, E., Demaurex, N., and Krause, K. H. (2000). A mammalian H^+ channel generated through alternative splicing of the NADPH oxidase homolog *NOH-1. Science* **287,** 138–142.

Bánfi, B., Molnar, G., Maturana, A., Steger, K., Hegedus, B., Demaurex, N., and Krause, K. H. (2001). A Ca^{2+}-activated NADPH oxidase in testis, spleen, and lymph nodes. *J. Biol. Chem.* **276,** 37594–37601.

Bánfi, B., Clark, R. A., Steger, K., and Krause, K. H. (2003). Two novel proteins activate superoxide generation by the NADPH oxidase NOX1. *J. Biol. Chem.* **278,** 3510–3513.

Bánfi, B., Malgrange, B., Knisz, J., Steger, K., Dubois-Dauphin, M., and Krause, K. H. (2004). NOX3, a superoxide-generating NADPH oxidase of the inner ear. *J. Biol. Chem.* **279,** 46065–46072.

Biberstine-Kinkade, K. J., Yu, L., Stull, N., Le Roy, B., Bennett, S., Cross, A., and Dinauer, M. C. (2002). Mutagenesis of p22^{phox} histidine 94: A histidine in this position is not required for flavocytochrome b_{558} function. *J. Biol. Chem.* **277,** 30368–30374.

Bokoch, G. M., and Knaus, U. G. (2003). NADPH oxidases: Not just for leukocytes anymore! *Trends Biochem. Sci.* **28,** 502–508.

Cheng, G., Cao, Z., Xu, X., van Meir, E. G., and Lambeth, J. D. (2001). Homologs of gp91^{phox}: Cloning and tissue expression of Nox3, Nox4, and Nox5. *Gene* **269,** 131–140.

Cheng, G., and Lambeth, J. D. (2004). NOXO1, regulation of lipid binding, localization, and activation of Nox1 by the Phox homology (PX) domain. *J. Biol. Chem.* **279,** 4737–4742.

Cheng, G., Ritsick, D., and Lambeth, J. D. (2004). Nox3 regulation by NOXO1, p47^{phox}, and p67^{phox}. *J. Biol. Chem.* **279,** 34250–34255.

Cross, A. R., and Segal, A. W. (2004). The NADPH oxidase of professional phagocytes—prototype of the NOX electron transport chain systems. *Biochim. Biophys. Acta* **1657,** 1–22.

De Deken, X., Wang, D., Many, M. C., Costagliola, S., Libert, F., Vassart, G., Dumont, J. E., and Miot, F. (2000). Cloning of two human thyroid cDNAs encoding new members of the NADPH oxidase family. *J. Biol. Chem.* **275,** 23227–23233.

de Mendez, I., and Leto, T. L. (1995). Functional reconstitution of the phagocyte NADPH oxidase by transfection of its multiple components in a heterologous system. *Blood* **85,** 1104–1110.

Dupuy, C., Ohayon, R., Valent, A., Noel-Hudson, M. S., Deme, D., and Virion, A. (1999). Purification of a novel flavoprotein involved in the thyroid NADPH oxidase. Cloning of the porcine and human cDNAs. *J. Biol. Chem.* **274,** 37265–37269.

Fortemaison, N., Miot, F., Dumont, J. E., and Dremier, S. (2005). Regulation of H_2O_2 generation in thyroid cells does not involve Rac1 activation. *Eur. J. Endocrinol.* **152,** 127–133.

Geiszt, M., Kopp, J. B., Varnai, P., and Leto, T. L. (2000). Identification of renox, an NAD(P)H oxidase in kidney. *Proc. Natl. Acad. Sci. USA* **97,** 8010–8014.

Geiszt, M., Lekstrom, K., Brenner, S., Hewitt, S. M., Dana, R., Malech, H. L., and Leto, T. L. (2003a). NAD(P)H oxidase 1, a product of differentiated colon epithelial cells, can partially replace glycoprotein 91^{phox} in the regulated production of superoxide by phagocytes. *J. Immunol.* **171,** 299–306.

Geiszt, M., Lekstrom, K., Witta, J., and Leto, T. L. (2003b). Proteins homologous to p47^{phox} and p67^{phox} support superoxide production by NAD(P)H oxidase 1 in colon epithelial cells. *J. Biol. Chem.* **278,** 20006–20012.

Geiszt, M., and Leto, T. L. (2004). The Nox family of NAD(P)H oxidases: Host defense and beyond. *J. Biol. Chem.* **279,** 51715–51718.

Griendling, K. K. (2004). Novel NAD(P)H oxidases in the cardiovascular system. *Heart* **90,** 491–493.

Kawahara, T., Kuwano, Y., Teshima-Kondo, S., Takeya, R., Sumimoto, H., Kishi, K., Tsunawaki, S., Hirayama, T., and Rokutan, K. (2004). Role of nicotinamide adenine dinucleotide phosphate oxidase 1 in oxidative burst response to Toll-like receptor 5 signaling in large intestinal epithelial cells. *J. Immunol.* **172,** 3051–3058.

Kawahara, T., Kohjima, M., Kuwano, Y., Mino, H., Teshima-Kondo, S., Takeya, R., Tsunawaki, S., Wada, A., Sumimoto, H., and Rokutan, K. (2005). Helicobacter pylori lipopolysaccharide activates Rac1 and transcription of NADPH oxidase Nox1 and its organizer NOXO1 in guinea pig gastric mucosal cells. *Am. J. Physiol. Cell. Physiol.* **288,** C450–C457.

Kikuchi, H., Hikage, M., Miyashita, H., and Fukumoto, M. (2000). NADPH oxidase subunit, gp91^{phox} homologue, preferentially expressed in human colon epithelial cells. *Gene* **254,** 237–243.

Kuribayashi, F., Nunoi, H., Wakamatsu, K., Tsunawaki, S., Sato, K., Ito, T., and Sumimoto, H. (2002). The adaptor protein p40^{phox} as a positive regulator of the superoxide-producing phagocyte oxidase. *EMBO J.* **21,** 6312–6320.

Koga, H., Terasawa, H., Nunoi, H., Takeshige, K., Inagaki, F., and Sumimoto, H. (1999). Tetratricopeptide repeat (TPR) motifs of p67^{phox} participate in interaction with the small GTPase Rac and activation of the phagocyte NADPH oxidase. *J. Biol. Chem.* **274,** 25051–25060.

Lambeth, J. D. (2004). NOX enzymes and the biology of reactive oxygen. *Nat. Rev. Immunol.* **4,** 181–189.

Lapouge, K., Smith, S. J., Walker, P. A., Gamblin, S. J., Smerdon, S. J., and Rittinger, K. (2000). Structure of the TPR domain of p67^{phox} in complex with Rac.GTP. *Mol. Cell* **6,** 899–907.

Moreno, J. C., Bikker, H., Kempers, M. J., van Trotsenburg, A. S., Baas, F., de Vijlder, J.J, Vulsma, T., and Ris-Stalpers, C. (2002). Inactivating mutations in the gene for thyroid oxidase 2 (*THOX2*) and congenital hypothyroidism. *N. Engl. J. Med.* **347,** 95–102.

Nauseef, W. M. (2004). Assembly of the phagocyte NADPH oxidase. *Histochem. Cell Biol.* **122,** 277–291.

Paffenholz, R., Bergstrom, R. A., Pasutto, F., Wabnitz, P., Munroe, R. J., Jagla, W., Heinzmann, U., Marquardt, A., Bareiss, A., Laufs, J., Russ, A., Stumm, G., Schimenti, J. C., and Bergstrom, D. E. (2004). Vestibular defects in head-tilt mice result from mutations in Nox3, encoding an NADPH oxidase. *Genes Dev.* **18,** 486–491.

Price, M. O., McPhail, L. C., Lambeth, J. D., Han, C. H., Knaus, U. G., and Dinauer, M. C. (2002). Creation of a genetic system for analysis of the phagocyte respiratory burst: High-level reconstitution of the NADPH oxidase in a nonhematopoietic system. *Blood* **99,** 2653–2661.

Quinn, M. T., and Gauss, K. A. (2004). Structure and regulation of the neutrophil respiratory burst oxidase: Comparison with nonphagocyte oxidases. *J. Leukoc. Biol.* **76,** 760–781.

Sarfstein, R., Gorzalczany, Y., Mizrahi, A., Berdichevsky, Y., Molshanski-Mor, S., Weinbaum, C., Hirshberg, M., Dagher, M. C., and Pick, E. (2004). Dual role of Rac in the assembly of NADPH oxidase, tethering to the membrane and activation of p67^{phox}: A study based on mutagenesis of p67^{phox} -Rac1 chimeras. *J. Biol. Chem.* **279,** 16007–16016.

Shiose, A., Kuroda, J., Tsuruya, K., Hirai, M., Hirakata, H., Naito, S., Hattori, M., Sakaki, Y., and Sumimoto, H. (2001). A novel superoxide-producing NAD(P)H oxidase in kidney. *J. Biol. Chem.* **276,** 1417–1423.

Suh, Y. A., Arnold, R. S., Lassegue, B., Shi, J., Xu, X., Sorescu, D., Chung, A. B., Griendling, K. K., and Lambeth, J. D. (1999). Cell transformation by the superoxide-generating oxidase Mox1. *Nature* **401,** 79–82.

Takeya, R., and Sumimoto, H. (2003). Molecular mechanism for activation of superoxide-producing NADPH oxidases. *Mol. Cells* **16,** 271–277.

Takeya, R., Ueno, N., Kami, K., Taura, M., Kohjima, M., Izaki, T., Nunoi, H., and Sumimoto, H. (2003). Novel human homologues of p47^{phox} and p67^{phox} participate in activation of superoxide-producing NADPH oxidases. *J. Biol. Chem.* **278,** 25234–25246.

Ueno, N., Takeya, R., Miyano, K., Kikuchi, H., and Sumimoto, H. (2005). The NADPH oxidase Nox3 constitutively produces superoxide in a p22^{phox}-dependent manner: Its regulation by oxidase organizers and activators. *J. Biol. Chem.* **280,** 23328–23339.

Yang, S., Madyastha, P., Bingel, S., Ries, W., and Key, L. (2001). A new superoxide-generating oxidase in murine osteoclasts. *J. Biol. Chem.* **276,** 5452–5458.

[35] Activation of MEKK1 by Rho GTPases

By ZHUI CHEN and MELANIE H. COBB

Abstract

Mammalian MAP/ERK kinase kinase 1 (MEKK1) is MAP kinase kinase kinase (MAP3K) that is a crucial regulator of many cellular signaling cascades. One of the most important physiological functions of MEKK1 is its ability to regulate cell migration, because MEKK1 null mice are defective in eyelid closure. MEKK1 exhibits its signaling activity through interaction with a large array of cellular factors, including several proteins that are known to play central roles in controlling cell movement and motility. We have recently identified an interaction between MEKK1 and RhoA. This interaction occurs between the GTP-bound, active form of RhoA and the amino terminal region of MEKK1 that harbors a PHD domain with E3 ubiquitin ligase activity. RhoA-GTP activates MEKK1 *in vitro* and in cells. Here we describe in detail the assay methods for RhoA activation of MEKK1, including preparation of recombinant proteins and proteins immunoprecipitated from cells, pretreatment of proteins, and assay conditions. We also briefly explain the methods and conditions we use to identify the interaction between MEKK1 and RhoA in yeast and in mammalian cells.

METHODS IN ENZYMOLOGY, VOL. 406
Copyright 2006, Elsevier Inc. All rights reserved.

0076-6879/06 $35.00
DOI: 10.1016/S0076-6879(06)06035-6

Introduction

Mammalian MAP/ERK kinase kinase 1 (MEKK1) is a MAP3K that plays central roles in controlling a diverse array of cellular signaling cascades in response to cellular stresses. MEKK1 was first discovered as a homolog of the yeast MAP3 kinase Ste11p (Lange-Carter *et al.*, 1993). Since then, it has been implicated in signal transduction pathways activated by numerous stimuli, including UV light, cold shock, osmotic stress, lysophosphatidic acid (LPA), cytokines, chemokines, and a variety of microtubule-interfering agents (Hagemann and Blank, 2001; Uhlik *et al.*, 2004; Yujiri *et al.*, 2000). Overexpression of MEKK1 enhances the activities of multiple MAP kinases (MAPKs), including the c-Jun N-terminal Kinase (JNK), ERK1/2, and to a lesser extent p38-MAPK (Lin *et al.*, 1995; Xu *et al.*, 1996; Yujiri *et al.*, 1998). In addition, MEKK1 is also involved in the regulation of transcriptional regulators such as AP-1 and NFκB (Cuevas *et al.*, 2003; Lee *et al.*, 1997, 1998). The most prominent physiological role of MEKK1 seems to be its function in cell migration. Gene disruption experiments in mice revealed that MEKK1-deficient mice were defective in eyelid closure resulting from impaired epithelial cell migration over the surface of the cornea (Xia *et al.*, 2000; Yujiri *et al.*, 2000). Embryonic stem cells and embryonic fibroblast cells generated from the mice indeed migrated significantly more slowly than wild-type cells.

MEKK1 is a large protein that can interact with numerous signaling proteins. Several binding partners have been identified ranging from components of MAPK pathways, including JNK, ERK2, MEK1, MEK7, Raf-1, the Nck-interacting kinase (NIK), and the germinal center kinase (GCK) to other proteins such as Ras, 14-3-3, Tax, and α-actinin (Gallagher *et al.*, 2002; Uhlik *et al.*, 2004). Thus, MEKK1 seems to function as a scaffold for signaling events, much like the yeast scaffold protein Ste5p (Choi *et al.*, 1994; Marcus *et al.*, 1994; Printen and Sprague, 1994). These interacting proteins bind to MEKK1 at different locations spanning almost the entire MEKK1 sequence, including the kinase domain that lies at its C terminus and the PHD domain near its N terminus. The PHD domain is an E3 ubiquitin ligase and can promote degradation of ERK2 (Lu *et al.*, 2002). Therefore, the specific interactions with individual partners are very likely to affect different aspects of the biological function of MEKK1.

Several of these defined interactors of MEKK1 may be key to its role in cell migration. For instance, MEKK1 can interact with all three major classes of the Rho-family of small GTPases, Rho, Rac, and Cdc42, which are central components in governing cytoskeleton reorganization and subsequent cell migration and motility (Bar-Sagi and Hall, 2000; Fanger *et al.*, 1997; Gallagher *et al.*, 2004). Overexpression of MEKK1 stimulates the formation of lamellipodia, a process regulated by Rac1 (Fanger *et al.*, 1997). MEKK1 has been shown to localize to actin stress fibers, focal adhesions, microtubules, and intermediate filaments (Christerson *et al.*, 1999) and bind directly to cytoskeletal proteins including α-actinin, vimentin, and focal adhesion kinase (FAK). In ovarian cancer cells, it has been reported that MEKK1 is activated by LPA in a Ras-dependent fashion and leads to the redistribution of FAK to focal contact regions, which are essential for cell migration (Bian *et al.*, 2004). Consistent with a major action on multiple cytoskeletal elements, several microtubule-interfering agents robustly activate MEKK1 (Gibson *et al.*, 1999; Yujiri *et al.*, 2000).

We have recently identified an interaction between MEKK1 and RhoA (Gallagher *et al.*, 2004). A negative regulator of RhoA, p115Rho-GAP, has also been shown to bind to MEKK1 (Christerson *et al.*, 2002). Through yeast two-hybrid and coimmunoprecipitation experiments we have confirmed that RhoA binding occurs on the N terminus of MEKK1 near its PHD domain. Rac and Cdc42 do not bind to this region. The interaction is abolished with mutation of essential cysteine residues in this region and may somehow be related to the ubiquitin ligase activity of MEKK1. RhoA-GTP, the active form of RhoA, dramatically stimulates the kinase activity of full-length MEKK1 but was not ubiquitinated by MEKK1. Such activation could be crucial, because a RhoA-MEKK1 pathway has been suggested to mediate activin-induced migration of keratinocytes (Zhang *et al.*, 2005).

In Vitro MEKK1 Kinase Assay with RhoA-GTP

Purification of MEKK1 from Sf9 Cells

The full-length MEKK1 sequence was first isolated from a rat cDNA library and encodes 1493 amino acids (Xu *et al.*, 1996). Recombinant HA-tagged rat MEKK1 is expressed as a 195-kDa protein in mammalian cells, the same size as the largest immunoreactive form of endogenous MEKK1 (Xu *et al.*, 1996).

To test the effect of RhoA on MEKK1 kinase activity in *in vitro* kinase assays, full-length MEKK1 is required, because the kinase domain resides at the C terminus of MEKK1, whereas the Rho-binding PHD domain exists at the N terminus of the protein. We have previously purified different fragments of MEKK1 proteins from bacteria that possess kinase activity when they contain the C-terminal kinase domain. However, the yield of tagged protein is usually low, although the purified proteins are very active. Because protein kinases as large as MEKK1 usually cannot be purified with high yield and activity from bacterial systems, we elected to use the insect *Spodoptera frugiperda* (Sf9) cell system to produce stable and full-length MEKK1 that is robustly active in *in vitro* kinase assays against a variety of substrate proteins.

Materials and Conditions

Recombinant baculovirus encoding His_6-MEKK1
Cell line: Sf9 insect cells
Medium: Grace's insect cell culture medium (Invitrogen)
Multiplicity of infection (m.o.i.): 2.5–5
PBS: phosphate-buffered saline, pH 7.4
Resin: Ni^{2+}-NTA agarose
Lysis buffer: 10 mM HEPES, pH 7.6, 1.5 mM $MgCl_2$, 10 mM NaCl, 1 mM EDTA, and 1 mM EGTA
Washing buffer: lysis buffer plus containing 50 mM sodium phosphate at pH 8.5
Elution buffer: lysis buffer without or with 0.5 M imidazole.

The MEKK1 cDNA sequence was amplified from the pCEP4-HA-MEKK1 plasmid and ligated into the baculovirus transfer vector pVL1393 (Invitrogen, Carlsbad, CA). Recombinant baculoviruses are generated by contransfection with linear wild-type baculoviral DNA into Sf9 cells. Production of MEKK1 proteins from Sf9 cells is carried out as previously described using an Invitrogen baculovirus expression system (Huang *et al.*, 2000). Sf9 cells are grown in suspension in Grace's insect cell culture medium with 5% fetal bovine serum until the cells reach ~1 million/ml. The cell culture is then infected with MEKK1 baculovirus (m.o.i., 2.5–5.0). The infected cells are harvested by centrifugation 48 h after infection. Cell pellets are washed three times with ice-cold PBS, pelleted again, and flash frozen in liquid nitrogen and stored at −80°. This freezing step increases the efficiency of lysis of the cells.

Purification of His_6-MEKK1 is carried out by resuspending the cell pellets in 4 ml of lysis buffer. Cells are further incubated on ice for 30 min and lysed using 30 or more strokes in a Dounce homogenizer. Cell debris is removed by centrifugation at 40,000g for 30 min at 4°. The supernatant is applied twice to a column containing ~1.5 ml Ni^{2+}-NTA agarose resin. The column is thoroughly washed with 30 ml of washing buffer and eluted with 40 ml of elution buffer (lysis buffer with a 0–0.5 M imidazole gradient). Protein is eluted in 1-ml fractions. The protein composition of each fraction is analyzed by SDS-PAGE, which is generally adequate to identify the 195-kDa MEKK1 band. Immunoblotting and protein kinase assays are optional. Fractions with MEKK1 are combined, aliquoted, and stored at −80° for future use.

Dephosphorylation of MEKK1

Recombinant MEKK1 purified from Sf9 cells typically has high basal activity caused by phosphorylation on several residues in Sf9 cells. The high basal activity dramatically reduces the stimulation that might be detected in kinase assays when MEKK1 is treated with other factors. Incubation of MEKK1 with phosphoprotein phosphatases 2A (PP2A) dephosphorylates the protein, significantly reducing the basal activity of MEKK1 generated from insect cells (Gallagher et al., 2004). Thus, purified MEKK1 is first incubated with ~0.5 μg of PP2A for 30–60 min before testing the effects of small GTPases.

Purification and GTP-Loading of RhoA

Like many other small GTPases, RhoA undergoes a series of posttranslational modification steps including prenylation. The prenyl group is essential for the correct intracellular distribution and functions of RhoA, as well as its appropriate interaction with other proteins. To produce prenylated RhoA proteins, RhoA is coexpressed with GST-tagged guanine nucleotide dissociation inhibitor in Sf9 cells as previously described (Wells et al., 2002). Cell pellets are flash frozen and thawed three times in standard lysis buffer containing protease inhibitors. Cell debris is removed by ultracentrifugation at 100,000g for 30 min at 4°. The resulting supernatants are applied to a glutathione-sepharose column and eluted with the same buffer containing 1% cholate to separate RhoA from GST-guanine nucleotide dissociation inhibitor. Eluted factions are concentrated, and cholate is removed by dilution followed by a final concentration by pressure filtration through an Amicon PM10 membrane (Wells et al., 2002).

RhoA cycles between a GTP-bound, active state and a GDP-bound, inactive state in cells. Purified RhoA will be isolated primarily in the GDP-bound state. In yeast two-hybrid tests and coimmunoprecipitation experiments, we have found that only a mutant of RhoA, RhoA L63, that remains predominantly in the GTP-bound form, efficiently associates with MEKK1 and activates it. Therefore, it is essential to pre-exchange GDP for GTP on purified RhoA to allow it to become fully active in MEKK1 kinase assays. A nonhydrolyzable form of GTP, GTPγS, is used in the preloading step to lock RhoA irreversibly in the active conformation. RhoA proteins are incubated with 17.5 μM GTPγS in 50 mM HEPES, pH 7.6, 5 mM EDTA, 1 mM dithiothreitol, and 50 mM NaCl at 30° for 30 sec. The GTPγS-loaded RhoA is purified through a G-25 column equilibrated with the loading buffer with the addition of 5 mM MgCl$_2$. The eluted RhoA-GTPγS is active and ready to be added into MEKK1 kinase assays (Kozasa *et al.*, 1998).

Kinase Assays

Multiple proteins have been identified as direct substrates of MEKK1, including several of the MAP2Ks such as MEK1, MEK2, and MEK4. MEK4 has been most frequently used in MEKK1 kinase assays in our hands. A kinase dead, K131M, mutant form of MEK4 is purified from bacteria using standard method. The lysine-to-methionine substitution abolishes the endogenous kinase activity of MEK4 and thus prevents its autophosphorylation during kinase reactions.

Buffers and Reaction Mixture

Reaction mixture (30 μl): 2 μl MEKK1 (0.1–0.5 mg/ml)
4 μl MEK4 K131M (1–2 mg/ml)
2 μl RhoA (0.1–0.5 mg/ml)
6 μl 5× kinase buffer
15 μl H$_2$O
1 μl 10 mCi/ml [γ-^{32}P] ATP
Reaction time: 30 min
Reaction temperature: 30°
5× kinase buffer: 100 mM HEPES, pH 7.8, 50 μM ATP, 50 mM MgCl$_2$, and 50 mM β-glycerophosphate.

In the standard MEKK1 kinase assay, 0.5–1 μg of PP2A-treated recombinant MEKK1 is mixed with 4–8 μg of purified MEK4 K131M along with 5 μCi of [γ-^{32}P] ATP in kinase buffer in the absence or presence of GDP or GTP-loaded RhoA. The reaction is carried out in a total volume of 30 μl for 30 min at 30°. The kinase reaction is terminated by the addition of 5× SDS-PAGE buffer and loaded on to a gel. Gels are dried and exposed to X-ray film to reveal the phosphorylated MEK4 signal. After determining the location of MEK4 K131M on the gel by comparing it to the autoradiograph, each MEK4 K131M band is excised from the gel and counted by liquid scintillation counting to quantify the incorporation of radioactive phosphate. MEKK1 activity is normally increased by RhoA-GTP 10-fold or more, but less than three fold with RhoA-GDP (Fig. 1).

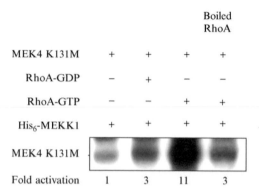

			Boiled RhoA
MEK4 K131M	+	+	+	+
RhoA-GDP	−	+	−	−
RhoA-GTP	−	−	+	+
His$_6$-MEKK1	+	+	+	+
MEK4 K131M				
Fold activation	1	3	11	3

FIG. 1. Stimulation of MEKK1 kinase activity by RhoA *in vitro*. MEKK1 and RhoA proteins were purified from Sf9 lysates. MEKK1 was pretreated with PP2a and used in kinase assays in the presence of Rho-GTP, Rho-GDP, or boiled RhoA with MEK4 K131M as substrate. MEK4 K131M phosphorylation was visualized by autoradiography of the gel. Fold activation is shown below the lanes. One of four similar experiments is shown. This figure is adapted from Gallagher *et al*.

Assay of MEKK1 Kinase Activity from Cells Expressing Active RhoA

Transient Transfection and Lysate Preparation

To test the effect of RhoA on MEKK1 in cells, MEKK1 is immuno-precipitated from mammalian cells assayed for its activity. Typically, 293 cells are grown in Dulbecco's modified Eagle's medium containing 10% fetal bovine serum and antibiotics. Cells are grown to 50–80% confluent in 60-mm plates and transfected with various RhoA constructs. After 48 h, transfected cells are washed three times in ice-cold PBS and harvested by scraping off the plates with 0.5 ml lysis buffer (50 mM Tris-Cl, pH 8.0, 150 mM NaCl, 1% Triton X-100, 1 mM sodium orthovanadate, 80 mM β-glycerophosphate, 1 mM phenylmethylsulfonyl fluoride, 10 μg/ml leupeptin, and 7 μg/ml aprotinin). Lysates are sedimented at 200g to remove nuclei and cell debris. The resulting supernatants are adjusted to a final concentration of 1 M NaCl and homogenized using a Dounce homogenizer. After 30 min on ice, the lysates are further clarified by ultracentrifugation for 30 min at 100,000g.

Immunoprecipitation and Kinase Assay

Lysates from 293 cells (0.5 ml at 3–4 mg/ml) are incubated in 1.5-ml microfuge tubes with 1–2 μg of an anti-MEKK1 antibody (C-22 from Santa Cruz, Santa Cruz, CA) for 1 h followed with 30 μl of protein A-sepharose at 4 C for 2 h to overnight with constant rotation. Beads and the bound proteins are sedimented in a microcentrifuge. The pellets are then washed for 1 h with 1 ml of lysis buffer three times. The long washing time is important to reduce the background signal in the following kinase assays.

Immunoprecipitates are used for kinase assays. The kinase assays are performed essentially the same as the *in vitro* assays. The amounts of MEKK1 should be examined by immunoblotting to ensure that nearly equal amounts of them are present in each sample. RhoA L63, the active mutant of RhoA, dramatically stimulated the activity of endogenous MEKK1 in these assays, whereas RhoA WT and RhoA N19, a dominant negative mutant of RhoA, showed little activation effect (Fig. 2).

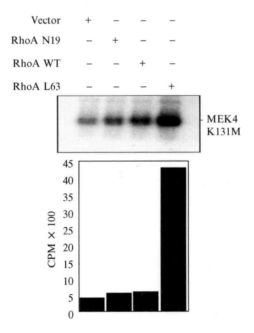

FIG. 2. Stimulation of MEKK1 kinase activities by overexpression of RhoA in cells. HEK 293 cells were transfected with pCMV vector and pCMV Myc-RhoA N19, Myc-RhoA WT, or Myc-RhoA L63. After 48 h, cells were lysed, and endogenous MEKK1 was immunoprecipitated from the lysates. The immunoprecipitates were used to perform kinase assays with MEK4 K131M as a substrate. MEK4 K131M phosphorylation by MEKK1 was visualized by autoradiography and quantitated by liquid scintillation counting. One of three similar experiments is shown. This figure is adapted from Gallagher *et al.*

Yeast Two-hybrid Binding of RhoA to MEKK1

The ability of RhoA to bind to MEKK1 was tested because we have previously found that MEKK1 can interact with p115 RhoGAP (Christerson *et al.*, 2002). The MEKK1–RhoA interaction was evaluated using the yeast two-hybrid system, and the interacting domain was confirmed to be at the N terminus of MEKK1. The strains, constructs, and strategies used are similar to standard two-hybrid assays previously described. (Bai and Elledge, 1996). Here we introduce the materials and conditions that we used to study the interaction between RhoA and MEKK1.

The *Saccharomyces cerevisiae* Y190 cells are transformed using lithium acetate/polyethylene glycol with a mouse T-cell library of GAL4 activation domain-cDNA fusions and the GAL4 DNA-binding domain fused to a series of MEKK1 fragments. The MEKK1 fragments incorporated into pAS1CYH2 vector includes 1–132, 149–347, 149–636, 630–772, 766–1173, 766–1493, 1–719, and 565–1174. Approximately 2.5 million transformants were plated onto synthetic complete medium (plus 50 mM 3-amino triazole, 20 μg/ml 5-bromo-4-chloro-3-indoyl-b-D-galactopyranoside, and lacking histidine, leucine, and tryptophan) and incubated at 30° for 5–7 days to allow colony formation and final growth selection.

As described in a recent report from our laboratory, the fragment of MEKK1 containing residues 1–719 interacts with wild-type RhoA in the two-hybrid assays but not with other known MEKK1 interactors such as H-Ras, Rac, or Cdc42. An even smaller fragment of MEKK1, MEKK1-(149-636), also associates with wild-type RhoA in yeast. The constitutively active mutant of RhoA, RhoAL63, also binds to residues 149–636, whereas the GDP-bound, inactive mutant RhoAN19 fails to bind any of the MEKK1 fragments (Gallagher *et al.*, 2004). Thus, the interaction between RhoA and MEKK1 seems to be activation-dependent (GTP-binding dependent) in yeast, although we cannot completely eliminate the possibility that Rho-GDP retains some ability to bind. It is also interesting to note that a triple mutation (G452C, R454C, N455D) or another single cysteine mutation (C433A) near or within the PHD domain of MEKK1 completely abolishes its interaction with RhoA. In support of yeast two-hybrid assays, formation of the MEKK1 and RhoA complex can be confirmed by coimmunoprecipitation experiment with endogenous MEKK1 and RhoA; the interaction can be specifically blocked by an antigenic peptide.

Acknowledgment

We thank Steve Stippec, William Singer, and Yu-Chi Juang for technical advice and helpful discussion, Bing-e Xu and Michael Lawrence for critical reading of the manuscript, and Dionne Ware for administrative assistance.

References

Bai, C., and Elledge, S. J. (1996). Gene identification using the yeast two-hybrid system. *Methods Enzymol.* **273,** 331–347.

Bar-Sagi, D., and Hall, A. (2000). Ras and Rho GTPases: A family reunion. *Cell* **103,** 227–238.

Bian, D., Su, S., Mahanivong, C., Cheng, R. K., Han, Q., Pan, Z. K., Sun, P., and Huang, S. (2004). Lysophosphatidic acid stimulates ovarian cancer cell migration via a Ras-MEK kinase 1 pathway. *Cancer Res.* **64,** 4209–4217.

Choi, K. Y., Satterberg, B., Lyons, D. M., and Elion, E. A. (1994). Ste5 tethers multiple protein kinases in the MAP kinase cascade required for mating in *S. cerevisiae. Cell* **78,** 499–512.

Christerson, L. B., Gallagher, E., Vanderbilt, C. A., Whitehurst, A. W., Wells, C., Kazempour, R., Sternweis, P. C., and Cobb, M. H. (2002). p115 Rho GTPase activating protein interacts with MEKK1. *J. Cell Physiol.* **192,** 200–208.

Christerson, L. B., Vanderbilt, C. A., and Cobb, M. H. (1999). MEKK1 interacts with alpha-actinin and localizes to stress fibers and focal adhesions. *Cell Motil. Cytoskeleton* **43,** 186–198.

Cuevas, B. D., Abell, A. N., Witowsky, J. A., Yujiri, T., Johnson, N. L., Kesavan, K., Ware, M., Jones, P. L., Weed, S. A., DeBiasi, R. L., Oka, Y., Tyler, K. L., and Johnson, G. L. (2003). MEKK1 regulates calpain-dependent proteolysis of focal adhesion proteins for rear-end detachment of migrating fibroblasts. *EMBO J.* **22,** 3346–3355.

Fanger, G. R., Johnson, N. L., and Johnson, G. L. (1997). MEK kinases are regulated by EGF and selectively interact with Rac/Cdc42. *EMBO J.* **16,** 4961–4972.

Gallagher, E. D., Gutowski, S., Sternweis, P. C., and Cobb, M. H. (2004). RhoA binds to the amino terminus of MEKK1 and regulates its kinase activity. *J. Biol. Chem.* **279,** 1872–1877.

Gallagher, E. D., Xu, S., Moomaw, C., Slaughter, C. A., and Cobb, M. H. (2002). Binding of JNK/SAPK to MEKK1 is regulated by phosphorylation. *J. Biol. Chem.* **277,** 45785–45792.

Gibson, S., Widmann, C., and Johnson, G. L. (1999). Differential involvement of MEK kinase 1 (MEKK1) in the induction of apoptosis in response to microtubule-targeted drugs versus DNA damaging agents. *J. Biol. Chem.* **274,** 10916–10922.

Hagemann, C., and Blank, J. L. (2001). The ups and downs of MEK kinase interactions. *Cell Signal* **13,** 863–875.

Huang, Y. W., Lu, M. L., Qi, H., and Lin, S. X. (2000). Membrane-bound human 3beta-hydroxysteroid dehydrogenase: Overexpression with His-tag using a baculovirus system and single-step purification. *Protein Expr. Purif.* **18,** 169–174.

Kozasa, T., Jiang, X., Hart, M. J., Sternweis, P. M., Singer, W. D., Gilman, A. G., Bollag, G., and Sternweis, P. C. (1998). p115 RhoGEF, a GTPase activating protein for Galpha12 and Galpha13. *Science* **280,** 2109–2111.

Lange-Carter, C. A., Pleiman, C. M., Gardner, A. M., Blumer, K. J., and Johnson, G. L. (1993). A divergence in the MAP kinase regulatory network defined by MEK kinase and Raf. *Science* **260,** 315–319.

Lee, F. S., Hagler, J., Chen, Z. J., and Maniatis, T. (1997). Activation of the IkappaB alpha kinase complex by MEKK1, a kinase of the JNK pathway. *Cell* **88,** 213–222.

Lee, F. S., Peters, R. T., Dang, L. C., and Maniatis, T. (1998). MEKK1 activates both IkappaB kinase alpha and IkappaB kinase beta. *Proc. Natl. Acad. Sci. USA* **95,** 9319–9324.

Lin, A., Minden, A., Martinetto, H., Claret, F. X., Lange-Carter, C., Mercurio, F., Johnson, G. L., and Karin, M. (1995). Identification of a dual specificity kinase that activates the Jun kinases and p38-Mpk2. *Science* **268**, 286–290.

Lu, Z., Xu, S., Joazeiro, C., Cobb, M. H., and Hunter, T. (2002). The PHD domain of MEKK1 acts as an E3 ubiquitin ligase and mediates ubiquitination and degradation of ERK1/2. *Mol. Cell* **9**, 945–956.

Marcus, S., Polverino, A., Barr, M., and Wigler, M. (1994). Complexes between STE5 and components of the pheromone-responsive mitogen-activated protein kinase module. *Proc. Natl. Acad. Sci. USA* **91**, 7762–7766.

Printen, J. A., and Sprague, G. F., Jr. (1994). Protein-protein interactions in the yeast pheromone response pathway: Ste5p interacts with all members of the MAP kinase cascade. *Genetics* **138**, 609–619.

Uhlik, M. T., Abell, A. N., Cuevas, B. D., Nakamura, K., and Johnson, G. L. (2004). Wiring diagrams of MAPK regulation by MEKK1, 2, and 3. *Biochem. Cell Biol.* **82**, 658–663.

Wells, C., Jiang, X., Gutowski, S., and Sternweis, P. C. (2002). Functional characterization of p115 RhoGEF. *Methods Enzymol.* **345**, 371–382.

Xia, Y., Makris, C., Su, B., Li, E., Yang, J., Nemerow, G. R., and Karin, M. (2000). MEK kinase 1 is critically required for c-Jun N-terminal kinase activation by proinflammatory stimuli and growth factor-induced cell migration. *Proc. Natl. Acad. Sci. USA* **97**, 5243–5248.

Xu, S., Robbins, D. J., Christerson, L. B., English, J. M., Vanderbilt, C. A., and Cobb, M. H. (1996). Cloning of rat MEK kinase 1 cDNA reveals an endogenous membrane-associated 195-kDa protein with a large regulatory domain. *Proc. Natl. Acad. Sci. USA* **93**, 5291–5295.

Yujiri, T., Sather, S., Fanger, G. R., and Johnson, G. L. (1998). Role of MEKK1 in cell survival and activation of JNK and ERK pathways defined by targeted gene disruption. *Science* **282**, 1911–1914.

Yujiri, T., Ware, M., Widmann, C., Oyer, R., Russell, D., Chan, E., Zaitsu, Y., Clarke, P., Tyler, K., Oka, Y., Fanger, G. R., Henson, P., and Johnson, G. L. (2000). MEK kinase 1 gene disruption alters cell migration and c-Jun NH2-terminal kinase regulation but does not cause a measurable defect in NF-kappa B activation. *Proc. Natl. Acad. Sci. USA* **97**, 7272–7277.

Zhang, L., Deng, M., Parthasarathy, R., Wang, L., Mongan, M., Molkentin, J. D., Zheng, Y., and Xia, Y. (2005). MEKK1 transduces activin signals in keratinocytes to induce actin stress fiber formation and migration. *Mol. Cell. Biol.* **25**, 60–65.

Further Reading

Xu, S., Robbins, D., Frost, J., Dang, A., Lange-Carter, C., and Cobb, M. H. (1995). MEKK1 phosphorylates MEK1 and MEK2 but does not cause activation of mitogen-activated protein kinase. *Proc. Natl. Acad. Sci. USA* **92**, 6808–6812.

Yujiri, T., Fanger, G. R., Garrington, T. P., Schlesinger, T. K., Gibson, S., and Johnson, G. L. (1999). MEK kinase 1 (MEKK1) transduces c-Jun NH2-terminal kinase activation in response to changes in the microtubule cytoskeleton. *J. Biol. Chem.* **274**, 12605–12610.

[36] Activation of the Apoptotic JNK Pathway Through the Rac1-Binding Scaffold Protein POSH

By ZHIHENG XU and LLOYD A. GREENE

Abstract

The JNKs (c-Jun N-terminal protein kinases) play important roles in a variety of physiological and pathological functions including induction of apoptosis. A major pathway by which JNKs are activated in response to apoptotic stimuli includes the GTP•Rac1-binding scaffold protein POSH (plenty of SH3s). POSH acts as a scaffold for binding and autoactivation of the MLK family of MKKK proteins, which in turn phosphorylate and activate the MKK family members MKK4 and 7, which in turn phosphorylate JNKs. In this chapter, we describe methods and techniques that have been successfully used to study the POSH-dependent apoptotic JNK pathway. Use of these techniques may lead to a better understanding of the components of this pathway and of how it is suppressed in viable cells and rapidly activated in response to apoptotic stimuli.

Introduction

The c-Jun-N-terminal protein kinases (JNKs 1–3), also known as stress-activated protein kinases (SAPKs), are members of the MAP kinase superfamily. They are activated in response to diverse extracellular signals, including growth factors, cytokines, apoptotic stimuli, and environmental stresses and play multiple roles in mammalian systems, including mediation of cellular proliferation, differentiation, development, inflammation, and apoptosis (Davis, 2000; Weston and Davis, 2003). Like other MAP kinase superfamily members, JNKs are activated by a sequential protein kinase cascade consisting of MAPK kinase kinases (also designated as MKKKs or MEKKs) and MAPK kinases (MKKs or MEKs). The appropriate activated MAPK kinases, in turn, phosphorylate and activate JNKs. Once activated, JNKs phosphorylate a variety of target substrates, including, in response to apoptotic stimuli, c-Jun and the BH3-only protein Bim, both of which play required roles in cell death. (Davis, 2000; Lei and Davis, 2003).

One area in which JNKs have been intensively studied, and that will be the focus of this chapter, is their involvement in apoptotic cell death. Apoptosis is a precisely orchestrated process essential for normal development and homeostatic control of cell numbers, and malfunctions of apoptosis are important factors in maladies such as neural degeneration,

METHODS IN ENZYMOLOGY, VOL. 406   0076-6879/06 $35.00
DOI: 10.1016/S0076-6879(06)06036-8

cancer, and autoimmune diseases. JNK activation plays major roles in a variety of apoptotic cell death paradigms, including those associated with DNA damage, heat shock, ischemia-reperfusion, oxidative stress, hyperosmolarity, mechanical stress, and loss of trophic support (Davis, 2000; Franzoso et al., 2003).

The development and implementation of experimental means to study the mechanisms by which JNKs are activated under apoptotic conditions revealed a pathway that is dependent on the G proteins Rac1 and Cdc42. Supporting this, expression of constitutively active forms of Cdc42/Rac1 induce cell death by JNK activation, whereas dominant-negative forms of these proteins inhibit death (Bazenet et al., 1998; Chuang et al., 1997; Xu et al., 2001). Subsequent studies revealed a sequential kinase-signaling cascade downstream of Rac1/Cdc42 consisting of the MLK family of MKKKs, MKK 4/7 and JNKs (Bazenet et al., 1998; Weston and Davis, 2002; Xia et al., et al., 2001). Knock-down, inhibition or interference with the activity of any of these components blocks death induced by various apoptotic stimuli. Insight as to how G proteins might interface with JNK activation was provided by Tapon et al. (1998), who used a two-hybrid screen to identify the multidomain molecule POSH as a partner for GTP-bound Rac1 fibroblasts, as well as of other cell types. Moreover, POSH was found to act as a scaffold that binds and promotes activation of the MLK family (Xu et al., 2003). In addition to POSH, a second set of scaffold proteins, the JIPs (JNK-interacting proteins), participate in JNK activation. JIPs are reported to associate with MLKs, MKKs, and JNKs (Weston and Davis, 2002). Recently, we found that POSH binds JIPs to organize components of the JNK pathway, including GTP-Rac1, MLKs, MKK4/7, and JNKs into a multimolecular proapoptotic complex that we have designated PJAC (POSH and JIP-associated complex; see Fig. 1 [Kukekov et al., in press]).

Our aim here is to provide experimental pointers for studying the apoptotic POSH-dependent JNK pathway. Use of these techniques may lead to a better understanding of the components of this pathway and of how it is suppressed in viable cells and rapidly activated in response to apoptotic stimuli. Although additional signaling pathways seem to activate JNKs under both apoptotic and nonapoptotic conditions (Davis, 2000; Weston and Davis, 2002), these are beyond the scope of this chapter and are not discussed here.

Immunoreagents

The following materials have been especially useful for studying the POSH-dependent JNK pathway. Antisera/monoclonal antibodies specific for JNK, phospho-JNK, MKK4, phospho-MKK4, MLK3, phospho-MLK3, and phosphotyrosine are available from Cell Signaling Technology, Beverly, MA. Anti-His tag monoclonal antibodies from Novagen, Madison,

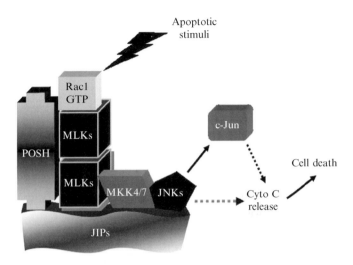

FIG. 1. Model for POSH-dependent interaction of JNK apoptotic pathway components.

MI; anti-HA tag antiserum from Clontech, Palo Alto, CA; anti-Myc tag, -ERK1, and -JNK1/2 antisera and JNK1/2 and GFP antibodies from Santa Cruz Biotechnology, Santa Cruz, CA; and anti-JIP1 monoclonal antibody and -MKK7 antiserum from BD Transduction Laboratories, San Diego, CA. Anti-Flag tag, -HA tag, and -Myc tag agarose-linked affinity gels are available from Sigma (St. Louis, MO). A POSH antiserum is currently available from Santa Cruz Biotechnology. However, although it recognizes the overexpressed protein, we have not successfully used it to identify endogenous POSH by Western blotting or immunostaining. We have generated a POSH antiserum against purified His-tagged POSH peptide (aa. 255–365 from mouse; His-tag cleaved off before immunization) that detects endogenous POSH (by blotting and immunostaining) in rodent cells after affinity purification with the peptide antigen (Xu *et al.*, 2005).

Reagents for Interfering with the POSH-Dependent JNK Pathway

A major means to study functional aspects of the POSH-dependent JNK pathway is to block the activities/expression of its various components. There are several alternative ways to achieve this.

Selective Small Molecule Inhibitors

CEP-1347 and its derivative CEP-11004 have been successfully used to block the activity of MLK family kinases both *in vitro* and *in vivo*. *In vitro* studies indicate good efficacy and relative specificity in the <200 n*M* range

(Maroney *et al.*, 2001). These inhibitors are not commercially available, but CEP-11004 can be requested for research purposes from Cephalon, Inc (http://www.Cephalon.com). SP600125 is marketed as a selective inhibitor of the JNKs (Calbiochem, San Diego, CA). Although widely used and a useful reagent, at high concentration ($>20 \mu M$) SP600125 inhibits activities of kinases in addition to JNKs (Bain *et al.*, 2003) and so, as with most "specific" inhibitors, must be used with caution.

Dominant-negative (D/N) Interfering Constructs

Because many JNK pathway components are protein kinases, constructs encoding kinase inactive mutant forms of those proteins can act as dominant-negatives. When overexpressed, these forms presumably compete with the corresponding wild-type endogenous proteins for access to substrates and activating partners such as POSH and, in this way, specifically interfere with phosphorylation of target proteins. D/N MLKs, MKK4 and 7, and JNKs have been described and are highly effective (Xu *et al.*, 2001). D/N forms of additional kinases that may affect JNK activation, such as ASK1 and MEKK1, are also reported (Nishitoh *et al.*, 1998). In the case of MLKs, competition for common substrates and binding partners such as POSH results in a situation in which D/N forms of any one MLK seem to interfere with actions of all other family members (Xu *et al.*, 2003). D/N forms of Rac1 and Cdc42 have also been described (Chuang *et al.*, 1997; Bazenet *et al.*, 1998). As detailed in the preceding references, D/N constructs can be prepared using PCR-based kits for site-directed mutagenesis.

Loss-Of-Expression Strategies

Although chemical inhibitors and D/N forms of kinases are very useful for inhibiting the POSH-dependent JNK pathway, they do have potential limitations. One is that they may not be entirely specific; chemical inhibitors are subject to possible nonspecificity as are D/N proteins. Also, for some pathway components, such as POSH, no inhibitor or D/N forms are currently available. An alternative strategy is to study cells with loss of expression of pathway components. Small interfering RNAs (siRNAs) have come to the forefront as an effective means to specifically and effectively downregulate expression of target genes (see Chapters 26 and 27 for detail). We have used both commercially synthesized double-stranded siRNAs and hairpin loop constructs that produce siRNAs to successfully target POSH (Xu *et al.*, 2003). siRNA strategies have also been used to knock-down both JNK1 and 2 (Alarcon-Vargas and Ronai, 2004). However, knock-down of JNK3 requires a separate construct. Because there are multiple MLKs with apparently overlapping functions, the siRNA approach is of limited value in this case. Another strategy is to exploit mice and cells cultured from them that are null for specific pathway components. To date, mice null for expression of JNKs, MKK4 and MKK7, and JIP1 have been reported (Brecht et al., 2005; Tournier et al., 2001).

Transfection of Cells for Studying the POSH-Dependent JNK Pathway

Certain biochemical studies involving the JNK pathway require high-efficiency transfections. This is particularly the case for components such as POSH and D/N forms of the MLK family that are very unstable and for use of siRNAs. Although many cell lines can be used to investigate the JNK pathway, HEK293 and HeLa cells are particularly useful for transfections. The CaPi method given in the following is a very efficient (as high as 99%) and economical way of transfecting HEK293, HeLa, and several other cell lines.

- Plate cells at $30,000/cm^2$ 20–24 h before transfection (50–70% confluent).
- The next day, change to fresh growth medium 3–4 h before transfection (10/3.3/1.5 ml per 10-/6-/3.5-cm-dish/well, respectively).
- Add 15/5/2.5 μg of plasmid DNA (per 10-/6-/3.5-cm-dish/well, respectively) to 500/168/83 μl of 250 mM $CaCl_2$ (per 10-/6-/3.5-cm-dish/well, respectively).
- Add the DNA/$CaCl_2$ mixture drop wise to 500/168/83 μl of 2XHBS (see following) while mixing by vortexing at low to medium speed.
- Add the mixture directly to the medium, mix well by shaking.
- Incubate for 4 h at $37°$, then remove the medium.
- Overlay cells with 10–15% DMSO-containing serum-free medium (3/1.5/1 ml per 10-/6-/3.5-cm-dish/well, respectively) for 1–2 min.
- Remove the DMSO-containing medium. Refeed the cells with 10/3/2 ml (per 10-/6-/3.5-cm-dish/well, respectively) of complete medium.

Solutions (100 ml): 250 mM $CaCl_2 \cdot 2$ H_2O (3.67 g/100 ml); 2XHBS buffer (pH = 7.10): 50 mM HEPES (1.19 g/100 ml), 280 mM NaCl (1.63 g/100 ml), 1.5 mM Na_2HPO_4 (0.02 g/100 ml).

Western Immunoblotting of POSH and MLKs

Because POSH and MLKs are often expressed at very low levels, particularly in viable cells, and the antisera against them are either expensive or in limited supply, we recommend the following modifications of standard Western immunoblotting procedures.

After protein transfer, the membranes are incubated in 5% BSA in TBST (which can be recycled many times). Primary antibodies are diluted in the same solution as well. Before incubation with appropriate secondary antibody, the membranes are rinsed twice with water then shaken and washed for 10 min with TBST buffer. Repeat the same rinse and wash procedure twice. After incubation with secondary antibody for 1 h, repeat the rinse and wash procedure three times. This method efficiently drops the background for Western immunoblotting.

Immunofluorescence Staining of Cultured Cells

Because of the low level expression of POSH and its associated proteins such as MLKs, it is recommended that cells are stained for a longer time than customary with primary antisera and extensively washed with PBS to reduce nonspecific staining.

Cells are fixed with 4% paraformaldehyde for 10 min, washed with PBS, and then permeabilized with 0.5% Triton X-100 in PBS for 10 min and blocked with 10% goat serum for 1 h at room temperature. Stain cells with antibody overnight at 4°, followed by three to four washes with PBS. Fluorescence-conjugated secondary antibody is applied for 1 h at room temperature. Cells are then washed with PBS once before staining with $1\mu g/ml$ Hoechst dye (Hoechst 33342; Sigma) in PBS to visualize nuclei, washed twice with PBS, and covered with mounting medium and a cover glass.

Immunoprecipitation (IP)/Co-IP

Because of the low cellular levels of endogenous POSH and associated proteins, especially in viable cells, IP is often required for their detection by Western immunoblotting. In addition, co-IP has proved to be especially useful for detecting both endogenous and cotransfected POSH-associated proteins.

1. For each IP/co-IP of relatively abundant (JNKs, JIPs) or tagged expressed proteins, at least 200 μg of total cell protein in 0.5 ml of cell lysis buffer (see following) is incubated for 2 h at 4° with the desired amount of antiserum and 15 μl of the appropriate suspended (50% v/v) sepharose conjugate (protein A-, protein G-, or protein A&G–sepharose mixture; Pharmacia). Before use, the beads are washed once and resuspended in lysis buffer. The choice of conjugate will depend on the affinity of each immunoreagent for either protein A or protein G (see details in "Research Applications and Support Products" section, Santa Cruz Biotechnology). For IP/co-IP of expressed tagged proteins, 15 μl of anti-c-Myc, Flag-, or HA-agarose conjugates (Sigma) is recommended. For endogenous proteins expressed at very low levels (POSH and MLKs), 1 μg antibody/ antiserum in 0.5 ml of cell lysis buffer is incubated and rotated for 2 h at 4° with the appropriate protein A– or protein G–conjugated sepharose beads. The beads are then centrifuged at 500 rpm for 20 sec and the supernatant

discarded. Up to several milligrams of total cell protein in lysis buffer (<1 ml) is then incubated and rotated with the antibody/antiserum-treated sepharose beads for 4 h to overnight at 4°.

2. The bead-associated IPs are washed four to five times with lysis buffer. The beads are centrifuged briefly (10 sec) at 5000 rpm after each wash, and most supernatant is removed. The residual buffer is removed with a 0.17-mm flat gel loading tip to avoid loss of beads.

3. The IPs are eluted with 2× sample buffer at room temperature by vortexing three times (3 sec each) and centrifuged briefly at high speed. The supernatant is transferred to a fresh tube with 0.17-mm flat gel loading tips, boiled for 4 min, and analyzed by Western immunoblotting. The advantage of eluting immunoprecipitates with 2× sample buffer at room temperature is that most of the IgG will not be eluted, yielding a lower background after immunoblotting.

NOTE: For IP/co-IP of multiple samples, it is recommended that the appropriate amount of beads is washed and resuspended in lysis buffer at 10 times the volume of beads. Mix well before transferring aliquots of beads for each IP/co-IP. In this way, the amount of beads added to each IP/co-IP is consistent.

In Vitro Binding Assays to Evaluate POSH-Interacting Proteins

We have used pGEX-2T to express GST fusion proteins in *E. coli* DH5α for pull-down assays involving POSH and other proteins, including constitutively active forms of Rac1 and Cdc42. Other vectors and *E. coli* strains, such as DH10B, can also be used for this purpose. Because expression of wt POSH is low in *E. coli* as in mammalian cells, we recommend that the POSH ring finger mutant (Xu *et al.*, 2003) be used if the study is unrelated to the function of the ringer finger. Purification of GST fusion proteins is as follows:

1. A single colony of transfected. *E. coli* is shaken overnight at 37° in LB medium containing suitable selection antibiotics.

2. The bacteria are diluted 1:100 into 200 ml of LB medium and shaken for another 2–3 h until the OD 560 reaches approximately 0.6–0.8. Isopropyl-β-D-thiogalactopyranoside (IPTG) is added to a final concentration of 0.2 mM.

3. After induction of protein expression with IPTG for 2–4 h, the bacteria are centrifuged down and resuspended in 20 ml of PBS containing 0.5% Tween, 16 μl of NaF (500 mM), 16 μl of Na$_3$VO$_4$ (500 mM), 100 μl PMSF (100 mM), 20 μl of dithiothreitol (DTT; 1 M), 400 μl of lysozyme (50 mg/ml), 30 μl of DNase I (10 mg/ml), and further disrupted by

sonification. Bacteria are sonicated on ice for 20 sec at highest power with additional 20 sec sonification intervals until the lysate clarifies.

4. The bacterial lysate is centrifuged at 12,000g for 20 min. The induced protein should be in the supernatant.

NOTE: To monitor the amounts of produced GST fusion protein, culture bacteria obtained before and after IPTG induction from steps 2 and 3 (40 and 20 μl, respectively); 10 μl of lysate and 10 μl of resuspended pellet (approximately 1 mm^3 of pellet in 100 μl 1× sample buffer) from step 4 should be analyzed by SDS-PAGE and staining with Coomassie blue. If most of the GST fusion protein is in the pellet in step 4, it indicates that the induced protein is within inclusion bodies. In this case, try to induce with 0.1 mM IPTG at OD 560 = 0.1 overnight at room temperature.

5. Take 1 ml of suspended bead-immobilized glutathione (50% v/v) (Amersham, Piscataway, NJ), centrifuge at 5000 rpm for 10 sec, discard the supernatant, and transfer the pellet to the preceding supernatant containing GST fusion protein. Incubate and rotate for 4 h to overnight to absorb the GST fusion protein.

6. After incubation, the beads are washed four to five times with cell lysis buffer and resuspended in 1 ml of the same buffer; 10 μl of the beads should be analyzed as previously. The remaining beads are aliquoted and stored at $-80°$.

7. cDNAs in pCMS.EGFP, pCDNA3, or other vectors are transcribed and translated *in vitro* using the TNT-coupled reticulocyte lysate system according to the manufacturer's protocol (Promega, Madison, WI) in the presence of [^{35}S]methionine.

A 10-μl aliquot of each *in vitro* translated polypeptide is incubated and rotated with glutathione beads carrying equal amounts of GST (as control for nonspecific binding to beads) or GST fusion proteins in 400 μl of lysis buffer at 4° for 3 h to overnight. The beads are washed six times with 600 μl of the same buffer, boiled in 25 μl of 2× SDS sample buffer, and the solubilized proteins are resolved by SDS-PAGE. Gels are fixed and stained with Coomassie blue, dried, and visualized by autoradiography. *In vitro* translated protein (approximately 1/6 of the total used in the binding assays) is also analyzed as standard.

Evaluation of JNK Pathway Kinase Activities

Activation of kinases in the JNK pathway requires phosphorylation at specific sites, and hence, antisera that differentially recognize these phosphorylated epitopes are convenient tools for monitoring whether the kinase in question has been activated. Reliable and sensitive antisera against activated/phosphorylated forms of JNK pathway kinases, such as JNKs, MKK4, and MLK3 (New England Biolabs, Beverly, MA) are presently available and can be used for Western immunoblotting or immunostaining. Overall activation of the JNK pathway can also be detected with antisera that recognize the phosphorylated form of c-Jun (phosphor-c-Jun [Ser63 or Ser73]; New England Biolabs, Beverly, MA), the direct substrate of JNKs. Despite the utility of phosphospecific immunoreagents, there are circumstances in which it is desirable to directly measure the activity of JNK pathway kinases using *in vitro* kinase assays (see Whitmarsh and Davis, 2001).

Detection of Phosphorylated Proteins

Many JNK pathway components, including those associated with POSH, are phosphorylated in response to apoptotic stimuli, including MLKs, MKKs, JNKs, POSH, and JIPs (Nihalani *et al.*, 2003; Tournier *et al.*, 2001; Xu *et al.*, 2001, 2005). There are reliable antisera available to detect specific phosphorylated forms of most of these, specifically at sites associated with activation (see "Immunoreagents"). However, for potential phosphorylation sites not known to be associated with catalytic activation as well as for POSH, suitable phosphospecific antisera are not presently available. An alternative approach relies on the observation that phosphorylation often decreases the mobility of proteins subjected to SDS-PAGE. To determine whether a given signal observed by Western blotting represents a phosphorylated form of the protein under study, phosphatase treatment can be adopted. This is performed by incubating aliquots of cell lysates (approximately 100 μg of protein in 50 μl) with 400 U of λ phosphatase (New England Biolabs) for 2 h at 30°. The reactions are terminated by adding 5× SDS sample buffer, and the samples are then analyzed by Western immunoblotting along with nonphosphatase-treated control material to detect whether phosphatase treatment increases the mobility of the corresponding protein.

Useful Buffers

Cell Lysis Buffer

10 mM Tris (pH 7.4), 1.0% Triton X-100, 0.5% Nonidet P-40, 150 mM NaCl, 20 mM sodium fluoride (NaF), 0.2 mM sodium orthovanadate (Na$_3$VO$_4$), 1.0 mM EDTA, 1.0 mM EGTA, 0.2 mM phenylmethylsulfonyl fluoride (PMSF).

NOTE: PMSF should be added while the buffer is being stirred to avoid precipitation. Alternately, protease inhibitor cocktail tablets (Roche) can be used. It is preferable to add NaF, PMSF, and Na_3VO_4 just before use, but the buffer can be stored at $-20°$. For samples to be subjected to kinase assay, Na_3VO_4 is omitted. NaF is omitted if samples are to be subjected to phosphatase treatment. The buffer (without NaF, PMSF, and Na4PPi) can also be used as washing buffer for IP/co-IPs, preparation of GST fusion proteins, and GST-pull down assays.

TBST Buffer

25 mM Tris-HCl (pH 7.4), 137 mM NaCl, 5 mM KCl, 0.7 mM $CaCl_2$, 0.1 mM $MgCl_2$, 0.1–0.2% (vol/vol) Tween 20.

2× Protein Sample Buffer

Mix 25 ml of 4×Tris-HCl/SDS, pH 6.8 (0.5 M Tris-HCl, 0.4% SDS), 20 ml of glycerol, 4 g of SDS, and 3 mg bromphenol blue in 2 ml of 2-mercaptoethanol. Bring to 100 ml with water. Store at $-20°$.

5× Protein Sample Buffer

Mix 60 ml of 4×Tris-HCl/SDS, pH 6.8 (0.5 M Tris-HCl, 0.4% SDS), 40 ml of glycerol, 10 g of SDS, 7.5 mg bromphenol blue, and 7.7 g DTT. Store at $-20°$.

References

Alarcon-Vargas, D., and Ronai, Z. (2004). c-Jun-NH2 kinase (JNK) contributes to the regulation of c-Myc protein stability. *J. Biol. Chem.* **279**, 5008–5016.

Bain, J., McLauchlan, H., Elliott, M., and Cohen, P. (2003). The specificities of protein kinase inhibitors: An update. *Biochem. J.* **371**, 199–204.

Brecht, S., *et al.* (2005). Specific pathophysiological functions of JNK isoforms in the brain. *Eur. J. Neurosci.* **21**, 363–377.

Bazenet, C. E., Mota, M. A., and Rubin, L. L. (1998). The small GTP-binding protein Cdc42 is required for nerve growth factor withdrawal-induced neuronal death. *Proc. Natl. Acad. Sci. USA* **95**, 3984–3989.

Chuang, T. H., Hahn, K. M., Lee, J. D., Danley, D. E., and Bokoch, G. M. (1997). The small GTPase Cdc42 initiates an apoptotic signaling pathway in Jurkat T lymphocytes. *Mol. Biol. Cell* **8**, 1687–1698.

Davis, R. J. (2000). Signal transduction by the JNK group of MAP kinases. *Cell* **103**, 239–252.

Franzoso, G., Zazzeroni, F., and Papa, S. (2003). JNK: A killer on a transcriptional leash. *Cell Death Differ.* **10**, 13–15.

Lei, K., and Davis, R. J. (2003). JNK phosphorylation of Bim-related members of the Bcl2 family induces Bax-dependent apoptosis. *Proc. Natl. Acad. Sci. USA* **100**, 2432–2437.

Maroney, A. C., Finn, J. P., Connors, T. J., Durkin, J. T., Angeles, T., Gessner, G., Xu, Z., Meyer, S. L., Savage, M. J., Greene, L. A., Scott, R. W., and Vaught, J. L. (2001). Cep-1347

(KT7515), a semisynthetic inhibitor of the mixed lineage kinase family. *J. Biol. Chem.* **276,** 25302–25308.

Nihalani, D., Wong, H. N., and Holzman, L. B. (2003). Recruitment of JNK to JIP1 and JNK-dependent JIP1 phosphorylation regulates JNK module dynamics and activation. *J. Biol. Chem.* **278,** 28694–28702.

Nishitoh, H., Saitoh, M., Mochida, Y., Takeda, K., Nakano, H., Rothe, M., Miyazono, K., and Ichijo, H. (1998). ASK1 is essential for JNK-SAPK activation by TRAF2. *Mol. Cell. Biol.* **2,** 389–395.

Tapon, N., Nagata, K., Lamarche, N., and Hall, A. (1998). A new rac target POSH is an SH3-containing scaffold protein involved in the JNK and NF-kappaB signalling pathways. *EMBO J.* **17,** 1395–1404.

Tournier, C., Dong, C., Turner, T. K., Jones, S. N., Flavell, R. A., and Davis, R. J. (2001). MKK7 is an essential component of the JNK signal transduction pathway activated by proinflammatory cytokines. *Genes Dev.* **15,** 1419–1426.

Weston, C. R., and Davis, R. J. (2002). The JNK signal transduction pathway. *Curr. Opin. Genet.* **12,** 14–21.

Xia, Z., Dickens, M., Raingeaud, J., Davis, R. J., and Greenberg, M. E. (1995). Opposing effects of ERK and JNK-p38 MAP kinases on apoptosis. *Science* **270,** 1326–1331.

Xu, Z., Kukekov, N. V., and Greene, L. A. (2003). POSH acts as a scaffold for a multiprotein complex that mediates JNK activation in apoptosis. *EMBO J.* **22,** 252–261.

Xu, Z., Kukekov, N. V., and Greene, L. A. (2005). Regulation of apoptopic N-terminal c-Jun kinase by a stabilization-based feed forward loop. *Mol. Cell. Biol.* **25** (in press).

Xu, Z., Maroney, A. C., Dobrzanski, P., Kukekov, N. V., and Greene, L. A. (2001). The MLK family mediates c-Jun N-terminal kinase activation in neuronal apoptosis. *Mol. Cell Biol.* **21,** 4713–4724.

[37]   Quantification of Isozyme-Specific Activation of Phospholipase C-β2 by Rac GTPases and Phospholipase C-ε by Rho GTPases in an Intact Cell Assay System

By David M. Bourdon, Michele R. Wing, Earl B. Edwards, John Sondek, and T. Kendall Harden

Abstract

Phospholipase C (PLC) catalyzes the hydrolysis of $PtdIns(4,5)P_2$, which results in both formation of the second messengers $Ins(1,4,5)P_3$ and diacylglycerol and alteration in the membrane association and/or activity of $PtdIns(4,5)P_2$-binding proteins. The existence of 13 different PLC isozymes suggests multiple mechanisms of regulation of inositol lipid signaling, and the recent realization that Rho-family GTPases directly bind and activate certain PLC isozymes has added to this potential diversity of

METHODS IN ENZYMOLOGY, VOL. 406
0076-6879/06 $35.00
DOI: 10.1016/S0076-6879(06)06037-X

inositol lipid–related signal transduction. With the goal of delineating a less labor-intensive method for quantification of intracellular inositol phosphate production, we have applied a commercially available yttrium silicate RNA binding resin selective for inositol phosphates to develop a high-throughput inositol phosphate scintillation proximity assay (SPA). We highlight the utility of this assay using COS-7 cells robotically transfected in a 96-well format. This method is readily applied to quantify activation of PLC by receptors and G proteins, and we illustrate here the selective activation of PLC-β2 by Rac but not by Rho GTPases and the selective activation of PLC-ε by Rho but not Rac GTPases.

Introduction

The central role of phospholipase C (PLC)–catalyzed breakdown of membrane phosphoinositides in the Ca^{2+}-mobilizing and protein-kinase C–activating action of a broad range of growth factors, hormones, neurotransmitters, and other extracellular stimuli was established in the 1980s (Berridge, 1987). A large family of PLC isozymes subsequently was identified, and this group of cell signaling proteins now includes PLC-β, -γ, -δ, -ε, -ζ, and -η isozymes (Rhee, 2001). All PLC-β isoforms are directly activated by Gα-subunits of the Gq family of heterotrimeric G proteins, and PLC-β2 and PLC-β3 additionally are regulated by G$\beta\gamma$-subunits. In contrast, PLC-γ1 and PLC-γ2 contain SH2 and SH3 domains and are regulated by receptor and nonreceptor tyrosine kinases.

Although early work from Wakelam et al. (1986) as well as Cantley and coworkers (Fleischman et al., 1986) suggested that Ras regulates inositol lipid hydrolysis, working models of hormonal regulation of phospholipase C–mediated cell signaling principally have included Gαq- and G$\beta\gamma$-mediated regulation of PLC-β isozymes and tyrosine kinase-regulated PLC-γ isozymes. This long-held view has proven to be an oversimplification, and the groundbreaking work of Gierschik and coworkers illustrating activation of PLC-β2 in neutrophils by Cdc42 and Rac presaged an increased appreciation of regulation of inositol lipid metabolism by Ras superfamily GTPases (Illenberger et al., 1998, 2002). This idea has been augmented by the recent discovery of PLC-ε as a Ras binding protein (Kelley et al., 2001; Lopez et al., 2001; Song et al., 2001).

Not only is PLC-β2 directly activated by Gαq and G$\beta\gamma$, but it also is directly activated by GTP-dependent binding of Rac1, Rac2, Rac3, and (to a lesser extent) Cdc42 (Illenberger et al., 2002). Interaction with the amino terminal PH domain entirely accounts for Rac binding to PLC-β2, as we initially illustrated by surface plasmon resonance (Snyder et al., 2003) and have more recently unambiguously confirmed in the crystal structure of a complex of GTP-Rac1 and PLC-β2 (Jezyk, M. *et al.*, unpublished observations).

The work of Kataoka (Song *et al.*, 2001) and Kelley, Smrcka and colleagues (Kelley *et al.*, 2001) established that Ras and Rap bind in a GTP-dependent manner to the RA domains present in the carboxy terminus of PLC-ε. Interestingly, PLC-ε also contains an amino terminal Cdc25 GEF domain and is an upstream regulator of Ras, Rap, and potentially other GTPases (Jin *et al.*, 2001; Lopez *et al.*, 2001). Although PLC-ε is not activated by Gαq, overexpression of Gα12 or Gα13 results in activation of this isozyme in cell-based assays of inositol lipid hydrolysis (Lopez *et al.*, 2001; Wing *et al.*, 2001). The fact that Gα12/13 activates RhoGEFs suggested to us that Rho also might be a direct regulator of PLC-ε. Indeed, RhoA, RhoB, and RhoC all activate PLC-ε *in vivo*, and this activity is independent of the Ras-binding RA domains in the PLC-ε carboxy terminus (Wing *et al.*, 2003). Our recent work with purified proteins has confirmed that the activation of PLC-ε by Rho GTPases is direct and occurs through interaction in the catalytic core of the isozyme (Seifert *et al.*, 2004).

The discovery that PLC-β2 and PLC-ε are effectors of Rho GTPases has made necessary the facile evaluation of the capacity of multiple Ras superfamily GTPases and mutants thereof to regulate inositol lipid hydrolysis in an intact cell context. The capacity of heterotrimeric G protein–coupled and tyrosine kinase receptors to stimulate PLC has been evaluated over the past two decades by preincorporation of [³H]inositol into phosphoinositide pools, incubation of cells or tissues in the presence of hormones and LiCl, and isolation of [³H]inositol phosphates by Dowex chromatography (Berridge *et al.*, 1982). We recently have developed a unified system that applies robotic transfection of COS-7 cells with expression vectors for Ras GTPases and PLC isozymes in a 96-well format, followed by quantification of [³H]inositol phosphates using scintillation proximity assay (SPA) beads that circumvents laborious separation of [³H]inositol phosphates using ion exchange columns. The methods underpinning this utility are described here as is the application of these techniques to illustrate the selective activation of PLC-β2 by Rac GTPases and PLC-ε by Rho GTPases.

Methods

Cell Culture

COS-7 cells are grown in Dulbecco's modified Eagle's medium (DMEM) supplemented with 10% fetal bovine serum, 4 mM L-glutamine, 200 U penicillin, and 0.2 mg/ml streptomycin (all obtained from Gibco,

Invitrogen Corp.) at 37° in a 10% CO_2/90% air-humidified atmosphere. COS-7 cells are maintained during routine passage in culture in 162-cm^2 (T-162) tissue culture flasks (Corning Costar). A confluent T-162 flask of COS-7 cells yields approximately 1×10^7 cells. To plate cells for assay in 96-well format, add 4 ml 0.25% trypsin/EDTA (Gibco Invitrogen Corp.) to a confluent T-162 flask and discard. This initial wash removes most FBS, thus improving subsequent cell detachment. Add 4 ml of trypsin/EDTA and incubate at 37° to allow cells to gently detach from the tissue culture flask surface. Add 10 ml of supplemented DMEM (as previously) and count cells using a hemacytometer. Add additional medium to bring cells to a density of between 50,000–80,000 cells/ml. Using a multichannel pipette, dispense 100 μl of the cell suspension to each well of a 96-well plate (Corning Costar 3595) and place in a tissue culture incubator maintained at 37° in a 10% CO_2/90% air-humidified atmosphere. Transfection of cells with expression vectors is usually carried out approximately 12 h subsequent to plating.

Transfection

Optimal transfection occurs with COS-7 cells at a cell surface density of 50–80%. DNA samples for transfection are prepared using FuGENE 6 (Roche Applied Science, Indianapolis, IN) according to the manufacturer's suggested guidelines. Specifically, 0.24 μl of FuGENE 6 is combined with 4 μl of serum-free DMEM to complex a total of 80 ng of plasmid DNA for transfection into a single well of a 96-well plate of COS-7 cells. First add the required amount of FuGENE to a polypropylene microfuge tube containing the corresponding amount of serum-free DMEM making certain that the FuGENE contacts the DMEM without touching the side of microfuge tube. Mix by inversion (no vortex) and incubate at room temperature for 5 min.

The plasmid dilutions needed for the automated transfection may be prepared before or during this incubation period. Individual expression plasmids are diluted to an initial concentration of 100 ng/μl in sterile water. DNAs are further diluted to the necessary concentrations with 100 ng/μl empty pcDNA3 (Invitrogen). We have found that transfection of 80 ng total DNA per well of a 96-well plate provides efficient and sufficient expression of various combinations of expression vectors for signaling proteins. Thus, all wells receive a total of 80 ng of DNA, which includes expression vectors for the PLC isozymes and GTPases, as well as empty vector DNA as needed. For example, to transfect 30 ng of PLC-β2 and 30 ng of Rac3 in a total of 80 ng DNA per well of COS-7 cells, dilute the 100 ng/μl stocks of PLC-β2 and Rac3 to 75 ng/μl with 100 ng/μl pcDNA3.

As a consequence, concentrations of individual DNAs are adjusted, but the overall DNA concentration remains at 100 ng/μl, and combination of 0.4 ul of PLC-β2 and 0.4 ul of Rac3 with 4 μl of serum-free DMEM and 0.24 μl of FuGENE provides the target amounts of 30 ng of PLC-β2, 30 ng of Rac3, and 20 ng of empty vector per well.

Dilution of the initial stock DNAs (e.g., PLC-β2 and Rac3) is carried out by hand in a V-bottom microtiter plate (Corning Costar 3896), whereas all replicates of DNA dilution combinations are accomplished with a Beckman Coulter BioMek FX fluid handling system (see later). To increase speed, reproducibility, and capacity to perform comprehensive screens of potential activators of phospholipases, an automated fluid handler such as the BioMek FX is used. Stock dilutions of all DNAs to be transfected are placed in the V-bottom microtiter plate. The BioMek FX fluid handler is programed to transfer the appropriate amount of each DNA from the stock microtiter plate into a given well of a new V-bottom microtiter plate. For each test condition (e.g., PLC-β2 and Rac3), the BioMek FX adds the appropriate amount of preincubated DMEM/FuGENE mix to the well and mixes, followed by a 15-min incubation period at room temperature for formation of DNA-lipid complexes. After 15 min, the previously plated COS-7 cells are removed from the 37° tissue culture incubator and placed on the BioMek FX deck for completion of the transfection process. In a final step, the fluid handler transfers 5 μl (containing 80 ng total DNA) of the DNA-FuGENE/DMEM complex to the 96-well plate containing 100 ul of medium followed by an automated mix step. The 96-well tissue culture plates are then returned to the 37° 10% CO_2/90% air-humidified atmosphere and incubated for at least 24 h.

Phosphoinositide Hydrolysis Detection by Scintillation Proximity Assay

Twenty-four hours after transfection, the medium is aspirated from the 96-well plates and replaced with serum-free inositol-free DMEM containing [^{3}H]*myo*-inositol. Each well of a 96-well plate is labeled with 1 μCi of [^{3}H]inositol by adding 1 μCi (1 μl) of [^{3}H]*myo*-inositol (American Radiolabeled Chemicals) in 40 μl serum-free, inositol-free DMEM (ICN Biomedical, Inc., Aurora, OH). Thus, for a 96-well plate combine 100 μCi of [^{3}H]inositol with 4 ml serum-free inositol-free DMEM and add 40 μl per well using a multichannel pipette. Return the 96-well plates to the tissue culture incubator in a 37° 10% CO_2/90% air-humidified atmosphere for 12–18 h.

Phospholipase C activity is measured 12–18 h after addition of [^{3}H] inositol. Remove the 96-well plates from the tissue culture incubator and add 10 μl of 50 mM LiCl to each well to give a final concentration of 10 mM

LiCl. Return the plate(s) to the tissue culture incubator for 1 h. LiCl inhibits inositol phosphate phosphatases and, therefore, allows quantification of [^{3}H]inositol phosphates that accumulate as a result of PLC-catalyzed hydrolysis of [^{3}H]PtdIns(4,5)P^2. Assays are terminated by aspiration of the labeling medium from the cells and addition of 30 μl of 50 mM formic acid followed by incubation of 96-well plates at 4° for 15 min. No pH neutralization of samples is required. Our laboratory has adapted a yttrium silicate (YSi) RNA binding resin developed by Amersham Biosciences (number RPNQ0013; Piscataway, NJ) as first described by Liu et al. (2003) for quantification of [^{3}H]inositol phosphate accumulation. The SPA beads bind [^{3}H]inositol phosphates (but do not bind [^{3}H]inositol) and scintillate on close proximity to radioactive material. Dilute the 100-mg/ml commercial suspension of SPA resin to 2.7 mg/ml, and add 75 μl (0.2 mg resin per well) to each well of a white opaque 96-well plate (OptiPlate; Packard). Use a multichannel pipette to add all 30 ul of the formic acid/cell lysate to the OptiPlate containing the SPA resin. Cover the Optiplate with clear adhesive microtiter plate covers (PerkinElmer TopSeal, #6005185) and mix at 300 rpm for 30 min. Allow the sample to settle at 4° for 4 h and quantify radioactivity using a microplate scintillation counter.

Application of Assay to Quantify Activation of PLC Isozymes by Rho Family GTPases

Our laboratory has routinely used Dowex ion exchange resin to separate [^{3}H]inositol phosphates from [^{3}H]inositol (Brown et al., 1991), as well as to resolve individual [^{3}H]inositol phosphates (Nakahata et al., 1987). This technology is both straightforward and highly effective, but it is labor intensive and is difficult to adapt to high-throughput assay protocols. Thus, the availability of SPA beads provides a potential means to automate quantification of [^{3}H]inositol phosphate accumulation. That is, as first reported by Liu and colleagues (2003), the scintillation proximity assay (SPA) beads bind [^{3}H]inositol phosphates but do not bind [^{3}H]inositol, and radioactivity is quantified by scintillation counting in an appropriate instrument without the need of the addition of scintillation fluid.

Quantification of [^{3}H]inositol phosphates was tested with SPA beads under the conditions of our assays, and the results were directly compared with results obtained with traditional separation of [^{3}H]inositol phosphates with Dowex ion exchange columns. SPA beads are not predicted to bind [^{3}H]inositol, and essentially no signal was obtained under conditions in which SPA beads were mixed with [^{3}H]inositol. The SPA bead technology was compared directly with quantification of [³H]inositol phosphates by Dowex column chromatography under several different experimental conditions. For example, the time course of [³H]inositol phosphate accumulation was measured in the absence or presence of carbachol in M1 muscarinic cholinergic receptor–expressing COS-7 cells. Essentially superimposable time courses were obtained with each method. Similarly, the fold stimulation of COS-7 cells expressing a GTPase-deficient mutant of Gαq was the same with each method. The absolute cpm obtained from [³H] inositol phosphates depends on the amount of SPA beads added (Fig. 1). However, as is also illustrated in Fig. 1, the signal obtained is linear over a wide range of SPA bead concentrations, and, therefore, low amounts of SPA beads added per well of a 96-well plate provide adequate sensitivity for measuring [³H]inositol phosphate responses. For example, the 0.2 mg amount of SPA beads routinely added per well in our studies enables

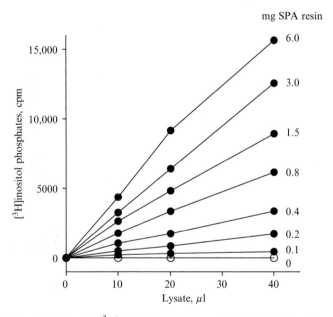

FIG. 1. Quantification of [³H]inositol phosphates with SPA beads. M1 muscarinic cholinergic receptor-expressing COS-7 cells grown in 96-well plates were labeled overnight with 1 μCi [³H]inositol/well as described in the "Methods." Incubations were carried out the following day in the presence of 100 μM carbachol and 10 mM LiCl for 60 min. [³H]inositol lipid hydrolysis was terminated, and [³H]inositol phosphate-containing medium was obtained by lysis of the cells with 30 μl of 50 mM formic acid per well. The indicated amounts of cell lysate were combined with the indicated amount of SPA resin in 96-well plates as described previously. Radioactivity was quantified using a microplate scintillation counter.

quantification of the amount of [³H]inositol phosphate present with high reliability, reproducibility, and linearity.

We have quantified activation of phospholipase C in intact cells using the robotic transfection of COS-7 cells with transfection vectors for Rho family GTPases and PLC isozymes and the assay system described previously for measurement of [³H]inositol phosphates. Transfection of COS-7 cells with PLC-β2, Rac1, Rac2, Rac3, or RhoA alone had little effect on [³H]inositol phosphate accumulation (Fig. 2). However, cotransfection of increasing amounts of expression vectors for Rac1, Rac2, or Rac3 along with 30 ng of PLC-β2 DNA resulted in a marked increase in [³H]inositol phosphate accumulation. This PLC-β2-dependent activation by Rac GTPases was dependent on the amount of GTPase expressed as detected by immunoblots (not shown), and expression of Rac GTPases had no effect on PLC-β2 expression (not shown). The fold stimulation of PLC-β2

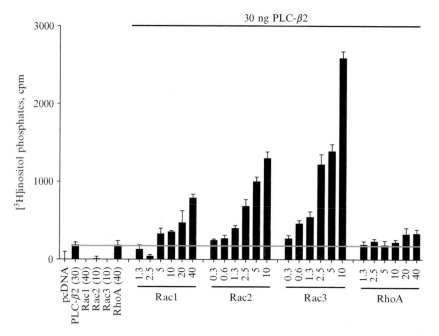

FIG. 2. GTPase selective activation of PLC-β2. COS-7 cells were transiently transfected (24 h) with pcDNA mammalian expression vectors encoding full-length PLC-β2 and GTPase-deficient mutant forms of Rac1, Rac2, Rac3, or RhoA. The values below each bar indicate the amount (ng) of DNA transfected. After an 18-h [³H]inositol labeling period and a 1-h LiCl treatment, cells were lysed and [³H]inositol phosphates quantified as described in the "Methods." Data shown are the mean ± S.E.M. from a single experiment representative of three replicate experiments.

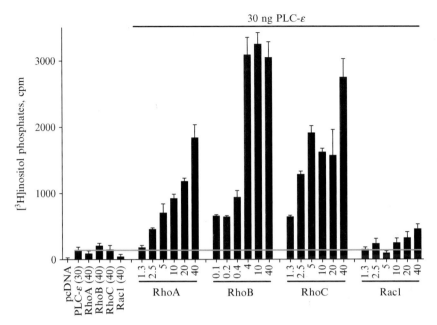

FIG. 3. GTPase selective activation of PLC-ε. COS-7 cells were transiently transfected (24 h) with pcDNA mammalian expression vectors encoding full-length PLC-ε and GTPase-deficient mutant forms of RhoA, RhoB, RhoC, or Rac1. The values below each bar indicate the amount (ng) of DNA transfected. After an 18-h [^{3}H]inositol labeling period and a 1-h LiCl treatment, cells were lysed and [^{3}H]inositol phosphates quantified as described in the "Methods." Data shown are the mean $\pm$ S.E.M. from a single experiment representative of three replicate experiments.

observed with maximally effective amounts of Rac GTPase expression was approximately 15-fold and was similar to that observed in other experiments in which [^{3}H]inositol phosphate accumulation was quantified using Dowex chromatography (Snyder *et al.*, 2003). Typical values for [^{3}H]inositol phosphates quantified under the conditions described were empty vector = 600 cpm, PLC-β2 = 800 cpm, maximally effective amount of Rac1 + PLC-β2 = 3700 cpm. In contrast, coexpression of PLC-β2 with RhoA (Fig. 2) or RhoB or RhoC (data not shown) had negligible effects on [^{3}H]inositol phosphate accumulation.

The robotic transfection/SPA bead system also was used to quantify activation of PLC-ε by Rho-family GTPases. Again, expression of PLC-ε, RhoA, RhoB, RhoC, or Rac1 alone had little effect on [^{3}H]inositol phosphate accumulation (Fig. 3). However, cotransfection of increasing amounts of expression vectors for RhoA, RhoB, or RhoC with PLC-ε

resulted in a marked increase in [^{3}H]inositol phosphate accumulation. In contrast, negligible PLC-ε-dependent stimulation of [^{3}H]inositol phosphate accumulation was observed with Rac1 (Fig. 3) or Rac2 or Rac3 (data not shown). The approximately 20-fold stimulation observed with PLC-ε and RhoA was similar to that previously observed in experiments in which [^{3}H] inositol phosphates were quantified using Dowex chromatography (Wing et al., 2003). Typical values for [^{3}H]inositol phosphates quantified under the conditions described were empty vector = 500 cpm, PLC-ε = 700 cpm, maximally effective amount of RhoA + PLC-ε = 3800 cpm.

Concluding Remarks

The existence of a broad family of at least 13 different mammalian PLC isozymes suggests multiple physiological roles of these signaling proteins, and PLC isozymes are important in many ways that transcend their historically considered function in hormone-regulated Ca^{2+} signaling. These enzymes process multiple forms of upstream regulation and coordinate multiple downstream responses unrelated to those initially associated with PLC action. The realization that Rho superfamily GTPases prominently regulate PLC isozymes in parallel with their activation by the more traditionally considered tyrosine phosphorylation or heterotrimeric G proteins adds remarkable and subtle complexities to the otherwise simple hydrolysis of PtdIns(4,5)P^2. Many biological responses are associated with Rho family GTPases, and it will be important to understand the roles played by PLC isozymes in Rac and Rho-regulated signaling. The method described here illustrates a simple and highly flexible approach for studying Rho GTPase–dependent activation of PLC isozymes at the cellular level.

Acknowledgments

This work was supported by National Institutes of Health grants GM29536, GM57391, and GM65533. D. B. recognizes support from the Pharmaceutical Research and Manufacturers of America Foundation. J. S. recognizes support from the Pew Charitable Trusts.

References

Berridge, M. J. (1987). Inositol trisphosphate and diacylglycerol: Two interacting second messengers. Annu. Rev. Biochem. 56, 159–193.
Berridge, M. J., Downes, C. P., and Hanley, M. R. (1982). Lithium amplifies agonist-dependent phosphatidylinositol responses in brain and salivary glands. Biochem. J. 206, 587–595.

Brown, H. A., Lazarowski, E. R., Boucher, R. C., and Harden, T. K. (1991). Evidence that UTP and ATP regulate phospholipase C through a common extracellular 5'-nucleotide receptor in human airway epithelial cells. *Mol. Pharmacol.* **40,** 648–655.

Fleischman, L. F., Chahwala, S. B., and Cantley, L. (1986). Ras-transformed cells: Altered levels of phosphatidylinositol-4,5-bisphosphate and catabolites. *Science* **231,** 407–410.

Illenberger, D., Schwald, F., Pimmer, D., Binder, W., Maier, G., Dietrich, A., and Gierschik, P. (1998). Stimulation of phospholipase C-β2 by the Rho GTPases Cdc42Hs and Rac1. *EMBO J.* **17,** 6241–6249.

Illenberger, D., Walliser, C., Nurnberg, B., Diaz, L. M., and Gierschik, P. (2002). Specificity and structural requirements of phospholipase C-β stimulation by Rho GTPases versus G protein βγ dimers. *J. Biol. Chem.* **278,** 3006–3014.

Jin, T. G., Satoh, T., Liao, Y., Song, C., Gao, X., Kariya, K., Hu, C. D., and Kataoka, T. (2001). Role of the CDC25 homology domain of phospholipase C-ε in amplification of Rap1-dependent signaling. *J. Biol. Chem.* **276,** 30301–30307.

Kelley, G. G., Reks, S. E., Ondrako, J. M., and Smrcka, A. V. (2001). Phospholipase C-ε: A novel Ras effector. *EMBO J.* **20,** 743–754.

Liu, J. J., Hartman, D. S., and Bostwick, J. R. (2003). An immobilized metal ion affinity adsorption and scintillation proximity assay for receptor-stimulated phosphoinositide hydrolysis. *Anal. Biochem.* **318,** 91–99.

Lopez, I., Mak, E. C., Ding, J., Hamm, H. E., and Lomasney, J. W. (2001). A novel bifunctional phospholipase C that is regulated by Gα12 and stimulates the Ras/mitogen-activated protein kinase pathway. *J. Biol. Chem.* **276,** 2758–2765.

Nakahata, N., and Harden, T. K. (1987). Regulation of inositol trisphosphate accumulation by muscarinic cholinergic and H1-histamine receptors on human astrocytoma cells. Differential induction of desensitization by agonists. *Biochem. J.* **241,** 337–344.

Rhee, S. G. (2001). Regulation of phosphoinositide-specific phospholipase C. *Annu. Rev. Biochem.* **70,** 281–312.

Seifert, J. P., Wing, M. R., Snyder, J. T., Gershburg, S., Sondek, J., and Harden, T. K. (2004). RhoA activates purified phospholipase C-ε by a guanine nucleotide-dependent mechanism. *J. Biol. Chem.* **279,** 47992–47997.

Snyder, J. T., Singer, A. U., Wing, M. R., Harden, T. K., and Sondek, J. (2003). The pleckstrin homology domain of phospholipase C-β2 as an effector site for Rac. *J. Biol. Chem.* **278,** 21099–21104.

Song, C., Hu, C. D., Masago, M., Kariyai, K., Yamawaki-Kataoka, Y., Shibatohge, M., Wu, D., Satoh, T., and Kataoka, T. (2001). Regulation of a novel human phospholipase C, PLC-ε, through membrane targeting by Ras. *J. Biol. Chem.* **276,** 2752–2757.

Wakelam, M. J., Davies, S. A., Houslay, M. D., McKay, I., Marshall, C. J., and Hall, A. (1986). Normal p21N-ras couples bombesin and other growth factor receptors to inositol phosphate production. *Nature* **323,** 173–176.

Wing, M. R., Houston, D., Kelley, G. G., Der, C. J., Siderovski, D. P., and Harden, T. K. (2001). Activation of phospholipase C-ε by heterotrimeric G protein βγ-subunits. *J. Biol. Chem.* **276,** 48257–48261.

Wing, M. R., Snyder, J. T., Sondek, J., and Harden, T. K. (2003). Direct activation of phospholipase C-ε by Rho. *J. Biol. Chem.* **278,** 41253–41258.

[38]   Activation of Rho and Rac by Wnt/Frizzled Signaling

By RAYMOND HABAS and XI HE

Abstract

Wnt/Frizzled (Fz) signaling controls developmental, physiological, and pathological processes by means of several distinct pathways. Wnt/Fz activation of the small GTPases Rho and Rac has emerged as a key mechanism that regulates cell polarity and movements during vertebrate gastrulation. In this chapter, we describe biochemical assays for the detection of Wnt/Fz activation of Rho and Rac in both mammalian cells and *Xenopus* embryo explants.

Introduction

The Wnt family of secreted signaling molecules has central roles in embryonic development and human diseases (Logan and Nusse, 2004). Although Wnt signaling by means of the canonical β-catenin pathway has been most intensively studied in cell fate determination, cell proliferation, and human cancers (Giles *et al.*, 2003; Logan and Nusse, 2004), Wnt activation of the Rho family of GTPases has received increased attention. This sometimes so-called non-canonical Wnt signaling pathway is essential for cell polarity and movements during vertebrate gastrulation and shares common components similar to those involved in the establishment of planar cell polarity (PCP) in *Drosophila*, thus it is also often referred to as the Wnt/PCP pathway (Mlodzik, 2002). In addition, Wnt proteins can also activate other non-canonical signaling pathways, including calcium calmodulin–dependent kinase 2 (CaMK2), protein kinase C (PKC) (Sheldahl *et al.*, 2003), and protein kinase A (PKA) (Chen *et al.*, 2005) (Fig. 1).

It is generally accepted that the Frizzled (Fz) family of serpentine receptors are cell surface receptors for Wnt proteins (Huang and Klein, 2004). There are 19 wnt genes and 10 fz genes in both the mouse and human genomes. These shear numbers and a general lack of purified soluble Wnt proteins (except for Wnt-3a; Willert *et al.*, 2003) have hindered the study of Wnt-Fz interaction and specificity, which is poorly understood in vertebrates. In addition to Fz, several other families of cell surface receptors, including LDL receptor–related proteins 5 and 6 (LRP5/6) (He *et al.*, 2004), Ryk (an atypical receptor tyrosine kinase) (Lu *et al.*,

METHODS IN ENZYMOLOGY, VOL. 406
0076-6879/06 $35.00
DOI: 10.1016/S0076-6879(06)06038-1

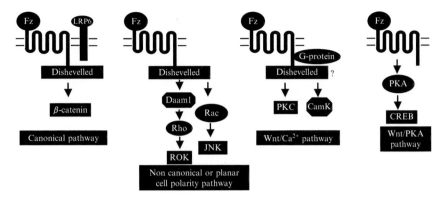

FIG. 1. A schematic representation of the several known Wnt signal transduction cascades. For the canonical pathway, signaling through the Fz receptor and LRP5/6 co-receptor complex and dishevelled induces the stabilization of β-catenin, which then translocates into the nucleus where it complexes with Lef/Tcf family members to mediate transcriptional induction of target genes. For non canonical or planar cell polarity (PCP) signaling, Wnt signaling is transduced through Fz independent of LRP5/6. This pathway through dishevelled mediates cytoskeletal changes though the small GTPases Rho and Rac, which in turn activates Rho kinase (ROK) and Jun N-terminal kinase (JNK), respectively. Trimeric G-proteins may be involved in both β-catenin and PCP pathways. For the Wnt/Ca^{2+} pathway, Wnt signaling by way of Fz mediates the activation of G-proteins to activate PLC, CaMK2, and PKC. There is disagreement about the involvement of Dsh in this pathway. The Wnt/PKA pathway stimulates PKA activation, which in turn activates CREB for gene induction events. It is not known whether dishevelled plays a role in this pathway.

2004), and some proteoglycans (Ohkawara et al., 2003; Topczewski et al., 2001) are also important for Wnt signal transduction, highlighting the complexity and versatility of the Wnt signaling system. Although LRP5/6 are most likely coreceptors for Fz proteins and required specifically for the Wnt/β-catenin pathway (Fig. 1) (He et al., 2004), proteoglycans seem to be involved in extracellular Wnt ligand transport and distribution, thus they potentially affect many Wnt pathways (Lin, 2004). The signaling pathway activated by Ryk and the relationship between Ryk and Fz proteins in Wnt signaling remains obscure (He, 2004).

Attempts have been made to classify Wnt molecules into either the canonical subfamily that includes Wnt-1, Wnt-3a, and Wnt-8 and that "only" activates the β-catenin pathway or the non canonical subfamily that activates β-catenin–independent pathways, such as Wnt-4, Wnt-5a, and Wnt-11. However, accumulating evidence suggests that such a classification may be oversimplified, and many Wnt proteins, including Wnt-1, Wnt-3a, Wnt-5a, and Wnt-11, can activate multiple pathways in different experimental contexts, likely depending on receptor complements (including Fz and other receptors) and other cofactors that these Wnt proteins may interact with (Endo *et al.*, 2005; Habas *et al.*, 2003; He *et al.*, 1997; Kishida *et al.*, 2004; Qiang *et al.*, 2003; Tao *et al.*, 2005). Thus, with perhaps a few exceptions, there are no clear rules yet that faithfully predict Wnt or Fz specificity for the activation of a particular pathway in vertebrates. Despite these caveats, it is now well established that activation of the Rho family of GTPases by Wnt/Fz signaling is critical for vertebrate development (Keller, 2002; Veeman *et al.*, 2003; Wallingford *et al.*, 2002).

The Rho family of GTPases plays important roles in regulating cytoskeletal architectures associated with cell polarity and motility. The prototypes of this family, Rho, Rac, and Cdc42, act as bimolecular switches cycling between the active GTP-bound form and the inactive GDP-bound form, with the GTP-bound form interacting with downstream effectors (Raftopoulou and Hall, 2004). During *Xenopus* and zebrafish embryogenesis, Wnt-11 activates both RhoA and Rac to regulate the so-called convergent extension (CE) movements that are a major driving force of gastrulation. During CE movements, cells in the dorsal mesoderm polarize, elongate, and align along the mediolateral axis and intercalate among one another, resulting in the mediolateral narrowing (convergence) and anteroposterior lengthening (extension) of the axis of the embryo (Keller, 2002; Wallingford *et al.*, 2002). Wnt-11 activation of RhoA and Rac can be demonstrated in dorsal embryo explants (Figs. 2 and 3), in which interfering with Wnt-11 or Fz function prevents RhoA and Rac activation (Habas *et al.*, 2001, 2003). Conversely, overexpression of Wnt-11 or *Xenopus* Fz7 (Xfz7) in the embryo ventral region, which neither exhibits CE movements nor expresses Wnt-11 and Xfz7, is sufficient to activate RhoA and Rac (Habas *et al.*, 2001, 2003).

Wnt/Fz activation of Rho and Rac has also been observed in several commonly used mammalian cell lines (Endo *et al.*, 2005; Habas *et al.*, 2001; Qiang *et al.*, 2003). In these experiments (Fig. 2), transfection of Wnt-1 or Wnt-3a cDNA, of certain (but not all) fz cDNA, or treatment with Wnt-1 or Wnt-3a conditioned medium results in RhoA and Rac activation. These observations in vertebrate embryos/mammalian cells followed and paralleled earlier genetic studies in *Drosophila*, which demonstrated that fz function in PCP relies on rhoA and rac gene function (Mlodzik, 2002). Thus, Wnt/Fz signaling to RhoA and Rac is conserved, although it remains unknown whether a Wnt ligand is required for PCP in flies.

Wnt/Fz activation of RhoA and Rac requires the cytoplasmic phosphoprotein Dishevelled (Dsh), which often acts downstream of Fz (Wharton, 2003). Indeed, Dsh is also essential for Wnt/Fz activation of the β-catenin pathway (Boutros and Mlodzik, 1999). There are three Dsh proteins in

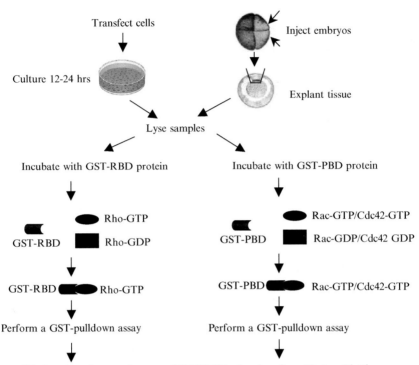

Rho, Rac/Cdc42 assay

FIG. 2. A schematic overview of the Rho and Rac/Cdc42 assays.

mammals (Dvl-1, 2, and 3), and all Dsh family members are composed of three highly conserved domains; an N-terminal DIX domain (for Dishevelled and Axin), a central PDZ domain (for postsynaptic density-95, discs-large, and zonula occludens-1), and a carboxyl DEP domain (for Dishevelled, Egl-10 Pleckstrin) (Wharton, 2003). The DIX domain is dispensable for PCP and for RhoA and Rac activation but is critical for β-catenin signaling (Veeman et al., 2003; Wharton, 2003). How Dsh channels Wnt/Fz signaling to different pathways is not understood. Downstream of Dsh, two independent and parallel pathways lead to the activation of RhoA and Rac, respectively (Fig. 1). The first pathway involves a molecule called Daam1 that binds to the PDZ domain of Dsh and can activate RhoA which in turn leads to the activation of the Rho-associated kinase Rock (Habas et al., 2003; Kishida et al., 2004; Marlow

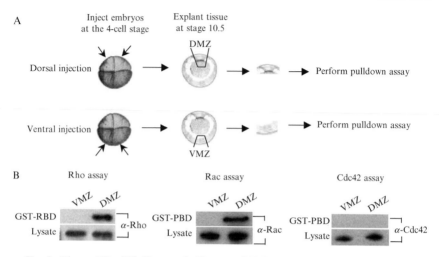

FIG. 3. Rho and Rac/Cdc42 assays in *Xenopus*. (A) Embryos are injected at the four-cell stage into the two dorsal or ventral cells and are allowed to develop to stage 10.5. At this point, the DMZ or VMZ are explanted and subjected to Rho or Rac/Cdc42 pull-down assays. (B) Examples of Rho, Rac, and Cdc42 assays performed with DMZs and VMZs. Rho, Rac, and Cdc42 activation detected from GST-RBD or GST-PBD pull-down samples show Rho and Rac but not Cdc42 is activated preferentially in the DMZ. Endogenous Rho, Rac, and Cdc42 levels are shown in the lysate samples.

et al., 2002). The second pathway requires the DEP domain of Dsh and activates Rac, which in turn stimulates Jun N-terminal kinase (JNK) activity (Boutros *et al.*, 1998; Habas *et al.*, 2003; Li *et al.*, 1999). Notably, the guanine nucleotide exchange factor (GEF) and the GTPase activating protein (GAP) for RhoA and Rac in Wnt/Fz signaling have not been identified.

It should be mentioned that a large number of genes, in addition to *fz*, *dsh*, *rhoA*, and *rac*, have been implicated in *Drosophila* PCP by genetic studies (Adler, 2002). Many of these gene products appear to act in feedback loops that result in polarized localization of Fz and Dsh and localized RhoA (and possibly Rac) activation, which are prerequisite for PCP establishment (Adler, 2002; Mlodzik, 2002). It is unknown whether localized Fz, Dsh and Rho/Rac activation occurs during Wnt-11 signaling. The vertebrate homologs of some of these genes, such as Strabismus and Prickle, have been shown to function in CE movements during gastrulation, but their mechanism of actions in relationship to Wnt/Fz signaling remains unclear (Veeman *et al.*, 2003).

Here we describe biochemical assays to investigate activation of Rho, Rac, and Cdc42 in mammalian culture cells and the *Xenopus* embryo. These assays use a glutathione S-transferase (GST)–pull-down strategy using fusion proteins that specifically bind to the activated or GTP-bound forms of Rho, Rac, and Cdc42. For the Rho assay, a Rho-binding fragment of the Rho effector Rhotekin is fused to GST and termed GST-RBD (Ren *et al.*, 1999) (Fig. 2). For the Rac/Cdc42 assay, the Rac/Cdc42 binding fragment of p21 (PAK) is fused to GST and termed GST-PBD (Akasaki *et al.*, 1999; Benard *et al.*, 1999) (Fig. 2). The GST-RBD and GST-PBD fusion proteins are produced in bacterial cells, purified, and incubated with cell lysates derived from either mammalian cells or *Xenopus* embryo explants (Fig. 2). GST-RBD and GST-PBD bind specifically to the GTP-bound forms of Rho and Rac or Cdc42, respectively, and are precipitated using GST-agarose beads and detected by conventional immunoblotting (Figs. 2 and 3). This assay provides an efficient means to access levels of activated Rho, Rac, and Cdc42.

Methods

Rho and Rac/Cdc42 Assays

Preparation of Recombinant GST-RBD Protein

1. Grow an overnight culture of a single colony of BL21 bacterial cell containing the GST-RBD plasmid in 20 ml LB-amp (100 mg/μl) at 30°.
2. Dilute the culture into 1 l of LB-amp (100 mg/μl) and grow at 30° until the optical density at 600 nm is 1.0. This takes 5–7 h, depending on starting optical density.
3. Induce the bacterial culture with 1 ml of 1 *M* IPTG and incubate for 3–4 h at 30°.
4. Aliquot the bacteria into 50-ml Falcon tubes and spin at 4000 rpm for 10 min to pellet the bacteria. Discard supernatant and flash freeze the pellets in liquid nitrogen.
5. Store the pellets at −80°. The pellets are stable for up to 1 year.

Preparation of Recombinant GST-PBD Fusion Protein

1. Grow an overnight culture of a single colony of BL21 bacterial cell containing the GST-RBD plasmid in 20 ml LB-amp (100 mg/μl) at 30°.

2. Dilute the culture into 1 l of LB-amp (100 mg/μl) and grow at 30°
until the optical density at 600 nm is 1.0. This takes 5–7 h, depending
on starting optical density.

3. Induce the bacterial culture with 1 ml of 1 M IPTG and incubate for
3–4 h at 30°.

4. Lyse the cells in 1× PBS and protease inhibitors using either
sonication or a French press.

5. Pellet the lysate by spinning at 15,000 rpm for 15 min and isolate the
supernatant.

6. Aliquot the supernatant into 1.5-ml Eppendorf tubes and flash
freeze in liquid nitrogen.

7. Store at −80°.

Extraction of GST-RBD and GST-PBD

GST-RBD

1. Prepare the glutathione sepharose beads by swelling approximately
100 μl with 1× PBS/10 mM DTT/1% Triton-X 100 for at least
30 min on ice, then wash three times with 500 μl of 1× PBS/10 mM
DTT/1% Triton-X 100. After the final wash, the beads can be stored
on ice as a 1× slurry. Do not spin beads higher than 3000 rpm during
the pelleting and washing stages, because this will damage the beads.
Also, prepare two tubes of beads, because you will have 2 ml of
lysate.

2. Thaw one aliquot of frozen GST-RBD pellet on ice and resuspend
in 2 ml 1× PBS.

3. Add 20 μl 1 M DTT, 20 μl protease inhibitor cocktail (Boehringer-
Mannheim), and 40 ml of 50 mg/μl lysozyme.

4. Vortex briefly to mix well and incubate on ice for 30 min.

5. Add 225 ml 10% Triton-X 100, 22.5 ml 1 M MgCl$_2$, and 22.5 ml of
10 mg/ml DNase1.

6. Vortex briefly to mix well and incubate on ice for 30 min.

7. Spin at 14,000 rpm at 4° for 2 min, and add 1 ml supernatant to each
of the two tubes of the preswollen beads.

8. Incubate on a Nutator at 4° for 45 min (do not exceed 1 h).

9. Spin and wash beads three times with 500 ml of 1× PBS/10 mM
DTT/1% Triton-X 100. After the final wash, store on ice in 1x slurry
with the final volume approximately 500 ml.

GST-PBD

1. Prepare the glutathione sepharose beads by swelling approximately 100 μl with 1$\times$ PBS/10 mM DTT/1% Triton-X 100 for at least 30 min on ice, then wash three times with 500 μl of 1$\times$ PBS/10 mM DTT/1% Triton-X 100, and after final wash, store on ice as a 1$\times$ slurry. Do not spin beads higher than 3000 rpm during the pelleting and washing stages, because this will damage the beads. Also prepare two tubes of beads, because you will have 1 ml of bacterial lysate.
2. Thaw one aliquot of frozen bacterial supernatant on ice.
3. Add 500 μl supernatant to each of the two tubes of the preswollen beads.
4. Incubate on a Nutator at 4° for 45 min and do not exceed 1 h.
5. Spin and wash beads three times with 500 μl of 1$\times$ PBS/10 mM DTT/1% Triton-X 100. After the final wash, store on ice in 1$\times$ slurry with the final volume approximately 500 μl.

Sample Preparations for the Pull-down Assays

Mammalian Cells

RHO ASSAY

1. Mammalian HEK293T cells are cultured in 6-well plates (30 mm) in 10% fetal bovine serum and DMEM media supplemented penicillin/streptomycin until 60% confluency.
2. 1 μg of cDNA is transfected into the cells using standard calcium-phosphate method.
3. Media is changed 12 h after transfection, and cells are incubated for 12–24 h.

RAC/CDC42 ASSAY

1. Mammalian HEK293T cells are cultured in 6-well plates (30 mm) in 10% fetal bovine serum and DMEM media supplemented penicillin/streptomycin until 60% confluency. At this point the cells are changed into 0.5% fetal bovine serum and DMEM media supplemented penicillin/streptomycin. This step helps to reduce the basal level of activated Rac/Cdc42.

2. Six hours after media is changed to 0.5% sera, 1 μg of cDNA is transfected into the cells using standard calcium-phosphate method.

3. Media is changed 12 h after transfection, and cells are incubated for an additional 12–24 h in 0.5% fetal bovine serum and DMEM media supplemented with penicillin/streptomycin.

Xenopus Embryos and Explants

1. *Xenopus* embryos are injected at the 4-cell stage into the two dorsal cells (for dorsal marginal zone [DMZ] explants) or into the two ventral cells (for ventral marginal zone [VMZ] explants) in 3% Ficoll/0.5× MMR.
2. Two hours after injections, embryos are changed into 0.1× MMR and cultured to stage 10.5.
3. The vitelline membrane is removed from the embryo, and DMZ or VMZ is explanted using forceps. Explants are pooled and stored on ice until they are lysed. All embryos are dissected on agarose-coated culture dishes in a solution of 0.5× MMR/1%BSA. *Xenopus* embryos and explants are handled as described elsewhere (Sive, 2000).

GST-RBD and GST-PBD Binding Assay

1. Lyse the cells in 500 μl of Rho or Rac lysis buffer and to each sample add 10 μl of 10 mg/ml DNase1 solution. Incubate on ice for 10 min and then spin samples at 14,000 rpm at 4°. For *Xenopus* explants, we typically use 50 explants (DMZ or VMZ) for each sample.

2. Remove 25 μl of supernatant and add 25 μl of 2× sample buffer, heat at 90° for 5 min and store. This is your whole cell lysate for control immunoblotting.

3. Remove the remaining 475 μl of supernatant and add to tubes containing approximately 50 μl of GST-beads coupled to the RBD or to PBD.

4. Incubate on a nutator at 4° for 1 h, and wash three times with Rho or Rac wash buffer. After final wash, resuspend in 50 μl of 2× sample buffer and heat at 90° for 5 min.

5. Perform Western Blotting.

Western Blotting

1. Resolve samples on a 12% SDS-PAGE gel and run until the bromophenol dye is approximately 1 inch from the bottom of the gel.
2. Transfer to nitrocellulose membrane, which is incubated for 1 h with 5% non-fat dry milk for 1 h at room temperature.
3. Wash twice for 5 min with 1× PBST.
4. Incubate with the primary antibody (Rho monoclonal, Santa Cruz, for mammalian cell extracts or Rho polyclonal, Santa Cruz, for

Xenopus explant extracts; Rac or Cdc42 monoclonal, Transduction Labs, for both mammalian cell and *Xenopus* explant extracts) at a 1/500 dilution for 1 h at room temperature or overnight at 4°.

5. Wash once for 15 min and four times for 5 min with 1× PBST.
6. Incubate with secondary antibody at 1/5000 solution for 1 h at room temperature.
7. Wash once for 15 min and four times for 5 min with 1× PBST.
8. Perform ECL reaction. We typically use SuperSignal PicoWest (Pierce).
9. The endogenous Rho/Rac/Cdc42 are detectable within 1 min in total lysates or 5 min in the GST-RBD/PBD pull-down samples for mammalian cell extracts. For *Xenopus* explant extracts, the endogenous Rho/Rac/Cdc42 are detectable within 3–5 min in lysates or 10 min in the GST-RBD/PBD pull-down samples.

Buffers

Rho Lysis Buffer. 50 mM Tris-HCl, pH 7.2, 500 mM NaCl, 1% Triton-X 100, 0.5% sodium deoxycholic acid, 0.1% SDS, 10 mM MgCl$_2$ and 1× protease inhibitors (added fresh each time for this and other buffers).

Rho Wash Buffer. 50 mM Tris-HCl, pH 7.2, 1% Triton-X 100, 150 mM NaCl, 10 mM MgCl$_2$ and 1× protease inhibitors.

Rac/CDC42 Lysis Buffer. 50 mM Tris, pH 7.5, 200 mM NaCl, 2% NP40, 10% glycerol, 10 mM MgCl$_2$ and 1× protease inhibitors.

Rac Wash Buffer. 25 mM Tris, pH 7.5, 40 mM NaCl, 1% NP$_4$0, 30 mM MgCl$_2$ and 1× protease inhibitors.

10x PBST. 995 ml of 10× PBS and 5 ml Tween-20 in 1 l.

10x MMR. 1 M NaCl, 20 mM KCl, 20 mM CaCl$_2$, 10 mM MgCl$_2$, 50 mM HEPES, pH to 7.6.

Notes

1. For each preparation of GST-RBD or GST-PBD bacteria, use one aliquot and follow the protocol to extract the fusion protein and resolve a fraction of the sample on a 12% SDS-PAGE gel and Coomassie blue stain. GST-RBD runs at approximately 35 kDa and GST-PBD approximately 37 kDa. If there is extensive degradation, redo the protein preparation.

2. Each 50-ml bacterial aliquot of GST-RBD yields approximately 200–300 μg of fusion protein, and each is enough for 10 samples for the GST-RBD assay. Each 1-ml aliquot of GST-PBD yields approximately 150–250 μg of protein, and each is enough for 10 samples for the GST-PBD assay.

3. Always keep samples on ice whenever possible and do not exceed the incubation times.

References

Adler, P. N. (2002). Planar signaling and morphogenesis in Drosophila. *Dev. Cell* **2**, 525–535.

Akasaki, T., Koga, H., and Sumimoto, H. (1999). Phosphoinositide 3-kinase-dependent and independent activation of the small GTPase Rac2 in human neutrophils. *J. Biol. Chem.* **274**, 18055–18059.

Benard, V., Bohl, B. P., and Bokoch, G. M. (1999). Characterization of rac and cdc42 activation in chemoattractant-stimulated human neutrophils using a novel assay for active GTPases. *J. Biol. Chem.* **274**, 13198–13204.

Boutros, M., and Mlodzik, M. (1999). Dishevelled: At the crossroads of divergent intracellular signaling pathways. *Mech. Dev.* **83**, 27–37.

Boutros, M., Paricio, N., Strutt, D. I., and Mlodzik, M. (1998). Dishevelled activates JNK and discriminates between JNK pathways in planar polarity and wingless signaling. *Cell* **94**, 109–118.

Chen, A. E., Ginty, D. D., and Fan, C. M. (2005). Protein kinase A signalling via CREB controls myogenesis induced by Wnt proteins. *Nature* **433**, 317–322.

Endo, Y., Wolf, V., Muraiso, K., Kamijo, K., Soon, L., Uren, A., Barshishat-Kupper, M., and Rubin, J. S. (2005). Wnt-3a-dependent cell motility involves RhoA activation and is specifically regulated by dishevelled-2. *J. Biol. Chem.* **280**, 777–786.

Giles, R. H., van Es, J. H., and Clevers, H. (2003). Caught up in a Wnt storm: Wnt signaling in cancer. *Biochim. Biophys. Acta* **1653**, 1–24.

Habas, R., Kato, Y., and He, X. (2001). Wnt/Frizzled activation of Rho regulates vertebrate gastrulation and requires a novel Formin homology protein Daam1. *Cell* **107**, 843–854.

Habas, R., Dawid, I. B., and He, X. (2003). Coactivation of Rac and Rho by Wnt/Frizzled signaling is required for vertebrate gastrulation. *Genes Dev.* **17**, 295–309.

He, X., Saint-Jeannet, J. P., Wang, Y., Nathans, J., Dawid, I., and Varmus, H. (1997). A member of the Frizzled protein family mediating axis induction by Wnt-5A. *Science* **275**, 1652–1654.

He, X. (2004). Wnt signaling went derailed again: A new track via the LIN-18 receptor? *Cell* **118**, 668–670.

He, X., Semenov, M., Tamai, K., and Zeng, X. (2004). LDL receptor-related proteins 5 and 6 in Wnt/beta-catenin signaling: Arrows point the way. *Development* **131**, 1663–1677.

Huang, H. C., and Klein, P. S. (2004). The Frizzled family: Receptors for multiple signal transduction pathways. *Genome Biol.* **5**, 234.

Keller, R. (2002). Shaping the vertebrate body plan by polarized embryonic cell movements. *Science* **298**, 1950–1954.

Kishida, S., Yamamoto, H., and Kikuchi, A. (2004). Wnt-3a and Dvl induce neurite retraction by activating Rho-associated kinase. *Mol. Cell Biol.* **24**, 4487–4501.

Li, L., Yuan, H., Xie, W., Mao, J., Caruso, A. M., McMahon, A., Sussman, D. J., and Wu, D. (1999). Dishevelled proteins lead to two signaling pathways. Regulation of LEF-1 and c-Jun N-terminal kinase in mammalian cells. *J. Biol. Chem.* **274**, 129–134.

Lin, X. (2004). Functions of heparan sulfate proteoglycans in cell signaling during development. *Development* **131**, 6009–6021.

Logan, C. Y., and Nusse, R. (2004). The Wnt Signaling Pathway in Development and Disease. *Annu. Rev. Cell Dev. Biol.* **20**, 781–810.

Lu, W., Yamamoto, V., Ortega, B., and Baltimore, D. (2004). Mammalian Ryk is a Wnt coreceptor required for stimulation of neurite outgrowth. *Cell* **119,** 97–108.

Marlow, F., Topczewski, J., Sepich, D., and Solnica-Krezel, L. (2002). Zebrafish Rho kinase 2 acts downstream of Wnt11 to mediate cell polarity and effective convergence and extension movements. *Curr. Biol.* **12,** 876–884.

Mlodzik, M. (2002). Planar cell polarization: Do the same mechanisms regulate Drosophila tissue polarity and vertebrate gastrulation? *Trends Genet.* **18,** 564–571.

Ohkawara, B., Yamamoto, T. S., Tada, M., and Ueno, N. (2003). Role of glypican 4 in the regulation of convergent extension movements during gastrulation in *Xenopus laevis.* *Development* **130,** 2129–2138.

Qiang, Y. W., Endo, Y., Rubin, J. S., and Rudikoff, S. (2003). Wnt signaling in B-cell neoplasia. *Oncogene* **22,** 1536–1545.

Raftopoulou, M., and Hall, A. (2004). Cell migration: Rho GTPases lead the way. *Dev. Biol.* **265,** 23–32.

Ren, X. D., Kiosses, W. B., and Schwartz, M. A. (1999). Regulation of the small GTP-binding protein Rho by cell adhesion and the cytoskeleton. *EMBO J.* **18,** 578–585.

Sheldahl, L. C., Slusarski, D. C., Pandur, P., Miller, J. R., Kuhl, M., and Moon, R. T. (2003). Dishevelled activates Ca^{2+} flux, PKC, and CamKII in vertebrate embryos. *J. Cell Biol.* **161,** 769–777.

Sive, H. L., Grainger, R. M., and Harland, R. M. (2000). "Early Development of *Xenopus laevis.* A laboratory manual." Cold Spring Harbor Press, Cold Spring Harbor, NY.

Tao, Q., Yokota, C., Puck, H., Kofron, M., Birsoy, B., Yan, D., Asashima, M., Wylie, C. C., Lin, X., and Heasman, J. (2005). Maternal Wnt11 activates the canonical Wnt signaling pathway required for axis formation in *Xenopus* embryos. *Cell* **120,** 857–871.

Topczewski, J., Sepich, D. S., Myers, D. C., Walker, C., Amores, A., Lele, Z., Hammerschmidt, M., Postlethwait, J., and Solnica-Krezel, L. (2001). The zebrafish glypican knypek controls cell polarity during gastrulation movements of convergent extension. *Dev. Cell* **1,** 251–264.

Veeman, M. T., Axelrod, J. D., and Moon, R. T. (2003). A second canon. Functions andmechanisms of beta-catenin-independent Wnt signaling. *Dev. Cell* **5,** 367–377.

Wallingford, J. B., Fraser, S. E., and Harland, R. M. (2002). Convergent extension: The molecular control of polarized cell movement during embryonic development. *Dev. Cell* **2,** 695–706.

Wharton, K. A., Jr. (2003). Runnin' with the Dvl: Proteins that associate with Dsh/Dvl and their significance to Wnt signal transduction. *Dev. Biol* **253,** 1–17.

Willert, K., Brown, J. D., Danenberg, E., Duncan, A. W., Weissman, I. L., Reya, T., Yates, J. R., 3rd, and Nusse, R. (2003). Wnt proteins are lipid-modified and can act as stem cell growth factors. *Nature* **423,** 448–452.

[39]   Fluorescent Assay of Cell-Permeable C3 Transferase Activity

By DANA LASKO and LISA McKERRACHER

Abstract

C3 transferase has been extensively used in the investigation of Rho signaling. It acts as a Rho antagonist by covalently attaching an ADP-ribose group to RhoA, B, and C proteins. Cell-permeable C3 fusion proteins have been developed so that lower doses of C3 can be used. We describe a simple and fast fluorescence-based assay to evaluate the enzymatic activity of cell-permeable C3 proteins purified from *Escherichia coli*. The assay measures glycohydrolase (GH) activity of C3 that cleaves NAD^+ into ADP-ribose and nicotinamide. Results from the GH activity correlate with other tests carried out in tissue culture cells such as neurite outgrowth or ADP ribosylation of RhoA. This method provides reliable measurements of the activity of permeable C3 proteins or other C3-related proteins.

Introduction

C3 transferase (called C3) is a bacterial protein that specifically ADP-ribosylates the active region of Rho A, B, and C proteins, and C3 is often used in cell biology as a Rho antagonist (Aktories *et al.*, 2004; Han *et al.*, 2001; Ménétrey *et al.*, 2002). C3 does not possess a natural cell-binding component that would allow it to efficiently enter cells. Methods that disrupt the cell membrane or microinjection are used to help aid the entry of C3 into cells. To overcome the limitation of delivery of C3, several groups have constructed cell-permeable forms of C3 (Aktories *et al.*, 2004; Sauzeav *et al.*, 2001; Winton *et al.*, 2002). We have compared several cell-permeable C3 proteins and selected C3-05, a cell permeable C3 with a proline-rich transport sequence, for further experiments. C3-05 was later modified to a version called C3-07 to permit more efficient purification (Bertrand *et al.*, 2005). Both C3-05 and C3-07 exhibit the same enzymatic activity, and both inactivate Rho.

Inside the cell, C3 catalyzes the transfer of an ADP-ribose group to Asn-41 (Sekine *et al.*, 1989) of RhoA, RhoB, or RhoC. Besides its Rho-specific ADP-ribosyltransferase activity, C3 exoenzyme has a glycohydrolase (GH) activity that cleaves NAD^+ into ADP-ribose and

METHODS IN ENZYMOLOGY, VOL. 406
Copyright 2006, Elsevier Inc. All rights reserved.

0076-6879/06 $35.00
DOI: 10.1016/S0076-6879(06)06039-3

nicotinamide in the absence of protein substrate. GH activity can be measured using radiolabeled NAD^+ (Ménétrey et al., 2002; Wilde et al., 2002). To determine the potency of cell-permeable C3 protein preparations, we developed an assay to measure the formation of fluorescent etheno-ADP-ribose (ε-ADP-ribose) in a 96-well microtiter plate format. GH activity in this assay correlated with two biological endpoints of C3 activity in cultured cells: (1) neurite outgrowth induced by C3 in culture of NG108 cells and (2) ADP-ribosylation of RhoA protein in homogenates of cells treated with C3. Therefore, the fluorescent GH assay provides a reliable quantitative measure of C3 activity.

Purification of Cell-Permeable C3 Proteins

C3-05 is fusion of C3 exoenzyme and a 19-mer peptide, a protein that demonstrated superior performance in neurite outgrowth bioassays (Winton et al, 2002). The protein was expressed in DH5α cells containing pGEX-4T-C3-05 by IPTG induction and purified by affinity chromatography (Dillon and Feig, 1995; Lehmann et al., 1999). As well, we used C3-07, a C3-05 derivative subcloned into the pET9a vector, containing a slight modification at the N-terminus, pET-9a-C3-07. Another construct used to validate the assay was pET9a-C3-07 Glu191Gln, which alters the active site conserved in the ARTT motif present of ADP-ribosyltransferases (Han et al., 2001; Saito et al., 1995). The C3-07 or C3-07 Q191E proteins were expressed in E. coli BL21 (DE3) cells and purified using column chromatography based on a method involving extraction in the presence of 0.15 M NaCl. We prepared the bacterial extract by sonication and used prepacked columns containing SPXL and Superdex 75 on an Äkta Explorer 100 FPLC (all from Amersham Pharmacia). A Q Sepharose column was used to reduce nucleic acid and endotoxin levels. With this method, protein was greater than 95% pure as determined by SDS-PAGE and contained less than 2 units/mg endotoxin as measured by Limulus Amebocyte Lysate assay (QCL-1000 kit, BioWhittaker). Protein concentration was measured by the method of Bradford or by UV spectrophotometry. Neurite outgrowth bioassay with NG-108 cells (Gringas et al., 2004) was in 96-well plates with incubated for 4 h with cell-permeable C3 (10 μg/ml). After the incubation, cells were fixed with 4% paraformaldehyde and 1% glutaraldehyde, then stained with 0.05% cresyl violet. Cells were examined by light microscopy, and neurite outgrowth was determined as the percentage of cells with at least one neurite longer than the cell body. Mobility shift of ADP-ribosylated RhoA was done with cell homogenates, as reported previously (Laplante et al., 2004).

Assay Method Overview

We investigated the use of glycohydrolase activity as a reporter for the mono-ADP-ribosyltransferase activity of cell-permeable C3 proteins. Barrio et al. (1972) synthesized ε-NAD$^+$, a fluorescent derivative, and found that ε-ADP-ribose had a 10-fold higher fluorescence intensity, an observation ascribed to the loss of quenching by the aromatic moiety of nicotinamide. Thus, no physical separation of reaction products from substrate is required to measure the conversion of ε-NAD$^+$ to ε-ADP-ribose, providing a simple, fluorescence-based assay for NADase activity. Pekala and Anderson (1978) used this approach to study a bovine glycohydrolase. Armstrong and Merrill (2001) included protein substrate and used ε-AMP (Secrist et al., 1972) for a standard curve to quantitate ε-ADP-ribose formation by Pseudomonas aeruginosa exotoxin A.

We adapted this assay to measure the GH activity of C3, using a 96-well format for fluorescence measurements. We estimated formation of ADP-ribose in the absence of a substrate Rho GTPase to eliminate a potential source of variability and expense. We used the assay for cell-permeable C3 at a stage in purification where residual NAD$^+$ from the host E. coli is negligible. The reaction followed is depicted in Fig. 1.

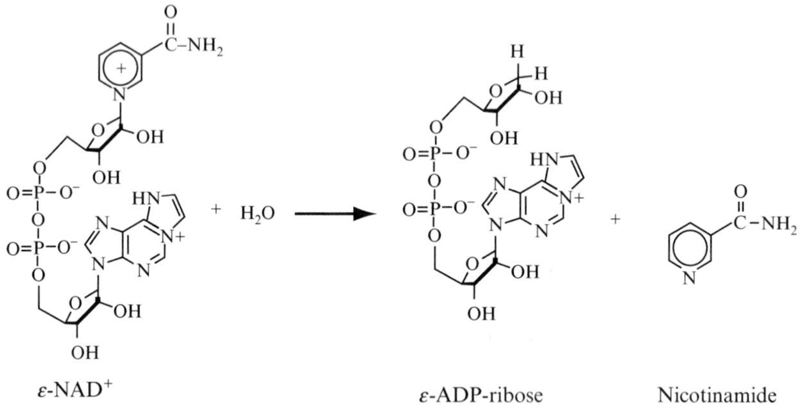

ε-NAD$^+$ ε-ADP-ribose   Nicotinamide

FIG. 1. Fluorescence-based glycohydrolase assay. The basis for measurement of product is the increase in fluorescence intensity of ε-ADP-ribose compared with ε-NAD$^+$ (drawn as double zwitterion). Amount of product formed is measured using a standard curve of the fluorescence intensity of ε-AMP solutions at concentrations of 0.25–10 μM. The assay is robust within the range of pH from 6–8 in several buffers.

Method for Measurement of NAD^+ Glycohydrolase Activity of Cell-Permeable C3 Proteins

A spectrofluorometer capable of excitation at 305 nm and emission at 410 nm is required for this assay. We used a Gemini EM (Molecular Devices) fitted with a 96-well microtiter plate adaptor. The method is carried out as follows:

1. Prepare stock solutions of 4 mM ε-NAD substrate (Sigma) and 50 μM ε-AMP (Sigma) in 10 mM HEPES-NaOH, pH 7.5. Store at $-20°$ protected from light.

2. Prepare reaction mixtures of 25 μl containing 10 mM HEPES-NaOH, pH 7.5, 2 mM MgCl$_2$, and 0.4 mM ε-NAD$^+$. This volume is for 0.5-ml polypropylene tubes. *Note*: Multiply volume accordingly for other formats.

3. Add C3 protein sample (6 μg) to be tested, and incubate for 90 min at 37°. Include negative (buffer blank) and positive controls in each set of assays.

4. During the incubation, dilute the ε-AMP stock solution in 10 mM HEPES-NaOH to make at least five standard solutions ranging from 0.25–10 μM ε-AMP. Transfer 125 μl of each solution (in duplicate) to a 96-well black fluorescent plate treated for tissue culture. These samples will be read in the spectrofluorometer at the same time as the reaction samples to prepare a standard curve.

5. Add 100 μl 10 mM HEPES-NaOH, pH 7.5, and 12.2 mM EDTA to each reaction tube. Place the test tube rack on ice.

6. Immediately transfer the contents of each tube to the 96-well black fluorescent plate containing the ε-AMP standard solutions.

7. Read the 96-well plate from the top using an endpoint fluorescence protocol, with excitation was at 305 nm and emission at 410 nm.

8. Subtract mean background fluorescence from the incubated buffer blank from the mean fluorescence of each sample to obtain net fluorescence in relative fluorescence units (RFU).

9. To prepare the standard curve, plot the concentration of standard solutions of ε-AMP on the x-axis and the observed RFU on the y-axis.

10. ε-ADP ribose is assumed to have similar fluorescence to ε-AMP. Estimate the amount of ADP-ribose formed in each GH reaction from the equation of the ε-AMP standard curve.

11. Calculate specific activity in units/mg protein. A unit is defined as nmol ADP-ribose formed in 30 min at 37°.

Results

Permeable C3 proteins prepared as described had >20 units/mg specific activity in the fluorescence-based glycohydrolase assay. The parameters that we evaluated included incubation time, concentration of permeable C3 protein, pH, and concentration of ε-NAD^{+}. A time course from 30–180 min revealed that ε-ADP-ribose increased in a linear fashion (Fig. 2). We chose an incubation time of 90 min for standard assay conditions. Similarly, the amount of product formed was linear with respect to concentration of C3 protein between 4–9 μM. When ε-NAD^{+} concentration was increased to 0.8 mM, ε-ADP-ribose formation reached a plateau. A concentration of 0.4 mM ε-NAD^{+} as a compromise to reduce consumption of substrate provided robust results over many assays (coefficient of variation <15%). The assay can be carried out in a number of buffers and within the range of pH 6–8. We have tested Tris-HCl, Tris-maleate, and sodium phosphate, as well as HEPES-NaOH. There was very little change in the amount of ε-ADP-ribose formed between pH 6 and 8. Limitations of this assay include requirement for microgram amounts of purified protein and buffer interference in the presence of high concentrations of Na^{+} or phosphate anion.

Measurements of GH Activity Using the Fluorescence-Based Assay

The simple GH assay is useful for determining the activity of cell-permeable C3, and when multiple experiments were performed, the relative standard deviation for wild-type protein was less than 10%

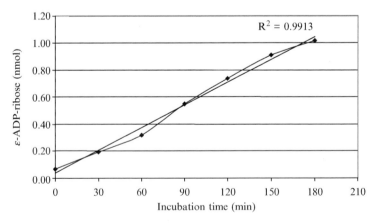

FIG. 2. Example of assay results—time course. Reactions containing 6 μg C3-07 protein were prepared and incubated at 37°. Time of incubation was plotted against amount of ε-ADP-ribose formed (nmol).

(Table I). It can also be used to characterize mutant proteins for GH activity. Point mutations of the glutamate residue in the ARTT turn of C3 transferase define the catalytic site (Aktories *et al.*, 2004; Han *et al.*, 2001; Saito *et al.*, 1995). Among the mutants that we chose to characterize is the Glu to Gln (E→Q) substitution at the corresponding position of permeable C3 proteins, which has been reported to abrogate both GH and ADP-ribosyltransferase activity (Aktories *et al.*, 2004). Table I shows that there is a substantial reduction in GH activity of the mutant permeable C3 protein as anticipated.

Purified mutant protein was mixed with wild-type protein, and the samples were assayed for GH activity under standard conditions. Figure 3 shows that specific activity decreases in linear fashion as the amount of mutant protein in the mixture is increased. This linear relationship is consistent with inactivation of the catalytic center of a monomer protein.

Correlation of Glycohydrolase Activity with Biological Activity

Two assays carried out in cultured cells provide complementary information on potency of cell-permeable C3. Bioassays demonstrated severely diminished induction of neurite outgrowth by Glu→Gln active site mutant protein (Table II). Results of mobility shift experiments in PC-12 cells that were incubated with cell-permeable C3 protein indicated the characteristic decrease in RhoA mobility ascribed to ADP-ribosylation (Table III). In contrast, RhoA mobility was normal when mutant protein or buffer was incubated with PC-12 cells.

Table III summarizes the correlation of glycohydrolase activity with assays in cultured cells. Possible applications of the GH assay include determination of potency of a batch of permeable C3 protein, comparison of protein preparations, characterization of mutant proteins, accelerated

TABLE I
GH ASSAY OF MUTANT PROTEIN

Sample	Specific activity	Percent relative standard deviation
C3-07 (positive control)	30.2 ± 1.96	6.49
E→Q mutant protein	1.02 ± 1.00	98
Vehicle control	0	0

Data are mean $\pm$ SD from three experiments each consisting of two separate assays in triplicate. Percent relative standard deviation for these experiments was calculated as 100* (SD/mean).

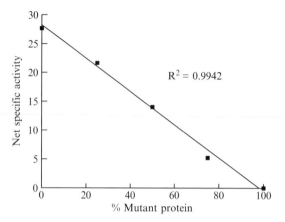

$R^2 = 0.9942$

FIG. 3. Mixing experiment with mutant protein. Active cell-permeable C3 was mixed in the proportions indicated with cell-permeable C3 containing a Glu mutation to Gln in the active site for glycohydrolase activity. Specific activity (units/mg) of the mixture was then determined, and the value for 100% mutant subtracted from each data point to generate net specific activity. A total of 6 μg of protein was present in each reaction.

TABLE II
NEURITE OUTGROWTH ACTIVITY OF ACTIVE SITE GLU→GLN MUTANT PROTEIN

Sample	Percent cells with neurites
Glu→Gln active site mutant	0.81 ± 0.02
Vehicle control	1.58 ± 0.77
C3–7 (Positive control)	21.51 ± 1.98

Representative experiment. NG-108-15 cells were treated with 10 μg/ml permeable C3 protein or Glu→Gln active site mutant.

TABLE III
COMPARISON OF GH ASSAY WITH BIOLOGICAL ASSAYS

Sample	GH activity (percentage of wild type)	Cells with neurites (percentage of wild type)	ADP-ribosylation
Glu→Gln mutant	3.37	3.77	−
Buffer control	0	7.34	−
C3-07 (positive control)	100	100	+

Comparison of representative experiments. ADP-ribosylation was measured by gel-shift assay, and the minus or plus sign indicates the presence or absence or RhoA with decreased mobility in SDS-polyacrylamide gels.

stability experiments such as heat inactivation, and real-time stability studies to determine shelf-life.

Permeable C3 proteins and mutant versions are tools for investigation of Rho GTPases *in vitro* or *in vivo*. The GH assay reported here provides a simple, fast, and accurate tool to measure enzymatic activity after purification. The combination of methods that includes GH activity, neurite outgrowth bioassay, and ADP ribosylation provides an accurate measure of C3 activity and intracellular penetration of cell-permeable C3 proteins. It is also important to characterize the potential confounding contaminants in preparations of permeable C3 proteins, particularly for *in vivo* experiments. For instance, endotoxin levels are high in preparations of permeable C3 protein purified from *E. coli* in one step by glutathione affinity chromatography.

The amount and complexity of information on signaling pathways involving Rho GTPases and cell physiology is increasing at a fast pace. Straightforward assays and well-characterized mutant proteins allow extension of cell-permeable C3 to new model systems with confidence.

Acknowledgments

We thank Véronique Boulanger and Drs. Arshad Jilani and Isabel Laplante for expertise in assay development.

References

Aktories, K., Wilde, C., and Vogelgesang, M. (2004). Rho-modifying C3-like ADP-ribosyltransferases. *Rev. Physiol. Biochem. Pharmacol.* **152,** 1–22.

Armstrong, S., and Merrill, A. R. (2001). Application of a fluorometric assay for characterization of the catalytic competency of a domain III fragment of *Pseudomonas aeruginosa* exotoxin A. *Anal. Biochem.* **292,** 26–33.

Barrio, J. R., Secrist, J. A., III, and Leonard, N. J. (1972). A fluorescent analog of nicotinamide adenine dinucleotide. *Proc. Natl. Acad. Sci. USA* **69,** 2039–2042.

Bertrand, J., Winton, M. J., Rodriguez-Hernandez, N., Campenot, R. B., and McKerracher, L. (2005). Application of Rho antagonist to neuronal cell bodies promotes neurite growth in compartmented cultures and regeneration of retinal ganglion cell axons in the optic nerve of adult rats. *J. Neurosci.* **25,** 1113–1121.

Dillon, S. T., and Feig, L. A. (1995). Purification and assay of recombinant C3 transferase. *Methods Enzymol.* **256,** 174–184.

Gringas, K., Avedissian, H., Thouin, E., Boulanger, V., Essagian, C., McKerracher, L., and Lubell, W. D. (2004). Synthesis and evaluation of 4-(1-aminoalkyl)-N-(4-pyridyl) cyclohexane carboxamides as Rho kinase inhibitors and neurite outgrowth promoters. *Bioorg. Med. Chem. Lett.* **14,** 4931–4934.

Han, S., Arval, A. S., Clancy, S. B., and Tainer, J. A. (2001). Crystal structure and novel recognition motif of Rho ADP-ribosylating C3 exoenzyme from *Clostridium botulinum*: Structural insights for recognition specificity and catalysis. *J. Mol. Biol.* **395,** 95–107.

Laplante, I., Beliveau, R., and Paquin, J. (2004). RhoA/ROCK and Cdc42 regulate cell-cell contact and N-cadherin protein level during neurodetermination of P19 embryonal stem cells. *J. Neurobiol.* **60,** 289–307.

Lehmann, M., Fournier, A., Selles-Navarro, I., Dergham, P., Sebok, A., Leclerc, N., Gigyi, G., and McKerracher, L. (1999). Inactivation of Rho signaling pathway promotes CNS axon regeneration. *J. Neurosci.* **19,** 7537–7547.

Ménétrey, J., Flatau, G., Stura, E. A., Charbonnier, J. B., Gas, F., Teulon, J. M., Le Du, M. H., Boquet, P., and Menez, A. (2002). NAD binding induces conformational changes in Rho ADP-ribosylating *Clostridium botulinum* C3 exoenzyme. *J. Biol. Chem.* **277,** 30950–30957.

Pekala, P. H., and Anderson, B. M. (1978). Studies of bovine erythrocyte NAD glycohydrolase. *J. Biol. Chem.* **253,** 7453–7459.

Saito, Y., Nemoto, Y., Ishizaki, T., Watanabe, N., Morii, N, and Narumiya, S. (1995). Identification of Glu^{173} as the critical amino acid residue for the ADP-ribosyltransferase activity of *Clostridium botulinum* C3 exoenzyme. *FEBS Lett.* **371,** 105–109.

Sauzeav, V., LeMellionnec, E., Bertoglio, J., Scalbert, E., Pacaud, P., and Loirand, G. (2001). Human urotensin-II–induced contraction and arterial smooth muscle cell proliferation are mediated by RhoA and Rho-kinase. *Circ. Res.* **88,** 1102–1104.

Sekine, A., Fujiwara, M., and Narumiya, S. (1989). Asparagine residue in the Rho gene product is the modification site for botulinum ADP-ribosyltransferase. *J. Biol. Chem.* **264,** 8602–8605.

Secrist, J. A., 3rd, Barrio, J. R., Leonard, N. J., and Weber, G. (1972). Fluorescent activation of adenosine-containing coenzymes, biological activities and spectroscopic properties. *Biochemistry* **11,** 3499–3506.

Wilde, C., Just, I., and Aktories, K. (2002). Structure-function analysis of the Rho-ADP-ribosylating exoenzyme C3stau2 from *Staphylococcus aureus*. *Biochemistry* **41,** 1539–1544.

Winton, M. J., Dubreuil, C. I., Lasko, D., Leclerc, N., and McKerracher, L. (2002). Characterization of new cell permeable C3-like proteins that inactivate Rho and stimulate neurite outgrowth on inhibitory substrates. *J. Biol. Chem.* **277,** 32820–32829.

Further Reading

Bourgeois, C., Okazaki, I., Cavanaugh, E., Nightingale, M., and Moss, J. (2003). Identification of regulatory domains in ADP-ribosyltransferase-1 that determine transferase and NAD glycohydrolase activities. *J. Biol. Chem.* **278,** 26351–26355.

Bradford, M. M. (1976). A rapid and sensitive method for the quantitation of microgram quantities of protein utilizing the principle of protein-dye binding. *Anal. Biochem.* **72,** 248–254.

[40] Use of TIRF Microscopy to Visualize Actin and Microtubules in Migrating Cells

By JEAN-BAPTISTE MANNEVILLE

Abstract

Total internal reflection fluorescence (TIRF) is the technique of choice to visualize and quantify cellular events localized at the basal plasma membrane of adherent cells. By selectively illuminating the first 200 nm

METHODS IN ENZYMOLOGY, VOL. 406   0076-6879/06 $35.00
                            DOI: 10.1016/S0076-6879(06)06040-X

above the basal membrane, it allows maximal resolution in the vertical z-axis. In this chapter, I describe a prism-based TIRF setup and the procedures to visualize the actin and microtubule cytoskeleton in migrating astrocytes. TIRF microscopy provides quantitative information on the organization of the cytoskeleton in both fixed and live migrating cells.

Introduction

Total internal reflection fluorescence (TIRF) microscopy is becoming increasingly popular among cell biologists. This technique allows visualization of cellular events localized at the plasma membrane of adherent cells in real time with unique nanometer axial resolution, low phototoxicity, and virtually no fluorescent background. After calibration, TIRF illumination can also provide quantitative data. Until the late 1990s, TIRF microscopy has mostly been used to study cell-substrate contacts during cell adhesion and locomotion. With the development of live cell imaging using fluorescent proteins or probes, TIRF has found new applications in fields such as membrane trafficking or cytoskeletal dynamics. A number of reviews have been published that cover technical advances in TIRF microscopy, as well as their applications in cell biology (Axelrod, 2001; Beaumont, 2003; Oheim et al., 1999; Sako and Uyemura, 2002; Schneckenburger, 2005; Steyer and Almers, 2001; Toomre and Manstein, 2001).

Key cellular events occur at the plasma membrane during directed migration of adherent cells. Most TIRF studies on migrating cells have focused on the role of the actin cytoskeleton and focal contacts in cell-substrate adhesion (Adams et al., 2004; Kawakami et al., 2001; Lanni et al., 1985; Lee and Jacobson, 1997). Visualization of actin dynamics in Dictyostelium cells has shown that initiation of actin polymerization can occur independently of membrane protrusion at the cell front (Bretschneider et al., 2004; Gerisch et al., 2004). TIRF microscopy has also been used to image polarized membrane trafficking in migrating cells and to show that both exocytic and endocytic events preferentially occur toward the leading edge (Rappoport and Simon, 2003; Schmoranzer et al., 2003). Interestingly, microtubule-dependent transport is required for polarized exocytosis during migration (Schmoranzer et al., 2003). Recently, more emphasis has been put on the role of the microtubule cytoskeleton. Imaging microtubule dynamics with TIRF shows that microtubule plus-ends approach within 50 nm of the basal plasma membrane and are specifically targeted to focal adhesions in fibroblasts (Krylyshkina et al., 2003). The microtubule minus-end–directed motor dynein localizes to the leading edge of migrating fibroblasts and appears as bright puncta in TIRF images (Dujardin et al., 2003), suggesting that interaction of microtubule plus-ends with the plasma membrane by means of the dynein/dynactin complex may play an important role in polarized cell migration.

The microtubule cytoskeleton has been shown to play a pivotal role during polarized migration of astrocytes (Etienne-Manneville and Hall, 2001, 2003). We have used TIRF microscopy to visualize the actin and microtubule cytoskeleton in astrocytes during wound-healing–induced migration. In this chapter, we first detail the tuning steps necessary to achieve total internal reflection on our homemade prism-based TIRF microscope. Although specific to our setup, adapting these basic steps to tune any prism-based TIRF setup is straightforward. We then describe our methods to image actin and microtubules both in living and fixed cells.

Dual-Color TIRF Microscopy

Principles of TIRF Microscopy

Total internal reflection is achieved when light (wavelength λ) is reflected at the interface between a high-refractive index medium (n_i) and a low-refractive index medium (n_t) with an incident angle θ above the so-called critical angle $\theta_c = \sin^{-1}(n_t/n_i)$. A fraction of the energy of the incident light is transmitted to the low-refractive index medium in the form of an evanescent wave. The intensity profile of the evanescent wave is exponentially decaying:

$$I(z) = I_0 \exp(-z/d_P) \tag{1}$$

where z is the vertical distance, $I_0 = I(z = 0)$ is the intensity at the interface, and d_P is the penetration depth and is given by:

$$d_P = \frac{\lambda}{4\pi\sqrt{n_t^2 \sin^2(\theta) - n_i^2}} \tag{2}$$

In TIRF microscopy, the evanescent wave is used to excite fluorophores at a glass/water interface. Only those fluorophores located within the evanescent field ($\sim 3 \times d_P$ distant from the interface) can be visualized, thus greatly reducing out of focus fluorescence. Also, after calibration of the evanescent field, quantitative information on the vertical position of the fluorophores can be obtained by inverting Eq. 1, $z = \ln(I/I_0)$. Note that vertical positions deduced in this way are not absolute but only relative to the $z = 0$ position. Variable-angle TIRF (VA-TIRF) setups have been developed to achieve precise absolute quantification of vertical distances (Burmeister *et al.*, 1994; Loerke *et al.*, 2000; Olveczky *et al.*, 1997; Rohrbach, 2000; Stock *et al.*, 2003).

TIRF Setup

Two types of TIRF setups have been developed: objective-based (or prismless) TIRF (commercially available) and prism-based TIRF (mostly homemade setups). The setup I describe in this article is a prism-based setup (Fig. 1; Manneville *et al.*, 2003), allowing dual-color TIRF imaging of fixed samples and single-color imaging of live samples. Total internal reflection at the glass slide/culture medium interface is obtained using a trapezoidal glass prism. The refractive index of the prism and culture medium are $n_i = n_{glass}$= 1.52 and $n_t = n_{culture\ medium} \approx 1.336$, respectively, giving a critical angle (θ_c) for total internal reflection of $\theta_c = 61.5°$. The angle of incidence θ of the excitation light can be adjusted (see below, step 3) but is fixed to 68–70° (above the critical angle) during an experiment. More sophisticated variable angle TIRF setups allow fast variation of the incident angle without change in spot position (Loerke *et al.*, 2000; Rohrbach, 2000; Stock *et al.*, 2003).

Materials. Experiments are carried out on an upright microscope (Axioplan, Zeiss, Oberkochen, Germany) using either an Argon ion laser (excitation $\lambda = 488$ nm, 25 mW, Melles-Griot, Carlsbad, CA or a Nd:YAG laser (excitation $\lambda = 532$ nm, 50 mW, CrystaLaser, Reno, NV). The penetration depth calculated from Eq. 2 for the argon ion laser is $d_P = 75$–85 nm and for the Nd:YAG laser is $d_P = 85$–95 nm. GFP or Cy2 (resp. rhodamine) fluorescence is excited by the argon ion (resp. Nd:YAG) laser. Differences in beam radius and divergence are corrected by additional lenses (beam expander and lens L_1 in Fig. 1) to get the same spot size and position on the sample. Light from TIRF images are passed through a dichroic filter (505DRLP02 for GFP and Cy2 fluorescence or 560DRLP for rhodamine fluorescence, Omega Optical, VT) and an emission filter (530DF30 or 565ALP). Standard epifluorescence is achieved using a 100-W mercury lamp (excitation filters 485DF22 or 525DF45). Temperature in the cell chamber is maintained at 35–37° by circulating temperature-controlled water around the prism; 5% CO_2 in O_2 is blown on the cell chamber through a collar around the objective to maintain the pH at 7.

Procedure for Setting up TIRF Illumination Mode

1. Turn on lasers, epifluorescence mercury lamp, and water bath 30 min before experiment to avoid thermal drifts.

2. First-time use only. Wear suitable eye protection. Remove mirror M_3 and project Ar and Nd:YAG laser beams on a wall at least 3 m away. Using standard laser alignment procedures, make the beams parallel to the optical table and superimpose Ar and Nd:YAG spots on the wall.

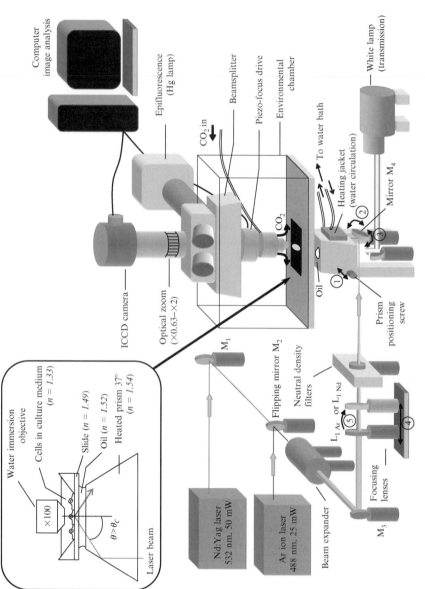

FIG. 1. Schematic diagram of the TIRF setup described in the text.

3. Turn laser power to minimum and neutral density to maximum. Directly monitor the position of the Ar laser spot at the top surface of the prism through the microscope oculars and a 20× air objective.

4. Adjust x-y position of the Ar spot with prism positioning screw (step 1, Fig. 1) and mirror M_4 (step 2, Fig. 1). Center the spot. Adjust the angle of the incident beam 5–10° above the critical angle with mirror M_4 (step 3, Fig. 1). Repeat steps 1 and 2 to center the spot.

5. Focus the Ar spot by moving the focusing lens ($L_{1\ Ar}$, step 4, Fig. 1).

6. Switch to the Nd:YAG beam by flipping mirror M_2 and changing $L_{1\ Ar}$ to $L_{1\ Nd}$ on the magnetic base (step 5, Fig. 1). Focus and adjust x-y position of the Nd:YAG spot by moving the focusing lens $L_{1\ Nd}$.

7. Switch back to Ar laser or keep Nd:YAG laser, depending on which fluorophore is observed. The resulting laser spot is slightly elongated in the direction of laser light propagation.

Sample Preparation

Confluent primary rat astrocytes are cultured, and monolayers are scratched as described in Chapter 44. Cells are observed or fixed 4–8 h after scratching.

Materials

For live experiments, astrocytes are electroporated with a nucleofector (Amaxa, Cologne, Germany) following the manufacturer instructions with tubulin-GFP. Cells are cultured in P50G-1.5–14-F glass-bottom dishes (MatTek Corp., Ashland, USA). Fixed cells are stained after standard immunofluorescence procedures. Primary antitubulin antibody used is rat monoclonal anti-α-tubulin diluted 1/200 (clone YL1/2 MCAP77, Serotec, Oxford, UK). Secondary antibodies are rhodamine (TRITC)-conjugated donkey anti-rat, Cy2-conjugated donkey anti-rabbit (Jackson ImmunoResearch Labs Inc., West Grove, PA). Actin is stained with rhodamine-phalloidin (Sigma) diluted 1/500 for 45 min. Cells are kept in PBS after staining. Note that cells must not be mounted after immunostaining, because TIRF imaging is not possible in mounting medium such as Mowiol because of a high refractive index.

Procedure

1. Clean the prism with ethanol and optical paper to remove traces of oil from the previous experiment.

2. Put one or two drops of clean microscope oil (Zeiss "Immersol" 518N) on top of the prism.

3. Place glass-bottom dish in holder on the x-y stage.

4. Couple prism to glass bottom dish by raising the prism assembly (screw at the back of the microscope).

5. Orient the scratch perpendicularly or parallel to the long axis of the TIRF spot, depending on the type of image that has to be acquired (see below and Fig. 2).

Live and Fixed Cells Observation and Image Acquisition

Migrating astrocytes develop very long protrusions (up to 100 μm in length) and a 40× objective is required to image whole cells. Higher magnification images of the leading edge are obtained with a 100× objective. Because cells migrate and elongate in the direction perpendicular to the scratch and because the TIRF spot is not circular but slightly elongated, the relative orientation of the scratch with the long axis of the spot has to be considered to maximize the cells area observed in TIRF (Fig. 2).

Materials

Live or fixed cells are observed with a 100× 1.0 n.a. water-immersion objective (Zeiss) or a 40× 0.75 n.a. water-immersion objective (Zeiss) driven by a piezoelectric focus drive (Physik Instrumente, Waldbronn, Germany). Fluorescence images are magnified by a 0.5–2× optical zoom (Zeiss), processed by an Argus 20 image processor (Hamamatsu Photonics K.K., Hamamatsu City, Japan) and collected with an intensified CCD camera (Remote Head Darkstar, S25 Intensifier, Photonics Science, UK). Images are digitized and stored in the memory of a Pentium III PC computer at a maximum rate of 25 frames/sec by a frame grabber (IC-PCI 4Mb (AMVS), Imaging Technology, MA) and then saved to disk. Live time-lapse images of GFP-labeled organelles are acquired at 0.5–2 frames/sec with online two frames averaging performed by the Argus-20 image processor. TIRF illumination is limited to 200 frames to minimize photobleaching and phototoxicity. Dual-color images of fixed cells are acquired with 256 frame averaging. The pixel size is 0.1092 μm (resp. 0.0870 μm) with a 1.6× (resp. 2×) optical zoom. The image size is typically 520 × 500 pixels.

Procedure

1. Choose 100× objective with scratch oriented perpendicular to the long axis of the TIRF spot (Fig. 2, upper left) for high-magnification images of the leading edge. Choose 40× objective for low-magnification

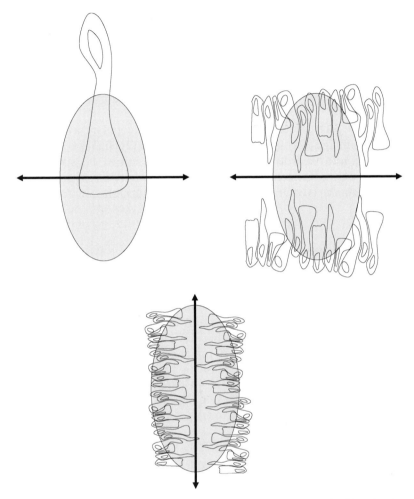

FIG. 2. Positioning of the cells in the TIRF spot to image details of the leading edge (upper left, high magnification, scratch oriented perpendicular to the long axis of the TIRF spot), whole cell protrusions (upper right, low magnification, scratch oriented perpendicular to the long axis of the TIRF spot), or the scratch edge (bottom, low magnification, scratch oriented parallel to the long axis of the TIRF spot).

images. Orient scratch perpendicular to the long axis of the TIRF spot to image whole protrusions (Fig. 2, upper right), or orient scratch parallel to the long axis of the TIRF spot to image several cells at the scratch edge (Fig. 2, bottom).

2. Use the 0.5–2× optical zoom to optimize image size.

3. Find GFP-tubulin expressing or stained cells in epifluorescence first, then switch to TIRF illumination mode.

4. Focus precisely the image with the piezo focus drive.

5. Live cells: lower Ar laser intensity and use neutral density filters to minimize photodamage, increase CCD gain, increase image contrast with Argus, acquire time-lapse image at 0.5–2 sec time interval and two frame online averaging.

6. Fixed cells: acquire images with 256 frame online averaging. For dual-color images of fixed cells, switch Ar and Nd:YAG lasers (see preceding), refocus precisely with the piezo focus drive, because the focus and evanescent wave penetration depth are slightly different depending on the wavelength. Acquire epifluorescence image with 256 frame online averaging.

Image Analysis and Quantification

To get a qualitative view of the morphology of the cytoskeleton close to the plasma membrane, the TIRF image (color-coded in green) of fixed cells stained for actin or tubulin can be superimposed with the corresponding epifluorescence (Epi, color-coded in red) image. Because of the exponential decay of the evanescent field, the TIRF image shows only cytoskeletal elements within about $3 \times d_P \sim 300$ nm from the basal plasma membrane, corresponding to a 100-fold decrease in light intensity, and greatly enhances the fluorescence close to the plasma membrane. Qualitatively, a mostly green feature in the TIRF-Epi merged image is thus within ~ 100 nm from the plasma membrane, whereas a red feature is >300 nm away from the membrane (Fig. 3).

To get a more quantitative three-dimensional view of the organization of the cytoskeleton, we use the theoretical relationship between the fluorescence intensity and the vertical position to determine the height profile ("z-profile") $z(r)$ of cytoskeletal elements according to:

$$z(r) = -d_P \ln \frac{I(r)}{I_{max}} \tag{3}$$

where $I(r)$ is the fluorescence intensity along the cytoskeletal element at a distance r from the leading edge of the cell, and I_{max} is the fluorescence of the brightest point in the image. Note again that vertical measurements are not absolute distances but relative to the lowest point in the image ($z = 0$ for $I = I_{max}$). This also assumes that tubulin (resp. actin) staining or tubulin-GFP (resp. actin-GFP) is homogeneously distributed along microtubules (resp. actin filaments) so that variations in fluorescence intensity

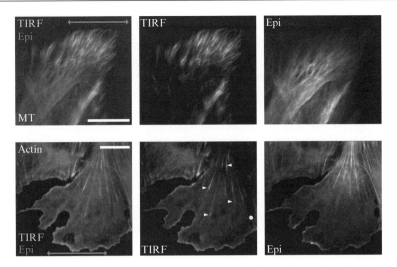

FIG. 3. High magnification images of the microtubule (upper images) and actin (lower images) cytoskeleton in fixed migrating astrocytes. Left images show merged TIRF (green) and epifluorescence (Epi, red). Note that microtubules come in close contact to the substrate only at the front of migrating cells, whereas actin is visible inside the protrusion. Focal contacts appear as bright spots in the actin TIRF image (arrowheads). The double red arrows indicate the direction of the scratch. Bars: 10 μm. (See color insert.)

reflect only variations in vertical position and not local variations in fluorescent probe concentration.

Microtubule plus-end dynamics can also be quantified from live time-lapse TIRF images, either by direct tracking of the plus-ends or by kymograph analysis. A kymograph is a position versus time plot along a line used to follow growth or shortening of cytoskeletal elements along their track (see, for example, Salmon *et al.* [2002], Vallotton *et al.* [2003], and references therein).

Materials

Image processing is carried out using Optimas 6.5 (Media Cybernetics L. P., Silver Spring, MD) and Adobe Photoshop. Customized macros (available upon request) written in the C^{++}-based language ALI (Analytical Language for Images) are run under Optimas 6.5. Kymograph analysis of microtubule dynamics is performed using the Metamorph software (Universal Imaging Corporation, Downingtown, USA).

TIRF/Epifluorescence Merged Images

1. During acquisition, adjust camera gain so that the fluorescence of the brightest point in the TIRF image is not saturated. In the case of a "bent" cytoskeletal element that spans the entire evanescent field, only the brightest part (i.e., the one in closer contact to the plasma membrane) is visible.

2. Open TIRF and corresponding epifluorescence images.

3. Color-code the 8-bit gray images in green and red for TIRF and epifluorescence, respectively.

4. Superimpose TIRF and epifluorescence image. No x-y adjustment of the two channels is needed.

z-Profiles of Microtubules or Actin Filaments (Custom C^{++} Macro Run under Optimas 6.5)

1. Open TIRF image of fixed cell.

2. Draw contour of leading edge.

3. Find brightest point in the image ($z = 0$ for $I = I_{max}$).

4. Input in the macro: I_{max}, d_P, and the w (width of the line for reducing noise by averaging data on w pixels, typically $w = 3$ pixels).

5. Draw a line along selected microtubule or actin filament.

6. Run the macro to measure distance from the microtubule plus-end or actin filament extremity to the leading edge, to measure fluorescence intensity $I(r)$ along the selected line of width w and to deduce the z-profile $z(r)$ according to Eq. 3.

Kymograph Analysis of Microtubule Dynamics (Metamorph Software)

1. Open TIRF time-lapse movie of live cell as a Metamorph stack.

2. Average time-lapse movie to visualize microtubule tracks ("Process/ Stack Arithmetic/Average" tool in Metamorph).

3. Select individual microtubule track by drawing a line in the averaged image ("line ROI" tool).

4. Transfer selected line to TIRF movie stack ("Regions/Transfer Regions" tool).

5. Use kymograph function in Metamorph to obtain kymograph plot ("Stack/Kymograph" tool).

6. Open an Excel file to transfer Metamorph data ("Log/Open Data Log" tool).

7. Kymograph analysis: find growth, pause, and shortening phases by eye or with quantitative criterion (e.g., threshold on velocity and duration

of each phase). Mark start and end points of each phase (corresponding to rescues and catastrophes) using "Apps/Measure XYZ distance" tool. Transfer data to Excel ("Log data" button). In Excel, convert pixel values to distances using image calibration. Deduce growth and shortening rates and catastrophe and rescue frequencies.

Acknowledgments

I thank Michael Ferenczi, Rod Chillingworth, and Ronnie Burns (NIMR, London, UK) for assistance with the TIRF setup.

References

Adams, M. C., Matov, A., Yarar, D., Gupton, S. L., Danuser, G., and Waterman-Storer, C. M. (2004). Signal analysis of total internal reflection fluorescent speckle microscopy (TIR-FSM) and wide-field epi-fluorescence FSM of the actin cytoskeleton and focal adhesions in living cells. *J. Microsc.* **216,** 138–152.

Axelrod, D. (2001). Total internal reflection fluorescence microscopy in cell biology. *Traffic* **2,** 764–774.

Beaumont, V. (2003). Visualizing membrane trafficking using total internal reflection fluorescence microscopy. *Biochem. Soc. Trans.* **31,** 819–823.

Bretschneider, T., Diez, S., Anderson, K., Heuser, J., Clarke, M., Muller-Taubenberger, A., Kohler, J., and Gerisch, G. (2004). Dynamic actin patterns and Arp2/3 assembly at the substrate-attached surface of motile cells. *Curr. Biol.* **14,** 1–10.

Burmeister, J. S., Truskey, G. A., and Reichert, W. M. (1994). Quantitative analysis of variable-angle total internal reflection fluorescence microscopy (VA-TIRFM) of cell/substrate contacts. *J. Microsc.* **173**(Pt 1), 39–51.

Dujardin, D. L., Barnhart, L. E., Stehman, S. A., Gomes, E. R., Gundersen, G. G., and Vallee, R. B. (2003). A role for cytoplasmic dynein and LIS1 in directed cell movement. *J. Cell Biol.* **163,** 1205–1211.

Etienne-Manneville, S., and Hall, A. (2001). Integrin-mediated Cdc42 activation controls cell polarity in migrating astrocytes through PKCz. *Cell* **106,** 489–498.

Etienne-Manneville, S., and Hall, A. (2003). Cdc42 regulates GSK3 and adenomatous polyposis coli (APC) to control cell polarity. *Nature* **420,** 629–635.

Gerisch, G., Bretschneider, T., Muller-Taubenberger, A., Simmeth, E., Ecke, M., Diez, S., and Anderson, K. (2004). Mobile actin clusters and traveling waves in cells recovering from actin depolymerization. *Biophys. J.* **87,** 3493–3503.

Kawakami, K., Tatsumi, H., and Sokabe, M. (2001). Dynamics of integrin clustering at focal contacts of endothelial cells studied by multimode imaging microscopy. *J. Cell Sci.* **114,** 3125–3135.

Krylyshkina, O., Anderson, K. I., Kaverina, I., Upmann, I., Manstein, D. J., Small, J. V., and Toomre, D. K. (2003). Nanometer targeting of microtubules to focal adhesions. *J. Cell Biol.* **161,** 853–859.

Lanni, F., Waggoner, A. S., and Taylor, D. L. (1985). Structural organization of interphase 3T3 fibroblasts studied by total internal reflection fluorescence microscopy. *J. Cell Biol.* **100,** 1091–1102.

Lee, J., and Jacobson, K. (1997). The composition and dynamics of cell-substratum adhesions in locomoting fish keratocytes. *J. Cell Sci.* **110**(Pt 22), 2833–2844.

Loerke, D., Preitz, B., Stuhmer, W., and Oheim, M. (2000). Super-resolution measurements with evanescent-wave fluorescence excitation using variable beam incidence. *J. Biomed. Opt.* **5**, 23–30.

Manneville, J. B., Etienne-Manneville, S., Skehel, P., Carter, T., Ogden, D., and Ferenczi, M. (2003). Interaction of the actin cytoskeleton with microtubules regulates secretory organelle movement near the plasma membrane in human endothelial cells. *J. Cell Sci.* **116**, 3927–3938.

Oheim, M., Loerke, D., Chow, R. H., and Stuhmer, W. (1999). Evanescent-wave microscopy: A new tool to gain insight into the control of transmitter release. *Philos. Trans. R. Soc. Lond. B Biol. Sci.* **354**, 307–318.

Olveczky, B. P., Periasamy, N., and Verkman, A. S. (1997). Mapping fluorophore distributions in three dimensions by quantitative multiple angle-total internal reflection fluorescence microscopy. *Biophys. J.* **73**, 2836–2847.

Rappoport, J. Z., and Simon, S. M. (2003). Real-time analysis of clathrin-mediated endocytosis during cell migration. *J. Cell Sci.* **116**, 847–855.

Rohrbach, A. (2000). Observing secretory granules with a multiangle evanescent wave microscope. *Biophys. J.* **78**, 2641–2654.

Sako, Y., and Uyemura, T. (2002). Total internal reflection fluorescence microscopy for single-molecule imaging in living cells. *Cell Struct. Funct.* **27**, 357–365.

Salmon, W. C., Adams, M. C., and Waterman-Storer, C. M. (2002). Dual-wavelength fluorescent speckle microscopy reveals coupling of microtubule and actin movements in migrating cells. *J. Cell Biol.* **158**, 31–37.

Schmoranzer, J., Kreitzer, G., and Simon, S. M. (2003). Migrating fibroblasts perform polarized, microtubule-dependent exocytosis towards the leading edge. *J. Cell Sci.* **116**, 4513–4519.

Schneckenburger, H. (2005). Total internal reflection fluorescence microscopy: Technical innovations and novel applications. *Curr. Opin. Biotechnol.* **16**, 13–18.

Steyer, J. A., and Almers, W. (2001). A real-time view of life within 100 nm of the plasma membrane. *Nature Rev. Mol. Cell Biol.* **2**, 268–276.

Stock, K., Sailer, R., Strauss, W. S., Lyttek, M., Steiner, R., and Schneckenburger, H. (2003). Variable-angle total internal reflection fluorescence microscopy (VA-TIRFM): Realization and application of a compact illumination device. *J. Microsc.* **211**, 19–29.

Toomre, D., and Manstein, D. (2001). Lighting up the cell surface with evanescent wave microscopy. *Trends Cell Biol.* **11**, 298–303.

Vallotton, P., Ponti, A., Waterman-Storer, C. M., Salmon, E. D., and Danuser, G. (2003). Recovery, visualization, and analysis of actin and tubulin polymer flow in live cells: A fluorescent speckle microscopy study. *Biophys. J.* **85**, 1289–1306.

[41]   Inhibition of ROCK by RhoE

By KIRSI RIENTO and ANNE J. RIDLEY

Abstract

RhoE belongs to the Rnd subfamily of small Rho-related GTP-binding proteins. Similar to other Rho proteins, RhoE regulates actin cytoskeleton dynamics. Expression of RhoE induces loss of actin stress fibers, and it also increases cell migration speed. In part, this is due to RhoE interaction with the RhoA effector ROCK I, a serine/threonine kinase that regulates the formation and contractility of stress fibers. Interestingly, RhoE does not interact with the highly homologous kinase ROCK II. RhoE binding inhibits ROCK I from phosphorylating its downstream target myosin light chain phosphatase, thus increasing the activity of the phosphatase to de-phosphorylate myosin II, which results in reduced actomyosin contractility. RhoE itself is phosphorylated by ROCK I, and this may enhance RhoE regulation of ROCK I function. This chapter describes the method used for studying ROCK inhibition by RhoE.

Introduction

The Rnd subfamily of Rho-related GTP-binding proteins consists of three proteins: Rnd1, Rnd2, and Rnd3/RhoE (Nobes *et al.*, 1998). Compared with other Rho proteins, Rnd proteins have no detectable intrinsic GTPase activity (Foster *et al.*, 1996; Guasch *et al.*, 1998; Nobes *et al.*, 1998). Consequently, they exist in a GTP-bound state, and their activity is not regulated by GTP hydrolysis. Rnd1 or RhoE expression in fibroblasts or epithelial cells induces loss of actin stress fibers, and, there-fore, they seem to function antagonistically to RhoA in actin cytoskeleton regulation (Guasch *et al.*, 1998; Nobes *et al.*, 1998). RhoE expression also increases cell migration speed (Guasch *et al.*, 1998) and blocks cell cycle progression (Villalonga *et al.*, 2004).

We have found that RhoE interacts specifically with the serine/threonine kinase ROCK I (Riento *et al.*, 2003) but not with ROCK II (Riento *et al.*, 2005). ROCKs are ubiquitously expressed protein kinases that regulate actin filament assembly and contractility (Riento and Ridley, 2003). Several proteins required for actin cytoskeleton dynamics are targets for ROCK-mediated phosphorylation. In particular, ROCKs can phosphorylate myosin light chain phosphatase (MLCP) and myosin light chain, thus resulting in

DOI: 10.1016/S0076-6879(06)06041-1

increased levels of phosphorylated myosin II and enhanced actomyosin contractility (Amano et al., 1996; Kimura et al., 1996). In addition, ROCKs increase actin filament assembly by phosphorylating LIM kinases, and this increases their ability to phosphorylate and inactivate the actin-depolymerizing protein cofilin (Maekawa et al., 1999). Other well-characterized ROCK substrates include adducin and ezrin/radixin/moesin proteins, which are required for membrane ruffling and protrusion formation, respectively (Fukata et al., 1999; Matsui et al., 1998). Actomyosin contractility regulates cell motility and contraction of smooth muscle. ROCK inhibitors, therefore, decrease vascular contraction and tumor cell dissemination in vivo, emphasizing their potential use in the treatment of hypertension and tumour cell metastasis (Itoh et al., 1999; Uehata et al., 1997).

In contrast to RhoA, which can stimulate ROCK kinase activity (Ishizaki et al., 1996; Matsui et al., 1996), RhoE inhibits ROCK I–induced actin filament formation and reduces ROCK I–mediated phosphorylation of MLCP (Riento et al., 2003). The RhoE-binding domain is located at the amino-terminus of ROCK I, which encompasses the kinase domain. Interestingly, this domain is different from the binding domain for RhoA. By directly binding to the kinase domain, RhoE may inhibit ROCK I from interacting/phosphorylating its downstream targets. RhoE is also a substrate for ROCK I kinase activity, and this may regulate the function of RhoE in ROCK I inhibition. Here, we describe the techniques we have used for studying RhoE binding to and inhibiting ROCK I and RhoE phosphorylation by ROCK I.

TABLE I

ROCK CONSTRUCTS USED IN THE BINDING ASSAY (THE MINIMUM RhoE BINDING DOMAIN IS AMINO ACIDS 1–420 OF ROCK I.)

Construct name	Amino acids	Binds to RhoE
myc-ROCK I/ pCAG	Full length	Yes
myc-ROCK II/ pSVneo	Full length	No
myc-Δ1ROCK I/pCAG[a]	1–1080	Yes
myc-Δ3ROCK I/pCAG[a]	1–727	Yes
myc-Δ5ROCK I/pCAG[a]	1–375	No
myc-CCROCK I/pCAG[b]	375–727	No
myc-R1ROCK I/pCAG[b]	1–420	Yes
myc-R2ROCK I/pCAG[b]	76–420	No

[a] Ishizaki et al. (1997).
[b] Riento et al. (2003).

RhoE Binding to ROCK I

GST pull-down assays are used to analyze the proteins that bind to RhoE. In this method, recombinant, glutathione S-transferase (GST)–tagged RhoE is produced in *Escherichia coli* and purified using glutathione-sepharose beads (Amersham Biosciences) that specifically bind to GST. Inducing protein expression in bacteria at 30° helped reducing cleavage of GST from RhoE and degradation of RhoE. GST-RhoE is used to pull down myc-tagged–ROCKs from COS7 cell lysates. We frequently use electroporation to transfect COS7 cells, because this method results in ~80% transfection efficiency. Different ROCK I forms lacking amino-terminal or carboxy-terminal domains were constructed for the assay (Table I).

Methods

Production of Recombinant GST-RhoE. E. coli (BL21 strain) containing the plasmid pGEX-2T GST-RhoE are streaked from a glycerol stock onto a Luria broth (LB)-agar plate containing 100 μg/ml ampicillin and grown overnight at 37°. A single colony is inoculated into 100 ml LB, 100 μg/ml ampicillin, and grown at 37° overnight. This culture is diluted into 900 ml LB, 50 μg/ml ampicillin, and grown at 37° for 1 h. Protein expression is induced with 0.2 mM isopropyl-β-D-1-thiogalactopyranoside (IPTG) for 3 h at 30°. Cells are centrifuged at 2000g for 10 min at 4°. The cells are resuspended in 6 ml of cold lysis buffer (50 mM Tris, pH 7.5, 50 mM NaCl, 5 mM MgCl$_2$, 1 mM dithiothreitol [DTT], 1 mM phenylmethanesulfonyl fluoride [PMSF]), and placed on ice. The bacteria are sonicated six times for 10 sec, with 1 min between each burst, keeping them on ice throughout the sonication. The cells are centrifuged at 10,000g for 10 min at 4°, and the supernatant is retained. Glutathione-sepharose bead suspension (500 μl) is centrifuged in a microcentrifuge, the supernatant removed, and the beads washed twice with lysis buffer. The beads are suspended in 500 μl lysis buffer, and the bacterial supernatant is added. The beads and supernatant are inverted for 60 min at 4°. The GST-RhoE is now bound to the beads. The beads are centrifuged, the supernatant removed, and the beads washed three times with 5 ml cold lysis buffer without PMSF. The beads are suspended in 400 μl lysis buffer, 10% glycerol (without PMSF), aliquoted, and stored at −80°. Before use, the beads are washed with lysis buffer required for the GST pull-down assay.

Electroporation of COS7 Cells. COS7 cells are grown on 14-cm plates in Dulbecco's modified Eagle's medium (DMEM) containing 10% heat-inactivated fetal calf serum (FCS), 100 U penicillin, and 10 μg/ml streptomycin until they are almost confluent. The cells are split 1:2 on 14-cm plates the day before transfection; one plate is required per transfection. The cells are detached from the plate, centrifuged, and washed with 5 ml of cold electroporation buffer (120 mM KCl, 10 mM K_2PO_4/KHPO$_4$, pH 7.6, 25 mM HEPES, pH 7.6, 2 mM MgCl$_2$, 0.5% Ficoll 400) that has been filter-sterilized. The cells are centrifuged again and suspended in 250 μl electroporation buffer per transfection. Cells (250 μl) are mixed with 5 μg DNA (expression vectors encoding various ROCK proteins) in a 0.4-cm electroporation cuvette and left on ice for 10 min. Cells are electroporated at 250 V, 975 μF. The cuvette is left on ice for 5 min then at room temperature for 5 min. Cells are added to 10 ml of growth medium on a 10-cm dish and incubated at 37°, 10% CO_2 for 24 h.

GST-RhoE Pull-down. COS7 cells expressing myc epitope–tagged ROCKs are washed twice with phosphate-buffered saline (PBS) containing 0.01% (w/v) Ca^{2+} and 0.1% (w/v) Mg^{2+}. Ice-cold lysis buffer (1% Triton-X 100, 20 mM Tris, pH 8, 130 mM NaCl, 1 mM DTT) is prepared and has the following inhibitors added just before use: 10 mM NaF, 1% aprotinin, 10 μg/ml leupeptin, 0.1 mM Na$_3$VO$_4$, and 1 mM PMSF. Lysis buffer (1 ml) is added to each plate and left on ice for 10 min. The cells are scraped into a microcentrifuge tube, and the insoluble material is removed by centrifugation at 13,000g for 10 min at 4°. The supernatant is transferred to a fresh tube, and 20 μl of the lysate is removed to analyze protein expression. Glutathione beads with GST-RhoE attached (20 μl) are added to the lysate and rotated at 4° for 1 h. The beads are centrifuged and washed four times with lysis buffer. The wash buffer is removed, and 20 μl sample loading buffer is added. The samples are boiled, and bound proteins are analyzed by immunoblotting.

RhoE and ROCK I Microinjection in Swiss 3T3 Cells

RhoE effects on the actin cytoskeleton are best observed in mouse Swiss 3T3 fibroblast cells, because serum starvation induces loss of stress fibers in these cells, and stress fibers are induced by treatment with reagents such as lysophosphatidic acid (Ridley and Hall, 1992). However, Swiss 3T3 cells are poorly transfectable, and the cDNAs encoding RhoE and/or ROCK I are, therefore, microinjected in the cells.

RhoE expression in growing fibroblasts (Fig. 1A) (Guasch *et al.*, 1998; Nobes *et al.*, 1998) and epithelial cells (Guasch *et al.*, 1998) induces loss of stress fibers. Mutation of the RhoE "effector" domain does not abolish this response to RhoE (Fig. 1A). However, mutation of threonine 37 to asparagine, which mimics the dominant negative mutation N19 in Rho, affects RhoE localization and ability to induce stress fiber disruption (Fig. 1A). Expression of ROCK I in starved Swiss 3T3 cells induces formation of thick

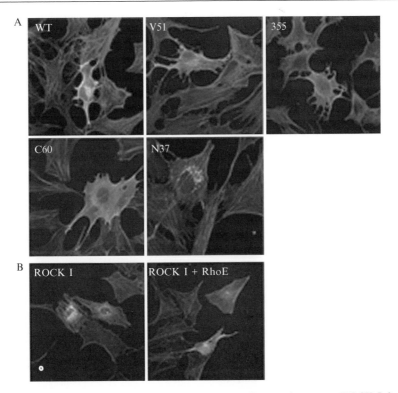

FIG. 1. Expression of RhoE induces loss of stress fibers and prevents ROCK I from inducing stress fibers in Swiss 3T3 cells. (A) Growing Swiss 3T3 cells were injected with wild-type or mutated FLAG-RhoE forms. RhoE was localized with anti-FLAG antibody (green) and F-actin with phalloidin (red). (B) ROCK I expression in starved Swiss 3T3 cells increases the formation of actin filaments. Coexpression of RhoE prevents ROCK I–induced stress fiber formation. Myc-ROCK I or FLAG-RhoE is shown in green, F-actin in red. (See color insert.)

actin bundles (Fig. 1B). Interestingly, microinjection of RhoE and ROCK I cDNAs together prevents ROCK I–induced actin filament formation, suggesting that RhoE inhibits ROCK I function in actin filaments.

Methods

Microinjection of Swiss 3T3 Cells. Swiss 3T3 cells are grown in DMEM containing 10% FCS, penicillin, and streptomycin. For microinjections, they are seeded on 13-mm diameter glass coverslips at between 1 and 2×10^4 cells/coverslip. When the cells are approximately 50% confluent, they are starved for 24 h before microinjection with expression vectors encoding FLAG-RhoE (50 μg/ml) and/or myc-ROCK I (25 μg/ml) expression vectors in the cell nuclei. After injection, the cells are incubated at 37° for 6 h, fixed with 4% paraformaldehyde for 20 min, permeabilized with 0.2% Triton X-100 for 5 min, blocked with 10% goat serum for 30 min, and incubated with primary antibodies followed by incubation with fluorophore-conjugated goat secondary antibodies. Cells are washed extensively with PBS between treatments. To stain for filamentous (F-)actin, the cells are incubated with fluorophore-conjugated phalloidin. The specimens are mounted in mounting medium (DakoCytomation), the mounting reagent is allowed to set, and images are generated with a confocal microscope.

RhoE Inhibits Phosphorylation of Myosin Light Chain Phosphatase by ROCK I

Expression vectors encoding full-length ROCK I or C terminally truncated ROCK I, ROCK Δ1 (both myc epitope–tagged at the N-terminus) are electroporated with or without FLAG-RhoE into COS7 cells as described previously. After 24 h, the level of phosphorylated myosin light chain phosphatase is analyzed in the cell lysates by immunoblotting (Fig. 2). Both full-length and ROCK Δ1 increase the level of MLCP phosphorylation. However, coexpression of RhoE reduces ROCK I–induced phosphorylation of MLCP. Even the constitutively active form of ROCK I, ROCK Δ1, induced much less MLCP phosphorylation when expressed together with RhoE, suggesting that by binding to the amino-terminal domain of ROCK I RhoE can block its kinase activity.

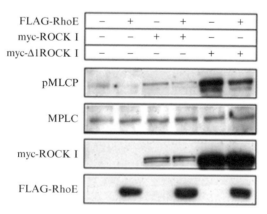

FIG. 2. RhoE inhibits ROCK I–induced phosphorylation of the myosin-binding subunit of MLCP. Full-length myc epitope-tagged ROCK I or C-terminally truncated, active ROCK I was expressed together with FLAG-RhoE. The cell lysates were analyzed by immunoblotting with anti-phosphoMLCP antibody (Thr696, from Upstate Biotechnology).

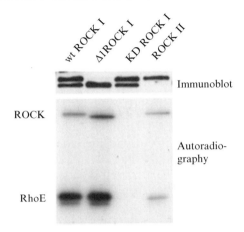

FIG. 3. RhoE is phosphorylated by ROCK I but not by ROCK II. RhoE was incubated with full-length ROCK I, Δ1ROCK I, kinase-dead ROCK I, or full-length ROCK II in an *in vitro* kinase assay. The presence of the kinases was verified by immunoblotting, and phosphorylation of the proteins was determined by SDS-PAGE and autoradiography. Reproduced from Riento *et al.*, 2005.

In Vitro Phosphorylation of RhoE by ROCK I

ROCKs can phosphorylate several proteins involved in regulating actin filament formation and contractility. RhoE was also found to be a substrate for ROCK I (Fig. 3) Recombinant RhoE is combined with immunoprecipitated myc-tagged ROCK I or ROCK II, and the complexes are subjected to *in vitro* kinase assay. ROCKs can phosphorylate their substrates even while bound to beads. By use of radioactive ATP and autoradiography, RhoE was found to be phosphorylated by ROCK I but not by ROCK II. Autophosphorylation of the kinases confirms that the enzymes are active.

Methods

Purification of RhoE. Recombinant GST-RhoE is produced in *E. coli* and purified as described previously. The GST-tag is cleaved off with thrombin by suspending the GST-RhoE beads in 500 μl thrombin buffer (50 mM Tris, pH 7.5, 2.5 mM CaCl$_2$, 100 mM NaCl, 5 mM MgCl$_2$, 1 mM DTT) and adding 5 units thrombin per 1 ml beads. The beads are inverted overnight at 4°. The supernatant is taken off and retained. The beads are resuspended in another 500 μl of thrombin buffer to wash off all RhoE and centrifuged. This second supernatant is combined with the first one, and 10 μl p-aminobenzamide–agarose bead suspension is added (Sigma) to remove thrombin. Incubate with inversion for 30 min at 4°. The beads are centrifuged and the supernatant taken off and retained. The supernatant is centrifuged again to ensure all beads are removed. The supernatant containing RhoE is dialyzed twice against dialysis buffer (50 mM Tris, pH 7.5, 100 mM NaCl, 5 mM MgCl$_2$, 1 mM DTT). Protein concentration is determined by BioRad assay, and the protein is concentrated if necessary.

Purification of ROCK I and ROCK II. Vectors encoding myc-ROCK I or myc-ROCK II are electroporated into COS7 cells as described previously. After incubation and cell lysis, the insoluble material is removed by centrifugation at 13,000g in a microcentrifuge for 10 min at 4°. The supernatant is transferred to a fresh tube, and 20 μl of anti-myc epitope-antibody covalently linked to agarose (Santa Cruz) is added. Samples are incubated with inversion for 2 h at 4°. The beads are washed four times with lysis buffer containing 0.5 M NaCl to remove any ROCK-binding proteins.

In Vitro *Kinase Assay.* The myc-ROCK beads are washed twice with kinase buffer (50 mM Tris, pH 7.5, 10 mM MgCl$_2$, 1 mM DTT). Purified recombinant RhoE (2 μg) is added to the beads, and the volume increased to 20 μl with kinase buffer containing 30 μM ATP, 2 μCi γ-^{32}P ATP. Samples are incubated at 30° for 30 min, then 7 μl 4× sample buffer is added. Samples are boiled and analyzed by SDS-PAGE. The polyacrylamide gel is fixed with 10% acetic acid for 30 min before drying and exposed to X-ray film.

References

Amano, M., Ito, M., Kimura, K., Fukata, Y., Chihara, K., Nakano, T., Matsuura, Y., and Kaibuchi, K. (1996). Phosphorylation and activation of myosin by Rho-associated kinase (Rho-kinase). *J. Biol. Chem.* **271**, 20246–20249.

Foster, R., Hu, K. Q., Lu, Y., Nolan, K. M., Thissen, J., and Settleman, J. (1996). Identification of a novel human Rho protein with unusual properties: GTPase deficiency and *in vivo* farnesylation. *Mol. Cell. Biol.* **16**, 2689–2699.

Fukata, Y., Oshiro, N., Kinoshita, N., Kawano, Y., Matsuoka, Y., Bennett, V., Matsuura, Y., and Kaibuchi, K. (1999). Phosphorylation of adducin by Rho-kinase plays a crucial role in cell motility. *J. Cell Biol.* **145**, 347–361.

Guasch, R. M., Scambler, P., Jones, G. E., and Ridley, A. J. (1998). RhoE regulates actin cytoskeleton organization and cell migration. *Mol. Cell. Biol.* **18**, 4761–4771.

Ishizaki, T., Maekawa, M., Fujisawa, K., Okawa, K., Iwamatsu, A., Fujita, A., Watanabe, N., Saito, Y., Kakizuka, A., Morii, N., and Narumiya, S. (1996). The small GTP-binding protein Rho binds to and activates a 160 kDa Ser/Thr protein kinase homologous to myotonic dystrophy kinase. *EMBO J.* **15**, 1885–1893.

Ishizaki, T., Naito, M., Fujisawa, K., Maekawa, M., Watanabe, N., Saito, Y., and Narumiya, S. (1997). p160ROCK, a Rho-associated coiled–coil forming protein kinase, works downstream of Rho and induces focal adhesions. *FEBS Lett.* **404**, 118–124.

Itoh, K., Yoshioka, K., Akedo, H., Uehata, M., Ishizaki, T., and Narumiya, S. (1999). An essential part for Rho-associated kinase in the transcellular invasion of tumor cells. *Nat. Med.* **5**, 221–225.

Kimura, K., Ito, M., Amano, M., Chihara, K., Fukata, Y., Nakafuku, M., Yamamori, B., Feng, J., Nakano, T., Okawa, K., Iwamatsu, A., and Kaibuchi, K. (1996). Regulation of myosin phosphatase by Rho and Rho-associated kinase (Rho-kinase). *Science* **273**, 245–248.

Maekawa, M., Ishizaki, T., Boku, S., Watanabe, N., Fujita, A., Iwamatsu, A., Obinata, T., Ohashi, K., Mizuno, K., and Narumiya, S. (1999). Signaling from Rho to the actin cytoskeleton through protein kinases ROCK and LIM-kinase. *Science* **285**, 895–898.

Matsui, T., Amano, M., Yamamoto, T., Chihara, K., Nakafuku, M., Ito, M., Nakano, T., Okawa, K., Iwamatsu, A., and Kaibuchi, K. (1996). Rho-associated kinase, a novel serine/threonine kinase, as a putative target for small GTP binding protein Rho. *EMBO J.* **15**, 2208–2216.

Matsui, T., Maeda, M., Doi, Y., Yonemura, S., Amano, M., Kaibuchi, K., and Tsukita, S. (1998). Rho-kinase phosphorylates COOH-terminal threonines of ezrin/radixin/moesin (ERM) proteins and regulates their head-to-tail association. *J. Cell Biol.* **140**, 647–657.

Nobes, C. D., Lauritzen, I., Mattei, M. G., Paris, S., Hall, A., and Chardin, P. (1998). A new member of the Rho family, Rnd1, promotes disassembly of actin filament structures and loss of cell adhesion. *J. Cell Biol.* **141**, 187–197.

Ridley, A. J., and Hall, A. (1992). The small GTP-binding protein rho regulates the assembly of focal adhesions and actin stress fibers in response to growth factors. *Cell* **70**, 389–399.

Riento, K., Guasch, R. M., Garg, R., Jin, B., and Ridley, A. J. (2003). RhoE binds to ROCK I and inhibits downstream signaling. *Mol. Cell. Biol.* **23**, 4219–4229.

Riento, K., and Ridley, A. J. (2003). Rocks: Multifunctional kinases in cell behaviour. *Nat. Rev. Mol. Cell Biol.* **4**, 446–456.

Riento, K., Totty, N., Garg, R., Villalonga, P., and Ridley, A. J. (2005). RhoE function is regulated by ROCK I-mediated phosphorylation. *EMBO J.* **24**, 1170–1180.

Uehata, M., Ishizaki, T., Satoh, H., Ono, T., Kawahara, T., Morishita, T., Tamakawa, H., Yamagami, K., Inui, J., Maekawa, M., and Narumiya, S. (1997). Calcium sensitization of smooth muscle mediated by a Rho-associated protein kinase in hypertension. *Nature* **389**, 990–994.

Villalonga, P., Guasch, R. M., Riento, K., and Ridley, A. J. (2004). RhoE inhibits cell cycle progression and Ras-induced transformation. *Mol. Cell. Biol.* **24**, 7829–7840.

[42]   Conditional Regulation of a ROCK-Estrogen Receptor Fusion Protein

By Daniel R. Croft and Michael F. Olson

Abstract

The Rho-associated kinases ROCK I and ROCK II are serine/threonine kinases that play central and critical roles in regulating the actin cytoskeleton. Activation of ROCK proteins contributes positively to the phosphorylation of myosin II light chains (MLC), myosin ATPase activity, stabilization of filamentous actin, and coupling of the actin-myosin filaments to the plasma membrane, thereby leading to the increased actin-myosin force generation and contractility. We have constructed a conditionally-activated form of ROCK II (called ROCK:ER) by fusing the ROCK II kinase domain to the estrogen receptor hormone–binding

METHODS IN ENZYMOLOGY, VOL. 406
Copyright 2006, Elsevier Inc. All rights reserved.

0076-6879/06 $35.00
DOI: 10.1016/S0076-6879(06)06042-3

domain. In this chapter, we describe the construction and characterization of this regulatable ROCK:ER fusion protein.

Introduction

Rho GTPases influence the architecture of the actin cytoskeleton, resulting in cell shape changes that affect adhesion, motility, growth, and cytokinesis (Coleman *et al.*, 2004; Etienne-Manneville and Hall, 2002). Within the Rho protein family are the RhoA and RhoC GTPases, which promote actin-myosin filament formation and the generation of contractile force. Acting downstream of RhoA and RhoC are the pivotal effectors ROCK I and ROCK II, serine/threonine kinases that influence actin-myosin cytoskeletal structures through phosphorylation of numerous protein substrates (Riento and Ridley, 2003). Phosphorylation of myosin II light chains (MLC), which enhances the velocity and force of actin-myosin contraction, results from direct phosphorylation by ROCK (Amano *et al.*, 1996; Totsukawa *et al.*, 2000) together with ROCK-mediated phosphorylation and concomitant inactivation of the myosin binding subunit (MYPT1) of the MLC phosphatase (Feng *et al.*, 1999; Kimura *et al.*, 1996). ROCK also promotes actin filament stabilization by phosphorylating and activating the LIM kinases (LIMK) 1 and 2, which in turn phosphorylate and inactivate the actin-severing protein cofilin (Maekawa *et al.*, 1999). There is considerable interest in the role of ROCK kinases in pathophysiological conditions, including cancer, hypertension, asthma, and glaucoma (Wettschureck *et al.*, 2002), at least partially because of studies in which potentially therapeutic ROCK inhibitors such as Y-27632 and fasudil have been used (Uehata *et al.*, 1997).

As a complementary approach to the use of ROCK inhibitors, many studies have transiently or stably overexpressed constitutively active ROCK I or ROCK II, either alone or in combination with inhibitors. Ectopic expression studies, however, do not allow for manipulation of ROCK signaling intensity or time course of activation.

Therefore, to study ROCK signaling in a more regulated manner and to complement the use of small-molecule ROCK inhibitors, we have developed a conditionally active form of ROCK II (ROCK:ER) by fusing the kinase domain to the hormone-binding domain of the estrogen receptor (ER). In this chapter, we characterize ROCK:ER fusion protein activation by the estrogen antagonist 4-hydroxytamoxifen (4-HT). Ultimately, ROCK:ER activation leads to phosphorylation of proteins involved in the regulation of the actin-myosin cytoskeleton, generation of pronounced actin stress fibers, and changes in cell morphology.

Methods

Generation of Constructs

To generate the ROCK:ER fusion protein, we joined the kinase domain along with part of the coiled–coil region of bovine ROCK II (residues 5–553) to the hormone-binding domain of the mouse estrogen receptor (hbER) at the carboxyl-terminus and enhanced green fluorescent protein (EGFP) at the amino terminus (ROCK:ER, Fig. 1). As a control, a kinase-dead version (KD:ER) in which lysine 121 in the ATP-binding pocket was changed to glycine (K121G) was used to generate KD:ER (Fig. 1). To achieve this, a pBABEpuro3 retroviral vector, which carries a puromycin-resistance selection marker, was initially made encoding EGFP:ER with *Bam*HI and *Eco*RI sites at the junction between the EGFP and ER moieties. EGFP was amplified by PCR using forward (GGGAGATCTGCCACCATGGTGAGCAAGGGC) and reverse oligos (CCCGGAATTCCCGGATCCCTTGTACAGCTCGTCCATCCC), digested with *Bgl*II and *Eco*RI, and ligated into *Bam*HI and *Eco*RI sites of pBABEpuro3 EGFP:ΔRaf-1:ER (Woods *et al.*, 1997). The kinase domains of bovine ROCK II or the kinase-dead (KD) mutant were removed from pGEX-2T using *Bam*HI and *Eco*RI and subcloned in frame into pBABEpuro3

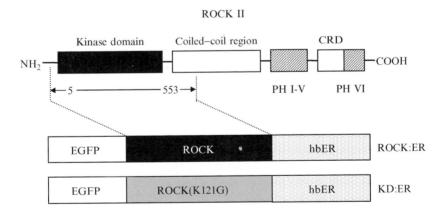

FIG. 1. Schematic representation of ROCK II and the conditionally regulated ROCK:ER. A diagram of ROCK II indicating the functional domains (PH, Pleckstrin homology domain, CRD, cysteine-rich domain). Amino acids 5–553 were joined in frame to enhanced green fluorescence protein (EGFP) and estrogen receptor hormone–binding domain (hbER) moieties to create the ROCK:ER fusion protein. A kinase-dead version (KD:ER) was created by changing lysine 121 to glycine (ROCK(K121G)).

EGFP:ER, thereby generating pBABEpuro 3 EGFP:ROCK:ER and pBA-BEpuro3 EGFP:KD:ER, respectively.

Retroviral Transduction

To package replication-defective retrovirus, pBABEpuro3 EGFP:ROCK:ER or pBABEpuro3 EGFP:KD:ER is transiently transfected into the ecotropic packaging cell line, BOSC 23 (Pear *et al.*, 1993). To do this, BOSC 23 cells are plated at 2.5×10^5 cells per well of a 6-well dish and 48 h later are transfected with retroviral constructs (0.5 μg) using Effectene (Qiagen). The following day, transfected cells are washed twice with phosphate-buffered saline (PBS) and then 1.5 ml of Dulbecco's modified Eagles medium (DMEM) (Invitrogen) supplemented with 10% (v/v) fetal bovine serum (FBS) (Invitrogen) and 100 U/ml penicillin-streptomycin (Invitrogen) is added to each well. Retrovirus-containing supernatant is collected approximately 40 h after transfection and any residual cells removed by centrifugation (3000g for 5 min). Retroviral supernatant can be used immediately or aliquoted and stored at −80°.

Here we describe retroviral transduction of mouse NIH 3T3 fibroblasts with ROCK:ER and KD:ER. We have also transduced a number of human cell lines (including LS174T and HCT116 colon carcinoma [Croft *et al.*, 2004], A375 melanoma, and A431 squamous carcinoma cell lines) that had been stably transfected with the retroviral ecotropic receptor (Scholz and Beato, 1996). Twenty-four hours before transduction, NIH 3T3 cells are plated into 6-well dishes at a density of 1×10^5 per well. To each well, 1 ml of retrovirus plus 4 μg/ml polybrene (Sigma) is added and then incubated for 5–6 h, after which 2 ml of DMEM containing 10% (v/v) donor calf serum (DCS) (Invitrogen) is added. The following day, retrovirus is removed and 3 ml of fresh 10% DCS/DMEM added. Cells are normally confluent after 2–3 days and then can be trypsinized, plated out in 15-cm dishes, and selected using 2.5 μg/ml puromycin (Sigma). Selection is maintained until colonies are large enough to isolate or pools of cells have grown out. In our laboratory, we normally generate a number of different pools or clones to match cell lines with similar levels of fusion protein expression.

Cell Culture and Western Blotting

We routinely maintain KD:ER and ROCK:ER-expressing cell lines in DMEM supplemented with 10% (v/v) DCS containing 100 U/ml penicillin-streptomycin and 2.5 μg/ml puromycin. For experiments, LS174T cells (1×10^6 per 10-cm dish) or NIH 3T3 cells (0.6×10^6 per 10-cm dish) are plated in 10% DCS/DMEM. The following day, cells are transferred to serum-free medium for 24 h. Subsequently, cells are treated with 4-HT

(Sigma), with or without 10 μM Y-27632 (Calbiochem) for 16 h. Dishes are placed on ice, rinsed with ice-cold PBS, and then 200–300 μl of RIPA buffer (10 mM Tris-HCl [pH 7.5], 5 mM EDTA, 1% [v/v] Nonidet P-40, 0.5% [w/v] sodium deoxycholate, 40 mM Na PP$_i$, 1 mM Na$_3$VO$_4$, 50 mM NaF, 1 mM phenylmethylsulfonyl fluoride [PMSF], 0.025% [w/v] SDS, 150 mM NaCl, and 1× complete protease inhibitors [Roche]) is added. The cells are scraped and the crude lysate centrifuged at 13,000g for 15 min. Equal 60-μg aliquots of each lysate are run on SDS-polyacrylamide gels, transferred to nitrocellulose, and blocked with 5% (w/v) dried milk in TBST (25 mM Tris, 144 mM NaCl, and 0.1% [v/v] Tween 20) before immunoblotting with anti-ERα (Santa Cruz Biotechnology, anti-β-tubulin [Clone Tub2.1; Sigma) or anti-ERK2 (C. J. Marshall, Institute of Cancer Research, London, UK) antibodies.

To determine the levels of LIMK1, phospho-LIMK, phospho-MLC, or MLC, we lyse cells directly in 3× Laemmli sample buffer. Samples are scraped from the plate, sonicated (15 sec at 30% amplitude) using a Branson Digital Sonifier, heated at 100° for 5 min, and centrifuged at 13,000g for 5 min. Samples are separated by SDS-PAGE and transferred to nitrocellulose membrane. Antibody sources are as follows: phospho-MLC (Ser19), phospho-MLC (Thr18/Ser19), phospho-LIMK1 (Thr508)/LIMK2 (Thr505), and LIMK1 are from Cell Signaling Technologies; MLC (Clone MY21) is from Sigma. Goat anti-mouse and goat anti-rabbit horseradish peroxidase antibodies are from Pierce.

In Vitro *Kinase Assays*

To determine the catalytic activity of the ROCK:ER fusion protein under differing 4-HT treatment regimens, cells are lysed in RIPA buffer, and samples prepared as described previously. For each sample, ER-fusion protein is immunoprecipitated from 500 μg of cleared lysate with 30 μl of protein G Sepharose (Sigma) coupled to 3 μg of anti-ERα (Clone HC20; Santa Cruz Biotechnology) for 90 min at 4°. Immunoprecipitates are washed three times with kinase-wash buffer (50 mM Tris [pH 7.5], 10 mM MgCl$_2$, 0.1 mM PMSF, 40 mM NaPP$_i$, 1 mM Na$_3$VO$_4$, 50 mM NaCl, 1 mM DTT, 10% [v/v] glycerol, and 0.03% [v/v] Brij35) and then twice in kinase-assay buffer) 50 mM Tris [pH 7.5], 1 mM EDTA [pH 8.0], 10 mM MgCl$_2$, 50 mM NaCl, 1 mM DTT, and 0.03% [v/v] Brij35), taking care to remove the last of the buffer with a gel-loading tip. Resuspend beads in 30 μl of kinase-assay buffer containing 10 μM ATP, 5 μCi [γ-^{32}P] ATP and 50 ng/μl histone H1. After incubating for 30 min at 30°, the reactions are terminated by adding 6 μl of 6× Laemmli buffer and then boiling at 100° for 5 min. Reaction products are separated on an 8%

SDS-polyacrylamide gel, transferred to nitrocellulose, and kinase activity is quantified using a STORM phosphorimager (Molecular Dynamics). Immunoprecipitation of ER-fusion proteins is confirmed by Western blotting with ERα antibody.

Immunofluorescence

To examine how activation of ROCK:ER affects the organization of the actin cytoskeleton, 2×10^4 NIH 3T3 ROCK:ER or KD:ER cells are plated onto glass coverslips (12-mm diameter) in a 24-well dish in 10% DCS/DMEM for 24 h. Cells are serum starved for 24 h before treatment with 1 μM 4-HT for 16 h. Cells are washed once with PBS and then fixed with 4% (w/v) *p*-formaldehyde in PBS for 10 min at room temperature. Cells are washed three times with PBS and then permeabilized in 0.5% (v/v) Triton X-100 in PBS for 10 min. Nonspecific antibody binding is blocked by incubating for 1 h at room temperature in 3% (w/v) bovine serum albumin (BSA) in PBS. Coverslips are incubated for 60 min with anti-phospho-MLC (Ser19) antibody (Cell Signaling Technologies) diluted 1 in 1000 with 3% (w/v) BSA in PBS, followed by PBS washes, and then a 60-min incubation with a FITC-labeled secondary antibody (Jackson Immunoresearch Laboratories). Actin structures are visualized by staining with 2 μg/ml Texas Red–conjugated phalloidin (Molecular Probes) in 3% (w/v) BSA in PBS. Coverslips are then washed three times with PBS and once with water before mounting onto glass slides with 10 μl Mowiol (Calbiochem) (containing 0.1% [w/v] *p*-phenylenediamine (Sigma) as an antiquenching agent). Confocal laser-scanning microscopy is performed using a Bio-Rad MRC 1024 microscope and LaserSharp software (Bio-Rad).

Results and Discussion

We examined the dose-dependence and time-dependence of ROCK:ER activation after treatment of LS174T cells with 4-HT. Immunoprecipitated ROCK:ER was assayed for its ability to phosphorylate purified histone H1 *in vitro*. Kinase activity was elevated with increasing 4-HT concentrations, reaching a maximum of eightfold activation at 1 μM 4-HT (Fig. 2A). The solubility of 4-HT in ethanol limited the maximum concentration tested; it is possible that further activation might occur at higher doses. In contrast, KD:ER had no kinase activity at 1 μM 4-HT (Fig. 2A). ROCK:ER kinase activity could be detected as early as 15 min after 4-HT addition, with activity continuing to increase over the 15-h time-course (Fig. 2B). Immunoprecipitated ROCK:ER protein levels did not change over time, which is

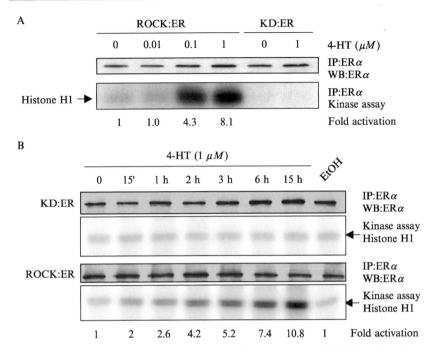

FIG. 2. 4-HT induces kinase activity in conditionally active ROCK:ER. (A) Dose-dependent increase in ROCK:ER kinase activity. LS174T cells expressing ROCK:ER or KD:ER were treated for 16 h in the absence or presence of various 4-HT concentrations. ROCK:ER or KD:ER were immunoprecipitated from equal quantities of whole cell lysate with anti-ERα antibody, followed by an *in vitro* kinase assay with histone H1 as substrate. Fold activation was calculated as the ratio of histone H1 phosphorylation in 4-HT–treated sample relative to untreated. (B) Time course of ROCK:ER activation. LS174T cells expressing KD:ER or ROCK:ER were treated for the indicated times with 1 μM 4-HT or with an equal volume of ethanol drug vehicle for 15 h and kinase activity of the fusion proteins determined as described in the text.

indicative of increased specific activity being responsible for elevated kinase activity. No KD:ER kinase activity was detected at any time, despite similar levels of ROCK:ER and KD:ER being purified from equivalent amounts of lysate (Fig. 2B). These data show that ROCK:ER can be specifically activated in a dose-dependent and time-dependent manner.

ROCK regulation of actin-myosin contractile force generation is mediated by means of the phosphorylation of several proteins (Riento and Ridley, 2003). We wished to determine whether these endogenously expressed proteins were phosphorylated after ROCK:ER activation. The effects of ROCK:ER activation were examined in NIH 3T3 fibroblasts using phospho-specific antibodies against MLC (Ser19 and Thr18/Ser19) and against the common epitope in LIMK 1 and 2 (LIMK1 Thr508, LIMK2 Thr505 [Amano *et al.*, 2001; Ohashi *et al.*, 2000]). Treatment with 1 μM 4-HT for 16 h resulted in increased phosphorylation of MLC and LIMK1/2 in ROCK:ER, but not KD:ER, expressing cells and could be blocked by Y-27632 (Fig. 3A). We next looked at phosphorylation kinetics after ROCK:ER activation in serum-starved NIH 3T3 ROCK:ER cells. Increased phospho-MLC levels were apparent 4 h after 4-HT addition

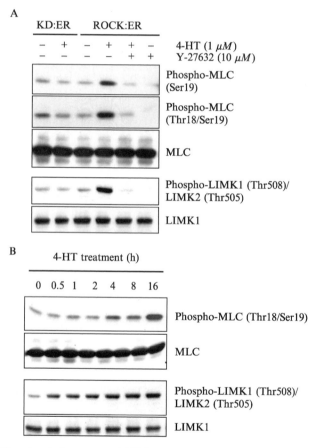

FIG. 3. Phosphorylation states of MLC and LIMK vary with time after ROCK:ER activation. (A) KD:ER and ROCK:ER-expressing NIH 3T3 fibroblasts were either left untreated or treated with 1 μM 4-HT, 1 μM 4-HT plus 10 μM Y-27632, or 10 μM Y-27632 as indicated for 16 h. Cell lysates were Western blotted for phospho-MLC (Ser19 or Thr18/Ser19), MLC, phospho-LIMK1 (Thr508)/LIMK2 (Thr505), or LIMK1. (B) Serum-starved NIH 3T3 ROCK:ER cells were treated with 1 μM 4-HT for the times indicated. Lysates were Western blotted as described previously.

and were maximal at 16 h (Fig. 3B). By contrast, increased phospho-LIMK1/2 levels were evident as early as 30 min and increased steadily with time, mirroring the time-dependent increase in *in vitro* kinase activity (Fig. 2B). These results indicate that the phosphorylation of MLC, unlike LIMK1/2, is influenced by factors in addition to ROCK. Negative feedback mechanisms may be activated to limit signaling duration and/or intensity.

Given that constitutively active ROCK induces cytoskeletal rearrangements (Amano *et al.*, 1997), we examined how ROCK:ER activation influenced actin structures. Serum-starved ROCK:ER-expressing cells treated with ethanol vehicle showed only cortical actin and weak stress fibers (Fig. 4A, left panel), whereas 100 nM 4-HT treated cells exhibited thick actin stress fibers (Fig. 4A, right panel), which were not seen in cells given Y-27632 together with 4-HT (Fig. 4A, lower left panel). Treatment of KD:ER-expressing cells with 100 nM 4-HT had no effect on actin structures (Fig. 4A, lower right panel). This suggests that 4-HT activation of ROCK:ER resulted in actin rearrangements that were dependent on ROCK catalytic activity.

We observed that in serum-starved ROCK:ER-expressing cells, weak phospho-MLC staining colocalized with cortical actin and disorganized stress fibers (Fig. 4B, upper panels). In contrast, ROCK:ER activation increased phospho-MLC staining, which colocalized to actin stress fibers, with no increased cortical actin–associated phospho-MLC (Fig. 4B, lower panels). These data are consistent with the proposal that ROCK phosphorylates MLC in the central region of cells, which contributes to actin stress fiber formation (Totsukawa *et al.*, 2000).

Given that ROCK:ER activation *in vitro* was dose-dependent, we examined the effects of increasing 4-HT concentrations on actin structures in ROCK:ER-expressing NIH 3T3 cells. Actin stress fibers were observed at 4-HT concentrations as low as 1 nM (Fig. 5A), although *in vitro* kinase activity was not found to be elevated at concentrations less than 100 nM (Fig. 2A). These results suggest that the *in vivo* biological responses are more sensitive than the *in vitro* biochemical assay to changes in ROCK activity. Concentrations of 4-HT greater than 1 nM gave rise to thicker actin bundles, with the formation of "stellate" stress fibers starting at 100 nM. In addition, we observed a dose-dependent increase in contracted and blebbing ROCK:ER-expressing cells up to a maximum 50% at 1 μM (Fig. 5A, lower right), an effect not observed in KD:ER-expressing cells at any concentration. The morphological effects did not result from the induction of apoptosis and caspase activity, because the caspase 3 substrate poly-ADP ribose polymerase (PARP) was not cleaved after treatment of ROCK:ER-expressing cells with 1 μM 4-HT for 16 h (Fig. 5A, graph inset panel). PARP cleavage, however, was evident after treatment with an

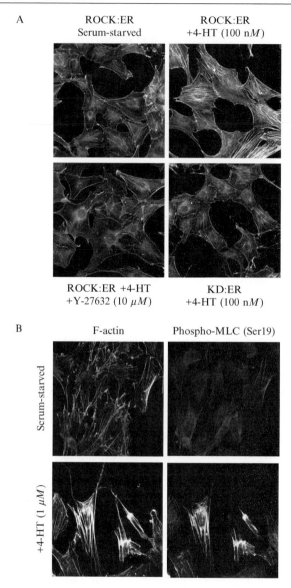

FIG. 4. Stress fiber induction and phospho-MLC colocalization after ROCK:ER activation. (A) Serum-starved ROCK:ER or KD:ER-expressing NIH 3T3 cells were treated with ethanol drug vehicle, 100 nM 4-HT, or 100 nM 4-HT plus 10 μM Y-27632 for 16 h. Cells were fixed and F-actin structures stained with Texas Red–conjugated phalloidin. (B) Serum-starved NIH 3T3 ROCK:ER cells were treated either with ethanol drug vehicle or with 1 μM 4-HT for 16 h, then were fixed and stained with Texas Red–conjugated phalloidin and with anti-phospho MLC (Ser19) antibody.

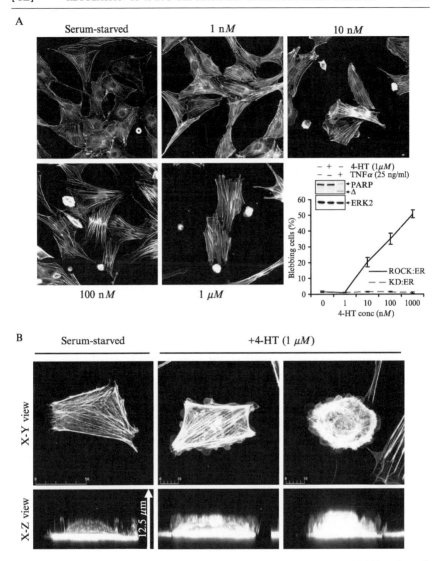

FIG. 5. ROCK:ER induction of cell rounding and dynamic membrane blebbing through contractile stress fibers. (A) Serum-starved ROCK:ER and KD:ER-expressing NIH 3T3 cells were treated with or without 4-HT at indicated concentrations for 16 h. Cells were fixed and stained for F-actin, and the percentage of contracted blebbing cells was determined at each concentration. Inset panel shows Western blot of PARP and ERK2 from ROCK:ER cells that were untreated, treated with 1 μM 4-HT for 16 h or 25 ng/ml TNFα plus 10 μg/ml cycloheximide for 2 h. (B) Serum-starved NIH 3T3 ROCK:ER cells were left untreated (left panel, scale bar = 50 μm) or treated (middle and right panels, scale bars = 10 μm) with 1 μM 4-HT for 16 h. A series of optical Z-sections were collected at 0.1-μm intervals, then reconstructed in X-Y (top panels) and X-Z views (lower panels) using LaserSharp software. X-Z views were stretched in the Z-dimension. Note that image height represents 12.5 μm.

apoptotic stimulus for 2 h. Video time-lapse microscopy revealed that blebbing cells sustained membrane blebbing for some hours (data not shown), in contrast to *bona fide* apoptotic blebbing that lasts for ~20–30 min in NIH 3T3 cells. These results are consistent with the morphological responses being a direct result of ROCK:ER signaling and not through the induction of apoptosis.

Reconstruction of serial Z-sections reveals that a serum-starved ROCK:ER-expressing NIH 3T3 cell, with some actin stress fibers when viewed "top-down" in an X-Y view (Fig. 5B, lower left panel), is relatively flat when viewed "side-on" in an X-Z view, with some central elevation above the nucleus and actin protrusions at attachment points (Fig. 5B, lower left panel). ROCK:ER activation results in thick actin stress fibers and cell contraction along the X-Y plane (Fig. 5B, middle and right panels). Accompanying X-Y contraction is an increase in cell height (Z-dimension) together with the generation of actin-rich membrane blebs (Fig. 5B, lower panels). Thus, as a consequence of ROCK:ER activation, contractile actin stress fibers form and pull centripetally, leading to cell rounding and membrane blebbing.

Concluding Remarks

We believe that conditionally active ROCK will be a valuable research tool to study the function of ROCK kinases in processes such as cell cycle control, apoptosis, and cell motility. We recently demonstrated the value of ROCK:ER in studying the role of ROCK in keratinocyte differentiation (McMullan *et al.*, 2003), tumor cell invasiveness and angiogenesis (Croft *et al.*, 2004), and in apoptotic nuclear breakdown (Croft *et al.*, 2005). In addition, because ROCK:ER-induced effects on actin filament assembly and MLC phosphorylation can be assessed by fluorescence microscopy, high-content, fluorescence-based cellular assays could be developed to facilitate the identification of novel potentially-therapeutic ROCK inhibitors.

Acknowledgments

This work was supported by Cancer Research UK and an NIH/NCI (R01 CA030721) grant to M. F. O. Thanks to Martin McMahon (UCSF) for the gift of pBABEpuro3 EGFP: ΔRaf-1:ER and to David Drechsel (UCL) for the gift of pGEX-2T ROCK II.

References

Amano, M., Chihara, K., Kimura, K., Fukata, Y., Nakamura, N., Matsuura, Y., and Kaibuchi, K. (1997). Formation of actin stress fibers and focal adhesions enhanced by Rho-kinase. *Science* **275**, 1308–13011.

Amano, M., Ito, M., Kimura, K., Fukata, Y., Chihara, K., Nakano, T., Matsuura, Y., and Kaibuchi, K. (1996). Phosphorylation and activation of myosin by Rho-associated kinase (Rho- kinase). *J. Biol. Chem.* **271,** 20246–20249.

Amano, T., Tanabe, K., Eto, T., Narumiya, S., and Mizuno, K. (2001). LIM-kinase 2 induces formation of stress fibres, focal adhesions and membrane blebs, dependent on its activation by Rho-associated kinase-catalysed phosphorylation at threonine-505. *Biochem. J.* **354,** 149–159.

Coleman, M. L., Marshall, C. J., and Olson, M. F. (2004). RAS and RHO GTPases in G1-phase cell-cycle regulation. *Nat. Rev. Mol. Cell Biol.* **5,** 355–366.

Croft, D. R., Coleman, M. L., Li, S., Robertson, D., Sullivan, T., Stewart, C. L., and Olson, M. F. (2005). Actin-myosin-based contraction is responsible for apoptotic nuclear disintegration. *J. Cell Biol.* **168,** 245–255.

Croft, D. R., Sahai, E., Mavria, G., Li, S., Tsai, J., Lee, W. M., Marshall, C. J., and Olson, M. F. (2004). Conditional ROCK activation *in vivo* induces tumor cell dissemination and angiogenesis. *Cancer Res.* **64,** 8994–9001.

Etienne-Manneville, S., and Hall, A. (2002). Rho GTPases in cell biology. *Nature* **420,** 629–635.

Feng, J., Ito, M., Ichikawa, K., Isaka, N., Nishikawa, M., Hartshorne, D. J., and Nakano, T. (1999). Inhibitory phosphorylation site for Rho-associated kinase on smooth muscle myosin phosphatase. *J. Biol. Chem.* **274,** 37385–37390.

Kimura, K., Ito, M., Amano, M., Chihara, K., Fukata, Y., Nakafuku, M., Yamamori, B., Feng, J., Nakano, T., Okawa, K., Iwamatsu, A., and Kaibuchi, K. (1996). Regulation of myosin phosphatase by Rho and Rho-associated kinase (Rho-kinase). *Science* **273,** 245–248.

Maekawa, M., Ishizaki, T., Boku, S., Watanabe, N., Fujita, A., Iwamatsu, A., Obinata, T., Ohashi, K., Mizuno, K., and Narumiya, S. (1999). Signaling from Rho to the actin cytoskeleton through protein kinases ROCK and LIM-kinase. *Science* **285,** 895–898.

McMullan, R., Lax, S., Robertson, V. H., Radford, D. J., Broad, S., Watt, F. M., Rowles, A., Croft, D. R., Olson, M. F., and Hotchin, N. A. (2003). Keratinocyte differentiation is regulated by the Rho and ROCK signaling pathway. *Curr. Biol.* **13,** 2185–2189.

Ohashi, K., Nagata, K., Maekawa, M., Ishizaki, T., Narumiya, S., and Mizuno, K. (2000). Rho-associated kinase ROCK activates LIM-kinase 1 by phosphorylation at threonine 508 within the activation loop. *J. Biol. Chem.* **275,** 3577–3582.

Pear, W. S., Nolan, G. P., Scott, M. L., and Baltimore, D. (1993). Production of high-titer helper-free retroviruses by transient transfection. *Proc. Natl. Acad. Sci. USA* **90,** 8392–8396.

Riento, K., and Ridley, A. J. (2003). Rocks: Multifunctional kinases in cell behaviour. *Nat. Rev. Mol. Cell. Biol.* **4,** 446–456.

Scholz, A., and Beato, M. (1996). Transient transfection of ecotropic retrovirus receptor permits stable gene transfer into non-rodent cells with murine retroviral vectors. *Nucleic Acids Res.* **24,** 979–980.

Totsukawa, G., Yamakita, Y., Yamashiro, S., Hartshorne, D. J., Sasaki, Y., and Matsumura, F. (2000). Distinct roles of ROCK (Rho-kinase) and MLCK in spatial regulation of MLC phosphorylation for assembly of stress fibers and focal adhesions in 3T3 fibroblasts. *J. Cell Biol.* **150,** 797–806.

Uehata, M., Ishizaki, T., Satoh, H., Ono, T., Kawahara, T., Morishita, T., Tamakawa, H., Yamagami, K., Inui, J., Maekawa, M., and Narumiya, S. (1997). Calcium sensitization of smooth muscle mediated by a Rho-associated protein kinase in hypertension. *Nature* **389,** 990–994.

Wettschureck, N., and Offermanns, S. (2002). Rho/Rho-kinase mediated signaling in physiology and pathophysiology. *J. Mol. Med.* **80,** 629–638.

Woods, D., Parry, D., Cherwinski, H., Bosch, E., Lees, E., and McMahon, M. (1997). Raf-induced proliferation or cell cycle arrest is determined by the level of Raf activity with arrest mediated by p21Cip1. *Mol. Cell. Biol.* **17,** 5598–5611.

[43] Rational Design and Applications of a Rac GTPase–Specific Small Molecule Inhibitor

By HUZOOR AKBAR, JOSE CANCELAS, DAVID A. WILLIAMS, JIE ZHENG, and YI ZHENG

Abstract

Rac GTPases are involved in the regulation of multiple cell functions and have been implicated in the pathology of certain human diseases. Dominant negative mutants of Rac have been the tool of choice in studying Rac function in cells. Given the difficulty of introducing high concentrations of the Rac mutants into primary cells and nonspecific effects of the mutants on Rho guanine nucleotide exchange factor (GEF) activities, it is desirable to develop small molecule inhibitors that could specifically inhibit Rac activities. Here we describe the rational design, characterization, and applications of a first-generation Rac-specific small molecule inhibitor. On the basis of the structure-function information of Rac interaction with GEFs, in a computer-based virtual screening we have identified NSC23766, a highly soluble and membrane permeable compound, as a specific inhibitor of a subset of GEF binding to Rac and, therefore, Rac activation by these GEFs. In fibroblast cells, NSC23766 inhibited Rac1 GTP-loading without affecting Cdc42 or RhoA activity and suppressed cell proliferation induced by a Rac GEF Tiam1. It has little effect on cell growth induced by a constitutively active Rac1 mutant. In addition, NSC23766 inhibited: (1) the anchorage-independent growth and invasion phenotypes of human prostate cancer PC-3 cells; (2) Rac activation and Rac-dependent aggregation of platelets stimulated by thrombin; and (3) Rac1 and Rac2 activities of hematopoietic stem/progenitor cells and induced their mobilization from mouse bone marrow to peripheral blood. Thus, NSC23766 is a lead small molecule inhibitor of Rac activity and could be useful for studying Rac-mediated cellular functions and for modulating pathological conditions in which Rac-deregulation may play a role.

Introduction

Rho-family GTPases are molecular switches that control signaling pathways regulating a myriad of cellular functions (Etienne-Manneville and Hall, 2002). They can be activated through interaction with the Dbl family guanine nucleotide exchange factors (GEFs) that catalyze the agonist-receptor–mediated GTP/GDP exchange (Schmidt and Hall, 2002; Zheng,

METHODS IN ENZYMOLOGY, VOL. 406   0076-6879/06 $35.00
DOI: 10.1016/S0076-6879(06)06043-5

2001). Because there is a sizable disparity between the numbers of Dbl family GEFs and Rho GTPase substrates (Zheng, 2001), the signaling specificity mediated by Rho proteins, in part, is governed by specific Rho-GEF interactions. One strategy to inhibit the activity of a Rho GTPase and the associated downstream signaling events might be to target the Rho GTPase activation step by blocking Rho GTPase interaction with specific Dbl family GEFs.

A wealth of structure-function information made available by X-ray crystallography and mutagenesis studies has laid out the foundation for derivation of a mechanism-based targeting strategy. Earlier structural mapping studies of Rac1 interaction with its GEFs have identified a groove formed by the switch I, switch II, and $\beta 1/\beta 2/\beta 3$ regions of Rac1 as the key area rendering GEF specification (Gao et al., 2001; Worthylake et al., 2000). A Trp[56] residue in this region has been pinpointed as the critical determinant of Rac1 for discrimination by a subset of Rac1-specific GEFs including Trio and Tiam1 (Gao et al., 2001; Karnoub et al., 2001). The requirement for the Trp[56] residue in Rac-GEF interaction and the strategy to target this structural site to achieve specific inhibition of the interaction was also highlighted by the observation that a polypeptide derived from Rac1 sequences containing Trp[56] specifically inhibits binding of Rac1 to its GEF (Gao et al., 2001). These analyses led us to consider the possibility that the structural features surrounding Trp[56] of Rac1 might be explored to design a chemical compound to specifically inhibit activation of Rac GTPases by preventing the Rac-GEF interaction. In this chapter, we describe the mechanism-based design, characterization, and applications of a Rac activation-specific, first-generation small molecule inhibitor.

Rational Design of Rac Activation-Specific Inhibitors

In the three-dimensional (3D) structure of Rac1-Tiam1 complex, the functionally important Trp[56] residue of Rac1 is buried in a pocket formed by residues His[1178], Ser1[184], Glu[1183], and Ile [1197] of Tiam1 and Lys[5], Val[7], Thr[58], and Ser[71] of Rac1 (Worthylake et al., 2000). To design Rac-specific inhibitors targeting this site, a putative inhibitor-binding pocket was created with residues of Rac1 within 6.5 angstroms of Trp[56] in the Rac1-Tiam1 monomer, including residues Lys[5], Val[7], Trp[56], and Ser[71] of Rac1. A 3D database containing the coordinates of more than 140,000 small chemical compounds available from the National Cancer Institute (http://www.dtp.nih.gov) was searched for compounds whose conformations would fit into this pocket. The UNITY program, whose Directed Tweak algorithm allows a conformationally flexible 3D search, was applied during the initial screening process to accommodate for the flexibility of the compounds (Hurst,

1994). The UNITY program yielded a number of small molecule hits. These were docked onto the predicted binding pocket of Rac1 containing Trp[56] by use of energy minimization modeling software, FlexX, which allows flexible docking to protein binding sites (Rarey et al., 1996). The compounds identified by this protocol were ranked on the basis of their predicted ability to interact with the binding pocket using Cscore, software that generates a relative, consensus score based on how well the individual scoring functions of the protein-ligand complex perform (Clark et al., 2002). Initially, 100 compounds with the highest consensus scores were chosen. Fifty-eight of these whose docking conformation did not seem to involve residue Trp[56] of Rac1 by visual inspection were eliminated from the list, and 15 of the remaining 42 compounds were chosen for further evaluation because of the availability and solubility issues.

In Vitro Screening and Examination

The ability of the chosen compounds to inhibit the Rac1 binding interaction with its GEF was assessed in a complex formation assay. The Dbl-homology and Pleckstrin homology domains of TrioN, which can specifically activate Rac1 but not Cdc42 (Gao et al., 2001) and Intersectin, a Cdc42-specific GEF (Karnoub et al., 2001), were used to assay the binding activities of Rac1 and Cdc42 in the presence of 1 mM of each individual compound. TrioN readily coprecipitated with GST-Rac1, but not with GST (Gao et al., 2001). Among the panel of candidates tested, NSC23766 was the only one that significantly inhibited TrioN binding to Rac1 (Fig. 1A) and did not interfere with the Cdc42 binding to Intersectin. The chemical structure of NSC23766 is depicted in Fig. 1B. In the simulated model, the binding site for NSC23766 on Rac1 appears to be a surface cleft formed by residues Lys[5], Asp[38], Asn[39], Trp[56], Asp[57], Thr[58], Leu[70], and Ser[71] (Fig. 1C), conforming to the previously predicted groove between the switch I, switch II, and $\beta1/\beta2/\beta3$ regions.

To characterize the inhibitory mechanism of NSC23766, the structural requirement, concentration-dependence, and specificity issues were further examined in vitro. Mutation of most of the residues in the predicted docking site of Rac1 impaired GEF binding and/or catalysis (Gao et al., 2001). In particular, Rac1L70A/S71A retained binding to TrioN but lost the inhibitory response to NSC23766 (Fig. 2A), supporting the possibility that the chemical may, indeed, use the predicted site for the interference of GEF recognition. NSC23766 inhibited the Rac1-TrioN interaction in a concentration-dependent manner with an inhibitory concentration-50 (IC$_{50}$) of ~50 μM in vitro (Fig. 2B). Increasing the amounts of TrioN diminished the inhibitory effect, suggesting that NSC23766 competitively

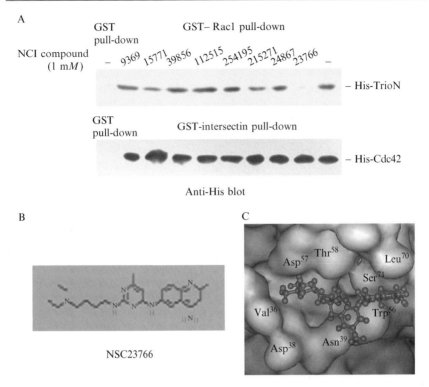

FIG. 1. Identification of NSC23766 as a Rac-activation specific inhibitor. (A) The inhibitory effect of a panel of compounds predicted by virtual screening on Rac1 interaction with TrioN was tested in a complex formation assay; 0.5 μg of (His)$_6$-tagged TrioN was incubated with GST alone or nucleotide-free GST-Rac1 (2 μg) in the presence or absence of 1 mM indicated NCI compound and 10 μl suspended glutathione-agarose beads. After an incubation at 4° for 30 min, the beads with associated (His)$_6$-TrioN were detected by anti-His Western blotting. Lower panel, the effect of the compounds on Cdc42 binding to Intersectin was determined similarly; ~1 μg of GST or GST-tagged Intersectin was incubated with the nucleotide-free (His)$_6$-tagged Cdc42 (0.25 μg) under similar conditions. (B) The chemical structure of NSC23766. (C) A simulated docking model of NSC23766 on the Rac1 surface. (See color insert.)

inhibits the Rac-GEF interaction. Investigations of the effects of NSC23766 on Rac1-Tiam1, RhoA-PDZ-RhoGEF, Rac1-BcrGAP, and Rac1-PAK1 interactions further revealed that whereas NSC23766 inhibits Tiam1 binding to Rac1, it has no detectable effect on RhoA binding to its GEF, PDZ-RhoGEF, or on Rac1 binding to BcrGAP or effector PAK1 at similar or higher concentrations (Gao et al., 2001). These data suggest that NSC23766 specifically inhibits the Rac1-GEF interaction.

Whether NSC23766 inhibits the GEF-stimulated guanine nucleotide exchange reaction of Rac1 by blocking the Rac-GEF interaction was

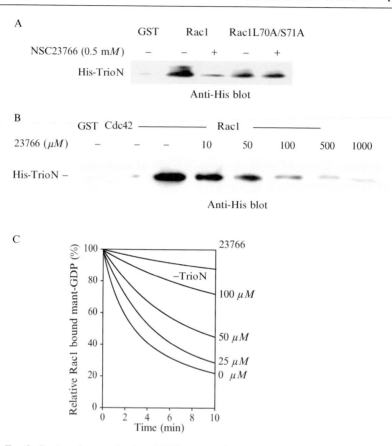

FIG. 2. *In vitro* characterization of NSC23766. (A) The effect of NSC23766 was examined with the RaclL70A/S71A mutant binding to TrioN. (B) Concentration-dependent inhibition of TrioN interaction with Racl by NSC23766. (C) NSC23766-inhibited Trio catalyzed GDP/GTP exchange of Racl in a concentration-dependent manner; 200 nM Racl loaded with mant-GDP was incubated at 25° in an exchange buffer containing 100 mM NaCl, 5 mM $MgCl_2$, 50 mM Tris-HCl (pH 7.6), and 0.5 mM GTP in the absence (top line) or presence of 100 nM TrioN. Increasing concentrations of NSC23766 were included in the exchange buffer as indicated.

assessed using a fluorescent mant–GDP/GTP exchange assay for Racl, Cdc42, and RhoA catalyzed by TrioN, Intersectin, and PDZ-RhoGEF, respectively. Figure 2C shows that NSC23766 blocked the nucleotide exchange of Racl catalyzed by TrioN in a concentration-dependent manner. The IC_{50} value for inhibiting GDP/GTP exchange was consistent with the IC_{50} for inhibiting the binding interaction. At up to 200 μM concentration, NSC23766 failed to detectably affect either the Intersectin-stimulated

Cdc42 exchange or the PDZ-RhoGEF-stimulated RhoA exchange (Gao *et al.*, 2001). These results indicate that NSC23766 specifically inhibits the activation of Rac1 by its GEF *in vitro*.

Effect of NSC23766 on Rac1 Activity in Fibroblast Cells

Rac-specific GEFs such as Tiam1 are involved in serum or PDGF-induced activation of Rac1 in fibroblasts (Hawkins *et al.*, 1995; Ridley *et al.*, 1992). The possibility that NSC23766 may inhibit Rac1 activity *in vitro* was investigated using NIH 3T3 cells. The cells grown in 10% calf serum were incubated with varying concentrations of NSC23766 overnight, and the endogenous activities of Rho GTPases were probed by an effector domain pull-down assay that specifically reveals Rac1-GTP, RhoA-GTP, and Cdc42-GTP contents. NSC23766 inhibited Rac1-GTP formation induced by serum but had no effect on activation of Cdc42 or RhoA (Fig. 3A). Furthermore, NSC23766 suppressed the PDGF (10 nM)-induced activation of Rac1 in serum-starved NIH 3T3 cells in a concentration-dependent manner (Fig. 3B), effectively inhibited the PDGF-stimulated lamellipodia formation (Gao *et al.*, 2004). These results show that NSC23766 is capable of specifically inhibiting Rac1 activity in cells.

Constitutively active Rac1 is known to increase fibroblast cell growth, whereas the dominant negative Rac1 can slow the cell growth rate (Khosravi-Far *et al.*, 1995). The efficacy of NSC23766 in inhibiting cell growth by blocking Rac1 activation was examined by use of wild-type NIH 3T3 cells and cells expressing the Rac-specific GEF, Tiam1, or the constitutively active Rac1 mutant, L61Rac1. Data in Fig. 3C show that NSC23766 slowed the growth of wild-type and Tiam1-expressing cells but had no effect on the growth rate of L61Rac1 cells. In addition, NSC23766 reduced the foci-forming activity of Tiam1, but not that of L61Rac1, and decreased the Tiam1-induced colony size on soft agar (Gao *et al.*, 2001). These data suggest that NSC23766 inhibits fibroblast growth by blocking the cellular Rac1 activity.

Applications of NSC23766 in Diverse Cellular Systems

Reversal of PC-3 Prostate Tumor Cell Phenotypes

Genetic alterations or mitogenic insults in cancer cells may lead to the hyperactivation of Rac GTPases that contributes to the transformation and/or invasion phenotypes of the cells. In the prostate adenocarcinoma PC-3 cells, the tumor suppressor PTEN is defective, resulting in elevated

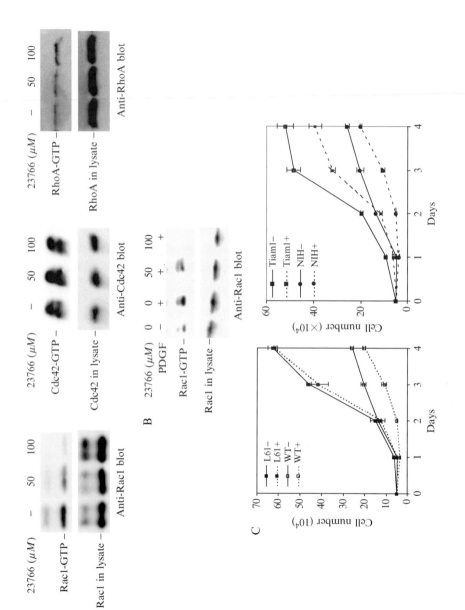

levels of PIP$_3$ and consequent Rac1 hyperactivation. As shown in Fig. 4A, incubation of PC-3 cells with NSC23766 led to a concentration-dependent inhibition of anchorage-independent growth. At similar concentrations of NSC23766, the normal prostate epithelial cells, RWPE-1, remained viable. In an *in vitro* invasion assay, NSC23766 inhibited PC-3 cell invasion through a layer of Matrigel by 85% at 25 μM concentration (Fig. 4B). These results show that NSC23766 can effectively reverse the proliferation and invasion phenotypes of PC-3 tumor cells by inhibiting the Rac1 GTPase.

Inhibition of Platelet Activation

Rac GTPases have been shown to be activated during thrombin-induced or collagen-induced platelet activation (Soulet *et al.*, 2001). Rac1-dependent activation of PAK has been reported to regulate lamellipodia induction by collagen in platelets (Vidal *et al.*, 2002). Gene targeting of Rac1 and Rac2 GTPases impairs platelet adhesion to fibrinogen, α-granule secretion, and aggregation and causes prolonged tail-bleeding times in the mice, implicating Rac GTPases as potential antithrombotic targets (Huzoor-Akbar *et al.*, unpublished observations). We investigated the possibility that inhibition of Rac GTPases by NSC23766 may block platelet aggregation. Platelets from human blood were isolated and washed, and aggregation studies were carried out as detailed earlier (Huzoor-Akbar *et al.*, 1993). As shown in Fig. 5, a 3-min preincubation of washed human platelets with NSC23766 effectively inhibited thrombin-induced activation of Rac1 and Rac2, as well as platelet aggregation. These results suggest that NSC23766 and/or its derivatives may serve as novel antithrombotic agents by targeting Rac GTPases in platelets.

Fig. 3. NSC23766 was effective in specifically inhibiting Rac1 activation in cells. (A) The activation states of endogenous Rac1, Cdc42, and RhoA in NIH3T3 cells with or without NSC23766 treatment were detected by the effector pull-down assays. At 80% confluency in the presence of 10% serum, NIH 3T3 cells in 100-mm dishes were treated with the indicated concentrations of NSC23766 for 12 h. Cell lysates containing similar amount of Rac1, Cdc42, or RhoA were incubated with the agarose immobilized GST-PAK1, GST-WASP, or GST-Rhotekin, and the coprecipitates were subjected to anti-Rac1, Cdc42 or RhoA Western blot analysis to reveal the amount of GTP-bound Rho proteins. (B) The inhibitory effect of NSC23766 on the PDGF-stimulated Rac1 activation was determined by the GST-PAK1 pull-down assay. Serum-starved NIH 3T3 cells in the DMEM medium with different concentrations of NSC23766 were treated with 10 nM PDGF for 2 min. (C) NSC23766 specifically inhibited Rac GEF--stimulated cell growth. Wild-type (WT) Tiam1 or L61Rac1 expressing NIH 3T3 cells was grown in 5% serum in the presence (---) or absence (—) of 100 μM NSC23766. The cells were split in triplicate in 6-well plates at a density of 5×10^4 cells per well for the growth measurements.

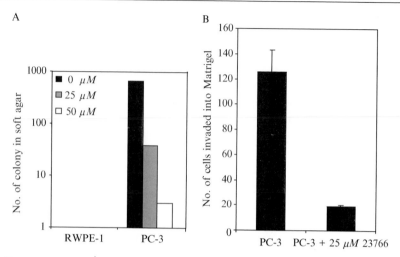

FIG. 4. NSC23766 inhibited anchorage-independent growth and invasion of PC-3 prostate cancer cells. (A) PC-3 and RWPE-1 prostate epithelial cells (1.25×10^3 per well) were grown in 0.3% agarose in the presence of different concentrations of NSC23766, and the number of colonies formed in soft agar was quantified 12 days after plating. (B) PC-3 cells were placed in an invasion chamber at 37° in the absence or presence of 25 μM NSC23766, and the number of penetrated cells into the Matrigel™ matrix after 24 h was quantified.

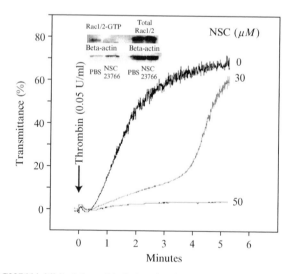

FIG. 5. NSC23766 inhibited thrombin-induced activation of Rac GTPases and aggregation of platelets. Washed human platelets were incubated with NSC23766 at indicated concentrations for 3 min before stimulation with thrombin (0.05 U/ml). The Rac-GTPase activity was determined by the GST-PAK1 pull-down assay (insert), and platelet aggregation induced by the thrombin was determined by a light-scattering–based absorption method. (See color insert.)

Induction of Hematopoietic Stem Cell Mobilization

In steady-state hematopoiesis, a small pool of circulating hematopoietic stem cells/progenitors (HSC/P) is in equilibrium with the large pool of bone marrow HSC/P. Mobilization of HSC/P from bone marrow to peripheral blood is routinely used in stem cell transplantation. Some patients show deficient mobilization by current methods for stem cell collection by apheresis. Rac GTPases have been reported to be important for HSC/P homing and migration. Deficiency of Rac2 induces a moderate mobilization (Yang *et al.*, 2001), whereas the combined deficiency of Rac1 and Rac2 induces a striking mobilization of HSC/P (Gu *et al.*, 2003). We investigated whether NSC23766 could inhibit Rac1 and Rac2 activities and induce mobilization of HSC/P (Cancelas *et al.*, 2005). The effect of NSC23766 on the endogenous Rac1 and Rac2 activities was quantified by incubating the bone marrow Lin-/c-kit+ HSC/P cells with SDF-1α in the presence or absence of NSC23766. Data in Fig. 6 show that NSC23766 at 10 or 25 μM significantly inhibited activation of both Rac1 and Rac2 by SDF-1α as assayed by the GST-PAK1 effector domain pull-down method. Intraperitoneal administration of NSC23766 (2.5 mg/kg) into the "poorly mobilizing" C57Bl/6 mouse strain led to a twofold increase in circulating HSC/P 6 h after

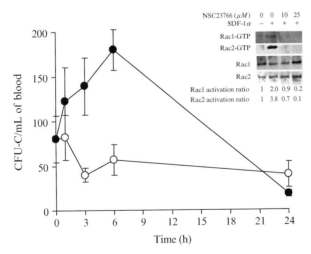

Fig. 6. NSC23766 induces Rac-GTPase–mediated HSC/P mobilization. Colony-forming-units (CFU-C) of the HSC/P in peripheral blood were determined after treatment with NSC26337 (-●-) or PBS control (-○-) in 8-week-old C57Bl/6J mice. Insert, the Rac1 and Rac2 activities before and after stimulation (100 ng/mL SDF-1α) of Lin-/c-kit+ HSC/P cells were measured by the GST-Pak1 effector domain pull-downs in cells preincubated with different concentrations of NSC23766. Normalized Rac1-GTP and Rac2-GTP were calculated as the normalized band density ratios between Rac1-GTP or Rac2-GTP and total Rac1 or Rac2.

injection. The level of circulating HSC/P returned to the normal level 24 h after NSC23766 administration (Fig. 6). These data suggest that NSC23766 may be useful for HSC/P mobilization in animals by down-regulating Rac1/2 activities.

Summary

In this Chapter, the mechanism-based rationale to design and develop NSC23766, a lead compound that specifically inhibit Rac GTPases by blocking a subset of Rac-GEF interactions, is described. NSC23766, by virtue of its ability to specifically target Rac GTPases, is found to be effective in modulating Rac-mediated functions in diverse cellular systems, including tumor cell transformation and invasion, platelet aggregation, and hematopoietic stem/progenitor cell mobilization. It is anticipated that NSC23766 and future improved derivatives may serve as valuable tools for studying Rac-mediated cellular functions and for downregulating aberrant Rac activities in therapeutics.

References

Cancelas, J. A., Lee, A. W., Prabhakar, R., Stringer, K., Zheng, Y., and Williams, D. A. (2005). Rac1 and Rac2 GTPases distinctly regulate hematopoietic stem cell localization and are novel targets for stem cell mobilization. *Nature Medicine* **11**, 886–891.

Clark, R. D., Strizhev, A., Leonard, J. M., Blake, J. F., and Matthew, J. B. (2002). Consensus scoring for ligand/protein interactions. *J. Mol. Graph Model* **20**, 281–295.

Etienne-Manneville, S., and Hall, A. (2002). Rho GTPases in cell biology. *Nature* **420**, 629–635.

Gao, Y., Xing, J., Streuli, M., Leto, T. L., and Zheng, Y. (2001). Trp(56) of rac1 specifies interaction with a subset of guanine nucleotide exchange factors. *J. Biol. Chem.* **276**, 47530–47541.

Gao, Y., Dickerson, J. B., Guo, F., Zheng, J., and Zheng, Y. (2004). Rational design and characterization of a Rac GTPase-specific small molecule inhibitor. *Proc. Natl. Acad. Sci. USA* **101**, 7618–7623.

Gu, Y., Filippi, M. D., Cancelas, J. A., Siefring, J. E., Williams, E. P., Jasti, A. C., Harris, C. E., Lee, A. W., Prabhakar, R., Atkinson, S. J., Kwiatkowski, D. J., and Williams, D. A. (2003). Hematopoietic cell regulation by Rac1 and Rac2 guanosine triphosphatases. *Science* **302**, 445–449.

Hawkins, P. T., Eguinoa, A., Qiu, R. G., Stokoe, D., Cooke, F. T., Walters, R., Wennstrom, S., Claesson-Welsh, L., Evans, T., and Symons, M., and Stephens, L. (1995). PDGF stimulates an increase in GTP-Rac via activation of phosphoinositide 3-kinase. *Curr. Biol.* **5**, 393–403.

Hurst, T. (1994). Flexible 3D searching: The directed tweak technique. *J. Chem. Inf. Comput. Sci.* **34**, 190–196.

Huzoor-Akbar, Wang, W., Kornhauser, R., Volker, C., and Stock, J. B. (1993). Protein prenylcysteine analog inhibits agonist-receptor-mediated signal transduction in human platelets. *Proc. Natl. Acad. Sci. USA* **90**, 868–872.

Karnoub, A. E., Worthylake, D. K., Rossman, K. L., Pruitt, W. M., Campbell, S. L., Sondek, J., and Der, C. J. (2001). Molecular basis for Rac1 recognition by guanine nucleotide exchange factors. *Nat. Struct. Biol.* **8**, 1037–1041.

Khosravi-Far, R., Solski, P. A., Clark, G. J., Kinch, M. S., and Der, C. J. (1995). Activation of Rac1, RhoA, and mitogen-activated protein kinases is required for Ras transformation. *Mol. Cell. Biol.* **15,** 6443–6453.

Rarey, M., Kramer, B., Lengauer, T., and Klebe, G. (1996). A fast flexible docking method using an incremental construction algorithm. *J. Mol. Biol.* **261,** 470–489.

Ridley, A. J., Paterson, H. F., Johnston, C. L., Diekmann, D., and Hall, A. (1992). The small GTP-binding protein rac regulates growth factor-induced membrane ruffling. *Cell* **70,** 401–410.

Schmidt, A., and Hall, A. (2002). Guanine nucleotide exchange factors for Rho GTPases: Turning on the switch. *Gene Dev.* **16,** 1587–1609.

Soulet, C., Gendreau, S., Missy, K., Benard, V., Plantavid, M., and Payrastre, B. (2001). Characterisation of Rac activation in thrombin- and collagen-stimulated human blood platelets. *FEBS Lett.* **507,** 253–258.

Vidal, C., Geny, B., Melle, J., Jandrot-Perrus, M., and Fontenay-Roupie, M. (2002). Cdc42/Rac1-dependent activation of the p21-activated kinase (PAK) regulates human platelet lamellipodia spreading: Implication of the cortical-actin binding protein cortactin. *Blood* **100,** 4462–4469.

Worthylake, D. K., Rossman, K. L., and Sondek, J. (2000). Crystal structure of Rac1 in complex with the guanine nucleotide exchange region of Tiam1. *Nature* **408,** 682–688.

Yang, F. C., Atkinson, S. J., Gu, Y., Borneo, J. B., Roberts, A. W., Zheng, Y., Pennington, J., and Williams, D. A. (2001). Rac and Cdc42 GTPases control hematopoietic stem cell shape, adhesion, migration, and mobilization. *Proc. Natl. Acad. Sci. USA* **98,** 5614–5618.

Zheng, Y. (2001). Dbl family guanine nucleotide exchange factors. *Trends Biochem. Sci.* **26,** 724–732.

[44] *In Vitro* Assay of Primary Astrocyte Migration as a Tool to Study Rho GTPase Function in Cell Polarization

By SANDRINE ETIENNE-MANNEVILLE

Abstract

Rho GTPases are key players in cell migration. The contribution of Rho, Rac, and Cdc42 to the regulation of the actin and microtubule cytoskeletons is essential for membrane protrusion and cell retraction (Etienne-Manneville and Hall, 2002). The polarization of these protrusive and retracting activities in a migrating cell is also under the control of Rho GTPases, in particular Cdc42 (Nobes and Hall, 1999). *In vitro* study of cell migration has shown that Cdc42 activity is required for polarized cell migration in several cell types, including fibroblasts, neutrophils, macrophages, and astrocytes (Allen *et al.*, 1998; Etienne-Manneville, 2004; Etienne-Manneville and Hall, 2001; Palazzo *et al.*, 2001; Srinivasan *et al.*, 2003). Using scratch-induced migration assay, we have previously used primary astrocytes as a tool to study the molecular mechanisms controlling

METHODS IN ENZYMOLOGY, VOL. 406   0076-6879/06 $35.00
Copyright 2006, Elsevier Inc. All rights reserved.   DOI: 10.1016/S0076-6879(06)06044-7

cell polarization at the onset of migration (Etienne-Manneville and Hall, 2001, 2003). On scratching of the monolayer, astrocytes polarize perpendicularly to the scratch to migrate and close the wound. Astrocyte polarization is characterized by the formation of a protrusion in the direction of migration, the elongation of the microtubules that fill the protrusion, and the reorientation of the centrosome, which serves as a microtubule-organizing center toward the direction of migration. This *in vitro* migration assay allows us to simultaneously investigate the mechanisms controlling cell migration, cell protrusion, and cell polarization. Primary astrocytes, although more constraining, provide a more physiological model than immortalized cell lines. Moreover, astrocyte culture can be obtained in a large number and, therefore, also allows biochemical analysis. Here I describe the procedure by which we can obtain and purify primary rat astrocytes and the different assays we have previously used to analyze the role of Rho GTPases and their downstream targets in cell migration and polarization.

Introduction

Astrocytes are the most abundant cells in the central nervous system (CNS) with intricate relationships with neurons, blood vessels, and meninges. The ratio of astrocytes to neurons has increased during evolution. Although astrocytes have been traditionally viewed as neuron-supporting cells, evidence is now accumulating that astrocytes play a wide range of functions critical for maintenance of a homeostatic neuronal environment and are also directly involved in neuronal development and functions (Haydon, 2000; Song et al., 2002; Ullian et al., 2001). Astrocytes are suspected to be involved in a wide range of CNS pathologies, including trauma, viral, or bacterial infections but also neurodegeneration (Mucke and Eddleston, 1993). In such situations, astrocytes undergo a reaction called astrogliosis with increased cell division and protein expression and a characteristic morphological change (the hypertrophy of their cellular processes) associated with an increased motility (Ridet et al., 1997). The poor ability of mammalian central nervous system axons to regenerate has been attributed, in part, to astrocyte behavior after axonal injury (Shearer and Fawcett, 2001; Sivron and Schwartz, 1995). This behavior is manifested by the limited ability of astrocytes to migrate and thus repopulate the site of injury. The slow repopulation of the injury site by astrocytes is in apparent correlation with the failure of the regrowing injured axons to traverse this site (Rhodes et al., 2003; Silver and Miller, 2004). Understanding why mammalian astrocytes migrate or fail to repopulate the site of injury and finding ways to overcome this failure may contribute significantly to the achievement of CNS regeneration.

Although connected by gap junctions, astrocytes do not form any defined structure amenable to experimental manipulation, and, therefore, astrocytes have been relatively difficult to study *in vivo*. The development of *in vitro* models of astrocyte migration has provided a useful tool to investigate the molecular mechanisms controlling astrocyte migration. Our experimental procedure was derived from ones used to follow gliosis in response to *in vitro* injury or the effects of wound-associated factors on astrocyte migratory response (Faber-Elman *et al.*, 1996) and recapitulate most of the major astrocyte responses to *in vivo* injury.

In addition to their physiological relevance, primary astrocytes have appeared as a beautiful cell biology model. Using this scratch-induced *in vitro* migration assay, we have observed that astrocytes undergo slow and polarized migration in the direction perpendicular to the scratch (Etienne-Manneville and Hall, 2001). Cells do not move individually but rather migrate as a sheet toward the other edge of the scratch. Migration stops when the two edges of the scratch meet. This assay has proved to be a very powerful tool to investigate the molecular mechanisms controlling cell migration and cell polarization (Etienne-Manneville and Hall, 2001, 2003). Confluent astrocytes are barely motile, and migration is initiated on scratching. Migrating astrocytes are morphologically different from immobile cells. They are extremely flat, firmly adherent, and with an elongated protrusion easily reaching 100 μm long. Indeed, on scratching, the leading edge protrudes continuously for at least 12 h, forming an elongated protrusion extending in the direction of migration and reminiscent of the large astrocytic process visible *in vivo*. The protrusion is filled with elongated microtubules. Protrusion formation, as well as migration, is dependent on microtubule dynamics, which makes it a valuable model to study the regulation of the microtubule network. As part of microtubule network rearrangement, the centrosome reorients in front of the nucleus in the direction of cell migration. Although not always required for migration, centrosome reorientation is a common theme in many migrating cells, including fibroblasts, epithelial cells, and neurons (Magdalena *et al.*, 2003; Palazzo *et al.*, 2001; Yvon *et al.*, 2002). In astrocytes, centrosome position seems to be a good indicator for the orientation of the protrusion and, therefore, reflects the orientation of the polarized cells. Therefore, protrusion formation and centrosome reorientation are two essential markers of polarization of migrating astrocytes, which can be used to investigate the pathways involved in cell polarization.

Primary Culture of Rat Cerebral Astrocytes

Reagents

> Dissection buffer (DB): PBS (without calcium and magnesium), 0.6% glucose.
> Astrocyte culture medium (ACM): DMEM, 1 g/l glucose, 2 m*M* L-glutamine, 100 mg/ml penicillin/streptomycin, 10% heat-inactivated fetal calf serum.
> Poly-L-ornithine hydrobromide (Sigma-Aldrich).

Dissection Procedure

To purify primary astrocytes we use a method derived from El-Etr *et al.* (1989) and Weiss *et al.* (1986)

1. An 18-day pregnant Sprague-Dawley female rat is put to death in a CO_2 chamber according to published recommendations (Close *et al.*, 1996a,b).
2. Embryos are collected in a Petri dish containing cold DB and placed on ice.
3. Embryos are decapitated and heads moved to a second Petri dish containing cold DB and placed on ice.
4. While steadying the embryo head with forceps, cut the skin and the skull from the midline of the head at the base of the skull to the eyes with microdissecting curved scissors.
5. If necessary, cut the skull at the midline fissure, without cutting into the brain tissue, to facilitate brain extraction.
6. Release the brain from the skull cavity. Apply slight pressure with microdissecting curved forceps, sliding along the length of the brain from the mid-eye area backward to the base of the skull.
7. Using small curved forceps, transfer each brain to a 60-mm Petri dish containing cold DB and placed on ice. Leave on ice approximatively 15 min. Use this time to coat tissue culture plates with polyornithine (see "Tissue Culture Plate Coating").
8. While steadying the brain with forceps, separate the cerebrum from the cerebellum and the brain stem.
9. Cut with fine microdissecting scissors on each sides of the midline fissure.
10. Gently peel the meninges from individual cortical lobes.
11. Dissect out each striatum and immediately transfer it to a 35-mm Petri dish containing 2 ml of cold serum-free ACM.

12. Once all the striata have been transferred to the 35-mm Petri dish, check the absence of remaining meninges.
13. Cut striata in pieces and transfer them with the 2 ml of serum-free ACM to a 15-ml tube.
14. Mechanically dissociate into a cell suspension with a narrow Pasteur pipette by pipetting up and down at least 15 times.
15. Add 2 ml of serum-free ACM and mix by gentle inversion.
16. Decant for 5 min.
17. Transfer the top 2 ml in a 50-ml tube containing 37° preheated ACM 37° and keep it in the 37° incubator.
18. Add 1 ml of serum-free ACM.
19. Repeat steps 4–6 using a narrower Pasteur pipette.
20. After decantation, pool the cell suspension with the previous one and discard the remaining cell agglomerates at the bottom of the tube. You should not have to discard more than 100–200 μl. If cells have not been sufficiently dissociated, go back to step 19.
21. Centrifuge at low speed (800 rpm) for 8 min.
22. Discard the supernatant and resuspend the pellet in ACM.
23. Count the cells and dilute them to $3 \cdot 10^5$ cells/ml in the appropriate final volume of preheated ACM.
24. Keep the cells at 37° while washing the tissue culture plates (see "Tissue Culture Plate Coating").
25. Plate 10 ml ($3 \cdot 10^6$ cells) of cells in poly-L-ornithine–coated 90-mm tissue culture dishes.
26. Place the tissue culture dishes at 37° in a moist 5% CO_2, 95% air atmosphere.

Tissue Culture Plate Coating

You should start the coating before having finished the dissection procedure, so that you do not have to keep the cells in suspension in the incubator before plating.

1. As a first estimation, you can coat one 90-mm tissue culture Petri dish per embryo. Add another three to four plates in case you get more cells at the end of the dissection.
2. Prepare a 1.5 μg/ml poly-L-ornithine solution in sterile distilled water.
3. Use 4 ml of poly-L-ornithine solution for each 90-mm tissue culture Petri dish.
4. Incubate the plates for 1 h at 37°.
5. Discard the poly-L-ornithine solution.

6. Wash the plates once with 10 ml of water.

Culture Purification and Differentiation

1. Keep the cells in the incubator for 7 days without changing the medium.
2. After 7 days, remove the medium and replace it with 10 ml PBS.
3. Seven days after dissection, only few astrocytes are visible; most cells are neural precursors.
4. Wash thoroughly the entire cell monolayer by flushing 10 times PBS directly onto the cells. This should separate remaining oligodendrocytes and microglial cells; remove all debris, dead cells, and aggregates from the plates.
5. Replace the PBS with 10 ml of fresh ACM. Return the plates to the 37° incubator for 48 h.
6. Change the medium every 2 days until cells form a confluent monolayer. Cells usually form a confluent monolayer 10–12 days after dissection. Confluent plates contain $4 \cdot 10^6$ to $5 \cdot 10^6$ cells.
7. After 3 weeks in culture, routinely more than 95% of the cells are positively stained for glial fibrillary acidic protein (GFAP).
8. Confluent cells can then be kept for 2 weeks (changing medium twice a week) or be passed once onto poly-L-ornithine–coated plates or glass coverslip for further enrichment.
9. For better results, astrocytes must be passaged during the first week after confluency.
10. Wash cells in calcium- and magnesium-free PBS.
11. Add 1 ml of trypsin 0.25%, EDTA 0.02% per 90-mm tissue culture dish and incubate for 5 min at 37°.
12. Detach the cells by flushing the trypsin solution on the monolayer, add the cell suspension in a 15-ml tube containing 14 ml of ACM and centrifuge for 8 min at 800 rpm.
13. Resuspend the cell pellet in the appropriate ACM volume, so that cell density/surface is divided three times. (One 90-mm dish will give rise to three 90-mm dishes of confluent astrocytes or to the equivalent surface on glass coverslips).
14. Secondary cultures are grown to confluency changing medium every 48 h and can be kept for a further 2–3 weeks changing medium twice a week.
15. We do not recommend a second passage.

Scratch-Induced Astrocyte Migration

Astrocytes like many other cell types will undergo migration after scratching of the cell monolayer (Fig. 1). Therefore, cell migration will occur whether you use fully differentiated astrocytes or not. However, astrocyte-specific responses, in particular astrocyte protrusion, will occur only if the cells are fully differentiated. *In vitro* differentiation of astrocytes can be monitored by the nature of the intermediate filaments. Differentiated astrocytes are characterized by GFAP expression (Eliasson *et al.*, 1999; Pekny and Pekna, 2004). Regular GFAP staining is recommended. Because the number of differentiated astrocytes increases while the cells are maintained in culture as confluent monolayers, the longer you wait the more differentiated they will be. So, we recommend waiting at least 7 days after confluence before performing a scratch assay.

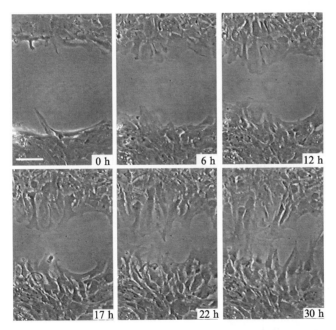

FIG. 1. Scratch-induced migration of primary rat astrocytes. Phase-contrast images from a video showing scratch-induced astrocyte migration. The time period after scratching is indicated at the bottom of each picture. Note the morphological changes (protrusion formation) occurring during the first 12 h and the nuclear movement only seen later. Bar, 100 μm.

Single Wound Procedure for Cell Biology Analysis

1. Use a confluent monolayer of astrocytes plated on a glass coverslip (diameter, 12–14 mm) placed in a 4-well plate.
2. Change the cell culture medium 16 or 24 h before scratching.
3. Make a single scratch in the monolayer by moving a flamed glass capillary at regular speed across the glass coverslip. Depending on the size of the glass capillaries, the width of the scratch will vary. Because astrocytes migrate relatively slowly, scratches 300-mm wide are sufficient for most experiments. However, for migration assays, use large glass capillaries or Pasteur pipette to make wide scratch (500 μm).
4. Do not change the medium and return the coverslip in the 4-well plate in the incubator for the desired time period.

Multiple Wound Procedure for Biochemical Analysis

To assess regulation of enzymatic activity (kinase assays), changes in protein content, GTP-loading of GTPases, or protein complex formation, you must first obtained a large number of migrating astrocytes. To scale up the number of scratch-activated cells, a large number of scratches is done on large plate of confluent cells.

1. Use confluent monolayers of astrocytes plated on 35-mm dishes or 90-mm dishes.
2. Change the cell culture medium 16 h before scratching.
3. Make multiple scratches in the monolayer so that at least 50% of the cells are on the edge of a scratch. We recommend the use of an eight-channel pipet with 1–10 μl tips (three tips are enough to scratch a 35-mm dish). Slide the eight tips altogether three times around the plate and 10 times across it turning the plate a 20-degree angle each time.
4. Incubate the plate for the required amount of time in a 37°, 5% CO_2 incubator.

Analysis of Scratch-Induced Astrocyte Responses

Procedure for Immunostaining

1. Wash cells twice with 1 ml PBS.
2. Fix cells with 250 μl PBS, paraformaldehyde 4%, (supplemented with glutaraldehyde 0.5% for microtubules or actin filament staining) for 10 min at room temperature.

3. Wash twice with 1 ml PBS.
4. Permeabilize the cells with PBS, Triton 0.2%, for 10 min at room temperature.
5. Wash twice with 1 ml PBS.
6. Quench for 10 min at room temperature with 2 mg/ml $NaBH_4$ in PBS.
7. Wash twice with 1 ml PBS.
8. Incubate with primary antibodies for 45 min at room temperature.
9. Wash 10 times in PBS.
10. Incubate with labeled-secondary antibodies for 30 min at room temperature.
11. Wash 10 times with PBS and twice more with H_2O.
12. Mount the coverslip in 5 μl of Mowiol (Calbiochem) or other mounting reagent. Let the Mowiol dry for 30 min at 37° or overnight at 4°.

Astrocyte Migration Assay

One must be careful not to assess cell protrusion instead of cell translocation. Indeed, the cell leading edge movement is not representative of the movement of the cell rear. Nuclear migration is barely visible during the first 6 h, and translocation of the cell body starts 8–12 h after scratching, while protrusions have already invaded the space left by the scratch (Fig. 2A). Therefore, we strongly recommend assessment of nuclear migration as a valuable sign of astrocyte migration rather than the closure of the wound, which reflects both migration and protrusion.

1. Make a large wound (approximately 500-mm wide) across the astrocyte monolayer plated on a 12- to 14-mm coverslip.
2. Immediately microinject pEGFP vector in the nucleus of cells on the edge of the scratch.
3. Place the cells back in the 37°, 5% CO_2 incubator and leave them to migrate for 48 h.
4. Wash cells twice with 1 ml PBS.
5. Incubate the cells with Hoechst solution for 10 min at room temperature. Hoechst staining is sufficient to assess astrocyte migration; however, staining for actin, microtubules, or any protein of interest can be performed on the same cells. In this case, use the immunostaining procedure accordingly.
6. Wash four times in PBS.
7. Mount the cells by inverting the glass coverslip on a microscope slide on which you will have put 5 μl Mowiol.

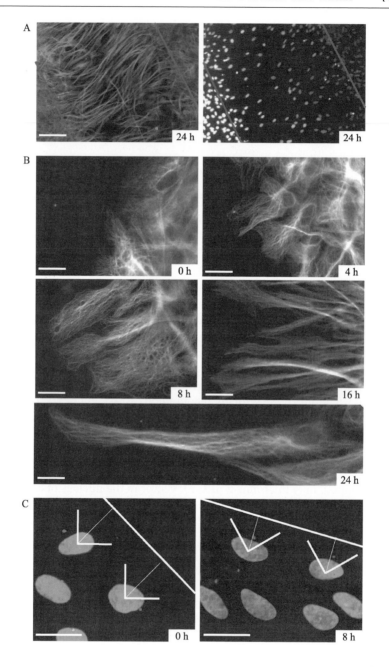

8. Observe under the microscope at low magnification ($10\times$ or $20\times$ objectives). Looking at the Hoechst staining, you will easily recognize the initial position of the scratch as a line with more numerous (corresponding to the accumulation of nonmigrating cells) and condensed nuclei (which mark dead cells left after scratching) (Fig. 2A, right panel). Find the GFP-expressing cells and among these score as migrating cells those cells whose nuclei have moved away from their original place. Do not expect all expressing cells to have migrated. Approximatively only 60% of expressing cells are migrating cells.

An alternative (and more fastidious) way to test astrocyte migration is to do a time-lapse movie for 36–48 h after scratching and to assess nuclear migration by tracking the nuclei of wound-edge cells during wound closure.

Quantification of Astrocyte Protrusion

We have defined astrocyte protrusion as a morphological characteristic of migrating astrocytes (Etienne-Manneville and Hall, 2001). Protrusion starts during the first hour after scratching (Fig. 2B). However, for the first 4–6 h, the protrusion is comparable in shape and length to a fibroblast lamellipodia. It is only after 8 h that the elongation of the protrusion becomes evident. The protrusion being very flat, we recommend the use of microtubule staining (Fig. 2B). Indeed, microtubules elongate in the entire protrusion and allow easy measurement of the cell length. Protrusion formation can, therefore, be assessed 8 h after wounding; however, for easier quantification, we generally score the number of protrusions formed 16 h after scratching. Except for the migration time, the protocol is identical in both cases.

1. Make a wound (approximately 300-mm wide) across the astrocyte monolayer plated on a 12- to 14-mm coverslip.

FIG. 2. Scratch-induced astrocyte responses. (A) Cell migration. Left panel shows tubulin (microtubules, green) and Hoechst (nuclei, blue) 24 h after scratching. From the Hoechst image shown on the right panel, the initial position of the scratch can be drawn (red line). Bar, 100 μm. (B) Protrusion formation and polarization of the microtubule network. Staining with antitubulin antibodies at indicated time after scratching shows the elongation of the microtubule network and its polarization along the axis of migration. Bar, 10 μm. (C) Centrosome reorientation. Concomitant Hoechst (nuclei, blue) and pericentrin (centrosome, red) staining allow the quantification of centrosome reorientation. In the left panel (0 h after scratching), the two cells of the front row are not polarized, their centrosome is not localized in the forward-facing quadrant. In the right panel (8 h after scratching), the centrosomes of the scratch edge cells are localized in the forward-facing quadrant. Bar, 10 μm. (See color insert.)

2. Immediately microinject pEGFP vector or your vector of interest in the nucleus of cells of the wound edge.
3. Place the cells back in the 37°, 5% CO_2 incubator and leave them to migrate for 16 h.
4. Wash cells twice with 1 ml PBS.
5. Use the immunostaining procedure with antibodies directed against tubulin.
6. Observe under the fluorescence microscope using (20× or 40× objective). Find the GFP/protein of interest expressing cells. Among these cells, looking at the microtubule staining, score as "protruding" cells with one major protrusion that appear at least four times longer than wide. Do not expect all expressing cells to have protrusions. Approximately 65% of GFP-expressing cells form protrusions. We score protruding cells regardless of the orientation of the protrusion. When polarization is inhibited in the entire monolayer (for instance with PKCξ or GSK3β inhibitors (Etienne-Manneville and Hall, 2001, 2003), the number of protruding cells remains normal; however, protrusions form in random directions, and smaller protrusions can be seen on the sides of the major protrusion.

Centrosome Reorientation Assay

The centrosome reorientation assay allows a precise quantification of cell orientation (Fig. 2C). In astrocytes, the centrosome also serves as the major microtubule organizing center, where most microtubule minus-ends are anchored. In a confluent astrocyte monolayer, the centrosome is localized in close proximity to the nuclear envelope in a random position around the nucleus. On scratching, the centrosome of most cells (about 75%, 16 h after scratching) localizes in front of the nucleus toward the front edge (Etienne-Manneville and Hall, 2001). Centrosome reorientation essentially occurs during the first 6–8 h after scratching and before cell translocation.

1. For centrosome reorientation assay, allow the cells to migrate for 8 h.
2. Wash cells twice with 1 ml PBS.
3. Use the immunostaining procedure with antibodies directed against pericentrin and Hoechst.
4. Observe under the microscope using 63× objective. Centrosome reorientation must be assessed in the cells located at the wound edge

(Fig. 2C). Eight hours after wounding, the centrosome of cells in the second and third rows of cells is not reoriented yet. To quantify centrosome reorientation, separate the space in four equal quadrants joining in the center of the nucleus of interest and placed so that one of the quadrant is facing the scratch (the median of each 90-degree angle being either parallel or perpendicular to the scratch) (Fig. 2C). Do not try to determine centrosome orientation along curved or irregular portions of the scratch. Centrosomes are scored as "polarized" or "reoriented" when the centrosome is placed in the quadrant facing the scratch. When cells have two centrosomes (i.e., cells in G2), centrosomes are scored polarized if one of the two centrosomes is correctly positioned. Note that in a nonpolarized population of cells, such as wound edge cells just after wounding, 25% of the centrosome will be positively scored; 25% of "polarized" centrosome, therefore, corresponds to a random cell orientation. Eight hours after wounding, you can expect approximately 65% of "polarized" centrosomes.

References

Allen, W. E., Zicha, D., Ridley, A. J., and Jones, G. E. (1998). A role for Cdc42 in macrophage chemotaxis. *J. Cell Biol.* **141,** 1147–1157.

Close, B., Banister, K., Baumans, V., Bernoth, E.-M., Bromage, N., Bunyan, J., Erhardt, W., Flecknell, P., Gregory, N., Hackbarth, H., Morton, D., and Warwick, C. (1996a). Recommendations for euthanasia of experimental animals: Part 1. *Lab. Animals* **30,** 293–316.

Close, B., Banister, K., Baumans, V., Bernoth, E.-M., Bromage, N., Bunyan, J., Erhardt, W., Flecknell, P., Gregory, N., Hackbarth, H., Morton, D., and Warwick, C. (1996b). Recommendations for euthanasia of experimental animals: Part 2. *Lab. Animals* **31,** 1–32.

El-Etr, M., Cordier, J., Glowinski, J., and Premont, J. (1989). A neuroglial cooperativity is required for the potentiation by 2-chloroadenosine of the muscarinic-sensitive phospholipase C in the striatum. *J. Neurosci.* **9,** 1473–1480.

Eliasson, C., Sahlgren, C., Berthold, C. H., Stakeberg, J., Celis, J. E., Betsholtz, C., Eriksson, J. E., and Pekny, M. (1999). Intermediate filament protein partnership in astrocytes. *J. Biol. Chem.* **274,** 23996–24006.

Etienne-Manneville, S. (2004). Cdc42 - the centre of polarity. *J. Cell Sci.* **117,** 1291–1300.

Etienne-Manneville, S., and Hall, A. (2001). Integrin-mediated Cdc42 activation controls cell polarity in migrating astrocytes through PKCz. *Cell* **106,** 489–498.

Etienne-Manneville, S., and Hall, A. (2002). Rho GTPases in cell biology. *Nature* **420,** 629–635.

Etienne-Manneville, S., and Hall, A. (2003). Cdc42 regulates GSK3 and adenomatous polyposis coli (APC) to control cell polarity. *Nature* **421,** 753–756.

Faber-Elman, A., Solomon, A., Abraham, J. A., Marikovsky, M., and Schwartz, M. (1996). Involvement of wound-associated factors in rat brain astrocyte migratory response to axonal injury: *In vitro* simulation. *J. Clin. Invest.* **97,** 162–171.

Haydon, P. G. (2000). Neuroglial networks: Neurons and glia talk to each other. *Curr. Biol.* **10,** R712–R714.

Magdalena, J., Millard, T. H., and Machesky, L. M. (2003). Microtubule involvement in NIH 3T3 Golgi and MTOC polarity establishment. *J. Cell Sci.* **116,** 743–756.

Mucke, L., and Eddleston, M. (1993). Astrocytes in infectious and immune-mediated diseases of the central nervous system. *FASEB J.* **13,** 1226–1232.

Nobes, C. D., and Hall, A. (1999). Rho GTPases control polarity, protrusion and adhesion during cell movement. *J. Cell Biol.* **144,** 1235–1244.

Palazzo, A. F., Joseph, H. L., Chen, Y. J., Dujardin, D. L., Alberts, A. S., Pfister, K. K., Vallee, R. B., and Gundersen, G. G. (2001). Cdc42, dynein, and dynactin regulate MTOC reorientation independent of Rho-regulated microtubule stabilization. *Curr. Biol.* **11,** 1536–1541.

Pekny, M., and Pekna, M. (2004). Astrocyte intermediate filaments in CNS pathologies and regeneration. *J. Pathol.* **204,** 428–437.

Rhodes, K. E., Moon, L. D., and Fawcett, J. W. (2003). Inhibiting cell proliferation during formation of the glial scar: Effects on axon regeneration in the CNS. *Neuroscience* **120,** 41–56.

Ridet, J. L., Malhotra, S. K., Privat, A., and Cage, F. H. (1997). Reactive astrocytes: Cellular and molecular cues to biological function. *Trends Neurosci.* **20,** 570–577.

Shearer, M. C., and Fawcett, J. W. (2001). The astrocyte/meningeal cell interface—a barrier to successful nerve regeneration? *Cell Tissue Res.* **305,** 267–273.

Silver, J., and Miller, J. H. (2004). Regeneration beyond the glial scar. *Nat. Rev. Neurosci.* **5,** 146–156.

Sivron, T., and Schwartz, M. (1995). Glial cell types, lineages and response to injury in rat and fish: Implication for regeneration. *Glia* **13,** 157–165.

Song, H., Stevens, C. F., and Gage, F. H. (2002). Astroglia induce neurogenesis from adult neural stem cells. *Nature* **417,** 39–44.

Srinivasan, S., Wang, F., Glavas, S., Ott, A., Hofmann, F., Aktories, K., Kalman, D., and Bourne, H. R. (2003). Rac and Cdc42 play distinct roles in regulating PI(3,4,5)P3 and polarity during neutrophil chemotaxis. *J. Cell Biol.* **160,** 375–385.

Ullian, E. M., Sapperstein, S. K., Christopherson, K. S., and Barres, B. A. (2001). Control of synapse number by glia. *Science* **291,** 657–660.

Weiss, S., Pin, J. P., Sebben, M., Kemp, D. E., Sladeczek, F., Gabrion, J., and Bockaert, J. (1986). Synaptogenesis of cultures striatal neurons in serum-free medium: A morphological and biochemical study. *Proc. Natl. Acad. Sci. USA* **83,** 2238–2242J.

Yvon, A. M., Walker, J. W., Danowski, B., Fagerstrom, C., Khodjakov, A., and Wadsworth, P. (2002). Centrosome reorientation in wound-edge cells is cell type specific. *Mol. Biol. Cell* **13,** 1871–1880.

[45]   Real-Time Centrosome Reorientation During Fibroblast Migration

By EDGAR R. GOMES and GREGG G. GUNDERSEN

Abstract

The centrosome is positioned between the nucleus and the leading edge of many types of migrating cells. Cdc42 regulates this centrosome reorientation through its effectors Par6 and MRCK. Using time-lapse microscopy of live cells, the mechanisms and kinetics of centrosome reorientation can be studied. In this chapter, we describe a modification in the standard wound healing assay that allows the study of signaling pathways involved in centrosome reorientation and other polarization events that occur before cell migration. We also describe a method for visualization of centrosome reorientation by time-lapse microscopy using NIH 3T3 fibroblasts stably transfected with GFP-tubulin.

Introduction

During migration of many cell types, the centrosome is reoriented to a position between the nucleus and the leading edge. The centrosome is the microtubule (MT) nucleating site, and in most migrating cells, the MTs remain attached to the centrosome, giving rise to a radial array of MTs. Because of the organizing capability of the centrosome, it is sometimes referred to as the MT organizing center (MTOC), and centrosome reorientation is also referred to as MTOC reorientation. Centrosome reorientation toward the leading edge has been described in migrating fibroblasts, neurons, astrocytes, endothelial cells, and macrophages (Etienne-Manneville and Hall, 2001; Euteneuer and Schliwa, 1992; Gotlieb *et al.*, 1981; Gregory *et al.*, 1988; Gundersen and Bulinski, 1988; Kupfer *et al.*, 1982; Nemere *et al.*, 1985). Centrosome reorientation occurs in some nonmigratory situations such as in endothelial cells *in situ* (Rogers *et al.*, 1985) or in response to shear stress in culture (Tzima *et al.*, 2003) and in T cells, where the centrosome is reoriented toward the target cell (Kupfer *et al.*, 1983). The reorientation of the centrosome and the associated Golgi apparatus may facilitate the delivery of components to the leading edge (Bergmann *et al.*, 1983; Prigozhina and Waterman-Storer, 2004; Schmoranzer *et al.*, 2003). Centrosome reorientation may also contribute to the appropriate geometry to ensure that the nucleus moves forward with

METHODS IN ENZYMOLOGY, VOL. 406
Copyright 2006, Elsevier Inc. All rights reserved.

the cell (Gomes *et al.*, 2005; Solecki *et al.*, 2004). The molecular mechanisms of centrosome reorientation are not known, however, in several systems, Cdc42 and its effectors, Par6 and MRCK, and dynein have been implicated in the regulation of centrosome reorientation (Etienne-Manneville, 2004; Gomes *et al.*, 2005).

Centrosome reorientation in migrating cells is easily quantified in fixed preparations stained with centrosomal and nuclear markers. A cell is determined to have a reoriented centrosome when the centrosome is located in the "pie slice" defined by the nucleus and the leading edge (Fig. 1). As this sector occupies ~1/3 of the cell, by chance, ~33% of the

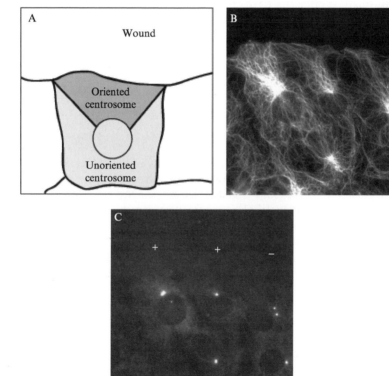

FIG. 1. (A) Diagram of criteria used to determined centrosome orientation. When the centrosome is localized in the sector between the nucleus and the leading edge, it is scored as oriented. Localization in other sectors is scored as non-oriented. (B, C) Wound edge cells with oriented and non-oriented centrosomes. Cells were treated with LPA, fixed, and stained for MTs (B) and pericentrin, a centrosomal protein (C). In both images, the nucleus is detected as an oval nonfluorescent area. Cells with (+) and without (−) reoriented centrosomes are indicated in C.

cells will have a "reoriented centrosome." In wounded monolayers, which are typically used to study centrosome reorientation, maximal centrosome reorientation reaches 70–90% and occurs ~2 h after wounding of endothelial and fibroblast monolayers (Gundersen and Bulinski, 1988) and ~8 h after wounding astrocyte monolayers (Etienne-Manneville and Hall, 2001).

Determination of centrosome reorientation in fixed-cell preparations is a simple assay. However, it does not provide information on the mechanism, time, and speed of centrosome reorientation or what actually moves during centrosome reorientation. A number of techniques have been used to visualize centrosomes in living cells so that these issues can be addressed. Stable cell lines expressing green fluorescent protein (GFP)-tagged γ-tubulin, a centrosomal protein, have characterized centrosome position during migration of CHO and Ptk cells (Yvon *et al.*, 2002). A transgenic mouse expressing GFP-centrin-2 has been described, and it will be a valuable tool to study centrosome reorientation by time-lapse microscopy (Higginbotham *et al.*, 2004). Centrosome position can also be determined by visualizing the focus of MTs (i.e., by the position of the MTOC) (Gomes *et al.*, 2005). MTs visualized by modulated polarization microscopy have been used to visualize centrosome reorientation in T cells (Kuhn and Poenie, 2002).

In this chapter, we describe the preparation and use of stably expressing GFP-α-tubulin NIH 3T3 cell line (3T3-GFPTub) to observe centrosome reorientation in wounded monolayers (Gomes *et al.*, 2005). As the GFP-tubulin is incorporated into both centriolar and cytoplasmic MTs, the dynamics of the centrosome and cytoplasmic MTs can be followed simultaneously. This allows movements of the centrosome to be related to changes in MT dynamics and/or interactions with cortical sites. Another advantage of using such a cell line is that the position of the nucleus can be determined, because the nucleus is easily identified in the GFP-tubulin–expressing cell as a nonfluorescent area. In our experience, transient expression of GFP-tubulin has a number of drawbacks: there is substantial cell–cell variability in expression levels, and high levels of GFP-tagged tubulin can alter MT dynamics and give rise to substantial diffuse cytoplasmic fluorescence. MTs can also be labeled by microinjection of fluorescently labeled tubulin (Waterman-Storer, 2002). This method requires a microinjection apparatus and can be more time consuming and technically demanding.

Generation of a Stable Cell Line Expressing GFP-Tubulin

We used a standard method for generating stable NIH-3T3 cell lines expressing GFP-tubulin (Sullivan and Satchwell, 2000). Cells are transfected with a commercially available enhanced GFP-α-tubulin vector containing a neomycin-resistance cassette (BD Biosciences Clontech).

Addition of neomycin or G418 (a neomycin analog) selects for cells that incorporate the vector. Resistant clones are screened for GFP-α-tubulin expression levels and checked for growth rates, migration, centrosome reorientation, and formation of stable MTs (Cook *et al.*, 1998; Gomes *et al.*, 2005).

Materials

NIH 3T3 cells (ATCC)
DMEM (Invitrogen)
Enhanced GFP-α-tubulin mammalian expressing vector (Clontech)
Lipofectamine (Invitrogen)
G418 (Calbiochem)
Cloning cylinders (Sigma)

NIH 3T3 fibroblasts are cultured in DMEM supplemented with 10% calf serum (Cook *et al.*, 1998; Gundersen *et al.*, 1994) and kept at subconfluent densities to prevent spontaneous transformation.

Transfect cells with GFP-α-tubulin vector using lipofectamine according to the protocol provided by the manufacture. Two days after transfection, plate cells at low density with DMEM medium supplemented with 1 mg/ml G418. Once colonies are visible by direct observation of the plates (5–7 days after transfection), select individual colonies using cloning cylinders and transfer them to larger plates for expansion. Reduce G418 concentration to 0.5 mg/ml at this stage. When clones have been expanded, check for GFP-α-tubulin expression by plating cells on acid-washed glass coverslips, fixing in −20° methanol, rehydrating, and observing by fluorescence microscopy. Select clones with fluorescent MTs that are in a normal radial arrays (Fig. 1B,C). Characterize positive clones by growth rate, migration into the wound, and centrosome reorientation (Fig. 2). We also check clones for formation of stabilized MTs, because this is another rearrangement of MTs that occurs in cells in response to wounding (Cook *et al.*, 1998; Gundersen and Bulinski, 1988). Select clones that behave similarly to parental cells.

The level of GFP-α-tubulin relative to endogenous α-tubulin is <5% in the selected clone (3T3-GFPTub) (Fig. 2). Multiple clones should be checked for use in cell assays to ensure that the behavior being studied is typical.

Wound Healing Assay

Wound healing assays are useful to study cell polarization and migration. When a confluent monolayer of cells is wounded, the cells at the wound edge synchronously polarize and begin migrating in the direction

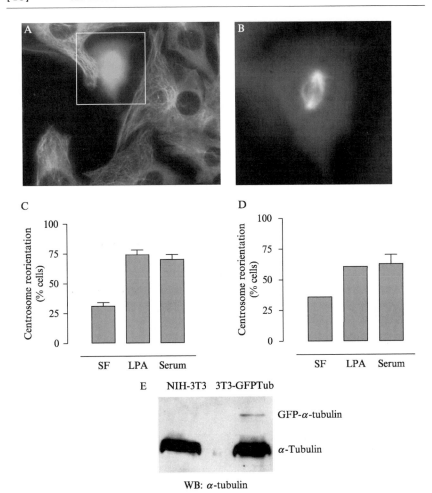

FIG. 2. Characterization of the NIH 3T3 cell line stably expressing GFP-α-tubulin (3T3 GFPTub). (A) GFP-tubulin fluorescence in 3T3-GFPTub cells. (B) Higher magnification of the inset in (A). The different focal plane shows the MTs of the mitotic spindle. Centrosome reorientation stimulated by LPA or serum is identical in parental NIH 3T3 (C) and 3T3 GFPTub (D) cells. Serum-starved cells were left in serum-free DMEM (SF) or treated with LPA or calf serum for 2 h. Cells were fixed with methanol, immunofluorescently stained for centrosomes and MTs, and the extent of centrosome reorientation determined. (E) Expression levels of GFP-α-tubulin in 3T3-GFPTub cells determined by Western blot using an anti-α-tubulin antibody. GFP-α-tubulin is expressed at <5% the level of endogenous α-tubulin.

of the wound. For cells maintained in serum, the polarization and migration responses are triggered as a result of wounding. We introduced a small variation in the wounding assay that allows us to trigger wound response with soluble factors rather than by wounding alone. If confluent monolayers of NIH 3T3 cells are serum starved before wounding, the cells at the edge of the wound do not polarize or migrate until the appropriate serum factors are added (Fig. 3) (Gundersen et al., 1994). The mitogenic lipid lysophosphatidic acid (LPA) is a major serum factor required for MT polarizations in response to wounding (Cook et al., 1998; Palazzo et al., 2001). LPA induces the same MT reorganizations as serum, yet LPA alone is not sufficient to stimulate cell migration (Gomes et al., 2005). Thus, addition of LPA only activates cell polarization and allows the study of polarization events independently from those of active cell migration. To induce both cell polarization and migration responses, serum can be added to the starved cells (Gomes et al., 2005; Gundersen and Bulinski, 1988; Gundersen et al., 1994).

Because serum factors are necessary for triggering the responses to wounding, serum starvation also allows a separation of the physical act of wounding from the triggering of polarization and migration responses. This has a number of advantages for studying cell polarization and migration, particularly in studying the very early polarization and migration responses. For example, purified proteins, antibodies, or expression plasmids can be introduced into wound edge cells by microinjection before stimulation, which is not possible with the regular wounding assay.

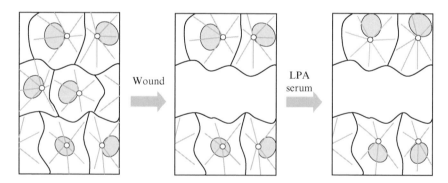

FIG. 3. Wound assay using serum-starved monolayers of NIH 3T3 fibroblasts. Schematic representation of a serum-starved monolayer with MTs, centrosome and nucleus. When the starved monolayer is wounded, centrosome reorientation does not occur. Centrosome reorientation is triggered only on the addition of LPA or serum. (See Palazzo et al. [2001] for details.)

Introduced proteins can also be tested for sufficiency in induction of cell polarization, which is not possible in a regular wounding assay (Gomes *et al.*, 2005; Palazzo *et al.*, 2001; Wen *et al.*, 2004). We have also found that serum-starved cells are flatter and provide better subjects for optical imaging. In the following, we describe the specific protocols for serum starvation and wounding of NIH 3T3 fibroblasts. We have found that similar protocols work well for other rodent fibroblast cell lines (Swiss 3T3, Balb 3T3, and NRK fibroblasts).

Materials

   Glass coverslips, washed with 1 N HCl
   Jeweler's screwdriver
   DMEM medium
   NIH-3T3 cells
   Viokase or trypsin/EDTA
   LPA (1 mM)

Preparation of LPA Stock Solution. Add 2.3 ml of a sterile 10 mM HEPES, pH 7.4, 10 mg/ml BSA (fatty acid free, Sigma), 100 mM NaCl solution to 1 mg of LPA (Sigma). Mix well. Aliquot and freeze at $-80°$. Frozen stocks are good for at least 2 months. Thawed aliquots remain usable for 24 h.

Preparation of Starved Cell Monolayers and Wounding. Cover the bottom of a 10-cm tissue culture dish with coverslips. Detach NIH 3T3 cells from a stock plate using trypsin/EDTA or Viokase and resuspend in DMEM with 10 % calf serum. Dilute cells into 10 ml of DMEM plus 10% calf serum and add 10 ml per 10-cm dish so that cells will be approximately $\sim$20% confluent. Allow cells to grow until just confluent (48 h). Fill three 6-cm dishes with 8 ml of DMEM without serum. Fill 35-mm dishes with 1 ml of DMEM without serum (or each well of a 6-well plate). Using a sterile forceps, transfer each coverslip (containing the confluent monolayer of cells) serially to the 6-cm dishes containing DMEM. Place the coverslip in the 35-mm dish and starve the cells for 48 h. Some cells will detach during starvation; however, the remaining cells should still form a confluent monolayer and are fully viable (Gundersen *et al.*, 1994). Using a sterile jeweler's screwdriver (tip width 1–2 mm), wound the monolayer with an even, continuous movement. Multiple wounds can be made on a single coverslip. Replace cells in incubator and let recover for 20 min. Treat cells with 2 μM LPA to induce centrosome reorientation and MT stabilization or add serum (1–5%) to stimulate migration. Centrosome reorientation, MT stabilization, and cell migration can be assessed with fixed cell assays (see Cook *et al.* [1998]; Palazzo *et al.* [2001]; Wen *et al.* [2004]), or 3T3-GFPTub cells can be used for live cell imaging (see the following).

Note

Wound edge cells can be microinjected after recovery from the wound. Proteins can be microinjected into the cytoplasm, and DNA can be injected into the nucleus. Typically, we find that DNA expression vectors based on CMV promoters allows for efficient protein expression in 1–2 h after microinjection. Because cells will still polarize and migrate if LPA or serum are added up to 8 h after wounding, this allows proteins to be expressed before triggering wound responses.

Imaging Centrosome Reorientation in Living Cells

For real-time imaging of centrosome reorientation, we use 3T3-GFPTub cells prepared as described previously. The preparation of the 3T3-GFPTub cell monolayers for wounding and serum starvation is similar to the preparation of NIH 3T3 cells for wound healing assays (see preceding) with a few modifications to allow for optimal imaging and maintenance of cells under the microscope. 3T3-GFPTub cells are plated in specific imaging chambers, and modified medium ("recording medium") is used for starvation and for maintaining the cells while imaging.

Materials

Inverted microscope equipped with epi-fluorescence (Nikon Eclipse TE300)

  60× Plan Apo (NA 1.4) objective
  40× Plan Fluor (NA 0.6) phase objective
  Computer-controlled shutters (Sutter and Uniblitz) on the epi-illumination light path and the bright-field illumination light path (for phase-contrast imaging)
    Neutral density filters (ND4 and ND8) on the epi-illumination light path
    GFP-optimized filter set (Chroma)
    Temperature-controlled chamber surrounding the microscope
    Motorized Z-focus (Nikon)
    Motorized XY stage (Prior)
    Cooled CCD camera (Coolsnap HQ)

Computer with software to control shutters, time-lapse acquisition, Z-focus, and XY stage (Metamorph, Universal Imaging Corporation)
   Recording medium
   Glass coverslip chambers
   Preparation of Recording Medium. HBSS (Gibco) containing essential and nonessential MEM amino acids (Gibco), 2.5 g/l D-glucose, 2 mM glutamine, 1 mM sodium pyruvate, and 10 mM HEPES, pH 7.4.
   Preparation of Glass Coverslip Imaging Chambers. 35-mm plastic tissue culture dishes are used. A hole (10–15 mm diameter) is punched in the bottom of the dish with a hand punch tool (Roper Whitney). An acid-washed glass coverslip (22 mm^2) is glued to the bottom of the dish, covering the hole, using Sylgard 184 Silicone Elastomer (Dow Corning) or VALAP (Vaseline/lanolin/paraffin = 1:1:1). Plates are sterilized under UV light overnight. We have also had good results with commercially prepared glass coverslip dishes (35-mm dish with 10-mm hole size) from Mattek.
   Live Cell Imaging of Centrosome Reorientation. Plate 3T3-GFPTub cells in imaging chambers at a 20% density (see preceding). Once cells reach confluency (after 48 h), wash the cells twice with recording medium and add 4 ml of recording medium. Starve cells in recording medium for 24 h. Equilibrate the microscope temperature chamber at 34–35° for at least 1 h before recording. Wound the monolayer and allow to recover for 20 min in the incubator. Transfer the imaging chamber to the microscope stage. To minimize photodamage (see following), insert the highest neutral density filter in the epi-illumination light path that still allows observation of the centrosome. Select cells with distinct centrosome and MTs. Minimize as much as possible the exposure of cells to epi-fluorescence illumination. Determine the minimum exposure time necessary for obtaining an image with a distinct centrosome (see following). Start image acquisition at a rate that does not induce photobleaching (see following) to characterize centrosome position before stimulation. To induce centrosome reorientation, add 100 μl of 80 μM LPA drop-wise (2 μM final concentration) or 40 μl of serum (1% final concentration). Restart image acquisition.
   We have successfully followed centrosome position up to 4 h after addition of LPA, acquiring 1 image per minute, or up to 18 h after addition of serum, acquiring 1 image every 10 min. By use of this protocol, we were able to observe that centrosome reorientation occurs by nuclear movement away from the leading edge, whereas the centrosome remains at the cell centroid (Fig. 4).
   If additional information about the nucleus or the cell boundaries are required (for example, to relate centrosome reorientation to leading edge

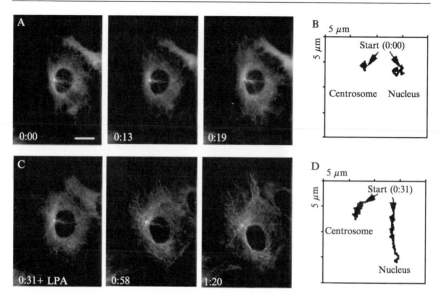

FIG. 4. High-magnification imaging of centrosome and nucleus position during LPA-stimulated centrosome reorientation. Selected frames from a time-lapse movie of a serum-starved 3T3-GFPTub cell before (A) and after (C) the addition of LPA at 0:26 (h:min). The centrosome is identified as a bright point at the center of the MT array. The wound is at the top of the panel. The nucleus is the nonfluorescent oval area. For each time-point, the position of the centrosome and the centroid of the nucleus were determined and plotted before (B) and after (D) the addition of LPA. Before the addition of LPA, both centrosome and LPA remain in the same position. After the addition of LPA, the nucleus moves away from the leading edge, whereas the centrosome remains stationary at the cell center. Thus, the rearward movement of the nucleus reorients the centrosome. Images were acquired with a 60× Plan Apo (NA 1.4) objective, ND8 density filter, camera binning of 2 × 2, and 1500 msec exposure at a rate of 1 frame/min. Previously published in *Cell* (Gomes *et al.*, 2005).

structures, such as lamellipodia or ruffles) sequential phase and fluorescent images can be acquired. To do this, the microscope needs to be equipped with phase or differential interference contrast (DIC) objectives and a shutter in the transmitted light path. The phase or DIC image also contains information regarding possible rotation of the nucleus by reference to the position of the nucleolus and other structures within the nucleus. Meta-morph and other commercially available software can be programmed to acquire both epifluorescence and phase images at each time point.

Notes

Chamber Temperature. Preheating the microscope incubator and increasing room temperature minimizes focal drift. We use a temperature of 34–35° instead of 37°, because it is easier to maintain this temperature and focal drift is less of a problem.

Recording Media. DMEM medium is not suitable for fluorescence microscopy because it contains riboflavin that interferes with GFP fluorescence microscopy. Riboflavin has an excitation peak at $\sim$450 nm and an emission peak at $\sim$530 nm, which is similar to enhanced GFP excitation and emission peaks (480 nm and 530 nm, respectively). We also leave out the phenol red that is commonly included in media, because it can absorb light, and this is important for fluorochromes such as rhodamine or DsRed. The recording medium is buffered with HEPES, not bicarbonate, because most microscope chambers are not equipped with a CO_2-enriched atmosphere. The imaging chamber must be covered, because evaporation is significant at 37°, especially during long recordings.

Starvation. Starvation time is reduced to 24 h when recording medium is used, because cells start to detach from the coverslip if they are starved for longer times in recording medium. Alternately, cells can be starved with DMEM for 48 h, wounded, transferred to the microscope, and then stimulated with recording medium containing LPA or serum. Starved cells are sensitive to changes in medium; therefore, if the starvation medium is replaced by recording medium without LPA or serum, cells can detaching rapidly (15–30 min). Imaging of the centrosome position before stimulation is not possible if cells are starved in DMEM.

Photodamage. Long exposure times can result in photodamage of the cells without detectable photobleaching of the GFP-tubulin. Photodamage is recognized when cells bleb, MTs depolymerize or, more subtly, do not respond to LPA or serum addition. To minimize photodamage, reduce light exposure as much as possible while searching for cells and during imaging. It is helpful to dark-adapt one's eyes by waiting $\sim$5 min in complete darkness before searching for cells under the microscope. Also, we have found that the use of neutral density filters in combination with longer exposure times generates less photodamage than short exposure times without neutral density filters.

Image Acquisition. Imaging centrosome reorientation can involve long (up to 4 h) recording times. To prevent photobleaching and photodamage, we acquired images at a rate between 1 frame per 1 min and 1 frame every 10 min. These acquisition rates are sufficient to characterize both centrosome and nucleus movements. The exposure used for image acquisition should be as short as possible without compromising the visualization of the centrosome. Setting the camera binning to 2 × 2, acquisition time can be shortened; however, image resolution is also reduced.

Triggering Centrosome Reorientation. Centrosome reorientation can be triggered either by LPA or serum, as described previously. The concentration of serum should not be higher that 1%, because serum contains fluorescent factors that increase the background fluorescence, decreasing image quality. This amount of serum is sufficient to induce cell polarization and subsequent migration (Fig. 5).

Objectives. We find that a 60× oil Plan Apo objective (NA 1.4) gives the best combination of resolution, sensitivity, and field size for imaging centrosome reorientation. With this objective, the centrosome appears distinct, and individual MTs can be observed (Fig. 4). With a higher magnification objective (e.g., 100×), only part of the cell is observed, and photobleaching occurs faster. When only the position of the centrosome

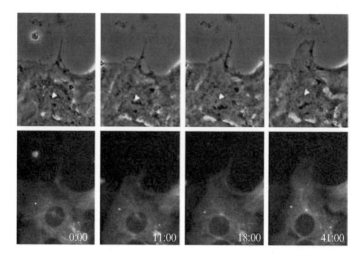

FIG. 5. Low-magnification imaging of centrosome and nucleus position during serum-stimulated centrosome reorientation. Selected frames from a time-lapse movie of a starved 3T3-GFPTub cell after the addition of calf serum at 0:00 (min:sec). In this example, a 40× phase objective was used, and both phase-contrast (top) and epifluorescence (bottom) images were acquired. Although it is not possible to observe individual microtubules in the fluorescence image, the centrosome is visible. The position of the centrosome is represented by a white triangle in the phase-contrast image. Images were acquired with a 40× Plan Fluor (NA 0.6) phase objective, ND8 density filter, camera binning of 1 × 1, and 500 msec exposure time at a rate of 1 frame every 3 min.

is to be determined, we obtained good results using a 40× Plan Fluor objective NA 0.6 (Fig. 5). One advantage of using a lower magnification objective is that higher numbers of cells can be imaged.

XY Motorized Stage. Using a motorized XY stage, we have been able to acquire multiple fields (up to 60) at each time point (one frame every 3 min), increasing the speed of data acquisition. A motorized Z-focus and an autofocus controller are necessary for the acquisition of in-focus images. Autofocusing can be software controlled. The autofocus software routine must be performed using phase-contrast illumination to prevent photobleaching of GFP; therefore, a shutter in the transmitted illumination path and phase condenser and objectives are required. We find that 40× dry objectives perform optimally in this scenario. Oil immersion objectives can also be used as long as the travel distances between each field are minimized (<1 cm).

Acknowledgments

We thank Jan Schmoranzer for comments on the manuscript. This work was supported by a postdoctoral fellowship from Fundação para a Ciência e Tecnologia, Portugal (to E. R. G.), and an NIH grant GM062938 (to G. G. G.).

References

Bergmann, J. E., Kupfer, A., and Singer, S. J. (1983). Membrane insertion at the leading edge of motile fibroblasts. *Proc. Natl. Acad. Sci. USA* **80,** 1367–1371.

Cook, T. A., Nagasaki, T., and Gundersen, G. G. (1998). Rho guanosine triphosphatase mediates the selective stabilization of microtubules induced by lysophosphatidic acid. *J. Cell Biol.* **141,** 175–185.

Etienne-Manneville, S. (2004). Cdc42 – the centre of polarity. *J. Cell Sci.* **117,** 1291–1300.

Etienne-Manneville, S., and Hall, A. (2001). Integrin-mediated activation of Cdc42 controls cell polarity in migrating astrocytes through PKCzeta. *Cell* **106,** 489–498.

Euteneuer, U., and Schliwa, M. (1992). Mechanism of centrosome positioning during the wound response in BSC-1 cells. *J. Cell Biol.* **116,** 1157–1166.

Gomes, E. R., Jani, S., and Gundersen, G. G. (2005). Nuclear movement regulated by Cdc42, MRCK, myosin and actin flow establishes MTOC polarization in migrating cells. *Cell* **121,** 451–463.

Gotlieb, A. I., May, L. M., Subrahmanyan, L., and Kalnins, V. I. (1981). Distribution of microtubule organizing centers in migrating sheets of endothelial cells. *J. Cell Biol.* **91,** 589–594.

Gregory, W. A., Edmondson, J. C., Hatten, M. E., and Mason, C. A. (1988). Cytology and neuron-glial apposition of migrating cerebellar granule cells *in vitro*. *J. Neurosci.* **8,** 1728–1738.

Gundersen, G. G., and Bulinski, J. C. (1988). Selective stabilization of microtubules oriented toward the direction of cell migration. *Proc. Natl. Acad. Sci. USA* **85,** 5946–5950.

Gundersen, G. G., Kim, I., and Chapin, C. J. (1994). Induction of stable microtubules in 3T3 fibroblasts by TGF-beta and serum. *J. Cell Sci.* **107**, 645–659.

Higginbotham, H., Bielas, S., Tanaka, T., and Gleeson, J. G. (2004). Transgenic mouse line with green-fluorescent protein-labeled Centrin 2 allows visualization of the centrosome in living cells. *Transgenic Res.* **13**, 155–164.

Kuhn, J. R., and Poenie, M. (2002). Dynamic polarization of the microtubule cytoskeleton during CTL-mediated killing. *Immunity* **16**, 111–121.

Kupfer, A., Louvard, D., and Singer, S. J. (1982). Polarization of the Golgi apparatus and the microtubule-organizing center in cultured fibroblasts at the edge of an experimental wound. *Proc. Natl. Acad. Sci. USA* **79**, 2603–2607.

Kupfer, A., Dennert, G., and Singer, S. J. (1983). Polarization of the Golgi apparatus and the microtubule-organizing center within cloned natural killer cells bound to their targets. *Proc. Natl. Acad. Sci. USA* **80**, 7224–7228.

Nemere, I., Kupfer, A., and Singer, S. J. (1985). Reorientation of the Golgi apparatus and the microtubule-organizing center inside macrophages subjected to a chemotactic gradient. *Cell Motil.* **5**, 17–29.

Palazzo, A. F., Joseph, H. L., Chen, Y. J., Dujardin, D. L., Alberts, A. S., Pfister, K. K., Vallee, R. B., and Gundersen, G. G. (2001). Cdc42, dynein, and dynactin regulate MTOC reorientation independent of Rho-regulated microtubule stabilization. *Curr. Biol.* **11**, 1536–1541.

Prigozhina, N. L., and Waterman-Storer, C. M. (2004). Protein kinase D-mediated anterograde membrane trafficking is required for fibroblast motility. *Curr. Biol.* **14**, 88–98.

Rogers, K. A., McKee, N. H., and Kalnins, V. I. (1985). Preferential orientation of centrioles toward the heart in endothelial cells of major blood vessels is reestablished after reversal of a segment. *Proc. Natl. Acad. Sci. USA* **82**, 3272–3276.

Schmoranzer, J., Kreitzer, G., and Simon, S. M. (2003). Migrating fibroblasts perform polarized, microtubule-dependent exocytosis towards the leading edge. *J. Cell Sci.* **116**, 4513–4519.

Solecki, D. J., Model, L., Gaetz, J., Kapoor, T. M., and Hatten, M. E. (2004). Par6α signaling controls glial-guided neuronal migration. *Nat. Neurosci.* **7**, 1195–1203.

Sullivan, J., and Satchwell, M. F. (2000). Development of stable cell lines expressing high levels of point mutants of human opsin for biochemical and biophysical studies. *Methods Enzymol.* **315**, 30–58.

Tzima, E., Kiosses, W. B., del Pozo, M. A., and Schwartz, M. A. (2003). Localized cdc42 activation, detected using a novel assay, mediates microtubule organizing center positioning in endothelial cells in response to fluid shear stress. *J. Biol. Chem.* **278**, 31020–31023.

Waterman-Storer, C. M. (2002). Fluorescent speckle microscopy (FSM) of microtubules and actin in living cell. *In* "Current Protocols in Cell Biology" (K. S. Morgan, ed.), pp. 4.10.11–14.10.26. John Wiley & Sons Inc., New York.

Wen, Y., Eng, C. H., Schmoranzer, J., Cabrera-Poch, N., Morris, E. J., Chen, M., Wallar, B. J., Alberts, A. S., and Gundersen, G. G. (2004). EB1 and APC bind to mDia to stabilize microtubules downstream of Rho and promote cell migration. *Nat. Cell Biol.* **6**, 820–830.

Yvon, A. M., Walker, J. W., Danowski, B., Fagerstrom, C., Khodjakov, A., and Wadsworth, P. (2002). Centrosome reorientation in wound-edge cells is cell type specific. *Mol. Biol. Cell* **13**, 1871–1880.

[46] Lentiviral Delivery of RNAi in Hippocampal Neurons

By Justyna Janas, Jacek Skowronski, and Linda Van Aelst

Abstract

The breakthrough discovery that double-stranded RNA of 21 nucleotides in length (referred to as short or small interfering RNA; siRNA) can trigger sequence-specific gene silencing in mammalian cells has led to the development of a powerful new approach to study gene function (Dillon *et al.*, 2005; Dykxhoorn *et al.*, 2003; Elbashir *et al.*, 2001; Hannon *et al.*, 2004). Effective delivery of siRNA molecules into target cells or tissues is critical for successful RNA interference (RNAi) application. Here, we describe the use of human immunodeficiency virus type 1 (HIV-1)–based lentiviral vectors for delivery of short hairpin RNA (shRNA), a precursor of siRNA, into primary neurons to suppress gene expression. Major advantages of lentiviral vectors are their ability to transduce nondividing cells and to confer long-term expression of transgenes. This chapter covers selection of short hairpin sequences, vector design, production of lentiviral supernatants, transduction of dissociated primary hippocampal neurons, and testing the effectiveness of shRNA-mediated silencing.

Introduction

Lentiviral vectors based on, for example, human immunodeficiency virus type 1 (HIV-1) have been successfully used for effective gene silencing in neurons using RNAi (Cottrell *et al.*, 2004; Dittgen *et al.*, 2004). The lentiviral system bypasses deficiencies of other techniques such as transient transfections of siRNAs and the use of plasmid-based mammalian expression vectors by allowing stable long-lasting expression of shRNAs, which are processed *in vivo* to siRNAs, in cells that are difficult to transfect. Furthermore, in contrast to retroviral vectors derived from murine retroviruses, lentiviral vectors can efficiently infect both actively dividing, and nondividing postmitotic and terminally differentiated cells, such as brain and muscle cells (Kafri *et al.*, 1997; Naldini *et al.*, 1996; Van den Haute *et al.*, 2003). We and others have found lentiviral vectors to be valuable tools for stable gene silencing in a wide range of cell types, including neurons at different stages of development.

METHODS IN ENZYMOLOGY, VOL. 406   0076-6879/06 $35.00
          DOI: 10.1016/S0076-6879(06)06046-0

Lentiviral Vectors for shRNA Delivery

Most commonly used lentiviral vectors are derived from human immunodeficiency virus type 1 (HIV-1). These vectors represent the most advanced lentiviral systems for gene delivery. Because they originate from viruses that are pathogenic for humans, the major emphasis in the construction of these vectors has been on their safety. The general strategy has been to use as few genetic elements of the original lentiviral genome as possible and to minimize the probability of generating a replication-competent virus through infrequent recombination events but at the same time to maintain high transduction efficiency of the vector and its ability to infect postmitotic cells. This approach yielded several successive generations of lentiviral vectors that offer safe and efficient means for gene delivery (Naldini, 1998; Romano et al., 2003).

Lentiviral vectors contain only viral cis-acting sequences that are essential for transcription of viral genetic material, its encapsidation, reverse transcription, and integration into the host genome. To generate viral particles, the vector's genomic transcript needs to be coexpressed with HIV-1 core proteins, enzymes, and a heterologous envelope protein. Commonly used second-generation packaging systems provide virion components (Gag and Pol) and regulatory proteins, such as Tat and Rev, which are both required for the efficient expression of the proviral genome from a single plasmid. In the third-generation packaging system, Rev is provided from a distinct plasmid. In this way, helper functions are divided, and the risk of generating replication-competent helper virus is further reduced (Barker and Planelles, 2003; Dull et al., 1998). Instead of using HIV-1 envelope glycoprotein, which confers a relatively narrow infection tropism, viral particles are usually pseudotyped with vesicular stomatitis virus G glycoprotein (VSV-G). VSV-G mediates entry into the target cells possibly by binding to ubiquitous phospholipids, such as phosphatidylcholine (PC) or phosphatidylserine (PS), which greatly extends the tropism of pseudotyped virions and also enhances their stability (Burns et al., 1993; Verhoeyen et al., 2004).

Lentiviral vectors for RNAi applications generally contain two expression cassettes. One cassette contains a hairpin sequence encoding siRNA precursor. The other cassette is a RNA polymerase II transcription unit directing expression of a marker protein, such as GFP, which facilitates detection of the infected cells. Lentiviral systems for shRNA delivery usually use the type III class of RNA polymerase III (pol III) promoter sequences to drive constitutive or inducible expression of the hairpin. The most frequently used pol III promoters are those derived from genes encoding the H1-RNA component of RNAse P and U6 small nuclear

RNA. For shRNA delivery to primary neurons, we have been using successfully TRIPΔU3-EF1α-EGFP (TRIP), a replication-defective self-inactivating (SIN) lentiviral vector (Gimeno *et al.*, 2004; Stove *et al.*, 2005; Zennou *et al.*, 2000) (Fig. 1). In SIN vectors, deletion within the U3 region of 3′ long terminal repeat (LTR) removes sequence elements crucial for viral transcription and thereby inactivates the viral promoter located in the 5′ LTR (Miyoshi *et al.*, 1998; Zufferey *et al.*, 1998).

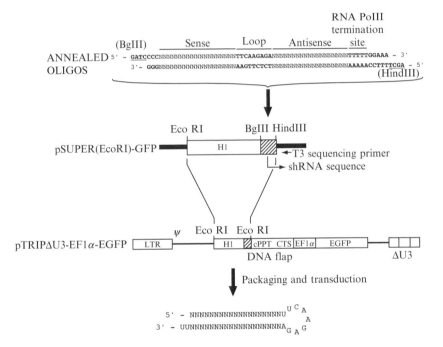

FIG. 1. Schematic representation of TRIPΔ U3-EF1α-EGFP containing H1-shRNA expression cassette. pTRIPΔ U3-EF1α-EGFP contains HIV-1–derived packaging signal required for encapsidation of genomic RNA and a central polypurine tract (cPPT) associated with a central termination sequence (CTS). CTS along with cPPT mediate a central strand displacement event during reverse transcription that results in a stable 99-nucleotide long plus-strand overlap in linear DNA molecules, known as DNA flap. The central DNA flap is the cis-acting determinant of HIV's genome nuclear import and influences the lentiviral vector's ability to transduce nondividing cells (Sirven *et al.*, 2000). Internal EF1α gene–derived promoter drives the expression of enhanced green fluorescent protein (EGFP) marker. H1-hairpin cassette is located upstream of EF1α-EGFP. The cloning of double-stranded hairpin sequences into pSUPER(EcoRI) is described in the text.

Design and Cloning of Hairpin Sequences

Selection of siRNA Target Sequences and Design of shRNA Templates

Identification of optimal target sequences for siRNA is critical to achieve potent and specific gene silencing. The effectiveness of gene silencing depends on the sequence-specific and thermodynamic properties of siRNA, as well as requires elaborate assessment of its specificity to minimize potential off-target effects. Generally, shRNA contains a 19–29 nucleotide (nt) duplex connected by a short loop sequence (Siolas *et al.*, 2005). Together they form a stem-loop structure ("hairpin") with a 3′ UU-overhang, necessary for efficient target cleavage, which is subsequently processed to siRNA. The loop characteristics seem to influence silencing (Brummelkamp *et al.*, 2002). However, there are no clear guidelines with regard to its design, and loops of different length (from 3–23nt) and sequences have been successfully used. For selection of siRNA target sequences, we have been using the Sfold software, available through http://sfold.wadsworth.org (other web sites that provide siRNA designing software include http://design.dharmacon.com/, http://www.cshl.org/ public/SCIENCE/hannon.html, http://jura.wi.mit.edu/ siRNAext, http://www.oligoengine.com/). Besides using conventional rules for siRNA target selection based on empirical design patterns (Elbashir *et al.*, 2002), Sfold also offers algorithms for RNA secondary structure prediction, target sequence accessibility, as well as includes rational design criteria recently described by Reynolds *et al.* (2004). The siRNA target sequence can be located either within the coding sequence or in the 3′ untranslated region (3′UTR) of the mRNA of interest. In the latter case, convenient rescue of the knockdown phenotype can be achieved by expressing the target gene from a construct that lacks the 3′UTR. This approach has been used to provide formal proof that the observed RNAi phenotype is target gene specific. The specificity of selected siRNAs should also be confirmed by blast-searches (http://www.ncbi.nlm.nih.gov/BLAST) against relevant EST libraries or mRNA sequence databases to reduce the probability of an off-target gene knockdown.

We construct shRNA templates containing a 19-nt long siRNA target sequence and its complement separated by a 9-nt spacer, followed by a pol III transcription termination signal consisting of five thymidines (TTTTT). Such a double-stranded fragment is flanked by 5′ *Bgl*II and 3′ *Hind*III restriction sites that enable cloning downstream of the H1-RNA promoter in an intermediate vector, pSUPER(EcoRI) (see Fig. 1). This vector is a derivative of pSUPER (Brummelkamp *et al.*, 2002) and contains a modified multiple cloning site and GFP marker gene. Notably, the choice of the promoter for shRNA expression has an additional constraint on the nucleotide sequence of shRNA. The first nucleotide for shRNA expressed from H1-RNA and U6 gene promoters should be A/G and G, respectively, to enable transcriptional initiation at the first shRNA nucleotide.

Cloning shRNA Expression Cassette into pTRIPΔU3-EF1α-EGFP Lentiviral Vector

Sense and antisense oligonucleotides encoding shRNA templates, as shown in Fig. 1, can be obtained from commercial suppliers. Dissolve lyophilized oligonucleotides in sterile H_2O at a concentration of 1 mg/ml.

1. Combine 0.5 μg of sense and antisense DNA template oligonucleotides (1 μl total).
2. Phosphorylate oligonucleotides by adding 3 units of T4 polynucleotide kinase (T4 PNK), 1 μl of 10× PNK buffer, 1 μl of 10 mM ATP stock, and H_2O to the final volume of 10 μl. Incubate for 30 min at 37°.
3. Inactivate T4 PNK and denature the oligonucleotides by incubating at 95° for 10 min, followed by 10 min at 70°.
4. Anneal the oligonucleotides by slowly cooling the sample to 4°. Annealed oligonucleotides can be stored at −20°.
5. Clone the annealed oligonucleotides into the BglII and HindIII digested pSUPER(EcoRI) intermediate vector.
6. Verify the presence of the hairpin insert by restriction enzyme analysis. Digestion of pSUPER(EcoRI) with EcoRI results in excision of an approximately 0.3-kb fragment corresponding to complete shRNA expression cassette containing the hairpin and H1 promoter sequences. Verify its sequence by sequencing using T3 promoter sequencing primer (3′ → 5′).
7. Isolate EcoRI fragment containing the H1 RNA promoter and the hairpin sequence from pSUPER(EcoRI) and subclone into EcoRI site in pTRIPΔU3-EF1α-EGFP (pTRIP) lentiviral vector.

Production and Titration of Lentiviral Supernatants

Production of Lentiviral Supernatants

Viral particles are produced by transient coexpression of the lentiviral vector along with either second- or third-generation packaging constructs in human embryonic kidney (HEK) 293T cells, using the calcium phosphate coprecipitation method. We have been producing lentiviral particles using a second-generation packaging system. The TRIP vector was pseudotyped with VSV-G encoded by pMD.G (Dull et al., 1998). Gag, Pol, and Tat were expressed from packaging construct pCMVΔR8.91 (Zufferey et al., 1997).

HEK 293T cells are cultured in Dulbecco's modified Eagle's medium (DMEM [Gibco/Invitrogen Cat#11995-065]) supplemented with 10% fetal bovine serum (FBS [HyClone Cat#SH30071.03]), 1% penicillin/streptomycin (Gibco/Invitrogen Cat#15070-063), in a $37°$ incubator with 5% CO_2.

1. Plate 2 to 3 × 10^6 HEK 293T cells in a 10-cm tissue culture dish with 10 ml of medium 24 h before transfection. Cells should be approximately 60–80% confluent at the time of transfection.
2. Prepare 1 ml of calcium phosphate–DNA precipitate for transfection of one 10-cm plate of HEK 293T cells as follows:
   a. Combine in sterile microcentrifuge tube 5 μg of pTRIP vector, 10 μg of pCMVΔR8.91, and 6 μg of pMD.G (VSVG). *Note:* For optimal results, the relative amounts of plasmids used for cotransfection should be determined empirically.
   b. Adjust the volume to 250 μl with 0.1× TE (1 mM Tris-HCl, pH 8.0, 0.1 mM EDTA).
   c. Add 250 μl of 2× $CaCl_2$ solution (0.5 M $CaCl_2$, 1 mM Tris-Base, 0.1 mM EDTA) and mix by vortexing.
   d. Place 500 μl of 2× HEPES buffered saline (HBS) (0.05 M HEPES-NaOH, pH 7.12, 0.28 M NaCl, 1.5 mM sodium phosphate buffer, pH 7.0) in a sterile tube (e.g., 2065 Falcon tube).
   e. Add drop wise calcium-DNA solution to 500 μl of 2× HBS while vortexing slowly. Let the suspension stand for 10 min to allow precipitate formation.

Note: 2× HBS should be stored at $4°$ and can be kept for up to 2 weeks. Aliquots of 2× $CaCl_2$ solution and 0.5 M HEPES-NaOH stock should be stored frozen at $-20°$. All solutions should be thawed at room temperature before use.

3. Distribute the precipitate evenly over the cells in the culture dish. Mix gently by rocking the dish back and forth, and return the plates to the incubator.
4. Replace medium 5–8 h after transfection with 10 ml fresh culture medium.
5. Collect the medium containing viral particles 48–60 h after transfection.
6. Centrifuge freshly collected supernatant at 2000g to remove cell debris.
7. Filter the supernatant through a 0.45-μm filtration unit. The filter should be prerinsed with cell culture medium before use.
8. Aliquot and store at $-80°$ or proceed to concentration step.

Concentration of Viral Particles by Pelleting Through Sucrose Cushion

The protocol described in the preceding yields high titer viral stocks. However, for some purposes (e.g., transduction of neurons in brain slices) further concentration of the supernatant may be required.

1. Place 4 ml of sterile 20% sucrose (wt/vol, in TNE buffer: 10 mM Tris-HCl, pH 7.4, 100 mM NaCl, 1 mM EDTA) in an Ultra-Clear tube for SW28 Rotor (Beckman Cat#344058).

2. Carefully fill up the tube with filtered supernatant containing the virus.

3. Centrifuge at 120,000g for 3 h at 4°.

4. Remove the supernatant and resuspend the viral pellet in an appropriate volume of either TNE buffer, Hanks balanced salt solution without calcium (Ca) and magnesium (Mg) (HBSS) (Gibco/Invitrogen Cat# 14185-052) or suitable culture medium.

5. Store the tube containing the concentrated viral suspension at 4° for a few hours to ensure complete pellet resuspension. Aliquot and store at −80°.

Titration of Lentiviral Supernatants

Although VSV-G confers a relatively broad infection tropism, the infectivity of viral particles will vary depending on the type of target cells and will differ for each preparation. Therefore, it is imperative to titer viral supernatants on the cell type under study. Lentiviruses generated using pTRIP carry an EGFP marker gene, and hence the infected cells can be distinguished by the presence of green fluorescence. If primary neurons are the target cells, virus titration can be accomplished in two ways. Cultures containing a known number of neurons can be transduced with serially diluted viral supernatants, and viral titers can be calculated from the percentage of EGFP positive cells, which is determined by fluorescence microscopy. Alternately, HeLa cells can be used for titration experiments, which enable reliable quantification of infected cells by flow cytometry. In this case, the relation between the viral titers determined on HeLa cells and on neurons will need to be established.

1. Plate 1 × 10^5 HeLa cells into wells of a 12-well tissue culture plate.

2. Infect cells with serial dilutions of viral suspension 12 h after plating. Incubate for 48 h.

3. Remove the medium, wash the cells with phosphate buffer saline (PBS), and detach with 200 μl of trypsin/EDTA for 1–2 min at 37°.

4. Fix the cells and inactivate viral particles by adding to each well 300 μl of 4% (w/v) paraformaldehyde in PBS. Incubate for 15 min at room temperature.

5. Determine the percentage of EGFP-positive cells by flow cytometry. Calculate the titer with the following equation:

$$\text{Titer[TU/ml]} = \frac{\text{Target cell number} \times \% \text{of EGFP-positive cells}}{\text{Volume of viral supernatant[ml]}/100}$$

Titrations should be performed at relatively low multiplicity of infection (MOI). MOI is the ratio of infectious virus particles to the number of cells being infected (see formula following). Infections at low MOI minimize the probability that a cell will become infected multiple times, which would result in underestimation of the viral titer. Therefore, we calculate viral titers only when less than 20% of the target cells are EGFP positive.

$$\text{MOI} = \frac{\text{Volume of viral supernatant[ml]} \times \text{Titer[TU/ml]}}{\text{Target cell number}}$$

Lentiviral Transduction of Dissociated Hippocampal Neurons

Hippocampal neurons are typically prepared from rat fetuses at embryonic day 18–19 (E18-E19) or from mouse fetuses at a comparable developmental stage. A number of protocols have been developed to isolate and culture hippocampal neurons (Banker et al., 1998). In the protocol we routinely use (see following), rat hippocampi are dissociated with trypsin. The dissociated cells are then plated at a medium density on poly-L-lysine or poly-L-lysine and laminin-treated substrates and maintained in serum-free culture medium (Brewer et al., 1993).

Isolation of Primary Hippocampal Neurons

1. Euthanize an E19 pregnant rat, separate the pups from the placenta, and place them in a large sterile tissue culture dish. Using scissors, collect heads of the fetuses into a beaker with ice-cold sterile modified HBSS (mHBSS: for 500 ml, add 50 ml 10× HBSS without Mg or Ca (Gibco/Invitrogen Cat#14185-052); 16.5 ml 0.3 M HEPES, pH 7.3 (Sigma Cat#H4034); 0.5 ml phenol red (Gibco/Invitrogen Cat#15100-043); 1.25 ml 200 mM L-glutamine; 6.3 ml 1 M CaCl$_2$; 2.5 ml 1 M MgCl$_2$; 2 ml 1 M MgSO$_4$; adjust pH to 7.3; add MilliQ water to the final volume of 500 ml; filter sterilize and store at 4°). Cut the skulls along the midline from the base of the neck to the nose, remove the brains using a spatula, and place them in a small beaker with ice-cold mHBSS.

2. Place the brains in a small dish containing ice-cold mHBSS. Separate hemispheres completely and remove the corpus callosum and cerebellum using a pair of Dupont forceps and a dissecting chisel (Fine Surgical Instruments Cat#10095-12). Remove membranes and meninges

using forceps. Dissect the hippocampi and place them in a 15-ml tube containing 12 ml of ice-cold mHBSS.

3. Rinse hippocampi twice with 12 ml of prewarmed mHBSS* (mHBSS without $CaCl_2$, $MgCl_2$, and $MgSO_4$). After the final rinse, add 5 ml mHBSS* containing 0.5 ml of 2.5% trypsin and incubate in a 37° water bath for 15 min.

4. Remove the trypsin solution and wash the hippocampi three times with 13 ml of mHBSS. Add 3 ml of mHBSS and dissociate the cells by pipetting up and down several times using first a regular Pasteur pipette and subsequently a fire-polished Pasteur pipette with approximately half the normal pore diameter.

5. Spin cells at 1000 rpm for 5 min and resuspend the cell pellet in 3 ml (for eight hippocampi) of prewarmed culture medium (culture medium: 97 ml Neurobasal medium (Gibco/Invitrogen cat#21103-049); 2 ml B27 Supplement (Gibco/Invitrogen Cat#17504-044); 250 μl of 200 mM glutamine; 125 μl of 10 mM glutamate; 1 ml penicillin/streptomycin (Gibco/Invitrogen Cat#15070-063, optional); no glutamate after day 4). Determine cell number with a hemacytometer. Each hippocampus should yield approximately 5×10^4 cells.

6. Plate 3 to 4×10^5 cells on poly-L-lysine (or poly-L-lysine and laminin)–coated 60-mm dishes in 5 ml culture medium, or plate 1 to 3×10^5 cells on coated coverslips (22-mm diameter) in a 6-well plate with 2 ml of culture medium. After 4 days, feed the cells by replacing half of the medium with fresh medium without glutamate.

Lentiviral Transduction of Hippocampal Neurons

1. To assess knockdown of a protein at early stages of neuronal development, plate the neurons at the density of 4×10^5 cells in a 60-mm culture dish or 2 to 3×10^5 cells on a coverslip in a 6-well plate. To assess knockdown at a later development stage, plate neurons at the density of 3×10^5 cells in a 60-mm culture dish or 1 to 2×10^5 cells on a coverslip in a 6-well plate.

2. Infect neurons at the desired stages of development at an appropriate MOI by adding to the culture medium a suitable amount of viral supernatant. For studying neurons at early stages of development, infection can be performed 30–60 min after plating, and in the case of mature neurons, infection can be done 12–14 days after plating. Return cells to the incubator. Importantly, half the media needs to be replaced with fresh culture media every 3–4 days in culture throughout the duration of the experiment. In our hands, infection at MOI of 5 results in more than 90% transduction efficiency. As an example, we found that ~90–95% of dissociated hippocampal neurons in culture become infected with TRIP

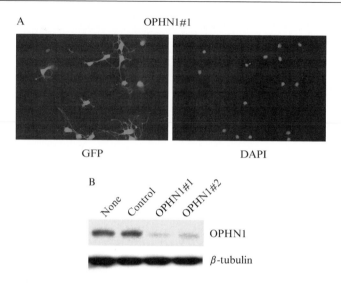

FIG. 2. OPHN1 knockdown in hippocampal neurons using a lentiviral-based system. Dissociated hippocampal neurons isolated from E18 rats were plated on coverslips placed in a 6-well plate and cultured for 4–6 days. The neurons were then infected with 25 μl of viral stocks containing TRIP vectors encoding two different shRNAs for *oligophrenin-1* (*OPHN1#1* or *OPHN1#2*) or a control TRIP vector with an unrelated hairpin. (A) After 30 h in culture, neurons were fixed in 4% paraformaldehyde and mounted in Vectashield containing DAPI (Vector Laboratories). Cells were analyzed for GFP expression and DAPI labeling using fluorescence microscopy. As shown for *OPHN1#1*, approximately 90–95% of the neurons were infected. (B) Lysates were prepared 30 h after infection and analyzed by Western blotting with antibodies to OPHN1 and to β-tubulin as a loading control. Cells infected with *OPHN1#1* or *OPHN1#2* showed a significant decrease in OPHN1 levels compared with cells infected with control viruses. (See color insert.)

encoding shRNAs for *oligophrenin-1* (*OPHN1*), which resulted in >90% reduction in OPHN1 protein level (Fig. 2). *OPHN1* encodes a Rho-GTPase–activating protein required for dendritic spine morphogenesis (Govek *et al.*, 2004).

Effectiveness and Specificity of shRNA-Mediated Silencing

It is necessary to test how effectively each hairpin reduces the level of expression of the target gene. The degree of silencing should be verified at protein and/or mRNA level. Reduction in protein levels can be assessed by Western blot analysis, immunohistochemical analysis, and/or immunofluorescence-based assays. Reduction of target mRNA levels can be evaluated by Northern blotting, RT-PCR, QT-PCR, and/or branched DNA (bDNA) assays (Gruber *et al.*, 2005).

It is also imperative to ensure that the reduction in the level of target gene expression is specific and does not reflect a global suppressive effect on gene expression, which could result from, for example, a high level of double-stranded RNA expression. Appropriate controls to address these possibilities need to be included. The use of controls, such as scrambled or unrelated shRNAs, will ensure that the observed changes in gene expression are not a result of the delivery method used. The use of two or more independent hairpins designed to knock down a single target gene should induce similar phenotypic and gene expression profile changes. However, the ultimate control involves rescue of the knockdown phenotype upon re-expression of the target gene product. The use of shRNAs directed against 3'UTR of the target sequence enables a straightforward rescue experiment by reintroduction of the coding sequence of the target gene lacking the UTR. Alternately, a coding sequence of the target gene with silent mutations that abolish its recognition by RNAi can be generated and used in rescue experiments.

Acknowledgments

We thank Bruno Verhasselt (Ghent University, Belgium) for providing the TRIPΔU3-EF1α-EGFP and pSUPER(EcoRI) vectors. We also thank members of Van Aelst's and Skowronski's laboratories for sharing their expertise and helpful discussions. Our research is supported by NIH grants to L. V. A. (CA096882, CA64593) and J. S. (AI-42561).

References

Banker, G., and Goslin, K. (eds.) (1998). "Culturing Nerve Cells," 2nd Ed. MIT Press, Cambridge.

Barker, E., and Planelles, V. (2003). Vectors derived from the human immunodeficiency virus, HIV-1. *Front. Biosci.* **8,** d491–d510.

Brewer, G. J., Torricelli, J. R., Evege, E. K., and Price, P. J. (1993). Optimized survival of hippocampal neurons in B27-supplemented Neurobasal, a new serum-free medium combination. *J. Neurosci. Res.* **35,** 567–576.

Brummelkamp, T. R., Bernards, R., and Agami, R. (2002). A system for stable expression of short interfering RNAs in mammalian cells. *Science* **296,** 550–553.

Burns, J. C., Friedmann, T., Driever, W., Burrascano, M., and Yee, J. K. (1993). Vesicular stomatitis virus G glycoprotein pseudotyped retroviral vectors: Concentration to very high titer and efficient gene transfer into mammalian and nonmammalian cells. *Proc. Natl. Acad. Sci. USA* **90,** 8033–8037.

Cottrell, J. R., Borok, E., Horvath, T. L., and Nedivi, E. (2004). CPG2: A brain- and synapse-specific protein that regulates the endocytosis of glutamate receptors. *Neuron* **44,** 677–690.

Dillon, C. P., Sandy, P., Nencioni, A., Kissler, S., Rubinson, D. A., and Van Parijs, L. (2005). Rnai as an experimental and therapeutic tool to study and regulate physiological and disease processes. *Annu. Rev. Physiol.* **67,** 147–173.

Dittgen, T., Nimmerjahn, A., Komai, S., Licznerski, P., Waters, J., Margrie, T. W., Helmchen, F., Denk, W., Brecht, M., and Osten, P. (2004). Lentivirus-based genetic manipulations of cortical neurons and their optical and electrophysiological monitoring *in vivo*. *Proc. Natl. Acad. Sci. USA* **101**, 18206–18211.

Dull, T., Zufferey, R., Kelly, M., Mandel, R. J., Nguyen, M., Trono, D., and Naldini, L. (1998). A third-generation lentivirus vector with a conditional packaging system. *J. Virol.* **72**, 8463–8471.

Dykxhoorn, D. M., Novina, C. D., and Sharp, P. A. (2003). Killing the messenger: Short RNAs that silence gene expression. *Nat. Rev. Mol. Cell. Biol.* **4**, 457–467.

Elbashir, S. M., Harborth, J., Lendeckel, W., Yalcin, A., Weber, K., and Tuschl, T. (2001). Duplexes of 21-nucleotide RNAs mediate RNA interference in cultured mammalian cells. *Nature* **411**, 494–498.

Elbashir, S. M., Harborth, J., Weber, K., and Tuschl, T. (2002). Analysis of gene function in somatic mammalian cells using small interfering RNAs. *Methods* **26**, 199–213.

Gimeno, R., Weijer, K., Voordouw, A., Uittenbogaart, C. H., Legrand, N., Alves, N. L., Wijnands, E., Blom, B., and Spits, H. (2004). Monitoring the effect of gene silencing by RNA interference in human CD34+ cells injected into newborn RAG2-/- gammac-/-mice: Functional inactivation of p53 in developing T cells. *Blood* **104**, 3886–3893.

Govek, E. E., Newey, S. E., Akerman, C. J., Cross, J. R., Van der Veken, L., and Van Aelst, L. (2004). The X-linked mental retardation protein oligophrenin-1 is required for dendritic spine morphogenesis. *Nat. Neurosci.* **7**, 364–372.

Gruber, J., Lampe, T., Osborn, M., and Weber, K. (2005). RNAi of FACE1 protease results in growth inhibition of human cells expressing lamin A: Implications for Hutchinson-Gilford progeria syndrome. *J. Cell Sci.* **118**, 689–696.

Hannon, G. J., and Rossi, J. J. (2004). Unlocking the potential of the human genome with RNA interference. *Nature* **431**, 371–378.

Kafri, T., Blomer, U., Peterson, D. A., Gage, F. H., and Verma, I. M. (1997). Sustained expression of genes delivered directly into liver and muscle by lentiviral vectors. *Nat. Genet.* **17**, 314–317.

Miyoshi, H., Blomer, U., Takahashi, M., Gage, F. H., and Verma, I. M. (1998). Development of a self-inactivating lentivirus vector. *J. Virol.* **72**, 8150–8157.

Naldini, L. (1998). Lentiviruses as gene transfer agents for delivery to non-dividing cells. *Curr. Opin. Biotechnol.* **9**, 457–463.

Naldini, L., Blomer, U., Gage, F. H., Trono, D., and Verma, I. M. (1996). Efficient transfer, integration, and sustained long-term expression of the transgene in adult rat brains injected with a lentiviral vector. *Proc. Natl. Acad. Sci. USA* **93**, 11382–11388.

Reynolds, A., Leake, D., Boese, Q., Scaringe, S., Marshall, W. S., and Khvorova, A. (2004). Rational siRNA design for RNA interference. *Nat. Biotechnol.* **22**, 326–330.

Romano, G., Claudio, P. P., Tonini, T., and Giordano, A. (2003). Human immunodeficiency virus type 1 (HIV-1) derived vectors: Safety considerations and controversy over therapeutic applications. *Eur. J. Dermatol.* **13**, 424–429.

Siolas, D., Lerner, C., Burchard, J., Ge, W., Linsley, P. S., Paddison, P. J., Hannon, G. J., and Cleary, M. A. (2005). Synthetic shRNAs as potent RNAi triggers. *Nat. Biotechnol.* **23**, 227–231.

Sirven, A., Pflumio, F., Zennou, V., Titeux, M., Vainchenker, W., Coulombel, L., Dubart Kupperschmitt, A., and Charneau, P. (2000). The human immunodeficiency virus type-1 central DNA flap is a crucial determinant for lentiviral vector nuclear import and gene transduction of human hematopoietic stem cells. *Blood* **96**, 4103–4110.

Stove, V., Van de Walle, I., Naessens, E., Coene, E., Stove, C., Plum, J., and Verhasselt, B. (2005). Human immunode Ficiency virus NeF induces rapid internalization of the T-cell coreceptor CD8$\alpha\beta$. *J. Virol.* **79**, 11422–11433.

Van den Haute, C., Eggermont, K., Nuttin, B., Debyser, Z., and Baekelandt, V. (2003). Lentiviral vector-mediated delivery of short hairpin RNA results in persistent knockdown of gene expression in mouse brain. *Hum. Gene Ther.* **14,** 1799–1807.

Verhoeyen, E., and Cosset, F. L. (2004). Surface-engineering of lentiviral vectors. *J. Gene Med.* **6**(Suppl 1), S83–S94.

Zennou, V., Petit, C., Guetard, D., Nerhbass, U., Montagnier, L., and Charneau, P. (2000). HIV-1 genome nuclear import is mediated by a central DNA flap. *Cell* **101,** 173–185.

Zufferey, R., Dull, T., Mandel, R. J., Bukovsky, A., Quiroz, D., Naldini, L., and Trono, D. (1998). Self-inactivating lentivirus vector for safe and efficient in vivo gene delivery. *J. Virol.* **72,** 9873–9880.

Zufferey, R., Nagel, D., Mandel, R, J., Mandel, R. J., Naldinl, L., and Trono, D. (1997). Multiply attenuated lentiviral vector achieves efficient gene delivery *in vivo. Nat. Biotechnol.* **15,** 871–875.

[47] Methods for Studying Neutrophil Chemotaxis

By XUEMEI DONG and DIANQING WU

Abstract

Chemotaxis is a fascinating biological process, through which a cell migrates along a shallow chemoattractant gradient that is less than 5% difference between the anterior and posterior of the cell. Chemotaxis is composed of two independent, but interrelated processes—motility and directionality—both of which are regulated by extracellular stimuli and chemoattractants; small GTPases have been shown to be involved. In this chapter, the methods that are used to prepare mouse neutrophils and study their chemotactic behaviors and morphological and biochemical changes in response to chemoattractant stimulation are described.

Introduction

Leukocyte chemotaxis underlies leukocyte recruitment, infiltration, homing, and trafficking, which are not only required for leukocyte function and development and normal immune responses but also responsible for initiation and progression of many human diseases that are associated with uncontrolled inflammatory reactions. Leukocyte chemotaxis is regulated by a large number of chemoattractants, including bacterial by-product fMLP, complement proteolytic fragment C5a, and the superfamily ($\sim$50) of small (8–10 kDa), inducible, secreted, proinflammatory cytokines called

METHODS IN ENZYMOLOGY, VOL. 406
0076-6879/06 $35.00
DOI: 10.1016/S0076-6879(06)06047-2

chemokines (Baggiolini *et al.*, 1997; Zlotnik and Yoshie, 2000). These chemoattractants bind to their specific cell surface receptors and mainly activate the Gi proteins in leukocytes (Murphy, 1994). The Gi proteins belong to the family of heterotrimeric G proteins, which consist of three subunits: α, β, and γ. The α subunits contain GTPase activity. Ligand-bound receptors promote the exchange of GDP with GTP on the α subunits, and the GTP-bound forms of Gα subunits are subsequently dissociated from the $\beta\gamma$ subunits. Although the α subunits of the Gi proteins inhibit adenylyl cyclases, the G$\beta\gamma$ subunits can regulate a number of effectors, including phospholipase (PLC) β2/3, phosphatidylinositide 3 (PI3K)γ, ion channels, G protein–coupled receptor kinases, and some subtypes of adenylyl cyclases (Neer, 1995). We recently added PAK1 to the list by demonstrating the interaction between G$\beta\gamma$ and PAK1, which is involved in chemoattractant-mediated activation of Cdc42 (Li *et al.*, 2003). Chemoattractants can also couple to G16 (Mao *et al.*, 1998a) and probably G12/13 (Xu *et al.*, 2003), which have been shown to activate small GTPase RhoA via a guanine nucleotide exchange factor (GEF) 115 (Hart *et al.*, 1998; Kozasa *et al.*, 1998; Mao *et al.*, 1998b).

Chemotaxis is a fundamental biological process in which a cell migrates following the direction of a spatial cue. This spatial cue is provided in a form of a gradient of chemoattractants. Neutrophils are specialized to possess the capability of sensing and processing the information conveyed by the chemical gradient, which allows telling the small difference in chemoattractant concentrations between its two ends and translating this small difference into a much greater intracellular biochemical gradient. For a neutrophil to move even in a random migration, the cell has to polarize by forming F-actin–rich lamellipodia at the direction of the migration, because continuous formation of F actin at the leading edge is required for cell locomotion. Chemoattractants augment and stabilize cell polarization by stimulating F actin polymerization particularly in the lamellipodia; thus, the cell becomes highly polarized. When a chemoattractant gradient is present, the cell can sense the gradient and align its polarity with the gradient to migrate directionally up the gradient. One of the characteristics of a highly polarized cell is its elongated shape with a more blunted leading edge and narrower posterior, resulting from the formation of F actin–rich lamellipodia at the leading edge and actomyosin structure at the posterior. In addition, in a polarized cell, many intracellular proteins, including F actin, lipids, and small G proteins, including RhoA, Cdc42, and Rac, show polarized distribution, which can readily detected by staining. Small GTPases, particularly the members of the Rho subfamily, have been shown to play important roles in regulation of neutrophil chemotaxis by regulating the localization and formation of F actin (Li *et al.*, 2005; Xu *et al.*, 2003;

Zhang *et al.*, 2003). In this chapter, we will discuss the methods that are used to prepare mouse neutrophils and study their chemotactic behaviors and morphological and biochemical changes in response to chemoattractant stimulation.

Preparation of Mouse Neutrophils

We routinely prepare primary mouse bone marrow neutrophils from femurs using a protocol initially described by Lowell *et al.* (1996) with some modifications. Mice that are 2–4 months old are euthanized by carbon dioxide asphyxiation. Femurs are removed from the mice, and bone marrow cells are flushed from the femurs using buffer A (Ca^{2+}/Mg^{2+}-free Hank's buffered saline solution [HBSS; Invitrogen, Grand Island, NY] containing 0.1% bovine serum albumin and 5 mM HEPES). Red blood cells are lysed by incubating the bone marrow cells with a RBC lysis buffer containing 0.15 M NH$_4$Cl, 10 mM KHCO$_3$, and 0.1 mM Na$_2$ EDTA for 4 min at room temperature. Then the cells are spun down, washed twice with buffer A, and resuspended in 3 ml of a HBSS (Ca^{2+}/Mg^{2+} -free) solution that contains 45% Percoll (Amersham Biosciences, Uppsala, Sweden). In a 15-ml polystyrene tube (Nalgene, Rochester, NY), a Percoll density gradient is prepared by layering successively 2 ml each of 62, 55, and 50% Percoll solutions made in Ca^{2+}/Mg^{2+} -free HBSS on the top of an 81% Percoll solution (3 ml) as depicted in Fig. 1. Percoll gradient is made from a stock solution vol by vol as follows: first, the stock Percoll solution is made by adding 1 ml of 1.08 M NaCl to per 7 ml Percoll; then, the stock solution is mixed with HBSS (Ca^{2+}/Mg^{2+} -free) vol by vol to make 81, 62, 55, 50, and 45% Percoll solutions. Finally, Percoll density gradient in the centrifuge tube is set by layering, successively, the 62, 55, and 50% Percoll solutions on the top of the 81% Percoll solution. Leukocytes in 45% Percoll solution are loaded on the top of this density gradient and centrifuged at 1600g for 30 min at room temperature in a swinging bucket rotor. It is important not to apply the brake, because it will disrupt the gradient. The mature neutrophils that are banded between the 81% and 62% layers can be harvested by inserting a Pasteur pipette on the top of the cell layer and gently sucking the cells into the pipette. The cells are immediately diluted into 10 ml of fresh buffer A, washed twice, and resuspended in HBSS containing Ca^{2+}/Mg^{2+}. The number of cells is determined by counting with a hemacytometer. We generally obtain 3–5 million cells per CD-1 mouse (different strain may affect the number of cells).

As quality control, we often examine whether the cells show polarized distribution of F actin in response to stimulation of chemoattractants. Twenty microliters of isolated mouse bone marrow neutrophils (2×10^6/ml)

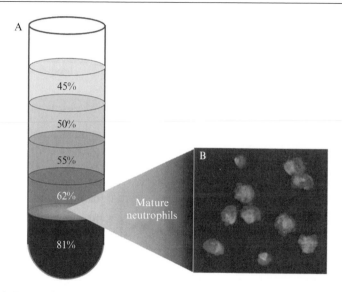

FIG. 1. Preparation of primary mouse bone marrow neutrophils by density gradient. (A) Schematic representation of the step Percoll gradient in a 15-ml centrifuge tube for neutrophil separation. The mature neutrophils are banded between the layers of 62% and 81%. (B) Neutrophils were stimulated with uniform fMLP as described in the text and stained with Alexa-633-phalloidin and FITC-conjugated anti-CD43. Anti-Cd43 is used to counterstain phalloidin, which primarily stains the leading in polarized cells. (See color insert.)

are placed on a coverslip for 5 min at room temperature, followed by adding 2 μl HBSS containing 100 μM fMLP for 5 min. These cells are subsequently fixed with 1 ml of ice cold 4% paraformaldehyde (Mediatech, Inc, Herndon, VA) in PBS at room temperature for 15 min. To stain F actin, the coverslip containing fixed cells is washed with PBS, and cells are permeabilized with a solution of 0.1% triton X-100 in PBS for 5 min at room temperature. After washing with PBS, the cells are incubated with a mixture of 200 μl of PBS containing 1% BSA and 5 μl of Alexa Fluor 633 phalloidin (Molecular Probes, Eugene, OR) for 30 min at room temperature. After this incubation, cells are washed, and the coverslip is mounted with a drop of the Antifade reagent (Molecular Probes). Figure 1B show an example of neutrophils stained for F actin. Although neutrophils can be identified by their distinct "donut"-shaped nuclei, we found that nuclear morphology may not be an accurate marker for distinguishing mature from less mature neutrophils. Although most of the cells in the band formed between the 55 and 62% Percoll layers have the typical "donuts"-shaped nuclei, they do not show highly polarized F actin stain as shown by cells in the band between the 62 and 81% layers (Fig. 1B).

Tracking Mouse Neutrophil Migration

Cell migratory behaviors can be recorded and analyzed in a shallow chemoattractant gradient using a Zigmond or Dunn chamber as described (Allen *et al.,* 1998; Zigmond, 1988). The Dunn chamber as shown in Fig. 2 is believed to maintain the gradient for a longer time than the Zigmond chamber. Here we only describe the procedures and our experience of using the Dunn chamber to monitor the migration of mouse neutrophils. First, a coverslip is seeded with 50 μl of isolated mouse bone marrow neutrophils (2×10^6/ml) in HBSS containing Ca^{2+} and Mg^{2+} for 5 min at room temperature. Then, the concentric well of the chemotaxin chamber is filled with HBSS, and the coverslip with cells is inverted and placed onto the chamber in an offset position leaving a narrow slit at one edge (Fig. 2). HBSS containing 10 μM fMLP (Sigma, St Louis, MO) or 100 nM C5a (Sigma) is carefully pipetted into the outer well. Finally, the coverslip is gently moved to cover the entire outer well and annular bridge. The assembled chamber is left at room temperature for 20 min before the cells are observed under a time-lapse video microscope using the Metamorph software (West Chester, PA). Images are taken at 5- to 15-sec intervals.

Although chemoattractant gradients are supposed to be formed across the annular bridge that separates the inner and outer wells of the chamber, we have noticed a considerable variation in the quality of gradient at different locations of the bridge. We found that the gradient at the left side of the annular bridge (if the filling slit is at the right side of the annular bridge, Fig. 2A) is relatively reproducible, and a consistent migration parameter in terms of both directionality and translocation rate can be obtained at this side of annular bridge. Therefore, we usually track cells at this specific location on the bridge as indicated in Fig. 2A.

Distribution plots are constructed for analysis of directionality of cell migration. Time-lapse video images are used to calculate the final position of a neutrophil relative to its starting position, and this is graphed using Excel to determine the directionality of the cell migration. On these graphs, a positive x distance reflects travel up the gradient. The absolute y values represent lateral deviation along the gradient. As shown in Fig. 2B, most of wild-type neutrophils migrate toward the source of the chemoattractant.

Another key parameter for cell migration is speed. The cell migration speed is calculated by tracking the center of the cell mass using Metamorph software, which then determines the speed as the distance covered by the centroid over the time of observation (normally 10–15 min). We have found that the duration in which cells are incubated in the chamber containing the gradient is an important factor that can affect the calculated speed of cell migration; the longer cells stay in the gradient, the faster they move. Therefore, to

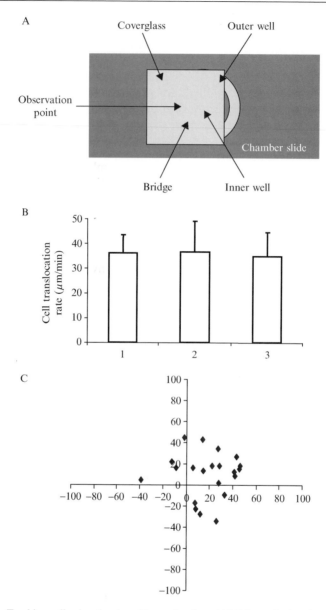

FIG. 2. Tracking cell migration in a Dunn chamber. (A) Schematic drawing of a Dunn Chamber. (B) Directionality plot constructed from tracking normal mouse neutrophil migration in a Dunn chamber containing an fMLP gradient. (C) Translocation rates of primary mouse bone marrow neutrophils from three independent experiments. (See color insert.)

accurately compare the migration speeds of two populations of cells, cell migration needs to be recorded with cells that are being incubated in the chamber for the same interval (for example, the first 10 min).

Because of the heterogeneity in primary neutrophil preparations, not all of the cells chemotax in response to chemoattractants. Arbitrary cutoff values are usually used to wean off the unresponsive cells and cells that do not translocate rather than waving around by forming pseudopods all over the cells. The latter can introduce significant errors in both directionality and speed calculations. We usually use cells that translocate more than 5 μm (a half of the average mouse neutrophil diameter) in the duration of the measurement for the calculations. Figure 2C shows the translocation rates of wild-type neutrophils in three independent experiments following the rules described previously.

Directional Sensing Detection

The ability of neutrophils to sense chemoattractant gradient can be detected by subjecting the cells to acute stimulation (30–45 sec) from a micropipette as depicted in Fig. 3A. Cell locator (Eppendorf, Westbury, NY) is used to locate the stimulated cell that is fixed and stained with fluorescence-labeled phalloidin for F actin. We found that F actin is preferentially formed and localized at the sides of cells that are proximal to the source (i.e., micropipette) of the chemoattractant. This preference can be quantified by arbitrarily dividing the cell image into three sectors as shown in Fig. 3A and comparing the pixel values of F actin staining. In the wild-type neutrophils, there is a significantly higher level of pixels in the sector that faces the micropipette than in the other two sectors (Fig. 3B).

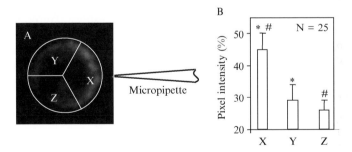

FIG. 3. Detection of directional sensing. (A) Schematic depiction of the stimulation from a point source delivered by a micropipette that is 40–50 μm from the cell. The cell was fixed and stained with phalloidin. For quantification, the cell was divided into three sectors, and pixel values were determined. (B) Pixel values of 25 neutrophils in each sector are compared. (See color insert.)

Localization of Active Small GTPases

As different small GTPases have their specific downstream target genes, it has been shown that some of the target protein contains sequences that specifically bind to active small GTPases. For instances, active RhoA and its close homologs specifically bind to the RhoA-binding domain (RBD) of Rhotekin (Alblas *et al.*, 2001), whereas active CDC42 binds to the CDC42-binding domain (CBD) of WASP (Symons *et al.*, 1996). We have shown that when neutrophils are stimulated by a uniform, point source (by a micropipette as shown in Fig. 2A) or gradient chemoattractant (as in a Dunn chamber) and stained with RBD or CBD, CBD preferentially stains the leading edges, whereas RBD preferentially stains the posterior (Li *et al.*, 2005; Zhang *et al.*, 2003). This suggests that distribution of small GTPases is polarized in stimulated neutrophils. To perform the staining, cells are fixed with cold 4% paraformaldehyde in PBS for 15 min and permeabilized with 0.1% triton X-100 in PBS for 5 min at room temperature. The cells on the coverslip are then washed with PBS and incubated with 1 μg purified GST-RBD or CBD-GST protein in 100 μl PBS containing 1% BSA at 4° overnight. After washing with PBS, the cells are incubated with Alexa Fluor 488 conjugate anti-GST antibody (1:100; Molecular Probes) at room temperature for 1 h, followed by washing and mounting with a drop of Antifade reagent (Molecular Probes). The RBD-GST and CBD GST proteins were prepared as previously described (Alblas *et al.*, 2001; Symons *et al.*, 1996).

Concluding Remarks

In this chapter, we discuss the methods for preparation of primary mouse neutrophils and analysis of neutrophil polarization, migration, and directional sensing. We have used these methods in the study of neutrophils from a number of gene-targeted mouse lines, which include those lacking PLC $\beta2$, PLC $\beta3$, PIXα, and PI3Kγ (Hannigan *et al.*, 2002; Jiang *et al.*, 1997; Li *et al.*, 2000, 2003, 2005). Although PLC $\beta2$- or $\beta3$-deficiency does not impair neutrophil chemotaxis, PIXα or PI3Kγ deficiency preferentially leads to defects in directionality.

References

Alblas, J., Ulfman, L., Hordijk, P., and Koenderman, L. (2001). Activation of Rhoa and ROCK are essential for detachment of migrating leukocytes. *Mol. Biol. Cell* **12,** 2137–2145.

Allen, W. E., Zicha, D., Ridley, A. J., and Jones, G. E. (1998). A role for Cdc42 in macrophage chemotaxis. *J. Cell Biol.* **141,** 1147–1157.

Baggiolini, M., Dewald, B., and Moser, B. (1997). Human chemokines: An update. *Annu. Rev. Immunol.* **15,** 675–705.

Hannigan, M., Zhan, L., Li, Z., Ai, Y., Wu, D., and Huang, C. K. (2002). Neutrophils lacking phosphoinositide 3-kinase gamma show loss of directionality during N-formyl-Met-Leu-Phe-induced chemotaxis. *Proc. Natl. Acad. Sci. USA* **99,** 3603–3608.

Hart, M. J., Jiang, X. J., Kozasa, T., Roscoe, W., Singer, W. D., Gilman, A. G., Sternweis, P. C., and Bollag, G. (1998). Direct stimulation of the guanine nucleotide exchange activity of P115 RhoGEF by Gα13. *Science* **280,** 2112–2114.

Jiang, H., Kuang, Y., Wu, Y., Xie, W., Simon, M. I., and Wu, D. (1997). Roles of phospholipase C β2 in chemoattractant-elicited responses. *Proc. Natl. Acad. Sci. USA* **94,** 7971–7975.

Kozasa, T., Jiang, X. J., Hart, M. J., Sternweis, P. M., Singer, W. D., Gilman, A. G., Bollag, G., and Sternweis, P. C. (1998). P115 RhoGEF, a GTPase activating protein for Gα12 and Gα13. *Science* **280,** 2109–2111.

Li, Z., Dong, X., Wang, Z., Liu, W., Deng, N., Ding, Y., Tang, L., Hla, T., Zeng, R., Li, L., and Wu, D. (2005). Regulation of PTEN by Rho small GTPases. *Nat. Cell Biol.* **7,** 399–404.

Li, Z., Hannigan, M., Mo, Z., Liu, B., Lu, W., Wu, Y., Smrcka, A. V., Wu, G., Li, L., Liu, M., Huang, C. K., and Wu, D. (2003). Directional sensing requires G beta gamma-mediated PAK1 and PIX alpha-dependent activation of Cdc42. *Cell* **114,** 215–227.

Li, Z., Jiang, H., Xie, W., Zhang, Z., Smrcka, A. V., and Wu, D. (2000). Roles of PLC-β2 and -β3 and PI3Kgamma in chemoattractant-mediated signal transduction. *Science* **287,** 1046–1049.

Lowell, C. A., Fumagalli, L., and Berton, G. (1996). Deficiency of Src family kinases p59/61hck and p58c-fgr results in defective adhesion-dependent neutrophil functions. *J. Cell Biol.* **133,** 895–910.

Mao, J., Xie, W., Yuan, H., Simon, M. I., Mano, H., and Wu, D. (1998a). Specific involvement of GEF115 in activation of Rho and SRF by Gα13. *Proc. Natl. Acad. Sci. USA* **95,** 12973–12976.

Mao, J., Yuan, H., Xie, W., and Wu, D. (1998b). Specific involvement of G proteins in regulation of SRF by receptors. *J. Biol. Chem.* **273,** 27118–27123.

Murphy, P. M. (1994). The molecular biology of leukocyte chemoattractant receptors. *Annu. Rev. Immunol.* **12,** 593–633.

Neer, E. J. (1995). Heterotrimeric G proteins: Organizers of transmembrane signals. *Cell* **80,** 249–257.

Symons, M., Derry, J. M., Karlak, B., Jiang, S., Lemahieu, V., McCormick, F., Francke, U., and Abo, A. (1996). Wiskott-Aldrich syndrome protein, a novel effector for the GTPase CDC42Hs, is implicated in actin polymerization. *Cell* **84,** 723–734.

Xu, J., Wang, F., Van Keymeulen, A., Herzmark, P., Straight, A., Kelly, K., Takuwa, Y., Sugimoto, N., Mitchison, T., and Bourne, H. R. (2003). Divergent signals and cytoskeletal assemblies regulate self-organizing polarity in neutrophils. *Cell* **114,** 201–214.

Zhang, Y., Hoon, M. A., Chandrashekar, J., Mueller, K. L., Cook, B., Wu, D., Zuker, C. S., and Ryba, N. J. (2003). Coding of sweet, bitter, and umami tastes: Different receptor cells sharing similar signaling pathways. *Cell* **112,** 293–301.

Zigmond, S. H. (1988). Orientation chamber in chemotaxis. *Methods Enzymol.* **162,** 65–72.

Zlotnik, A., and Yoshie, O. (2000). Chemokines: A new classification system and their role in immunity. *Immunity* **12,** 121–127.

[48] Measurement of Epidermal Growth Factor Receptor Turnover and Effects of Cdc42

By QIONG LIN, WANNIAN YANG, and RICHARD A. CERIONE

Abstract

Ligand-induced degradation represents an essential component of the overall regulation of EGF receptor (EGFR)–coupled signal transduction. Following activation, EGFRs are monoubiquitinated, subsequently sorted by ubiquitin-interaction–based sorting machinery, and transported to multivesicular bodies (MVBs) and lysosomes for degradation. The Rho-family small G-protein, Cdc42, has been implicated in the regulation of EGFR degradation. Here we describe routine methods for assaying EGFR endocytosis and degradation. In addition, we have introduced procedures for determining the effects of Cdc42 and its downstream targets, in particular, ACK (Activated Cdc42-associated Kinase) and p85Cool-1 (Cloned out of library)/Pix (for Pak-interactive exchange factor), on EGFR degradation.

Introduction

The epidermal growth factor receptor (EGFR) is a member of the ErbB family of receptor tyrosine kinases and is activated by multiple ligands, including EGF, TGFα, heparin-binding EGF-like growth factor, amphiregulin, β-cellulin, and epiregulin (Kim and Muller, 1999). On ligand binding, EGFRs form homodimers or heterodimers with other members of the ErbB family, including ErbB2/Neu and ErbB3. EGFR dimerization is accompanied by the trans-phosphorylation of tyrosine residues within the cytoplasmic domain, which then serve as docking sites for downstream adaptors and signaling molecules (Hackel *et al.*, 1999; Moghal and Sternberg, 1999; Schlessinger, 2000, 2002).

EGFRs are internalized and degraded promptly on EGF stimulation (Sorkin, 2001). The sorting process for EGFR degradation is signaled by ubiquitination as catalyzed by the adaptor protein Cbl, which functions as an ubiquitin E3 ligase. Cbl interacts by way of an SH2-like domain with a phosphotyrosine residue (tyrosine 1045) on the EGFR. The EGF-dependent phosphorylation of Cbl, at tyrosine 371, has been suggested to activate its E3 ligase activity, resulting in the monoubiquitination of the EGFR (Levkowitz *et al.*, 1998, 1999). Various studies indicate that monoubiquitination of the EGFR serves as a lysosomal degradation–sorting signal (Haglund *et al.*, 2003; Mosesson *et al.*, 2003). Cbl-interacting proteins

METHODS IN ENZYMOLOGY, VOL. 406
0076-6879/06 $35.00
DOI: 10.1016/S0076-6879(06)06048-4

including CIN85, endophilin, Grb2, and Sprouty have been reported to form complexes that influence the ubiquitination of EGFRs (Rubin *et al.*, 2003; Schmidt *et al.*, 2003; Soubeyran *et al.*, 2002; Waterman *et al.*, 2002).

The endosomal sorting of EGFRs is organized by the hepatocyte growth factor (HGF)–regulated tyrosine phosphorylation substrate (Hrs). Monoubiquitinated EGFRs are recognized by the UIM domain of Hrs (Urbe *et al.*, 2003). Hrs interacts with the signal-transducing adaptor molecule (STAM) proteins and brings EGFR-loaded clathrin-coated vesicles to multivesicular bodies (MVBs). The latter occurs through an interaction with Tsg101, which is part of the endosomal-sorting complex required for transport (ESCRT-I) (Bache *et al.*, 2003a,b; Lu *et al.*, 2003). The EGFRs in MVBs are eventually transported to lysosomes for degradation.

The small GTP-binding protein, Cdc42, has been suggested to influence EGFR degradation in two ways. First, activated forms of Cdc42 negatively regulate the binding of Cbl to the EGFR (Wu *et al.*, 2003). This occurs through the interaction of activated Cdc42 with the p85Cool-1 (for *C*loned *o*ut *o*f *l*ibrary)/β-Pix (*P*ak-*i*nteractive *ex*change factor) protein, which in turn binds to Cbl. When Cbl is part of this ternary complex, it is not able to bind to the EGFR. Dissociation of Cdc42 from the complex allows Cbl to bind to the EGFR and catalyze its ubiquitination.

A second way in which Cdc42 can influence EGFR degradation is through its interactions with a specific target/effector called ACK (*A*ctivated *C*dc42-associated *K*inase). Two forms of ACK have been identified and characterized, designated as ACK1 (Manser *et al.*, 1993) and ACK2 (Yang and Cerione, 1997). Both ACK1 and ACK2 have been found to bind clathrin (Teo *et al.*, 2001; Yang *et al.*, 2001), and ACK2 has been shown to interact with the sorting nexin SH3PX1 (SNX9) (Lin *et al.*, 2002). Moreover, ACK2 can be coimmunoprecipitated with the EGFR, indicating that ACK2 is associated with EGFR complexes and participates in EGFR-coupled signaling (Lin *et al.*, 2002; Yang and Cerione, 1997). EGFR degradation is facilitated by coexpression of SH3PX1 with wild-type but not kinase-dead ACK2, suggesting that ACK2 complexes with and phosphorylates SH3PX1 to regulate receptor degradation (Lin *et al.*, 2002).

In the following sections, we will describe the procedures used to examine different aspects of EGFR down-regulation and how Cdc42 and its target/effectors influence these events.

Approaches for Examining EGFR Endocytosis and Degradation

Three common methods are used for assaying EGFR endocytosis and degradation. The first involves detecting EGFR levels in total cell lysates by immunoblotting with anti-EGFR antibodies. The second method involves the use of fluorescent microscopy to visualize the endocytic Texas red–conjugated EGF/EGFR complexes in cells or immunofluorescent-stained endocytic EGFRs. The third approach is a radioactive isotope–based assay that detects the endocytic ^{125}I-EGF/EGFR complexes in cells.

A number of different cell types including COS7, NIH3T3, and HEK293 cells are routinely used for monitoring the degradation of both endogenous and ectopically expressed EGFRs, whereas CHO cells are typically used when assaying ectopically expressed receptors. COS7 cells express high levels of endogenous EGFR, whereas NIH3T3 and HEK293 cells express relatively low levels of receptors. The endogenous EGFRs can be easily detected in COS7 cells and reasonably well in NIH3T3 cells when immunoblotting cell lysates; however, endogenous EGFRs are only detectable in HEK293 cells when immunoprecipitated with anti-EGFR antibodies. For investigation of endocytosis and degradation of ectopically expressed EGFRs, CHO cells have proven to be a good model system, because they express few endogenous EGFRs but contain all of the appropriate machinery for the endocytosis and degradation of EGFRs.

Detection of EGFR Endocytosis Using Fluorescent Microscopic Methods

Materials

COS7 cells are typically used for these experiments. Common reagents include EGF-conjugated with Texas-red (Molecular Probes), anti-EGFR antibody (mAb528) (Santa Cruz Biotech, Inc.), Texas red–conjugated anti-mouse IgG (Molecular Probes), and acidic stripping solution: 0.1 M acetic acid, pH 2.7, and 0.15 M NaCl.

EGF Receptor Endocytosis Assays Using Texas-Red–Conjugated EGF

COS7 cells are cultured on glass coverslips to 50–90% confluence and serum-starved for 12 h. The cells are then chilled on ice for 10 min, and Texas red–conjugated EGF (20 ng/ml) is added to the medium for 60 min. To initiate endocytosis, the cells are rinsed with serum-free DMEM and incubated at 37° for the desired time. The cells are then chilled on ice to stop endocytosis. The remaining cell surface–bound EGF is stripped by incubating the cells in ice-cold stripping solution for 5 min. The cells are again briefly washed with the stripping solution and rinsed with pre-cooled PBS (1×), followed by fixation with 3.7% paraformaldehyde at 25° for 10 min. The cells are then washed in PBS (3×) and the endocytosed, Texas-red–conjugated EGF is monitored under a fluorescent microscope.

EGF Receptor Endocytosis Assays as Monitored by Immunofluorescent Staining of the Receptors

COS7 cells are cultured on glass coverslips to 50–90% confluence and serum-starved for 12 h. Receptor endocytosis is assayed at 37° and is initiated by the addition of EGF (100 ng/ml) to the medium at different time points. The cells are washed with chilled PBS (1×), followed by fixation with 3.7% paraformaldehyde in PBS at 25° for 10 min. The cells are then permeabilized by incubation with 0.1% Triton X-100 in PBS at 25° for 10 min. After washing with PBS (3×), the cells are incubated with anti-EGFR (mAb528, 4 μg/ml) at 25° for 60 mins. For secondary antibody incubation, the cells are washed with PBS once, and then incubated with Texas red–conjugated anti-mouse IgG (1:250) for 60 min at 25°. To remove unbound secondary antibody, the cells are washed with PBS (3×). Immunofluorescent staining of EGFRs is then observed under a fluorescent microscope.

An example of EGFR endocytosis assays performed using Texas red–conjugated EGF and anti-EGFR immunofluorescent staining is shown in Fig. 1. When cells are not stimulated with EGF, the EGFRs are located on the plasma membrane of the cells (Fig. 1A,C). When cells are stimulated with EGF for 30 min, EGFRs are endocytosed and localized in endosomes (Fig. 1B,D).

Detection of EGFR Endocytosis by [^{125}I]-EGF Labeling

Materials

COS7 cells are typically used for these assays. Reagents include [^{125}I]-EGF (750 Ci/mmol, 5 μCi) (Amersham); acidic stripping solution: 0.1 M acetic acid, pH 2.7, 0.15 M NaCl; and mammalian cell lysis buffer (MCL): 40 mM HEPES, pH 7.4, 1% Triton X-100, 100 mM NaCl, 1 mM EDTA, 25 mM β-glycerolphosphate, 1 mM Na-orthovanadate, 10 μg/ml leupeptin and aprotinin.

Procedures

After 12 h of serum starvation, the cells are chilled on ice for 10 min, and then [^{125}I]-EGF (0.1 μCi) along with unlabeled EGF (final concentration = 20 ng/ml) is added to the culture medium. The cells are incubated on ice for 60 min and then rinsed with DMEM to remove the unbound ligand and rapidly warmed to 37°. Endocytosis is assayed for 2, 5, 10, 30, 60 90, and 120 min and stopped by the addition of chilled acidic stripping solution. EGF that has not been taken up into the cell as an outcome of EGFR

0 min   30 min

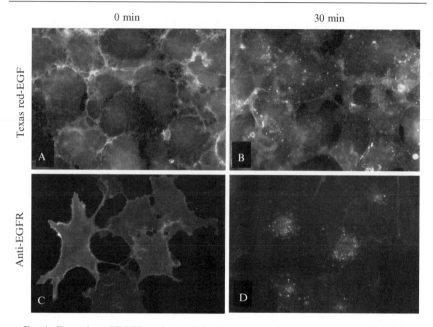

FIG. 1. Detection of EGFR endocytosis by fluorescent microscopic methods. (A, B) Texas red–conjugated EGF was used as a fluorescent dye to track EGFR endocytosis after the cells were stimulated by Texas red-EGF (20 ng/ml) for 0 and 30 min. (C, D) EGFR was detected by immunofluorescent staining with anti-EGFR antibody (mAb528) after the cells were stimulated by EGF (100 ng/ml) for 0 and 30 min. The punctated fluorescence in B and D represents endocytosed EGFRs.

endocytosis is stripped by incubating the cells in stripping solution for 5 min. The cells are then rinsed with acidic stripping solution and lysed with MCL buffer. The radioactivity of both the stripping solution and the lysates is measured using a γ-ray counter. The counts in the lysates represent the relative amount of internalized EGFRs.

Detection of EGFR Degradation by Immunoblotting

Materials

The cell line most typically used is COS7. The cells are cultured in DMEM plus 10% FBS at 37° in 5% CO_2. The cell lysis buffer (MCL) contains 40 mM HEPES, pH 7.4, 1% Triton X-100, 100 mM NaCl, 1 mM EDTA, 25 mM β-glycerolphosphate, 1 mM Na-orthovanadate, 10 μg/ml leupeptin and 10 μg/ml aprotinin. The EGF stock solution (100 μg/ml) is prepared in phosphate-buffered saline (PBS; i.e., 20 mM phosphate, pH

7.4, and 150 mM NaCl). The (2×) SDS-PAGE sample buffer contains 125 mM Tris-Cl, pH 6.8, 4% SDS, 5% β-mercaptoethanol, 20% glycerol, and 0.005% bromophenol blue.

Procedures

COS7 cells are cultured in 6-well plates in DMEM plus 10% FBS, to 90% confluence followed by serum-starvation overnight (12 h). For EGF stimulation, the EGF stock solution (100 μg/ml) is directly added to the culture medium at a final concentration of 100 ng/ml for 0.5, 1, 2, 3, and 4 h. The cells are then briefly rinsed with PBS once and lysed by the addition of 250 μl of chilled MCL buffer, with rocking for 20 min at 4°. The cell lysates are collected into an Eppendorf tube and cleared by centrifugation at 14,000 rpm for 4 min in a Microfuge. To prepare the lysate samples for SDS-PAGE, aliquots (20 μl) of the cleared cell lysates are mixed with 20 μl of 2× SDS-PAGE sample buffer. The lysate samples are loaded onto an 8% SDS-polyacrylamide gel. After transfer to a PVDF membrane, the EGFRs in the lysate samples are detected by immunoblotting with an anti-EGFR antibody (1005) (Santa Cruz), at 1:1000 dilution, using the ECL (enhanced chemiluminescence) assay kit (Amersham). A representative result is shown in Fig. 2. The degradation of EGFRs in COS7 cells reaches a maximum level after 4 h of EGF stimulation.

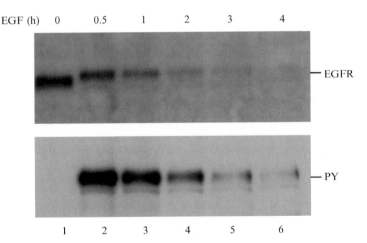

Fig. 2. Detection of EGFR degradation by immunoblotting assays. COS7 cells were cultured in DMEM plus 10% FBS to 90% confluence and then serum-starved for 12 h. The cells were stimulated by EGF (100 ng/ml) for 0, 0.5, 1, 2, 3, and 4 h. Receptor degradation was detected by immunoblotting the cell lysates with anti-EGFR antibody (Santa Cruz) (upper panel). The tyrosine phosphorylation of EGFRs was detected by immunoblotting the cell lysates with antiphosphotyrosine (4G10, UBI) (lower panel).

Additional Comments

Among the four members of the ErbB tyrosine kinase family, only the EGFR undergoes ligand-induced degradation (Baulida *et al.*, 1996). It also has been suggested that the heterodimerization between EGFRs and other members of ErbB family (e.g., Neu/ErbB2) blocks the ligand-induced degradation of EGFRs (Wang *et al.*, 1999). Thus, endogenous expression levels of ErbB2, ErbB3, or ErbB4 may affect ligand-induced EGFR degradation. The degradation rates for EGFRs may vary in different cell lines. For example, in COS7 cells, completion of EGFR degradation occurs 3–4 h after EGF stimulation, whereas in 293 cells, EGFR degradation occurs in 1–2 h. Therefore, the time scale for EGF treatment and for visualizing EGFR degradation should be adjusted to the particular cell lines used.

When the expression levels of EGFRs in cells are low, the immunoprecipitation of EGFRs using an anti-EGFR antibody (e.g., mAb528) should be included. For example, when HEK293 cells are used as the cell system, the endogenous EGFRs are difficult to visualize when immunoblotting the cell lysate samples. In this case, immunoprecipitation of the EGFRs is necessary for monitoring EGFR degradation.

Effects of Cdc42 and ACK on EGFR Degradation

The ACKs are Cdc42-specific target/effectors that are capable of binding to clathrin and phosphorylating SH3PX1, a sorting nexin (Manser *et al.*, 1993; Yang and Cerione, 1997). ACK2 has been shown to associate with EGFRs and facilitate EGFR degradation by phosphorylating SH3PX1 in an EGF-dependent manner (Lin *et al.*, 2002). The detailed procedures that have been used to examine the effects of ACK2 and SH3PX1 on EGFR degradation are as follows.

Materials

COS7 cells are typically used and transfected with plasmids encoding ACK2 and SH3PX1 using Lipofectamine (Invitrogen). These plasmids include pcDNA3, pcDNA3-Myc-ACK2 (wild-type), pcDNA3-Myc-ACK2-(K158R) (kinase-defective), and pcDNA3-HA-SH3PX1 (described in Lin *et al.*, 2002). The antibodies used include anti-Myc (9E10, Covance), anti-HA (HA.11, Covance), anti-phosphotyrosine (4G10, UBI), and anti-EGFR (Santa Cruz). Cell lysis is performed in a buffer containing 40 mM HEPES, pH 7.4, 1% Triton X-100, 100 mM NaCl, 1 mM EDTA, 25 mM β-glycerolphosphate, 1 mM Na-orthovanadate, 10 μg/ml leupeptin and 10 μg/ml aprotinin.

Procedures

COS7 cells are cultured in 6-well plates to 90% confluence before transfection. Transfections are performed using LipofectAmine, following the instructions of the manufacturer. Commonly used constructs include pcDNA3 (1 μg/well), pcDNA3-Myc-ACK2 (1 μg/well), or pcDNA3-Myc-ACK2-(K158R) (1 μg/well), individually or cotransfected together with pcDNA3-HA-SH3PX1 (1 μg/well). After transfection for 36 h, the cells are serum-starved for 12 h and then stimulated by the addition of EGF (final concentration 100 ng/ml) for different time periods. The cells are rinsed with PBS once, and transferred on ice. Lysis of the cells is performed by addition of chilled MCL (250 μl/well), followed by incubation at 4° with rocking for 20 min. The cell lysates are transferred to Eppendorf tubes and cleared by centrifugation at 14,000 rpm for 4 min in a Microfuge. The cleared lysates are mixed with 2× SDS-PAGE sample buffer, denatured at 100° for 2 min, and loaded onto an 8% SDS polyacrylamide gel.

The gel is transferred onto a PVDF membrane. The transferred PVDF membrane is incubated with anti-EGFR (1:1000), anti-Myc (1:1000), and anti-HA (1:1000) at 25° for 1–3 h, followed by incubation with horseradish peroxidase–conjugated secondary antibody (1:5000 dilution) at 25°for 1 h. The immunoreactive protein bands are visualized on X-ray film by ECL (enhanced chemiluminescence).

An example is shown in Fig. 3. Expression of SH3PX1 alone (lanes 9–12), or together with kinase-defective ACK2(K158R) (lanes 5–8), did not significantly change the degradation rate for EGFRs compared with vector-transfected samples (lanes 13–16). However, coexpression of wild-type ACK2 with SH3PX1 significantly enhanced EGFR degradation (lanes 1–4).

Effects of Cdc42 and Cool-1/β-Pix on Cbl-Catalyzed EGFR Ubiquitination

Materials

CHO and HEK293 cells are typically used in these experiments. The cells are transfected using LipofectAmine (Invitrogen) with HA-tagged c-Cbl and HA-tagged ubiquitin, together with different Cdc42 and p85Cool-1/β-Pix constructs. Cell lysis is performed as described in the preceding sections.

Procedures

HEK293 cells are transiently cotransfected with plasmids encoding HA-tagged ubiquitin, different forms of Myc-tagged Cdc42 (either wild-type Cdc42, GTPase-defective Cdc42[Q61L], or constitutively active

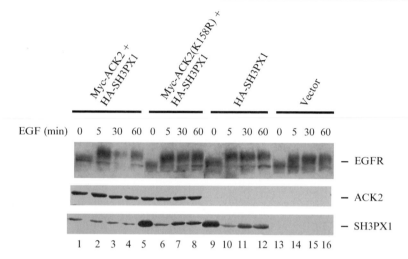

FIG. 3. Examination of the effects of the Cdc42-specific target, ACK2, on EGFR degradation. The HA-tagged SH3PX1 was transfected or cotransfected with Myc-ACK2 or the kinase-dead mutant ACK2(K158R) mutant into COS7 cells for 36 h. The cells were then serum-starved for 12 h and treated with EGF for 0, 5, 30, and 60 min. The amount of EGFRs in whole lysates was detected by immunoblotting with an anti-EGF receptor antibody (top panel). The expression level of Myc-ACK2 or the kinase-dead mutant ACK2(K158R) was detected by immunoblotting the lysates with an anti-Myc antibody (middle panel). The expression levels of HA-SH3PX1 were detected by immunoblotting the lysates with an anti-HA antibody (bottom panel). Data from Lin *et al.* (2002).

Cdc42[F28L], described in Wu *et al.*, 2003), and different forms of Myc-tagged p85Cool-1/β-Pix (either wild-type or the Cbl-binding-defective p85Cool-1[W43K]) (also, see Wu *et al.*, 2003). The cells are cultured in 6-well plates for 24 h and then serum-starved for an additional 12 h. At that point, the cells are treated with EGF (100 ng/ml) for 10 min at 37° or are untreated. The cells are lysed, and endogenous EGFRs are immunoprecipitated using an anti-EGFR antibody (Gibco BRL), and the ubiquitinated EGFRs are detected by Western blotting with anti-HA (HA.11, Covance) antibody. When using CHO cells, the cells are transfected with an expression plasmid encoding the EGFR, together with plasmids encoding HA-tagged ubiquitin, HA-tagged c-Cbl, and different Myc-tagged Cdc42 constructs. Cell lysates are subjected to immunoprecipitation with an anti-EGFR antibody, and the levels of ubiquitinated EGFRs are determined by immunoblotting with anti-HA antibody. Typically, activated forms of Cdc42 will cause a significant reduction in the levels of EGFR ubiquitination, as readout by a decrease in the incorporation of HA-ubiquitin into the receptors, and an accumulation of total cellular EGFRs as indicated by

Western blotting total cell lysates with an anti-EGFR antibody. However, in all cases, it is important to perform a number of controls. These include verifying that there is no ubiquitination of the EGFRs in the absence of added EGF and that equivalent levels of EGFR and HA-tagged c-Cbl are being compared for the different experimental conditions.

Discussion

At present, the detailed mechanisms underlying EGFR down-regulation and degradation are far from being completely understood. A good deal remains to be determined regarding the role of Cbl in different aspects of EGFR degradation, particularly its relative importance in endocytosis versus endosomal sorting (Haglund et al., 2003; Mosesson et al., 2003; Soubeyran et al., 2002). The exact roles played by members of the sorting nexin family also need to be better established. For example, what specific function does sorting nexin 1 play (Kurten et al., 1996) relative to other members of its family (i.e., sorting nexin 3/SH3PX1)? We also have much to learn regarding how Cdc42 influences different aspects of EGFR degradation. The regulatory effects imparted by Cdc42 on the Cbl-catalyzed ubiquitination of EGFRs are likely to be finite in duration, because GTP hydrolysis by Cdc42 should allow it to dissociate from p85Cool-1 and enable Cbl to bind to EGFRs and signal their degradation. Thus, it is attractive to think of Cdc42 as a timing device that ensures EGFR-Cbl interactions do not occur either too quickly or too slowly. However, how are the actions of Cdc42 on EGFR-Cbl interactions linked to its regulation of ACK-SH3PX1 interactions and EGFR sorting, and are these regulatory influences by Cdc42 occurring on a broad scale or are they limited to specific cell types? Many additional questions also await future studies, including how the degradation of other members of the EGFR subfamily is mediated, if at all, and what kinds of regulatory mechanisms are used by other receptor tyrosine kinases (e.g., PDGF receptor, nerve growth factor receptors) to ensure their proper cellular homeostasis?

References

Bache, K. G., Brech, A., Mehlum, A., and Stenmark, H. (2003a). Hrs regulates multivesicular body formation via ESCRT recruitment to endosomes. J. Cell Biol. 162, 435–442.

Bache, K. G., Raiborg, C., Mehlum, A., and Stenmark, H. (2003b). STAM and Hrs are subunits of a multivalent ubiquitin-binding complex on early endosomes. J. Biol. Chem. 278, 12513–12521.

Baulida, J., Kraus, M. H., Alimandi, M., Di Fiore, P. P., and Carpenter, G. (1996). All ErbB receptors other than the epidermal growth factor receptor are endocytosis impaired. J. Biol. Chem. 271, 5251–5257.

Hackel, P. O., Zwick, E., Prenzel, N., and Ullrich, A. (1999). Epidermal growth factor receptors: Critical mediators of multiple receptor pathways. *Curr. Opin. Cell Biol.* **11,** 184–189.

Haglund, K., Sigismund, S., Polo, S., Szymkiewicz, I., Di Fiore, P. P., and Dikic, I. (2003). Multiple monoubiquitination of RTKs is sufficient for their endocytosis and degradation. *Nat. Cell Biol.* **5,** 461–466.

Kim, H., and Muller, W. J. (1999). The role of the epidermal growth factor receptor family in mammary tumorigenesis and metastasis. *Exp. Cell Res.* **253,** 78–87.

Kurten, R. C., Cadena, D. L., and Gill, G. N. (1996). Enhanced degradation of EGF receptors by a sorting nexin, SNX1. *Science* **272,** 1008–1010.

Levkowitz, G., Waterman, H., Ettenberg, S. A., Katz, M., Tsygankov, A. Y., Alroy, I., Lavi, S., Iwai, K., Reiss, Y., Ciechanover, A., Lipkowitz, S., and Yarden, Y. (1999). Ubiquitin ligase activity and tyrosine phosphorylation underlie suppression of growth factor signaling by c-Cbl/Sli-1. *Mol. Cell* **4,** 1029–1040.

Levkowitz, G., Waterman, H., Zamir, E., Kam, Z., Oved, S., Langdon, W. Y., Beguinot, L., Geiger, B., and Yarden, Y. (1998). c-Cbl/Sli-1 regulates endocytic sorting and ubiquitination of the epidermal growth factor receptor. *Genes Dev.* **12,** 3663–3674.

Lin, Q., Lo, C. G., Cerione, R. A., and Yang, W. (2002). The Cdc42 target ACK2 interacts with sorting nexin 9 (SH3PX1) to regulate epidermal growth factor receptor degradation. *J. Biol. Chem.* **277,** 10134–10138.

Lu, Q., Hope, L. W., Brasch, M., Reinhard, C., and Cohen, S. N. (2003). TSG101 interaction with HRS mediates endosomal trafficking and receptor down-regulation. *Proc. Natl. Acad. Sci. USA* **100,** 7626–7631.

Manser, E., Leung, T., Salihuddin, H., Tan, L., and Lim, L. (1993). A non-receptor tyrosine kinase that inhibits the GTPase activity of p21cdc42. *Nature* **363,** 364–367.

Moghal, N., and Sternberg, P. W. (1999). Multiple positive and negative regulators of signaling by the EGF-receptor. *Curr. Opin. Cell Biol.* **11,** 190–196.

Mosesson, Y., Shtiegman, K., Katz, M., Zwang, Y., Vereb, G., Szollosi, J., and Yarden, Y. (2003). Endocytosis of receptor tyrosine kinases is driven by monoubiquitylation, not polyubiquitylation. *J. Biol. Chem.* **278,** 21323–21326.

Rubin, C., Litvak, V., Medvedovsky, H., Zwang, Y., Lev, S., and Yarden, Y. (2003). Sprouty fine-tunes EGF signaling through interlinked positive and negative feedback loops. *Curr. Biol.* **13,** 297–307.

Schlessinger, J. (2000). Cell signaling by receptor tyrosine kinases. *Cell* **103,** 211–225.

Schlessinger, J. (2002). Ligand-induced, receptor-mediated dimerization and activation of EGF receptor. *Cell* **110,** 669–672.

Schmidt, M. H., Furnari, F. B., Cavenee, W. K., and Bogler, O. (2003). Epidermal growth factor receptor signaling intensity determines intracellular protein interactions, ubiquitination, and internalization. *Proc. Natl. Acad. Sci. USA* **100,** 6505–6510.

Sorkin, A. (2001). Internalization of the epidermal growth factor receptor: Role in signalling. *Biochem. Soc. Trans.* **29,** 480–484.

Soubeyran, P., Kowanetz, K., Szymkiewicz, I., Langdon, W. Y., and Dikic, I. (2002). Cbl-CIN85-endophilin complex mediates ligand-induced downregulation of EGF receptors. *Nature* **416,** 183–187.

Teo, M., Tan, L., Lim, L., and Manser, E. (2001). The tyrosine kinase ACK1 associates with clathrin-coated vesicles through a binding motif shared by arrestin and other adaptors. *J. Biol. Chem.* **276,** 18392–18398.

Urbe, S., Sachse, M., Row, P. E., Preisinger, C., Barr, F. A., Strous, G., Klumperman, J., and Clague, M. J. (2003). The UIM domain of Hrs couples receptor sorting to vesicle formation. *J. Cell Sci.* **116,** 4169–4179.

Wang, Z., Zhang, L., Yeung, T. K., and Chen, X. (1999). Endocytosis deficiency of epidermal growth factor (EGF) receptor-ErbB2 heterodimers in response to EGF stimulation. *Mol. Biol. Cell* **10**, 1621–1636.

Waterman, H., Katz, M., Rubin, C., Shtiegman, K., Lavi, S., Elson, A., Jovin, T., and Yarden, Y. (2002). A mutant EGF-receptor defective in ubiquitylation and endocytosis unveils a role for Grb2 in negative signaling. *EMBO J.* **21**, 303–313.

Wu, W. J., Tu, S., and Cerione, R. A. (2003). Activated Cdc42 sequesters c-Cbl and prevents EGF receptor degradation. *Cell* **114**, 715–725.

Yang, W., and Cerione, R. A. (1997). Cloning and characterization of a novel Cdc42-associated tyrosine kinase, ACK-2, from bovine brain. *J. Biol. Chem.* **272**, 24819–24824.

Yang, W., Lo, C. G., Dispenza, T., and Cerione, R. A. (2001). The Cdc42 target ACK2 directly interacts with clathrin and influences clathrin assembly. *J. Biol. Chem.* **276**, 17468–17473.

[49] Tumor Cell Migration in Three Dimensions

By STEVEN HOOPER, JOHN F. MARSHALL, and ERIK SAHAI

Abstract

In almost all physiological and pathological situations, cells migrate through three-dimensional environments, yet most studies of cell motility have used two-dimensional substrates. It is clear that two-dimensional substrates do not mimic the *in vivo* environment accurately, and recent work using three-dimensional environments has revealed many different mechanisms of cell migration (Abbott, 2003; Sahai and Marshall, 2003; Wolf *et al.,* 2003). This chapter will describe methods for generating three-dimensional matrices suitable for studying cell motility, methods for imaging the morphology of motile cells *in situ,* and methods for quantifying cell migration through three-dimensional environments.

Introduction

Precisely regulated cell movement is crucial for many processes in multicellular organisms, including embryogenesis, mobilization of immune cells, and wound healing. In contrast, inappropriate cell movement contributes to pathologic conditions such as tumor metastasis and a range of inflammatory diseases.

For many years, the study of cell motility has used two-dimensional substrates; however, cells move in three-dimensional environments *in vivo.* Recent advances in imaging technologies and developments in producing extra-cellular matrix component have enabled cell movement to be studied

METHODS IN ENZYMOLOGY, VOL. 406   0076-6879/06 $35.00
          DOI: 10.1016/S0076-6879(06)06049-6

in detail in artificially generated three-dimensional matrices. These studies have revealed different modes of cell movement and altered sensitivity to pharmacological agents in three-dimensional environments (Sahai and Marshall, 2003; Wolf *et al.*, 2003). It will therefore be necessary to re-evaluate much of what we have learned from studying cell motility in two-dimensional contexts in three-dimensional environments. Here we describe simple methods to generate three-dimensional matrices and to image invasion of cells into these matrices. We present analysis of the invasive properties of tumor cells, stromal fibroblasts, and macrophages.

Choice and Generation of Three-Dimensional Matrices

Cells within tissues are surrounded by a heterogeneous mixture of complex polysaccharides and proteins; collectively this milieu is called the extracellular matrix (ECM). The most common polysaccharides in the ECM are glycosaminoglycans (GAGs) that form hydrated gels because of their very hydrophilic nature (Alberts *et al.*; Bosman and Stamenkovic, 2003). Most GAGs are linked to core proteins to form higher order structures called proteoglycans. The most common ECM proteins belong to the collagen and laminin families. Collagens form trimeric coiled-coil complexes that are usually organized into higher order fibrous structures (Bornstein and Sage, 1980; Linsenmayer, 1991). Collagen fibers, particularly collagens I and III, are abundant in the ECM of connective tissues and the dermis. In contrast, several members of the laminin family and collagen IV are organized into a meshlike network that forms a specialized ECM structure called a basement membrane, which defines epithelial tissue boundaries (Geiger *et al.*, 2001; Hudson *et al.*, 1993). Epithelial tumors are not considered invasive until they have crossed the basement membrane that delineates the tissue from which they originate. Fibronectin and Tenascins are also widely distributed ECM proteins; however, they will not be discussed in this chapter.

To study cell migration in three-dimensional environments, it is necessary to generate a matrix that resembles the ECM found in real tissues. This section will discuss some of the commonly used matrices and provide protocols to generate some of these matrices.

Collagen I

Analysis of tumors has revealed that they are rich in collagen fibers. The most widely used collagen for *in vitro* studies is collagen I, which forms insoluble fibers *in vivo*. Collagen I can be rendered soluble by a combination of acidic pH and partial pepsin digestion (Bornstein, 1958). This form

A

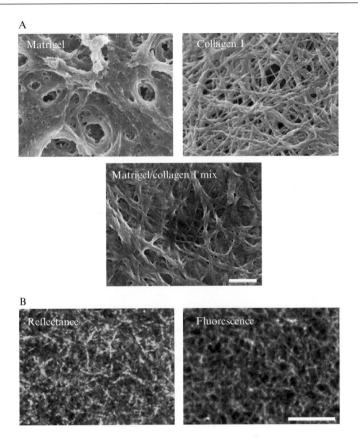

FIG. 1. Images of different types of reconstituted ECM. (A) Scanning electron micrographs of 3 mg/ml matrigel gel, 3 mg/ml collagen I gel, and 5 mg/ml matrigel/collagen I gel (1:1 mix). Bar is 1 μm. (B) Laser-scanning micrographs of 5 mg/ml collagen I gel; left-hand panel shows a reflectance image, and right-hand panel shows a fluorescence image. Bar is 10 μm.

of solubilized collagen I is widely available and can be induced to form a fibrous network by returning the solution to neutral pH. Collagen I fibers formed in this way are of similar dimension to those found *in vivo*, and even some higher order collagen I bundles can be formed (Fig. 1A). However, the pepsin digestion used to prepare the collagen means that these fibers are not identical to native collagen I. Nonetheless, the ease of preparing these matrices means that they are very widely used and have proved very useful for modeling a number of biological systems. Some native preparations of collagen I are commercially available, but these are harder to manipulate and will not be discussed here.

Procedure for Generating Collagen Gels

REAGENTS. Collagen I (pepsin-solubilized; concentration may vary between suppliers from 1–0 mg/ml; BD Bioscience Rat Tail collagen I cat #354236), 5× cell culture media, 1 *M* HEPES, pH 7.5 (Sigma H3375), dH$_2$O.

1. Decide what final concentration of collagen I you require: 2–8 mg/ml collagen I gels are relatively easy to work with.
2. Prepare the collagen I mix with chilled solutions on ice: 20% 5× cell culture media, 2% 1 *M* HEPES, pH 7.5, up to 78% collagen I, depending on the concentration required. If less than 78% collagen I is required, replace the missing volume with d H$_2$0. Keep sterile.
3. Pipette collagen I mix onto or into relevant dishes or wells.
4. Place in cell culture incubator for 60–90 min. When gel has set, it will become translucent.
5. Cover gels with cell culture media.

Notes

Solutions of collagen can be very viscous; therefore, it is recommended to make 20% more collagen mix than you actually require to allow for wastage/difficulties associated with pipetting a viscous solution. When making 5× cell culture media, remember to include appropriate amounts of any supplements and NaHCO$_3$ if it is required. We routinely fix collagen I gels with 4% paraformaldehyde.

Visualization of Collagen I in Gels

There are several methods for visualizing collagen fibers in the gels. Collagen I gels usually have a fibrous appearance when viewed using conventional phase contrast microscopy, although only the larger bundles of fibers can be observed clearly. Laser-scanning (confocal) microscopy can reveal much more detail of the collagen fibers and can also resolve three-dimensional aspects of the sample more readily. The simplest method relies on capturing the reflected light from the specimen. The microscope should be set up exactly as for fluorescence imaging except that it should capture an image of the same wavelength light that is being used to illuminate the specimen (any wavelength can be used to illuminate the specimen). Figure 1B shows a reflectance image of a collagen I gel. This method can be combined with fluorescence imaging on almost all laser-scanning microscopes and does not rely on any labeling of the collagen. It should be remembered that any reflective material would be imaged using reflectance imaging, not just collagen fibers.

Collagen fibers will also generate second harmonic signals when excited by high-intensity beams of near-infrared laser light (Zoumi *et al.*, 2002). Laser-scanning microscopes equipped with Ti-sapphire lasers and multi-photon imaging capability can be used to image these second harmonic signals. Wavelengths between 700 and 900 nm can be used to excite the collagen fibers that will then produce light at half the wavelength used for excitation. This technique is not widely used because of the high cost of the equipment needed.

An alternative is to use fluorescently labeled collagen I, which can be purchased from commercial suppliers. A small amount of fluorescent collagen can then be added to the collagen I mixture while it is still soluble; 0.1 mg/ml of fluorescent collagen is sufficient to efficiently label collagen fibers (Sigma C4361). As before, laser-scanning microscopy is the best way to visualize the fibers, because conventional epifluorescence cannot deal easily with the three-dimensional nature of the gels. Figure 1B shows that both fluorescence and reflectance imaging of collagen I gels give similar results, but neither has the resolution to visualize the individual fibers apparent by electron microscopy. Figure 2A shows a three-dimensional reconstruction of an invading cell surrounded by collagen I fibers.

Matrigel™

Matrigel is a commercially available ECM preparation that is purified from the ECM produced by a mouse sarcoma cell line grown *in vivo*. Matrigel™ is a heterogeneous mixture of proteins composed primarily of collagen IV and laminins (Kleinman *et al.*, 1982; McGuire and Seeds, 1989), a composition that is similar to that of basement membranes. Matrigel™ is much less fibrous and more gel-like than collagen I (Fig. 1A). The increased complexity of Matrigel™ relative to collagen I has the potential advantage that it may more closely resemble the ECM in tumors. However, the increased complexity means that not all the components of Matrigel™ are defined, and it contains several growth factors and proteases that may complicate the interpretation of experiments. To counteract this problem, a growth factor–reduced version of Matrigel™ is available, although this still contains low levels of many growth factors.

Basic Procedure for Generating Matrigel™ Gels

REAGENTS. Matrigel™ (normal or growth factor reduced; BD Bioscience #354234 and #356230, respectively), Cell culture media.

1. Decide what final concentration of Matrigel™ is required: 2.5–8mg/ml Matrigel™ gels are relatively easy to work with. Below 2.5 mg/ml, Matrigel™ gels lack sufficient rigidity to generate a workable three-

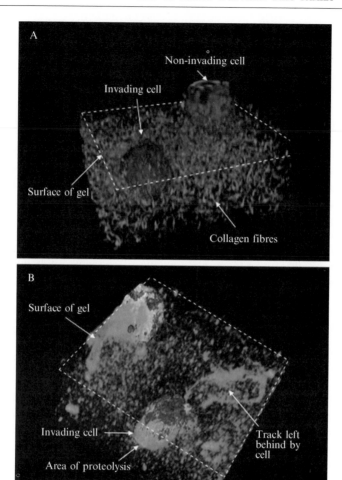

FIG. 2. Three-dimensional images of A375 cells invading labeled collagen gels. (A) F-actin staining of A375 cells in red and FITC collagen I fibers in green. Laser-scanning microscope z-sections were reconstructed using Volocity software (Improvision). Dashed lines marking the surfaces of the gel are 80 μm. (B) F-actin staining of A375 cells in red and DQ collagen I in green. Laser-scanning microscope z-sections were reconstructed using Volocity software (Improvision). Dashed lines marking the surfaces of the gel are 80 μm. (See color insert.)

dimensional matrix. Dilute preparations of Matrigel™ are still useful for coating surfaces.

2. On ice, dilute Matrigel™ with chilled cell culture media appropriate for your cells. Keep sterile.

3. Pipette Matrigel™ onto or in relevant dishes or wells. If working with small volumes, it is beneficial to use chilled pipette tips.
4. Place in cell culture incubator for 30–60 min.
5. Cover with cell culture media.

Notes

Matrigel™ is only a solution at low temperature (roughly between 0 and 10°); it is crucial to keep solutions of matrigel on ice to prevent them gelling inappropriately. Matrigel™ is supplied frozen; to avoid repeated freeze–thaw cycles, matrigel should be aliquoted and stored at −80°. Matrigel™ can routinely be fixed with 4% paraformaldehyde; if it is being used at a low concentration, the gels become quite fragile; this can be partly overcome by using 4% paraformaldehyde/0.2% glutaraldehyde. Matrigel™ and collagen I matrices can be combined to provide a matrix with both fibrous and gel-like properties (shown in Fig. 1A).

Visualizing Matrix Degradation

Proteolysis of the ECM is a key feature of tumor cell invasion; this can be visualized *in situ* by the use of labeled ECM components whose fluorescent properties change on cleavage. Labeling of either collagen I, IV, or gelatin with fluorophores at very high stoichiometry leads to auto-quenching of fluorescence (called DQ collagen I, IV, or gelatin available from Molecular Probes). Proteolytic cleavage can relieve the autoquench-ing and lead to an increase in fluorescence. For most applications, 0.1 mg/ml of commercially available DQ collagen I is sufficient. This increase in fluorescence can either be imaged live or it can be fixed with parafor-maldehyde. Figure 2B shows matrix degradation surrounding an invading cell and its track. Care should be taken interpreting experiments using DQ collagen/gelatin, because many cells are able to locally concentrate ECM components (Grinnell, 2003); this may lead to an apparent increase in DQ collagen fluorescence that is not associated with proteolysis. To eliminate this possibility, it is sensible to perform the same experiments in parallel using normal fluorescently labeled collagen/gelatin. If similar changes in fluorescence are observed in both cases, proteolysis is not the cause of the changes in DQ collagen/gelatin fluorescence. A comparison of Fig. 2A and B shows that normally labeled collagen has a relatively uniform distribution throughout the gel, whereas the signal from DQ collagen is concentrated near the invading cell and cell tracks. This difference gives one confidence that the DQ collagen signal is the result of proteolysis.

Types of Three-Dimensional Migration Assays

Many different assays have been developed that aim to investigate the migration of cells in or through three-dimensional environments. This section will discuss the pros and cons of these assays and provide protocols for some of the more widely used assays. A good three-dimensional assay needs to be quantifiable and to mimic the *in vivo* environment more accurately than more conventional two-dimensional assays.

Invasion Assays Without Gradients

Figure 3 shows various configurations for invasion assays; A and B are the simplest to set up. Configuration A involves overlaying a matrix of choice on top of cells that are already adherent. Use at least 50 μl of matrix per cm^2; this will theoretically generate a matrix 500-μm thick. The most suitable dishes for this type of experiment have a base of coverslip glass that is suitable for imaging; these can be purchased from the MatTek Corporation (http://www.glass-bottom-dishes.com/index.html). One of the

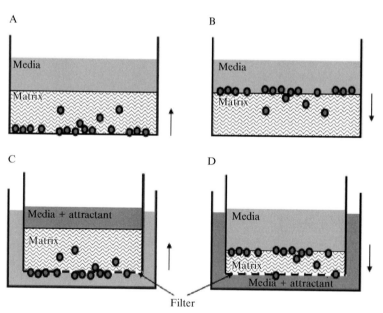

Fig. 3. Different configurations of invasion assays. (A) Simple overlay assay, cells move up into matrix from the bottom of a dish. (B) Cells plated on top of matrix and invade down into matrix. (C) Inverted transwell assay. Cells are plated on the underside of a transwell and cross filter and then invade up into the matrix. (D) Conventional transwell assay. Cells are plated in a transwell and invade down into the matrix to the other side of the filter. Arrows indicate the direction of cell invasion.

main benefits of this type of assay is that the cells start at a clearly defined place; after a suitable period of time, the assay can be fixed and stained (see later). The assay can then be quantified using a laser-scanning microscope, and qualitative data can be captured using a standard epifluorescence microscope (see later). A major complication with this assay is that the potentially invasive cells are initially placed in a competitive situation. Some cell types may adhere very strongly to the base of the dish/coverslip, and these adhesions may be stronger than those adhesions that can be made with the overlying matrix. If this is the case, cells with an inherent invasive potential will not invade into the matrix.

Protocol for Setting up Configuration an Invasion Assay

DAY 1

1. Prepare 35-mm Mat Tek dishes with 10-mm wells by washing with 1 N HCl for 15 min then washing twice with PBS, twice with 70% ethanol, and once with media.
2. Trypsinize cells as normal; plate 50–200,000 cells per dish.

DAY 2

3. Prepare 100 μl of collagen I or matrigel mix per dish (see protocols in "Visualization of collagen I in Gels").
4. Carefully aspirate as much media as possible from cells, then overlay 80 μl of collagen mix.
5. Incubate at 37° for 60–90 min.
6. Overlay with 2 ml media.

Configuration B is also very easy to set up. Prepare the matrix of your choice and spread onto the bottom of a cell culture dish. Use at least 30 μl of matrix per cm^2; this will theoretically generate a matrix 300-μm thick. If the matrix becomes too thin, it may present cells with an environment that is effectively two-dimensional. This assay is well suited to qualitative assessment of tumor cell invasion; high-quality images can easily be produced using an upright microscope fitted with a dipping objective or with inverted microscopes. The main drawback is the difficulty in quantifying the extent of invasion, because it is very difficult to define the position at which the cells start.

Invasion Assays with Gradients

Cell motility is often directed by a gradient of a chemotactic factor. Neither of the two experimental setups outlined previously enable a chemotactic gradient to be established within the three-dimensional matrix.

The experimental setups shown in Fig. 3C and D allow for gradients of chemotactic factors to be generated. Both systems use "transwell" inserts that form a well with a filter within a regular cell culture dish. These inserts are commercially available with a range of different pore sizes in the filters; 8-μm pores are the most suitable for cell invasion assay. A matrix is prepared within the well, and the ability of cells to invade into or through the matrix is assayed. These assays have the benefit of having either a defined starting point (configuration C) or a defined endpoint (configuration D). By adding a chemoattractant in either the upper chamber (transwell) or lower chamber (reservoir) of the assay, a gradient can be established across the matrix into which the cells are invading. To determine whether cell invasion is chemotactic, it is necessary to compare the extent of invasion in the presence and absence of a gradient; only if invasion is increased in the presence of gradient can it be described as chemotactic.

Protocol for Setting up Configuration C Invasion Assay

1. Prepare 60 μl of collagen I or Matrigel™ mix per transwell (from Costar cat #3422). Protocols for preparing collagen I or matrigel are in "Visualization of Collagen in Cells".
2. Pipette 50 μl of matrix solution into the upper well of the transwell.
3. Incubate at 37° for 60–90 min.
4. Trypsinize cells as normal. Turn plate containing transwells upside-down. Lift base of plate up to leave inverted transwells resting on the lid of the plate. Pipette 50–100 μl of cell suspension onto inverted transwell (cell suspension should contain 5000–25,000 cells per well). Replace the base of the plate so that it covers the inverted transwells.
5. Leave in incubator for 2–4 h until cells have adhered.
6. Turn plate back up the "right" way. Add 1 ml media to lower chamber and 200 μl media plus chemoattractant to the upper chamber. Return to incubator.

Notes

The length of time for which invasion assays should be left needs to be determined empirically, typically 1–3 days is sufficient. To set up configuration D, adapt the preceding protocol: reduce the amount of matrix mix used (we have successfully used 9 μl spread using a chilled glass rod with a rounded end). The thickness and uniformity of the gel can be estimated by including some fluorescently labeled collagen and imaging by laser-scanning microscopy. Once the matrix has set, plate cells in the upper chamber in a volume of 200 μl and add 1 ml to the lower chamber.

Stability of the Gradient. To investigate the stability of gradients generated using transwell invasion systems, we monitored the diffusion of a dye from the lower chamber into the upper chamber. The lower chamber was filled with a 5 mM solution of bromophenol blue (Mr 669) in PBS, and the upper chamber was filled PBS and incubated at 37°. Figure 4 shows the concentration of dye in the upper chamber at various time intervals in a transwell system with no matrix or with 50 μl of collagen I gel above the filter. In both cases, the dye had almost reached equilibrium after 20 h. The presence of a collagen gel slows the diffusion of the dye such that a significant concentration difference (at least threefold) persists between the chambers for 3–4 h. However, given that most invasion assays take 24 h or more, a gradient of small molecules (Mr < 1000) would only be present for the first few hours of the assay. Gradients of larger molecules such as polypeptide growth factors will persist for longer but are likely to have significantly decreased after 24 h. Although it is impossible to accurately generalize from this simple analysis, it should be noted that gradients used in invasion assays are not stable and, in the case of small molecules, are unlikely to persist for a significant amount of time relative to the length of the assay. This problem can be partly counteracted by replacing the media in the upper and lower chambers at regular intervals.

The ability of small molecules (Mr < 1000) to diffuse through collagen gels within a few hours means that small molecule inhibitors can be used in invasion assays. For most experiments, it is sufficient to add inhibitor at the same time as the cells, because it is likely that the inhibitor will diffuse faster than the cells can move. If this is a particular concern or one wishes to use blocking antibodies that will diffuse much slower, the inhibitor or antibody can be included in the collagen I or matrigel mix as well as the cell culture media.

Protocols for Fixation and Staining of Invasion Assays

1. Remove media (from both upper and lower chamber if using configurations C or D). Rinse gently once with PBS.
2. Fix using 4% paraformaldehyde in PBS (for matrigel gels <4 mg/ml also include 0.25% glutaraldehyde for 30 min. Use sufficient volume to cover the gel (if using configurations C or D put fix in both chambers).
3. Remove fixing solution and permeabilize using 0.2% Triton X100 in PBS for 30 min (if using configurations C or D, put TritonX100/PBS in both chambers).
4. Rinse gently twice with PBS.

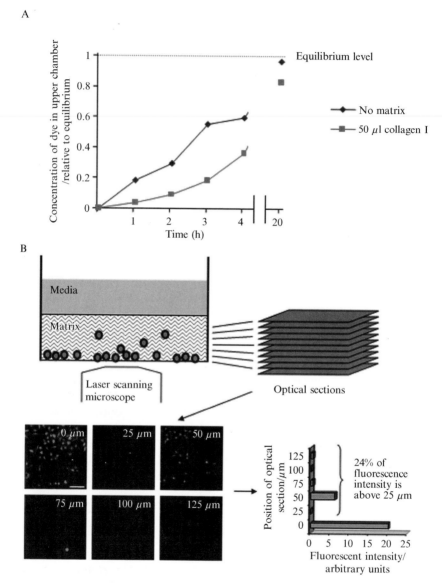

FIG. 4. Characterization and quantification of invasion assays. (A) Stability of a small molecule gradient in transwell assay. Graph shows the relative concentration of bromophenol blue in the upper chamber of a transwell assay using a filter with 8-μm pores with or without a 50-μm collagen I gel. Initially, dye was placed in lower chamber; only then assay was incubated at 37°; 1 is the concentration of bromophenol blue at equilibrium. (B) Quantification of an overlay invasion assay. Top left shows a schematic of experimental setup; top right shows representation of the optical section; bottom left shows representative optical sections from an assay; and bottom right shows quantification of the images in bottom left. (See color insert.)

5. Stain using 0.2 μg/ml TRITC Phalloidin (Sigma P1951) and 2 μg/ml DAPI (Sigma D9542) for 2–3 h in the dark.
6. Rinse gently three times with PBS, allow 15 min per rinse.
7. Samples are now ready to be imaged; they can be stored at 4° for 1–2 weeks.

Notes

Invasion assays should be imaged without being mounted, because mounting can lead to considerable spatial distortion, and most mounting media do not readily diffuse into the gel. Indirect immune-fluorescence labeling of invading cells is possible, particularly when using configuration B. Follow the same staining procedure for regular indirect immune fluorescence but increase the incubation times of both the primary and secondary antibodies.

Procedure for Imaging and Quantification Using a Laser-Scanning Microscope

1. Choose an objective with a relatively long working distance (>0.25 mm is good). $10\times$, $20\times$, or $25\times$ objectives are suitable for quantification. For high-resolution images, we used a Zeiss LSM 510 with a $63\times$ C-Apochromat water correction lens with working distance 0.24 mm and NA 1.2; for quantification we used a $25\times$ Plan-Neofluor multicorrection lens (used with water) with working distance 0.22 mm and NA 0.8.
2. Place specimen on microscope and focus on cells that remain adherent to the coverglass at the base of the dish. Switch to laser-scanning mode.
3. Start scanning and recheck focal plane.
4. Set up microscope to capture a z-series starting at cells that remain adherent to the coverglass at the base of the dish and moving into the overlying gel. For quantification, z intervals between 2.5 and 5 μm are suitable (for simplicity, Fig. B only shows z sections 25 μm apart), whereas for high-resolution images, z intervals between 0.5 and 1 μm are best. Check that the pinhole is adjusted so that the thickness of optical section is less than the z interval being used. Ensure that laser power and gain are adjusted so that images captured are within the linear range of photodetectors.
5. Begin optical sectioning.
6. Quantify the total DAPI fluorescence in each optical section (subtract background if necessary). Propidium iodide or other nucleic acid–binding fluorophores can be used in place of DAPI. Alternately, manually count the number of cells in each section. Plot histograms of fluorescent intensity or number of cells against optical section number (i.e., physical position). These data can be presented in number of ways; it is

often convenient to determine the proportion of cells that have invaded further than a threshold distance; in the example in Fig. 4B, 24% of the fluorescent intensity is at or above 25 μm. For cells that span the threshold distance, this method scores the part of the cell at or beyond the threshold.

7. Laser-scanning microscope z-sections can be reconstructed to generate three-dimensional images of invading cells. Figure 5 shows the difference between single confocal sections and three-dimensional reconstructions of z-series.

Notes

Generally, inverted microscopes are more convenient for imaging invasion assays, but upright microscopes can be used as well. Phase-contrast and epifluorescence microscopes will provide qualitative information about the extent of invasion and morphology of invading cells. However, laser-scanning microscopes are most suitable for imaging invasion assays, because they can provide accurate optical sectioning of the three-dimensional environment. Assays using configuration C and D can be imaged on an inverted microscope by placing the transwell in a MatTek dish containing PBS (http://www.glass-bottom-dishes.com/index.html). If access to laser-scanning microscopes is not possible, assays can be fixed as described previously, mounted in paraffin, sectioned, and stained using hematoxylin–eosin or immunohistochemical methods and analyzed using a bright-field microscope. It is important to score the proportion of invading cells and not the total number of invading cells. Methods that involve wiping all the cells from one side of the filter and then staining and scoring the absolute number of invading cells will be misleading if experimental manipulations affect the proliferation or survival of cells.

Comparison of Different Cell Lines and Different Motility Assays

To compare the morphology of different cell lines, we performed the invasion assays (as shown in Fig. 3B); cells were allowed to invade into the collagen I/matrigel mix for 24 h and were then fixed and stained. High-resolution imaging of the actin cytoskeleton revealed that different cells exhibit a diverse range of cell morphologies while invading three-dimensional matrices. Figure 5 shows the various morphologies displayed by cells invading a three-dimensional matrix compared with cells cultured on coverslips. BAC macrophages have a somewhat elongated shape with numerous small actin-rich protrusions; tumor-derived fibroblasts are also elongated but have a smoother surface with few protrusions. A375 melanoma cells adopt a very rounded shape with small bleblike membrane

2D culture   Invading in 3D

Confocal section   Confocal section   3D reconstruction

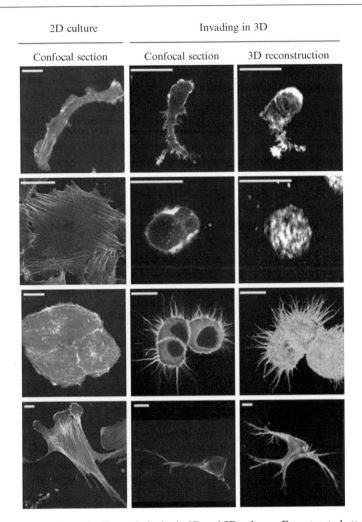

Fig. 5. Comparison of cell morphologies in 2D and 3D cultures. From top to bottom, the left-hand panels show F-actin staining of BAC macrophages, A375 melanoma cells, A431 squamous carcinoma cells, and tumor fibroblasts cultured on a coverslip. Middle panels show single laser-scanning microscope sections of BAC macrophages, A375 melanoma cells, A431 squamous carcinoma cells, and tumor fibroblasts invading into a 5-mg/ml collagen I/matrigel mix. Right-hand panels show three-dimensional reconstructions (using Volocity software) of BAC macrophages, A375 melanoma cells, A431 squamous carcinoma cells, and tumor fibroblasts invading into a 5-mg/ml collagen I/matrigel mix. Bar is 20 μm.

TABLE I

MOTILITY OF DIFFERENT CELL LINES IN INVASION ASSAYS

Cell line	Config D No matrix	Config D 60–80 μm matrix	Config C
A375 melanoma	19%	1.9%	1.7%
A431 squamous ca.	33%	4.3%	11%
BAC macrophage	7.9%	4.9%	2.6%
Tumour fibroblasts	53%	20%	10%

The percentage of invading cells is shown in a range of cell motility assays. For A375 and A431 cells DMEM +10%FCS + 10 nM EGF was used as a chemo-attractant, for BAC macrophages DMEM + 10%FCS + 3000 units/ml CSF-1 was used as a chemo-attractant, for tumour fibroblasts DMEM + 10%FCS was used as a chemo-attractant. DMEM was placed in the upper chamber in Configuration D and in the lower chamber in Configuration C. The percentage of invading cells was calculated by manually counting the number of invading and the total number of cells in z-series (numbers are average of two experiments).

protrusions, whereas A431 squamous cell carcinoma cells have numerous large actin-rich spikes. In some cases, there is a clear relation between the morphology of cells in two- and three-dimensional environments (for example BAC macrophages). However, the dramatic difference between the A375 and tumor fibroblasts could not be predicted from their morphology in two-dimensional cultures. These differences between two- and three-dimensional systems highlight the potentially significant information that can be overlooked when using two-dimensional environments.

We next compared the ability of these different cell lines to invade in variations of the invasion assays shown in Fig. 3C and D (configuration D is modified from Hennigan et al. [1994]). The variations on the assays were as follows: configuration D but with no matrix added for 24 h, configuration D with a 60- to 80-μm-thick matrigel mix coated above the filter for 36 h, configuration C with a 1-mm thick matrigel mix coated above the filter for 36 h. For the assays using configuration D, the proportion of cells that had moved through the filter and were adherent to its lower surface was scored. For the assay using configuration C, the proportion of cells that had invaded more than 8 μm into the matrigel in the upper chamber was scored. Unsurprisingly, all the cell lines showed the greatest amount of cell motility when there was no matrix barrier present. Fibroblasts were the most invasive cells in both types of invasion assay, whereas the proportion of invasive A375 melanoma cells was low. This is consistent with a previous report that cells that invade with a rounded morphology are relatively inefficient compared with those that invade with an elongated "mesenchymal" motility (Sahai and Marshall, 2003). The distance invaded by cells with a rounded morphology is typically 5–15 μm/day, whereas cells with a mesenchymal morphology can invade 5–75 μm/day, depending on the cell line.

Inhibition of ROCK Has Different Effects Depending on the Assay

Given that cells adopt very different morphologies when moving in two- and three-dimensional environments (Fig. 5), one might expect differing molecular mechanisms of cell motility depending on the cellular environment. We have previously demonstrated that ROCK (Rho-kinase) function is required for the rounded cell invasion of A375 cells (Sahai and Marshall, 2003). Table II shows that this requirement for ROCK function is only observed when A375 cells are moving with a rounded morphology in a three-dimensional environment. Inhibition of ROCK using Y-27632 actually enhanced the movement of A375 cells in the absence of a three-dimensional matrix or a very thin coating of matrigel. Only when A375 cells had to move through a relatively thick layer of ECM did they require ROCK function. Furthermore, previous studies have shown that the rounded morphology adopted by A375 cells in a three-dimensional environment is dependent on ROCK function (Sahai and Marshall, 2003).

Perspective

The cell invasion assays described here are by no means a comprehensive list; many more specialised systems have been developed for the study of particular of tissues that more accurately recreate the tissue microenvironment. For example, the invasive behavior of squamous cell carcinomas can be accurately modeled by culturing cells on a matrix containing fibroblasts and exposed to the air (Parenteau et al., 1992; Stenback et al., 1999), whereas the ability of cells to invade chick-heart explants has also been used successfully (Noe et al., 1999). Breast epithelial cells can be induced to grow in acinar structures (Debnath et al., 2003), whereas kidney epithelial cells will form tubules when cultured in the appropriate three-dimensional

TABLE II
EFFECT OF ROCK INHIBITOR IN DIFFERENT CELL MOTILITY ASSAYS

	Config D no matrix	Config D 60–80 μm matrix	Config C
Cell motility in presence of ROCK inhibitor (relative to control)	79% increase	88% decrease	84% decrease

The percentage change in cell motility caused by treatment with 10 μM Y-27632 is stated (average of two experiments). Y-27632 was added to both upper and lower chambers of motility assays.

matrix (Zegers *et al.*, 2003). The formation of these structures is disrupted by the activation of oncogenic signaling pathways. Because of space limitations and the highly specialized nature of some of these systems, they have not been described in detail here. The protocols described here are a starting point for the exploration of the very complex behavior of tumor cells in three-dimensional environments.

Materials and Methods

Cell Lines

A375 (subclone M2) was a gift from R. Hynes (MIT, Boston), A431 cells were obtained from the ATCC, BAC1.2F5 cells were a gift from E. R. Stanley (Albert Einstein College of Medicine, New York), tumor fibroblasts were a gift from K. Harrington (ICR, London).

Miscellaneous

Y-27632 was purchased from Tocris UK (cat# 1254), Bromophenol Blue (Sigma-Aldrich #11,439-1). Three-dimensional image reconstruction was performed using Volocity software (www.improvision.com).

Acknowledgments

We thank Steve Gschmeissner for electron microscopy services and Sophie Pinner for proofreading. S. H., J. M., and E. S. are funded by Cancer Research UK.

References

Abbott, A. (2003). Cell culture: Biology's new dimension. *Nature* **424**, 870–872.

Alberts, B., *et al.* "Molecular Biology of the Cell." Edition IV.

Bornstein, M. B. (1958). Reconstituted rattail collagen used as substrate for tissue cultures on coverslips in Maximow slides and roller tubes. *Lab. Invest.* **7**, 134–137.

Bornstein, P., and Sage, H. (1980). Structurally distinct collagen types. *Annu. Rev. Biochem.* **49**, 957–1003.

Bosman, F. T., and Stamenkovic, I. (2003). Functional structure and composition of the extracellular matrix. *J. Pathol.* **200**, 423–428.

Debnath, J., Muthuswamy, S. K., and Brugge, J. S. (2003). Morphogenesis and oncogenesis of MCF-10A mammary epithelial acini grown in three-dimensional basement membrane cultures. *Methods Enzymol.* **30**, 256–268.

Geiger, B., Bershadsky, A., Pankov, R., and Yamada, K. M. (2001). Transmembrane crosstalk between the extracellular matrix? Cytoskeleton crosstalk. *Nat. Rev. Mol. Cell Biol.* **2**, 793–805.

Grinnell, F. (2003). Fibroblast biology in three-dimensional collagen matrices. *Trends Cell Biol.* **13**, 264–269.

Hennigan, R. F., Hawker, K. L., and Ozanne, B. W. (1994). Fos-transformation activates genes associated with invasion. *Oncogene* **9**, 3591–3600.

Hudson, B. G., Reeders, S. T., and Tryggvason, K. (1993). Type IV collagen: Structure, gene organization, and role in human diseases. Molecular basis of Goodpasture and Alport syndromes and diffuse leiomyomatosis. *J. Biol. Chem.* **268,** 26033–26036.

Kleinman, H. K., McGarvey, M. L., Liotta, L. A., Robey, P. G., Tryggvason, K., and Martin, G. R. (1982). Isolation and characterization of type IV procollagen, laminin, and heparan sulfate proteoglycan from the EHS sarcoma. *Biochemistry* **21,** 6188–6193.

Linsenmayer, T. F. (1991). Collagen. *In* "Cell Biology of Extracellular Matrix" (E. D. Hay, ed.), pp. 5–37.

McGuire, P. G., and Seeds, N. W. (1989). The interaction of plasminogen activator with a reconstituted basement membrane matrix and extracellular macromolecules produced by cultured epithelial cells. *J. Cell Biochem.* **40,** 215–227.

Noe, V., Willems, J., Vandekerckhove, J., Roy, F. V., Bruyneel, E., and Mareel, M. (1999). Inhibition of adhesion and induction of epithelial cell invasion by HAV-containing E-cadherin-specific peptides. *J. Cell Sci.* **112**(Pt. 1), 127–135.

Parenteau, N. L., Bilbo, P., Nolte, C. J., Mason, V. S., and Rosenberg, M. (1992). The organotypic culture of human skin keratinocytes and fibroblasts to achieve form and function. *Cytotechnology* **9,** 163–171.

Sahai, E., and Marshall, C. J. (2003). Differing modes of tumour cell invasion have distinct requirements for Rho/ROCK signalling and extracellular proteolysis. *Nat. Cell Biol.* **5,** 711–719.

Stenback, F., Makinen, M. J., Jussila, T., Kauppila, S., Risteli, J., Talve, L., and Risteli, L. (1999). The extracellular matrix in skin tumor development–a morphological study. *J. Cutan. Pathol.* **26,** 327–338.

Wolf, K., Mazo, I., Leung, H., Engelke, K., Von Andrian, U. H., Deryugina, E. I., Strongin, A. Y., Brocker, E. B., and Friedl, P. (2003). Compensation mechanism in tumor cell migration: Mesenchymal-amoeboid transition after blocking of pericellular proteolysis. *J. Cell Biol.* **160,** 267–277.

Zegers, M. M., O'Brien, L. E., Yu, W., Datta, A., and Mostov, K. E. (2003). "Epithelial polarity and tubulogenesis *in vitro*." *Trends Cell Biol.* **13**(4), 169–176.

Zoumi, A., Yeh, A., and Tromberg, B. J. (2002). Imaging cells and extracellular matrix *in vivo* by using second-harmonic generation and two-photon excited fluorescence. *Proc. Natl. Acad. Sci. USA* **99,** 11014–11019.

[50] An *In Vitro* Model to Study the Role of Endothelial Rho GTPases During Leukocyte Transendothelial Migration

By JAIME MILLÁN, LYNN WILLIAMS, and ANNE J. RIDLEY

Abstract

Tissue injury induces release of cytokines that stimulate the expression of adhesion receptors in the endothelial wall of neighboring vessels. These endothelial receptors recruit leukocytes from the bloodstream and facilitate their transendothelial migration (TEM) toward the inflamed area. The molecules involved in leukocyte–endothelium interaction and TEM have been

METHODS IN ENZYMOLOGY, VOL. 406
0076-6879/06 $35.00
DOI: 10.1016/S0076-6879(06)06050-2

studied *in vivo* and *in vitro* over the past 20 years. Human umbilical vein endothelial cells (HUVECs) are a popular *in vitro* model to analyze TEM, and have been used to investigate the role of Rho GTPases in this process. Here we describe methods to activate HUVECs, to investigate Rho GTPase-activation by the endothelial adhesion receptor ICAM-1, and to inhibit or activate Rho GTPases using C3 transferase or adenoviruses coding for dominant negative or constitutive active Rho GTPases. Finally, we describe how to image and quantitate leukocyte TEM by digital time-lapse microscopy.

Introduction

Endothelial cells form a monolayer that lines blood vessels and regulates the passage of cells and macromolecules between the blood and the surrounding tissues (Wojciak-Stothard and Ridley, 2002). Pathological alterations to the endothelium contribute to arteriosclerosis and chronic inflammatory or autoimmune diseases (Davignon and Ganz, 2004). In addition, endothelial cells lead to the formation of new vessels in response to proangiogenic stimuli, which is often essential for the growth of solid tumors (Jain, 2003). Studying the mechanisms that regulate endothelial cell morphology, motility, and proliferation are, therefore, highly important to biomedicine.

RhoGTPases have been shown to regulate contraction, motility, and proliferation in endothelial cells (Connolly *et al.,* 2002; Mettouchi *et al.,* 2001; Wojciak-Stothard and Ridley, 2002). At the early stages of inflammation, cytokines induce the expression of endothelial adhesion receptors of the immunoglobulin superfamily such as ICAM-1 or VCAM-1 in the surrounding vessels. These receptors, along with many others, facilitate the adhesion and transendothelial migration (TEM), also called diapedesis, of leukocytes from the bloodstream toward the inflamed site (Butcher, 1991). Leukocytes adhere to ICAM-1 and VCAM-1 by means of integrins Mac-1 ($\alpha M\beta 2$) or LFA-1 ($\alpha L\beta 2$), and VLA-4 ($\alpha 4\beta 1$), respectively (Shimizu *et al.,* 1991). In addition to their role in leukocyte capture, ICAM-1 and VCAM-1 induce signaling on integrin ligation, including activation of Rho GTPases, which mediate endothelial remodeling. This can lead to the opening of intercellular gaps, which may facilitate leukocyte TEM (Millan and Ridley, 2005; Thompson *et al.,* 2002; van Wetering *et al.,* 2003). Rho GTPases may be important for leukocyte TEM *in vivo*, because statins, which inhibit leukocyte TEM in mouse models of arteriosclerosis and multiple sclerosis (Davignon, 2004), act, at least in part, by inhibiting Rho GTPase prenylation and thereby their activity (Greenwood *et al.,* 2003). Here we describe methods and protocols to set up an *in vitro* model to analyze leukocyte TEM and to study the role of Rho GTPases in signaling downstream of the endothelial adhesion receptor ICAM-1.

HUVEC Culture and TNF-α Stimulation

Human umbilical vein endothelial cells (HUVECs) are often used for *in vitro* studies on leukocyte TEM. Endothelial cells *in vivo* are very heterogeneous in morphology, receptor expression, and function, depending on the nature and location of the vessel (Aird, 2003). Cultured HUVECs are flat, polygonal, and have cell–cell junctions similar to those of endothelial cells from large veins. Methods for isolating HUVECs from the umbilical cord have been reviewed elsewhere (Larrivee and Karsan, 2005). Nowadays, several companies sell HUVECs.

Materials

Vials of HUVECs from pooled donors (5×10^5 cells) and culture medium (EBM-2) plus growth factors and serum are obtained from Cambrex Bio Science Wokingham, Ltd (Wokingham, UK), human plasma fibronectin is obtained from Sigma-Aldrich (Gillingham, UK), and recombinant human tumor necrosis factor (TNF)-α is from Serotec Ltd. (Kidlington, UK).

Methods

Growth of HUVECs. Success in growing endothelial cells depends on plating them on a proper extracellular matrix (ECM) component to promote their proliferation. We have obtained optimal results with human fibronectin, but other ECM proteins such as collagen have been used by others. Plastic dishes and flasks are coated with 10 μg/ml fibronectin in PBS for 30 min at 37°. The cells are thawed following the manufacturer's instructions, adding the cells to 15 ml of EBM-2 medium containing growth factors and 2% FCS (EGM-2) in a fibronectin-coated 75-cm^2 flask. The cells are not centrifuged to remove the DMSO, because this reduces viability. The medium is replaced after 24 h. HUVECs grow optimally at 37° and 5% CO_2.

Once HUVECs reach confluency in the 75-cm^2 flask (between 4 and 5 million cells), they are trypsinized and passaged by diluting 1:3 onto plastic dishes precoated with fibronectin. The cells are used between passages 2 and 4. At passage 4 at least 3.5×10^7 cells are obtained. In further passages, cells start to age, and between passages 14 and 18, cells stop growing. If HUVECs are going to be plated on glass coverslips, for immunofluorescence experiments for example, the coverslips are incubated with 10 μg/ml fibronectin for a minimum of 6 h or preferably overnight at 37° before adding the cells. HUVECs grow better on plastic than on glass.

To analyze leukocyte TEM, it is necessary to generate a confluent monolayer with intact intercellular junctional complexes. Time-course analyses of adherens junction formation in our laboratory have shown that HUVECs require at least 24 h at confluency to generate mature adherens junctions (data not shown). Optimal junctional integrity is obtained after 2–3 days at confluency with daily replacement of medium.

Induction of Endothelial Adhesion Receptors by TNF-α. Some endothelial receptors involved in leukocyte TEM such as JAMs and PECAM-1 are constitutively expressed in HUVECs (Muller, 2003). However, those involved in the early steps of leukocyte–endothelial interactions are induced by proinflammatory stimuli such as TNF-α. The kinetics of expression varies depending on the receptor, and, therefore, the length of stimulation must be chosen depending on the molecule being analyzed. For example, expression of E-selectin, which is involved in leukocyte rolling (Tedder *et al.,* 1995), peaks between 2 and 4 h and decreases dramatically by 16 h after TNF-α stimulation (data not shown). On the other hand, ICAM-1 is highly expressed at the cell surface between 4 and 16 h after TNF-α stimulation. At earlier time points (1–2 h), ICAM-1 is expressed but most is localized intracellularly (data not shown).

Starvation of HUVECs. EGM-2 media contains several growth factors to enhance endothelial cell growth with a minimal requirement for serum. To get reproducible induction of adhesion molecules after TNF-α stimulation, it is preferable to stimulate cells in EBM-2 medium with no growth factor addition. Typically, a confluent monolayer of HUVECs is starved for 4 h in EBM-2 containing 1% FCS (starving medium) at 37°. Cells can be starved for up to 24 h, but for longer times, EBM-2 containing 2% FCS is used. If the aim is to reduce ICAM-1 levels to the minimum before the stimulation, at least 24 h of starvation is necessary. Cells are then stimulated with 10 ng/ml TNF-α in starving medium. Concentrations of 1–100 ng/ml TNF-α are used in the literature, but 10 ng/ml is sufficient to induce endothelial adhesion receptors for TEM.

Morphological Changes Induced by TNF-α. HUVECs change their morphology dramatically after prolonged TNF-α stimulation. TNF-α induces some stress fiber formation within the first minutes of stimulation (Wojciak-Stothard *et al.,* 1998 and data not shown), and over several hours, endothelial cells switch from a cobblestone morphology to an elongated shape. These morphological changes reflect molecular alterations in cell–cell and cell–substrate adhesive complexes, and this must be taken into account in studies of leukocyte TEM. We have monitored changes in cell shape in response to TNF-α by time-lapse microscopy and find that at least 6 h of stimulation is necessary to induce significant cell elongation (Fig. 1), although at 24 h of stimulation, shape changes are much more dramatic.

Analysis of ICAM-1–Induced RhoA Activation

Leukocytes induce signaling in endothelial cells on binding to the apical (luminal) endothelial surface. Leukocyte adhesion has been reported

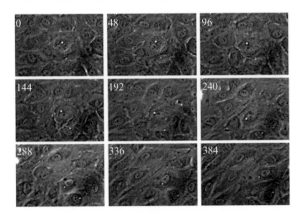

FIG. 1. Analysis of endothelial cell shape changes on TNF-α stimulation in HUVECs by time-lapse digital microscopy. HUVECs were starved for 4 h, stimulated with 10 ng/ml of TNF-α in starving medium, and observed by time-lapse phase-contrast microscopy; images were taken at 4-min intervals. Time of stimulation is shown in minutes. Early changes in cell shape started to be detected 240 min after stimulation, whereas at 384 min (>6 h) HUVECs appeared significantly elongated.

to induce transient endothelial intracellular calcium release, stress fiber formation, and contraction by a RhoA/ROCK-dependent pathway (Ehringer *et al.*, 1998; Lorenzon *et al.*, 1998). Analysis of signal transduction pathways activated by individual endothelial receptors has been carried out either by using function-blocking antibodies or by crosslinking each receptor with antibodies. ICAM-1, in particular, has been shown to be an endothelial receptor essential not only for leukocyte adhesion but also for TEM (Barreiro *et al.*, 2002; Lyck *et al.*, 2003). Blocking antibodies against ICAM-1 abrogate leukocyte-induced calcium release (Pfau *et al.*, 1995) and, in conjunction with anti-ICAM-2 antibodies, alter the pattern of leukocyte locomotion before TEM (Schenkel *et al.*, 2004). ICAM-1 engagement induces signaling by means of RhoA/ROCK and stress fiber assembly (Thompson *et al.*, 2002).

Materials

Mouse anti-ICAM-1 antibody (clone BBIG-I1, number BBA4) used for crosslinking, immunofluorescence and function blocking is from R&D Systems. Rabbit anti-ICAM-1 antibody (H-1080; ref. sc-7891) used for Western blotting is from Santa Cruz Biotechnology. Secondary antibody goat anti-mouse antibodies (ref. 115-005-003) and goat anti-mouse antibodies conjugated to different fluorophores are from Jackson Immunoresearch.

Methods

Anti-ICAM-1 antibodies can be used to analyze signaling pathways induced by ICAM-1 ligation. A detailed protocol is described as follows:

1. HUVECs are grown, starved, and stimulated for 18–24 h with 10 ng/ml TNF-α in starving medium, as described previously, to induce ICAM-1 expression.

2. During the TNF-α stimulation, a flask with enough starving medium to perform the antibody crosslinking is prewarmed in the cell incubator at 37° and 5% CO_2. The minimum volume required is 5 volumes of the medium in each dish.

3. The anti-ICAM-1–specific antibody is added at between 1 and 10 μg/ml to the cells without changing the medium, and the cells are incubated at 37° and 5% CO_2 for 60 min.

4. The cells are washed twice to remove excess anti-ICAM-1 antibody with one volume of the prewarmed medium. During the washing, the dishes are not left without medium for longer that a few seconds, and the medium is added gently on the wall of the dish. The way the cells are rinsed in this experiment is crucial, because endothelial cells are very sensitive to osmotic or shear stress, and careless washes may induce stress fiber–mediated cell contraction.

5. Anti-mouse secondary antibody (1–10 μg/ml) is added in one volume of prewarmed starving medium for between 30 and 60 min at 37° and 5% CO_2. This is the time frame when an increase of GTP-loaded RhoA and/or stress fiber formation is observed. The cells are rinsed twice with one volume of preequilibrated medium and then fixed to detect F-actin with fluorophore-conjugated phalloidin (Ridley, 1995) or harvested to perform biochemical analysis of Rho-GTP levels with GST-RBD in pull-down assays as described elsewhere (Ren and Schwartz, 2000; Thompson *et al.*, 2002).

6. Routinely, controls in this experiment are cells with no primary antibody, and primary antibody or secondary antibody alone, at each time point.

Analysis of Rho GTPase Function in TNF-α–Stimulated HUVECs

Materials

AD-293 cells (Stratagene, La Jolla, CA) are cultured in DMEM containing 10% FCS at 37°C, 5% CO_2, PD-10 columns are from Amersham Biosciences (Amersham, UK), caesium chloride from BD Biosciences (Oxford, UK), Microsol Solution and Wipes from Anachem Ltd. (Luton,

UK) and AdenoX Rapid Titre Kit from BD Clontech (Mountain View, CA). C3 transferase is purified from *E. coli* as previously described (Dillon and Feig, 1995).

Methods

Treatment of HUVECs with C3 Transferase. C3 transferase, an exoenzyme from *Clostridium botulinum*, inactivates the closely related RhoA, RhoB, and RhoC proteins by ADP-ribosylation at amino acid 41 (Aktories and Barbieri, 2005). C3 transferase can be introduced into single cells by microinjection, and this affects the adhesion of monocytes to HUVECs (Wojciak-Stothard *et al.*, 1999).

Alternately, C3 transferase can be added directly to HUVECs. Purified recombinant C3 transferase is diluted in starving medium to a final concentration of 15 μg/ml. Cells are stimulated with TNF-α for 1 h then incubated with this C3 transferase solution for 3 h at 37°. This method allows the effect of Rho inhibition to be studied, for example, on monocyte adhesion (Wojciak-Stothard *et al.*, 1999).

Preparation of High-titer Purified Adenoviruses Expressing Rho GTPase Mutants. Dominant negative forms of Rho proteins are created by a single amino acid substitution, most frequently from Thr to Asn at amino acid 17 in Rac and Cdc42 or at position 19 in RhoA (Feig, 1999; Ridley, 2001). These mutants compete for exchange factors with their respective endogenous proteins, inhibiting their function. Constitutively active mutants of Rho proteins can be obtained by a single amino acid mutation from Gly to Val at position 12 (Rac, Cdc42) or 14 (RhoA) (Diekmann *et al.*, 1991; Paterson *et al.*, 1990). To overcome the low efficiency of transfection in primary cells and in HUVECs in particular, we and others have generated adenoviruses to express some of these Rho GTPase mutants.

All adenoviruses used have their E1 and E3 genes deleted and cannot propagate unless they are grown in AD-293 human embryonic kidney cell lines, which stably express the adenoviral E1A protein, thereby allowing viral replication (He *et al.*, 1998). Adenoviruses in our laboratory were generated by subcloning cDNAs coding for myc-tagged N19RhoA, N17Cdc42, N17Rac1, and V12Rac1 into the admid transfer vectors pCR259 and pCR244 (Wojciak-Stothard *et al.*, 1999). The admid transfer vectors were transformed into an *E. coli* strain containing a vector encoding a transposase and an Ad5-based adenovirus vector with the E1 and E3 genes deleted (Ad5 ΔE1AE3). Transposition of the Rho cDNAs from the admid vector into Ad5 ΔE1AE3 created the recombinant vectors coding for Rho GTPase mutants under the control a CMV promoter. The method for purifying adenoviruses is described here.

1. An archive of virus containing crude viral lysate is diluted in 9 ml serum-free DMEM and used to infect a 75-cm^2 flask of confluent 293 cells for 1 h, then 0.5 ml of FCS is added. After 3 days, viral infection will cause the cells to detach and undergo cell lysis and full cytopathic effect is observed (CPE). The medium is removed and frozen/thawed twice to ensure a complete release of viral particles. This is then used to amplify the viral preparation.

2. The medium with the viral particles is diluted into 200 ml of serum-free DMEM and used to infect 10 162-cm^2 flasks of 90% confluent 293 cells for 1 h; 1 ml of FCS is added, and cells are incubated until they begin to detach (36–48 h). Cells are then tapped off the flask, and the cell suspension is collected. This cell suspension is centrifuged at 250g for 5 min; all cells pooled and centrifuged again for 5 min. Cells are resuspended in 0.1 M Tris, pH 8.0, and frozen/thawed three times before being passed through a blunt 19-gauge needle three to four times to shear chromatin. Finally, lysed cells are centrifuged at 500g for 5 min.

3. This supernatant is saved and made up to 11.4 ml with 0.1 M Tris solution; 6.6 ml of saturated CsCl$_2$ is added, and the solution is carefully placed into two ultracentrifuge tubes and centrifuged for 16–18 h at 120,000g at 4°. The band containing the virus is drawn off and placed into a new ultracentrifuge tube. This tube is topped up with 1.34 g/ml CsCl$_2$ in PBS containing 0.01% (w/v) calcium and 0.1% (w/v) magnesium (PBS2$^+$) and centrifuged again overnight at 120,000g at 4°. The band containing the virus particles is recovered from the ultracentrifuge tube in a volume less than 2.5 ml. CsCl$_2$ is then removed by passing through a PD10 column previously equilibrated in PBS2$^+$. The adenovirus particles are eluted with a further addition of 3.5 ml of PBS2$^+$ and collected into a bijou tube containing 0.5 ml of glycerol, mixed, and filtered with a sterile 0.45-μm filter. Finally, the purified virus is aliquoted and stored at −70°.

4. To ensure that the recombinant adenoviruses are not replication-competent, 10 μl of purified virus is used to infect one 25-cm^2 flask of slow-growing adherent cells such as human skin fibroblasts or HUVECs. The virus is left to infect the cells for several days, and the cells are monitored for any cytopathic effects.

Titration of Adenovirus. We routinely use the Adeno-X Rapid titer kits to titer viruses (BD Clontech) and follow the manufacturer's instructions. AD-293 cells are infected with a succession of serial dilutions of purified adenoviral stock. After an incubation period, which allows the virally infected cells to produce hexon proteins, the cells are fixed and treated with an anti-hexon protein-specific antibody. A secondary, HRP-conjugate antibody is then added and allowed to develop with DAB substrate. Any positive cells will turn brown/black and can be counted on a microscope using a 20× objective. The infectious units per milliliter (ifu/ml) can then be calculated as the average number of positive cells per dilution.

Infection of HUVECs. More than 80% of HUVECs can be infected using adenoviruses at a multiplicity of infection (m.o.i.) of 500 (500 ifu/cell). The viral stock is diluted in EGM-2 medium and incubated with HUVECs for 90 min at 37°. The volume of medium must be reduced during the viral incubation as much as possible without compromising cell viability. Cells are then washed once and incubated with starving medium containing TNF-α for 18 h as described previously. Different assays, such as the transmigration assays described in the following, can then be performed. Infection efficiency is monitored by immunofluorescence with an anti-myc antibody to detect myc-tagged Rho GTPases and staining with F-actin to follow changes in the actin cytoskeleton caused by expression of the Rho mutant (Fig. 2).

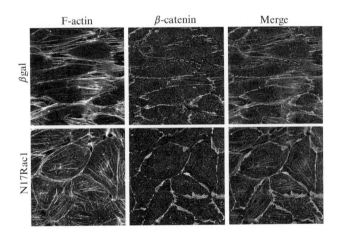

FIG. 2. Effect of adenoviral N17Rac1 on HUVECs. HUVECs were incubated with an adenovirus coding for β-galactosidase (βgal) or N17Rac1 (500 m.o.i.) for 90 min in EGM-2 medium at 37°. Cells were washed with PBS and stimulated with 10 ng/ml of TNF-α in starving medium for 18 h. Cells were then fixed with 4% paraformaldehyde, permeabilized with 0.2% TX-100 in PBS, and stained with TRITC-conjugated phalloidin for F-actin detection and anti-β-catenin rabbit antibody followed by a FITC-conjugated secondary anti-rabbit antibody and analyzed by confocal microscopy. Uninfected cells (not shown) or βgal-expressing cells showed bundles of long stress fibers running almost parallel to the axes of the elongated cells, ending at the intercellular junctions (β-catenin staining). However, N17Rac1-expressing cells are not elongated, and their stress fibers are short and not connected to the intercellular junctions. (See color insert.)

Analysis of Leukocyte Transendothelial Migration

Different strategies are used to analyze leukocyte TEM. One of the most popular is transmigration through filters of a particular pore size that retains the endothelial cells but allows the passage of smaller leukocytes (Millan et al., 2002). This method is suitable for the study of leukocyte motility in cytokine gradients or for long-term analysis of endothelial processes, because cells transmigrated across the filter are detected in the bottom chamber after a few hours. However, to study endothelial remodeling during TEM and leukocyte parameters such as cell migration speed and preferred sites of TEM, TEM is monitored in live cells by time-lapse microscopy. This enables analysis of leukocyte TEM under flow conditions if a parallel-plate flow chamber is fitted into the microscope (Cinamon and Alon, 2004).

Materials

Monocytes are obtained by elutriation (Williams and Ridley, 2000). Human T lymphoblasts are prepared from peripheral blood mononuclear cells by treatment with 0.2% phytohemagglutinin (Amersham Pharmacia Biotech) for 48 h. Cells are washed and cultured in RPMI 1640 medium containing 10% human serum and 10 U/ml interleukin-2 (IL-2) at 37°, 5% CO_2. T lymphoblasts are typically used after culture for 7–15 days.

Methods

Time-lapse Digital Microscopy. HUVECs are grown in 35-mm dishes and starved of growth factors then stimulated with TNF-α as described previously, then they are affixed to the stage of an Axiovert inverted microscope (Zeiss) and maintained at 37° in an atmosphere containing 5% CO_2. Once dishes are on the stage, between 5×10^5 and 10^6 leukocytes are added to the HUVEC monolayer. Sequences of time-lapse images are collected through a 20× objective and projected onto KPM1E CCD cameras (Hitachi Denshi Ltd.) using a 10%/90% pellicle beam splitter (Melles Griot). The acquisition of the image data and synchronization of the illumination are controlled by Tempus Meteor software (Kinetic Imaging Ltd, Liverpool, UK), with a frame repetition rate of 1 min for monocytes and of 15–30 sec for memory T lymphoblasts over a 60-min period. During this period, transmigration is detected when phase-bright monocytes or T lymphoblasts on the apical surface became phase-dark (Fig. 3). For quantitation purposes, leukocytes are tracked using Motion Analysis software (Kinetic Imaging Ltd, Liverpool, UK) as previously described (Jones

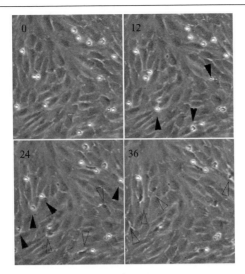

FIG. 3. Detection of monocyte transmigration in TNF-α–stimulated HUVECs by time-lapse digital microscopy. HUVECs were starved for 4 h and stimulated with 10 ng/ml of TNF-α in starving medium for 16 h. HUVECs were cocultured with 6 × 10⁵ primary human monocytes and analyzed by phase-contrast time-lapse microscopy; images were taken at intervals of 1 min. Numbers represent time of coculture in minutes. Black arrows point out cells before TEM, open arrows show the same cells after TEM. Note the change of phase brightness and shape between monocytes above and below the HUVEC monolayer.

et al., 2000). In addition to transmigration, tracking allows us to analyze other important leukocyte parameters such as apical locomotion and the subsequent basal locomotion under the endothelial monolayer, directional changes, and preferential areas of TEM.

References

Aird, W. C. (2003). Endothelial cell heterogeneity. *Crit. Care Med.* **31**, S221–S230.

Aktories, K., and Barbieri, J. T. (2005). Bacterial cytotoxins: Targeting eukaryotic switches. *Nat. Rev. Microbiol.* **3**, 397–410.

Barreiro, O., Yanez-Mo, M., Serrador, J. M., Montoya, M. C., Vicente-Manzanares, M., Tejedor, R., Furthmayr, H., and Sanchez-Madrid, F. (2002). Dynamic interaction of VCAM-1 and ICAM-1 with moesin and ezrin in a novel endothelial docking structure for adherent leukocytes. *J. Cell Biol.* **157**, 1233–1245.

Butcher, E. C. (1991). Leukocyte-endothelial cell recognition: Three (or more) steps to specificity and diversity. *Cell* **67**, 1033–1036.

Cinamon, G., and Alon, R. (2004). Real-time *in vitro* assay for studying chemoattractant-triggered leukocyte transendothelial migration under physiological flow conditions. *Methods Mol. Biol.* **239,** 233–242.

Connolly, J. O., Simpson, N., Hewlett, L., and Hall, A. (2002). Rac regulates endothelial morphogenesis and capillary assembly. *Mol. Biol. Cell* **13,** 2474–2485.

Davignon, J. (2004). Beneficial cardiovascular pleiotropic effects of statins. *Circulation* **109,** III39–III43.

Davignon, J., and Ganz, P. (2004). Role of endothelial dysfunction in atherosclerosis. *Circulation* **109,** III27–III32.

Diekmann, D., Brill, S., Garrett, M. D., Totty, N., Hsuan, J., Monfries, C., Hall, C., Lim, L., and Hall, A. (1991). Bcr encodes a GTPase-activating protein for p21rac. *Nature* **351,** 400–402.

Dillon, S. T., and Feig, L. A. (1995). Purification and assay of recombinant C3 transferase. *Methods Enzymol.* **256,** 174–184.

Ehringer, W. D., Edwards, M. J., Wintergerst, K. A., Cox, A., and Miller, F. N. (1998). An increase in endothelial intracellular calcium and F-actin precedes the extravasation of interleukin-2-activated lymphocytes. *Microcirculation* **5,** 71–80.

Feig, L. A. (1999). Tools of the trade: Use of dominant-inhibitory mutants of Ras-family GTPases. *Nat. Cell Biol.* **1,** E25–E27.

Greenwood, J., Walters, C. E., Pryce, G., Kanuga, N., Beraud, E., Baker, D., and Adamson, P. (2003). Lovastatin inhibits brain endothelial cell Rho-mediated lymphocyte migration and attenuates experimental autoimmune encephalomyelitis. *FASEB J.* **17,** 905–907.

He, T. C., Zhou, S., da Costa, L. T., Yu, J., Kinzler, K. W., and Vogelstein, B. (1998). A simplified system for generating recombinant adenoviruses. *Proc. Natl. Acad. Sci. USA* **95,** 2509–2514.

Jain, R. K. (2003). Molecular regulation of vessel maturation. *Nat. Med.* **9,** 685–693.

Jones, G. E., Ridley, A. J., and Zicha, D. (2000). Rho GTPases and cell migration: Measurement of macrophage chemotaxis. *Methods Enzymol.* **325,** 449–462.

Larrivee, B., and Karsan, A. (2005). Isolation and culture of primary endothelial cells. *Methods Mol. Biol.* **290,** 315–329.

Lorenzon, P., Vecile, E., Nardon, E., Ferrero, E., Harlan, J. M., Tedesco, F., and Dobrina, A. (1998). Endothelial cell E- and P-selectin and vascular cell adhesion molecule-1 function as signaling receptors. *J. Cell Biol.* **142,** 1381–1391.

Lyck, R., Reiss, Y., Gerwin, N., Greenwood, J., Adamson, P., and Engelhardt, B. (2003). T-cell interaction with ICAM-1/ICAM-2 double-deficient brain endothelium *in vitro*: The cytoplasmic tail of endothelial ICAM-1 is necessary for transendothelial migration of T cells. *Blood* **102,** 3675–3683.

Mettouchi, A., Klein, S., Guo, W., Lopez-Lago, M., Lemichez, E., Westwick, J. K., and Giancotti, F. G. (2001). Integrin-specific activation of Rac controls progression through the G(1) phase of the cell cycle. *Mol. Cell* **8,** 115–127.

Millan, J., and Ridley, A. J. (2005). Rho GTPases and leucocyte-induced endothelial remodelling. *Biochem. J.* **385,** 329–337.

Millan, J., Montoya, M. C., Sancho, D., Sanchez-Madrid, F., and Alonso, M. A. (2002). Lipid rafts mediate biosynthetic transport to the T lymphocyte uropod subdomain and are necessary for uropod integrity and function. *Blood* **99,** 978–984.

Muller, W. A. (2003). Leukocyte-endothelial-cell interactions in leukocyte transmigration and the inflammatory response. *Trends Immunol.* **24,** 327–334.

Paterson, H. F., Self, A. J., Garrett, M. D., Just, I., Aktories, K., and Hall, A. (1990). Microinjection of recombinant p21rho induces rapid changes in cell morphology. *J. Cell Biol.* **111,** 1001–1007.

Pfau, S., Leitenberg, D., Rinder, H., Smith, B. R., Pardi, R., and Bender, J. R. (1995). Lymphocyte adhesion-dependent calcium signaling in human endothelial cells. *J. Cell Biol.* **128,** 969–978.

Ren, X. D., and Schwartz, M. A. (2000). Determination of GTP loading on Rho. *Methods Enzymol.* **325,** 264–272.

Ridley, A. J. (1995). Growth factor-induced actin reorganization in Swiss 3T3 cells. *Methods Enzymol.* **256,** 306–313.

Ridley, A. J. (2001). Rho proteins: Linking signaling with membrane trafficking. *Traffic* **2,** 303–310.

Schenkel, A. R., Mamdouh, Z., and Muller, W. A. (2004). Locomotion of monocytes on endothelium is a critical step during extravasation. *Nat. Immunol.* **5,** 393–400.

Shimizu, Y., Newman, W., Gopal, T. V., Horgan, K. J., Graber, N., Beall, L. D., van Seventer, G. A., and Shaw, S. (1991). Four molecular pathways of T cell adhesion to endothelial cells: Roles of LFA-1, VCAM-1, and ELAM-1 and changes in pathway hierarchy under different activation conditions. *J. Cell Biol.* **113,** 1203–1212.

Tedder, T. F., Steeber, D. A., Chen, A., and Engel, P. (1995). The selectins: Vascular adhesion molecules. *FASEB J.* **9,** 866–873.

Thompson, P. W., Randi, A. M., and Ridley, A. J. (2002). Intercellular adhesion molecule (ICAM)-1, but not ICAM-2, activates RhoA and stimulates c-fos and rhoA transcription in endothelial cells. *J. Immunol.* **169,** 1007–1013.

Van Wetering, S., van den Berk, N., van Buul, J. D., Mul, F. P., Lommerse, I., Mous, R., ten Klooster, J. P., Zwaginga, J. J., and Hordijk, P. L. (2003). VCAM-1-mediated Rac signaling controls endothelial cell-cell contacts and leukocyte transmigration. *Am. J. Physiol. Cell Physiol.* **285,** C343–C352.

Williams, L. M., and Ridley, A. J. (2000). Lipopolysaccharide induces actin reorganization and tyrosine phosphorylation of Pyk2 and paxillin in monocytes and macrophages. *J. Immunol.* **164,** 2028–2036.

Wojciak-Stothard, B., Entwistle, A., Garg, R., and Ridley, A. J. (1998). Regulation of TNF-alpha-induced reorganization of the actin cytoskeleton and cell-cell junctions by Rho, Rac, and Cdc42 in human endothelial cells. *J. Cell Physiol.* **176,** 150–165.

Wojciak-Stothard, B., and Ridley, A. J. (2002). Rho GTPases and the regulation of endothelial permeability. *Vasc. Pharmacol.* **39,** 187–199.

Wojciak-Stothard, B., Williams, L., and Ridley, A. J. (1999). Monocyte adhesion and spreading on human endothelial cells is dependent on Rho-regulated receptor clustering. *J. Cell Biol.* **145,** 1293–1307.

[51] Analysis of a Mitotic Role of Cdc42

By SHINGO YASUDA and SHUH NARUMIYA

Abstract

Rho GTPases including Rho, Rac, and Cdc42 determine the cell shape by regulating the actin and microtubule dynamics. Through these actions on cytoskeleton, Rho GTPases also regulate cell cycle progression. Specifically, Rho, Rac, and Cdc42 regulate G1-S progression, and Rho regulates cytokinesis. However, involvement of these GTPases in nuclear division has not been definitely shown. This seems to be due to the lack of standard procedures examining mitosis-specific functions of these GTPases. Recently, we have found a mitosis-specific role of Cdc42 by enrichment of cells in mitosis by cell cycle synchronization and interfering with functions of Rho GTPases. This chapter describes the procedures we used in these studies.

Introduction

Rho GTPases, including Rho, Rac, and Cdc42, regulate reorganization of the actin cytoskeleton and modulate local dynamics of microtubules to generate cell shapes and to contribute to cell functions such as migration and proliferation (Etienne-Manneville and Hall, 2002). Although functions of these GTPases in interphase cells have been extensively studied, little is known about their roles in cell division except that of Rho in cytokinesis (Narumiya *et al.*, 2004). This seems to be due to the lack of standard procedures examining mitosis-specific functions of these GTPases. Recently, we found a mitosis-specific role of Cdc42 (Oceguera-Yanez *et al.*, 2005; Yasuda *et al.*, 2004) by enrichment of cells in mitosis by cell cycle synchronization and interfering with functions of Rho GTPases by three different methods. The methods we used for functional interference include treatment with *Clostridium difficile* toxin B, expression of dominant active or dominant negative Rho GTPase mutants, and depletion of endogenous Rho GTPases by RNA interference. Effects of such interference on mitosis were examined by immunofluorescence of fixed cells and live cell imaging of mitosis. We describe in the following the procedures we used in these studies.

METHODS IN ENZYMOLOGY, VOL. 406
0076-6879/06 $35.00
DOI: 10.1016/S0076-6879(06)06051-4

Mitosis-Specific Treatment of HeLa Cells with
   Clostridium difficile Toxin B

   Clostridium difficile toxin B glycosylates a critical threonine residue of
Rho GTPases, Thr37 of Rho and Thr35 of Rac and Cdc42, and thereby
inhibits their functions (Just *et al.*, 1995). Because botulinum C3 exoenzyme
specifically inactivates the Rho subfamily of GTPases such as RhoA, B,
and C, this toxin, used in combination with C3 exoenzyme, is an appropriate tool to analyze functions of Rho GTPases other than the Rho subfamily. However, when added to randomly growing cells, toxin B induces
growth arrest at various points of cell cycle. For analysis of the mitotic role
of Rho GTPases by toxin B, cells, therefore, need to be enriched in mitosis.

Procedure

   HeLa S3 cells are plated onto a 10-cm dish at a density of 1×10^6 cells
per dish and maintained in the DMEM (Invitrogen) supplemented with
10% fetal calf serum (FCS) and antibiotics at 37° with an atmosphere
containing 10% CO_2. After overnight culture, the cells are subjected to
double thymidine block (Bostock *et al.*, 1971) (Fig. 1). The medium is
replaced with DMEM containing 10 mM thymidine and 5% FCS, and the
cells are cultured for 24 h. After the thymidine-containing medium is
removed, the cells are washed three times with phosphate-buffered saline
without Ca^{++} and Mg^{++} ions (PBS(-)). The cells are further incubated in
DMEM supplemented with 10% FCS for 10 h, and then subjected to the
second thymidine block for 14 h. After thymidine-containing medium is
removed, the cells are washed three times with PBS(-) and cultured in the
DMEM containing 10% FCS. After 8 h, nocodazole is added to the
medium at the final concentration of 40 ng/ml, and the culture is continued
for another 6 h. Round mitotic cells are then purified by a shaken-off
procedure. Collected cells are suspended in DMEM containing 10% FCS,
40 ng/ml nocodazole, and toxin B (10 ng/ml) for 2 h in a 50-ml Falcon tube
in a CO_2 incubator. The cells are collected by centrifugation, washed with
PBS(-) three times, and suspended in DMEM containing 10% FCS to be
released from the nocodazole arrest and removed of toxin B. The cell are
plated on a coverglass (12 × 12 mm) coated with poly-L-lysine in a 6-well
plate, and cells are fixed at 45, 60, 90, and 180 min for prometaphase,
metaphase, telophase, and G1 phase, respectively (Kimura *et al.*, 2000),
with cold methanol chilled at −20° for 5 min. For Mad2 staining, cells are
permeabilized by 0.005% digitonin in a solution containing 110 mM potassium acetate, 20 mM HEPES (pH 7.3), 2 mM magnesium acetate, 0.5 mM
EGTA, 2 mM dithiothreitol, and the protease inhibitor complete mimi
(Roche) for 5 min at room temperature before fixation by cold methanol.

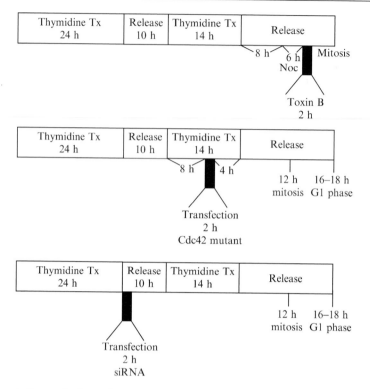

FIG. 1. Protocols for cell cycle synchronization and treatment. Protocols of cell synchronization used for toxin B treatment (upper), expression of Rho GTPase mutants (middle), and RNA interference (lower) are shown.

Mitosis-Specific Expression of Rho GTPase Mutants

Rho GTPase mutants (dominant active mutants: V12-Cdc42, V12-Rac1, V14-RhoA; dominant negative mutants: N17-Cdc42, N17-Rac1, N19-RhoA) are used (Yasuda *et al.*, 2004). HeLa S3 cells are seeded on a coverglass in a 35-mm dish at a density of 1×10^5 cells per dish and are subjected to double thymidine block after overnight culture as described previously (Fig. 1). From 8–10 h after the beginning of the second thymidine block, HeLa cells are transfected with pEGFP plasmid harboring each Rho GTPase mutant using Lipofectamine 2000 (Invitrogen). A total of 1 μg DNA is dissolved in 100 μl Opti-MEM (Invitrogen) containing 10 mM thymidine. In parallel, 3 μl of Lipofectamine 2000 is diluted in a final volume of 100 μl of thymidine-containing Opti-MEM and incubated for 5 min at room temperature. After 5 min incubation, the DNA solution is mixed gently with the diluted solution of Lipofectamine 2000 (a total volume, 200 μl) and incubated for 15 min at room temperature. The mixture is added to the cells in a 35-mm dish that have been washed and incubated in 1.8 ml thymidine-containing Opti-MEM. The cells are left for 2 h at 37° in a CO_2 incubator. The medium is then aspirated, and the cells are cultured in DMEM containing 10 mM thymidine and 5% FCS. After 4 h, the thymidine-containing DMEM is replaced with DMEM containing 10% FCS as described previously. According to this procedure, most cells enter mitosis 12 h after thymidine removal and enter G1 phase after another 4 h. The cells are fixed by 3.7% formaldehyde and subjected to immunofluorescence. For Mad2 staining, cells are treated and fixed as described previously.

Knockdown of Rho GTPase Expression with RNA Interference

Plasmids and siRNA Preparation

The pGEX-Cdc42, pGEX-Rac1, and pGEX-RhoA plasmids were described previously (Watanabe *et al.*, 1999). Complementary DNAs for Cdc42, Rac1, and RhoA are amplified by polymerase chain reaction using the primers listed below from pGEX-Cdc42, pGEX-Rac1, and pGEX-RhoA, respectively, and subcloned into pEGFP vector to produce pEGFP-Cdc42, pEGFP-Rac1, and pEGFP-RhoA, respectively.

    Cdc42; (M57298)
    forward primer ATGCAGACAATTAAGTGTGTTGTTGTGGGC-
      GA
    reverse primer TCATAGCAGCACACACCTGCGGCTCTTCTT
    RhoA; (L25080)
    forward primer ATGGCTGCCATCCGGAAGA
    reverse primer TCACAAGACAAGGCAACCAG
    Rac1; (NM_006908)
    forward primer ATGCAGGCCATCAAGTGTGTGG
    reverse primer TTACAACAGCAGGCATTTTCTC

To generate siRNA against Cdc42, Rac1, and RhoA, we used the BLOCK-iT RNAi-TOPO Transcription and the BLOCK-iT Dicer RNAi kits (Invitrogen). The full coding sequences of Rho GTPases are amplified by using the preceding primers. The PCR fragments obtained are ligated to the T7 promoter. Using a combination of T7 primer and the gene-specific forward or reverse primer, T7-linked PCR products are amplified to produce sense and antisense DNA templates. We use the sense and antisense DNA templates in an *in vitro* transcription reaction to produce sense and antisense RNA transcripts, respectively. Sense and antisense RNA transcripts are purified by using the RNA purification regents supplied in the kit and annealed to form double-strand RNA. The double-strand RNA obtained are cleaved into 21–23 nucleotide fragments with Dicer enzyme and purified by using the RNA purification regents supplied in the kit. As a control, we use siRNA for the *E. coli* LacZ gene as provided by the manufacturer.

Cell Culture, Synchronization, and Transfection

HeLa S3 cells are seeded at a density of 2×10^5 per dish onto 35-mm dishes and are maintained in DMEM supplemented with 10% FCS and antibiotics at 37° with 10% CO_2. After overnight culture, the cells are subjected to double thymidine block as described (Fig. 1C.) Just after the first thymidine block, the cells are transfected with appropriate siRNA for Rho GTPases using Lipofectamine 2000 (Invitrogen). A total of 0.3 μg of siRNA together with 0.5 μg of pEGFP-histoneH2BK (Yasuda *et al.*, 2004) is dissolved in a final volume of 100 μl of Opti-MEM (Invitrogen). In parallel, 2.4 μl of Lipofectamine 2000 is diluted in a final volume of 100 μl of Opti-MEM and incubated for 5 min at room temperature. After 5 min incubation, the DNA solution is mixed with the diluted Lipofectamine 2000 (a total volume, 200 μl) and incubated for 15 min at room temperature. The mixture is added to the cells in a 35-mm dish that have been washed and incubated in 1.8 ml Opti-MEM. The cells are left for 2 h at 37° in an atmosphere containing 10% CO_2. After 2 h incubation, the medium is aspirated and replaced with DMEM containing 10% FCS. The cells are further incubated in DMEM supplemented with 10% FCS for 8 h and then subjected to the second thymidine block for 14 h. After thymidine-containing medium is removed, the cells are washed three times with PBS(-) and cultured in the DMEM containing 10% FCS. Most cells enter mitosis 12 h after thymidine removal and enter interphase (G1 phase) after another 4 h. The cells are fixed with 3.7% formaldehyde and subjected to immunofluorescence.

Characterization of Knockdown of Cdc42

HeLa cells are cotransfected with siRNA for LacZ (control), Cdc42, Rac1, or RhoA, and pEGFP. The cells are collected 24 h or 48 h after transfection and subjected to immunoblotting by using antibody to Cdc42, Rac1, or RhoA, respectively (Fig. 2). RNAi for Cdc42, Rac1, or RhoA effectively suppresses expression of respective endogenous protein to <10% of that of control cells. On the contrary, such depletion is not observed in cells subjected to RNAi for LacZ.

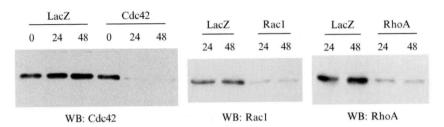

FIG. 2. Depletion of endogenous Cdc42, Rac1, and RhoA by RNAi. HeLa cells are transfected with siRNA for respective Rho GTPase or LacZ. After 24 and 48 h, the cells are lysed and subjected to immunoblot with the respective antibody (Cdc42, Rac1, and RhoA).

Immunofluorescence

After blocking with PBS(-) containing 1% BSA, the fixed cells are incubated for 90 min at room temperature in PBS(-) with primary antibodies to EB1 (1:200, mouse, Transduction Laboratories), Mad2 (1:100, rabbit, Covance), or CREST serum (1:1000, human) to visualize spindle microtubules, Mad2 or CENP-A, respectively. An antibody to β-tubulin (1:200, mouse, Sigma) can be used for microtubule staining instead of that to EB1. After washing with PBST (PBS(-) supplemented with 0.1% Tween20) three times, the samples are incubated for 30 min at room temperature in PBS(-) with secondary antibodies to mouse (1:500, Alexa 488 or 647, Molecular Probe), rabbit (1:500, Alexa 488 or 647, Molecular Probe), and human (1:500, Cy3, Jackson Lab.) IgG. DNA is stained with TOPRO-3 (Molecular Probe) or Hoechst 33342 during the secondary antibody incubation. Cells are examined with a Zeiss LSM510 confocal imaging system or a Leica MDW microscope system.

Effect of Toxin B Treatment and Expression of Rho GTPase Mutants on Mitosis

An example for mitotic cells treated with toxin B or those expressing N17Cdc42 is shown in Fig. 3A. The cells are fixed 45–90 min after the nocodazole release in the former, and the cells are fixed 12 h after the release from the second thymidine block in the latter. DNA staining shows that chromosomes are distributed widely in the cytoplasm and do not congress at the metaphase plate, providing a picture of the prometaphase arrest. As previously reported (Yasuda et al., 2004), some cells treated with toxin B or expressing N17 Cdc42 exit to G1 phase without distinct anaphase. An example of these cells is shown in Fig. 3B. DNA staining illustrates that

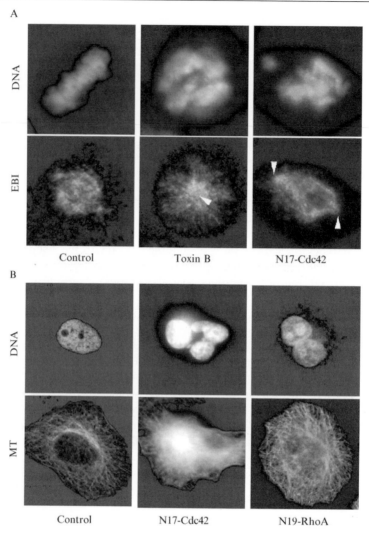

FIG. 3. Immunofluorescence of cells treated with toxin B or expressing Rho GTPase mutants. (A) Misaligned chromosome induced by toxin B treatment (middle) and N17-Cdc42 expression (right). HeLa cells are stained by anti-EB1 antibody for mitotic spindle and by TOPRO-3 for DNA. Misaligned chromosomes are seen in a toxin B–treated cell and a N17-Cdc42–expressing cell. Centrosomes are indicated by arrowheads. (B) Nuclear phenotypes of cells expressing N17-Cdc42 (middle) and N19-RhoA (right). HeLa cells are stained by anti-β-tubulin antibody for microtubules and by TOPRO-3 for DNA. The cells expressing N17-Cdc42 exhibit tetraploidy with micronuclei and macronuclei as a consequence of impaired chromosome segregation, whereas the cells expressing N19-RhoA show the binucleate phenotype.

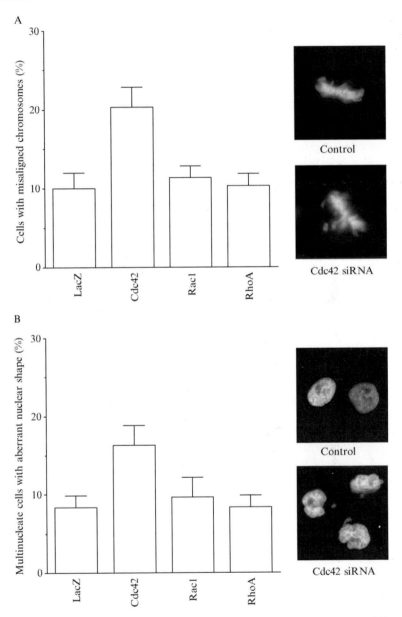

FIG. 4. Effect of RNAi for Cdc42 on mitosis. Misaligned chromosomes (A) and multinucleated cells with aberrant nuclear shape (B) induced by RNAi for Cdc42. HeLa cells are transfected with RNAi for LacZ (control) or respective Rho GTPase together with pEGFP–HistoneH2BK after the first thymidine block; 12 h (A) and 16 h (B) after the second

cells expressing N17-Cdc42 exhibit tetraploidy with micronuclei and macro-nuclei, whereas the cells expressing N19RhoA show the binucleate pheno-type, having two nuclei of an equal size.

Effect of Rho GTPases Knockdown on Mitosis

In RNAi experiments, we cotransfect GFP-Histone H2BK to visualize nuclear morphology and to identify transfection positive cells and monitor chromosome segregation. Most cells enter mitosis 12 h after thymidine removal. Only cells treated with siRNA for Cdc42 exhibit misaligned chromosome (Fig. 4A). On quantitative analysis, the percentage of cells showing misaligned chromosome is significantly high in the Cdc42-depleted cell population compared with the control cell population or cells depleted of either Rac 1 or RhoA. In the protocol described previously, most cells enter interphase16 h after thymidine removal. At this time, a significantly higher number of multinucleate cells containing macronuclei and micro-nuclei and nuclear bridges is found in the population of cells subjected to RNAi for Cdc42 than those subjected to control RNAi or RNAi for Rac 1 or RhoA (Fig. 4B).

Time-Lapse Live Imaging

HeLa S3 cells are plated onto a 35-mm glass bottom dish at a density of 2×10^5 per dish and maintained in the DMEM supplemented with 10% FCS and antibiotics at 37° with an atmosphere containing 10% CO_2. After overnight culture, the cells are transfected with 0.25 μg of pEGFP-EB1 and 0.25 μg of pDsRed2-histone H2Bk, together with 0.3 μg of siRNA for appropriate Rho GTPases using Lipofectamine 2000 as described previ-ously. Forty-eight hours after transfection, the dish is placed on a tempera-ture-controlled stage in a chamber maintained at 37° with an atmosphere containing 10% CO_2. Time-lapse live imaging is performed at an appro-priate interval by an inverted microscopy (model DMIRE2: Leica) with $63\times$ NA 1.3 lenses equipped with a xenon lamp. For an example of time-lapse live imaging experiment, see Yasuda et al. (2004) or Oceguera-Yanez et al. (2005).

---

thymidine block, cells are fixed by 3.7% formaldehyde and subjected to immunofluorescence. Graph shows the percentage of cells with misaligned chromosome (A) or multinucleated cells with aberrant nuclear shape (B) among cells expressing GFP-HistoneH2BK. Data are means $\pm$ S.E.M. of 100 cells for each experiment ($n = 3$). The right panel (A,B) shows control RNAi (upper) and Cdc42 RNAi (lower) cells. Cdc42 RNAi cell exhibits misaligned chromosome (A) and micronuclei (B).

References

Bostock, C. J., Prescott, D. M., and Kirkpatrick, J. B. (1971). An evaluation of the double thymidine block for synchronizing mammalian cells at the G1-S border. *Exp. Cell Res.* **68,** 163–168.

Etienne-Manneville, S., and Hall, A. (2002). Rho GTPases in cell biology. *Nature* **420,** 629–635.

Just, I., Selzer, J., Wilm, M., von Eichel-Streiber, C., Mann, M., and Aktories, K. (1995). Glucosylation of Rho proteins by *Clostridium difficile* toxin B. *Nature* **375,** 500–503.

Kimura, K., Tsuji, T., Takada, Y., Miki, T., and Narumiya, S. (2000). Accumulation of GTP-bound RhoA during cytokinesis and a critical role of ECT2 in this accumulation. *J. Biol. Chem.* **275,** 17233–17236.

Narumiya, S., Oceguera-Yanez, F., and Yasuda, S. (2004). A new look at Rho GTPases in cell cycle: Role in kinetochore-microtubule attachment. *Cell Cycle* **3,** 855–857.

Oceguera-Yanez, F., Kimura, K., Yasuda, S., Higashida, C., Kitamura, T., Hiraoka, Y., Haraguchi, T., and Narumiya, S. (2005). Ect2 and MgcRacGAP regulate the activation and function of Cdc42 in mitosis. *J. Cell Biol.* **168,** 221–232.

Watanabe, N., Kato, T., Fujit, A., Ishizaki, T., and Narumiya, S. (1999). Cooperation between mDia1 and ROCK in Rho-induced actin reorganization. *Nat. Cell Biol.* **1,** 136–143.

Yasuda, S., Oceguera-Yanez, F., Kato, T., Okamoto, M., Yonemura, S., Terada, Y., Ishizaki, T., and Narumiya, S. (2004). Cdc42 and mDia3 regulate microtubule attachment to kinetochores. *Nature* **428,** 767–771.

[52]   Plexin-Induced Collapse Assay in COS Cells

By LAURA J. TURNER and ALAN HALL

Abstract

Semaphorins are a family of growth cone guidance molecules. When associated with their receptors and coreceptors, plexins and neuropilins, they act as chemorepellents for an extensive range of neuronal populations. The prototypic semaphorin, Sema3A, has a potent inhibitory effect on sensory axons emanating from dorsal root ganglia. This has formed the basis of the most famous assay for semaphorin activity, the chick dorsal root ganglia collapse assay. Recently, a heterologous, highly tractable assay has been used to investigate semaphorin signaling. In this system, the binding of recombinant semaphorins to COS cells expressing plexins and neuropilins induces a morphological collapse that may correlate with growth cone collapse. This chapter describes the optimization of this assay and outlines the subtle differences required to enable Sema3A-Fc and Sema4D-Fc to induce identical collapse phenotypes in COS cells expressing Plexin-A1 and neuropilin-1, or Plexin-B1, respectively.

METHODS IN ENZYMOLOGY, VOL. 406
0076-6879/06 $35.00
DOI: 10.1016/S0076-6879(06)06052-6

Introduction

Plexins are a family of transmembrane receptors. In association with neuropilins, plexins act as functional receptors for members of the semaphorin family of growth cone guidance molecules (He and Tessier-Lavigne, 1997; Rohm et al., 2000b; Takahashi et al., 1999; Tamagnone et al., 1999; Winberg et al., 1998). Semaphorins have traditionally been regarded as modulators of neural behavior, acting as chemorepellents for an extensive range of neuronal populations. The prototypic semaphorin, Sema3A, has a potent inhibitory effect on sensory axons emanating from dorsal root ganglia. This has formed the basis of the most famous assay for semaphorin activity, the chick dorsal root ganglia (DRG) collapse assay (Luo et al., 1993). Recently, a heterologous system has also been used to investigate semaphorin signaling. In this system, the binding of Sema3A to COS-7 cells expressing the semaphorin receptors Plexin-A1 and neuropilin-1 (NP-1) induces a morphological "collapse" that may correlate with growth cone collapse (Takahashi et al., 1999).

Although the chick DRG collapse assay offers the obvious advantage of studying endogenous plexins, manipulations of primary neurons in vitro often requires the use of complex trituration or viral-mediated gene transfer techniques. In addition, although cell-permeable drugs may be used to investigate DRG collapse, many of these drugs alter basal neuronal outgrowth and are thus difficult to use over the necessary time-periods. As an alternative, the COS collapse assay provides a highly reproducible, tractable system that can be used to study the signals that may both mediate and regulate semaphorin activity. Because large numbers of transfected cells are easily available, this system provides the additional advantage that biochemical investigations may be performed in parallel with morphological studies.

This chapter describes the preparation of recombinant semaphorins and outlines the methods by which these semaphorins may be used to induce collapse in COS-7 cells overexpressing Plexin-A1 or Plexin-B1.

Materials

Antibodies/Probes

Mouse anti-vsv-G (clone P5D4) was from Roche Diagnostics Inc. (Indianapolis, IN), goat anti-neuropilin-1 (clone C-19) was from Santa Cruz Biotechnology Inc. (Santa Cruz, CA), rat anti-tubulin was from Serotec (Oxford, UK), goat anti-human IgG, Fcγ fragment specific (goat anti-hFc) was from Jackson ImmunoResearch Inc. (West Grove, PA), and rhodamine-conjugated phalloidin was from Sigma (Dorset, UK).

DNA Constructs

vsv-Plexin-A1 in pBK-CMV was provided by Dr. A. Puschel (Westfalische Wilhelms Universitat, Munster, Germany; Rohm *et al.*, 2000a). NP-1 in pMT21 was provided by Dr. M. Tessier-Lavigne (Howard Hughes Institute, Stanford University, Stanford, CA; He and Tessier-Lavigne, 1997). Sema3A-Fc in pIG I and pIG I-Fc empty vector were provided by Dr. B. Eickholt (Kings College, London, UK; Eickholt *et al.*, 1997). Plexin-B1 in pCDNA3.1 was provided by Dr. M. Driessens (The Netherlands Cancer Institute, Amsterdam, The Netherlands; Driessens *et al.*, 2001). Sema4D-Fc in pEF-Fc was provided by Prof. H. Kikutani (Osaka University, Osaka, Japan; Kumanogoh *et al.*, 2000).

Cell Lines and Culture Conditions

COS-7 cells were cultured in Dulbecco's modified Eagle's medium (DMEM, Gibco-BRL, Gaithersburg, MD) supplemented with 10% complete foetal calf serum (FCS, PAA Laboratories), 100 U/ml penicillin, and 100 μg/ml streptomycin (pen/strep, Gibco-BRL). The Chinese hamster ovary cell line CHO-K1 was cultured in DMEM-F10 plus L-glutamate (Gibco-BRL) supplemented with 10% FCS and pen/strep. During semaphorin preparation, FCS was substituted with gamma globulin–free FCS (Ultrapure FCS, Sigma, St Louis, MO) to minimize the copurification of IgGs. Cell lines were grown at $37°$ in 10% CO_2 and passaged at 80–90% confluency.

Methods

Preparation of Recombinant Semaphorins

Recombinant Sema3A and Sema4D (ligands for Plexin-A1 and B1 respectively) and a "mock" empty vector negative control preparation were prepared by protein-A sepharose affinity chromatography from the conditioned medium of stably transfected CHO-K1 cells (Sema3A-Fc::pIG I, Eickholt *et al.*, 1997) or transiently transfected COS-7 cells (Sema4D-Fc:: pEF or mock::pIG I-Fc). In each case, the semaphorins were fused to the Fc fragment of human IgG1 (Fawcett *et al.*, 1994). This enabled the proteins to be efficiently purified using protein-A and ensured that they were present in suspension as dimers. This was necessary because dimerization is critical for the biological activity of semaphorins (Eickholt *et al.*, 1997).

For each protein preparation, 100 ml of conditioned medium was harvested 72 h after passaging (CHO-K1) or transfection (COS-7) and stored at $-20°$ until required. Conditioned medium was thawed on ice and then centrifuged at 3000 rpm for 10 min to remove cell debris. The supernatants were added to 2 ml of a 50% slurry of protein-A Sepharose (Sigma) and rotated overnight at $4°$. The suspension was placed in a disposable column (Falcon) at $4°$ and washed with 30 ml cold PBS (phosphate buffered saline)-A. Bound semaphorins were eluted using 5 ml of 100 mM glycine, pH 2.7. The eluates were neutralized using 500 μl of 1 M Tris, pH 9.0, and placed on ice. Semaphorins were concentrated using centrifugal concentrators with a molecular weight cutoff of 10,000 (Millipore, Billerica, MA) to a final volume of approximately 200 μl, aliquoted, and snap frozen in liquid nitrogen before being stored at $-80°$.

Protein concentration and purity were analyzed by Western blotting and SDS-PAGE (see Fig. 1). Coomassie staining revealed Sema3A-Fc as a single band of approximately 145 kDa and Sema4D-Fc as a doublet of approximately 140 and 160 kDa. The doublet is likely to represent the precursor and proteolytically cleaved forms of Sema4D-Fc. In addition, each semaphorin preparation contained two bands with molecular masses of approximately 65 and 75 kDa. These proteins were also present in the mock preparation and are therefore nonspecific interactors of protein-A and not degradation products of semaphorins. The concentration of recombinant semaphorin in each preparation was estimated by comparison of the band intensities of known volumes of semaphorins (single band only for Sema4D-Fc) with known quantities of BSA standards. In Fig. 1, Sema3A-Fc is approximately 200 μg/ml, and Sema4D-Fc is approximately 100 μg/ml.

Sema3A-Fc–Induced Collapse

Sema3A is unable to interact directly with its receptor, Plexin-A1. Instead, NP-1 acts as a bridge, connecting Sema3A to Plexin-A1. In this receptor complex, Plexin-A1 and NP-1 are present in a stoichiometric ratio of 1:1. For COS-7 cells to collapse in response to Sema3A-Fc, they must, therefore, express approximately equal levels of Plexin-A1 and NP-1. To achieve this, cells were transiently transfected with varying ratios of vsv-Plexin-A1 and NP-1 cDNA (1.5 μg total cDNA/transfection). Cells were seeded in 6-well plates (1.5 × 10^5 cells per well) and transfected 16–24 h later using the Genejuice reagent according to the manufacturers instructions (Novagen). Alternative transfection reagents may be used, but cells should be transfected in an adherent rather than suspended state. This prevents problems with cell spreading and enhances the distinction between collapsed and noncollapsed cells. Receptor expression levels were

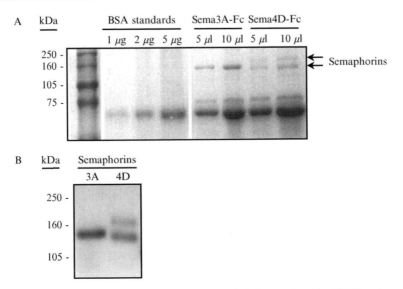

FIG. 1. Preparation of recombinant semaphorins. (A) Coomassie-stained 7.5% polyacrylamide gel showing 5- and 10-μl aliquots of each semaphorin preparation and 1-, 2-, and 5-μg aliquots of BSA standards. (B) Western blot of 5-μl aliquots of purified Sema3A-Fc and Sema4D-Fc probed with goat anti-hFc.

assessed 24 h after transfection by Western blot analysis (see Fig. 2A). A cDNA ratio of 1:2 of vsv-Plexin-A1:NP-1 provided the best expression levels of each receptor and was used in all subsequent collapse assays.

To enable useful conclusions to be drawn from COS collapse assays, it is important to demonstrate that ectopically expressed plexins and neuropilins interact only with the appropriate recombinant semaphorins. To do this, parallel immunofluorescence and Western blot analyses were performed. COS-7 cells were transfected with vsv-Plexin-A1 and NP-1 and incubated for 24 h. Cells were then stimulated for 30 min with 1 μg/ml of Sema3A-Fc, Sema4D-Fc or a mock preparation. After stimulation, cells on coverslips were washed twice in PBS-A then fixed and stained for vsv-Plexin-A1 (mouse anti-vsv) and semaphorins (goat anti-hFc). Of the cells expressing vsv-Plexin-A1, 95% were strongly positive for Sema3A-Fc staining. In contrast, only 8% of expressing cells were positive for Sema4D-Fc staining (see Fig. 2B). This demonstrates that Sema3A-Fc, but not Sema4D-Fc, interacts with Plexin-A1 and NP-1. This was confirmed by Western blot analysis. After semaphorin stimulation, cells were washed twice in PBS-A, then lysates were prepared and semaphorins that

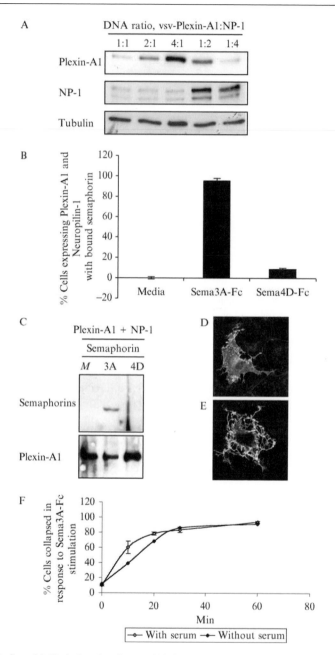

FIG. 2. Sema3A-Fc–induced collapse. (A) Lysates from cells transfected with varying ratios of vsv-Plexin-A1 and NP-1 were probed using mouse anti-vsv and goat anti-NP-1. Blots

remained associated with the cells were detected using the goat anti-hFc antibody (see Fig. 2C). Once again, the vsv-Plexin-A1/NP-1 receptor complex interacted with Sema3A-Fc and not Sema4D-Fc.

Having confirmed the specificity of the semaphorin/plexin interaction, the ability of Sema3A-Fc to alter the morphology of COS-7 cells was assessed. Cells overexpressing either vsv-Plexin-A1 or NP-1 alone, or both receptors together, were stimulated 24 h after transfection for 1 h at 37° with fresh culture medium supplemented with 0, 0.5, 1, 2, or 5 μg/ml of Sema3A-Fc. After stimulation, cells were washed twice in PBS-A, fixed in 4% paraformaldehyde in PBS-A for 20 min at room temperature, and stained with rhodamine-conjugated phalloidin and either mouse anti-vsv or goat anti-hFc to detect F-actin, Plexin-A1, and bound semaphorins to reveal the overall cell morphology. Cells were classified as collapsed (see Fig. 2E) or noncollapsed (see Fig. 2D) on the basis of visual changes, although alterations in cell surface area can also be accurately quantified using digital analysis software (Takahashi et al., 1999).

Stimulation with Sema3A-Fc triggered a striking collapse in 80% of vsv-Plexin-A1/NP-1–expressing cells with bound Sema3A-Fc when concentrations of 2 or 5 μg/ml of Sema3A-Fc were used. In contrast, the morphology of Sema3A-Fc–stimulated cells expressing either vsv-Plexin-A1 or NP-1 alone was unaltered in most cases, and no changes were apparent in cells expressing vsv-Plexin-A1 and NP-1 stimulated with a mock preparation. Collapse did not result in complete cell rounding, because F-actin–rich protrusions persisted to the edges of the original cell boundaries. Concentrations of 0, 0.5, or 1 μg/ml failed to induce any morphological changes. As a result, 2 μg/ml was chosen as an appropriate concentration of Sema3A-Fc for use in collapse assays. Because collapse was highly concentration dependent and semaphorin concentrations determined by

---

were also probed with rat anti-tubulin to enable total protein levels in lysates to be compared. (B and C) Sema3A-Fc, but not Sema4D-Fc, binds to COS-7 cells expressing vsv-Plexin-A1 and NP-1. To detect bound semaphorins, cells were either fixed and stained using goat anti-hFc and mouse anti-vsv (B), or lysates were prepared and probed using goat anti-hFc (C). (D and E) Sema3A-Fc induces collapse of COS-7 cells expressing vsv-Plexin-A1 and NP-1. (D) Unstimulated cells expressing vsv-Plexin-A1 and NP-1 stained with mouse anti-vsv-G. (E) Cells expressing vsv-Plexin-A1 and NP-1 stimulated for 30 min with 2 μg/ml Sema3A-Fc and stained with goat anti-hFc. (F) Cells expressing vsv-Plexin-A1 and NP-1 were stimulated with 2 μg/ml Sema3A-Fc for varying times, in the presence or absence of serum. "With serum" indicates that cells were cultured and stimulated in serum at all times. "Without serum" indicates that cells were starved of serum 16 h before stimulation.

polyacrylamide gel electrophoresis are only approximates, each new semaphorin preparation was also titrated in comparison with previous functional preparations.

In cells overexpressing vsv-Plexin-A1 and NP-1, Sema3A-Fc–induced collapse progressed according to a stereotypical time course. In the presence of serum, the percentage of collapsed cells increased rapidly until approximately 20 min, when a plateau was reached. At this point, 79% of cells with bound semaphorins had collapsed. In comparison, in the absence of serum, the initial increase in the percentage of collapsed cells occurred at a slightly slower rate.

Despite this, a plateau was still observed when the percentage of collapsed cells reached approximately 85%, this time at 30 min. As a result, 30 min was chosen as an appropriate length of semaphorin incubation for future collapse assays.

Sema4D-Fc–Induced Collapse

Unlike Plexin-A1, it has previously been reported that COS cells overexpressing Plexin-B1 do not collapse in response to stimulation with their ligand, Sema4D. Instead, Oinuma et al. (2003) report that Sema4D-induced collapse only occurs if Rnd1 or Rnd3 are overexpressed alongside Plexin-B1.

In contrast, we find that Sema4D-Fc induces an identical collapse phenotype in COS-7 cells expressing Plexin-B1 (see Fig. 3A), but the conditions necessary for this to occur differ from those required for Sema3A-Fc–induced collapse. When COS-7 cells were transfected with Plexin-B1 (1.5 μg cDNA/transfection) and stimulated 24 h after transfection with 2 μg/ml Sema4D-Fc for 30 min, only 2% of cells with bound semaphorins collapsed (see Fig. 3B). On this occasion, increasing the concentration of Sema4D-Fc to 5 μg/ml had no effect on the level of collapse (data not shown). Instead, the length of the incubation period between transfection and stimulation proved critical for the ability of Sema4D-Fc to induce a collapse response. When cells overexpressing Plexin-B1 were stimulated 48 h after transfection with 2 μg/ml of Sema4D-Fc for 30 min, the percentage of collapsed cells rose to 58%. Interestingly, in this instance reducing the concentration of Sema4D-Fc to 1 μg/ml had little effect on these changes, because 54% of cells with bound Sema4D-Fc still underwent a collapse response (see Fig. 3B).

Although the percentage of cells collapsing in response to Sema4D-Fc is lower than the 80% of cells that collapse in response to Sema3A-Fc stimulation, it is equivalent to the levels of collapse achieved by Oinuma et al.

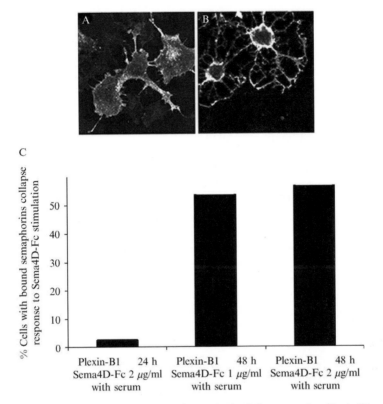

FIG. 3. Sema4D-Fc–induced collapse. (A and B) Cells expressing Plexin-B1 were stimulated either 24 h (A) or 48 h (B) after transfection for 30 min with 2 μg/ml Sema4D-Fc. Cells were probed with goat anti-hFc. (C) Cells expressing Plexin-B1 were stimulated either 24 or 48 h after transfection for 30 min with either 1 or 2 μg/ml Sema4D-Fc.

when co-expressing Plexin-B1 and Rnd1 (Oinuma *et al.*, 2003). Two main differences in method may account for the disparity observed between these results and those of Oinuma *et al.* First, the incubation time between transfection of Plexin-B1 and stimulation with Sema4D-Fc is critical for collapse. Because Oinuma *et al.* stimulate their cells 16 h after transfection, morphological changes might not be expected to occur. Second, Oinuma *et al.* incubate cells expressing Plexin-B1 with Sema4D-Fc for 5 min before assessing collapse. In light of the time course of Sema3A-Fc–induced morphological changes, after 5 min only 30% of cells with bound semaphorins might be predicted to collapse.

Discussion

The COS collapse assay provides a highly reproducible, tractable system with which to investigate semaphorin activity. To date, it has only been used in studies involving Plexin-A1, NP-1, and Plexin-B1. In the future, this assay may prove useful in comparing the mechanisms of collapse mediated by different plexin subfamily members.

The ease of manipulation of COS cells makes this assay extremely attractive. To test the involvement of specific molecules in plexin-mediated signal transduction, mutants of putative mediators of semaphorin responses can be cotransfected with plexins and neuropilins before semaphorin stimulation to assess their effects on collapse. In addition, cell-permeable, small molecular inhibitors of interesting molecules can be used to treat cells before semaphorin stimulation and their ability to inhibit collapse can be assessed, and biochemical experiments in COS cells can be performed in parallel with morphological studies to investigate changes in the activation status of relevant proteins. Finally, mutagenesis studies can be used to highlight sequences within plexins that are critical for mediating semaphorin-induced responses.

The COS collapse assay has already proved useful in developing our understanding of the role of small GTPases in plexin signaling. Using this assay, Sema3A-Fc–induced collapse has been shown to require Rac1 and Cdc42 but not RhoA (Turner et al., 2004); the functional requirement for a putative GTPase activating protein domain in the cytoplasmic tail of Plexin-A1 has been demonstrated (Rohm et al., 2000b), and recruitment of Rnd1 to Plexin-A1 has been shown to induce collapse in the absence of semaphorin stimulation (Zanata et al., 2002). Similar approaches in the future may allow us to determine which guanine nucleotide exchange factors and GTPase effector proteins and pathways are responsible for mediating the collapse responses. Ultimately, this assay may help us to explain the molecular mechanisms that underlie the semaphorin-induced collapse response.

Finally, when using the COS collapse assay, we must remain aware of its limitations. As well as being chemorepellents, semaphorins are believed to function as chemoattractants (Wong et al., 1999). Using this assay, it will be not possible to distinguish between molecules that inhibit collapse and those that convert semaphorin-induced collapse (i.e., repulsion) into attraction. Second, in vivo, growth cone turning in response to a repulsive cue is an asymmetrical event that involves a localized collapse in the region nearest to the repellent. Total cell collapse is an experimental outcome resulting from a generalized application of a chemorepellent. It is not currently possible to use this assay to study the polarized nature of sema-phorin-induced responses. Third, in cases in which the semaphorin being investigated is a transmembrane protein (e.g., Sema4D), the recombinant protein used in this assay is a secretable form of the extracellular domain of this protein. It is possible that the responses stimulated by these recombinant semaphorins may differ slightly from those induced by endogenous, transmembrane versions. In addition, no information will be gleaned about reverse signaling that may occur *in vivo* in the semaphorin-expressing cell in response to the interaction with a plexin receptor. Finally, because this is a heterologous assay, we must remember that the results obtained provide clues to the mechanisms that may induce growth cone collapse, but these observations should ultimately be confirmed in neuronal systems where possible.

Acknowledgments

We thank Dr. B. Eickholt for the kind gift of the Sema3A-Fc expressing CHO-K1 cell line, Sema3A-Fc cDNA, and protocols for purifying Sema3A-Fc. We also thank Dr. A. Puschel, Dr. M. Tessier-Lavigne, Dr. M. Driessens, and Prof. H. Kikutani for the gift of cDNAs. This work was generously supported by a programme grant from Cancer Research UK. L. J. T. was the recipient of an MRC PhD fellowship.

References

Driessens, M. H., Hu, H., Nobes, C. D., Self, A., Jordens, I., Goodman, C. S., and Hall, A. (2001). Plexin-B semaphorin receptors interact directly with active Rac and regulate the actin cytoskeleton by activating Rho. *Curr. Biol.* **11**, 339–344.

Eickholt, B. J., Morrow, R., Walsh, F. S., and Doherty, P. (1997). Structural features of collapsing required for biological activity and distribution of binding sites in the developing chick. *Mol. Cell Neurosci.* **9**, 358–371.

Fawcett, J., Holness, C. L. L., and Simmons, D. L. (1994). Molecular tools for the analysis of cell adhesion. *Cell Ad. Comm.* **2**, 275–285.

He, Z., and Tessier-Lavigne, M. (1997). Neuropilin is a receptor for the axonal chemorepellent Semaphorin III. *Cell* **90**, 739–751.

Kumanogoh, A., Watanabe, C., Lee, I., Wang, X., Shi, W., Araki, H., Hirata, H., Iwahori, K., Uchida, J., Yasui, T., Matsumoto, M., Yoshida, K., Yakura, H., Pan, C., Parnes, J. R., and Kikutani, H. (2000). Identification of CD72 as a lymphocyte receptor for the class IV semaphorin CD100: A novel mechanism for regulating B cell signalling. *Immunity* **13**, 621–631.

Luo, Y., Raible, D., and Raper, J. A. (1993). Collapsin: A protein in brain that induces the collapse and paralysis of neuronal growth cones. *Cell* **75**, 217–227.

Oinuma, I., Katoh, H., Harada, A., and Negishi, M. (2003). Direct interaction of Rnd1 with Plexin-B1 regulates PDZ-RhoGEF-mediated Rho activation by Plexin-B1 and induces cell contraction in COS cells. *J. Biol. Chem.* **278**, 25671–25677.

Rohm, B., Ottemeyer, A., Lohrum, M., and Puschel, A. W. (2000a). Plexin/neuropilincomplexes mediate repulsion by the axonal guidance signal semaphorin 3A. *Mech. Dev.* **93,** 95–104.

Rohm, B., Rahim, B., Kleiber, B., Hovatta, I., and Puschel, A. W. (2000b). The semaphorin 3A receptor may directly regulate the activity of small GTPases. *FEBS Lett.* **486,** 68–72.

Takahashi, T., Fournier, A., Nakamura, F., Wang, L. H., Murakami, Y., Kalb, R. G., Fujisawa, H., and Strittmatter, S. M. (1999). Plexin-neuropilin-1 complexes form functional semaphorin-3A receptors. *Cell* **99,** 59–69.

Tamagnone, L., Artigiani, S., Chen, H., He, Z., Ming, G. I., Song, H., Chedotal, A., Winberg, M. L., Goodman, C. S., Poo, M.-M., Tessier-Lavigne, M., and Comoglio, P. M. (1999). Plexins are a large family of receptors for transmembrane, secreted, and GPI-anchored semaphorins in vertebrates. *Cell* **99,** 71–80.

Turner, L. J., Nicholls, S., and Hall, A. (2004). The activity of the Plexin-A1 receptor is regulated by Rac. *J. Biol. Chem.* **279,** 33199–33205.

Winberg, M. L., Noordermeer, J. N., Tamagnone, L., Comoglio, P. M., Spriggs, M. K., Tessier-Lavigne, M., and Goodman, C. (1998). Plexin A is a neuronal semaphorin receptor that controls axon guidance. *Cell* **95,** 903–916.

Wong, J. T., Wong, S. T., and O' Connor, T. P. (1999). Ectopic semaphorin 1a functions as an attractive guidance cue for developing peripheral neurons. *Nat. Neurosci.* **2,** 798–803.

Zanata, S. M., Hovatta, I., Rohm, B., and Puschel, A. W. (2002). Antagonistic effects of Rnd1 and RhoD GTPases regulate receptor activity in semaphorin 3A-induced cytoskeletal collapse. *J. Neurosci.* **22,** 471–477.

[53] Morphological and Biochemical Analysis of Rac1 in Three-Dimensional Epithelial Cell Cultures

By Lucy Erin O'Brien, Wei Yu, Kitty Tang, Tzuu-Shuh Jou, Mirjam M. P. Zegers, and Keith E. Mostov

Abstract

Rho GTPases are critical regulators of epithelial morphogenesis. A powerful means to investigate their function is three-dimensional (3D) cell culture, which mimics the architecture of epithelia *in vivo*. However, the nature of 3D culture requires specialized techniques for morphological and biochemical analyses. Here, we describe protocols for 3D culture studies with Madin-Darby Canine Kidney (MDCK) epithelial cells: establishing cultures, immunostaining, and expressing, detecting, and assaying Rho proteins. These protocols enable the regulation of epithelial morphogenesis to be explored at a detailed molecular level.

METHODS IN ENZYMOLOGY, VOL. 406   0076-6879/06 $35.00
Copyright 2006, Elsevier Inc. All rights reserved.   DOI: 10.1016/S0076-6879(06)06053-8

Introduction

Epithelial tissues are coherent sheets of cells that form a barrier between the body interior and the outside world. These sheets line internal organs such as the kidney, lung, and mammary gland. They are commonly one cell thick and enclose a central lumen, thereby creating the spherical cysts and cylindrical tubules that are the architectural hallmarks of internal epithelial organs. The formation of cysts and tubules during embryogenesis requires coordinated regulation of cellular behaviors such as migration, differentiation, proliferation, and polarization. Studies with epithelial cells grown as monolayers on Petri dishes or permeable filter supports have established critical roles for Rho GTPases in these processes, and studies *in vivo* have demonstrated the necessity of these factors during epithelial organogenesis (Burridge and Wennerberg, 2004; Van Aelst and Symons, 2002). Between these two experimental approaches, a mechanistic understanding of how Rho GTPases function to coordinate cell behavior during morphogenesis of multicellular tissue architecture is largely lacking.

Three-dimensional cell culture is a powerful tool for bridging this gap (Debnath and Brugge, 2005; O'Brien *et al.*, 2002). When embedded inside a 3D extracellular matrix (ECM), many epithelial cell lines reproduce a tissue-like architecture (Walpita and Hay, 2002). For instance, when grown in a matrix of collagen I, MDCK cells form structures that resemble epithelial organ rudiments. Furthermore, MDCK cysts develop branching tubules after treatment with mesenchymally derived hepatocyte growth factor (HGF), a response reminiscent of the epithelial–mesenchymal interactions that induce tubulogenesis *in vivo* (Montesano *et al.*, 1991a,b).

The relative simplicity and tractability of 3D cultures enables detailed exploration of how Rho GTPases specify the organization of multicellular epithelia. Here, we delineate the methods we have used to investigate the role of the Rac1 GTPase during MDCK development in 3D. The methods are divided into techniques for morphological analysis and biochemical analysis. Combining these methods can shed significant insight on the mechanisms that regulate epithelial morphogenesis.

Morphological Analysis of 3D MDCK Cultures

Background

We have found *in situ* confocal analysis of immunostained, whole-mounted 3D cultures to be a powerful tool for understanding MDCK cyst and tubule morphogenesis (O'Brien *et al.*, 2001; Pollack *et al.*, 1998). Staining and imaging intact gels involves considerably less time than cutting sections, enables a large number of structures to be examined without changing slides, and allows 3D reconstruction with imaging software. Phalloidin staining of F-actin is sufficient to reveal detail at the cellular and, in some cases, subcellular level. This resolution allows the morphogenesis of these 3D structures to be examined in a depth not possible with bright-field microscopy.

Growing MDCK Cysts for Immunofluorescence

Reagents (All Sterile)

1. L-Glutamine stock, 29.2 mg/ml in H_2O.
2. $NaHCO_3$ stock, 2.35 g/100 ml in H_2O.
3. 10× MEM without glutamine (Invitrogen 11430-030).
4. Purified bovine collagen I (Vitrogen 100, 3 mg/ml, Angiotech Biomaterials FXP-019).
5. Nunc Anapore membrane inserts, 0.2-μm pore, 10-mm diameter (Nalge Nunc 136935).

Preparation of Cells. There are numerous clones of MDCK cells, which can differ in their ability to form cysts and tubules in 3D culture. We use MDCK strain II cells, derived from a clone isolated by Louvard (Louvard, 1980). These cells are maintained in MEM supplemented with 10% FCS, 100 IU/ml penicillin, and 100 mg/ml streptomycin in a humidified 37° incubator with 5% CO_2. MDCK cells optimally form cysts if they are actively proliferating at the time they are plated in collagen I. An easy way of assuring that the cells are rapidly dividing is by splitting a confluent 10-cm Petri dish of cells 1:10 1 day before plating in collagen I.

Plating in Collagen I

1. Prepare the collagen I solution by combining the following: 750 μl glutamine stock, 625 μl $NaHCO_3$ stock, 625 μl 10× MEM without glutamine, 125 μl 1 M Hepes (pH 7.6), and 4.13 ml Vitrogen. The final concentration of collagen I in this mixture is about 2 mg/ml. Slowly pipet up and down to mix, avoiding bubbles. Store this solution at 4° while preparing the cells to avoid premature polymerization.

2. Place the Nunc filter inserts individually in the wells of a standard 24-well tissue culture plate.

3. Trypsinize cells and resuspend in about 10 ml of standard culture media.

4. Pellet cells, remove the supernatant, and resuspend in 2–3 ml standard culture media. Triturate cells thoroughly to attain a single cell suspension.

This step is critical, because single cells tend to form spherical cysts with single lumens, whereas clumps of cells tend to form irregular cysts with multiple lumens. Determine the number of cells/ml using a hemocytometer.

5. Add cells to the collagen I solution for a final concentration between 2×10^4 and 5×10^4 cells/ml, typically 50–200 μl of resuspended cells. If the volume of the cell solution exceeds roughly 10% of the collagen I solution, the ability of the collagen I to polymerize will be diminished. Gently pipet to mix, avoiding bubbles.

6. Add 150 μl of the cell–collagen I mixture to each filter insert. Gently tap the tray to spread the mixture over the entire surface of each filter.

7. Place tray in a 37° oven (not a CO_2 incubator) for 10–20 min to harden. Because the collagen I solution is bicarbonate-buffered, it has a higher pH in room air than in 5% CO_2, and the collagen I sets faster and more reliably at this higher pH. The collagen I is polymerized when gentle taps generate no ripples.

8. Once the collagen I has solidified, add 0.5 ml standard culture media to the top of filter and 1 ml to the bottom. Place in a 37° CO_2 incubator for development.

Plating in Other ECMs. MDCK cells are competent to form cysts in other types of ECMs, and comparing cyst morphogenesis in different matrices can be revealing. One alternative matrix is Matrigel™ (BD Biosciences 356234), reconstituted basement membrane purified from the Engelbreth-Holm-Swarm mouse sarcomaline. Plating cysts in Matrigel™ is largely similar to plating cysts in collagen I. The differences reflect the fact that Matrigel™ solidifies very rapidly above 4°; thus, all plastic ware that contacts the liquid Matrigel™ must be diligently cooled to 4° before use, and once plated the cell/Matrigel™ mixture polymerizes within 5 min. Matrigel™-grown MDCK cysts do not tubulate in response to HGF (Santos and Nigam, 1993).

MDCK cells also form cysts in collagen I mixed with ECM components such as laminin-1, collagen IV, and fibronectin (BD Biosciences). These other components are unable to form a solid gel alone, so mixing them with collagen I provides structural support. To enable polymerization, care must be taken not to dilute the collagen I below a final concentration of 1 mg/ml. Purified laminin may be added directly to the collagen I solution. Purified collagen IV and fibronectin are stored in acid; to remove the acid, we dialyze these components overnight against PBS$^+$ at 4° before adding the collagen I solution.

Maintaining Cyst Cultures. Maintain the growing cysts by replacing media every 2–3 days. In our hands, lumens are visible by bright-field microscopy after 4–5 days, and cysts are terminally developed by 10–12 days.

Conditionally Expressing Rho-family GTPases in Cyst Culture

Background. Rho GTPases are critical regulators of epithelial polarity and morphology. Conditional expression of dominant-negative and dominant-active Rho GTPases enables detailed investigation of their function during cyst morphogenesis. We have extensively studied the role of Rac1 during cyst development with MDCK cell lines that inducibly express the dominant negative N17 or dominant active V12 alleles of Rac1 under the tetracycline-repressible system (Jou and Nelson, 1998; Jou *et al.*, 1998; O'Brien *et al.*, 2001). Inducible expression of these mutant proteins confers the ability to modulate expression levels, control for clonal variation, and determine whether effects of mutant protein expression are reversible.

In our hands, expression of exogenous Rho GTPases is induced efficiently only when cells are actively proliferating. We believe this characteristic may be intrinsic to the parental tet-off MDCK cell line (Clonetech 630913). Inducing exogenous Rac1 expression in mature cysts is thus difficult, because cell proliferation diminishes to replacement levels once cysts complete development. Because tubules are derived from mature cysts and Rac1 is required for cyst formation, this barrier has stymied our efforts to use the tet-off system during tubulogenesis.

Expressing Mutant Rac1 GTPases During Cyst Development

1. To ensure uniform expression and minimize cell line drift, we freeze a large number of vials of low-passage cells and thaw a fresh vial for each experiment. Plate the thawed cells in a 10-mm Petri dish in the presence of 20 ng/ml doxycycline to suppress expression completely.

2. The next day, wash cells thrice with PBS⁻ to remove doxycycline and induce expression. Cells can then be plated in collagen I in the complete absence of doxycycline. For the N17Rac1 cell line we use, this results in expression levels approximately sixfold above endogenous Rac1. Expression can be modulated by varying doxycycline concentrations; with 200 pg/ml doxycycline results in N17Rac1 expression comparable to endogenous levels.

Inducing Tubulogenesis with HGF

Background. During embryogenesis, epithelial tubulogenesis is often initiated in response to signals from mesenchymal growth factors (Hogan and Kolodziej, 2002). Tubulogenesis of MDCK cysts in response to mesenchymal HGF is a cell-culture analogy of this paradigm. Montesano *et al.* (Montesano *et al.*, 1991a,b) detailed three ways in which MDCK cells could be grown for assaying tubulogenesis: as a monolayer on top of a cell-free collagen I layer, as a single-cell suspension in collagen I, and as a culture of mature cysts. We have had best results with this last approach. With mature cysts, tubulogenesis proceeds with relative synchrony and completes within 3–4 days. When starting with a single cell suspension, tubule development is highly asynchronous and substantially slower (10–14 days).

MDCK Strain tubulogenesis involves two distinct stages (O'Brien et al., 2004), although other MOCK lines may tubulate via a slightly different process (Williams and Clark, 2003). The initial stage is a partial epithelial–mesenchymal transition (p-EMT). The extensions that mark the first manifestation of p-EMT are visible at 6 h. These extensions develop into single-file chains of cells that are apparent by 24 h. Cells in the chains have lost apicobasolateral polarity and are migratory and spindle shaped. The second stage of tubule development is epithelial redifferentiation. At approximately 48 h, cells in chains cease migration, repolarize, and become cuboidal, forming multilayered cords in the process. Nascent lumens develop between these cell layers. By 72–96 h, tubules reach maturity. Repolarization is complete, and lumens are continuous (Pollack et al., 1998). With 20× phase magnification, we are able to identify extensions and chains and distinguish each of these from redifferentiating tubules. We require confocal analysis to distinguish cords from mature tubules.

Whenever possible, we use recombinant, purified HGF. Media conditioned by fibroblasts such as MRC-5 cells contain HGF and can be used to induce tubulogenesis at a dilution ranging from 1:2–1:8 with standard MDCK culture medium. However, it is important to confirm results obtained with conditioned media using purified HGF.

HGF Treatment

1. Add 10 ng/ml HGF or diluted, conditioned medium to a culture of 10-day MDCK cysts in collagen I.
2. Replenish HGF-containing media daily until samples are harvested.

Immunostaining 3D MDCK Cultures In Situ

Reagents

1. Collagenase-1 (Type CVII, Sigma C-2799). Use of high-purity collagenase is critical. We use aliquots of a stock of 1000 U/ml in PBS$^+$, flash frozen and stored at 80°. Thaw and dilute aliquots 1:10 in PBS$^+$ before use.
2. Paraformaldehyde (PFA), 4%.
3. Quenching solution. Made fresh for each experiment from two stock solutions, 1 M NH$_4$Cl and 1 M glycine (pH 8). Combine 1.5 ml

NH$_4$Cl stock and 0.4 ml glycine stock, then add PBS$^+$ to a final volume of 20 ml.

4. Saponin stock solution. 10% saponin (Calbiochem 558255) in H$_2$O, filter sterilized, and stored in aliquots at 4° because prone to contamination.

5. Permeabilization solution. Made fresh for each experiment from 3.5 g fish skin gelatin (Sigma G-7765) and 1.25 ml saponin stock brought up to 500 ml with PBS$^+$. Triton X-100 at a final concentration of 0.5% can be substituted for saponin and is preferable for some antigens. The permeabilization solution can be stored at 4° over the course of a single experiment with the addition of 0.02% NaN$_3$.

6. RNase A (if nuclei are to be stained). Made fresh for each experiment by diluting RNase A stock solution (10 mg/ml in H$_2$O) 1:100 with permeabilization solution.

7. ProLong mounting media (Molecular Probes P-7481), prepared according to manufacturer's directions.

Collagenase Treatment. Brief treatment of 3D cultures with collagenase before fixation significantly increases the ability of antibodies to permeate the collagen I gel without detectably compromising extracellular antigens.

1. Leaving the gels in the filters, quickly wash the filters three times with room-temperature PBS$^+$.
2. Add 1× collagenase solution to the filters (approximately 0.25 ml on top, 0.5 ml on bottom) and incubate 8–10 min at 37°.

Fixation and Quenching

1. Wash three times quickly with room-temperature PBS$^+$.
2. Add 4% PFA and shake samples slowly 30 min.
3. Wash five times with PBS$^+$, three quick washes and two washes shaking slowly for 5 min each.
4. Samples may be stored up to 2 weeks in 0.1% PFA at 4°.

Permeabilization and Antibody Treatment

1. Incubate with permeabilization solution 30 min at room temperature or overnight at 4°, shaking slowly.

2. If nuclei are to be stained, incubate with 1× RNase A solution 60 min at 37°.

3. Dilute primary antibodies in permeabilization solution, 0.5 ml per sample. Add 0.25 ml of diluted antibodies to top of filter and 0.25 ml below filter. Incubate 2–3 h at room temperature or overnight at 4°, shaking slowly.

4. Wash filters with permeabilization solution, once quickly and three times for 10 min each, shaking slowly at room temperature.

5. Dilute fluorophore-conjugated secondary antibodies in permeabilization solution, 0.5 ml/filter. Because of the thickness of the gel, best results are obtained with bright fluorophores such as the Alexa dyes from Molecular Probes, which we use at a dilution of 1:400. Add 0.25 ml to the top and bottom of filter. Incubate 2–3 h at room temperature or overnight at 4°, shaking slowly.

6. Wash with permeabilization solution, once quickly and three times for 10 min each time, shaking slowly at room temperature.

7. Wash twice with PBS$^+$ for 5 min each.

Postfixation, Nuclear Staining, and Mounting

1. Add 4% PFA to samples and incubate 30 min at room temperature, shaking slowly.

2. Wash five times with PBS$^+$, three times quickly and twice for 5 min each.

3. If nuclear staining is desired, the far red nucleic acid dye TO-PRO3 (Molecular Probes T-3605) works well. Dilute TO-PRO3 1:100 in PBS$^+$, add to samples, and incubate 60 min at 37°. Do not wash samples before mounting; unbound dye does not fluoresce, and washing significantly diminishes the nuclear signal. Hoechst 33342 at a dilution of 1:1,000 can also be used, but include the Hoechst stain with the secondary antibody incubation, not after postfixation.

4. Aspirate PBS$^+$ or TO-PRO$_3$ and add a drop of ProLong mounting media to the top of the filter.

5. Remove the gel from the filter. This can be done with a filter punchout tool (Nalge Nunc 139586) or by gently pinching opposite sides of the gel with fine forceps and lifting the gel from the filter.

6. Place a generous drop of ProLong on a microscope slide and place the gel on the drop. To facilitate confocal analysis, make sure that the top of the gel is facing up. Gently put a 25-mm diameter circular coverslip on top of the gel. The gel is less prone to slip out with a circular coverslip.

7. Wait approximately 5 min to allow the ProLong to spread underneath the entire surface of the coverslip. Do not press down on the coverslip during this time.

8. Carefully aspirate any excess ProLong from the edges of the coverslip.

9. Place slides in a dark area at room temperature and allow the ProLong to harden overnight.

10. Slides can be stored at 4° or room temperature. They remain suitable for confocal analysis for 3–6 months.

Biochemical Analysis of 3D MDCK Cultures

Background

Analyzing protein function directly in 3D cultures, rather than in monolayers, is imperative if at all possible. Changing the extracellular environment can substantially alter gene expression and signaling (O'Brien *et al.*, 2001, 2004; Wang *et al.*, 1998; Weaver *et al.*, 2002). Thus, extrapolating biochemical results obtained with monolayer cultures to cyst developmental phenotypes can be tenuous. However, biochemical analysis is inherently more difficult in 3D than in 2D. Because cells are embedded in ECM, accessibility to the basolateral surface is reduced, and accessibility to the apical surface is extremely limited. Furthermore, the density of cells in 3D culture is about two orders of magnitude less than in 2D, making biochemical analysis of low-abundance proteins problematic.

We have developed several strategies for analyzing proteins in 3D, including two methods for generating cyst and tubule lysates. The first of these involves direct and rapid solubilization of 3D cultures *in situ*. Note that for secreted and transmembrane proteins, the solubilized lysate contains intracellular, membrane-bound, and/or secreted populations. Because of its rapidity, we use this technique to analyze the activation state of signaling molecules such as Rac1 and extracellular-regulated kinase (ERK). The disadvantage of solubilization is that the resulting lysates are substantially diluted. Immunoprecipitation of the protein of interest can sometimes overcome this difficulty. The second strategy involves isolating intact cysts and tubules by digesting the collagen I gel. This technique allows populations of intracellular, membrane-bound, and soluble proteins to be distinguished. It also enables the cells to be lysed in a small volume, resulting in a concentrated lysate. The major disadvantage of isolation is the substantial time needed to proteolyze the gel. During this period, cyst and tubule architecture begins to disintegrate, and thus analysis of the activation state of dynamic signaling molecules is not advisable.

Growing MDCK Cysts for Biochemical Analysis

Two modifications of the cyst plating protocol described previously facilitate biochemical analysis of 3D MDCK cultures. The first involves cell density, whereas an initial plating density of 2×10^4 cells/ml produces a convenient density of cysts for morphological analysis, it can be difficult to obtain sufficient cellular material for biochemical analysis. We have grown cysts at an initial plating density as high as 2×10^5 cells/ml to generate lysates. This higher density does not apparently alter cyst morphogenesis. However, an important caveat is that cysts at this concentration do not tubulate robustly; thus, we use the standard 2×10^4 density for tubulogenesis studies.

The second modification involves a double layer of collagen I. During the time required for the collagen I mixture to polymerize, some cells sink to the bottom and adhere to the filter. These cells do not proliferate into cysts but rather form a pseudostratified layer on the filter interface. This layer does not interfere with morphological analysis, because one can simply adjust the microscope's plane of focus. However, it could arguably interfere with biochemical analysis, because cells in the pseudostratified layer would comprise a portion of the lysate. To avoid this scenario, we first add 100 μl of cell-free collagen I liquid to the filter and allow it to set. We then plate the cell/collagen I mixture on top of this cell-free collagen I layer. In this manner, cells that sink to the bottom are still completely surrounded by collagen I and form true 3D structures instead of pseudostratified layers. To generate cyst lysates, we harvest both layers as many cells adhere to the bottom layer when the two are separated.

Solubilizing 3D Cultures

Reagents

1. SDS lysis buffer. 0.5% SDS, 100 mM NaCl, 50 mM triethanolamine (TEA)-Cl, pH 8.1, 5 mM EDTA, pH 8.0, 0.2% NaN$_3$, 1× protease inhibitor cocktail (below).
2. Protease inhibitor cocktail, 100× stock. 5 mg/ml pepstatin, 10 mg/ml chymostatin, 5 mg/ml leupeptin, 50 mg/ml antipain, 500 mM benzamidine, 10 IU/ml aprotinin, and 1 mM PMSF.

Solubilization of 3D Cultures

1. Wash gels four times quickly in PBS$^+$.
2. Remove gels from the filters by pinching opposite edges of the gel together with forceps. Place gels in a Microfuge tube containing SDS lysis buffer. We generally combine two duplicate samples together in a single tube with 0.8 ml SDS lysis buffer.
3. Boil gels for 15 min. Shake tubes a couple of times during this incubation to facilitate matrix denaturation. The final volume of two solubilized gels in 0.8 ml lysis buffer is about 1 ml.
4. Add protease inhibitor cocktail to 1×.

Isolating Cysts from Collagen I

Reagents

1. MEM buffered with 20 mM HEPES (pH 7.6) and including 4% BSA and 1% antibiotics.
2. Collagenase-1 solution, 0.5 ml/filter. Low-purity collagenase contains significant quantities of contaminating proteases. Thus, use of high-purity collagenase (for instance, Sigma C-0773 or Worthington CLSPA) is crucial if the protein of interest is extracellular. Considerable lot-to-lot variability exists. Data on the purity of Worthington collagenase is available online (http://www.worthington-biochem.com/CLS/clssamp.html), and we often sample several lots before purchasing a large quantity. Protease inhibitor cocktail can also be included; these inhibitors do not affect collagenase-1 but do inhibit some common contaminants. Lower purity collagenase such as Worthington CLS-4 may be used for examining intracellular proteins. Prepare a solution of 4000 U collagenase/ml and 0.05 mg/ml DNase (Boehringer 104159) in MEM. The DNase prevents the DNA of any prematurely lysed cells from causing cell clumping.
3. Trypsin stock solution. To isolate the intracellular population of a secreted or transmembrane protein, trypsin (Worthington TRL3) can be included in the collagenase solution. Make a solution of 50 mg/ml trypsin in PBS$^+$.
4. Ultra Low Cluster 24-well plates (Corning 3473) to minimize cell adhesion.
5. TLCK   (L-1-chloro-3-(4-tosylamido)-7-amino-2-heptanone-hydrochloride) stock solution. Prepare a solution of 10 mg/ml TLCK (Boehringer 874485) in 1 mM HCl. Stock can be stored for a month at room temperature.
6. Protease inhibitor cocktail as described in the solubilization protocol previously.

Collagen I Digestion

1. Wash filters three times quickly with PBS$^+$.
2. Aliquot 0.5 ml collagenase solution into the wells of a 24-well low cluster plate, one well per gel. If digestion of extracellular proteins is desired, add 100 μl of trypsin stock to each well. The MEM will turn yellowish with trypsin addition. Prewarm the plate to 37°.
3. Remove gels from filters and place one gel in each well. Gently create four to five tears in the gel with forceps to facilitate digestion. Rotate at 37° 45 min or until gels are completely digested.
4. Add protease inhibitor cocktail to digested samples. If trypsin was included, add TLCK stock to 1×.

5. Transfer digested samples to Microfuge tubes. Even with low-cluster plates, we find that significant numbers of cells often adhere to the wells. It may be helpful to visually examine the wells at low magnification to assess whether additional washing to remove the cells is warranted. To wash, add 400 μl MEM to well and pipette up and down, then add the wash to the rest of the sample in the Microfuge tube.

6. Pellet cells by centrifuging 5 min at 3200g. A small pellet should be visible.

7. Remove the supernatant. This fraction contains digested matrix, as well as factors in the matrix that were not strongly cell associated. If desired, it can be set aside for immunoprecipitation of proteins of interest.

8. Wash pellets twice with ice-cold PBS$^+$, gently resuspending and then repelleting cells with each wash.

9. If sample normalization is desired, a small, concentrated fraction of the pellet should be set aside at this point. To do this, we resuspend each pellet in a volume of 50 μl (corresponding to one filter) and set aside 7.5 μl to measure for protein concentration using a bicinchoninic assay (BCA) (Sigma-Aldrich BCA-1). Because of the difficulty of repelleting the remaining 42.5 μl, we dilute the remaining suspension with about 1 ml PBS$^+$, then repellet the cells and remove the supernatant.

10. At this point, the cell pellet can be either lysed directly in sample buffer for SDS-PAGE or processed for immunoprecipitation (see the following).

Immunoprecipitation

Reagents

1. Triton dilution buffer. 5% Triton X-100, 100 mM TEA-Cl (pH 8.1), 100 mM NaCl, 5 mM EDTA, 1% Trasylol, 0.1% NaN$_3$.
2. SDS lysis buffer. Described previously in the solubilization protocol.
3. Superose 6 beads, "prep grade" (Amersham Pharmacia 17-0489-01).

Preparing Samples for Immunoprecipitation

1. If the starting material is heat-solubilized lysates already in SDS lysis buffer, add 500 μl Triton dilution buffer per 1 ml lysate.

2. If the starting material is a cell pellet, add 1 ml SDS lysis buffer containing 1× protease inhibitor cocktail, boil for 10 min, then add 500 μl Triton dilution buffer.

3. Preclear lysates by rotating with 20 μl Superose 6 beads at least 20 min at room temperature. Spin 5 min at 14,000g and retain the supernatant.

4. Add primary antibody for immunoprecipitation and continue as for conventional lysates (Lipschutz *et al.*, 2001).

Assaying Rac1 Activation

Background. The activation state of Rho GTPases in the context of 3D morphogenesis can be mechanistically revealing. In 2D culture, pull-down assays have been a common strategy to determine the extent of activated Rho proteins (Ren and Schwartz, 2000). Such assays take advantage of the ability of activated, GTP-bound Rho proteins to specifically interact with particular effector domains. For instance, the conserved cdc42/Rac1 interacting binding (CRIB) domain of Pak kinases 1–3 strongly binds only the GTP-forms of Rac1 and cdc42. Conjugating a recombinant gluta- thione S-transferase (GST) CRIB domain fusion protein to glutathione beads thus provides a means to uniquely isolate, or "pull down", the activated population of Rac1 or cdc42. This standard protocol requires that cells be lysed rapidly at 4° (to minimally perturb GTPase activation state) and without denaturation (to preserve the GTPases' CRIB-binding competence). The two methods of 3D culture lysis that were described here are, therefore, incompatible with the standard pull-down assay; colla- genase isolation requires a lengthy incubation at 37°, and heat-mediated solubilization causes denaturation.

Recently, we have developed a lysis technique compatible with the requirements of the pull-down assay (Yu *et al.*, 2005). We have used this modified protocol to examine Rac1-GTP in 3D cultures. Because GTP- bound Rac1 is inherently unstable, the entire assay must be performed on ice within an hour of lysis. When performed with cyst and tubule lysates, the assay measures steady-state activation levels. We also describe an alternative 3D culture technique of collagen I overlay, reported by Schwimmer and Ojakian (Schwimmer and Ojakian, 1995), that can be used to determine the kinetics of β1-integrin–mediated Rac1 activation in response to cell adhesion to collagen.

Reagents

1. 24-mm transwell filters (Costar 3412) for collagen I overlay.

2. 2× lysis buffer containing 2% Triton X-100, 40 mM Tris (pH 8.0), 1 M NaCl, 20 mM MgCl$_2$, 30% glycerol, 1 mM DTT, and EDTA-free protease inhibitors (Roche).

3. GST-CRIB beads coupled to glutathione, prepared as described in Ren and Schwartz (2000) using a recombinant GST-CRIB fusion protein of Pak3 (Bagrodia et al., 1995).

Overlaying with Collagen I and Lysing Cells. Overlaying filter-grown MDCK monolayers with collagen I initiates the formation of tubulocysts through a mechanism dependent on Rac1 and $\beta 1$ integrin (Schwimmer and Ojakian, 1995; Yu et al., 2005). During tubulocyst formation, cells in the monolayer first proliferate to form a stratified multilayer and then create lumens between the stratified layers. Many of the changes in cell shape and apicobasolateral polarity that occur during this process are reminiscent of cyst and tubule development, and in at least some of these cases similar signaling pathways are involved.

1. Trypsinize a confluent plate of MDCK cells. Split cells and plate onto 24-mm transwell filters, one filter per time point. Allow at least 4 days for cells to form a polarized monolayer. Remove old media and replace with fresh media every 2–3 days, including the day before the experiment.

2. Make collagen I solution as for cyst plating.

3. Wash filters once with PBS⁻ then once with 0.4 ml collagen I solution.

4. Remove PBS⁻ from both top and bottom chambers and wash the apical surface of the cells (top chamber) once quickly with 0.4 ml collagen I solution.

5. Add 0.4 ml collagen I solution on top of each filter and 2 ml growth media to the bottom chamber. Incubate at 37°. If doing a time course, stagger the overlay step as appropriate. Include a filter with cells but no collagen I overlay (as the zero time point) and a filter with collagen I but no cells (as a negative control).

Lysing Collagen I Cultures

1. Wash cyst cultures or the bottom chamber of overlaid filters once with ice-cold PBS⁺.

2. Peel off the gel and place into a Microfuge tube containing ice-cold 2× lysis buffer. For overlay cultures, cut out the filter with a scalpel and add this to the tube. At time points earlier than 20 min, the collagen is likely not polymerized. In this case, transfer the collagen to the tube using a 1-ml pipette tip whose opening has been widened by cutting a few millimeters off the tip end. The volume of 2× lysis buffer should be such that the final concentration of the lysis buffer will be 1× after solubilization. This volume is 0.5 ml for collagen I–overlaid monolayers on a 24-mm filter. Therefore, to correct for the collagen present in the assay, we add 0.4 ml of gelled collagen solution to the tube containing the

filter that has not been overlaid with collagen. If starting with a different-sized filter or a different volume of collagen I, the amount of 2× lysis buffer to add for a final concentration of 1× can be estimated by weighing the filter and collagen I and using 1 ml of 2× lysis buffer per gram of weight.

3. Rotate the tubes at 4° for 30 min. After this incubation, over 95% of the collagen should be solubilized.

4. Spin samples at 12,000 rpm in a Microfuge at 4° for 5 min to remove traces of undissolved collagen gel.

5. Set aside 50 μl of the supernatant as a total Rac1 loading control. Add 50 μl sample buffer to the loading control and boil 5 min.

Pulling Down Rac1-GTP

1. Add 800 μl of the remaining supernatant to Microfuge tubes containing 20 μl CRIB beads and rotate at 4° for 30 min.

2. Spin at 2000 rpm for 2 min to pull down Rac1-GTP bound to the CRIB beads. Discard the supernatant.

3. Wash beads twice with 1× lysis buffer. Completely remove the last wash solution using a Hamilton syringe.

4. Add 12 μl sample buffer to beads. Boil 5 min. Pellet beads at 2000 rpm for 2 min. The supernatant contains the Rac1-GTP fraction and is ready for SDS-PAGE.

Acknowledgments

L. E. O. was supported by an American Heart Association predoctoral fellowship and an NIH training grant. W. Y. is supported by fellowships from the American Heart Association, Western Affiliate (0325013Y and 0120084Y), and the National Kidney Foundation. T.-S. J. was supported by a research grant from the National Taiwan University Hospital (NTUH 92-S014). M. Z. was supported by a fellowship of the American Cancer Society, California Division (2–7-99) and start-up funds from the Department of Surgery, University of Cincinnati. This work was supported by NIH grants to K. E. M.

References

Bagrodia, S., Taylor, S. J., Creasy, C. L., Chernoff, J., and Cerione, R. A. (1995). Identification of a mouse p21Cdc42/Rac activated kinase. *J. Biol. Chem.* **270,** 22731–22737.
Burridge, K., and Wennerberg, K. (2004). Rho and Rac take center stage. *Cell* **116,** 167–179.
Debnath, J., and Bragge, J. S. (2005). Modelling glandular epithelial cancers in three-dimensional cultures. *Nat. Rev. Cancer.* **5,** 675–688.
Hogan, B. L., and Kolodziej, P. A. (2002). Organogenesis: Molecular mechanisms of tubulogenesis. *Nat. Rev. Genet.* **3,** 513–523.
Jou, T. S., and Nelson, W. J. (1998). Effects of regulated expression of mutant RhoA and Rac1 small GTPases on the development of epithelial (MDCK) cell polarity. *J. Cell Biol.* **142,** 85–100.

Jou, T. S., Schneeberger, E. E., and Nelson, W. J. (1998). Structural and functional regulation of tight junctions by rhoA and rac1 small GTPases. *J. Cell Biol.* **142,** 101–115.

Lipschutz, J. H., O'Brien, L. E., Altshuler, Y., Avrahami, D., Nguyen, Y., Tang, K., and Mostov, K. E. (2001). Analysis of MEMBRANE TRAFFIC IN POLARIZED EPITHELIAL CELLS. *In* "Current Protocols in Cell Biology" (J. S. Bonifacano, M. Dasso, J. B. Harford, J. Lippincott-Schwartz, and K. M. Yamada, eds.), Vol. Unit 15.5. John Wiley & Sons, Indianapolis.

Louvard, D. (1980). Apical membrane aminopeptidase appears at site of cell-cell contact in cultured kidney epithelial cells. *Proc. Natl. Acad. Sci. USA* **77,** 4132–4136.

Montesano, R., Matsumoto, K., Nakamura, T., and Orci, L. (1991a). Identification of a fibroblast-derived epithelial morphogen as hepatocyte growth factor. *Cell* **67,** 901–908.

Montesano, R., Schaller, G., and Orci, L. (1991b). Induction of epithelial tubular morphogenesis *in vitro* by fibroblast-derived soluble factors. *Cell* **66,** 697–711.

O'Brien, L. E., Zegers, M. M., and Mostov, K. E. (2002). Opinion: Building epithelial architecture: Insights from three-dimensional culture models. *Nat. Rev. Mol. Cell. Biol.* **3,** 531–537.

O'Brien, L. E., Jou, T. S., Hansen, S. H., Pollack, A. L., Zhang, Q., Yurchenco, P. D., and Mostov, K. E. (2001). Rac1 orientates epithelial apical polarity through effects on basolateral laminin assembly. *Nat. Cell Biol.* **3,** 831–838.

O'Brien, L. E., Tang, K., Kats, E. S., Schutz-Geschwender, A., Lipschutz, J. H., and Mostov, K. E. (2004). ERK and MMPs sequentially regulate distinct stages of epithelial tubule development. *Dev. Cell* **7,** 21–32.

Pollack, A. L., Runyan, R. B., and Mostov, K. E. (1998). Morphogenetic mechanisms of epithelial tubulogenesis: MDCK cell polarity is transiently rearranged without loss of cell-cell contact during scatter factor/hepatocyte growth factor-induced tubulogenesis. *Dev. Biol.* **204,** 64–79.

Ren, X. D., and Schwartz, M. A. (2000). Determination of GTP loading on Rho. *Methods Enzymol.* **325,** 264–272.

Santos, O. F., and Nigam, S. K. (1993). HGF-induced tubulogenesis and branching of epithelial cells is modulated by extracellular matrix and TGF-β. *Dev. Biol.* **160,** 293–302.

Schwimmer, R., and Ojakian, G. K. (1995). The $\alpha 2\beta 1$ integrin regulates collagen-mediated MDCK epithelial membrane remodeling and tubule formation. *J. Cell Sci.* **108,** 2487–2498.

Van Aelst, L., and Symons, M. (2002). Role of Rho family GTPases in epithelial morphogenesis. *Genes Dev.* **16,** 1032–1054.

Walpita, D., and Hay, E. (2002). Studying actin-dependent processes in tissue culture. *Nat. Rev. Mol. Cell. Biol.* **3,** 137–141.

Wang, F., Weaver, V. M., Petersen, O. W., Larabell, C. A., Dedhar, S., Briand, P., Lupu, R., and Bissell, M. J. (1998). Reciprocal interactions between $\beta 1$-integrin and epidermal growth factor receptor in three-dimensional basement membrane breast cultures: A different perspective in epithelial biology. *Proc. Natl. Acad. Sci. USA* **95,** 14821–14826.

Weaver, V. M., Lelievre, S., Lakins, J. N., Chrenek, M. A., Jones, J. C., Giancotti, F., Werb, Z., and Bissell, M. J. (2002). $\beta 4$ integrin-dependent formation of polarized three-dimensional architecture confers resistance to apoptosis in normal and malignant mammary epithelium. *Cancer Cell* **2,** 205–216.

Williams, M. J., and Clark, P. (2003). Microscopic analysis of the cellular events during scatter factor/hepatocyte growth factor-induced epithelial tubulogenesis. *J. Anat.* **203,** 483–503.

Yu, W., Datta, A., Leroy, P., O'Brien, L. E., Mak, G., Jou, T. S., Matlin, K. S., Mostov, K. E., and Zegers, M. M. (2005). $\beta 1$-integrin orients epithelial polarity via Rac1 and laminin. *Mol. Biol. Cell* **16,** 433–445.

[54] Using Three-Dimensional Acinar Structures for Molecular and Cell Biological Assays

By BIN XIANG and SENTHIL K. MUTHUSWAMY

Abstract

Epithelial cells grown in three-dimensional matrix allows one to investigate biological processes in a setting that is closer to *in vivo* conditions than those obtained by growing cells on plastic culture dishes. Here we outline procedures that allow one to investigate molecular mechanisms that regulate formation and disruption of three-dimensional epithelial structures.

Introduction

Epithelial cells lining glandular structures such as the breast exist in a three-dimensional (3D) space and assume a 3D architecture. Several studies in the past have demonstrated that the 3D organization is critical for normal physiological functions of mammary epithelial cells. Although *in vitro* culture has armed us with powerful tools to investigate molecular mechanism of development and disease, cells cultured on plastic dishes lose their 3D organization and, hence, limits our ability to conduct physiologically relevant investigations.

Several recent studies demonstrate that mammary epithelial cells undergo a morphogenetic program that results in formation of 3D structures reminiscent of acinar structures *in vivo*, when grown under 3D culture conditions (Bissell *et al.*, 2003; Muthuswamy *et al.*, 2001). These studies have generated a new wave of interest in using modified *in vitro* culture conditions to investigate the molecular mechanism that regulates normal morphogenesis and mechanism by which oncogenes disrupt them.

Several reports have highlighted the fact that epithelial cells grown as 3D structures respond to oncogenic stimuli and promote uncontrolled proliferation, prevent apoptosis, and induce invasion, three properties of the cancer phenotype (Bissell *et al.*, 2003; Muthuswamy *et al.*, 2001; Seton-Rogers *et al.*, 2004; Weaver *et al.*, 2002). We have recently outlined a detailed method for culturing human mammary epithelial cells on a 3D matrix and analyzing them using cell biological tools (Debnath *et al.*, 2003). Here we provide a basic protocol for growing cells in 3D matrix and

METHODS IN ENZYMOLOGY, VOL. 406
0076-6879/06 $35.00
DOI: 10.1016/S0076-6879(06)06054-X

provide detailed methods for using the 3D structures to perform molecular and biochemical analyses of cell cycle, apoptosis and invasion.

Methods

Setting-Up 3D Assay

Materials. Growth factor–reduced Matrigel™ (BD Biosciences, Cat# 354230); chamber slides (Falcon Cat# 354108), assay media (DMEM/F12 [Invitrogen Cat# 11965-118], insulin [10 μg/ml], hydrocortisone [0.5 μg/ml], cholera toxin [100 ng/ml], 2% horse serum [HS], 1× Pen/Strep [Invitrogen, Cat # 15070-063]).

1. Thaw Matrigel™ on ice (usually takes a long time and will depend on the size of the aliquot). Note: Matrigel™ will remain as liquid on ice but will solidify at room temperature.

2. Add 40 μl of Matrigel™ to each well of 8-well glass chamber slide (at 50 μl/cm^2) and spread the Matrigel™ evenly using a P-200 tip. Take care not to generate a meniscus or air bubbles. You may want to do this step on ice until you get used to the procedure.

3. Place the slides in the cell culture incubator and allow the Matrigel™ to solidify (takes 15–20 min).

4. Trypsinize cells and resuspend in 1.0 ml of media containing 20% HS (no other additives). Add additional 2.0 ml of 20% HS containing media and incubate the cells for 1–2 min at RT (this step ensures inactivation of trypsin).

5. Spin the cells at 900 rpm for 3 min.

6. Resuspend the cells in 10 ml of assay medium.

7. Count cells and aliquot 100,000 cells in to a fresh 15-ml conical tube. Bring up the volume to 4.0 ml using assay media (label this tube as "A"). Final concentration will be 2500 cells/100 μl.

8. Prepare assay media containing 5% Matrigel™ and 10 ng/ml EGF (label this tube "B"). Use media that is cooled on ice—does not have to be ice cold! Because the EGF stock solution is stored at 100 μg/ml, it is advisable to prepare a working dilution of 10 μg/ml using assay media. This diluted stock may not be stored for extended use, because EGF will not remain stable when diluted. Attempting to aliquot very small volumes directly from the 100 μg/ml stock will be a likely source of experimental error.

9. Mix tube A and tube B 1:1 and plate 400 μl of the Matrigel™/cell mix onto the solidified Matrigel™ (Step 3).

10. Gently transfer the chamber slide into the incubator.

11. Change media once every 4 days with assay media containing 2.5% Matrigel™ and 5.0 ng/ml EGF. MCF-10A cells take about 8–12 days to form acini-like structures.

Indirect Immunofluorescence

All liquid aspirations should be performed with a mild vacuum. If you are not sure, use a P-1000 pipette to aspirate off solutions. All washes should be performed on a gentle rocker. Keep in mind that you are likely to loose the structures if you aspirate using a powerful vacuum manifold or if you add wash buffers too fast or if the rocker is set to high speed.

Materials. 10% Formalin (Sigma Catalog # HT50-1-4); dilute 1:1 in $1\times$ PBS to make 5% solution.

Goat serum (Sigma Cat# G 6767); F(ab)2, (Jackson Immunochemicals Catalog #M 35200). Secondary antibodies are from Molecular Probes, TOPRO-3 (Molecular Probes), DAPI (Boehringer Mannheim, Catalog # 236 276), prolong antifade, (Molecular Probe, Catalog # P-7481).

$10\times$ PBS/glycine: Dissolve 38.00 g NaCl; 9.38 g Na_2HPO_4; 2.07 g NaH_2PO_4; 37.50 g glycine in 500 ml of distilled deionized water, pH 7.4 and filter sterilize.

$10\times$ IF-wash: Dissolve 38.00 g NaCl; 9.38 g Na_2HPO_4; 2.07 g NaH_2PO_4; 2.5 g NaN_3; 5.0 g BSA (fraction V); 10.00 ml Triton X-100; 2.5 ml Tween-20 in 500 ml of distilled deionized water, pH 7.4, and filter sterilize.

1. Aspirate media.
2. Quick wash with 400 μl of $1\times$ PBS
3. Fix cells with 5% formalin (diluted in $1\times$ PBS) for 30 min at room temperature (RT) OR use 1:1 mix of methanol/acetone and fix at $-20°$ for 10 min.
4. Rinse three times, 10 minutes each, with 400 μl $1\times$ PBS/glycine
5. If fixing with formalin, permeabilize using 400 μl of in $1\times$ PBS + 0.5% Triton X-100 at RT for 5 min.
6. Rinse three times, 10 minutes each, with $1\times$ IF wash at RT.
7. Incubate with 400 μl of primary block ($1\times$ IF + 10% goat serum) for 1–1.5 h at RT.
8. Aspirate the primary block and incubate with 200 μl secondary block [IF + 10% goat serum + 1:50 dilution of F(ab)2 (1.0 mg/ml stock)] for 30–40 minutes at RT.
9. Aspirate the secondary block and incubate with 150 μl of primary antibody diluted in secondary block for 1–2 h at RT on a gentle rocker. If the structures were fixed using PFA you can incubate the primary antibody overnight at RT or at 4°. If incubation is carried out at 4°, there is a possibility that the Matrigel™ will liquefy and you may loose the structures.

10. Rinse three times, 20 min each, with 400 μl of 1× IF wash.

11. Incubate with 150 μl of secondary antibody diluted in primary block for 50–60 min at RT. Almost all of our secondary antibodies are Alexa-conjugated from Molecular Probes and is used at 1:100 dilution. Experiment proceeds in dark (wrapped in aluminum foil).

12. Rinse three times, 20 min each, with IF wash at RT.

13. Incubate with 1× PBS containing 1:100 dilution of TOPRO-3 for 15 min at RT.

14. Incubate with 1× PBS containing 0.5 ng/ml DAPI for 5–10 min at RT.

15. Mount with freshly prepared Prolong (anti-fade) (Molecular Probes) and allow it to dry overnight at RT. Once dry, the slides can be stored at RT for 2–2 weeks. However, it is suggested that you plan to analyze the slides as soon as possible. You may experience dehydration of slides when stored at −20°.

Preparation of Protein Extracts

Yield: Two chamber slides of acinar structures can yield 0.5–1.0 mg of protein.

1. Wash the acini twice with 800 μl of 1× PBS.

2. Add 250 μl of trypsin-EDTA (10×, Invitrogen, Catalog #15400-054) to each well and incubate at 37° for about 25 min until the acinar structures detach from Matrigel™, but will remain as loose cell clumps. Collect the structures using a P-1000 pipette tip and transfer the cells to a Microfuge tube containing 1.0 ml of resuspension media (DMEM/F12 + 20% horse serum).

3. Spin down cells at 3000 rpm for 2 min. Wash the pellet twice with 1.0 ml of 1× PBS. The 1× PBS wash is critical to minimize Matrigel™ contamination.

4. Lyse cells with 200.00 μl of RIPA lysis buffer on ice for 30 min.

5. Clear the lysate by centrifuging at 13,000 rpm for 15 min at 4°.

6. Transfer the supernatant to a new Microfuge tube and measure protein concentration. In many instances, we notice that the optical density measurements are not reliable because of traces of Matrigel™ present in lysate. To control for Matrigel™ contamination, we routinely perform a second-tier normalization by resolving 10 μg of lysates and immunoblotting with performing beta-actin antibodies.

7. The protein lysates can be used for immunoblot and immunoprecipitation assays.

Isolation of RNA

Yield: One chamber slide can yield 20–50 μg of total RNA.

1. Lyse 3D acini using TRIzol (Invitrogen, Catalog #15596-018) 125 μl/well of 8-well chamber slide. If using 2- or 4-well chamber slides, the amount of TRIzol needed for each well should be adjusted to keep the total volume per chamber slide to 1.0 ml.
2. Leave the slide at RT for 10 min.
3. Collect everything in each well, including Matrigel™. The TRIzol-lysate can be store at −80° and can be used to isolate RNA at a later time.
4. Isolate total RNA from the TRIzol lysates following manufacturer's protocol.
5. Store isolated RNA at −80°.

Preparing Cell for FACS

Yield: You need at least one chamber slide for each condition.

1. Trypsinize and collect the acinar structures as described under "Preparation of Protein Extracts." Wash the cells twice with 1× PBS.
2. Resuspend the cells in 0.5 ml of 1× PBS and transfer the cells to a 15-ml conical tube containing 4.5 ml ice-cold 70 % ethanol. To prevent formation of cell clumps, the cells should be added as drops while vortexing the 70% ethanol in the 15.0-ml tube.
3. Fix the cells by incubating on ice for 4 hours. (Fixed cells can be stored at 4° for several months).
4. To stain with propidium iodide, spin down cells at 700 rpm for 3 min at RT and wash with 1× PBS once. Resuspend the cells in 0.5 ml of PI staining buffer (0.1% (v/v) Triton X-100, 0.2 mg/ml DNase-free RNase A, and 20 μg/ml propidium iodide in PBS) and incubate at RT for 30 min.
5. Analyze cells with flow cytometer.

Applications

Cell Cycle

Antibodies: ki67 (Oncogene, Catalog #NA59), Phospho-Histone3 (Upstate, Catalog #05–817), MCM (Mendez and Stillman, 2000), Cdk2 (Santa Cruz, Catalog #sc-163), Cdk4 (Santa Cruz, Catalog #sc-601), cyclin D1 (Santa Cruz, Catalog# sc-718), cyclin D2 (BD Pharmingen, Catalog

#554200), cyclin D3 (BD Transduction Laboratories, Catalog #610280), cyclin E (Santa Cruz, Catalog #sc-247), cyclin E2 (Santa Cruz, Catalog #sc-9566), p27 (BD Transduction Laboratories, Catalog #K25020).

Indirect Immunofluorescence. Cell cycle status can be monitored using immunofluorescence approaches. This approach is particularly useful for monitoring cell cycle status in individual acinus and in conditions where the phenotype is not highly penetrant and only a fraction of the acini display cell cycle deregulation. A brief outline of the protocol is provided in "Indirect Immunofluorescence" and additional details can be found elsewhere (Debnath *et al.*, 2003).

FACS. Propidium iodide stained–cells can be analyzed by FACS to monitor changes in cell cycle status. Our studies suggest that MCF-10A cells undergo a G1 arrest during morphogenesis (unpublished observations).

Changes in Protein Expression. Protein lysates collected at different stages of acinar morphogenesis can be used to analyze expression of cell cycle regulators by immunoblots. As shown in Fig. 1, significant changes in expression of cell cycle inhibitor p27 and changes in levels of phosho-Rb can be readily observed during morphogenesis.

Changes in Kinase Activity

CDK2 KINASE ASSAYS

1. Lyse cells with cdk2 lysis buffer (50 mM Tris-HCl, pH 7.4, 150 mM NaCl, 1% NP-40, 50 mM NaF, 1 mM PMSF, 10 μg/ml leupeptin, 10 μg/ml aprotinin).
2. Immunoprecipitate cdk2 using 0.5 mg of protein lysate.
3. Wash the immunoprecipitates twice with ice-cold lysis buffer.
4. Resuspended in 45 μl of cdk2 kinase buffer (20 mM Tris-HCl, pH 7.4, 7.5 mM MgCl$_2$, 1 mM DTT, 32 μM ATP, 1 μg histone H1, 10 μCi γ-^{32}P ATP at 37° for 30 min.
5. Stop the reaction by adding 9 μl 6× SDS sample buffer.
6. Boil the samples at 95° for 2 min and resolve on a 12% SDS-PAGE.
7. Transfer proteins to PVDF membrane.
8. Expose PVDF membrane with film or PhosphoImager screen. Scan and analyze image.
9. The PVDF membrane can be inmmunoblotted with anti-cdk2 antibody to serve as control.

CDK4 KINASE ASSAY

1. Lyse cells with cdk4 lysis buffer (50 mM HEPES, pH 7.5, 150 mM NaCl, 1 mM EDTA, 2.5 mM EGTA, 0.1% Tween-20, 10% glycerol,

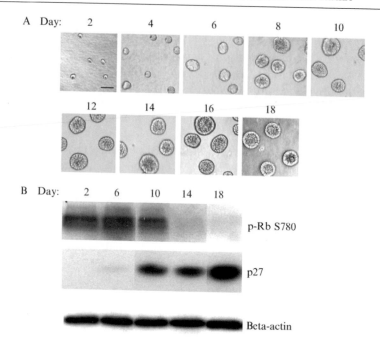

FIG. 1. (A) Phase images of morphogenesis. MCF-10A cells are seeded on Matrigel™ in chamber slide at 5000 cells per well density. Medium is changed every 4 days. Phase images are taken every 2 days. Scale bar, 50 μm. (B) Expression of p-Rb, p27 during 3D acini morphogenesis. The 3D acinar structures are trypsinized at different growth stages and lysed. An equal amount of protein is loaded onto SDS-PAGE gel, and phospho-Rb and p27 levels are analyzed by immunoblotting with the specific antibodies. Beta-actin blot serves as loading control.

10 mM β-glycerophosphate, 1 mM NaF, 0.1 mM orthovanadate, 10 μg/ml leupeptin, 10 μg/ml aprotinin, 1 mM DTT, 0.1 mM PMSF).

2. Immunoprecipitate cdk4 using 2.0 mg of protein lysate.

3. Wash the immunoprecipitates twice with ice-cold lysis buffer and three times with HEPES-DTT buffer (50 mM HEPES pH 7.5, 1 mM DTT).

4. Resuspend the immunoprecipitate in 30 μl of cdk4 kinase buffer (100 mM HEPES. pH 7.5, 2 mM DTT, 5 mM EGTA, 20 mM MgCl$_2$, 40 μM ATP, 0.2 mM orthovanadate, 2 mM NaF, 20 mM β-glycerophosphate, 10 μg Rb, 10 μCi γ-^{32}P ATP at 30° for 30 min.

5. Stop the reaction by adding 6 μl 6× SDS sample buffer and proceed as outlined previously.

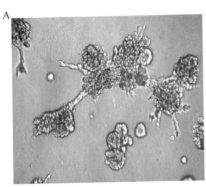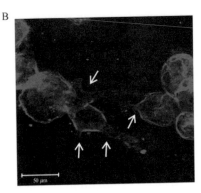

FIG. 2. Invasion phenotype of MCF-10A cells in 3D acini. (A) Phase morphology of invasion. MCF-10A cells expressing ErbB2 and ErbB3 chimera plasmids are cultured on Matrigel™/collagen mixture for 6 days and stimulated with synthetic dimerizer for 2 days. (B) Loss of intact laminin in invasive structures. The structure from A is fixed and stained with anti-laminin 5 antibody. Arrows show loss of laminin 5 at the basement membrane of invasive structures. Red, laminin 5. (See color insert).

Apoptosis

Indirect Immunofluorescence. There are two ways to monitor changes in apoptosis, monitoring EtBr uptake or immunofluorescence for cleaved caspase-3. Both these approaches have been outlined elsewhere and will not be discussed here (Debnath *et al.*, 2003).

Invasion

Acinar structures grown in Matrigel™ can be monitored for invasive potential. We have observed development of multicellular protrusions and development of interacinar bridges as indicators of invasive behavior (Fig. 2A).

In many instances, using pure Matrigel™ does not allow the cells to display an invasive behavior. Recent observations have demonstrated that viscoelastic properties of the matrix can regulate branching morphogenesis of mammary epithelial cells (Paszek and Weaver, 2004). To increase the matrix stiffness, we plate cells on a 1:1 mixture of Matrigel™ and collagen.

Matrigel™/Collagen Invasion Assay. All the following steps should be performed on ice.

1. Tube A: Mix 1.0 ml of collagen (Vitrogen, Cohesion Technologies Catalog #FXP-019), 125 μl of 10× PBS, 125 μl of 0.1 M NaOH, and 25 μl of 0.1 M HCl.

2. Tube B: 1.0 ml of Matrigel™.

3. Mix 1.0 ml from tube A with the Matrigel™ in tube B using a P1000 pipette.

4. Coat each well of 8-well chamber slide with 50 μl of Matrigel™/ collagen I mixture. Let it solidify by placing the slide in cell culture incubator for 30 min.

5. Plate cells on this Matrigel™/collagen mixture as outlined under "setting up 3D assay."

6. Change medium every 4 days. Invasive structures can be observed by 6–12 days in culture.

Gelatin Zymography. Conditioned media from 3D cultures can be analyzed for presence of secreted, activated forms of matrix metalloproteins (MMPs). Details for gelatin zymography can be found elsewhere (Pruijt *et al.*, 1999).

Indirect Immunofluorescence. Polarized epithelial cells are known to secrete and assemble basement membrane matrix in their basal surface. Non-invasive 3D acinar structures possess an intact layer of basement membrane that can be visualized by indirect immunofluorescence using anti-laminin 5 (Chemico, Catalog #MAB19562) antibodies as outlined in "Indirect Immunofluorescence." Loss of an intact laminin layer can be monitored to determine the invasive potential of the cells grown in 3D. As shown in Fig. 2B, invasive regions of the structures lack an intact layer of laminin.

References

Bissell, M. J., Rizki, A., and Mian, I. S. (2003). Tissue architecture: The ultimate regulator of breast epithelial function. *Curr. Opin. Cell Biol.* **15**, 753–762.

Debnath, J., Muthuswamy, S. K., and Brugge, J. S. (2003). Morphogenesis and oncogenesis of MCF-10A mammary epithelial acini grown in three-dimensional basement membrane cultures. *Methods* **30**, 256–268.

Mendez, J., and Stillman, B. (2000). Chromatin association of human origin recognition complex, cdc6, and minichromosome maintenance proteins during the cell cycle: Assembly of prereplication complexes in late mitosis. *Mol. Cell. Biol.* **20**, 8602–8612.

Muthuswamy, S. K., Li, D., Lelievre, S., Bissell, M. J., and Brugge, J. S. (2001). ErbB2, but not ErbB1, reinitiates proliferation and induces luminal repopulation in epithelial acini. *Nat. Cell Biol.* **3**, 785–792.

Paszek, M. J., and Weaver, V. M. (2004). The tension mounts: Mechanics meets morphogenesis and malignancy. *J. Mammary Gland Biol. Neoplasia* **9**, 325–342.

Pruijt, J. F., Fibbe, W. E., Laterveer, L., Pieters, R. A., Lindley, I. J., Paemen, L., Masure, S., Willemze, R., and Opdenakker, G. (1999). Prevention of interleukin-8-induced mobilization of hematopoietic progenitor cells in rhesus monkeys by inhibitory antibodies against the metalloproteinase gelatinase B (MMP-9). *Proc. Natl. Acad. Sci. USA* **96**, 10863–10868.

Seton-Rogers, S. E., Lu, Y., Hines, L. M., Koundinya, M., LaBaer, J., Muthuswamy, S. K., and Brugge, J. S. (2004). Cooperation of the ErbB2 receptor and transforming growth factor beta in induction of migration and invasion in mammary epithelial cells. *Proc. Natl. Acad. Sci. USA* **101**, 1257–1262.

Weaver, V. M., Lelievre, S., Lakins, J. N., Chrenek, M. A., Jones, J. C., Giancotti, F., Werb, Z., and Bissell, M. J. (2002). Beta4 integrin-dependent formation of polarized three-dimensional architecture confers resistance to apoptosis in normal and malignant mammary epithelium. *Cancer Cell* **2**, 205–216.

Further Reading

Debnath, J., Mills, K. R., Collins, N. L., Reginato, M. J., Muthuswamy, S. K., and Brugge, J. S. (2002). The role of apoptosis in creating and maintaining luminal space within normal and oncogene-expressing mammary acini. *Cell* **111**, 29–40.

[55] TC10 and Insulin-Stimulated Glucose Transport

By Shian-Huey Chiang, Louise Chang, and Alan R. Saltiel

Abstract

Insulin stimulates glucose uptake in insulin-responsive tissues by means of the translocation of the glucose transporter GLUT4 from intracellular sites to the plasma membrane. Two pathways are required, one involving activation of a phosphatidylinositol 3-kinase (PI 3-kinase) and downstream protein kinases, and one involving activation of the Rho-family GTPase TC10. TC10 activation by insulin is catalyzed by the exchange factor C3G, which is translocated to lipid rafts along with its binding partner CrkII as a consequence of Cbl tyrosine phosphorylation by the insulin receptor. This activation of TC10 is dependent on localization of TC10 in the lipid raft subdomains of the plasma membrane. We describe experimental approaches using the insulin-responsive cell line 3T3-L1 adipocytes to study the role of TC10 in insulin-stimulated glucose transport.

Introduction

Rho-family small GTPases are molecular switches in signal transduction pathways that link cell surface receptors to a variety of intracellular responses. These small GTP-binding proteins are critical for intracellular vesicle trafficking events including phagocytosis, endocytosis, mast cell degranulation, and exocytosis (Symons and Rusk, 2003). TC10 is an unusual member of the Rho family that has been recently identified as an important proximal protein in the control of insulin action

METHODS IN ENZYMOLOGY, VOL. 406   0076-6879/06 $35.00
        DOI: 10.1016/S0076-6879(06)06055-1

in adipocytes (Chiang *et al.*, 2001; Khan and Pessin, 2002; Saltiel and Pessin, 2002, 2003).

Insulin increases glucose uptake in skeletal muscle and adipose tissue by means of the glucose transporter GLUT4. In the basal state, GLUT4 cycles slowly between the plasma membrane and intracellular vesicles, with most of the transporters residing within cells. Insulin triggers a large increase in the rate of GLUT4 vesicle exocytosis and a smaller decrease in GLUT4 endocytosis, resulting in its accumulation on the cell surface and a concomitant increase in glucose uptake (Bryant *et al.*, 2002; Khan and Pessin, 2002; Mora and Pessin, 2002; Watson and Pessin, 2001). The G protein TC10 is highly expressed in insulin-sensitive tissues, including fat and muscle (Neudauer *et al.*, 1998). TC10 is activated by insulin, and, in turn, plays a part in regulating insulin-stimulated Glut4 trafficking, docking, and fusion with the plasma membrane in adipocytes (Chang *et al.*, 2002; Inoue *et al.*, 2003). TC10 localizes to caveolin-enriched lipid microdomains in the plasma membrane (Chiang *et al.*, 2002; Watson *et al.*, 2001). The compartmentalization of TC10 within lipid raft microdomains is essential for insulin-dependent activation of TC10 and GLUT4 translocation (Watson *et al.*, 2001). The TC10 pathway functions in parallel with phosphatidylinositol 3-kinase activation pathway to stimulate GLUT4 translocation in response to insulin (Chiang *et al.*, 2001). In this chapter, we will discuss methods to study the activity and role of TC10 in insulin-stimulated GLUT4 translocation in the model adipocyte cell line 3T3-L1.

Evaluation of TC10 Activation State *In Vitro*

Even though TC10 is most similar to Cdc42 among proteins in the Rho family, differences between these two proteins have been noted, including tissue distribution, downstream signaling effects, upstream activating proteins, downstream inactivating proteins, interactions with effectors, and intrinsic GTPase activity (Chang and Saltiel, unpublished results, 2005; Chiang *et al.*, 2003; Neudauer *et al.*, 1998). In addition, TC10 can also be differentiated from other Rho family proteins on the basis of its requirement for magnesium ions during GTP loading (Chang and Saltiel, unpublished results, 2005).

Preparation of GST-Fusion Proteins

Plasmids pGEX-2T expressing glutathione S-transferase (GST), GST-TC10, or GST-Cdc42 proteins are transformed into BL21(DE3)pLysS bacteria and inoculated into 100 ml LB containing 100 μg/ml ampicillin.

When bacteria reach an $OD_{600} = 0.7$, IPTG is added to a final 0.1 mM concentration, and then incubated at room temperature overnight with constant agitation. For high protein expression of small GTPases, it is best to induce bacteria with a low concentration of IPTG at a low temperature for an extended period of time. Inducing protein expression of GTPases in bacteria at higher temperatures or with high concentrations of IPTG tends to be toxic to the bacteria, decreasing protein expression. The cells are pelleted at 5000×g for 10 min. The bacteria are then resuspended in ice-cold 4 ml TEN (50 mM Tris, pH 7.5, 0.5 mM EDTA, 0.3 M NaCl). 40 μl 100 mg/ml lysozyme is added. After 15 min incubation on ice, 80 μl 10% NP-40 is added for 10 min. Cells are snap-frozen in liquid nitrogen, and allowed to thaw slowly at room temperature. Two Complete Mini Protease inhibitor cocktail tablets (Roche Diagnostics) are dissolved in 6 ml ice-cold NaCl-Mg (1.5 M NaCl, 12 mM MgCl$_2$) and then added to cell lysate; 200 μl 10 mg/ml DNase I is then added. To break up DNA, cells are intermittently pipetted up and down several times using a 10-ml pipette until the DNA is completely digested, when the solution in no longer viscous. The lysate is then centrifuged for 20 min at 15,000×g, and the supernatant is stored in 1-ml aliquots at −70°.

To couple GST or GST-fusion proteins to glutathione-coupled agarose (Amersham Biosciences), 1 ml lysates are incubated with 200 μl 1:1 slurry glutathione-agarose in PBS (phosphate buffered saline, pH 7.4) at 4° for 1 h with constant inversion. The glutathione-agarose bound GST fusion proteins are washed three times with 1 ml PBS, pH 7.4, and resuspended in an equal volume of PBS, pH 7.4, making a 1:1 slurry. Glutathione-agarose bound GST fusion proteins are stored in 25- or 50-μl aliquots at −70°. To determine protein concentrations, 5-μl aliquots of GST or GST-fusion proteins bound to glutathione agarose are diluted with 5 μl 2× SDS-PAGE sample buffer and compared with known concentrations of BSA analyzed by SDS-PAGE and Coomassie blue staining.

GTP Loading of TC10

One microgram of GST-fusion protein coupled to glutathione-agarose is equilibrated in 200 μl binding buffer (25 mM Tris, pH 7.5, 40 mM NaCl, 1.0 mM DTT, and 0.5% NP-40) containing the desired concentration of MgCl$_2$ (0–100 mM), and incubated at room temperature for 10 min; 0.2 pmol [^{35}S] GTPγS at a specific activity of 1000 cpm/fmol is diluted in 50 μl binding buffer plus the indicated MgCl$_2$ concentration and is added to the beads. The samples are incubated with constant inversion for 1 h. The beads are washed three times with 1.0 ml binding buffer containing the corresponding MgCl$_2$ concentration. Washed beads are resuspended

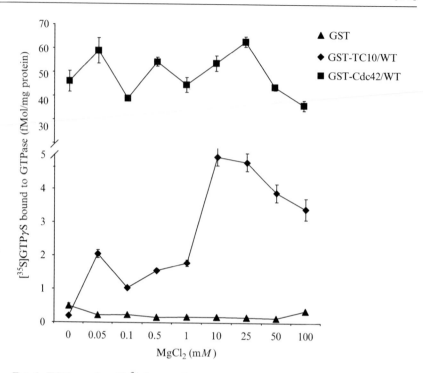

FIG. 1. TC10 requires Mg^{2+} for guanine nucleotide binding. GST, GST-TC10, or GST-Cdc42 proteins were expressed in bacteria, purified, and coupled to glutathione-agarose. One microgram of GST, GST-TC10, or GST-Cdc42 coupled to glutathione-agarose was resuspended in binding buffer containing the indicated amount of $MgCl_2$ incubated at room temperature for 10 min., and then 0.2 pMol [^{35}S]GTPγS was added. Samples were incubated at room temperature with constant inversion for 1 h and washed three times with binding buffer containing the indicated amount of $MgCl_2$. The beads were then resuspended in 100 μl binding buffer, added to 5-ml scintillation cocktail, and then counted. Bound [^{35}S]GTPγS was expressed in fMol/mg protein. All assays were done in triplicate, and the results are the mean ± standard error. The results are representative of five independent experiments.

in 100 μl binding buffer and added to 5 ml scintillation cocktail for counting using a Beckman Coulter LS-6500 multipurpose scintillation counter. Bound [^{35}S] GTPγS is expressed as fmol/mg protein. All assays are done in triplicate, and results are the mean ± standard error. GST is included in the assay as a negative control to confirm specificity of GTP binding to the GST-fusion protein.

As shown in Fig. 1, Cdc42 does not require Mg^{4+} for binding to guanine nucleotides. However, TC10 does require Mg^{2+} for guanine nucleotide loading. The optimal Mg^{2+} concentration for GTP binding to TC10 is between 10 and 25 mM (Fig. 1) (Chang and Saltiel, unpublished results, 2005).

TC10 Activation in Response to Insulin in 3T3-L1 Adipocytes

TC10 undergoes activation in response to insulin in 3T3-L1 adipocytes (Chiang et al., 2001, 2002). This effect of insulin is rapid, dose-dependent and not blocked by pharmacological inhibitors of the PI 3-kinase or MAP kinase pathways (Chiang et al., 2001).

Differentiation of 3T3-L1 Adipocytes

3T3-L1 fibroblasts in 150-mm dishes are cultured in 25 ml of high-glucose DMEM supplemented with 25 mM glucose and 10% newborn calf serum at 37° in 10% CO_2 incubator. For differentiation into adipocytes, fibroblasts are cultured for 3–5 days after reaching confluence with changes of medium every 4 days, after which the cells are cultured for 3 days in DMEM containing 10% fetal bovine serum, 1 μg/ml insulin (Sigma #I-5523), 0.25 μM dexamethasone (Sigma #D-1756), and 0.5 mM methylisobutylxanthine (Sigma #I-5879) and for 2 days in DMEM containing 10% fetal bovine serum and 1 μg/ml insulin. After differentiation, adipocytes are maintained in DMEM supplemented with 10% fetal bovine serum. Adipocytes are used for electroporation 7–8 days after the onset of differentiation.

DNA Preparation

DNA for electroporation is prepared by double-CsCl banding for best efficiency. One liter of overnight bacterial culture is resuspended in 20 ml of GTE buffer (50 mM glucose, 25 mM Tris, pH 8, 10 mM EDTA). Cells are then lysed in 40 ml of alkali lysis solution 0.2 N NaOH, 1% SDS at room temperature for 5 min, followed by addition of 30 ml of NaOAc, pH 4.8, and incubation for 10 min. After centrifugation in JA-10 rotor at 8000 rpm ($7000 \times g$) for 20 min, supernatant containing plasmid DNA is precipitated with 0.7 volume of isopropanol. Crude plasmid DNA is pelleted by JA-10 rotor for 20 min at 8000 rpm ($7000 \times g$). DNA is resuspended in 6 ml of distilled water; 10 g of CsCl and ethidium bromide (0.2 mg/ml) is added to the solution to final 10.5 ml before loaded into two ultracentrifuge tubes. Samples are spun in a NVT90 rotor at 60,000 rpm ($250,000 \times g$) for 5 h. After the first spin, the upper band is withdrawn with an 18-gauge needle and reloaded to the ultracentrifuge tube for second CsCl banding. After the second spin, the ethidium bromide bound to the DNA is extracted with water-saturated n-butanol until clear. DNA is then twice dialyzed overnight with 2 l of TE buffer. After dialysis, DNA is precipitated with three volumes of absolute ethanol and 1/10 volume of 3 M NaOAc, pH 7. Plasmid DNA is resuspended in TE buffer to a final concentration of 5–10 mg/ml for electroporation.

Electroporation

On the day of electroporation, adipocytes are placed into suspension by mild trypsinization, followed by two rinses with PBS buffer. Cell suspensions are pelleted by centrifugation for 5 min at 1000 rpm ($200 \times g$). Final washed pellets are resuspended to a final volume of 0.9 ml in PBS per 150-mm dish of cells; 450 μl of adipocytes are mixed with 50 μg of pKH3-TC10 mutants to a total volume of less than 600 μl in a 0.4-cm Gene Pulser cuvette (BioRad). Electroporation is performed under low-voltage conditions (160V, 950 μF) with a Gene Pulser II (BioRad). The time constant is usually between 21–23 msec. Electroporated cells are recovered immediately by adding 1 ml of medium into cuvettes and transferred into 15-ml conical tubes in the presence of 5 ml of medium for 10 min. Floating cell debris is removed from the tube before replating. Adipocytes from two electroporations are combined and plated onto five wells of a 6-well plate and maintained for 40–45 h. Medium is changed twice to remove cell debris during this period of time.

GST-Pak1 Pull-Down Assay to Determine Activity State of TC10

Transfected adipocytes are starved in low-glucose (5 mM) DMEM containing 0.5% fetal bovine serum for 3 h before incubation in the absence or presence of 100 nM insulin. After insulin stimulation as indicated, adipocytes are immediately washed with cold PBS three times and lysed in 0.3 ml of NP40-buffer (25 mM Tris, pH 7.5, 137 mM NaCl, 1% NP40, 5 mM MgCl$_2$, 10% glycerol) in the presence of 1 mM sodium orthovanadate and protease inhibitor cocktail (Roche) for 30 min at 4°. Fat cake (on the top) and unbroken cells (at the pellet) are separated from lysates after centrifugation at 13,000 rpm for 20 min. Protein concentration is measured by BioRad Protein Assay; 240 μg of cell extract in 100 μl of lysis buffer supplemented with 200 μl of binding buffer (25 mM Tris, pH 7.5, 40 mM NaCl, 30 mM MgCl$_2$, 1 mM dithiothreitol, 0.5% NP-40) are pre-cleared with 5 μg of GST-agarose beads for 30 min, after which lysates are recovered by centrifugation at 3000 rpm ($800 \times g$) for 2 min.

To determine the activity state of TC10, its interaction with an immobilized effector molecule is evaluated. The effector of choice is the protein kinase PAK, which contains a p21 binding domain (PBD) that interacts directly with TC10 (Neudauer et al., 1998). Pre-cleared lysates

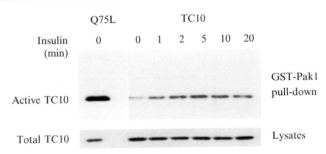

FIG. 2. Insulin activates TC10. Differentiated 3T3-L1 adipocytes were electroporated with HA-tagged human TC10 or its constitutively active mutant Q75L. Cells were stimulated with 100 nM insulin for 0, 1, 2, 5, 10, or 20 min. Fifty micrograms of lysates was subjected to SDS PAGE and immunoblotted with an anti-HA antibody (shown in the lower panel). GTP-bound TC10 pulled down by GST-Pak1 is shown in the upper panel.

are incubated with 5 μg of GST PAK-PBD agarose beads (Cat#PAK02, Cytoskeleton, Inc.) for 1 h at 4°. Beads are then washed three times with 1 ml of binding buffer. After the final wash, beads are resuspended in 2× SDS-PAGE sample buffer and are boiled for 10 min before loading. Samples are separated by 4–20% Tris-glycine SDS-PAGE gels, transferred to nitrocellulose membrane and blotted with monoclonal HA antibody (Santa Cruz Biotech.) to detect active TC10. An HA immunoblot is shown in Fig 2. Immunoblots are detected with chemiluminescence reagents from PerkinElmer Life Sciences. Activated TC10 is normalized with total TC10 in cell lysates. As a positive control for effector interaction, constitutively active mutant of TC10 (TC10/Q75L) is expressed in 3T3-L1 adipocytes. As shown in Fig. 2, constitutively active TC10 binds efficiently to GST-PAK-PBD.

Wortmannin Treatment

Wortmannin is a specific inhibitor for the class I PtdIns3-kinase, the enzyme that produces PtdInsP3 from PtdIns(4,5)P2. The inhibitory effect on PtdInsP3 formation in intact cells is dose-dependent, with an IC_{50} of approximately 5 nM. The lipid kinase activity immunoprecipitated by antibodies against the p85 regulatory subunit is totally inhibited when assayed *in vitro* at concentrations of 10–100 nM ($IC_{50} = 1$ nM). To test the effect of PI 3-kinase pathway on TC10 activation in response to insulin, adipocytes electroporated with HA-TC10 are starved with low-glucose DMEM supplemented with 0.5% FBS for 2.5 h before wortmannin treatment. A 1000× stock solution of wortmannin at the concentration of

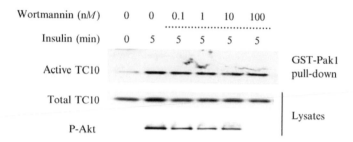

FIG. 3. Insulin activates TC10 through a PI(3)K-independent pathway in 3T3-L1 adipocytes. 3T3-L1 adipocytes electroporated with HA-tagged TC10 were incubated for 30 min with increasing doses of wortmannin (WM; 0–100 nM) and treated with insulin for 5 min. Top, GST–Pak1 pull-down assay probed with anti-HA. Middle, Anti-HA immunoblot with lysates. Bottom, Phospho-Akt (ser473) blot with the same lysates.

100 μM is prepared in dimethyl sulfoxide, stored at $-20°$ in the dark for stability and diluted with distilled water immediately before addition to cells. Wortmannin (at doses up to 100 nM) is added in 6-well plates for 30 min, followed by 100 nM insulin stimulation for 5 min. After insulin treatment, cells are lysed for the GST-Pak1 pull-down assay as described previously.

As a positive control for the wortmannin effect, lysates are blotted with antibody against phospho (serine 473)-AKT (Cell Signaling), a kinase downstream of the PI 3-kinase pathway; 100 nM wortmannin completely inhibits AKT phosphorylation stimulated in response to insulin. As shown in Fig. 3, wortmannin at concentrations as high as 100 nM does not have any effect on TC10 activation in response to insulin, indicating that TC10 activation is not regulated by the PI 3-kinase pathway.

Effect of TC10 on GLUT4 Translocation

Electroporation

Differentiated 3T3L1 adipocytes are electroporated with 50 μg of an expression plasmid encoding a fusion protein with GLUT4 and enhanced green fluorescence protein (GLUT4-EGFP) (Min *et al.*, 1999) and 200 μg of empty vector, TC10/WT, TC10/T31N, TC10/Q75L, Cdc42/WT, Cdc42/Q61L or Cdc42/T17N (Neudauer *et al.*, 1998). After replating on 6-well plates with coverslips, cells are allowed to recover 40–45 h and then stimulated with or without 100 nM insulin for 30 min.

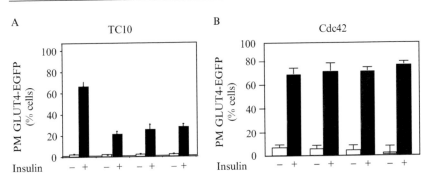

Fig. 4. TC10 overexpression blocks insulin-stimulated GLUT4 translocation. (A) Differentiated 3T3-L1 adipocytes were electroporated with 50 μg of an expression plasmid encoding GLUT4-EGFP and 200 μg of empty vector (sample 1), TC10/WT (sample 2), TC10/ T31N (sample 3), or TC10/Q75L (sample 4). (B) Differentiated 3T3-L1 adipocytes were electroporated with 50 μg of an expression plasmid encoding GLUT4-EGFP and 200 μg of empty vector (sample 1), Cdc42/WT (sample 2), Cdc42/Q61L (sample 3), or Cdc42/T17N (sample 4).

Staining and Counting

After treatment with insulin, cells are immediately washed twice in cold PBS, fixed for 10 min with 10% buffered neutral formalin at room temperature, and rinsed with PBS, followed by permeabilization with 0.5% triton X-100/ PBS for 5 min. Cells are then quenched by 100 mM glycine/PBS for 10 min. Coverslips are removed from the 6-well plate, and samples are mounted on glass slides with a drop of Vectashield (Vector Laboratories). EGFP fluorescence is analyzed by confocal immunofluorescence microscopy. Representative images should be taken from 5–8 independent experiments in which 100–200 individual cells are examined.

As shown in Fig. 4, coexpression of GLUT4-EGFP with TC10/WT, TC10/Q75L, or TC10/T31N cDNAs resulted in a marked decrease in the number of the cells responding to insulin. In contrast, overexpression of cDNAs encoding for Cdc42/WT, Cdc42/Q61L, or Cdc42/T17N had no effect on insulin-stimulated GLUT4-EGFP translocation.

2-Deoxyglucose Uptake

Cells in 12-well dishes are serum-deprived in low glucose DMEM supplemented with 0.5%FBS for 3 h. Cells are washed twice with PBS and incubated in 0.45 ml of KRBH (30 mM HEPES, pH 7.4, 10 mM NaHCO$_3$, 120 mM NaCl, 4 mM KH$_2$PO$_3$, 1 mM MgSO$_4$, and 1 mM CaCl$_2$ without glucose or BSA) in the presence or absence of 100 nM insulin for 30 min at 37°. Assay of glucose uptake was initiated by the addition of 50 μl

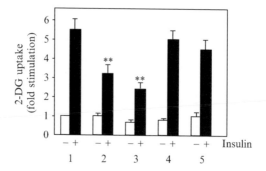

FIG. 5. Expression of TC10 inhibits insulin-stimulated glucose uptake and GLUT4 translocation in 3T3-L1 adipocytes. Differentiated 3T3-L1 adipocytes were electroporated with TC10 and Cdc42 mutants. The cells were left untreated or were stimulated with 100 nM insulin for 30 min. DNA transfected: 1, vector; 2: HA-TC10/WT; 3:HA-TC10/T31N; 4:HA-Cdc42/WT; 5: HA-Cdc42/T17N. The rate of [^{14}C] 2-deoxyglucose (2-DG) uptake was determined.

of fresh reaction mixture containing 0.1 μCi of 2-[U-^{14}C] deoxyglucose (2-DG) (NEN) and 0.05 μl of 200 mM cold 2-deoxyglucose (final 20 μM) in KRBH buffer. After incubation for 5 min at room temperature, the reaction is stopped by the addition of 50 μl of 200 mM 2-DG into each well. Wells are washed four times with ice-cold PBS on ice, and then solubilized in 0.5 ml of 0.1% SDS in distilled water at room temperature for 10 min; 10 μl of each sample is removed for measurement of protein concentration. Samples are then assessed for radioactivity by scintillation counting in Ready Gel (Beckman) (see Figure 5). Results are the mean $\pm$ SE of three to five independent experiments from individual experiments performed in triplicate with a transfection efficiency of ~50%.

Localization of TC10 in Adipocytes

Immunofluorescence and Image Analysis

3T3-L1 adipocytes are electroporated with 400 μg of HA-TC10. Cells are plated on coverslips and recovered for 4–5 days. Transfected adipocytes on coverslips are washed in ice-cold PBS, fixed, and permeabilized as described previously. Cells are blocked in 2% goat serum plus 1% BSA and 1% ovalbumin (Sigma-Aldrich) in PBS for 1 h. To detect HA-TC10, cells are stained with anti-HA monoclonal antibody. Caveolin, which localizes in the caveolae of the plasma membrane, is used as a plasma membrane marker. Anti-caveolin polyclonal (Transduction Laboratories) and anti-HA monoclonal (Santa Cruz Biotechnology) antibodies

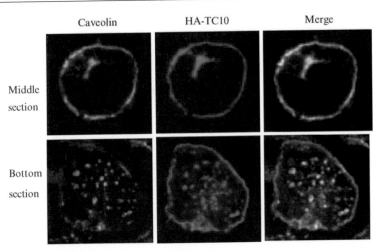

Fig. 6. Subcellular localization of TC10. 3T3-L1 adipocytes are electroporated with 400 μg of HA-TC10. Cells are plated on coverslips and recovered for 4–5 days. Cells were then immunostained with anti-caveolin and anti-HA antibodies. Images from different sections of the cells are taken as indicated. (See color insert.)

are diluted 1:200 in blocking buffer. After incubation with primary antibodies, cells are incubated with Alexa[488] or Alexa[594] goat anti-rabbit or anti-mouse immunoglobulin (IgG), obtained from Molecular Probes. Samples are mounted on glass slides with Vectashield (Vector Laboratories). Images are captured by using an Olympus FV300 confocal laser–scanning microscope. Images are then imported into Adobe Photoshop (Adobe Systems, Inc.) for processing and composite files are generated (see Figure 6).

Sucrose Density Gradient Centrifugation

3T3-L1 adipocytes are electroporated with 50 μg of HA-TC10 and replated in 100-mm dishes; 40–45 h after electroporation, cells are washed with ice-cold HES buffer (20 mM HEPES, pH 7.5, 250 mM sucrose, 1 mM EDTA) containing protease inhibitors (Roche) three times, then are homogenized in a Dounce homogenizer with 20 strokes. Homogenized cells are centrifuged at 800×g for 10 min to remove the unbroken cells, and the supernatants are centrifuged at 19,000×g to produce crude plasma membrane pellets. The resulting pellets are resuspended in 1.5 ml of 40% sucrose/MES buffer. This fraction is overlaid with 0.5 ml of 30, 20, 10, and 5% sucrose. The gradient is then centrifuged at 39,000 rpm (144,000×g) for 16 h with a SW55 rotor; 440-μl fractions are collected from the top

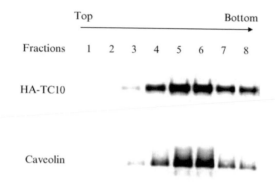

FIG. 7. TC10 localizes with caveolae/lipid rafts. Plasma membrane of transfected cells is fractionated by sucrose gradient centrifugation. Aliquots of each fraction were immunoblotted with anticaveolin and HA antibody to detect caveolin and TC10 protein, respectively.

of each gradient. Aliquots of each fraction are collected and subjected to SDS-PAGE followed by immunoblotting with anti-caveolin and HA antibody (see Figure 7).

Functional Studies of TC10 Downstream Effectors

Electroporation

3T3-L1 adipocytes are electroporated with eGFP-Caveolin (50 μg), myc-CIP4/2 (100 μg), HA-TC10/wt (50 μg), or HA-TC10/Q75L (50 μg). Electroporated cells are replated on glass coverslips, allowed to recover for 48 h, serum-starved with low glucose (5 mM) containing DMEM plus 0.5% FBS, treated with or without 100 nM insulin, and then processed for plasma membrane sheets.

Preparation of Plasma Membrane Sheets

To prepare plasma membrane sheets, cells are placed on ice after treatment with or without insulin, washed twice with ice-cold PBS, and then incubated in 0.2% Triton-X100/MES buffer (2 mM MES, pH 6.0, 130 mM NaCl, 0.5 mM MgCl$_2$, 1 mM CaCl$_2$, 10 mM NaF, 10 mM Na$_4$P$_2$O$_7$, 1 mM Na$_3$VO$_4$, and 1% Triton-X100) for 5 min on ice. The cells are next washed twice with ice-cold PBS, fixed in 4% paraformaldehyde (Electron Microscopy Sciences) in PBS, pH 7.4, and then processed for immunofluorescence.

Immunofluorescence

Cells are stained with anti-myc mAb and anti-HA pAb (Santa Cruz Biotechnologies) at 1 μg/ml, followed by Alexa[568] goat anti-mouse IgG (1 μg/ml; Molecular Probes) and Alexa[633] goat anti-rabbit IgG (1 μg/mL;

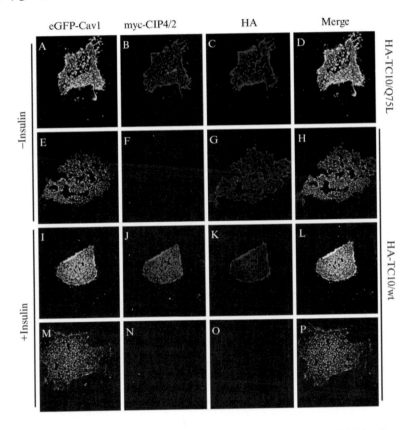

FIG. 8. TC10 recruits CIP4/2 to lipid rafts after insulin stimulation. 3T3-L1 adipocytes were electroporated with eGFP-Caveolin (50 μg), myc-CIP4/2 (100 μg), HA-TC10/wt (50 μg), or HA-TC10/Q75L (50 μg). The electroporated cells were plated on glass coverslips, and plasma sheets were prepared. Cells were stained with anti-myc mAb and anti-HA pAb (Santa Cruz Biotechnologies) at 1 μg/ml, followed by Alexa[568] goat anti-mouse IgG (1 μg/ml; Molecular Probes) and Alexa[633] goat anti-rabbit IgG (1 μg/ml; Molecular Probes) to visualize myc-CIP4/2 (red) and HA-TC10 (blue), respectively. Fluorescence was visualized by using an Olympus Fluoview 300 confocal laser scanning microscope, and images are representative of three independent determinations. In the presence of TC10/WT after insulin stimulation, myc-CIP4/2 was detected in the plasma membrane sheets (J). Consistent with these data, the constitutively active mutant TC10/Q75L also recruited myc-CIP4/2 to the plasma membrane (D). In the absence of TC10 expression, myc-CIP4/2 is not detected in the plasma membrane sheets after insulin stimulation (N and O). (See color insert.)

Molecular Probes) to visualize myc-CIP4/2 (red) and HA-TC10 (blue), respectively (Fig. 8). Fluorescence is visualized by using an Olympus Fluoview 300 confocal laser–scanning microscope, and images are representative of three independent determinations.

Acknowledgments

This work was supported by National Institutes of Health Grant RO1 DK60591 to A.R.S.

References

Bryant, N. J., Govers, R., and James, D. E. (2002). Regulated transport of the glucose transporter GLUT4. *Nat. Rev. Mol. Cell Biol.* **3**, 267–277.

Chang, L., Adams, R. D., and Saltiel, A. R. (2002). The TC10-interacting protein CIP4/2 is required for insulin-stimulated Glut4 translocation in 3T3L1 adipocytes. *Proc. Natl. Acad. Sci. USA* **99**, 12835–12840.

Chiang, S. H., Baumann, C. A., Kanzaki, M., Thurmond, D. C., Watson, R. T., Neudauer, C. L., Macara, I. G., Pessin, J. E., and Saltiel, A. R. (2001). Insulin-stimulated GLUT4 translocation requires the CAP-dependent activation of TC10. *Nature* **410**, 944–948.

Chiang, S. H., Hou, J. C., Hwang, J., Pessin, J. E., and Saltiel, A. R. (2002). Cloning and functional characterization of related TC10 isoforms, a subfamily of Rho proteins involved in insulin-stimulated glucose transport. *J. Biol. Chem.* **277**, 13067–13073.

Chiang, S. H., Hwang, J., Legendre, M., Zhang, M., Kimura, A., and Saltiel, A. R. (2003). TCGAP, a multidomain Rho GTPase-activating protein involved in insulin-stimulated glucose transport. *EMBO J.* **22**, 2679–2691.

Inoue, M., Chang, L., Hwang, J., Chiang, S. H., and Saltiel, A. R. (2003). The exocyst complex is required for targeting of Glut4 to the plasma membrane by insulin. *Nature* **422**, 629–633.

Khan, A. H., and Pessin, J. E. (2002). Insulin regulation of glucose uptake: A complex interplay of intracellular signalling pathways. *Diabetologia* **45**, 1475–1483.

Min, J., Okada, S., Kanzaki, M., Elmendorf, J. S., Coker, K. J., Ceresa, B. P., Syu, L. J., Noda, Y., Saltiel, A. R., and Pessin, J. E. (1999). Synip: A novel insulin-regulated syntaxin 4-binding protein mediating GLUT4 translocation in adipocytes. *Mol. Cell* **3**, 751–760.

Mora, S., and Pessin, J. E. (2002). An adipocentric view of signaling and intracellular trafficking. *Diabetes Metab. Res. Rev.* **18**, 345–356.

Neudauer, C. L., Joberty, G., Tatsis, N., and Macara, I. G. (1998). Distinct cellular effects and interactions of the Rho-family GTPase TC10. *Curr. Biol.* **8**, 1151–1160.

Saltiel, A. R., and Pessin, J. E. (2002). Insulin signaling pathways in time and space. *Trends Cell Biol.* **12**, 65–71.

Saltiel, A. R., and Pessin, J. E. (2003). Insulin signaling in microdomains of the plasma membrane. *Traffic* **4**, 711–716.

Symons, M., and Rusk, N. (2003). Control of vesicular trafficking by Rho GTPases. *Curr. Biol.* **13**, R409–R418.

Watson, R. T., and Pessin, J. E. (2001). Subcellular compartmentalization and trafficking of the insulin-responsive glucose transporter, GLUT4. *Exp. Cell Res.* **271**, 75–83.

Watson, R. T., Shigematsu, S., Chiang, S. H., Mora, S., Kanzaki, M., Macara, I. G., Saltiel, A. R., and Pessin, J. E. (2001). Lipid raft microdomain compartmentalization of TC10 is required for insulin signaling and GLUT4 translocation. *J. Cell Biol.* **154**, 829–840.

[56]   GTPases and the Control of Neuronal Polarity

By JENS C. SCHWAMBORN, YINGHUA LI, and ANDREAS W. PÜSCHEL

Abstract

Neurons are probably the most highly polarized cell type and typically develop a single axon and several dendrites. The establishment of a polarized morphology and the functional specialization of axonal and dendritic compartments are essential steps in the differentiation of neurons. Primary cultures of dissociated hippocampal neurons are a widely used system to study the development of neuronal differentiation. In this article, we will describe gain-of-function and loss-of-function approaches that allow us to analyze the role of GTPases in neuronal differentiation.

Introduction: The Development of Neuronal Polarity

Neurons are probably the most highly polarized cell type and typically develop a single axon and several often highly branched dendrites. The establishment of a polarized morphology and the functional specialization of axonal and dendritic compartments are essential steps in the differentiation of neurons. Primary cultures of dissociated neurons are a well-established system to study the development of neuronal polarity (Bradke and Dotti, 2000; Da Silva and Dotti, 2002), and the differentiation of hippocampal neurons in culture has been described in detail (Dotti *et al.*, 1988). After attaching to the substrate, neurons initially develop lamellipodia (stage 1; Fig. 1A). Within a few hours, several undifferentiated processes of similar length (minor neurites) are formed (stage 2). At 1–2 days *in vitro* (d.i.v.; depending on the culture conditions), one of these neurites is selected to become the axon and extends rapidly (stage 3), whereas the length of the remaining neurites does not change significantly (Fig. 1B). Thus, the transition form stage 2 to stage 3 is the crucial step in neuronal differentiation when neurons acquire a polar morphology (Bradke and Dotti, 1999, 2000). On 3 d.i.v., minor neurites begin to elongate and differentiate into dendrites (stage 4; Fig. 1B) followed by the maturation of axonal and dendritic processes and the formation of functional synapses (stage 5). The signaling pathways that lead to the establishment of neuronal polarity are still poorly understood. The use of primary cultures has facilitated the identification of several signaling molecules essential for neuronal polarity, with GTPases assuming a central role (Da Silva and Dotti, 2002; Govek *et al.*, 2005; Schwamborn and Püschel,

METHODS IN ENZYMOLOGY, VOL. 406
0076-6879/06 $35.00
DOI: 10.1016/S0076-6879(06)06056-3

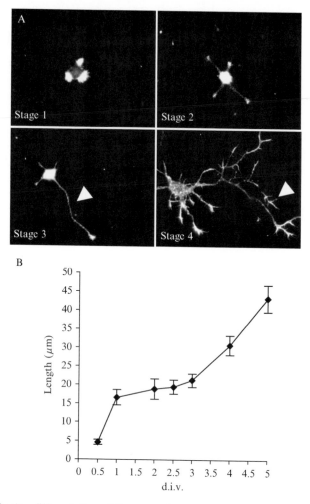

FIG. 1. *In vitro* differentiation of hippocampal neurons. (A) Hippocampal neurons were prepared as described, fixed at the indicated stages, and F-actin stained with rhodamine-phalloidin. Axons are marked by arrowheads. (B) The length of minor neurites/dendrites was determined between 0.5 and 5 d.i.v., using the UTHSCSA ImageTool v. 3.00 software.

2004). Several factors (Table I) including the evolutionarily conserved Par complex (consisting of Par3, Par6, and aPKCs), phosphatidylinositol 3-kinase (PI3K), GSK3β, and CRMP2 were implicated in the establishment of neuronal polarity (Inagaki *et al.*, 2001; Jiang *et al.*, 2005; Schwamborn and Püschel, 2004; Shi *et al.*, 2003, 2004; Yoshimura *et al.*, 2005). Their specific localization to the axon also makes them useful markers to verify

TABLE I
MARKERS FOR NEURONAL POLARITY

Marker	Stage 2	Stage 3	Ref.
MAP2	uniform	minor neurites	Caceres *et al.*, 1984
Tau-1	uniform	axon	Mandell and Banker, 1996
P-Akt (P-S473)	all neurites	axon	Shi *et al.*, 2003
Par3, Par6	all neurites	axon	Shi *et al.*, 2004
P-PKCζ/λ (P-T410/403)	—	axon	Shi *et al.*, 2003
P-GSK3βP-S9	all neurites	axon	Shi *et al.*, 2004
APC	all neurites	axon	Shi *et al.*, 2004
CRMP2	all neurites	axon	Inagaki *et al.*, 2001
Cdc42	several neurites	axon	Schwamborn and Püschel, 2004
Rap1B	single neurite	axon	Schwamborn and Püschel, 2004

the axonal identity of neurites. None of these factors, however, is restricted to a single neurite at stage 2 before polarity is detectable morphologically. Rap1B is the earliest marker identified so far that shows an asymmetrical distribution in stage 2 neurons (Schwamborn and Püschel, 2004). The sequential activity of Rap1B and Cdc42 directs the establishment of neuronal polarity in hippocampal neurons, and both are necessary and sufficient to specify axonal identity. Thus, the restriction of Rap1B to a single neurite is a decisive step in determining which neurite becomes the axon. Subsequently, Cdc42 and Par3 are localized to the Rap1B-positive growth cone, and active aPKC and GSK3β phosphorylated at Ser 9 become detectable at the tip of the axon in stage 3 neurons.

Preparation and Culture of Hippocampal Neurons

To isolate the neurons, the hippocampus is dissected from rat embryos at embryonic day 18 in ice-cold HBSS (without $CaCl_2$ and $MgCl_2$; Invitrogen). The hippocampi from several embryos are collected and washed three times in 10 ml ice-cold HBSS. The explants are incubated in 5 ml phosphate-buffered saline ($1 \times$ PBS) containing 10 mM glucose, 3 mg papain (Sigma, P-4762), and 0.05 mg DNase I (Sigma, D-7691) for 15 min at 37° to dissociate the cells. The degradation of DNA released from damaged cells by DNase I reduces the viscosity of the cell suspension and supports the dissociation of the tissue. After the treatment with papain, the cells still form large aggregates and are washed three times with 10 ml DMEM (supplemented with 10% fetal calf serum [FCS], 0.5 mM glutamine, and 100 U/ml penicillin/streptomycin; all Invitrogen). After the final wash, the cells are resuspended in 2 ml DMEM (supplemented with 10% FCS, 0.5

mM glutamine, and 100 U/ml penicillin/streptomycin), dissociated by repeated pipetting (using first a Gilson P1000 and subsequently a P200 pipette), and plated onto coverslips dispensed into 24-well plates. The coverslips are coated with polyornithine (15 μg/ml) or other adhesive substrates like fibronectin or laminin. For coating, the coverslips are incubated for at least 1 h at RT with 15 μg/ml polyornithine (Sigma) and washed twice with HBSS. The choice of the substrate can affect the time course of *in vitro* differentiation. Neurons grown on laminin-coated coverslips tend to show a polarized morphology at earlier time points than when cultured on polyornithine as a substrate. After dissociation, the neurons are diluted to the desired cell density with DMEM (supplemented with 10% FCS, 0.5 mM glutamine, and 100 U/ml penicillin/streptomycin), and 500-μl cell suspension is seeded per well. The number of cells plated per well depends on the experiment planned. To analyze the distribution of proteins by immunofluorescence, a density of 40,000–60,000 cells per well plate is recommended to obtain isolated neurons. A higher density is advisable for transfecting the neurons to achieve a sufficient transfection efficiency (see below); 4–6 h after plating, the medium is replaced by Neurobasal medium (supplemented with 2% B27, 0.5 mM glutamine and 100 U/ml penicillin/streptomycin; all Invitrogen). In this step, it is very important to minimize the time when the cells are not covered with medium by changing the medium one well at a time.

Transfection of Hippocampal Neurons

Several methods that differ in cost and efficiency are available to transfect hippocampal neurons. It should also be taken into account that the transfection efficiency depends on cell density and the developmental stage at the time of transfection. The transfection of hippocampal neurons shortly after plating has proven to be particularly difficult. In general, a higher density increases the transfection efficiency but makes it more difficult to unambiguously identify the neurites of transfected cells and determine the number and length of axons and dendrites. Methods using cationic lipids for DNA delivery are more cost-intensive than calcium phosphate coprecipitation but often give better transfection rates. Methods are also available for the electroporation of primary neurons (i.e., Amaxa nucleofector) that allow high transfection efficiency but are also more cost-intensive.

Lipofection

The cationic lipid Lipofectamine2000 (Invitrogen) gave the best results in our hands for the lipofection of hippocampal neurons directly after preparation (0 d.i.v.). To increase the efficiency of transfection, the cells are first plated at a very high density of approximately 800,000 cells per well in a 24-well plate. After plating, the cells are incubated for 2 h at $37°$ and 5% CO_2. Afterwards, the medium is replaced by 300 μl Opti-MEM (Invitrogen) per well. For one transfection, 0.5–1 μl Lipofectamine2000 and 1–2 μg DNA are each mixed with 12.5 μl Opti-MEM in two separate Eppendorf tubes. Both solutions are combined after incubation for 5 min at RT. After an additional 20 min incubation at RT to allow the formation of lipid/DNA complexes, the transfection mixture is added to one well. After an incubation for 2 h, the transfection medium is replaced by Neurobasal medium (supplemented with B27, glutamine and penicillin/streptomycin). The cells are detached after the transfection by moderate pipetting and replated onto new coverslips at a lower density (40,000–60,000 cells per coverslip in a 24-well plate; one transfection is sufficient for 12–20 coverslips). The neurons should not be cultured with the lipid/DNA complexes for more than 2 h; otherwise an increase in cell death can be observed. Expression from the transfected vectors can be first observed after several hours and reaches maximal levels 24–48 h after transfection.

Calcium Phosphate Coprecipitation

The calcium phosphate coprecipitation method can be used to transfect cultured hippocampal neurons at different stages of differentiation. Dissociated neurons are prepared as described previously, plated at 40,000 cells per well in a 24-well plate, and incubated at $37°$, 5% CO_2; 3–5 h after plating, the medium is replaced by Neurobasal medium (supplemented with 2% B27, 0.5 mM glutamine and 100 U/ml penicillin/streptomycin) and the culture continued for another 3 h. For the transfection of four wells, 8 μg DNA is mixed with 37.5 μl ddH$_2$O and 12,5 μl 1 M CaCl$_2$. To this solution, 50 μl prewarmed ($37°$) 2× BBS (50 mM BES, 280 mM NaCl, 1.5 mM Na$_2$HP0$_4$ and 1.5 mM Na$_2$HP0$_4$, pH 6.95) is added drop wise to prepare the DNA/calcium phosphate coprecipitate. The culture medium is removed, stored for later reuse at $37°$, and immediately replaced by 300 μl of prewarmed Opti-MEM; 25 μl of transfection mixture is added per well, and the cells incubated for 45–60 min at $37°$ and 5% CO_2. After removal of the transfection medium, the cells are washed once with serum-free DMEM. The medium saved at the previous step is added back the to cells and the culture continued at $37°$ and 5% CO_2. This protocol is optimized for the transfection of neurons at 0 d.i.v., and approximately 10–20 transfected neurons per coverslip can be typically found that are sufficient for analysis of neuronal morphology. Incubation with the transfection mixture for more than 1 h is not recommended at this stage to avoid increased cell death. A higher transfection efficiency can be achieved when transfections are done at 3–7 d.i.v., and later stages of neuronal differentiation are analyzed.

Analysis of Neuronal Differentiation

Immunofluorescence

The morphology of neurons can be easily visualized with anti-α-tubulin antibodies (Sigma, Clone DM 1A, T-9026) or by staining actin filaments with fluorescently labeled phalloidin (Molecular Probes) (Fig. 1). In addition to morphology, molecular markers should be used to verify the axonal and dendritic character of processes (Table I). The most widely used markers are the microtubule-associated proteins, MAP2, for dendrites (Caceres *et al.*, 1984) and a dephospho isoform of tau detected by the Tau-1 antibody (Mandell and Banker, 1996) for axons (Fig. 2). Anti-MAP2 antibodies label all minor processes in stage 3 neurons and often the proximal part of the axon, whereas the Tau-1 antibody stains the distal part of the axon. Establishment of the right conditions for fixation and staining is critical to obtain unambiguous results with anti-MAP2 and Tau-1 antibodies, because axonal and dendritic compartments are not fully segregated in stage 3 neurons (Da Silva and Dotti, 2002) making it more difficult than at later stages to specifically label axons and dendrites; 48–65 h after transfection, the cells are fixed and processed for indirect immunofluorescence. For most antibodies (including the anti-MAP2 and Tau-1 antibodies), the following procedure gives good results. The cultures

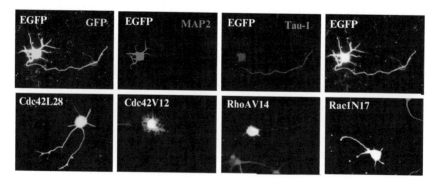

FIG. 2. Effect of GTPases on neuronal polarity. Hippocampal neurons were transfected at 0.5 d.i.v. with expression vectors for EGFP and EGFP fusion proteins of the indicated GTPase mutants, fixed at 2.5 d.i.v., and stained with Tau-1 (blue), anti-MAP2 (red), and anti-EGFP antibodies (green). (Modified after Schwamborn and Püschel, 2004). (See color insert.)

are fixed for 20 min at 4° with ice-cold 4% paraformaldehyde in 1× PBS containing 15% sucrose and washed three times for 5 min at RT with 1× PBS. After permeabilization with 0.1% Triton X-100, 0.1% sodium citrate in 1× PBS for 3 min at 4°, the cells are washed three times for 5 min in 1× PBS at RT, and nonspecific binding is blocked with 10% fetal calf serum in 1× PBS (PBSF) for 1 h at RT. The cells are incubated with the primary antibody (anti-MAP2 at 1:500, Tau-1 at 1:200; both from Chemicon) diluted in PBSF overnight at RT. After washing three times for 5 min with 1× PBS, Alexa Fluor–conjugated secondary antibodies (Molecular Probes) diluted in PBSF are added for 90 min at RT. Finally, the cells are washed three times for 5 min with 1× PBS and once with distilled water and mounted in an aqueous mounting medium.

Analysis of Neuronal Polarity

The polarization of neurons becomes detectable morphologically when one of the initially indistinguishable neurites is selected to become the axon. When analyzing the development of neuronal polarity, it is essential to clearly define the criteria that are used to categorize neuronal morphology. The stage of neuronal differentiation is usually determined by the criteria of Dotti et al. (1988). Axons are identified as processes that are at least twice as long as minor processes and show Tau-1 immunoreactivity but no MAP2 staining in their distal segments. MAP2-positive neurites longer than one cell diameter are counted as minor neurites. This definition excludes very short processes like filopodia from the quantification. Neurons that do not extend an axon are classified as unpolarized, and neurons with a single Tau-1 positive axon are classified as polarized. Neurons that extend several axons should be considered hyperpolarized. It is important to note that cultures of hippocampal neurons are heterogeneous, and different stages of differentiation can be found at the same time. For example, 4 h after plating, most of the cells will be stage 1 neurons, but there will also be some stage 2 and 3 neurons. The earliest indication of an asymmetry between neurites reported so far is Rap1B (Schwamborn and Püschel, 2004). Rap1B is initially found in all processes of early stage 2 neurons but becomes restricted to a single neurite in late stage 2 neurons. By contrast, Cdc42 initially is present at the tips of all neurites but becomes gradually restricted to the axon of stage 3 neurons. It has been reported (Da Silva et al., 2004) that Cdc42 shows a uniform distribution in polarized hippocampal neurons. However, the polyclonal antibody used in this study (anti-Cdc42Hs P1, Santa Cruz Biotechnology, sc-87) shows a strong nonspecific staining and a pattern distinct from that of a widely used monoclonal anti-Cdc42 antibody (BD Biosciences, #610928) when the conditions for fixation and blocking nonspecific binding are not tightly controlled. At suboptimal conditions, the signal obtained with this polyclonal antibody is not diminished by RNAi directed against Cdc42, whereas that of the monoclonal antibody is almost completely abolished (Fig. 4; Schwamborn and Püschel, 2004, unpublished results).

Functional Analysis of GTPases

Expression of GTPase Mutants

The most frequently used approach to investigate the function of GTPases during neuronal development is the expression of constitutively active or dominant-negative mutants (Luo, 2000). In most cases, mutations analogous to RasV12 and RasN17 are used (Govek et al., 2005; Luo, 2000). However, one has to be aware that dominant-negative mutants act by blocking guanine nucleotide exchange factors and may cause phenotypes that are not equivalent to a loss of the corresponding endogenous GTPase, especially after overexpression.

In some cases, the use of mutants that show an autonomous cycling between the GTP-bound and GDP-bound states (like Cdc42L28) can result in phenotypes different from those with constitutively active proteins that are locked in the GTP-bound state (Schwamborn and Püschel, 2004) (Fig. 2). The stage at which their function is impaired is an additional factor important in analyzing the role of GTPases in neuronal polarity. For example, expression of Rap1BV12 induces multiple axons only when neurons are transfected before polarity is established, whereas expression at 3 d.i.v. does not affect axon specification but stimulates axon branching (Schwamborn and Püschel, 2004, unpublished results). When investigating the function of a GTPase in the early steps of polarization it is, therefore, necessary to transfect the neurons directly after plating. Control neurons transfected with an expression vector for EGFP usually develop three to six minor processes and a single axon until 2.5 d.i.v (stage 3; Fig. 2). For analyzing the effects on neuronal polarity, it is necessary to determine the number of axons and dendrites per cell and the average length of axonal and dendritic processes (i.e., using the UTHSCSA Image Tool; http://ddsdx.uthscsa.edu/dig/itdesc.html) to be able to distinguish effects on polarity from more general effects on neurite extension. This is best illustrated by the phenotype of neurons expressing the fast cycling mutant Cdc42L28, constitutively active RhoAV14, and dominant-negative Rac1N17 (Fig. 2). Expression of Cdc42L28 induces the extension of multiple axons, whereas the number of dendrites per cell is not changed significantly (Fig. 2; Schwamborn and Püschel, 2004). By contrast, Cdc42V12 does not influence axon specification but actually prevents neurite formation when expressed at this stage (Fig. 2). Expression of constitutively active RhoA (RhoAV14) leads to a reduction in the number and length of processes irrespective of their identity, indicating that it does not affect polarity but neurite extension (see also Da Silva *et al.*, 2003). Like RhoAV14, dominant-negative Rac1 does not show an effect on the development of polarity but neurite elongation. Cells expressing Rac1N17 develop one axon and several dendrites, but both types of neurites are significantly shorter than in the controls with EGFP.

Knock-Down by RNA Interference

Loss-of-function experiments to show the involvement of GTPases in a specific cellular process have become much easier to accomplish with the demonstration that a knock-down by RNA interference (RNAi) is possible in mammalian cells if the double-stranded RNA (dsRNA) is shorter than 30 nucleotides (Elbashir *et al.*, 2001). The limit of 30 nucleotides results from a nonspecific response induced by longer dsRNAs that leads to a general inhibition of translation (Hannon, 2002). Suppressing the expression of proteins by RNAi is, at the moment, the most convenient method available and is much faster than for example the creation of knockout animals. As an additional advantage, RNAi allows the acute suppression of proteins, thereby reducing the possibility of compensatory effects that may occur when the protein is absent throughout development. A summary describing the mechanism and possible applications of RNAi can be found in numerous recent reviews (for example Hannon [2002]; He and Hannon [2004]; and Mittal [2004]). There are two possibilities to introduce small interfering RNAs (siRNAs) into target cells. The first method relies on chemically synthesized or enzymatically generated dsRNAs and transfection reagents that were developed specifically for siRNAs and mediate the uptake of RNA with a high efficiency. The second is based on expression vectors that contain a RNA polymerase III promotor (i.e., from the human U6 snRNA gene) and a transcription termination signal of five consecutive thymidine nucleotides. This expression vector produces a double-stranded short hairpin RNA (shRNA) that is processed to the siRNA, which mediates the knock-down of a specific transcript (Fig. 3). To identify transfected cells, an EGFP expression plasmid is cotransfected with the shRNA vector.

In our hands, the pSHAG vector (Paddison *et al.*, 2002) usually allows an efficient knock-down of the targeted protein. The gene-specific sequence for the shRNA is inserted into pSHAG by cloning oligonucleotides containing an inverted repeat of the target sequence separated by eight nucleotides that encode the loop of the hairpin. Detailed information,

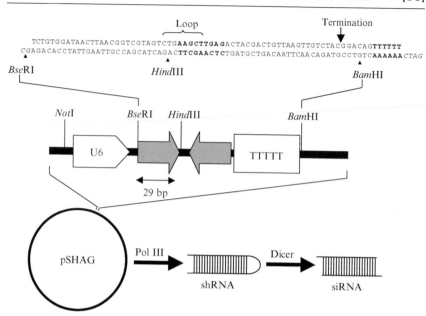

FIG. 3. Construction of an RNAi vector directed against Cdc42. A schematic representation of the pSHAG expression vector (Paddison *et al.*, 2002) for an shRNA directed against rat Cdc42 is shown (Schwamborn and Püschel, 2004). RNA polymerase III initiates transcription at the U6 promoter and terminates in a run of five consecutive thymidine nucleotides. The eight nucleotides that separate the gene-specific sequences give rise to the loop of the shRNA and are unrelated to the targeted transcript. The shRNA is processed to a siRNA that mediates the suppression of Cdc42 by RNA interference. The *Not*I site is located upstream of the U6 promoter, whereas the *Hind*III site is introduced into the vector by the oligonucleotides.

especially for designing oligonucleotides, can be found on the RNAiCentral web site (katahdin.cshl.org:9331/RNAi_web/scripts/main2.pl). Unfortunately, the prediction that sequence gives rise to functional RNAi constructs is difficult, and often several constructs have to be tested before a sequence is found that allows an efficient knock-down. Guidelines to improve the efficiency and specificity of RNAi in mammalian cells can also be found in the review by Mittal (2004). It is important to check whether the selected sequence shows homology to other genes in the rat genome. If the homology exceeds 18 nucleotides, it is advisable to select a different sequence; otherwise, the knock-down may not be limited to the targeted transcript. In our hands, one to five sequences have to be tested to obtain a construct that allows an efficient suppression. As an example, the effect of suppressing Cdc42 by RNAi is shown Figs. 3 and 4. Oligonucleotides encoding an shRNA directed against rat Cdc42 were cloned into the

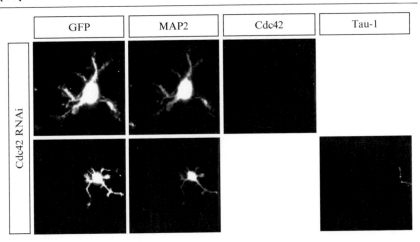

FIG. 4. Suppression of Cdc42 by RNAi blocks neuronal polarity. Neurons were cotransfected with expression vectors for an anti-Cdc42 shRNA and EGFP 12 h after plating and stained with the anti-MAP2, Tau-1, or anti-Cdc42 antibodies at 2.5 d.i.v. (Modified after Schwamborn and Püschel, 2004).

pSHAG vector (Schwamborn and Püschel, 2004). They were first phosphorylated by incubation of 500 pmol of each oligonucleotide with polynucleotide kinase (PNK), PNK buffer, and 100 μM ATP for 2 h at 37°. To anneal the complementary oligonucleotides, they are combined, incubated for 5 min at 100° and slowly cooled to RT. The pSHAG vector is digested with *Bse*RI and *Bam*HI and ligated with the annealed oligonucleotides overnight at 16°. The resulting clones are screened by digestion with *Not*I and *Hind*III, which produces a diagnostic fragment of 300 base pairs (Fig. 3). The effectiveness of the shRNA vector in mediating a knock-down can be tested by Western blot or immunofluorescence. Because of the low transfection efficiency of hippocampal neurons achieved with most transfection methods, the confirmation of a knock-down by immunofluorescence (Fig. 4) will usually be easier to perform. Alternately, a knock-down can be validated by Western blot of rat cell lines that express the targeted protein endogenously and allow an efficient transfection (i.e., REF52 cells) or by cotransfecting a heterologous cell line with expression vectors for the targeted rat protein and the RNAi vector. As control for the specificity of the shRNAs, the use vectors for shRNAs that contain 2 -3 mismatches to the target sequence is recommended (Huppi *et al.*, 2005). A small number of mismatches abolish the ability of the siRNA to suppress expression of the targeted mRNA (Mittal, 2004). In addition, a rescue experiment can serve to verify that the loss of the targeted protein is responsible for the observed phenotype. To achieve a rescue, the shRNA is expressed together with a transcript that contains silent mutations in the target sequence recognized by the siRNA (Huppi *et al.*, 2005). A few mismatches are usually sufficient to protect the transcript from a knock-down. However, this approach can prove problematic when the level or subcellular localization of the expressed protein is critical for its function. This is especially relevant for proteins involved in the establishment of neuronal polarity such as the Cdc42-interacting protein Par6 that blocks axon specification after overexpression (Schwamborn and Püschel, 2004; Shi *et al.*, 2003). A knock-down of Cdc42 leads to a complete loss of polarity as indicated by the absence of axons, whereas the number and length of minor processes is not affected (Fig. 4; Schwamborn and Püschel, 2004). The vector-based system described previously allows us to easily combine suppression by RNAi and overexpression of GTPase mutants. Using this approach, it was possible to show that Rap1B and Cdc42 act sequentially to establish neuronal polarity (Schwamborn and Püschel, 2004). The combination of mutants, RNAi, and pharmacological inhibitors allows it to establish the order in which GTPases, their regulators, and effectors act during neuronal differentiation. Thus, a functional analysis and the establishment of epistatic relationships are possible at the cellular level with genes that perform essential functions during development and would result in lethal phenotypes when inactivated by introducing mutations into the genome.

References

Bradke, F., and Dotti, C. G. (1999). The role of local actin instability in axon formation. *Science* **283**, 1931–1934.

Bradke, F., and Dotti, C. G. (2000). Establishment of neuronal polarity: Lessons from cultured hippocampal neurons. *Curr. Opin. Neurobiol.* **10**, 574–581.

Caceres, A., Banker, G., Steward, O., Binder, L., and Payne, M. (1984). MAP2 is localized to the dendrites of hippocampal neurons that develop in culture. *Brain Res.* **315**, 314–318.

Da Silva, J. S., and Dotti, C. G. (2002). Breaking the neuronal sphere: Regulation of the actin cytoskeleton in neuritogenesis. *Nat. Rev. Neurosci.* **3**, 694–704.

Da Silva, J. S., Schubert, V., and Dotti, C. G. (2004). RhoA, Rac1, and cdc42 intracellular distribution shift during hippocampal neuron development. *Mol. Cell. Neurosci.* **27**, 1–7.

Da Silva, J. S., Medina, M., Zuliani, C., Di Nardo, A., Witke, W., and Dotti, C. G. (2003). RhoA/ROCK regulation of neuritogenesis via profilin IIa-mediated control of actin stability. *J. Cell. Biol.* **162**, 1267–1279.

Dotti, C. G., Sullivan, C. A., and Banker, G. A. (1988). The establishment of polarity by hippocampal neurons in culture. *J. Neurosci.* **8**, 1454–1468.

Elbashir, S. M., Harborth, J., Lendeckel, W., Yalcin, A., Weber, K., and Tuschl, T. (2001). Duplexes of 21-nucleotide RNAs mediate RNA interference in cultured mammalian cells. *Nature* **411**, 494–498.

Govek, E. E., Newey, S. E., and Van Aelst, L. (2005). The role of the Rho GTPases in neuronal development. *Genes Dev.* **19**, 1–49.

Hannon, G. J. (2002). RNA interference. *Nature* **418,** 244–251.

He, L., and Hannon, G. J. (2004). MicroRNAs: Small RNAs with a big role in gene regulation. *Nat. Rev. Genet.* **5,** 522–531.

Huppi, K., Martin, S. E., and Caplen, N. J. (2005). Defining and assaying RNAi in mammalian cells. *Mol. Cell* **17,** 1–10.

Inagaki, N., Chihara, K., Arimura, N., Menager, C., Kawano, Y., Matsuo, N., Nishimura, T., Amano, M., and Kaibuchi, K. (2001). CRMP-2 induces axons in cultured hippocampal neurons. *Nat. Neurosci.* **4,** 781–782.

Jiang, H., Guo, W., Liang, X., and Rao, Y. (2005). Both the establishment and the maintenance of neuronal polarity require active mechanisms: Critical roles of GSK-3beta and its upstream regulators. *Cell* **120,** 123–135.

Luo, L. (2000). Rho GTPases in neuronal morphogenesis. *Nat. Rev. Neurosci.* **1,** 173–180.

Mandell, J. W., and Banker, G. A. (1996). A spatial gradient of tau protein phosphorylation in nascent axons. *J. Neurosci.* **16,** 5727–5740.

Mittal, V. (2004). Improving the efficiency of RNA interference in mammals. *Nat. Rev. Genet.* **5,** 355–365.

Paddison, P. J., Caudy, A. A., Bernstein, E., Hannon, G. J., and Conklin, D. S. (2002). Short hairpin RNAs (shRNAs) induce sequence-specific silencing in mammalian cells. *Genes Dev.* **16,** 948–958.

Schwamborn, J. C., and Püschel, A. W. (2004). The sequential activity of the GTPases Rap1B and Cdc42 determines neuronal polarity. *Nat. Neurosci.* **7,** 923–929.

Shi, S. H., Jan, L. Y., and Jan, Y. N. (2003). Hippocampal neuronal polarity specified by spatially localized mPar3/mPar6 and PI 3-kinase activity. *Cell* **112,** 63–75.

Shi, S. H., Cheng, T., Jan, L. Y., and Jan, Y. N. (2004). APC and GSK-3beta are involved in mPar3 targeting to the nascent axon and establishment of neuronal polarity. *Curr. Biol.* **14,** 2025–2032.

Yoshimura, T., Kawano, Y., Arimura, N., Kawabata, S., Kikuchi, A., and Kaibuchi, K. (2005). GSK-3beta regulates phosphorylation of CRMP-2 and neuronal polarity. *Cell* **120,** 137–149.

[57] *In Vitro* Assembly of Filopodia-Like Bundles

By Danijela Vignjevic, John Peloquin, and Gary G. Borisy

Abstract

A breakthrough in understanding the mechanism of lamellipodial protrusion came from development of an *in vitro* model system, namely the rocketing movement of microbes and activated beads driven by actin comet tails (Cameron *et al.*, 1999, 2000; Loisel *et al.*, 1999; Theriot *et al.*, 1994). As a model for investigation of the other major protrusive organelle, the filopodium, we developed *in vitro* systems for producing filopodia-like bundles (Vignjevic *et al.*, 2003), one of which uses cytoplasmic extracts and another that reconstitutes like–like bundles from purified proteins.

METHODS IN ENZYMOLOGY, VOL. 406   0076-6879/06 $35.00
DOI: 10.1016/S0076-6879(06)06057-5

Beads coated with Arp2/3-activating proteins can induce two distinct types of actin organization in cytoplasmic extracts: (1) comet tails or clouds displaying a dendritic array of actin filaments and (2) stars with filament bundles radiating from the bead. Actin filaments in star bundles, like those in filopodia, are long, unbranched, aligned, uniformly polar, and grow at the barbed end. Like filopodia, star bundles are enriched in fascin and lack Arp2/3 complex and capping protein. Similar to cells, the transition from a dendritic (lamellipodial) to a bundled (filopodial) organization is induced by depletion of capping protein, and add-back of this protein restores the dendritic mode. By use of purified proteins, a small number of components are sufficient for the assembly of filopodia-like bundles: WASP-coated beads, actin, Arp2/3 complex, and fascin. On the basis of analysis of this system, we proposed a model for filopodial formation in which actin filaments of a preexisting dendritic network are elongated by inhibition of capping and subsequently cross-linked into bundles by fascin.

Introduction

Lamellipodia and filopodia are both actin polymer–based protrusions of a crawling cell but are structurally and functionally distinct. Lamellipodia are sheetlike protrusions that contain a two-dimensional network of actin filaments and are a major locomotory organelle of the cell. Filopodia are rodlike extensions filled with tight bundles of actin filaments and are believed to be the cell's sensory and guiding organelles. Models have been proposed to explain how these distinct morphologies are constructed and how they produce force.

The dendritic nucleation/array treadmilling model (Borisy and Svitkina, 2000; Mullins et al., 1998) for lamellipodial protrusion proposes that activation of the Arp2/3 complex nucleates actin filaments on preexisting filaments (Higgs and Pollard, 2001). Repeated nucleation generates a branched array of filaments, as found at the leading edge of cells (Svitkina and Borisy, 1999; Svitkina et al., 1997). Capping protein limits filament growth spatially and to a length that is effective for pushing (Cooper and Schafer, 2000).

The convergent-elongation model for filopodia formation (Svitkina et al., 2003; Vignjevic et al., 2003) specifies that filopodia are initiated from the lamellipodial dendritic network by action of a complex that protects barbed ends against capping, allowing for their elongation and association. Barbed end complexes converge to form the tip complex that initiates cross-linking by fascin. This permits bundling to keep pace with elongation to produce efficient pushing. Initiated filopodia elongate and attain steady state by filament treadmilling. Filopodia contain an unbranched bundle of actin filaments that are aligned axially, packed tightly together, and have uniform polarity (Lewis and Bridgman, 1992; Small *et al.*, 1978). Arp2/3 complex is excluded from filopodia (Svitkina and Borisy, 1999), and filaments in filopodia are relatively long and do not turn over rapidly (Mallavarapu and Mitchison, 1999).

Model systems have proved valuable in explaining the mechanisms of formation and dynamics of lamellipodia and filopodia. The comet tail structure that propels bacteria or activated beads is structurally and functionally similar to actin assembly in lamellipodia, where the surface of the bacterium or bead is analogous to a fragment of the leading edge of the cell. Comet tails like lamellipodia consist of a dendritic network of actin containing similar actin-binding proteins and display comparable turnover (Cameron *et al.*, 2000). Filament depolymerization requires ADF/cofilin and occurs at a constant rate throughout the tail as in the lamellipodium (Carlier *et al.*, 1997). Filopodia-like bundles formed on activated beads, like filopodia, consist of actin filaments that are long, unbranched, and aligned (Vignjevic *et al.*, 2003). Both display uniform polarity with growth at barbed ends. The mechanism of formation of filopodial-like bundles *in vitro* is remarkably similar to the pathway of filopodia initiation *in vivo*, and they share essential actin-binding proteins (Svitkina *et al.*, 2003). Formation of filopodia-like bundles in extracts is dependent on depletion of capping protein. Cells depleted of capping protein by siRNA knock-down display similar structures (ventral stars) and an increased number of filopodia (Mejillano *et al.*, 2004).

Despite strong structural and compositional similarity, filopodia-like bundles differ dynamically and geometrically from filopodia. These bundles do not treadmill. The filament pointed ends may remain capped, perhaps not dissociating from the Arp2/3 complex that nucleated them, thus preventing depolymerization by ADF/cofilin that mediates treadmilling. Also in this system the bead is not analogous to the leading edge of the cell as in the comet tail motility system, because the barbed ends of the bundles point away from the bead and do not propel it forward as they grow. However, the bead would be analogous to a cluster of Arp2/3 activators at the leading edge. This mechanism may be used by cells to supply a high local concentration of barbed ends for filopodia formation. The availability of *in vitro* systems for construction of filopodia-like bundles allows facile manipulation of parameters involved in filopodial generation, organization, and growth. This provides means to dissect the biochemical and structural details of filopodial formation. Filopodia-like bundles *in vitro* could be assembled on coated beads in two systems: cell extracts or a biochemically defined system.

Preparation of Coated Beads

Reagents

- Brain buffer (BB): 10 mM Tris-Cl, 10 mM EGTA, 2 mM MgCl$_2$ and 10 μg/ml of protein inhibitor cocktail (Roche), pH 7.5.
- *Xenopus* buffer (XB): 100 mM KCl, 0.1 mM CaCl$_2$, 2 mM MgCl$_2$, 5 mM EGTA, 10 mM K HEPES, pH 7.7.
- WASP, purified as described by Vignjevic *et al.* (2003).
- ActA, purified as described by Cameron *et al.* (1999).
- Carboxylated polystyrene beads (Polysciences) ranging in size from 0.2–0.9 μm.

Procedure

To coat beads with ActA, we use the method of Cameron *et al.* (1999).

1. We generally use 0.5-μm beads coated with 37.5% ActA. Mix 1 μl beads, 1.87 μl of ActA (2 mg/ml), and 10.63 μl of 2 mg/ml ovalbumin.
2. Incubate mixture for 1 h at RT.
3. Pellet by centrifugation at 5000 rpm for 5 min.
4. Wash beads twice with the appropriate buffer and resuspend in 15 μl of buffer.

Coating beads with WASP/Scar proteins:

1. Mix 15 μl of 0.5 μM WASP or Scar protein with 15 μl of BB or XB and add 0.5 μl of 0.5-μm beads.
2. Incubate at RT for 1 h.
3. Pellet by centrifugation at 5000 rpm for 5 min.
4. Wash beads twice for 5 min each and resuspend in 10 μl of buffer. If stored at 4°, beads can be used for 1 week. For longer storage, supplement beads with 50% glycerol and store at −80°.

Assembly of Stars in Cytoplasmic Extracts

Reagents

The cytoplasmic extract assembly system requires either mouse brain or *Xenopus* oocyte extract, ActA, or WASP-coated beads and tetramethyl rhodamine (TMR)-actin.

- Mouse brain extract is prepared as described by Vignjevic *et al.* (2003).

• Metaphase *Xenopus* oocyte extract is prepared by the method of Murray *et al.* (1991). Immediately before use, centrifuge the extract at 100,000g for 1 h at 4° and use the supernatant for experiments.

• Dilution of extracts (from 50–80%) with appropriate buffers before use increases the number of assembled stars.

• TMR-actin. Actin was purified from rabbit muscle as described by Spudich and Watt (1971). TMR-actin is prepared by labeling F-actin with 5-(and 6) carboxytetramethylrhodamine succinimidyl ester (Molecular Probes) by a modification of the method of Isambert *et al.* (1995).

1. Dialyze 2.5 ml of G-actin at 2–3 mg/ml against 200 ml of labeling buffer (50 mM PIPES, pH 6.8, 50 mM KCl, 0.1 mM CaCl$_2$, 0.2 mM ATP) ON at 4° to produce F-actin.

2. Add with rapid mixing 5-(and 6)-carboxytetramethylrhodamine succinimidyl ester (TMR-SE, Molecular Probes, Eugene, Oregon) to 0.3 mM from a stock solution of 30 mM (16 mg/ml) in dry dimethyl formamide. The stock can be stored at −80° for years.

3. Incubate for 45 min at RT.

4. Stop the reaction by adding 50 mM lysine from a stock of 1 M lysine.

5. Centrifuge at 100,000g (Beckman TLA 100.3 rotor) for 40 min at 4°.

6. Resuspend the F-actin pellet in 1.0 ml G buffer using a Teflon-glass homogenizer.

7. Dialyze ON against 500 ml of G Buffer at 4°.

8. Centrifuge at 100,000g for 40 min at 4°.

9. To eliminate free dye, apply the supernatant to a 5-ml Excellulose desalting column (Pierce, Rockford, IL) equilibrated with buffer A. Collect 0.5-ml fractions. The sample will separate into two distinct red bands with the excluded TMR-actin eluting before free dye. Pool the TMR-actin peak, adjust it to 5 mM Tris, pH 8.0, 1 mM DTT, 1 mM ATP.

10. Drop-freeze 20-μl aliqouts in liquid nitrogen and store at −80°.

Before each experiment, we recycle TMR-actin to select for actin that is competent to polymerize, thus reducing fluorescent background because of inactive actin.

1. Polymerize a 20-μl aliquot of TMR-actin for 2 h on ice in the presence of 50 mM KCl, 2 mM MgCl$_2$ and 1 mM ATP.

2. Pellet polymerized actin by high-speed centrifugation (100,000g for 1.5 h) at 4°.

3. Resuspend the actin in 40 μl cold G buffer to a final concentration of 2 mg/ml.

4. Dialyze overnight at 4° against G buffer using microdialysis buttons (Hampton Research) and dialysis tubing (Pierce).

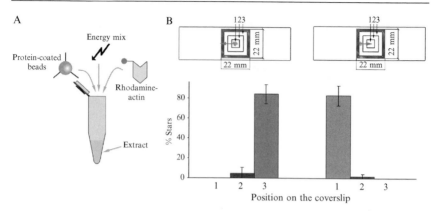

FIG. 1. Assay for filopodia-like bundle assembly *in vitro*. (A) Beads coated with Arp2/3-activators (ActA, WASP) were introduced into cell extract supplemented with energy mix and TMR-actin. (B) Formation of stars depended on position on the coverslip. (A) A drop of assay mix was placed on a glass slide, and a coverslip was applied such that the drop was at its center. After squashing, the drop spread toward the coverslip edges. (B) A drop of assay mix was placed on a glass slide, and a coverslip was applied such that the drop was at its edge. After squashing, the drop spread toward the coverslip center. Quantification of stars relative to their position on coverslip. 1, center of the coverslip; 2, transitional zone; 3, edge of the coverslip.

Light Microscopy

For light microscopic observation of star assembly, we use the following basic procedure (Fig. 1A), which is modified for specific needs.

1. Introduce 0.5 μl of coated beads into 10 μl of cell extract supplemented with energy mix (15 mM creatine phosphate, 2 mM ATP, and 2 mM MgCl$_2$) and 1.25 μM TMR-actin.

2. Take a 1-μl sample, place on a microscope slide, add a 22-mm square coverslip and press firmly using a second glass slide to provide even pressure to avoid cracking the coverslip. Firm pressure by hand creates a chamber about 5-mm thick. A thicker chamber will leave many filaments not immobilized; these floating filaments will contribute to fluorescence background and imaging will be difficult.

3. Seal the chamber with VALAP (Vaseline/lanolin/paraffin, at 1:1:1). It is vital that chambers are sealed completely; otherwise, a stream of fluid through a gap will prevent imaging of actin structures. The fluid flow can cause disruption of stars and drying of the sample in long-term observation.

4. Incubate samples at RT for 15 min.

5. Observe under the microscope taking images every 5 min for 2 h. We use an inverted microscope (Eclipse TE2000) equipped with epifluorescence optics with Plan 100× 1.3 NA, Plan 60×, and 20× objectives and a back-illuminated cooled CCD camera (CH250, Roper Scientific) driven by Metamorph imaging software (Universal Imaging Corp). Darken the room to increase sensitivity. Actin bundles of stars undergo Brownian fluctuations requiring a short exposure time (e.g., 300 msec) to obtain clear images.

Depending on the method of slide preparation, stars can be produced either at the edge or in the center of the coverslip. This is a consequence of star formation being dependent on depletion of capping protein (possibly other proteins as well, e.g., cofilin) by adsorption to the glass during sample spreading (Fig. 1B). For example, if a 1μl-drop of assay mix is placed on a glass slide and a coverslip applied such that the drop spread outward from the coverslip's center, stars are predominantly found at the edge of the coverslip and are absent from the center (Fig. 2B). In contrast, if the coverslip is applied such that one edge contacts the drop of assay mix that is then forced to spread toward the coverslip's center, stars are not observed at the

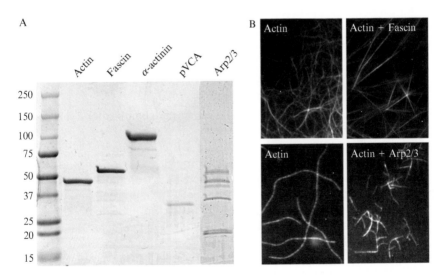

FIG. 2. Proteins used for biochemically defined filopodia-like bundle assembly. (A) SDS-PAGE pattern of proteins used in star reconstitution; 4–20% acrylamide gel. *Lane 1,* TMR-actin (7.5 μg); *lane 2,* recombinant fascin (6.8 μg); *lane 3,* FITC-labeled β-actinin (5 μg); *lane 4,* recombinant pVCA (profilin rich–verprolin homology–connecting–acidic region of WASP protein) (about 0.5 μg); *lane 5,* porcine brain Arp2/3 complex (8 μg). (B) Microscopic assays of protein functions.

initial contacting edge of the coverslip, whereas in the center of the cover-slip and at the opposite edge, stars are abundant (Fig. 1B). Pretreatment of the coverslip with 1% BSA blocks star formation, but other actin structures (tails and clouds) still form.

Stars may also be produced by (1) dilution of the extract with appro-priate buffer to reduce the concentration of capping protein or (2) pre-adsorption with glass:

1. Mix 30 μl of brain extract with 70 mg of ground coverslip glass in a mini filtration tube (0.65 μm, UltraFree-MC, Millipore).
2. Centrifuge twice for 1 min at 11,000g to collect the unadsorbed fraction.
3. As stars assembled in a tube are three dimensional, transferring the sample by pipetting through an arrow orifice leads to severe damage and loss of star structure. Use of tips with their ends cut off alleviates damage.

Procedure for Immunofluorescent Staining

Immunofluorescent staining of stars cannot be performed in the same chambers used for microscopic observation, because stars would be dam-aged by the shear force produced by removal of the coverslip for staining. Therefore we use perfusion chambers.

1. Take 4 μl of assay sample and sandwich it between a glass slide and a coverslip (22 × 22 mm) separated by two strips of Teflon tape.
2. Use VALAP to seal the Teflon tape sides.
3. Allow star formation by incubation of the chamber for 2 h at RT in a humidified container.
4. Perfuse the chamber with 10 μl of BB containing 0.2% Triton X-100 and 2 μM phalloidin by applying the solution on one side of chamber with a pipette and withdrawing it from the other side with filter paper.
5. Perfuse 10 μl of 2 μM phalloidin in BB.
6. Fix stars with 0.2% glutaraldehyde for 20 min at RT. If your antibody requires methanol fixation, fix with methanol for 10 min at $-20°$. Then continue with the procedure from step 9.
7. Wash two times with 10 μl PBS.
8. Quench for 20 min with 2 mg/ml of NaBH4 in PBS supplemented with 0.1% Tween-20. Tween-20 prevents bubble formation during reduction with NaBH4 that would damage star structure.
9. Wash two times with 10 μl PBS.

10. Perfuse with primary antibody and incubate at RT for 40 min.
11. Wash two times with 10 μl PBS.
12. Perfuse with labeled secondary antibody and incubate at RT for 40 min.
13. Wash two times with 10 μl PBS.
14. Seal chamber completely with VALAP and observe.

Electron Microscopy

1. Take 2 μl of sample and sandwich it between two coverslips (22 × 22 mm and 22 × 18 mm) that are separated by two strips (2-mm wide) of Teflon tape attached near the edges of the larger coverslip. Preparing samples for electron microscopy ultimately requires separation of coverslips resulting in destructive shear force that must be reduced by using thicker chambers than for light microscopy.

2. Incubate chambers for 2 h at RT in a humid environment.

3. Open the chambers under a solution containing 0.2% Triton X-100 and 2 μM phalloidin in BB.

4. Wash coverslips with 2 μM phalloidin in BB.

5. Fix with 2% glutaraldehyde.

6. Process for electron microscopy as described by Svitkina and Borisy (1998).

During this procedure, many stars are washed away or damaged (comet tails are more resistant). We tried several approaches to improve the preservation and adhesion of stars from a sandwich made of two coverslips as well as applied to a single coverslip. Even though none of these measures improved the yield of attached stars, we will list what was attempted so that others will not needlessly reproduce our negative results. We tried to clean the coverslips of grime or detergent. Coverslips were subjected to a series of washes with laboratory glass detergent, 1 mM EDTA, 70% ethanol, and 100% ethanol. Between each change of wash solution, coverslips were sonicated for 30 min. Also coverslips were treated with nitric acid or cleaning solution (chromic-sulfuric acid, Fisher) or floated on the surface of 5% hydrofluoric acid (Fisher) for 30 sec to 2 min. Coating coverslips with either 1% BSA or with 1 mg/ml polylysine was also tested. We tried to centrifuge stars through a sucrose cushion such that centrifugal force increase attachment to the coverslip. In this case, stars were assembled in a tube, stabilized with phalloidin and overlaid on a 25% sucrose cushion made in BB supplemented with phalloidin. Glass coverslips were placed on a rubber support at the bottom of the tube. After centrifugation for 20 min at 16,000g, the cushion was aspirated, samples were fixed with 2% glutaraldehyde for 20 min, and processed for EM. This procedure improved the binding of actin structures to the coverslips, but star structures were severely disrupted. Using freshly cleaved "mica" instead of glass also improved the number of adherent actin structures, but they were generally deformed and flattened. Although stars may be sparse on coverslips prepared using the preceding enumerated protocol, if you are patient, you will collect enough data.

Assembly of Filopodia-Like Bundles in the Reconstitution System

The reconstitution system requires: ActA or WASP-coated beads, TMR-labeled actin, Arp2/3, and fascin. SDS-PAGE of the purified proteins is shown on Fig. 2A.

Reagents

- Actin polymerization buffer ($1\times$ KME): 50 mM KCl, 1 mM MgCl$_2$, 1 mM EGTA, 10 mM imidazole, pH 7.
- Arp2/3: To prepare Arp2/3 and VCA (C-terminal domain of WASP), we used the method described by Laurent *et al.* (1999). The activity of Arp2/3 could be monitored microscopically using the ability of Arp2/3 to activate the branched polymerization of actin filaments in the presence of VCA. Incubation of 0.1 μM of Arp2/3 complex with 4 μM TMR-actin and 0.34 μM VCA in KME buffer when observed microscopically resolves individual filaments with branches (Fig. 2B). Alternately, the activity of Arp2/3 can be assayed based on the ability of Arp2/3 to initiate the formation of actin clouds at the surface of ActA coated beads incubated in 4 μM TMR-actin.
- Fascin: To purify recombinant human fascin:

1. Grow *E. coli* strain carrying the pGEX-4T-3-fascin plasmid in LB medium containing 50 μg/ml ampicillin at 37° until the OD600 reaches 0.6.
2. Induce protein expression at 20° for 4 h by adding 0.1 mM IPTG (fascin is mostly insoluble if induced at 37°).
3. Harvest bacteria by centrifugation and extract with B-PER, bacterial protein extraction reagent in phosphate buffer (Pierce) supplemented with 1 mM PMSF and 1 mM DTT. Use 10 ml of B-PER per 250 ml of bacterial culture.
4. Centrifuge the lysate at 48,000g for 15 min.
5. Mix the supernatant for 1 h at RT with 2 ml glutathione-Sepharose 4B beads (Pharmacia Biotech Inc.) equilibrated with PBS plus 1 mM DTT.
6. Pour the glutathione-Sepharose into a column and wash with 20 ml of PBS plus 1 mM DTT.

7. Add 80 units (80 μl) of thrombin (Pharmacia Biotech) and allow digestion to proceed ON at 4°.

8. Collect flow-through fractions (0.5 ml) in tubes containing enough PMSF to give a final concentration of 2 mM.

9. Concentrate by Centricon 10 (Amicon).

10. The activity of fascin to bundle actin can be assayed microscopically or by a centrifugation assay. For the centrifugation assay, we mix 12 μM F-actin with 1.2 μM fascin in KME buffer. After incubation at RT for 30 min, centrifuge the mixture at 14,000 rpm (in Microfuge) for 15 min, which will pellet bundled actin filaments, but not F-actin. Carefully remove the supernatant and resuspend pellets in the original volume. Pellets and supernatants (with or without fascin) are analyzed by SDS-PAGE. Formation of bundles can also be assayed microscopically, after incubation of 3 μM fascin with 7 μM TMR-actin in KME buffer for 2 h (Fig. 2B).

Procedure

In Vignjevic *et al.* (2003), we defined a minimal set of components necessary for star assembly. Because data obtained with extracts indicated that the presence of Arp2/3 complex and depletion of capping protein were both essential, initial reconstitution experiments were carried out with WASP-coated beads, actin, and Arp2/3 complex. Concentrations of proteins were based on published data for reconstitution of comet tails (Loisel *et al.*, 1999). These components were found to be insufficient to induce stars unless a bundling protein (fascin or alpha-actinin) was added (see Fig. 3).

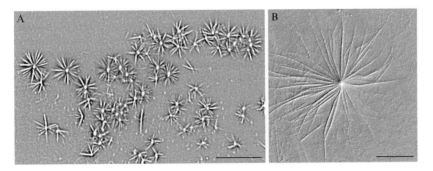

Fig. 3. Reconstitution of filopodia-like bundles from pure proteins. WASP-coated beads preincubated with TMR-actin, Arp2/3 complex, and fascin. Fluorescent images were processed with Metamorph to achieve a DIC-like edge enhancement effect for greater clarity in visualization. (A) Low magnification. Scale bar, 10 μm. (B) High magnification. Scale bar, 5 mm.

1. Add 0.5 μl of WASP-coated beads to 9.5 μl 1× KME buffer supplemented with 0.7 μM Arp2/3complex (at 0.7 μM Arp2/3 complex and above, actin polymerization at the bead surface produced clouds of actin filaments within 15 min. Lower concentrations of Arp2/3 failed to generate clouds in this time frame).

2. Incubate for 5 min at RT (this increases the amount of Arp2/3 recruited to the beads).

3. Add 6.9 μM TMR-actin to the mixture and incubate on ice for 20 min.

4. Add 3.1 μM fascin. (If fascin is added to the mixture simultaneously with actin, large bundles not connected to beads will form. Increasing the concentration of fascin promotes bundle formation both on beads and free in solution. At 3.1 μM fascin, greater than 95% of the beads showing actin polymerization displayed stars with straight, needle-like rays).

5. Take 1 μl of sample, place on a microscope slide, and press tightly between a slide and a 22-mm^2 coverslip. Coverslips and slides must be coated with 1% BSA to prevent protein absorption.

6. Incubate 15 min at RT and observe.

References

Borisy, G. G., and Svitkina, T. M. (2000). Actin machinery: Pushing the envelope. *Curr. Opin. Cell Biol.* **12**, 104–112.

Cameron, L. A., Footer, M. J., van Oudenaarden, A., and Theriot, J. A. (1999). Motility of ActA protein-coated microspheres driven by actin polymerization. *Proc. Natl. Acad. Sci. USA* **96**, 4908–4913.

Cameron, L. A., Giardini, P. A., Soo, F. S., and Theriot, J. A. (2000). Secrets of actin-based motility revealed by a bacterial pathogen. *Nat. Rev. Mol. Cell. Biol.* **1**, 110–119.

Carlier, M. F., Laurent, V., Santolini, J., Melki, R., Didry, D., Xia, G. X., Hong, Y., Chua, N. H., and Pantaloni, D. (1997). Actin depolymerizing factor (ADF/cofilin) enhances the rate of filament turnover: Implication in actin-based motility [see comments]. *J. Cell Biol.* **136**, 1307–1322.

Cooper, J. A., and Schafer, D. A. (2000). Control of actin assembly and disassembly at filament ends. *Curr. Opin. Cell Biol.* **12**, 97–103.

Higgs, H. N., and Pollard, T. D. (2001). Regulation of actin filament network formation through ARP2/3 complex: Activation by a diverse array of proteins. *Annu. Rev. Biochem.* **70**, 649–676.

Isambert, H., Venier, P., Maggs, A. C., Fattoum, A., Kassab, R., Pantaloni, D., and Carlier, M. F. (1995). Flexibility of actin filaments derived from thermal fluctuations. Effect of bound nucleotide, phalloidin, and muscle regulatory proteins. *J. Biol. Chem.* **270**, 11437–11444.

Laurent, V., Loisel, T. P., Harbeck, B., Wehman, A., Grobe, L., Jockusch, B. M., Wehland, J., Gertler, F. B., and Carlier, M. F. (1999). Role of proteins of the Ena/VASP family in actin-based motility of *Listeria monocytogenes*. *J. Cell Biol.* **144**, 1245–1258.

Lewis, A. K., and Bridgman, P. C. (1992). Nerve growth cone lamellipodia contain two populations of actin filaments that differ in organization and polarity. *J. Cell Biol.* **119**, 1219–1243.

Loisel, T. P., Boujemaa, R., Pantaloni, D., and Carlier, M. F. (1999). Reconstitution of actin-based motility of Listeria and Shigella using pure proteins [see comments]. *Nature* **401**, 613–616.

Mallavarapu, A., and Mitchison, T. (1999). Regulated actin cytoskeleton assembly at filopodium tips controls their extension and retraction. *J. Cell Biol.* **146**, 1097–1106.

Mejillano, M. R., Kojima, S., Applewhite, D. A., Gertler, F. B., Svitkina, T. M., and Borisy, G. G. (2004). Lamellipodial versus filopodial mode of the actin nanomachinery: Pivotal role of the filament barbed end. *Cell* **118**, 363–373.

Mullins, R. D., Heuser, J. A., and Pollard, T. D. (1998). The interaction of Arp2/3 complex with actin: Nucleation, high affinity pointed end capping, and formation of branching networks of filaments. *Proc. Natl. Acad. Sci. USA* **95**, 6181–6186.

Murray, M. T., Krohne, G., and Franke, W. W. (1991). Different forms of soluble cytoplasmic mRNA binding proteins and particles in *Xenopus laevis* oocytes and embryos. *J. Cell Biol.* **112**, 1–11.

Small, J. V., Isenberg, G., and Celis, J. E. (1978). Polarity of actin at the leading edge of cultured cells. *Nature* **272**, 638–639.

Spudich, J. A., and Watt, S. (1971). The regulation of rabbit skeletal muscle contraction. I. Biochemical studies of the interaction of the tropomyosin-troponin complex with actin and the proteolytic fragments of myosin. *J. Biol. Chem.* **246**, 4866–4871.

Svitkina, T. M., and Borisy, G. G. (1998). Correlative light and electron microscopy of the cytoskeleton of cultured cells. *Methods Enzymol.* **298**, 570–592.

Svitkina, T. M., and Borisy, G. G. (1999). Arp2/3 complex and actin depolymerizing factor/cofilin in dendritic organization and treadmilling of actin filament array in lamellipodia. *J. Cell Biol.* **145**, 1009–1026.

Svitkina, T. M., Verkhovsky, A. B., McQuade, K. M., and Borisy, G. G. (1997). Analysis of the actin-myosin II system in fish epidermal keratocytes: Mechanism of cell body translocation. *J. Cell Biol.* **139**, 397–415.

Svitkina, T. M., Bulanova, E. A., Chaga, O. Y., Vignjevic, D. M., Kojima, S., Vasiliev, J. M., and Borisy, G. G. (2003). Mechanism of filopodia initiation by reorganization of a dendritic network. *J. Cell Biol.* **160**, 409–421.

Vignjevic, D., Yarar, D., Welch, M. D., Peloquin, J., Svitkina, T., and Borisy, G. G. (2003). Formation of filopodia-like bundles *in vitro* from a dendritic network. *J. Cell Biol.* **160**, 951–962.

Author Index

Subject Index

A

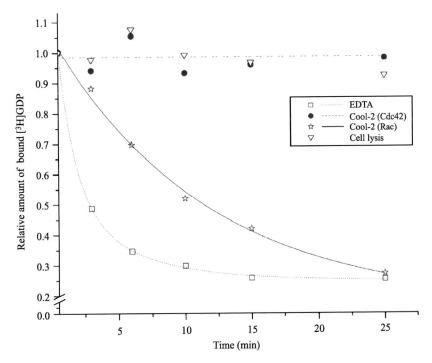

BAIRD *ET AL.*, CHAPTER 5, FIG. 1. The abilities of wild-type Cool-2 to exhibit GEF activity toward Cdc42 and Rac. Cool-2 stimulates the dissociation of [³H]GDP from Rac but not Cdc42. COS-7 cells were transiently transfected with a plasmid expressing Myc-tagged Cool-2, and the Cool-2 proteins were immunoprecipitated using an anti-Myc antibody. Recombinant (*E. coli*) Cdc42 or Rac was preloaded with [³H]GDP. The exchange of bound [³H]GDP for GTPγS, as monitored by the dissociation of the radiolabeled nucleotide from Cdc42 or Rac, was assayed in the presence of immunoprecipitated Myc-Cool-2 or in the presence of lysates from cells that were not transfected with the cDNA encoding Myc-Cool-2 (i.e., control cells). The data shown are representative of three experiments and were taken from Feng *et al.* (2004).

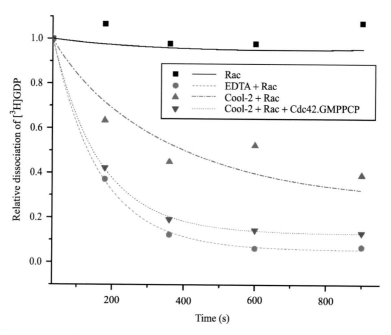

BAIRD *ET AL.*, CHAPTER 5, FIG. 2. GMP-PCP–bound Cdc42 stimulates the Rac-GEF activity of Cool-2/α-Pix. The GEF activity of Cool-2 toward Rac is specifically enhanced by GMP-PCP–bound Cdc42. Relative dissociation of preloaded [³H]GDP from Rac1 or Cdc42 in the presence of excess GTPγS and insect cell expressed His-Cool-2 was measured as outlined in this chapter. The data were taken from Baird *et al.* (2005).

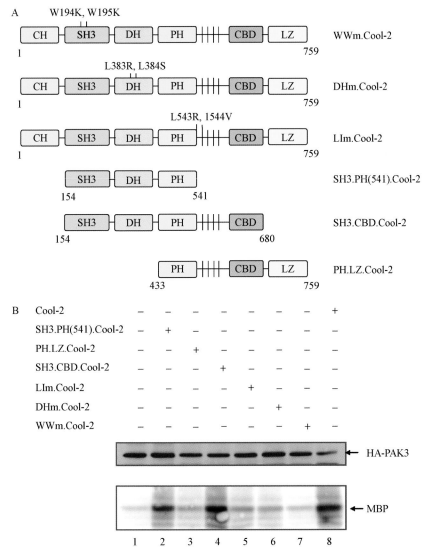

BAIRD *ET AL.*, CHAPTER 5, FIG. 3. The abilities of different Cool/Pix constructs to activate Pak *in vivo*. (A) Depiction of different Cool/Pix constructs used in the experiment. (B) SH3. PH.Cool-2 (lane 2), SH3.CBD.Cool-2 (lane 4), and wild-type Cool-2 (lane 8) can activate Pak *in vivo*, whereas PH.LZ.Cool-2 (lane 3), LIm.Cool-2 (lane 5), DHm.Cool-2 (lane 6), and WWm.Cool-2 (lane 7) are not able to activate Pak. The Pak assays (shown in the lower panel) were performed using myelin basic protein (MBP) as a phosphosubstrate. Upper panel shows the relative amounts of PAK3 assayed under the different conditions. Data were taken from Feng *et al.* (2004).

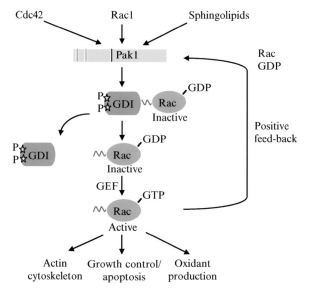

CELINE *ET AL.*, CHAPTER 7, FIG. 1. Model for Pak1-mediated positive feedback activation of Rac GTPase. The interaction of active Cdc42, Rac1, or sphingolipids with Pak1 stimulates the phosphorylation of Rac-RhoGDI complexes at Ser101 and Ser174. This induces the dissociation of Rac1GDP, which is then activated to Rac1GTP through interaction with GEF (s). Active Rac1 feeds back to further activate Pak1, leading to more free Rac1 in a positive feedback cycle that enhances subsequent biological activities.

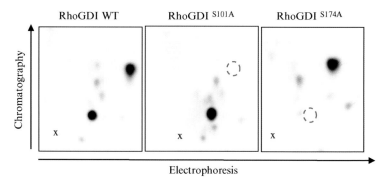

CELINE *ET AL.*, CHAPTER 7, FIG. 3. Analysis of Pak1 phosphorylation sites on RhoGDI. 293T cells expressing active Pak1 and RhoGDI constructs were metabolically labeled with $[\gamma\text{-}^{32}\text{P}]$ orthophosphoric acid and RhoGDI immunoprecipitated from the whole cell lysates and subjected to phosphopeptide mapping, as in text. RhoGDI constructs that were used are indicated at the top of each map, and the application point of each sample is indicated in the lower left-hand corner by a cross. An arrow shows the direction of electrophoresis and chromatography migration. On the electrophoresis axis, the cathode is on the right. Two distinct radiolabeled phosphopeptide spots were detected with RhoGDI wild type (left panel). Missing phosphopeptides in RhoGDI mutants are circled by dots (middle and right panel).

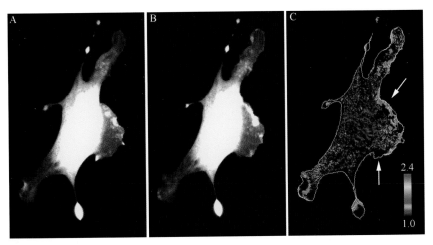

HODGSON *ET AL.*, CHAPTER 12, FIG. 1. Effects of motion artifacts on ratiometric measurements. (A and B) EGFP images taken before and after I-SO acquisition. Dividing the second EGFP image by the first shows where motion has occurred during the acquisition sequence. In this example, (B)—second EGFP acquisition—was divided by (A)—first EGFP acquisition—to produce ratiometric (C). Areas that underwent protrusion generated artificial high ratios (upper white arrow) and low ratios (lower white arrow). The outline of the cell is shown in white in (C).

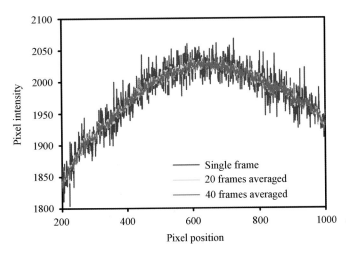

HODGSON *ET AL.*, CHAPTER 12, FIG. 3. The effect of averaging on shading correction images. A sample-free field of view was imaged under identical illumination conditions once (blue), 20 times (yellow), and 40 times (red) and averaged. Averaging of many frames reduces the stochastic noise in the image.

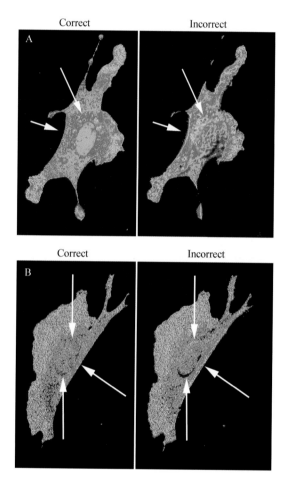

Hᴏᴅɢsᴏɴ ᴇᴛ ᴀʟ., Cʜᴀᴘᴛᴇʀ 12, Fɪɢ. 4. Image registration artifacts. (A) Example of misregistration producing artifactual edge effects and spurious intracellular features (white arrows). The images were misaligned by 5 whole pixels in both the x and y directions. (B) More subtle effects of misalignment resulting in edge effects around the nucleus (white arrows) plus an acute, high ratio region at the cell edge. In this case, the shifts were less than 1 pixel (0.1 pixel in the x direction, −0.8 pixel in the y direction). A telltale sign of misalignment is high values on one side of the misaligned object and low values on the opposite side.

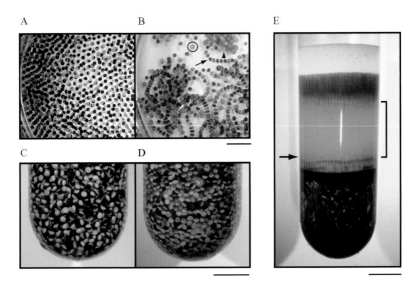

LEBENSOHN *ET AL.*, CHAPTER 13, FIG. 1. *Xenopus laevis* eggs. (A) Good-quality eggs are round and uniform. (B) Poor-quality eggs are often attached together forming strings (arrow) or clusters (arrowhead) or may be large and pale (circled). (C) Before dejellying, eggs are separated by their transparent jelly coats. (D) Dejellied eggs pack as tight spheres without any visible separation. (E) After centrifugation, three major layers result: lipid, cytoplasm, and yolk going from top to bottom. Cytoplasmic extract (marked by the bracket) is collected by puncturing the side of the tube near the bottom of the middle layer (at the site indicated by the arrow). Scale bar = 10 mm.

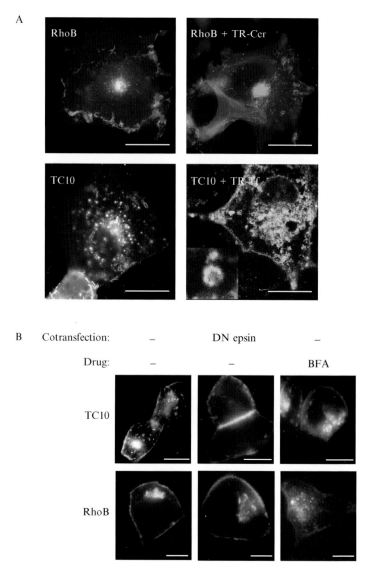

MICHAELSON AND PHILIPS, CHAPTER 22, FIG. 3. Identification of subcellular compartments in live cells. (A) Cos-1 cells were transfected as described with GFP-RhoB or GFP-TC10. The Golgi apparatus and endosomes were marked by incubating the cells for 30 min before imaging with, respectively, Texas Red–conjugated BODIPY ceramide (10 μg/ml, Molecular Probes) or transferrin (100 μg/ml, Molecular Probes). Excess label was washed away immediately before imaging. The inset on the lower right panel shows an enlargement of an endosome that reveals that the colocalization is on the limiting membrane. Bars indicate 10 μm. (B) MDCK cells were transfected with GFP-TC10 or GFP-RhoB with or without dominant negative epsin and imaged alive 16 h later. Cells transfected with GFP-TC10 or GFP-RhoB were treated with brefeldin A (BFA) for 5 h before imaging, causing dispersal of the Golgi. Bars indicate 10 μm.

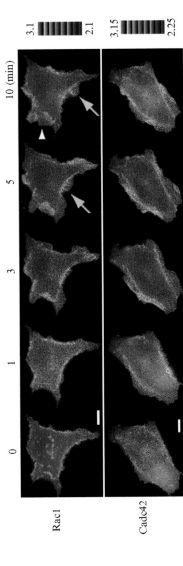

NAKAMURA *ET AL.*, CHAPTER 23, FIG. 3. Time-lapse experiments of Rac1 and Cdc42 activations upon EGF stimulation in Cos1 cells. Serum-starved Cos1 cells expressing Raichu-Rac1 or Raichu-Cdc42 were treated with EGF and imaged every 30 sec. FRET images at the indicated time points after EGF addition are shown in the IMD mode (see text). The yellow arrows and white arrowheads indicate nascent and retracting lamellipodia, respectively. The upper and lower limits of the ratio range are shown on the right. Bars indicate 10 μm. Reproduced with permission from Kurokawa *et al.* (2004).

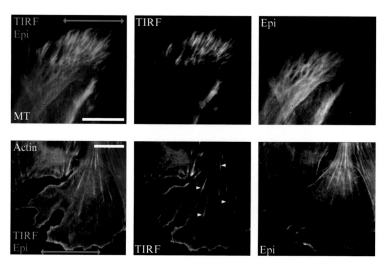

MANNEVILLE, CHAPTER 40, FIG. 3. High magnification images of the microtubule (upper images) and actin (lower images) cytoskeleton in fixed migrating astrocytes. Left images show merged TIRF (green) and epifluorescence (Epi, red). Note that microtubules come in close contact to the substrate only at the front of migrating cells, whereas actin is visible inside the protrusion. Focal contacts appear as bright spots in the actin TIRF image (arrowheads). The double red arrows indicate the direction of the scratch. Bars: 10 μm.

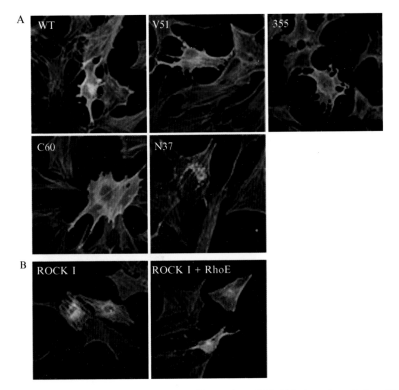

RIENTO AND RIDLEY, CHAPTER 41, FIG. 1. Expression of RhoE induces loss of stress fibers and prevents ROCK I from inducing stress fibers in Swiss 3T3 cells. (A) Growing Swiss 3T3 cells were injected with wild-type or mutated FLAG-RhoE forms. RhoE was localized with anti-FLAG antibody (green) and F-actin with phalloidin (red). (B) ROCK I expression in starved Swiss 3T3 cells increases the formation of actin filaments. Coexpression of RhoE prevents ROCK I–induced stress fiber formation. Myc-ROCK I or FLAG-RhoE is shown in green, F-actin in red.

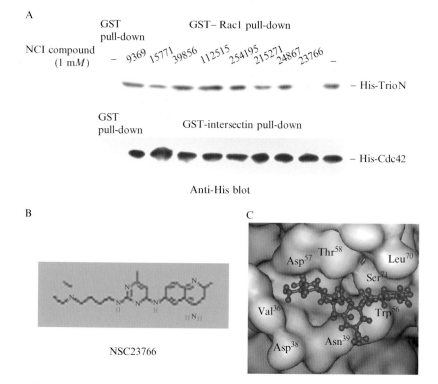

A

GST
pull-down   GST– Rac1 pull-down

NCI compound   – 9369 15771 39856 112515 254195 215271 24867 23766   –
(1 mM)

– His-TrioN

GST
pull-down   GST-intersectin pull-down

– His-Cdc42

Anti-His blot

B   C

NSC23766

Asp^{57} Thr^{58} Leu^{70}
                              Ser^{71}
Val^{36} Trp^{56}
            Asp^{38} Asn^{39}

Huzoor-AKBAR ET AL., CHAPTER 43, FIG. 1. Identification of NSC23766 as a Rac-activation specific inhibitor. (A) The inhibitory effect of a panel of compounds predicted by virtual screening on Rac1 interaction with TrioN was tested in a complex formation assay; 0.5 μg of (His)$_6$-tagged TrioN was incubated with GST alone or nucleotide-free GST-Rac1 (2 μg) in the presence or absence of 1 mM indicated NCI compound and 10 μl suspended glutathione-agarose beads. After an incubation at 4° for 30 min, the beads with associated (His)$_6$-TrioN were detected by anti-His Western blotting. Lower panel, the effect of the compounds on Cdc42 binding to Intersectin was determined similarly; ~1 μg of GST or GST-tagged Intersectin was incubated with the nucleotide-free (His)$_6$-tagged Cdc42 (0.25 μg) under similar conditions. (B) The chemical structure of NSC23766. (C) A simulated docking model of NSC23766 on the Rac1 surface.

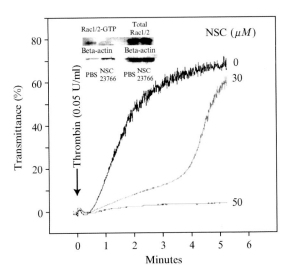

Huzoor-Akbar et al., Chapter 43, Fig. 5. NSC23766 inhibited thrombin-induced activation of Rac GTPases and aggregation of platelets. Washed human platelets were incubated with NSC23766 at indicated concentrations for 3 min before stimulation with thrombin (0.05 U/ml). The Rac-GTPase activity was determined by the GST-PAK1 pull-down assay (insert), and platelet aggregation induced by the thrombin was determined by a light-scattering–based absorption method.

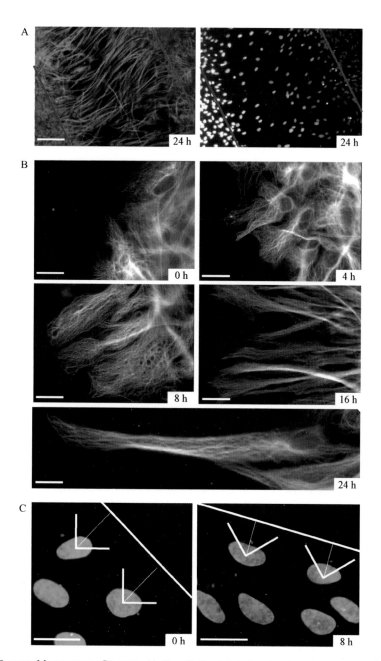

ETIENNE-MANNEVILLE, CHAPTER 44, FIG. 2. Scratch-induced astrocyte responses. (A) Cell migration. Left panel shows tubulin (microtubules, green) and Hoechst (nuclei, blue) 24 h after scratching. From the Hoechst image shown on the right panel, the initial position of the scratch can be drawn (red line). Bar, 100 μm. (B) Protrusion formation and polarization of the microtubule network. Staining with antitubulin antibodies at indicated time after scratching shows the elongation of the microtubule network and its polarization along the axis of migration. Bar, 10 μm. (C) Centrosome reorientation. Concomitant Hoechst (nuclei, blue) and pericentrin (centrosome, red) staining allow the quantification of centrosome reorientation. In the left panel (0 h after scratching), the two cells of the front row are not polarized, their centrosome is not localized in the forward-facing quadrant. In the right panel (8 h after scratching), the centrosomes of the scratch edge cells are localized in the forward-facing quadrant. Bar, 10 μm.

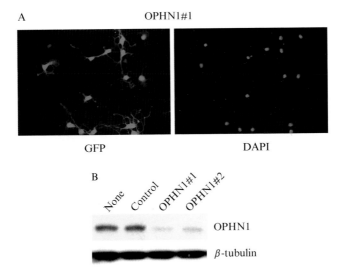

JANAS, CHAPTER 46, FIG. 2. OPHN1 knockdown in hippocampal neurons using a lentiviral-based system. Dissociated hippocampal neurons isolated from E18 rats were plated on coverslips placed in a 6-well plate and cultured for 4–6 days. The neurons were then infected with 25 μl of viral stocks containing TRIP vectors encoding two different shRNAs for *oligophrenin-1* (*OPHN1#1* or *OPHN1#2*) or a control TRIP vector with an unrelated hairpin. (A) After 30 h in culture, neurons were fixed in 4% paraformaldehyde and mounted in Vectashield containing DAPI (Vector Laboratories). Cells were analyzed for GFP expression and DAPI labeling using fluorescence microscopy. As shown for *OPHN1#1*, approximately 90–95% of the neurons were infected. (B) Lysates were prepared 30 h after infection and analyzed by Western blotting with antibodies to OPHN1 and to b-tubulin as a loading control. Cells infected with *OPHN1#1* or *OPHN1#2* showed a significant decrease in OPHN1 levels compared with cells infected with control viruses.

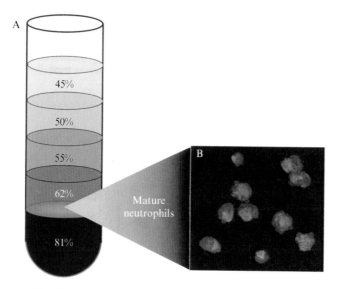

Dong and Wu, Chapter 47, Fig. 1. Preparation of primary mouse bone marrow neutrophils by density gradient. (A) Schematic representation of the step Percoll gradient in a 15-ml centrifuge tube for neutrophil separation. The mature neutrophils are banded between the layers of 62% and 81%. (B) Neutrophils were stimulated with uniform fMLP as described in the text and stained with Alexa-633-phalloidin and FITC-conjugated anti-CD43. Anti-Cd43 is used to counterstain phalloidin, which primarily stains the leading in polarized cells.

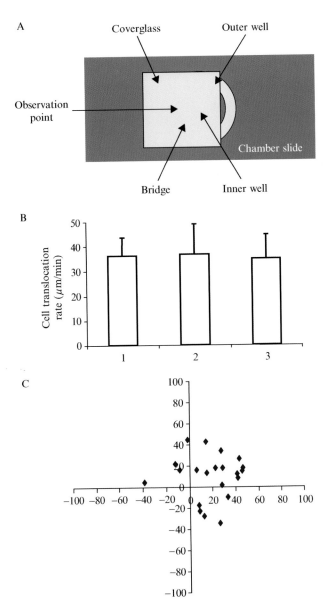

DONG AND WU, CHAPTER 47, FIG. 2. Tracking cell migration in a Dunn chamber. (A) Schematic drawing of a Dunn Chamber. (B) Directionality plot constructed from tracking normal mouse neutrophil migration in a Dunn chamber containing an fMLP gradient. (C) Translocation rates of primary mouse bone marrow neutrophils from three independent experiments.

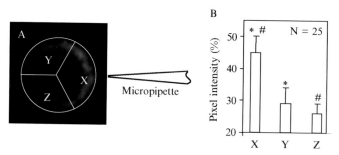

DONG AND WU, CHAPTER 47, FIG. 3. Detection of directional sensing. (A) Schematic depiction of the stimulation from a point source delivered by a micropipette that is 40–50 μm from the cell. The cell was fixed and stained with phalloidin. For quantification, the cell was divided into three sectors, and pixel values were determined. (B) Pixel values of 25 neutrophils in each sector are compared.

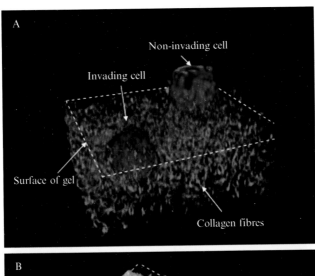

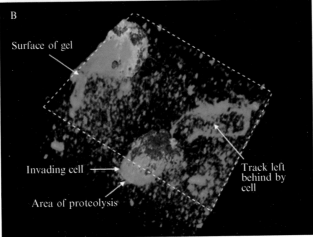

Hooper *et al.*, Chapter 49, Fig. 2. Three-dimensional images of A375 cells invading labeled collagen gels. (A) F-actin staining of A375 cells in red and FITC collagen I fibers in green. Laser-scanning microscope z-sections were reconstructed using Volocity software (Improvision). Dashed lines marking the surfaces of the gel are 80 μm. (B) F-actin staining of A375 cells in red and DQ collagen I in green. Laser-scanning microscope z-sections were reconstructed using Volocity software (Improvision). Dashed lines marking the surfaces of the gel are 80 μm.

A

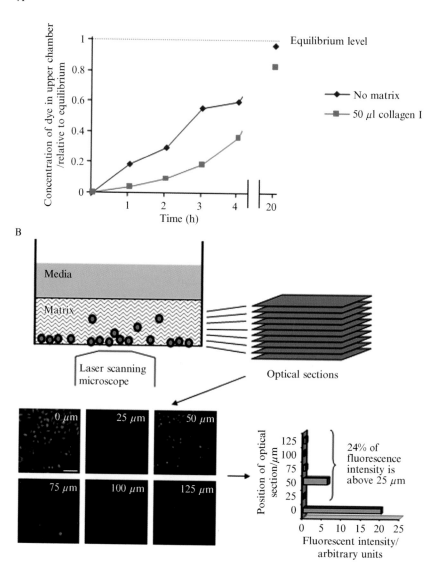

HOOPER *ET AL.*, CHAPTER 49, FIG. 4. Characterization and quantification of invasion assays. (A) Stability of a small molecule gradient in transwell assay. Graph shows the relative concentration of bromophenol blue in the upper chamber of a transwell assay using a filter with 8-μm pores with or without a 50-μm collagen I gel. Initially, dye was placed in lower chamber; only then assay was incubated at 37°; 1 is the concentration of bromophenol blue at equilibrium. (B) Quantification of an overlay invasion assay. Top left shows a schematic of experimental setup; top right shows representation of the optical section; bottom left shows representative optical sections from an assay; and bottom right shows quantification of the images in bottom left.

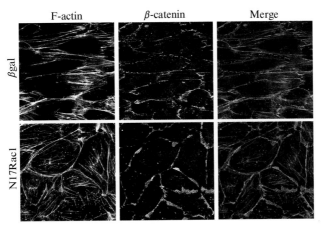

MILLÁN *ET AL.*, CHAPTER 50, FIG. 2. Effect of adenoviral N17Rac1 on HUVECs. HUVECs were incubated with an adenovirus coding for β-galactosidase (βgal) or N17Rac1 (500 m.o.i.) for 90 min in EGM-2 medium at 37°. Cells were washed with PBS and stimulated with 10 ng/ ml of TNF-α in starving medium for 18 h. Cells were then fixed with 4% paraformaldehyde, permeabilized with 0.2% TX-100 in PBS, and stained with TRITC-conjugated phalloidin for F-actin detection and anti-β-catenin rabbit antibody followed by a FITC-conjugated secondary anti-rabbit antibody and analyzed by confocal microscopy. Uninfected cells (not shown) or βgal-expressing cells showed bundles of long stress fibers running almost parallel to the axes of the elongated cells, ending at the intercellular junctions (β-catenin staining). However, N17Rac1-expressing cells are not elongated, and their stress fibers are short and not connected to the intercellular junctions.

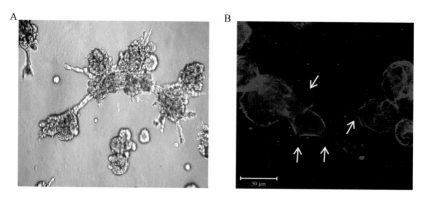

Xɪᴀɴɢ ᴀɴᴅ Mᴜᴛʜᴜsᴡᴀᴍʏ, Cʜᴀᴘᴛᴇʀ 54, Fɪɢ. 2. Invasion phenotype of MCF-10A cells in 3D acini. (A) Phase morphology of invasion. MCF-10A cells expressing ErbB2 and ErbB3 chimera plasmids are cultured on matrigel/collagen mixture for 6 days and stimulated with synthetic dimerizer for 2 days. (B) Loss of intact laminin in invasive structures. The structure from A is fixed and stained with anti-laminin 5 antibody. Arrows show the loss of laminin 5 at the basement membrane of invasive structures. Red, laminin 5.

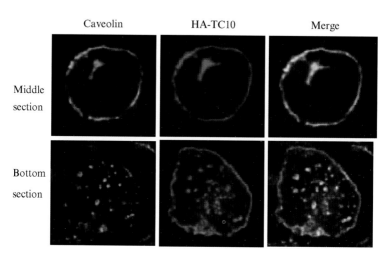

Cʜɪᴀɴɢ ᴇᴛ ᴀʟ., Cʜᴀᴘᴛᴇʀ 55, Fɪɢ. 6. Subcellular localization of TC10. 3T3-L1 adipocytes are electroporated with 400 μg of HA-TC10. Cells are plated on coverslips and recovered for 4–5 days. Cells were then immunostained with anti-caveolin and anti-HA antibodies. Images from different sections of the cells are taken as indicated.

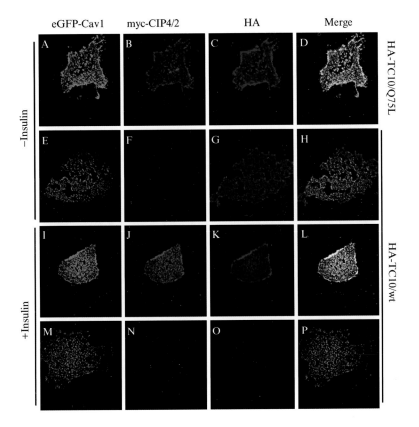

CHIANG ET AL., CHAPTER 55, FIG. 8. TC10 recruits CIP4/2 to lipid rafts after insulin stimulation. 3T3-L1 adipocytes were electroporated with eGFP-Caveolin (50 μg), myc-CIP4/2 (100 μg), HA-TC10/wt (50 μg), or HA-TC10/Q75L (50 μg). The electroporated cells were plated on glass coverslips, and plasma sheets were prepared. Cells were stained with anti-myc mAb and anti-HA pAb (Santa Cruz Biotechnologies) at 1 μg/ml, followed by Alexa[568] goat anti-mouse IgG (1 μg/ml; Molecular Probes) and Alexa[633] goat anti-rabbit IgG (1 μg/ml; Molecular Probes) to visualize myc-CIP4/2 (red) and HA-TC10 (blue), respectively. Fluorescence was visualized by using an Olympus Fluoview 300 confocal laser scanning microscope, and images are representative of three independent determinations. In the presence of TC10/WT after insulin stimulation, myc-CIP4/2 was detected in the plasma membrane sheets (j). Consistent with these data, the constitutively active mutant TC10/Q75L also recruited myc-CIP4/2 to the plasma membrane (d). In the absence of TC10 expression, myc-CIP4/2 is not detected in the plasma membrane sheets after insulin stimulation (n and o).

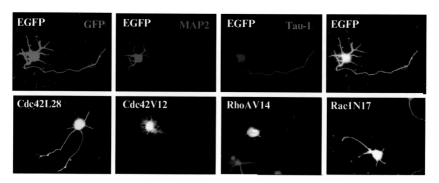

SCHWAMBORN *ET AL.*, CHAPTER 56, FIG. 2. Effect of GTPases on neuronal polarity. Hippocampal neurons were transfected at 0.5 d.i.v. with expression vectors for EGFP and EGFP fusion proteins of the indicated GTPase mutants, fixed at 2.5 d.i.v., and stained with Tau-1 (blue), anti-MAP2 (red), and anti-EGFP antibodies (green). (Modified after Schwamborn and Püschel, 2004).

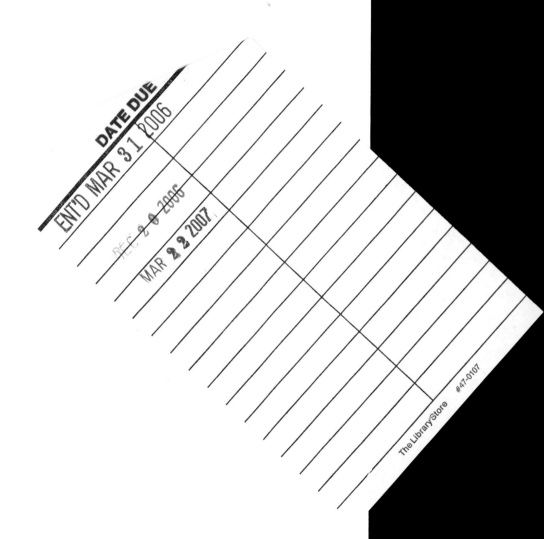